Life

The Science of Biology

FIFTH EDITION

Life

The Science of Biology

William K. Purves
Harvey Mudd College
Claremont, California

Gordon H. Orians
The University of Washington
Seattle, Washington

H. Craig Heller
Stanford University
Stanford, California

David Sadava
The Claremont Colleges
Claremont, California

SINAUER ASSOCIATES, INC.

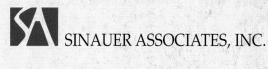
W. H. FREEMAN AND COMPANY

The Cover

Grizzly bears (*Ursus arctos*) take many years to mature reproductively, and cubs remain with their mothers for several years. If their populations are to persist, adult bears must have access to rich food resources. Grizzly bears that live in coastal Alaska depend on salmon that swim up the rivers to spawn. Given that abundant, high-quality food, they grow to become the world's largest carnivorous mammal. Photograph by Michio Hoshino/Minden Pictures.

The Frontispiece

A sunset scene with nesting painted storks (*Mycteria leucocephalus*) taken in Bhakatpur, India. Photograph by Mike Powles/Woodfall Wild Images.

LIFE: *The Science of Biology,* Fifth Edition

Copyright © 1998 by Sinauer Associates, Inc. All rights reserved.
This book may not be reproduced in whole or in part without permission.

Address editorial correspondence to Sinauer Associates, Inc., 23 Plumtree Road, Sunderland, Massachusetts 01375 U.S.A.

www.sinauer.com

Address orders to W. H. Freeman and Co. Distribution Center, 4419 West 1980 South, Salt Lake City, Utah 84104 U.S.A.

Examination copy information: 1-800-446-8923
Orders: 1-800-877-5351

www.whfreeman.com

Library of Congress Cataloging-in-Publication Data

Life, the science of biology / William K. Purves ... [et al.] —
   5th ed.
            p.   cm.
    Rev. ed. of: Life, the science of biology / William K. Purves,
Gordon H. Orians, H. Craig Heller. 4th ed. c1995.
     Includes bibliographical references and index.
     ISBN 0-7167-2869-9 (hardcover)
      1. Biology   I. Purves, William K. (William Kirkwood), 1934–   .
II. Purves, William K. (William Kirkwood), 1934– Life, the science
of biology

QH308.2.L565   1997   79-34772
579—dc21   CIP

Printed in U.S.A.
First printing 1998 RRD

To Jean, Betty, Renu, and Angeline

About the Authors

Bill Purves is Professor Emeritus of Biology as well as founder and former chair of the Department of Biology at Harvey Mudd College in Claremont, California. He received his Ph.D. from Yale University in 1959 under Arthur Galston. A fellow of the American Association for the Advancement of Science, Professor Purves has served as head of the Life Sciences Group at the University of Connecticut, Storrs, and as chair of the Department of Biological Sciences, University of California, Santa Barbara, where he won the Harold J. Plous Award for teaching excellence. His research  interests focus on the chemical and physical regulation of plant growth and flowering. Professor Purves elected early retirement in 1995, after teaching introductory biology for 34 consecutive years, in order to turn his skills to writing and producing multimedia for introductory biology students.

Craig Heller is the Lorry Lokey/  Business Wire Professor of Biological Sciences and Human Biology at Stanford University. He has served as Director of the popular interdisciplinary undergraduate program in Human Biology and is now Chairman of Biological Sciences. He is a fellow of the American Association for the Advancement of Science and received the Walter J. Gores Award for Excellence in Teaching. Dr. Heller received his Ph.D. from Yale University in 1970 and did postdoctoral work at Scripps Institution of Oceanography on how the brain regulates body temperature of mammals. His current research is on the neurobiology of sleep and circadian rhythms. Over the years Dr. Heller has done research on systems ranging from sleeping college students to diving seals to hibernating bears to meditating yogis. He teaches courses on animal and human physiology and neurobiology in Stanford's introductory core curriculum.

Gordon Orians is Professor Emeritus of Zoology at the University of Washington. He received his Ph.D. from the University of California, Berkeley, in 1960 under Frank Pitelka. Professor Orians has been elected to the National Academy of Sciences, the American Academy of Arts and Sciences, and is a Foreign Fellow of the Royal Netherlands Academy of Arts and Sciences. He was President of the Organization for Tropical Studies, 1988–1994, and President of the Ecological Society of America, 1995–1996. He is a recipient of the Distinguished Service Award of the American Institute of Biological Sciences. Professor  Orians is a leading authority in ecology and evolution, with research experience in behavioral ecology, plant–herbivore interactions, community structure, the biology of rare species, and environmental policy. He elected early retirement to be able to devote more time to writing and environmental policy activities.

David Sadava is the Pritzker Family  Foundation Professor of Biology at Claremont McKenna, Pitzer, and Scripps, three of the Claremont Colleges. He received his Ph.D. from the University of California, San Diego in 1972 and has been at Claremont ever since. The author of textbooks on cell biology and on plants, genes and agriculture, Professor Sadava has done research in many areas of cell biology and biochemistry, ranging from developmental biology, to human diseases, to pharmacology. His current research concerns human lung cancer and its resistance to chemotherapy. Virtually all of the research articles he has published have undergraduates as coauthors. Professor Sadava teaches introductory biology and has recently developed a new course on the biology of cancer. For the last 15 years, Dr. Sadava has been a visiting professor in the Department of Molecular, Cellular, and Developmental Biology at the University of Colorado, Boulder.

Preface

This is an exciting time to be a biologist: our knowledge of living systems is expanding rapidly and our technologies for research improve daily. This fifth edition of *Life: The Science of Biology* has been an opportunity for us to communicate to students the excitement of modern biology by expanding and refining our coverage, by finding new ways to make important concepts more understandable and memorable, and by conveying the sense of adventure in biological research.

Our overriding goal continues to be to stimulate students' interests in biology. We have tried to do this by making underlying concepts clear and easy to grasp and showing their relevance to medical, agricultural, and environmental issues. Also, we want students to appreciate *how* we know rather than just *what* we know. To that end, we discuss scientific methods and show how experiments, field observations, and comparative methods help biologists formulate and test hypotheses. In the preparation of this edition, we have tried to introduce opportunities for students to think about concepts rather than just learning facts.

Themes and approaches that characterize the new edition

Throughout the book, we use several themes to link chapters and provide continuity. These themes, which are introduced in Chapter 1, include evolution, the experimental foundations of our knowledge, the flow of energy in the living world, the application and influence of molecular techniques, and human health considerations

One of our approaches is to show how basic principles presented in earlier chapters apply in later chapters. For example, programmed cell death, also called apoptosis, has been a major focus of biological research in the past few years. This process is first presented in the context of cell reproduction (Chapter 9). Then we show its applications in development (Chapter 15), cancer (Chapter 17), and the immune system (Chapter 18). Another example is cladistics, introduced first in Chapter 22, and applied in subsequent chapters to show how evolutionary relationships help us understand a wide variety of biological problems.

A new organization enhances accessibility

In chapter after chapter, we have concentrated on making the descriptions, explanations, and applications more accessible to student readers. We have rewritten obscure or difficult passages, deleted some details, simplified the writing and illustrations, and shortened both paragraphs and sections. We have tried to tighten connections, improve transitions, and sharpen the focus. Many changes have been made in how information is distributed among the text, captions, figure labels, and a new feature of the illustrations—"balloon captions."

We have also taken a new approach to headings. We have tried to offer the reader more guidance in identifying, understanding, and interrelating key topics. We use two levels of heads (although occasionally a third level is introduced). Major heads divide the chapters into discrete topics, and second-level heads, now full sentences, identify the explicit focus of each subsection. In addition to providing a clear outline and introduction to covered topics, these "sentence heads" are useful to students for study and review.

To further guide the reader, we have provided explicit forecasts of concepts about to be discussed, both as part of the introduction to each chapter and as part of the introductions to most of the major sections within each chapter. This forecasting allows students to read with expectation and direction, better equipped to appreciate the implications of early topics and to see relationships among topics across the entire chapter.

Different students have different learning styles: some are more image-focused, others more text-focused. Line drawings and photographs have the advantages of directness, emphasis, and drama; on the other hand, text explanations provide explicit information and better describe events that occur through time. We have combined the strengths of both text and graphics through the abundant use of what we're calling "balloon captions." These brief statements are incorporated directly into the graphics and go beyond mere labeling to describe, define, or explain graphic elements. Thus, text becomes more intimately related to graphic representations and the graphics take on more significance. Balloon captions, sometimes numbered to clarify a sequence, guide the reader through the inevitable complexities of some figures; in other figures, balloons emphasize the most important features. This new feature has drawn extensive praise during the development of this edition, and we believe that students will find them highly effective aids to their learning.

A new format for the chapter summaries emphasizes the chapter outline, using major heads to distinguish and identify summary statements. The summary emphasizes major points but also includes specific references to key figures and tables where supporting details are found.

The seven parts: Content, changes, and themes

Each section of the book has undergone important changes. In Part One, The Cell, we eliminated some details and advanced topics, notably in Chapter 6 (Energy, Enzymes, and Metabolism), allowing us to develop certain key concepts such as allostery and cooperativity more clearly. New developments in such areas as protein folding are now introduced in a broad context so the student can relate them to other topics. When appropriate, we have tried to link biochemical and cellular phenomena to specific conditions and diseases that affect human health and well-being.

In Part Two, Information and Heredity, the first six chapters (Chapters 9–14) describe what we know and how we have gained some of this knowledge, and the final four (Chapters 15–18) describe its biological applications. The expression of DNA is dealt with separately in prokaryotes (Chapter 13) and eukaryotes (Chapter 14), and these principles are then used to describe the molecular analysis of development (Chapter 15), the manufacture of useful products via biotechnology (Chapter 16), the diagnosis and treatment of human genetic diseases (Chapter 17), and the production of antibodies (18). Because of its centrality to genetics and molecular biology, we now devote separate chapters to the structure and the role of DNA (Chapters 11 and 12, respectively).

The chapter on development (Chapter 15) in Part Two now concentrates entirely on molecular and genetic aspects of development; the cellular and tissue aspects of embryology are presented in Chapter 40. In addition to applying the principles of molecular biology to recombinant DNA technologies, Chapter 16 emphasizes how these technologies are being used in agriculture and medicine. The "molecular revolution" that is just beginning in medicine, including the Human Genome Project, is the subject of an extensively updated chapter (Chapter 17).

In Part Three, Evolutionary Processes, we have expanded the treatment of cladistic methods to assess evolutionary relationships and show how cladograms are constructed and why knowing evolutionary relationships helps us better understand a wide array of biological problems, including human health problems. With this background, we are able to use phylogenetic trees in subsequent chapters to illustrate evolutionary patterns that range from individual molecules to phyla.

Part Three also includes an entirely new chapter (Chapter 23) on molecular evolution, one of the most exciting and vigorous fields in contemporary biology. Contributed by Peg Riley of Yale University, this chapter emphasizes both detailed molecular comparisons among species and their implications as to why and how molecules change over evolutionary time as organisms encounter and survive environmental challenges.

The results of molecular evolutionary studies have led us to a new emphasis on lineages in Part Four, The Evolution of Diversity, especially in our treatment of bacteria, archaea, and protists. Systematics is in ferment, and we try to impart some sense of current controversies in the field in Chapters 25 and 26. We explicitly treat today's diversity of organisms as the product of evolution.

In Part Five of the fourth edition, we introduced a new chapter, The Biology of Flowering Plants, on plant responses to environmental challenges. It was so well received that we have enriched it with an up-to-date treatment of plant–pathogen interactions. This topic and others continue to emphasize the theme of evolution. Part Five also includes new findings on multiple phytochromes and on developmental mutants in *Arabidopsis*.

In response to requests from instructors, Part Six, The Biology of Animals, now features a chapter (Chapter 40) on animal embryology, which follows the chapter on animal reproduction. The coverage of neurobiology (Chapters 41–44) has been redesigned and expanded to include a new chapter (Chapter 43) on the organization and higher functions of the mammalian brain.

Our theme of human health concerns is manifest throughout Part Six. Chapter 47, on animal nutrition, includes new material on environmental toxicology, an emerging discipline we feel will be of increasing importance to the well-being of our planet.

In Part Seven, Ecology and Biogeography, we have further expanded our coverage of the role of experiments in helping biologists understand the complex interactions among organisms that structure ecological systems. New materials illustrate the role of phylogenetic analyses in behavioral ecology and biogeography. In Chapter 54 we have designed an original graphic method of displaying material on Earth's biomes. This new and striking presentation enables students to visualize and quantify the differences and similarities in the dominant features of Earth's major biomes.

We wish to thank a lot of people

We were all students and teachers long before we were textbook authors, and we want to help students in every way possible. In the next section, "To the Student," we offer some advice that many of our own students have told us they found helpful.

Again, we have been fortunate to receive cogent and significant advice from the more than 60 colleagues who reviewed chapters or whole sections of the book. Their names are listed after this Preface. Their reviews helped to shape many of the changes described above, ranging from the addition of new chapters to the many ways in which we worked to sharpen our story. We thank them all, and hope this new edition measures up to their expectations.

We were already indebted to J/B Woolsey Associates for the elegance and effectiveness of the art programs they developed for the third and fourth editions of this textbook. They have, of course, produced many new illustrations for this edition. However, rather than limiting ourselves to incremental changes in the existing art program, we have taken a major step forward this time with the introduction of the balloon captions. The success of this approach is the result of many factors. James Funston worked with authors and illustrators, offering input to virtually every pixel in the entire art program. John Woolsey and a dedicated team of artists led by Michael Demaray turned our ideas and suggestions into exciting new art.

James Funston, the developmental editor we chose to work with us on this edition, paid close attention to clarity and pedagogical focus. Stephanie Hiebert provided rigorous copy editing from beginning to end. Her sharp eye extended to the illustrations, and her polishing of and additions to the balloon copy often enhanced the clarity of the presentation. Carol Wigg once again coordinated and checked every change made by editors, artists and authors—indeed, she coordinated the entire preproduction process, and she applied her knowledge and talent to writing captions that tightly link the illustations to key points in the text. We owe her more than we can say for her patience, persistence, and skill. Jane Potter, as photo researcher, found many new and exciting photographs to enhance the learning experience and enliven the appearance of the book as a whole.

We wish to thank the dedicated professionals in W. H. Freeman's marketing and sales group. Their efficiency and enthusiasm has helped bring *Life* to a wider audience. We appreciate their constant support and valuable marketing feedback. A large share of *Life*'s success is due to their efforts in this publishing partnership.

Sinauer Associates provided the best publishing environment we can imagine. Their years of success in publishing biology books at the introductory, intermediate, and advanced levels result from their ability to envision a product and to guide, assist, and motivate authors through the long, demanding process. Remarkably, Andy Sinauer never ceases to extend helpful, and, above all, warm support to his authors.

Bill Purves   *Gordon Orians*   *Craig Heller*   *David Sadava*

November, 1997

Reviewers for the Fifth Edition

Henry W. Art, Williams College

Carla Barnwell, University of Illinois

Judith L. Bronstein, University of Arizona

Robert J. Brooker, University of Minnesota

Steven B. Carroll, Northeast Missouri State University

James J. Champoux, University of Washington

William A. Clemens, University of California/Berkeley

Frederick M. Cohan, Wesleyan University

Newton Copp, The Claremont Colleges

D. Andrew Crain, University of Florida

Joe W. Crim, University of Georgia

Rowland H. Davis, University of California/Irvine

Patrick E. Elvander, University of California/Santa Cruz

Wayne R. Fagerberg, University of New Hampshire

Michael Feldgarden, Yale University

Rachel D. Fink, Mt. Holyoke College

Barbara Fishel, University of Arizona

William Fixsen, Harvard University

Cecil H. Fox, Molecular Histology, Inc.

Stephen A. George, Amherst College

Wayne Goodey, University of British Columbia

Deborah Gordon, Stanford University

David M. Green, McGill University

Adrian Hayday, Yale University

Joseph Heilig, University of Colorado

Walter S. Judd, University of Florida

Mark V. Lomolino, University of Oklahoma

Michael A. Lydan, University of Toronto/Erindale

Denis H. Lynn, University of Guelph

Laura MacIntosh, Stanford University

James Manser, Harvey Mudd College

John M. Matter, Juniata College

Larry R. McEdward, University of Florida

Michael Meighan, University of California/Berkeley

Melissa Michael, University of Illinois

Charles W. Mims, University of Georgia

Anthony G. Moss, Auburn University

Shahid Naeem, University of Minnesota

Peter Nonacs, University of California/Los Angeles

Barry M. O'Connor, University of Michigan

Ron O'Dor, Dalhousie University

Richard Olmstead, University of Washington

Laura J. Olsen, University of Michigan

Judith A. Owen, Haverford College

Randall W. Phillis, University of Massachusetts

Lorraine Pillus, University of Colorado

Ellen Porzig, Stanford University

Thomas L. Poulson, University of Illinois at Chicago

Loren Reiseberg, Indiana University

Wayne C. Rosing, Middle Tennessee State University

Albert Ruesink, Indiana University

C. Thomas Settlemire, Bowdoin College

Joan Sharp, Simon Fraser University

Esther Siegfried, Pennsylvania State University

Anne Simon, University of Massachusetts

Mitchell L. Sogin, Marine Biological Laboratory, Woods Hole

Collette St. Mary, University of Florida

Millard Susman, University of Wisconsin

Elizabeth Vallen, Swarthmore College

Elizabeth Van Volkenburgh, University of Washington

Gary Wagenbach, Carleton College, Minnesota

Bruce Walsh, University of Arizona

Mark Wheelis, University of California/Davis

Brian White, Massachusetts Institute of Technology

Fred Wilt, University of California/Berkeley

Gregory A. Wray, State University of New York at Stony Brook

To the Student

Welcome to the study of life! In our student days—and ever since—we have enjoyed studying the fascinating and fast-changing field of biology, and we hope that you will, too.

There are a few things you can do to help you get the most from this book and from your course. For openers, read the book actively—don't just read passively, but do things that force you to think as you read. If we pose questions, stop and think about them. If a passage reminds you of something that has gone before, think about that, or even check back to refresh your memory. Ask questions of the text as you go. Do you understand what is being said? Does it relate to something you already know? Is it supported by experimental or other evidence? Does that evidence convince you? How does this passage fit into the chapter as a whole? Annotate the book—write down comments in the margins about things you don't understand, or about how one part relates to another, or even when you find an idea particularly interesting. The point of doing these things is that they will help you learn. People remember things they think about much better than they remember things they have read passively. Highlighting is passive; copying is drudge work; questioning and commenting are active and well worthwhile.

"Read" the illustrations actively too. You will find the balloon captions in the illustrations especially useful—they are there to guide you through the complexities of some topics and to highlight the major points.

The chapter summaries will help you quickly review the high points of what you have read. A summary identifies particular illustrations that you should study to help organize the material in your mind. It is essential that you study the cited illustrations and their captions as you review because important information that is covered in illustrations has been left out of the summary statements. Add concepts and details to the framework by reviewing the text. A way to review the material in slightly more detail after reading the chapter is to go back and look at the boldfaced terms. You can use the boldfaced terms to pose questions—and see if you can answer those questions. The boldfacing will probably be more useful on a second reading than on the first.

Use the self-quizzes and "Applying Concepts" questions at the end of each chapter. The self-quizzes are meant to help you understand some of the more detailed material and to help you sort out the information we have laid before you. Answers to all self-quizzes are in the back of the book. The concept questions, on the other hand, are often fairly open-ended and are intended to cause you to reflect on the material.

Two parts of a textbook that are, unfortunately, often underused or even ignored are the glossary and the index. Both can help you a great deal. When you are uncertain of the meaning of a term, check the glossary first—there are more than 1,500 definitions in it. If you don't find a term in the glossary, or if you want a more thorough discussion of the term, use the index to find where it's discussed.

What if you'd like to pursue some of the topics in greater detail? At the end of each chapter there is a short, annotated list of supplemental readings. We have tried to choose readings from books and magazines, especially *Scientific American*, that should be available in your college library.

To provide another kind of help for students, we commissioned a CD-ROM (*Life 5.0*) covering the subject matter of Parts One and Two of this textbook. *Life 5.0* introduces and illustrates (often with unique animations) over 1700 key terms and concepts. You can access this information in several ways: via *Life* chapter reviews; via minicourses such as "Molecular Structure," "The Cell Cycle," and "DNA Replication"; or via a hyperlinked index. There are also several hundred self-quiz items and dozens of thought problems. You may have a copy of the disk inside the front cover of this book; if not, and if you would like to purchase one, contact **www.mona-group.com**. If you use the disk, explore its contents to see which of its tools best correspond to your needs.

Most students occasionally have difficulty in courses, including biology courses. If you find that you are slipping behind in the course, or if a particular topic is giving you an unreasonable amount of trouble, here are some useful steps you might take. First, the basics: attend class, take careful lecture notes, and read the textbook assignments. Second, note that one of the most important roles of studying is to discover what you don't know, so that you can do something about it. Use the index, the glossary, the chapter summaries, and the text itself to try to answer any questions you have and to help you organize the material. Make a habit of looking over your lecture notes within 24 hours of when you take them—find out right away what points are unclear, and get them straightened out in your mind. The CD-ROM can help by providing a different perspective.

If none of these self-help remedies does the trick, get help! Other students are often a good source of help, because they are dealing with the material at the same level as you are. Study groups can be very useful, as long as the participants are all committed to learning the material. Tutors are almost always helpful and useful, as are faculty members. The main thing is to get help when you need it. It is not a good idea to be strong and silent and drift into a low grade.

But don't make the grade the point of this or any other course. You are in college to learn, to pursue interesting subjects, and to enjoy the subjects you are pursuing. We hope you'll enjoy the pursuit of biology.

Bill Purves   Gordon Orians   Craig Heller   David Sadava

Contents in Brief

Contents

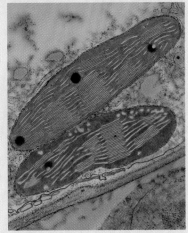

Part One

The Cell

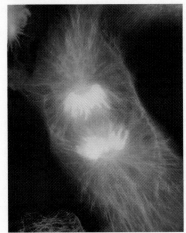

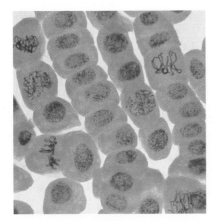

Part Three

Evolutionary Processes

Chapter 1

An Evolutionary Framework for Biology

We live on an ancient planet. People in some cultures believe that Earth has always existed—that it is eternal. In the Western world, however, people have long believed that Earth had a beginning, and a relatively recent one. In 1650 Irish Archbishop James Ussher, estimating from his close study of the Bible, calculated that Earth was created in 4004 B.C. Although not everyone agreed with his calculations, until the nineteenth century most people in the Western world shared Bishop Ussher's view that Earth was relatively young and that its entire history was chronicled in ancient texts.

During the nineteenth century, geologists and biologists accumulated evidence that Earth was much older, although they could not say exactly how old. Their evidence for an ancient Earth came primarily from the remains of organisms found in sedimentary rocks. The geologists' guiding concepts were simple: Rocks form slowly by the piling up of sediments, and younger rocks are deposited on top of older ones. A great canyon carved into sedimentary rocks may have a visible record of more than a billion years.

Preserved within some rocks were **fossils**—the remains of organisms that lived while the sediments were accumulating. When they compared older rocks with younger ones, geologists could detect slight but significant differences among similar fossil organisms. Furthermore, they found fossils of similar organisms at widely separated locations. By assuming that rocks at different locations containing the same type of fossil were of approximately the same age, early geologists determined the general order of events in the history of life on Earth. Although they could establish a sequence, these geologists had no method for determining the absolute ages of fossils.

One of the triumphs of twentieth-century science has been the development of methods to date materials formed in the past. The discovery that unstable forms (radioactive isotopes) of familiar atoms such as carbon and phosphorus decay at constant rates made it possible to date materials. Radioactive isotopes are incorporated into rocks and fossils in proportion to their presence in the environment when the rock solidified. Each type of radioactive isotope then begins to decay at its own constant rate, eventually becoming stable. Scientists can calculate the absolute ages of rocks from the proportions of radioactive and stable isotopes they contain.

Meanwhile, scientists in the fields of astronomy and physics, using data from the powerful telescopes and space probes that became available in the latter half of the twentieth century, have come to believe that our planet formed approximately 4 billion years ago. The earliest known fossils have been dated, using radioisotopes, as being 3.8 billion years old, so we know that life arose early in the history of Earth.

Evolutionary Milestones

The fullness of time is difficult for people to grasp. We all understand time spans measured in seconds, minutes, hours, days, years, and decades, but we find it difficult to comprehend millions, much less billions, of years. The following overview of the major evolutionary milestones is intended to provide a framework that presents life's characteristics as they will be covered in this book, and an overview of how these characteristics evolved during the history of life on Earth.

Life arose from nonlife

The first life must have come from nonlife. All matter, living and nonliving, is made up of chemicals. The smallest chemical units are atoms, which bond together into molecules (the properties of these molecules are the subject of Chapter 2). We think that the processes leading to life began nearly 4 billion years ago with interactions among small molecules that stored information in easy-to-copy sequences.

Chemical information became more complex when the information stored in these simple sequences resulted in the synthesis of larger molecules with complex but relatively stable shapes. Because they were both complex and stable, these molecules could participate in increasing numbers and kinds of chemical reactions. Certain types of large molecules—carbohydrates, lipids, proteins, and nucleic acids—are formed only by living systems, and they are found in all living systems. The properties and functioning of these complex *organic molecules* are the subject of Chapter 3.

About 3.8 billion years ago interacting systems of molecules came to be enclosed in compartments surrounded by *membranes*. Within these membrane-enclosed units, or *cells*, control was exerted over the entrance and retention of molecules, the chemical reactions taking place within the cell, and the exit of molecules. Cells and membranes are the subjects of Chapters 4 and 5.

Cells are so effective at capturing energy and replicating themselves—two fundamental characteristics of life—that since they evolved, cells have apparently outcompeted any noncellular life. The cell is the unit on which all life has been built.

An Englishman, Robert Hooke, built a simple microscope in 1665 and was the first person to observe cells. The cells he saw were those of cork, wood, and other dead plant materials, and they were empty. Living organisms that were fully contained in a single cell were first observed a few years later by the Dutch naturalist Antoni van Leeuwenhoek.

By 1839, microscopes had improved and enough living material had been observed that the German physiologist Theodor Schwann could assert that *all organisms consist of cells*. In 1858, the German physician Rudolf Virchow suggested that *all cells come from preexisting cells*. Experiments by the French chemist and microbiologist Louis Pasteur between 1859 and 1861 convinced most scientists that cells do not arise from noncellular material, but must come from other cells. In the modern world, life no longer arises from nonlife.

The first organisms were single cells

For 2 billion years all cells were small. They lived mostly autonomous lives, each separate from the other. Their lives were confined to the oceans, where they were shielded from lethal ultraviolet light. The relatively small amounts of genetic information that allowed these **prokaryotic cells** to replicate themselves and the biochemical machinery by which they obtained energy floated loose within an outer membrane. Some prokaryotes living today are similar to those that existed early in the evolution of living cells, several billion years ago (Figure 1.1).

To maintain themselves, to grow, and to reproduce, all organisms—whether they consist of one cell or tril-

1.1 Early Life May Have Resembled These Cells These "rock-eating" bacteria, appearing red in the artificially colored micrograph, were discovered in pools of water trapped between layers of rock more than 1,000 meters below Earth's surface. Deriving chemical nutrients from the rocks and living in an environment devoid of oxygen, they may resemble some of the earliest prokaryotic cells.

lions of cells—must obtain raw materials and energy from the environment. These raw materials are chemicals that are digested; the products are used to synthesize large carbon-based molecules. The energy obtained from chemical digestion is used to power the synthetic reactions. These conversions of matter and energy are called **metabolism**.

All organisms can be viewed as devices to capture, process, and convert matter and energy from one form to another; these conversions are the subjects of Chapters 6 and 7. *A major theme in the evolution of life is the development of increasingly diverse ways of capturing external energy and using it to drive biologically useful reactions.*

The earliest cells derived their energy from simple chemical compounds because complex molecules were scarce in their environment. On early Earth, volcanoes poured large quantities of methane and hydrogen sulfide into the atmosphere. Early prokaryotes evolved the ability to ingest these molecules and use them as sources of energy.

Photosynthesis and sex changed the course of evolution

Two powerful evolutionary events took place in the first billion years. One, the evolution of photosynthesis, created new metabolic pathways. The other, the evolution of sex, stimulated the evolution of the almost unimaginable diversity of organisms on Earth.

PHOTOSYNTHESIS CHANGED EARTH'S ENVIRONMENT. About 2.5 billion years ago, some prokaryotes evolved the ability to use the energy of sunlight to power their metabolism. Although raw chemicals were still taken up from the environment, the energy used to metabolize these chemicals came directly from the sun.

The early photosynthetic prokaryotes were probably similar to present-day cyanobacteria (Figure 1.2). The energy-capturing process they used—**photosynthesis**—is the basis of nearly all life on Earth today. As you will learn in Chapter 8, photosynthesis is a complex process made up of many chemical reactions. The ability to perform the photosynthetic reactions probably accumulated gradually during the first billion years or so of evolution, but once the ability had evolved, the effects of photosynthesis were dramatic.

Photosynthetic prokaryotes were so successful that they released vast quantities of oxygen gas (O_2) into the atmosphere. The presence of oxygen opened up new avenues of evolution. Metabolic pathways based on O_2—*aerobic metabolism*—came to be used by most organisms on Earth. The air we breathe today would not exist without photosynthesis.

Over a much longer time frame, the vast quantities of oxygen liberated by photosynthesis had another effect. A form of oxygen we call ozone (O_3) began to accumulate along with the O_2 in the atmosphere. The ozone slowly formed a dense layer that acted as a shield, intercepting much of the sun's deadly ultraviolet radiation. Eventually (although only within the last 800 million years of evolution) the presence of this shield allowed organisms to leave the protection of the ocean and find new lifestyles on Earth's land surfaces.

SEX CHANGED EVOLUTIONARY RATES. The earliest unicellular organisms reproduced by dividing. Progeny cells were identical to parent cells. But **sexual recombination**—the combining of genes from two cells in one cell—appeared early during the evolution of life. Sex is advantageous because an organism that receives genetic information from another individual produces

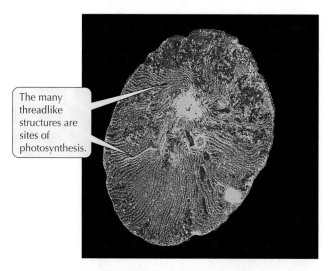

The many threadlike structures are sites of photosynthesis.

1.2 Oxygen Produced by Prokaryotes Changed Earth's Atmosphere This modern cyanobacterium is probably very similar to early photosynthetic prokaryotes.

offspring that are more variable. *Reproduction with variation is a major characteristic of life.* Because environments continuously vary, individuals that produce variable offspring rather than genetically identical "clones" are more likely to produce at least some offspring that *adapted* to changes in the environment.

Adaptation to environmental change is one of life's most distinctive features. An organism is adapted to its environment when it possesses features that enhance its survival and ability to reproduce in a given environment. Sex is so adaptive that today nearly all organisms on Earth engage in sex at least occasionally. By creating increased variation, sexual recombination increased the rate of evolutionary change.

Early prokaryotes engaged in sex (exchanging genetic material) and reproduction (cell division) at different times. Even today in many unicellular organisms, sex and reproduction occur at different times (Figure 1.3). But a different kind of organism evolved that would require a more complicated sex life.

Eukaryotes are "cells within cells"

As the ages passed, some prokaryotic cells became large enough to attack and consume smaller cells, becoming the first *predators*. Usually the smaller cells were destroyed within the predators' cells, but some of these smaller cells survived and became permanently integrated into the operation of their hosts' cells. In this manner, cells with complex internal compartments arose. We call these cells **eukaryotic cells**. Their appearance slightly more than 1.5 billion years ago opened more new evolutionary pathways.

Prokaryotic cells—including all the early bacteria and archaea—have only one membrane, the one that surrounds them. Eukaryotic cells, on the other hand, are filled with membrane-enclosed compartments. In eukaryotic cells, genetic material—*genes* and *chromo-*

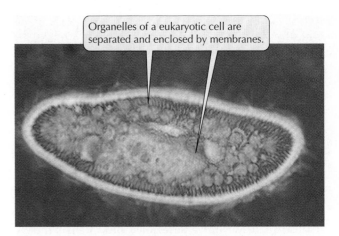

Organelles of a eukaryotic cell are separated and enclosed by membranes.

1.4 Multiple Compartments Characterize Eukaryotic Cells The nucleus and other specialized compartments (known as organelles) probably evolved from small prokaryotes that were ingested by a larger prokaryotic cell.

somes—became contained within a discrete **nucleus** and became increasingly complex. Other compartments became specialized for other purposes, such as photosynthesis (Figure 1.4). We refer to these specialized compartments as **organelles**.

Cells evolved the ability to change their structures and specialize

Until slightly more than 1 billion years ago, only unicellular organisms existed. Two key developments made the evolution of **multicellular organisms**—organisms consisting of more than one cell—possible. One was the ability of a cell to change its structure and functioning to meet the challenges of a changing environment. Prokaryotes accomplished this when they evolved the ability to change from rapidly growing cells into resting *spores* that could survive harsh environmental conditions. The second development allowed cells to stick together after they divided, forming a multicellular organism.

Once organisms began to be composed of many cells, it became possible for the cells to specialize. Certain cells, for example, could be specialized to perform photosynthesis. Other cells might become specialized to transport chemical raw materials, such as oxygen, from one part of an organism to another. Very early in the evolution of multicellular life, certain cells began to be specialized for sex. As multicellular life evolved, sex and reproduction became linked. In almost all present-day multicellular organisms, sex and reproduction occur together.

With more complicated and specialized sex cells, sex itself became more complicated. Simple cell division, which we know as **mitosis**, was and is sufficient for the needs of most cells. But a whole new method of cell division—**meiosis**—evolved that opened up new

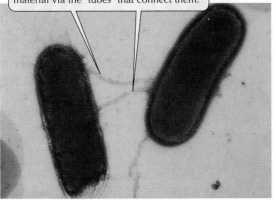

These two bacteria are exchanging genetic material via the "tubes" that connect them.

1.3 Sex Between Prokaryotes Genetic exchange produces variation that leads to adaptive evolution.

1.5 Organisms May Change Dramatically during Their Lives The caterpillar, pupa, and adult are all stages in the life cycle of a monarch butterfly. The transition from one stage to another is triggered by internal signals.

realms of recombination possibilities for the specialized sex cells, or *gametes*. Mitosis and meiosis are explained and compared in Chapter 9.

The cells of an organism are constantly adjusting

Both the emergence of multicellular life and the changes in Earth's atmosphere that allowed life to move out of the oceans and exploit the environments of the land masses quickened the pace of evolution. Photosynthetic green plants colonized the land and provided a rich source of energy for a vast array of organisms that consumed them. But whether it is made up one cell or many, an organism must respond appropriately to many signals emanating from its external and internal environments.

The external environment can change rapidly and unpredictably in ways that are outside of the organism's control. An organism can remain healthy only if its internal environment remains within a given range of physical and chemical conditions. Organisms maintain a relatively constant internal environment by adjusting their metabolism in response to external and internal signals indicating such things as a change in temperature, the presence or absence of sunlight, the presence or absence of specific chemicals, the need for nutrients (food) and water, or the presence of a foreign agent inside the organism's body. Maintenance of a relatively stable internal condition is called **homeostasis**.

The adjustments that organisms make to maintain constant internal conditions are usually minor; they are not obvious, because nothing appears to change. However, at some time during their lives many organisms respond to signals not by maintaining their status, but by undergoing major physical reorganization. We mentioned in the previous section the ability of prokaryotes to change from rapidly growing cells into dormant spores in response to environmental stresses. A striking example that evolved much later is *metamorphosis*, seen in many modern insects, such as butterflies. In response to internal chemical signals, a caterpillar changes into a pupa and then into an adult butterfly (Figure 1.5).

A major theme in the evolution of life is the development of increasingly complicated systems for responding to sig-

nals and maintaining homeostasis. Indeed, some animals exhibit a widespread and important biological process called *learning*, in which important changes in the internal environment result from responses to external signals (such as this textbook).

Multicellular organisms develop and grow

Multicellular organisms cannot achieve their adult shapes or function effectively unless their growth is carefully regulated. Uncontrolled growth, one example of which is cancer, ultimately destroys life. The functioning of a multicellular organism requires a sequence of events leading from a single cell to a multicellular adult. *A vital characteristic of living organisms is regulated growth.*

The activation of information within cells, and the exchange of information among many cells, produce the well-timed events that are required by the transition from single cell to adult form. Genes control the metabolic processes necessary for life. The astounding nature of the genetic material that controls these lifelong events has been understood only within the twentieth century; it is the story to which much of Part Two of this book is devoted.

Altering the timing of developmental processes can produce striking changes. Just a few genes can control processes that result in dramatically different adult organisms. Chimpanzees and humans share more than 97 percent of their genes, but the differences between

1.6 Genetically Similar but Quite Distinct By looking at the two, you would never guess that chimpanzees and humans share more than 97 percent of their genes.

the two in form and in behavioral abilities, most notably speech, are dramatic (Figure 1.6). When we realize how little information it sometimes takes to create major transformations, the still-mysterious process of *speciation* becomes a little less of a mystery.

Speciation has resulted in the diversity of life

All organisms on Earth today are the descendants of a unicellular organism that lived almost 4 billion years ago. The preceding sections of this chapter described the changes that led to the diversity present in life today. The course of this evolution has been accompanied by the storage of larger and larger quantities of information and increasingly complex mechanisms for using it. But if that were the entire story, only one kind of organism might exist on Earth today. Instead, Earth is populated by many millions of kinds of organisms that are genetically different from one another. We call these genetically independent groups **species**.

As long as individuals within a population mate at random and reproduce, structural and functional changes may occur, but only one species will exist. However, if a population becomes divided into two or more groups, individuals can mate only with individuals in their own group. When this happens, differences may accumulate with time, and the groups may evolve into different species.

The splitting of groups of organisms into separate species has resulted in the great richness and variety of life found on Earth today, as described in Chapter 19. How species form is explained in Chapters 20 and 21. From a single ancestor, many species may arise as a result of the repeated splitting of populations. How bi-

ologists determine which species have descended from a particular ancestor is discussed in Chapter 22.

Sometimes humans refer to species as "primitive" or "advanced." These and similar terms, such as "lower" and "higher," are best avoided because they imply that some organisms function better than others. The abundance of prokaryotes—all of which are relatively simple—readily demonstrates that they are highly functional, despite their relative simplicity. Therefore, in this book, we usually use the terms "ancestral" and "derived" to describe characteristics that appeared earlier and later in the evolution of lineages of life, respectively, recognizing that all organisms that have survived are successfully adapted to their environments. The wings that allow a bird to fly or the structures that allow green plants to survive in environments where water is either scarce or overabundant are examples of the rich array of adaptations found among organisms (Figure 1.7).

Biological Diversity: Domains and Kingdoms

As many as 30 million species of organisms inhabit Earth today. Many times that number lived in the past but are now extinct. To help us understand the past and current diversity of organisms, biologists use classification systems that group organisms according to their evolutionary relationships.

In the classification system used by most modern biologists, organisms are grouped into three **domains** and six **kingdoms** (Figure 1.8). Organisms belonging to a particular domain have been evolving separately from organisms in other domains for more than a billion years. Organisms in the domains **Archaea** and **Bacteria** are prokaryotes, single cells that lack a nucleus and other internal compartments found in the cells from other kingdoms.

Archaea and Bacteria differ so fundamentally from one another in the chemical pathways by which they function and in the products they produce that they are believed to have separated into distinct evolutionary lineages very early during the evolution of life. Each of these domains consists of a single kingdom, Archaebacteria and Eubacteria, respectively. These kingdoms are covered in Chapter 25.

Members of the other domain—**Eukarya**—have eukaryotic cells with nuclei and complex cellular compartments called *organelles*. The Eukarya are divided into four kingdoms—Protista, Plantae, Fungi, and Animalia. The kingdom **Protista** (protists), the subject of Chapter 26, contains mostly single-celled organisms. The remaining three kingdoms, nearly all of whose members are multicellular, are believed to have arisen from ancestral protists.

Most members of the kingdom **Plantae** (plants) convert light energy to chemical energy by photosynthesis.

1.7 Adaptations to the Environment   (a) The long, pointed wings of the peregrine falcon allow it to accelerate rapidly as it dives on its prey. (b) The wings of an Arctic tern allow it to hover above the water while searching for fish. (c) In a water-limited environment, this saguaro cactus stores water in its fleshy trunk. Its roots spread broadly to quickly extract water immediately after it rains. (d) The above-ground root system of red mangroves is a modification that allows them to thrive while inundated by water—an environment that would kill most terrestrial plants.

The biological molecules that they and some protists synthesize are the primary food for nearly all other living organisms. This kingdom is covered in Chapter 27.

The kingdom **Fungi**, the subject of Chapter 28, includes molds, mushrooms, yeasts, and other similar organisms, all of which are *heterotrophs*—that is, they require a food source of energy-rich molecules synthesized by other organisms. Fungi absorb food substances from their surroundings and digest them within their cells. Many are important as decomposers of the dead bodies of other organisms.

Members of the kingdom **Animalia** (animals) are heterotrophs that ingest their food source, break down (digest) the food outside their cells, and then absorb the products. Animals eat other forms of life for their raw materials and energy. Perhaps because we are animals ourselves, we are often drawn to study members of this kingdom, which is covered in this book in Chapters 29 and 30.

Biologists recognized that organisms were adapted for life in differing environments long before they understood how adaptation came about. Nearly 150 years ago, Charles Darwin and Alfred Russel Wallace proposed the first scientifically testable theory about adaptation. Their suggestion—that *adaptation is the result of evolution by natural selection*—has guided biological investigations ever since.

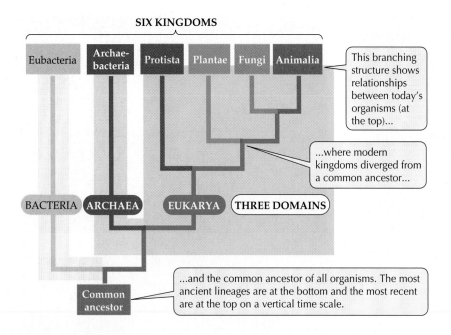

SIX KINGDOMS

Eubacteria | Archae-bacteria | Protista | Plantae | Fungi | Animalia

This branching structure shows relationships between today's organisms (at the top)...

...where modern kingdoms diverged from a common ancestor...

BACTERIA | ARCHAEA | EUKARYA | THREE DOMAINS

Common ancestor

...and the common ancestor of all organisms. The most ancient lineages are at the bottom and the most recent are at the top on a vertical time scale.

1.8 Domains and Kingdoms In the classification system used in this book, Earth's organisms are divided into three domains and six kingdoms.

The World Into Which Darwin Led Us

Long before scientists understood how biological evolution happened, they suspected that living organisms had evolved from organisms no longer alive on Earth. In the 1760s, the French naturalist Count George-Louis Leclerc de Buffon (1707–1788) wrote his *Natural History of Animals*, which contained a clear statement of the possibility of evolution.

Buffon originally believed that each species had been divinely created for different ways of life, but as he studied animal anatomy, doubts arose. He observed that the limb bones of all mammals, no matter what their way of life, were remarkably similar in many details (Figure 1.9). Buffon also noticed that the legs of certain animals, such as pigs, have toes that never touch the ground and appear to be of no use. He found it difficult to explain the presence of these seemingly useless small toes by special creation.

However, both these troubling facts could be explained if mammals had not been specially created in their present forms but had been modified from a common ancestor. Buffon suggested that the limb bones of mammals might all have been inherited, and that pigs might have functionless toes because they inherited them from ancestors with fully formed and functional toes. This was an early statement of evolution (descent with modification), although Buffon did not attempt to explain how such changes took place.

Buffon's student Jean Baptiste de Lamarck (1744–1829) wrote extensively about evolution and was the first person to propose a mechanism of evolutionary change. Lamarck suggested that lineages of organisms may change gradually over many generations as

offspring inherit structures that have become larger and more highly developed as a result of continued use or, conversely, have become smaller and less developed as a result of disuse.

For example, Lamarck suggested that aquatic birds extend their toes while swimming, stretching the skin between them. This stretched condition, he thought, could be inherited by the offspring, who would further stretch their skin during their lifetimes and would also pass this condition along to their offspring. According to Lamarck, birds with webbed feet would thereby evolve over a number of generations. He explained many other examples of adaptations in a similar way.

Today scientists do not believe that evolutionary changes are produced by inheritance resulting from use and disuse, as Lamarck suggested. But Lamarck did realize that species evolve with time, and his ideas

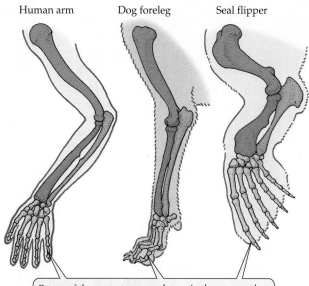

Human arm | Dog foreleg | Seal flipper

Bones of the same type are shown in the same color.

1.9 All Mammals Have Similar Limb Bones Mammalian forelimbs have different purposes: Humans use theirs for manipulating objects, dogs use theirs for walking, and seals use theirs for swimming. But the number and type of their bones are similar, indicating that they have been modified over time from a common ancestor.

deserved more attention than they received from his contemporaries, most of whom believed in a relatively young and unchanging universe. After Lamarck, other naturalists, scientists, and thinkers speculated that with time species change.

By 1858, the climate of opinion (among biologists, at least) was receptive to a theory of evolutionary processes proposed independently by Charles Darwin and Alfred Russel Wallace. By then geologists had shown that Earth had changed over millions of years, not merely a few thousand years. Thus, the presentation in the latter half of the nineteenth century of a well-documented and thoroughly scientific argument for evolution triggered a transformation of biology.

Darwin initiated the scientific study of evolution

Charles Darwin (1809–1882) based his approach to evolution on the following hypotheses:

1. Earth is very old, and organisms have been changing steadily throughout the history of life.
2. All organisms are descendants of a single common ancestor.
3. Species multiply by splitting into daughter species; such speciation has resulted in the great diversity of life found on Earth.
4. Evolution proceeds via gradual changes in populations, not by the sudden production of individuals of dramatically different types.
5. The major agent of evolutionary change is natural selection.

These five hypotheses have all been supported by the mass of research that has been conducted since Darwin published his book *The Origin of Species* in 1859.

Darwin's major insight was to perceive the significance of facts that were familiar to most of his fellow biologists. He understood that populations of all species have the potential for exponential increases in numbers. To illustrate this point, Darwin used the following example:

Suppose ... there are eight pairs of birds, and that only four pairs of them annually ... rear only four young, and that these go on rearing their young at the same rate, then at the end of seven years (a short life, excluding violent deaths for any bird) there will be 2048 birds instead of the original sixteen.

Charles Darwin

1.10 Many Organisms Have High Birth and Death Rates
This pair of sergeant majors is guarding the thousands of reddish eggs the female laid in a compact mat. If all the offspring of sergeant majors grew to adulthood and reproduced, the world's population would be overwhelming within a few years. However, most of these eggs and the fish that hatch from them will not survive.

Yet such rates of increase are rarely achieved in nature; the numbers of individuals of most species are relatively stable through time. Therefore, death rates in nature must be high (Figure 1.10). Without high death rates, even the most slowly reproducing species would quickly reach enormous population sizes.

Darwin also observed that, although offspring tend to resemble their parents, the offspring of most organisms are not identical to each other or to their parents (Figure 1.11). He suggested that slight variations among individuals significantly affect the chance that a given individual will survive and reproduce. He called this differential reproductive success of individuals **natural selection**.

Darwin probably used the words "natural selection" because he was familiar with the artificial selection practices of animal and plant breeders. Many of Darwin's observations on the nature of variation came from domesticated plants and animals. Darwin himself was a pigeon breeder, and he knew firsthand the astonishing diversity in color, size, form, and behavior that could be achieved by artifical human selection of which pigeons to mate (Figure 1.12). He recognized close parallels between artificial selection by breeders and selection in nature.

1.11 Offspring Differ from Their Parents These ducklings are members of a brood that hatched from a single clutch of eggs laid by this patterned female and fathered by a single male. Genetic variability among offspring of two parents is the norm.

Darwin states his case for natural selection

In *The Origin of Species* Darwin argued his case for natural selection:

> How can it be doubted, from the struggle each individual has to obtain subsistence, that any minute variation in structure, habits or instincts, adapting that individual better to the new conditions, would tell upon its vigour and health? In the struggle it would have a better chance of surviving; and those of its offspring which inherited the variation, be it ever so slight, would have a better chance.

That statement, written more than 100 years ago, still stands as a good expression of the idea of evolution by natural selection. Since Darwin wrote these words, biologists have developed a much deeper understanding of the genetic basis of evolutionary change and have assembled a rich array of examples of how natural selection acts.

Biology began a major conceptual shift a little more than a century ago with the general acceptance of long-term evolutionary change and the recognition that natural selection is the primary agent that adapts organisms to their environments. The shift took a long time because it required abandoning many components of an earlier worldview. The pre-Darwinian view held that the world was young, and that organisms had been created in their current forms. In the Darwinian view, the world is ancient, and both Earth and its inhabitants have been changing from forms very different from the ones they now have.

Accepting this new world view means accepting not only the processes of evolution, but also the view that the living world is constantly evolving, but without any "goals." The idea that evolutionary change is not directed toward a final goal or state has been more difficult for some people to accept than the process of evolution itself.

Asking the Questions "How?" and "Why?"

Biological processes and products can be viewed from two different but complementary perspectives. We ask functional questions: How does it work? We also ask adaptive questions: Why has it evolved to work in that way?

Suppose, for example, some marine biologists out on their research vessel are suddenly surrounded by dolphins leaping completely out of the ocean (Figure 1.13). Two obvious questions to ask are, *How* do these marine mammals achieve such a jump?, and *Why* do they do it? An answer to the how question would deal with the molecular mechanisms underlying muscular contraction, nerve and muscle interactions, and the receipt of stimuli by the dolphins' brains.

An investigation to answer the why question would attempt to determine why leaping out of the water is

1.12 Many Types of Pigeons Have Been Produced by Artificial Selection
Charles Darwin, who raised pigeons as a hobby, saw similar forces at work in artificial and natural selection.

1.13 How and Why do Dolphins Leap? Scientists from different disciplines focus on only one of these questions, and their answers are certain to be very different.

adaptive—that is, why it improves the survival and reproductive success of dolphins. For this particular instance, the why question is more obscure, and scientists still have no clear-cut answer. Some animals, such as frogs, evolved jumping behavior because it increased their chances of escape from predators; for others, such as mountain goats, jumping allowed them to cross obstacles like ravines and ridges. Neither of these explanations seems to apply to dolphins, who do not encounter barriers in the ocean and who jump when no predators are chasing them. Jumping may allow them to see better, or it may help dislodge parasites from their bodies; but neither of these possible benefits is established. Perhaps *you* can come up with an explanation of the behavior; in the meantime, scientists (along with most other humans) will enjoy watching it.

Is either of the two types of question more basic or important than the other? Is any one of the answers more fundamental or more important than the others? Not really. The richness of possible answers to apparently simple questions makes biology a complex science, but also an exciting field. Whether we're talking about molecules bonding, cells dividing, blood flowing, dolphins leaping, or forests growing, we are constantly posing both how and why questions. To answer these questions, scientists generate hypotheses that can be tested.

Hypothesis testing guides scientific research

Underlying all scientific research is the **hypothetico-deductive approach** with which scientists ask ques-

tions and test answers. This method allows scientists to modify and correct their beliefs as new observations and information become available. The method has five stages:

1. Making observations.
2. Asking questions.
3. Forming **hypotheses**, which are tentative answers to the questions.
4. Making predictions based on the hypotheses.
5. Testing the predictions by making additional observations or conducting experiments. The additional data gained may support or contradict the predictions being tested.

If the data support the hypothesis, it is subjected to additional predictions and tests. If they continue to support it, confidence in its correctness increases and the hypothesis comes to be considered a theory. If the data do not support the hypothesis, it is abandoned or modified in accordance with the new information. Then new predictions are made, and more tests are conducted. An example will illustrate this process.

APPLYING THE HYPOTHETICO-DEDUCTIVE METHOD. Biologists have long known that some caterpillars are conspicuously colored but that other caterpillars blend in with their backgrounds (Figure 1.14). Conspicuously colored caterpillars often live in groups but are seldom attacked by birds. These initial observations suggested questions that were used to develop hypotheses, make predictions, and devise an experiment to test the predictions. Let's examine how this was done.

GENERATING A HYPOTHESIS. The bright colors of some caterpillars, together with the observation that potential predators usually avoid brightly colored caterpillars, suggested a question that became a hypothesis: The bright color patterns of these caterpillars signal to potential predators that the caterpillars are distasteful or toxic. A companion hypothesis is that inconspicuous caterpillars are good to eat (palatable), and their coloration thus reduces the chance that predators will discover and eat them.

For each hypothesis of an effect there is a corresponding **null hypothesis** asserting that the proposed effect is absent. The null hypothesis for the hypotheses we have just stated is that there is no difference in palatability between colorful and camouflaged caterpillars.

Notice that these hypotheses depend on certain assumptions or on previous knowledge. For example, we assume that birds have color vision and can learn about the qualities of their prey by encountering and tasting them. If such assumptions are uncertain, they should be tested before other experiments are performed.

1.14 Caterpillars Can Be Easy or Hard to See (a) Many caterpillars blend into their surroundings, like these catocala moth larvae camouflaged by the bark of an oak tree. (b) This colorful butterfly larva, a stinging rose, contrasts with its leafy environment. Its name implies that its effect on a predator is unpleasant.

MAKING AND TESTING PREDICTIONS. The hypotheses about colorful and inconspicuous caterpillars led to predictions that were tested by an experiment. Captive birds, blue jays, were presented with both brightly colored monarch butterfly caterpillars and caterpillars that blended into their backgrounds. The blue jays were first deprived of food long enough to make them hungry, so they readily attacked the caterpillars. Ingesting even part of one monarch caterpillar caused a blue jay to vomit.

Because the birds were housed individually, the experimenters knew which ones had previously tasted monarchs and which ones had not. They found that a single experience with a monarch caterpillar was enough to cause a blue jay to reject all other monarch caterpillars presented to it. The camouflaged caterpillars, on the other hand, were readily attacked and eaten, and the jays continued to eat these caterpillars without showing signs of sickness or discontent.

These results supported the palatability hypothesis. The null hypothesis, that there is no difference in palatability between colorful and camouflaged caterpillars, was thus rejected.

Experiments are powerful tools

Scientists use a variety of methods to test predictions from their hypotheses. Among these are laboratory and field experiments and carefully focused observations. Each method has its strengths and weaknesses. The key feature of **experimentation** is the control of most factors so that the influence of a single factor can be seen clearly. In the experiments with blue jays, all birds were equally hungry and they were presented with caterpillars in the same way. By controlling these conditions, the experimenters could reject alternative explanations. Their results, for example, could not have been due to lack of hunger on the part of the birds or to the birds' failure to see the caterpillars.

The advantage of working in a laboratory is that control of the environment is easier. Field experiments are more difficult because it is usually impossible to control more than a small part of the total environment. But field experiments have one important advantage: Their results are more readily applicable to what happens where the organisms actually live and evolve. Just because an organism does something in the laboratory does not mean that it behaves the same way in nature. Because biologists usually wish to explain nature, not processes in the laboratory, combinations of laboratory and field experiments are needed to test most hypotheses about what organisms do (Figure 1.15).

A single piece of evidence supporting a hypothesis rarely leads to widespread acceptance of the hypothesis. Similarly, a single contrary result rarely leads to abandonment of a hypothesis. Negative results can be obtained for many reasons, only one of which is that the hypothesis is wrong. For example, the error may be that incorrect predictions were made from a correct hypothesis. A negative result can result from poor experimental design or because an inappropriate organism was chosen for the test. For example, a predator lacking color vision, or one that uses primarily its sense of smell, would not be appropriate for testing hypotheses about the colors of caterpillars.

A general textbook like this one presents hypotheses and theories that have been extensively tested and that are generally accepted. When possible in this text, we illustrate hypotheses and theories with observations and experiments that support them, but we cannot, because of space constraints, detail all the evidence. Remember as you read that statements of biological "fact" are mixtures of observations, predictions, and interpretations.

SOMETIMES ORGANISMS MUST BE SACRIFICED. Obtaining answers to many of the questions posed by biologists requires manipulating and sometimes sacrificing living organisms. To study the antipredator adaptations of caterpillars, the investigators had to keep blue jays in cages, make them hungry by depriving them of food, and then feed them caterpillars. This procedure resulted in the deaths of some caterpillars and temporary stress for some of the birds. To study their

(a)

(b)

1.15 Experimentation Is Essential in Biology Research and experimentation in biology are carried out in the field and in laboratories. (a) Biologists who study the canopies of rainforest trees use special climbing equipment that allows them to collect data and carry out vital studies in the field. (b) Many scientific experiments take place within the laboratory. Work with cells and many tiny organisms requires the use of microscopes.

detailed structure and functioning, scientists must often kill organisms.

No amount of observation without intervention could possibly substitute for experimental manipulation. However, this does not mean that scientists are insensitive to the welfare of the organisms with which they work. Most scientists who work with animals are continually alert to finding ways of getting answers that use the smallest number of experimental subjects and that cause the subjects the least pain and suffering.

Not all forms of inquiry are scientific

If you understand the methods of science, you can distinguish science from nonscience. Recently some people have claimed that "creation science," sometimes called "scientific creationism," is a legitimate science that deserves to be taught in schools together with the evolutionary view of the world presented in this book. In spite of these claims, creation science is not science.

Science begins with observations and the formulation of testable hypotheses that can be rejected by contrary evidence. Creation "science" begins with the unsubstantiated assertion that Earth is only about 4,000 years old and that all species of organisms were created in approximately their present forms. This assertion is not presented as a hypothesis from which testable predictions are derived. Advocates of creation science do not believe that tests are needed, because they assume the assertion to be true, nor do they suggest what evidence would refute it.

In this chapter we have outlined the hypothesis that Earth is about 4 billion years old, that today's living organisms evolved from single-celled ancestors, and that many organisms dramatically different from those we see today lived on Earth in the remote past. The rest of this book will provide evidence supporting this scenario. To reject this view of Earth's history, a person must reject not only evolutionary biology, but also modern geology, astronomy, chemistry, and physics. All of this extensive scientific evidence is rejected or misinterpreted by proponents of creation "science" in favor of a religious belief held by a very small proportion of the world's people.

Evidence gathered by scientific procedures does not diminish the value of religious accounts of creation. Religious beliefs are based on faith—not on falsifiable hypotheses, as science is. They serve different purposes, giving meaning and spiritual guidance to human lives. They form the basis for establishing values—something science cannot do. The legitimacy and value of both religion and science is undermined when a religious belief is called science.

Life's Emergent Properties

Biologists study structures and processes ranging from the simple to the complex and from the small to the large. They study the structure and functioning of small parts of cells, as well as the interactions among the hundreds or thousands of different types of organisms that live together in a particular region. Biology can be visualized as ordered into a hierarchy in which the units, from the smallest to the largest, are molecules, cells, tissues, organs, organisms, populations, communities, and biomes (Figure 1.16).

We have already identified the cell as the fundamental unit of life. A **tissue** is a group of cells with similar and coordinated functions. Several different tissues are usually joined to form larger structures called **organs**, such as hearts, brains, kidneys, and lungs. Organs often are joined to form **organ systems**, such as the nervous system of animals and the vascular system of plants. Organs and organ systems perform distinct functions for the **organism** in which they are found.

The organization of living systems extends beyond the individual organism. Organisms living in the same area that are capable of interbreeding with one another form a **population**. All of the populations of a particular kind of organism, whether or not they live in the same area, constitute a species. Individuals of many different species typically live together and interact to form an ecological **community**. Ecological communities are grouped by their distinctive vegetation into **biomes**. All the biomes on Earth constitute the **biosphere**.

The organism is the central unit of study in biology, and Parts Five and Six of this book discuss organismal biology in detail. But to understand organisms, biologists must study life at all its levels of organization. Biologists must study molecules, chemical reactions, and cells to understand the operations of tissues and organs. They study organs and organ systems to determine how whole organisms function and maintain internal homeostasis. At higher levels in the hierarchy, biologists study how organisms interact with one another to form social systems, populations, ecological communities, and biomes, which are the subjects of Part Seven of this text.

Each level of biological organization has properties, called **emergent properties**, that are not found at lower levels. For example, cells and multicellular organisms have processes and characteristics that are not shown by the molecules of which they are composed. Emergent properties arise in two ways.

First, many emergent properties of systems result from interactions among their parts. For example, at the organismal level, developmental interactions of cells result in a multicellular organism whose adult features are vastly richer than those of the single cell from which it developed. Other examples of properties that emerge through complex interactions are memory, learning, consciousness, and emotions such as hate, fear, envy, anger, and love. These properties result from interactions in the human brain among the 10^{12} (trillion) cells with 10^{15} (a quadrillion) connections. No single cell, or even small group of cells, possesses them.

Second, emergent properties arise because aggregations have collective properties that their individual

units lack. For example, individuals are born and they die—they have a life span. An individual does not have a birth rate or a death rate, but a population does. Death rate is an emergent property of a population. Other emergent properties of populations include age distribution and density. Evolution is an emergent property of populations that depends on differences in birth and death rates of individuals. Ecological communities possess emergent properties such as species richness.

Emergent properties do not violate principles that operate at lower levels of organization. However, emergent properties usually cannot be detected or even suspected by studying lower levels. Biologists could never discover the existence of human emotions by studying single nerve cells, even though they may eventually be able to explain those emotions in terms of interactions among many nerve cells.

Curiosity Motivates Scientists

The most important motivator of most biologists is curiosity. People are fascinated by the richness and diversity of life and want to learn more about organisms and how they function. Curiosity is probably an adaptive trait. Humans who were motivated to learn about their surroundings are likely to have survived and reproduced better, on average, than their less curious relatives. We hope this book will help you share the excitement biologists feel as they develop and test hypotheses. There are vast numbers of how and why questions for which we do not have answers, and new discoveries usually engender questions no one thought to ask before. Perhaps *your* curiosity will lead to an important new idea.

Summary of "An Evolutionary Framework for Biology"

- Earth is an ancient planet.
- Geologists in the nineteenth century provided the first evidence of Earth's antiquity and determined the sequence of events in life's evolution.
- The discovery of radioisotopes in the twentieth century enabled evolutionary events to be dated accurately.

Evolutionary Milestones

- Life arose from nonlife about 3.8 billion years ago when interacting systems of molecules became enclosed in membranes to form cells.
- All living organisms contain the same types of large molecules—carbohydrates, lipids, proteins, and nucleic acids. These organic molecules are formed only by living systems.
- All organisms consist of cells, and all cells come from preexisting cells. Life no longer arises from nonlife.

1.16 From Molecules to the Biosphere: The Hierarchy of Life

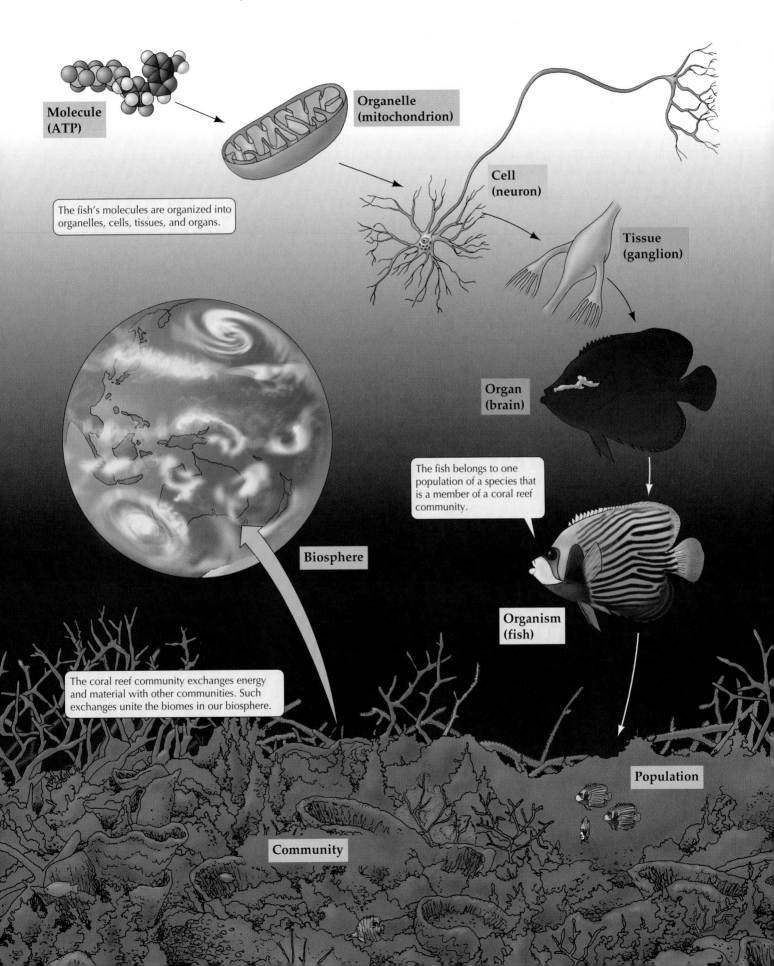

Molecule
(ATP)

Organelle
(mitochondrion)

Cell
(neuron)

The fish's molecules are organized into organelles, cells, tissues, and organs.

Tissue
(ganglion)

Organ
(brain)

The fish belongs to one population of a species that is a member of a coral reef community.

Biosphere

Organism
(fish)

The coral reef community exchanges energy and material with other communities. Such exchanges unite the biomes in our biosphere.

Population

Community

• A major theme in the evolution of life is the development of increasingly diverse ways of capturing external energy and using it to drive biologically useful reactions.

• Photosynthetic single-celled organisms released large amounts of O_2 into Earth's atmosphere, making possible the oxygen-based metabolism of large cells and, eventually, multicellular organisms.

• Reproduction with variation is a major characteristic of life. The evolution of sex greatly increased the rate of life's evolution.

• Complex eukaryotic cells evolved and were able to "stick together" after they divided, forming multicellular organisms. The individual cells of multicellular organisms became modified for specific functions within the organism.

• A major theme in the evolution of life is the development of increasingly complicated systems for responding to signals from the internal and external environments and for maintaining homeostasis.

• Regulated growth is a vital characteristic of life.

• Speciation resulted in the millions of species living on Earth today.

• Adaptation to environmental change is one of life's most distinctive features and is the result of evolution by natural selection. This principle guides virtually all biological investigation today.

Biological Diversity: Domains and Kingdoms

• Species are classified into three domains: Archaea, Bacteria, and Eukarya. The domains Archaea and Bacteria consist of prokaryotic cells and each contains one kingdom, Archaebacteria and Eubacteria, respectively. The domain Eukarya contains the kingdoms Protista, Plantae, Fungi, and Animalia, all of which have eukaryotic cells. **Review Figure 1.8**

The World Into Which Darwin Led Us

• Evolution is the theme that unites all of biology. The idea and evidence for evolution existed before Darwin.

• Darwin based his approach to evolution on five major hypotheses: (1) Earth is very old, (2) all organisms are descendants of a single common ancestor, (3) species multiply by splitting into daughter species, (4) evolution proceeds via gradual changes in populations, and (5) the major agent of evolutionary change is natural selection.

Asking the Questions "How?" and "Why?"

• Biologists ask two kinds of questions. Functional questions concern *how* organisms work. Adaptive questions concern *why* they evolved to work in that way.

• Both how and why questions are usually answered using a hypothetico-deductive approach. Hypotheses are tentative answers to the questions. Predictions are made on the basis of a hypothesis; then the predictions are tested by observations and experiments, which may support or refute the hypothesis.

• Science is based on the formulation of testable hypotheses that can be rejected by contrary evidence. The acceptance on faith of already refuted, untested, or untestable assumptions is not science.

Life's Emergent Properties

• Biology is organized into a hierarchy from molecules to the biosphere. Each level has emergent properties that are not found at the lower levels. **Review Figure 1.16**

Curiosity Motivates Scientists

• The most important motivator of most biologists is curiosity. There are vast numbers of unanswered—and in many cases, unasked—questions still to be explored.

Applying Concepts

1. According to the theory of evolution by natural selection, an organism evolves certain features because they improve the chances that it will survive and reproduce. There is no evidence, however, that evolutionary mechanisms have foresight or that organisms can anticipate future conditions. What do biologists mean when they say, for example, that wings are "for flying"?

2. Why is it so important in science that we design and perform tests capable of rejecting a hypothesis?

3. One hypothesis about the conspicuous coloration of caterpillars was described in this chapter, and some tests were mentioned. Suggest some other plausible hypotheses for conspicuous coloration in these animals. Develop some critical tests for one of these alternatives. What are the appropriate associated null hypotheses?

4. Some philosophers and scientists believe that it is impossible to prove any scientific hypothesis—that we can only fail to find a cause to reject it. Evaluate this view. Can you think of reasons why we can be more certain about rejecting a hypothesis than about accepting it?

Readings

Cziko, G. 1995. *Without Miracles: Universal Selection Theory and the Second Darwinian Revolution.* MIT Press, Cambridge, MA. An in-depth analysis of how an evolutionary view of the world can account for all of the features of life, including human self-awareness.

Darwin, C. 1859. *The Origin of Species by Means of Natural Selection.* John Murray, London. The book that set the world to thinking about evolution; still well worth reading. Many reprinted versions are available.

Futuyma, D. J. 1995. *Science on Trial: The Case for Evolution,* Revised Edition. Sinauer Associates, Sunderland, MA. A thorough presentation, for a general audience, of the scientific arguments that support evolution and its status as the single most fundamental principle in biology.

Margulis, L. and K. V. Schwartz. 1998. *Five Kingdoms: An Illustrated Guide to the Phyla of Life on Earth,* 3rd Edition. W. H. Freeman, New York. A good introduction to the kingdoms of organisms, in which the two kingdoms of prokaryotes are united into one. Excellent examples and illustrations.

Mayr, E. 1991. *One Long Argument: Charles Darwin and the Genesis of Modern Evolutionary Thought.* Harvard University Press, Cambridge, MA. An excellent account of the history of evolutionary thinking during the past century, written by a prominent exponent of the modern, or neo-Darwinian, synthesis of evolution.

Part Three
Evolutionary Processes

Chapter 19

The History of Life on Earth

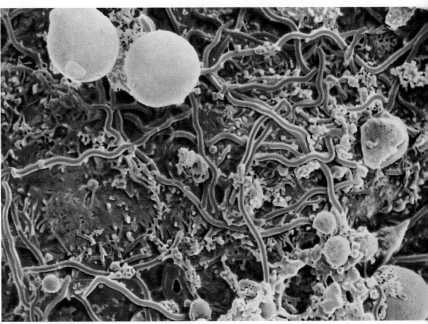

Borrelia bugdorferi

The Era of New Bacterial Diseases
The bacterium shown here causes Lyme disease. The disease has increased rapidly in North America since 1975.

*I*n 1967, the U.S. Surgeon General, William H. Stewart, announced that the time had come "to close the book on infectious diseases." Stewart and other world health officials were so confident of the power of their medical arsenal that they believed they were on the threshold of totally eradicating infectious diseases. However, the future did not conform to their prediction. In 1975, an era of new bacterial diseases began with legionnaires' disease, followed by Lyme disease, Brazilian purpuric fever, epithelioid angiomatosis, toxic shock syndrome, and others.

Since 1967, several hundred newly described bacteria have been associated with human disease, and dozens of other bacteria species that were thought to be harmless have been found to cause human disease. In addition, some historical diseases that had been nearly wiped out have resurged—some to epidemic proportions. Infectious diseases killed more than 16.5 million people in 1993, more than were killed by noninfectious diseases such as cancer and heart disease.

Health officials erred in their predictions because they did not understand the evolutionary capabilities of disease-causing organisms. Hippocrates described the symptoms of malaria, mumps, diphtheria, tuberculosis, and influenza 2,300 years ago, and human fossils show that our species has been plagued by infectious diseases throughout prerecorded history. Science, by making possible the rapid increase of Earth's human population, has helped create ideal conditions for the evolution of new pathogens.

Evolutionary changes are taking place all around us, and they have powerful implications for human welfare. Our attempts to control populations of undesir-

431

able species and increase populations of desirable species make us powerful agents of evolutionary change. In addition to producing the results we desire, we often cause undesirable outcomes, such as the evolution of resistance to medicines and pesticides and the evolution of greater virulence on the part of human pathogens. Medicine and agriculture can respond creatively to the evolutionary changes they are causing only if we understand why those changes happen.

What is biological evolution? **Biological evolution** is a change over time in the genetic composition of members of a population. Changes that take effect over a small number of generations constitute **microevolution**. Changes that take centuries, millennia, or longer to be completed are called **macroevolution**. The fossil record documents macroevolutionary changes among organisms. Many of these changes are dramatic. The goals of this part of the book are to document the history of life on Earth and to describe the processes of evolutionary change and the agents that cause them.

We begin with an overview of the history of life on Earth. Because the genetic material and basic cellular metabolic pathways are very similar or identical in all organisms, biologists believe that all living organisms have descended from a single ancestral lineage. How biologists determine the evolutionary histories of organisms, how they study the mechanisms of evolutionary change, and how the millions of species that live today (as well as those that became extinct) formed from a single common ancestor are addressed in subsequent chapters. Finally, in Chapter 24 we examine how life probably arose from nonliving matter several billion years ago.

To understand the long-term patterns of evolutionary change that we will document in this chapter, we must think in time frames spanning many millions of years and imagine events and conditions very different from those we now observe. The Earth of the distant past is, to us, a foreign planet inhabited by strange organisms. The continents were not where they are today, and climates were different. One of the remark-

TABLE 19.1   Earth's Geological History

EON	ERA	PERIOD	ONSET[a]	MAJOR PHYSICAL CHANGES ON EARTH
	Cenozoic	Quaternary	2 mya	Icehouse climate; repeated glaciations
		Tertiary	66 mya	Continents near current positions; climate cools
	Mesozoic	Cretaceous	138 mya	Northern continents attached; Gondwana begins to drift apart; meteorite strikes Yucatán Peninsula
		Jurassic	195 mya	Two large continents form: Laurasia (north) and Gondwana (south); climate warm
		Triassic	245 mya	Pangaea slowly begins to drift apart; hothouse climate
	Paleozoic	Permian	290 mya	Continents aggregate into Pangaea; large glaciers form; dry climates form in interior of Pangaea
		Carboniferous	345 mya	Climate cools; marked latitudinal climate gradients
		Devonian	400 mya	Continents collide at end of period; asteroid probably collides with Earth
		Silurian	440 mya	Sea levels rise; two large continents form; hothouse climate
		Ordovician	500 mya	Gondwana moves over South Pole; massive glaciation, sea level drops 50 m
		Cambrian	540 mya	O_2 levels approach current levels
Phanerozoic			600 mya	O_2 level at >5% of current level
Proterozoic			2.5 bya	O_2 level at >1% of current level
Archean			3.8 bya	O_2 first appears in atmosphere
Hadean			4.5 bya	

[a] In this chapter, figures that depict change over evolutionary time are shown with the most recent events at the top and the oldest at the bottom. mya, million years ago; bya, billion years ago.

able achievements of modern science has been the development of techniques that enable us to infer past conditions and to date them with some precision.

In this chapter we first examine how events in the distant past can be dated. Then we review the major changes in physical conditions on Earth during the past 4 billion years, look at how those changes affected life, and discuss the major patterns in the evolution of life.

Determining How Earth Has Changed

Geologists divide Earth's history into four **eons**: the Hadean eon, the Archean eon, the Proterozoic eon, and the Phanerozoic eon. The Phanerozoic eon is subdivided into **eras**, which are further subdivided into **periods** (Table 19.1). The boundaries between these divisions are based on major differences in the fossils contained in successive layers of rocks. The fossils of organisms, not the rocks themselves, provided the information geologists first used to order the sequence of events, because evolving life provided a record that can be ordered through time. There is no directional evolution of rocks; a rock of a particular type can be formed at any time. However, scientists do have ways of determining the date a rock was formed.

Radioactivity provides a way to date rocks

Radioactivity can be used to date rocks precisely because in successive, equal periods of time, an equal fraction of the remaining radioactive material of any radioisotope decays, becoming the corresponding stable isotope. For example, in 14.3 days, one-half of any sample of phosphorus-32 (^{32}P), a radioactive isotope of phosphorus, decays. During the next 14.3 days, one-half of the remaining half decays, leaving one-fourth of the original sample of ^{32}P. After 42.9 days, three half-lives have passed, so one-eighth (that is, $\frac{1}{2} \times \frac{1}{2} \times \frac{1}{2}$) of the original radioactive material remains, and so forth.

Each radioisotope has a characteristic half-life. Tritium (^{3}H) has a half-life of 12.3 years, and carbon-14 (^{14}C) has a half-life of about 5,700 years. Some radioisotopes have much longer half-lives: The half-life of potassium-40 (^{40}K) is 1.3 billion years; that of uranium-238 (^{238}U) is about 10 billion years. Which isotope is used to estimate the ages of some ancient material depends on how old the material is thought to be. The decay of potassium-40 to argon-40 has been used to date most of the ancient events in the evolution of life that we will describe in this chapter.

To use radioisotopes to date past events, we must know the concentrations of the isotopes at the time of those events. In the case of carbon, we know the initial amounts of ^{14}C because the production of new ^{14}C in the upper atmosphere (by the reaction of neutrons with ^{14}N) just balances the natural radioactive decay of ^{14}C. Therefore a steady state exists.

The ratio of radioactive ^{14}C to nonradioactive ^{12}C in a living creature is always the same as that in the environment because carbon is constantly being exchanged between the environment and organisms. However, as soon as a tree or any other living thing dies, it ceases to equilibrate its carbon compounds with the rest of the world. Its decaying ^{14}C is not replenished from outside, and the ratio of ^{14}C to ^{12}C decreases. By measuring the fraction of ^{14}C in a carbon specimen, we can easily calculate how much time has elapsed since it died, but ^{14}C can be used to date events only within the last 30,000 years. Some radiocarbon dates of archaeological objects are shown in Figure 19.1.

Armed with information on the ages of mileposts in Earth's history, we can analyze what caused the changes in types of organisms preserved in the rocks that led geologists to identify boundaries between geological periods. The major physical events we will describe are listed in Table 19.1, along with the most important events in the history of life.

MAJOR EVENTS IN THE HISTORY OF LIFE
Humans evolve; large mammals become extinct
Radiation of birds, mammals, flowering plants, and insects
Dinosaurs continue to radiate; flowering plants and mammals diversify. **Mass Extinction** at end of period (≈76% of species disappear)
Diverse dinosaurs; first birds; two minor extinctions
Early dinosaurs; first mammals; marine invertebrates diversify. **Mass Extinction** at end of period (≈76% of species disappear)
Reptiles radiate; amphibians decline; **Mass Extinction** at end of period (≈96% of species disappear)
Extensive forests; first reptiles; insects radiate
Fishes diversify; first insects and amphibians. **Mass Extinction** at end of period (≈82% of species disappear)
Jawless fishes diversify; first bony fishes; plants and animals colonize land
Mass Extinction at end of period (≈85% of species disappear)
Most animal phyla present; diverse algae
Ediacaran fauna
Eukaryotes evolve; several animal phyla appear
Origin of life; prokaryotes flourish

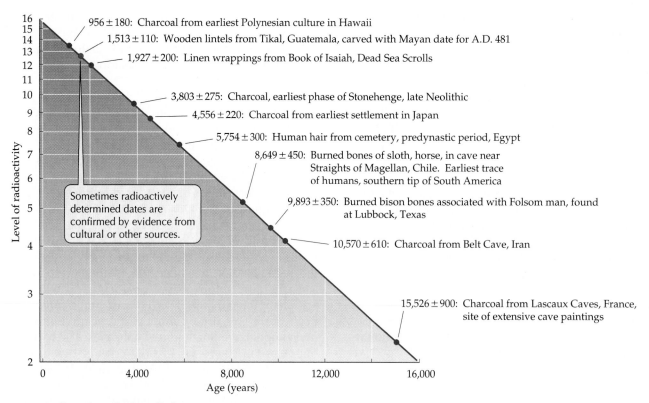

956 ± 180: Charcoal from earliest Polynesian culture in Hawaii

1,513 ± 110: Wooden lintels from Tikal, Guatemala, carved with Mayan date for A.D. 481

1,927 ± 200: Linen wrappings from Book of Isaiah, Dead Sea Scrolls

3,803 ± 275: Charcoal, earliest phase of Stonehenge, late Neolithic

4,556 ± 220: Charcoal from earliest settlement in Japan

5,754 ± 300: Human hair from cemetery, predynastic period, Egypt

8,649 ± 450: Burned bones of sloth, horse, in cave near Straights of Magellan, Chile. Earliest trace of humans, southern tip of South America

9,893 ± 350: Burned bison bones associated with Folsom man, found at Lubbock, Texas

10,570 ± 610: Charcoal from Belt Cave, Iran

15,526 ± 900: Charcoal from Lascaux Caves, France, site of extensive cave paintings

Sometimes radioactively determined dates are confirmed by evidence from cultural or other sources.

Level of radioactivity (y-axis)
Age (years) (x-axis)

19.1 Radioactive Clocks Tell Time By measuring the fraction of radioactive ^{14}C in any remains that contain carbon, archeologists and biologists can determine the approximate age of specimens such as those described here.

Unidirectional Changes in Earth's Atmosphere

The atmosphere of early Earth was a reducing one; that is, it lacked free oxygen. Perhaps the most important environmental change since Earth cooled enough for water to condense on its surface is the largely unidirectional increase in atmospheric O_2 concentrations that began in the Phanerozoic eon. The oxygen concentration increased because certain sulfur bacteria evolved the ability to use water as the source of hydrogen during photosynthesis. The cyanobacteria that evolved from these sulfur bacteria became very abundant. They liberated enough O_2 to open the way for the evolution of oxidation reactions as the energy source for the synthesis of ATP.

An oxygenated atmosphere also made possible larger cells and more complicated organisms. Small, unicellular aquatic organisms can obtain enough O_2 by simple diffusion even when O_2 concentrations are very low. Larger unicellular organisms have lower surface area-to-volume ratios. In order to obtain enough O_2 by simple diffusion, they must live in an environment where concentrations of O_2 are higher than those that can support small prokaryotic cells. Bacteria can thrive on 1 percent of current atmospheric O_2

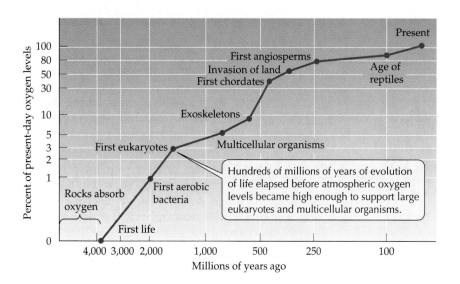

Present

First angiosperms

Invasion of land

First chordates

Age of reptiles

Exoskeletons

Multicellular organisms

First eukaryotes

Hundreds of millions of years of evolution of life elapsed before atmospheric oxygen levels became high enough to support large eukaryotes and multicellular organisms.

Rocks absorb oxygen

First aerobic bacteria

First life

Percent of present-day oxygen levels (y-axis)
Millions of years ago (x-axis)

19.2 Large Cells Need More Oxygen Although aerobic prokaryotes flourish with less, larger eukaryotic cells with lower surface area-to-volume ratios require oxygen levels that are at least 2 to 3 percent of current atmospheric O_2 concentrations. (Both axes of the graph are on logarithmic scales.)

levels, but eukaryotic cells require oxygen levels that are at least 2 to 3 percent of current atmospheric concentrations.

About 1,500 million years ago (mya), O_2 concentrations became high enough for large eukaryotic cells to flourish and diversify (Figure 19.2). Further increases in atmospheric O_2 levels some 700 to 570 mya enabled multicellular organisms to evolve. The fact that it took millions of years for Earth to develop an oxygenated atmosphere probably explains why only unicellular prokaryotes lived on Earth for more than a billion years.

Processes of Major Change on Earth

Unlike the largely unidirectional change in O_2 concentrations in Earth's atmosphere, most major changes on Earth have been characterized by irregular oscillations in the planet's internal processes, such as the activity of volcanoes and the shifting and colliding of the continents. External events, such as collision with meteorites, have also left their mark, sometimes causing major disruptions in the history of life.

The continents have changed position

The maps and globes that adorn our walls, shelves, and books give an impression of a static Earth. It is easy for us to believe that the continents have always been where they are, but this conclusion would be quite incorrect. Earth's crust consists of solid plates approximately 40 km thick that float on a fluid mantle. The mantle fluid circulates because heat produced by radioactive decay sets up convection cells. The plates move because the seafloor spreads along ocean ridges where material from the mantle rises and pushes the plates aside. Where plates come together, either they move sideways or one plate moves under the other, creating mountain ranges. The movement of the plates and the continents they contain—a process known as **continental drift**—has had enormous effects on climate, sea levels, and the distribution of organisms.

At times, the drifting of the plates has brought the continents together; at other times the continents have drifted apart. The positions and sizes of the continents influence ocean circulation patterns and sea levels. Mass extinctions of species, particularly marine organisms, have usually accompanied major drops in sea level (Figure 19.3). Later in this chapter we will discuss how the positions and movements of the continents, vulcanism, and large meteorites influenced the major events in the evolution of life.

Earth's climate shifts between hothouse and icehouse conditions

Through much of its history, Earth's climate was considerably warmer than it is today, and temperatures

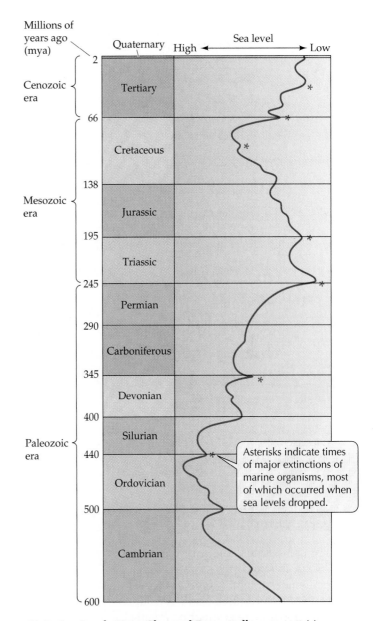

19.3 Sea Levels Have Changed Repeatedly As in Table 19.1, the most recent events are indicated at the top of the chart, the oldest at the bottom.

decreased more slowly toward the poles. At other times, however, Earth was colder than it is today. Large areas were covered with glaciers during the late Proterozoic, the Carboniferous, the Permian, and the Quaternary, but these cold periods were separated by long periods of milder climates (Figure 19.4). Because we live in one of the colder periods in the history of Earth, it is difficult for us to imagine the mild climates that were found at high latitudes during much of the history of life.

Usually climates change slowly, but major climatic shifts have taken place over periods as short as 5,000 to 10,000 years, primarily as a result of changes in

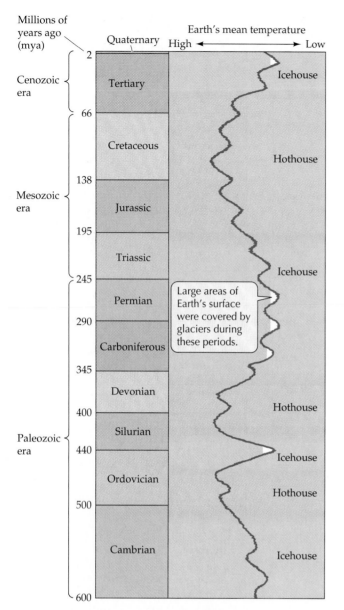

19.4 Hothouse and Icehouse Conditions Have Alternated Throughout Earth's history, periods of cold climates and glaciations have been separated by long periods of milder climates.

Earth's orbit around the sun. A few shifts appear to have been even more rapid. For example, during one interglacial period, the Antarctic Ocean changed from being ice-covered to being nearly ice-free in less than 100 years. Such rapid changes are usually caused by sudden shifts in ocean currents. Climates have sometimes changed rapidly enough that extinctions caused by them appear "instantaneous" in the fossil record.

Volcanoes disrupt the evolution of life

On the morning of August 27, 1883, Krakatau, an island the size of Manhattan located in the Sunda Strait between Sumatra and Java, was devastated by a series of volcanic eruptions. Tidal waves caused by the eruption hit the shores of Java and Sumatra, demolishing towns and villages and killing 40,000 people. The 1883 explosion at Krakatau was not the largest in recorded history. The 1815 eruption of Tambora on the Indonesian island of Sumbawa was five times greater. An even larger eruption on Sumatra 75,000 years ago created the 65-km-long Lake Toba. Impressive though these eruptions were, their effects were either local or short-lived. They did not cause major changes in patterns of the evolution of life. But the even larger volcanic eruptions that occurred several times during the history of Earth, did have major consequences for life on Earth.

The collision of continents during the Permian period (260 to 250 mya) to form a single, gigantic land mass called Pangaea, caused massive volcanic eruptions. The ash the volcanoes ejected into Earth's atmosphere reduced the penetration of sunlight to Earth's surface, lowering temperatures and triggering the massive glaciation of that time. Massive volcanic eruptions also occurred as the continents drifted apart during the late Triassic period and at the end of the Cretaceous.

External events have triggered other changes on Earth

At least 30 meteorites between the sizes of baseballs and soccer balls hit Earth each year, but collisions with large meteorites are rare. A large meteorite, weighing 5,000 kg, fell in Norton County, Kansas, on February 18, 1948. A prehistoric gigantic meteorite formed Canyon Diablo in Arizona.

The hypothesis that the mass extinction of life at the end of the Cretaceous period, about 65 mya, might have been caused by the collision of Earth with a large meteorite was proposed in 1980 by Luis Alvarez and several of his colleages at the University of California, Berkeley. These scientists based their hypothesis on the finding of abnormally high concentrations of the element iridium in a thin layer separating rocks deposited during the Cretaceous from those of the Tertiary (Figure 19.5). Iridium is abundant in some meterorites but is exceedingly rare on Earth's surface.

To account for the estimated amount of iridium in this layer, Alvarez postulated that a meteorite 10 km in diameter collided with Earth at a speed of 72,000 km per hour. The force of such an impact would have ignited massive fires, created great tidal waves, and sent up an immense dust cloud that encircled Earth, blocking the sun and cooling the planet. The settling dust would have formed the iridium-rich layer.

This hypothesis generated a great deal of controversy, which continues today. Many paleontologists (scientists who study fossils) believe that the fossil record shows that the mass extinction took place over

19.5 Evidence of a Meteorite Collision Iridium is a metal common in some meteorites but rare on Earth. Its presence has been cited as evidence of a meteorite collision that may have been the cause of the mass extinction at the end of the Cretaceous.

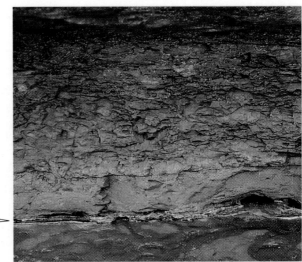

A thin band rich in iridium marks the boundary between rocks deposited in the Cretaceous and Tertiary periods.

a period of millions of years, not instantaneously as the meteorite theory demands. Also, because some volcanoes emit substantial quantities of iridium, the massive vulcanism of the time could have generated the iridium layer.

This controversy has stimulated much activity. Some scientists have tried to locate the site of impact of the supposed meteorite. Others have worked to improve the precision with which events of that age could be dated. Still others have tried to determine more exactly the speed with which extinctions occurred at the Cretaceous–Tertiary boundary. Progress on all three fronts has tended to support the meteorite theory.

The theory was supported by the discovery of a circular crater 180 km in diameter buried beneath the north coast of the Yucatán Peninsula, Mexico, thought to have been formed by an impact 65 mya. Some new fossil evidence also suggests that there may have been a sudden extinction of organisms 65 mya, as required by the meteorite theory. Therefore, many scientists accept that the collision of Earth with a large meteorite contributed importantly to the mass extinctions at the boundary between the Cretaceous and Tertiary periods, but concensus has not yet been reached.

The Fossil Record of Life

Geological evidence is a major source of information about changes on Earth during the remote past. But the preserved remains of organisms that lived in the past, not the rocks themselves, are what have enabled geologists to order those events in time. What are these remains and what do they tell us about the influence of physical events on the evolution of life on Earth? After examining the conditions that preserve remains, we will consider the completeness of the fossil record, and how that record reveals patterns in life's history.

Much of what we know about the history of life is derived from **fossils**—the preserved remains of organisms or impressions of organisms in materials that formed rocks. An organism is most likely to be preserved if it dies or is deposited in an environment that lacks O_2. However, most organisms live in oxygenated environments and therefore decompose when they die. Thus many fossil assemblages are collections of organisms that were transported by wind or water to

sites that lacked O_2. Occasionally, however, organisms are preserved where they lived. In such cases—especially if the environment in question was a cool, anaerobic swamp, where conditions for preservation were excellent—we obtain a picture of communities of organisms that lived together.

How complete is the fossil record?

About 300,000 species of fossil organisms have been described, and the number is growing steadily. However, this number is only a tiny fraction of the species that have ever lived. We do not know how many species lived in the past, but we have ways of making reasonable estimates. Of the present-day **biota**—that is, the species in all kingdoms (bacteria, archaea, protists, plants, fungi, and animals)—approximately 1.5 million species have already been named. The actual number of living species is probably at least 10 million (and possibly as high as 50 million) because most species of insects, the richest animal group, have not yet been described. Thus the number of known fossil species is less than 2 percent of the probable total of living species.

Because life has existed on Earth for at least 3.5 billion years, and because species last, on average, less than 10 million years, Earth's biota must have turned over many times during geological history. The total number of species over evolutionary time greatly exceeds the number living today.

The sample of fossils, although small in relation to the total number of extinct species, is better for some groups than for others. The record is especially good for marine animals that have hard skeletons. Among the nine major animal groups that have hard-shelled members (see Chapters 29 and 30), approximately 200,000 species have been described from fossils, roughly twice the number of living marine species in

19.6 A Fossil Spider Trapped in the sap of a tree in what is now Arkansas about 50 mya, this spider is exquisitely preserved in the amber that formed from the sap. The details of its external anatomy are clearly visible.

these same groups. Paleontologists lean heavily on these groups in their interpretations of the evolution of life in the past. Insects, although much rarer as fossils, are also relatively well represented in the fossil record (Figure 19.6).

The fossil record demonstrates several patterns

Despite its incompleteness, the fossil record reveals several patterns that are unlikely to be altered by future discoveries. First, great regularity exists: Organisms are not mixed together at random, but rather appear sequentially. Second, as we pass from ancient periods of geological time toward the present, fossils increasingly resemble species living today. The fossil record also tells us that extinction is the eventual fate of all species.

The fossil record contains many good series that demonstrate gradual change in lineages of organisms over time. A good example is the series of fossils showing the pathway by which whales evolved from hoofed terrestrial mammals, beginning about 50 mya. The intermediate fossils illustrate the major changes by which ancestors of whales became adapted for aquatic existence and lost their hind limbs (Figure 19.7). Interestingly, whales retain the genetic potential for developing legs; occasionally living whales have been found with small hind legs that extend outside their bodies. The claim made repeatedly by creationists that the fossil record does not contain examples of intermediates is false. Intermediates abound, and more and more of them are being discovered.

Nevertheless, the incompleteness of the fossil record can mislead our interpretations of what happened. Most described fossils come from a relatively small number of sites. Since any organism that was evolving elsewhere would have been absent at these sites, many organisms may be entirely unrepresented. Moreover, when a species that has evolved in one place appears among the fossils of another site, it gives the false impression that it evolved very rapidly from one of the species that already lived there.

For example, horses evolved slowly over millions of years in North America. Many different lineages arose and died out (Figure 19.8). Ancestors of horses crossed the Bering land bridge into Asia several times,

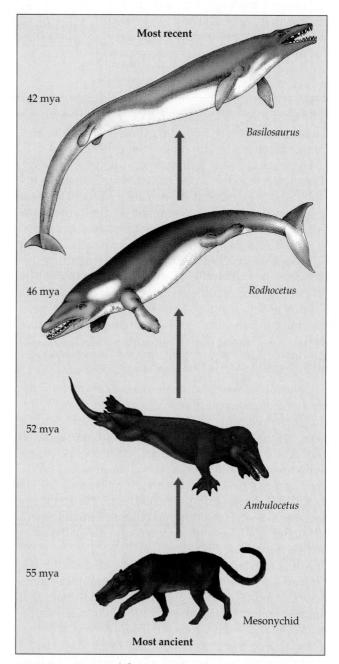

19.7 From Terrestrial to Aquatic Life An artist's reconstruction, based on fossil skeletons, of four ancestors that represent different stages in the evolution of modern whales from an early terrestrial mammal.

Early Tertiary (50 mya)

Mid Tertiary (20 mya)

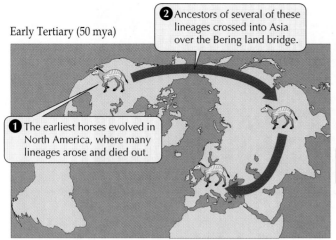

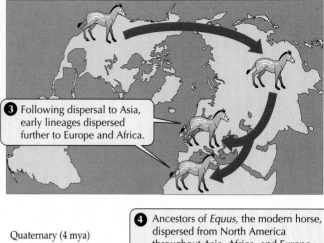

❷ Ancestors of several of these lineages crossed into Asia over the Bering land bridge.

❶ The earliest horses evolved in North America, where many lineages arose and died out.

❸ Following dispersal to Asia, early lineages dispersed further to Europe and Africa.

19.8 Horses Have a Complex Evolutionary History
Ancestors of horses crossed the Bering land bridge into Asia several times, the last one only a few million years ago (panel 4). If we lacked the earlier fossil evidence of horse evolution in North America, we might reach the false conclusion that horses evolved very rapidly somewhere in Asia.

Quaternary (4 mya)

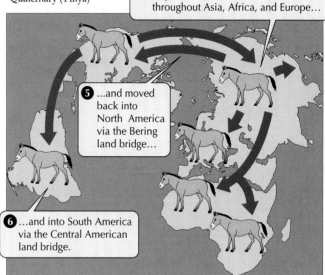

❹ Ancestors of *Equus*, the modern horse, dispersed from North America throughout Asia, Africa, and Europe...

❺ ...and moved back into North America via the Bering land bridge...

❻ ...and into South America via the Central American land bridge.

the last one only several million years ago. Evidence of each crossing appears suddenly in the Asian fossil record as a major new type of horse. If we lacked fossil evidence of horse evolution in North America, we might conclude that horses evolved very rapidly somewhere in Asia. On the other hand, an incomplete fossil record can also hide rapid changes.

Scientists have collected enough data to combine information of physical events during Earth's history with the course of evolution to compose pictures of what Earth and its inhabitants looked like at different times. We know in general where the continents were and how life changed, but many of the details are poorly known, especially for events in the more remote past.

Life in the Remote Past

The positions of the continents during the **Archean**, **Proterozoic**, and **Phanerozoic** eons are not well known. The major kingdoms of eukaryotic organisms evolved during the mid-Proterozoic eon. The fossil record shows that the volume of organisms increased dramatically in late Precambrian times, about 650 mya. The shallow Precambrian seas teemed with life. Protists and small multicellular animals fed on floating algae. Living plankton and plankton remains were devoured by animals that filtered food items from the water or ingested surface sediments and digested the organic remains in them. During the Phanerozoic, the continents were drifting apart.

The fossil record for Precambrian times is fragmentary. The best fossil assemblage of Precambrian animals, all soft-bodied invertebrates, is known as the Ediacaran fauna,* named after the Australian site where it was discovered (Figure 19.9). The Ediacaran animals are very different from any animals living today. Some of them may represent separate animal lineages that have no living descendants. In the discussions that follow in this section of the chapter, we will examine the changes that occurred in the Paleozoic, Mesozoic, and Cenozoic eras following the enormous proliferation of life in the Cambrian period.

Life exploded during the Cambrian period

During the **Cambrian period** (540 to 500 mya), O_2 levels in Earth's atmosphere approached their current concentration, and drifting of the plates resulted in several continental masses, the largest of which was

---

*The term *fauna* refers to all of the species of animals living in a particular area. The corresponding term for plants is *flora*.

Spriggina floundersi

Mawsonites

19.9 Ediacaran Animals   Fossils of soft-bodied invertebrates excavated at Ediacara in southern Australia are 600 million years old and illustrate the diversity of life that evolved in Precambrian times.

Gondwana (Figure 19.10*a*). Thus, O_2 concentration was no longer a constraint on the evolution of large, multicellular organisms. All animal phyla that have species living today had already evolved, as revealed by the exceptionally well preserved fossils in the Burgess Shale in British Columbia (Figure 19.10*b*), which were deposited in an equatorial sea. The evolution of hard skeletons in representatives of so many phyla about 540 mya suggests that predation became very intense during this period when life "exploded."

The Paleozoic era was a time of major changes

THE ORDOVICIAN.   During the **Ordovician period** (500 to 440 mya) the continents were located primarily in the Southern Hemisphere (Figure 19.11). No evidence exists of land or shallow marine environments in the Northern Hemisphere north of the Tropics during the Ordovician. Early during the period, the number of kinds of animals that filter small prey from the water, including brachiopods and mollusks, increased greatly. Floating graptolites, members of a now extinct phylum, were abundant. Ancestors of club mosses and horsetails colonized wet terrestrial environments, but they were still relatively small. At the end of the Ordovician, sea levels dropped about 50 m as massive glaciers formed over Gondwana (see Figure 19.3). Much of the continental shelf was exposed, and ocean temperatures dropped. About 85 percent of the species of marine animals became extinct, probably because of these major environmental changes.

19.10 Cambrian Continents and Animals (540–500 mya)
(a) Positions of the continents during mid-Cambrian times. *(b)* The Burgess Shale has yielded fossils of Cambrian animals, some of which may not be members of any lineage that survived to the present.

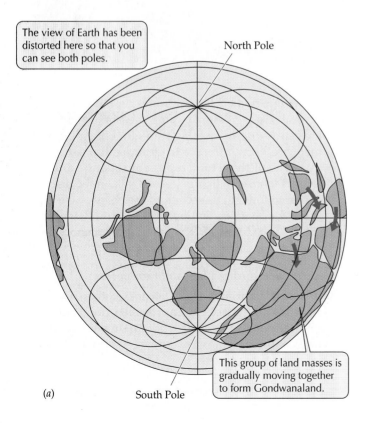

The view of Earth has been distorted here so that you can see both poles.

North Pole

This group of land masses is gradually moving together to form Gondwanaland.

(a)   South Pole

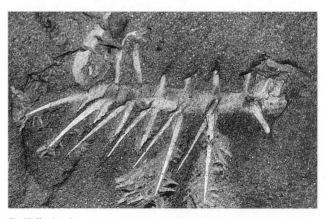

(b) Hallucigenia

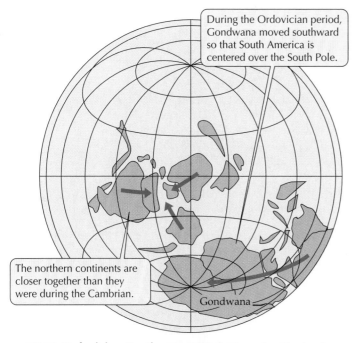

19.11 Ordovician Continents (500–440 mya)   The land mass of this period was still concentrated mainly in the Southern Hemisphere.

THE SILURIAN.   During the **Silurian period** (440 to 400 mya), the northern continents, which had been relatively close to one another during the Ordovician, coalesced, but the general positions of the continents did not change much. Marine life rebounded from the massive extinction at the end of the Ordovician, but no major new groups of marine organisms evolved. The tropical sea was uninterrupted by land barriers, and most marine genera were widely distributed. More significant changes happened on land where the first terrestrial arthropods—scorpions and millipedes—appeared.

THE DEVONIAN.   Rates of evolutionary change accelerated in many groups of organisms during the **Devonian period** (400 to 345 mya). Earth continued to be divided into northern and southern land masses, both of which were slowly moving northward (Figure 19.12a). There was a great evolutionary radiation of corals and shelled squidlike cephalopods (Figure 19.12b). (See Chapter 20 for a discussion of evolutionary radiation.) Fishes diversified as jawed forms replaced jawless ones, and the heavy armor that had characterized most earlier fishes gave way to the less rigid outer coverings of modern fishes.

19.12 Devonian Continents and Marine Communities (400–345 mya)   The reconstruction in (b) shows how a Devonian reef may have appeared.

Terrestrial communities also changed markedly during the Devonian period. Land plants became common, and some reached the size of trees. Most of them were club mosses and horsetails, along with some tree ferns. Toward the end of the period the first gymnosperms appeared. Distinct floras evolved on the two land masses. The first known fossils of centipedes, spiders, pseudoscorpions, mites, and insects date to this period, and the first fishlike amphibians began to occupy the land.

The end of the Devonian was marked by the extinction of about 80 percent of all marine species. Paleontologists disagree on the cause of this major extinction. Some believe that it was triggered by the col-

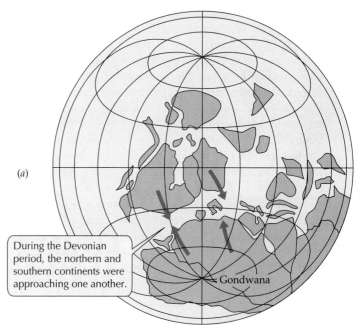

(a)

(b)

lision of the two continents, which destroyed much of the existing shallow, warm-water marine environment. This hypothesis is supported by the fact that extinction rates were much higher among tropical than among cold-water species. Other paleontologists believe that the extinction was caused by collision of a large asteroid with Earth.

THE CARBONIFEROUS. Extensive forests grew on the tropical continents during the **Carboniferous period** (345 to 290 mya) (see Figure 27.12). The compressed remains of trees that grew in swampy forests where they fell into deep, anaerobic mud that preserved them from biological degradation are the coal we now mine for energy. Carboniferous beds are rich in fossils, many of which retain traces of the fine details of their structure (Figure 19.13). The diversity of terrestrial animals increased greatly in the Carboniferous period. Snails, scorpions, centipedes, and insects were present in great abundance and variety. Insects evolved wings, which enabled them to move readily among tall and structurally complex plants. Many Carboniferous plant fossils show evidence of damage by feeding insects. Amphibians became better adapted to terrestrial existence. Some of them were large animals more than 5 m long, quite unlike any surviving today. From one amphibian stock, the first reptiles evolved late in the period.

THE PERMIAN. Deposits from the **Permian period** (290 to 245 mya) contain representatives of most modern groups of insects, including dragonflies with wingspreads that measured 2 feet, the largest insects that ever lived. By the end of the period, reptiles greatly outnumbered amphibians. These reptiles included a

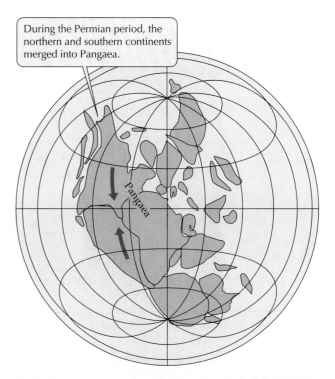

During the Permian period, the northern and southern continents merged into Pangaea.

Pangaea

19.14 Pangaea Formed in the Permian Period (290–250 mya) The interior of this supercontinent experienced very harsh climates. The largest glaciers in Earth's history formed during this period.

variety of terrestrial forms, as well as species that were major predators in marine and freshwater environments. The lineage leading to mammals diverged from one reptilian lineage late in the period. In fresh waters, the Permian period was a time of extensive radiation of bony fishes.

At the end of the Permian period, about 90 percent of all species, both terrestrial and marine, became extinct. What caused this mass extinction? There was massive volcanic activity in Siberia at that time, but there was no collision of a large meteorite with Earth. The Permian extinctions may have happened slowly, perhaps over more than 10 million years, but the time frame is uncertain. The most probable cause was the coalescing of the continents into the supercontinent Pangaea (Figure 19.14). The interior of Pangaea, far removed from the oceans, experienced very harsh climates. The largest glaciers in Earth's history formed, causing sea levels to drop, drying out many of the shallow seas where most marine organisms lived (see Figure 19.3).

Geographic differentiation increased in the Mesozoic era

At the start of the **Mesozoic era,** the few surviving organisms found themselves in a relatively empty world. As Pangaea slowly separated into individual

Alethopteris serlii

19.13 A Carboniferous Fossil (345–290 mya) A fossilized seed fern excavated near Washington, D.C. Like the fossil fuels we burn today, this fossil is a remnant of the massive forests of the Carboniferous period.

19.15 Mesozoic Dinosaurs   Dinosaurs of the Mesozoic era continue to capture our imagination. This painting illustrates some of the large species from the Jurassic period (195–138 mya).

continents, the glaciers melted and the oceans rose and reflooded the continental shelves, forming huge, shallow inland seas. Life again diversified, but lineages different from those of the Permian period came to dominate Earth. The trees that dominated the great coal-forming forests were replaced by cycadeoids, cycads, ginkgos, and conifers.

Earth's biota, which until that time had been relatively homogeneous, became increasingly **provincialized**; that is, distinctive terrestrial floras and faunas evolved on each continent, and the biotas of shallow waters bordering the continents also diverged from one another. The provincialization that began during the Mesozoic continues to influence the geography of life today (see Chapter 54).

THE TRIASSIC.   During the **Triassic period** (245 to 195 mya) many invertebrate lineages became more diverse, and many groups that previously had been restricted to living on surfaces of bottom sediments evolved burrowing forms. On land, gymnosperms and seed ferns became the dominant woody plants. A great radiation of reptiles began, which eventually gave rise to dinosaurs, crocodilians, and birds.

THE JURASSIC.   The transition from the Triassic period to the **Jurassic period** (195 to 138 mya) was marked by a mass extinction that eliminated about 75 percent of species on Earth. Why they went extinct is not known, but meteor impact is suspected. The mass extinction was followed by another period of evolutionary diversification during the Jurassic. Bony fishes began the great radiation that culminated in their dominance of the seas. Frogs, salamanders, and lizards first appeared. Flying reptiles evolved, and dinosaur lineages evolved into bipedal predators and large quadrupedal herbivores (Figure 19.15). Several groups of mammals also evolved.

THE CRETACEOUS.   By the early **Cretaceous period** (138 to 66 mya), the northern continents had completely separated from the southern ones and a continuous sea encircled the Tropics (Figure 19.16). Sea levels were high, and Earth existed in a hothouse state. Biological production was high both on land and in the oceans. Marine invertebrates increased in variety and number of species. On land, dinosaurs continued to diversify. The first snakes appeared during the Cretaceous, but their lineage did not radiate until much later.

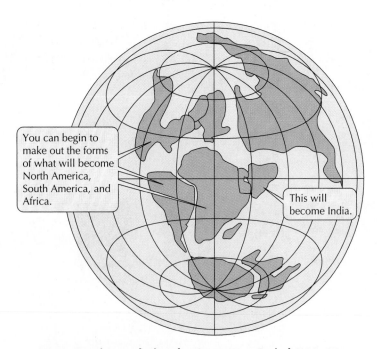

19.16 Continents during the Cretaceous Period (138–65 mya) Many groups of small mammals evolved during this period, and flowering plants began the radiation leading to their modern-day dominance on land.

By the end of the period, many groups of mammals had evolved, but these mammals were generally small. Early in the Cretaceous period, possibly somewhat earlier, flowering plants evolved from gymnosperm ancestors and began the radiation that led to their current dominance on land. And insects added angiosperms to the list of plants they ate.

Another mass extinction took place at the end of the Cretaceous period. On land, all vertebrates larger than about 25 kg in body weight apparently became extinct. In the seas, many planktonic organisms and bottom-dwelling invertebrates became extinct. As we discussed earlier, this extinction may have been caused by a large meteorite that collided with Earth off the Yucatán Peninsula.

The modern biota evolved during the Cenozoic era

By the early **Cenozoic era** (66 mya), the positions of the continents had begun to resemble those of today, but Australia was still attached to Antarctica, the Atlantic Ocean was much narrower, and the northern continents were

connected (Figure 19.17). The Cenozoic era was characterized by an extensive radiation of mammals, but other taxa were also undergoing important changes. Flowering plants diversified extensively and dominated world forests, except in cool regions.

THE TERTIARY. During the **Tertiary period** (66 to 2 mya), Australia began its northward drift, and by 20 mya it had nearly reached its current position. The map of the world during the Tertiary for the first time looks familiar to us.

In the middle of the Tertiary, when the climate became considerably drier and cooler, many lineages of flowering plants evolved herbaceous (nonwoody) forms; grasslands, with and without scattered trees, spread over much of Earth. By the beginning of the Cenozoic era, invertebrate faunas were already modern in most respects. It is among the vertebrates that evolutionary change during the Tertiary period was most rapid. Living groups of reptiles, such as snakes and lizards, underwent extensive radiations during this period, as did birds and mammals.

THE QUATERNARY. The present geological period, the **Quaternary period,** began with the Pleistocene epoch about 2 mya. The Pleistocene was a time of drastic cooling and climatic fluctuations. During four major and about 20 minor glacial episodes, Earth became much cooler and distributions of animals and plants shifted toward the equator. The last glaciers retreated from temperate latitudes less than 10,000 years ago.

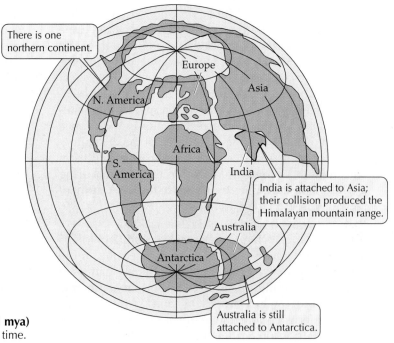

19.17 Continents during the early Tertiary Period (65 mya)
The continents approached their modern positions by this time.

Although Earth was relatively poor in species at the time of the other two evolutionary explosions as well, the species that were already present included a wide array of body plans and ways of life. Therefore, new major innovations were less likely to evolve at these times than in the Cambrian period.

The size and complexity of organisms have increased

The earliest organisms were small prokaryotes. A modest increase in size and structural complexity accompanied the evolution of eukaryotes 2.5 billion years ago. Since then, the maximum sizes of organisms in many lineages have increased, irregularly to be sure. E. D. Cope noted in 1885 that sizes of vertebrates often increase with time within lineages, a trend that has since been verified in many other groups. The most striking exception to this trend is insects, which have remained relatively small throughout their evolutionary history, although some were much larger during the Carboniferous.

The overall increase in body size is the result of two opposing forces. Within a species, selection often favors larger size because larger individuals can dominate smaller ones. But larger species on average survive for less time than small species do, which is one reason why Earth is not populated primarily by large organisms.

Predators have become more efficient

During the Cretaceous period (138 to 66 mya) many species of crabs with powerful claws evolved, and carnivorous marine snails able to drill holes in shells began to fill the seas. Skates, rays, and bony fishes with powerful teeth capable of crushing mollusk shells also evolved, and large, powerful marine reptiles—the placodonts—fed heavily on clams. The increasing thickness and narrowing openings of snail shells during the Cretaceous is evidence that predation rates intensified (Figure 19.20). Other evidence of heavy predation pressure is the increase in the percentage of fossil shells that show signs of having been repaired following an attack that did not kill its owner.

Although shell thickness provided some protection from predators, predators were so effective that clams disappeared from the surfaces of most marine sediments. The survivors in those environments were species that burrowed into the substrate, where they were more difficult to capture.

The Speed of Evolutionary Change

Following each mass extinction, the diversity of life rebounded. How fast did evolution proceed during those times? Were rates too high to be accounted for by the speed at which current evolutionary changes are happening? Why did some lineages evolve rapidly

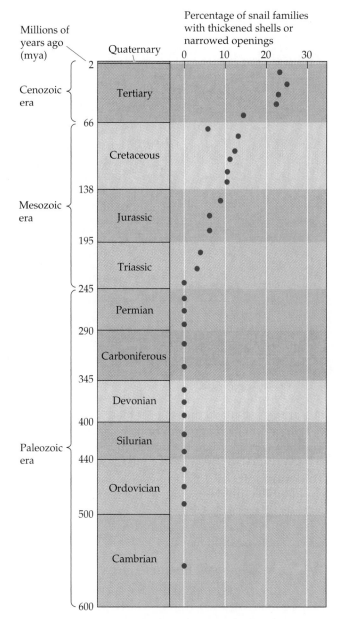

19.20 Snail Shells Have Thickened over Time   The percentage of subfamilies of snails that have internally thickened or narrowed openings of their shells has increased with evolutionary time—evidence that predation on shelled animals intensified.

while others did not? Scientists have made enough progress in studying evolution to give at least tentative answers to these questions.

Fossil evidence can be used to estimate rates of change of size and shape in many lineages of organisms. One of the fastest known rates of evolution has been in the skeletons of house sparrows that were introduced into the United States from Europe in the nineteenth century, but that rate is many times slower than rates achieved by scientists performing evolutionary experiments in the laboratory. Nonetheless, house sparrows diverged from their European ances-

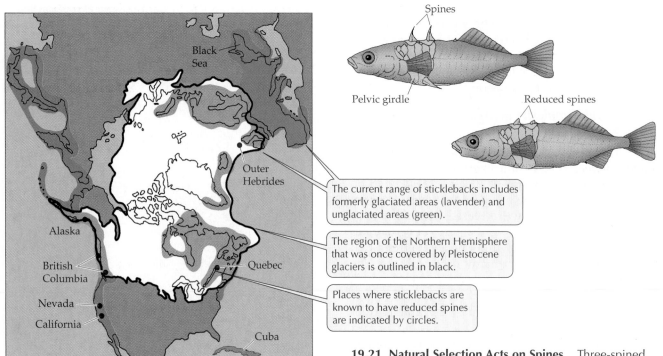

The current range of sticklebacks includes formerly glaciated areas (lavender) and unglaciated areas (green).

The region of the Northern Hemisphere that was once covered by Pleistocene glaciers is outlined in black.

Places where sticklebacks are known to have reduced spines are indicated by circles.

19.21 Natural Selection Acts on Spines Three-spined stickleback populations with reduced spines are found primarily in young lakes that were covered by ice during the last glaciation. These lakes lack large predatory fishes but do have predatory insects.

tors at rates about 20 times faster than the highest rate of change that has been measured in characters among rapidly evolving fossil mammals. Rates of increase in sizes of dinosaurs were ten times slower than that. Thus, the most rapid rates of evolutionary change are compatible with the microevolutionary mechanisms we will discuss in Chapter 20. In considering the speed of evolution, we will look at variations in the rates of both evolution and extinctions.

Why are evolutionary rates uneven?

The fossil record shows that many species have experienced times of **stasis**, during which they change very little over long periods. For example, many marine lineages evolved very slowly during the Silurian period. The horseshoe crabs that lived 300 mya are almost identical in appearance to those living today (see Figure 29.34). The chambered nautiluses of the late Cretaceous are indistinguishable from living species.

"Living fossils" are found today in environments that have changed relatively little for millennia. These environments tend to be harsh in one way or another. The sandy coastlines where horseshoe crabs spawn have extremes in temperature and in salt concentration that are lethal to many organisms. Chambered nautiluses spend their days in deep, dark ocean waters, ascending to feed in food-rich surface waters only under the protective cover of darkness. Their intricate shells provide little protection against today's visually hunting fishes.

Periods of stasis may be broken by times during which changes, either in the physical or biological envi-

ronment, create conditions that favor new traits. How new conditions favor rapid evolutionary changes is illustrated by the sizes of spines on the three-spined stickleback (*Gasterosteus aculeatus*). This widespread marine fish has repeatedly invaded fresh water throughout its long evolutionary history (Figure 19.21). Sticklebacks are quite tiny (usually less than 10 cm long); all marine and most freshwater populations have well-developed pelvic girdles with prominent spines that make it difficult for other fishes to swallow them. However, large predatory insects can readily grasp the stickleback's spines. These insects prey selectively on stickleback individuals with the largest spines. When sticklebacks invade freshwater habitats where predatory fish are absent but predatory insects are present, the lineages rapidly evolve smaller spines. Populations with reduced spines are found primarily in young lakes that were covered by ice during the last glaciation. These lakes do not have large predatory fishes.

The extensive fossil record of sticklebacks shows that spine reduction evolved many times in different populations that invaded fresh water. In addition, molecular data reveal that each freshwater population is most closely related to an adjacent marine population, not to other freshwater populations. Therefore, spine reduction has evolved rapidly many times in different places in response to the same ecological situation: the absence of predatory fish.

Extinction rates vary over time

More than 99 percent of the species that have ever lived are extinct. However, we know relatively little about causes of extinctions, except for species that have become extinct in historical times. Species have become extinct throughout the history of life, but extinction rates have fluctuated dramatically. In some periods, some groups have had high extinction rates while others were proliferating. Paleontologists distinguish between normal or background extinction rates and rates during mass extinctions (see Table 19.1).

Each mass extinction changed the flora and fauna of the next period by selectively eliminating some types of organisms, thereby increasing the relative abundance of others. For example, among planktonic foraminifera, important marine protists (see Figure 26.7), the only survivors of mass extinction were relatively simple species with broad geographic ranges. Among the seashells of the Atlantic coastal plain of North America, species with a broad geographic range were less likely to become extinct during normal periods than were species with a small geographic range.

On the other hand, during the mass extinction of the late Cretaceous, *groups of closely related species* with large geographic ranges survived better than those with small ranges, even if the *individual species* had small ranges. Similar patterns are found in other molluscan groups elsewhere, suggesting that traits favoring long-term survival during normal times are often different from those that favor survival during times of mass extinctions.

At the end of the Cretaceous period, extinction rates on land were much higher among large vertebrates than among small ones. The same was true during the Pleistocene mass extinction, when extinction rates were high only among large mammals and large birds. In addition, as we have seen, during some mass extinctions marine organisms were heavily hit while terrestrial organisms survived well. Other extinctions affected organisms living in both environments. These differences are not surprising, given that major changes on land and in the oceans did not always coincide.

The Future of Evolution

The agents of evolution are operating today as they have been since life first appeared on Earth, but major changes are under way as a result of the dramatic increase of Earth's human population. The loss of large vertebrates that dominated human-caused extinctions until recently is being supplemented by increasing extinctions of small species, driven primarily by changes in Earth's vegetation. Deliberately or inadvertently, people are moving thousands of species around the globe, reversing the provincialization of Earth's biota that evolved during the Cenozoic era.

Humans have taken charge of the evolution of certain valuable species by means of artificial selection. Our ability to modify species has been enhanced by modern molecular methods that enable us to move genes at will among species, even among distantly related ones. In short, humans have become the dominant evolutionary agent on Earth today. How we handle our massive influence will powerfully affect the future of life on Earth.

Summary of "The History of Life on Earth"

• Biological evolution is change over time in the genetic composition of members of a population.

Determining How Earth Has Changed

• Microevolutionary changes take effect over a small number of generations, macroevolutionary changes over centuries, millennia, or longer.
• The relative ages of rock layers in Earth's crust were determined from their embedded fossils. The eons during which the rock layers were laid down are divided into the eras and periods of Earth's geological history. The boundaries between these units are based on differences between their fossil biotas. **Review Table 19.1**
• Radioisotopes supplied the key for assigning absolute ages to the boundaries between geological time units. **Review Figure 19.1**

Unidirectional Changes in Earth's Atmosphere

• The early atmosphere was a reducing atmosphere; it lacked free oxygen. Oxygen accumulated after prokaryotes evolved the ability to use water as their source of hydrogen ions in photosynthesis. Increasing concentrations of atmospheric oxygen made possible the evolution of eukaryotes and multicellular organisms. **Review Figure 19.2**

Processes of Major Change on Earth

• Physical conditions have changed dramatically and repeatedly during Earth's history. **Review Table 19.1**
• Throughout Earth's history continents have drifted about, sometimes separating from one another, at other times colliding. Collisions typically have led to periods of massive vulcanism, glaciations, and major shifts in sea levels and ocean currents. **Review Figures 19.3, 19.4, 19.10, 19.11, 19.12, 19.14, 19.16, 19.17**
• External events, such as collisions with meteorites, also have changed conditions on Earth. A meteorite may have caused the abrupt mass extinction at the end of the Cretaceous period. Most other mass extinctions were apparently caused by events originating on Earth.

The Fossil Record of Life

• Much of what we know about the history of life on Earth comes from the study of fossils.
• The fossil record, although incomplete, reveals broad patterns in the evolution of life. About 300,000 fossil species

have been described. The best record is that of hard-shelled animals fossilized in marine sediments.

• Fossils show that many evolutionary changes are gradual, but an incomplete record can falsely suggest or conceal times of rapid change. **Review Figures 19.7, 19.8**

Life in the Remote Past

• The fossil record for Precambrian times is fragmentary, but fossils from Australia show that many lineages that evolved then may not have left living descendants.

Patterns of Evolutionary Change

• Truly novel features of organisms have evolved infrequently. Most evolutionay changes are the result of modifications of already existing structures. Striking changes in form can be caused by simple genetic changes that alter the rates of growth of different body parts. **Review Figure 19.18**

• Extinctions of major groups opened evolutionary opportunities for other groups of organisms. **Review Figure 19.19**

• Throughout evolution, organisms have increased in size and complexity. Predation rates have also increased, resulting in the evolution of better armor among prey species. **Review Figure 19.20**

The Speed of Evolutionary Change

• After each mass extinction, the diversity of life rebounded within 7 million years, but the groups of organisms that dominated the new biotas differed markedly from those characteristic of earlier biotas.

• Rates of evolutionary change have been very uneven, but even the fastest rates are slow enough to have been caused by known evolutionary agents.

• Periods of rapid evolution have followed times of mass extinction. Rapid evolution also has been stimulated by the provincialization of biotas when continents drifted apart. **Review Figure 19.21**

The Future of Evolution

• The agents of evolution continue to operate today, but human intervention, whether deliberate or inadvertent, now plays an unprecedented role in the history of life.

Self-Quiz

1. The number of species of fossil organisms that has been described is about
   a. 50,000.
   b. 100,000.
   c. 200,000.
   d. 300,000.
   e. 500,000.

2. Radioactive carbon can be used to date the ages of fossil organisms because
   a. all organisms contain many carbon compounds.
   b. radioactive carbon has a regular rate of decay to non-radioactive carbon.
   c. the ratio of radioactive to nonradioactive carbon in living organisms is always the same as that in the atmosphere.
   d. the production of new radioactive carbon in the atmosphere just balances the natural radioactive decay of ^{14}C.
   e. all of the above

3. The total of all species of organisms living in a region is known as its
   a. biota.
   b. flora.
   c. fauna.
   d. flora and fauna.
   e. diversity.

4. The coal beds we now mine for energy are the remains of
   a. trees that grew in swamps during the Carboniferous period.
   b. trees that grew in swamps during the Devonian period.
   c. trees that grew in swamps during the Permian period.
   d. herbaceous plants that grew in swamps during the Carboniferous period.
   e. none of the above

5. The cause of mass extinctions at the end of the Ordivician was
   a. the collision of Earth with a large meteorite.
   b. massive vulcanism.
   c. massive glaciation in Gondwana.
   d. the uniting of all continents to form Pangaea.
   e. changes in Earth's orbit.

6. The cause of the mass extinction at the end of the Mesozoic era probably was
   a. continental drift.
   b. the collision of Earth with a large meteorite.
   c. changes in Earth's orbit.
   d. massive glaciation.
   e. changes in the salt concentration of the oceans.

7. The times during the history of life when many new evolutionary lineages appeared were the
   a. Precambrian, Cambrian, and Triassic.
   b. Precambrian, Cambrian, and Tertiary.
   c. Cambrian, Paleozoic, and Triassic.
   d. Cambrian, Triassic, and Devonian.
   e. Paleozoic, Triassic, and Tertiary.

8. Many scientists believe that the collision of Earth with a large meteorite was a major contributor to the mass extinction at the boundary between the Cretaceous and Tertiary periods, because
   a. there is an iridium-rich layer at the boundary of rocks from these two periods.
   b. a crater that may be the site of the collision has been found off the Yucatán Peninsula.
   c. the mass extinction at the end of the Cretaceous may have been very sudden.
   d. new methods have allowed scientists to date the iridium layer very precisely.
   e. all of the above

9. We know that organisms can evolve very rapidly, because
   a. the fossil record reveals periods of very rapid evolutionary change.
   b. theoretical models of evolutionary change show that rapid change can be produced by natural selection.
   c. rapid evolutionary changes have been produced under artificial selection.
   d. rapid evolutionary changes have been measured in natural populations of organisms during the past century.
   e. all of the above

10. At which of the following times was there *no* mass extinction?
    a. The end of the Cambrian period
    b. The end of the Devonian period
    c. The end of the Permian period
    d. The end of the Triassic period
    e. The end of the Silurian period

Applying Concepts

1. Some lineages of organisms have evolved to contain large numbers of species, whereas other lineages have produced only a few species. Is it meaningful to consider the former as more successful than the latter? What does the word "success" mean in evolution? How does your answer influence your thinking about *Homo sapiens*, the only surviving respresentative of the Hominidae—a family that never had many species in it?

2. If extinction rates in groups of organisms are relatively constant over long time spans, as they sometimes are, then recently evolved species should be no better adapted to their environments than older species are. Does this observation contradict the belief that natural selection adapts organisms to their environments?

3. Why is it useful to be able to date past events absolutely as well as relatively?

4. In a study of lungfish evolution, fossil lungfishes and modern specimens were assigned scores on the basis of the number of derived versus ancestral traits they possessed. On this scale, an organism that has only ancestral traits receives a score of 0. An organism that has only derived traits receives a score of 100. The rate of appearance of these traits in lungfishes is plotted in the two graphs that folllow. What does this pattern suggest about evolutionary radiation among lungfishes?

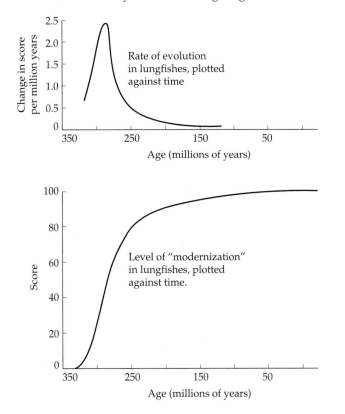

5. What factors favor increases in body size? How could *average* body size among species in a lineage decrease even if natural selection is favoring large body sizes in most of the species in the lineage?

Readings

Bonner, J. T. 1988. *The Evolution of Complexity by Means of Natural Selection.* Princeton University Press, Princeton, NJ. An excellent treatment of broad patterns in the evolutionary record and the developmental and physiological bases for them.

Burney, D. A. 1993. "Recent Animal Extinctions: Recipe for Disaster." *American Scientist*, vol. 81, pages 530–541. This account shows that late prehistoric extinctions were caused by factors, such as climate and vegetation change, human hunting, and the arrival of exotic animals, that are also serious today.

Dalziel, I. W. D. 1995. "Earth before Pangaea." *Scientific American*, January. An account of how geologists search for clues to the early wanderings of the continents.

Erwin, D. H. 1996. "The Mother of Mass Extinctions." *Scientific American*, July. A description of the mass extinction at the end of the Permian period.

Fenchel, T. and B. J. Finlay. 1994. "The Evolution of Life without Oxygen." *American Scientist*, vol. 82, pages 22–28. A discussion of clues about the evolution of early eukaryotes.

Futuyma, D. J. 1998. *Evolutionary Biology*, 3rd Edition. Sinauer Associates, Sunderland, MA. The most complete general treatment of evolution and its mechanisms.

Gates, D. M. 1993. *Climate Change and Its Biological Consequences.* Sinauer Associates, Sunderland, MA. An accessible book with extensive discussions of Earth's past climates and the methods scientists use to recreate climatic history.

Gordon, M. S. and E. C. Olson. 1995. *Invasions of the Land.* Columbia University Press, New York. A thorough account of the ways in which different lineages of organisms adapted to and colonized terrestrial environments.

Gould, S. J. 1989. *Wonderful Life: The Burgess Shale and the Nature of History.* W. W. Norton, New York. An engaging account of the remarkable fauna, containing representatives of phyla of animals that left no modern descendants, in the Burgess Shale. Explores the implications of this fauna for our view of the history of life.

Gould, S. J. 1994. "The Evolution of Life on Earth." *Scientific American*, October. A discussion of how catastrophes and random events have shaped the course of evolution.

Grimaldi, D. A. 1996. "Captured in Amber." *Scientific American*, April. This account shows how insects preserved in amber reveal aspects of genetic evolution.

Simpson, G. G. 1944. *Tempo and Mode in Evolution.* Columbia University Press, New York. A dated but classic book providing clear pictures of the long-term patterns of evolutionary change.

Stanley, S. M. 1987. *Extinction.* Scientific American Library, New York. A beautifully illustrated account of the fossil record and the phenomenon of extinction.

Ward, P. D. 1992. *On Methuselah's Trail: Living Fossils and the Great Extinctions.* W. H. Freeman, New York. A delightful personal account by a paleontologist who has studied both "living fossils" and true fossils embedded in rocks that formed at the times of the great extinctions.

Chapter 20

The Mechanisms of Evolution

Tuberculosis Is Again a Threat
In recent years there has been a significant increase in cases of tuberculosis, even in children under the age of 5. The bacterium responsible for the disease has evolved resistance to the medications humans use to combat it.

By the 1980s, tuberculosis had almost disappeared from the United States. Doctors and public health officials, believing that they had won the war against the disease, turned their attention to other diseases. But while they looked away, the incidence of tuberculosis began to increase. It increased by nearly 20 percent between 1985 and 1992. The number of cases among children under 5 years old rose 30 percent between 1987 and 1990 alone.

Tuberculosis returned because populations of *Mycobacterium tuberculosis*, the bacterium that causes tuberculosis, evolved resistance to isoniazid, the principal drug used to combat the disease. The bacterium became resistant to the drug when it dropped from its chromosome a gene called *katG*, which codes for the production of two enzymes, catalase and peroxidase. Bacteria lacking this gene produce almost no catalase and peroxidase and thus are resistant to isoniazid. Proof that the lack of *katG* is responsible for the resistance came when investigators reinserted the missing gene into resistant bacteria and the bacteria subsequently were killed by isoniazid.

Tuberculosis is just one of many "nearly cured" diseases whose incidences are increasing. Most of these diseases are resurging because the microorganisms that cause them have developed resistance to drugs that formerly killed them. Many major medical and agricultural problems today are due to the rapid evolution of organisms that we have tried to control or eliminate. Why have so many disease-causing organisms evolved so rapidly? What can we learn by studying these examples of "evolution in action"? Several times in the previous chapter we referred to evolutionary agents, but we did not identify them. In this chapter we examine the processes that drive evolutionary changes and describe the agents that cause them. But first we need to define biological evolution.

452

Biological evolution is the change over time in the genetic composition of members of a population. Changes that occur over a small number of generations constitute **microevolution**. Changes that happen over centuries, millennia, and longer are called **macroevolution**. Genetic changes result when an evolutionary agent acts on genetic variability within the population. In the mid-1800s Charles Darwin identified natural selection as the key evolutionary agent. At about the same time, Gregor Mendel discovered the genetic basis of the variation on which evolutionary agents act (see Chapter 10). Mendel's and Darwin's insights form the basis for most current evolutionary hypotheses and the studies designed to test them.

In this chapter we will discuss genetically based variation and how it is measured. We will describe the agents of evolution and the short-term studies designed to investigate them. And we will show how genetic variation is maintained in populations over space and time. When you understand these processes, you will understand the mechanisms of evolution and why humans are powerful evolutionary agents.

Variation in Populations

Because biological evolution is a change over time in the genetic composition of members of a population, biologists studying evolution attempt to measure genetic variability and how it changes. The appropriate unit for defining and measuring genetic variation is a **population**: a group of organisms of the same species occupying a particular geographic region. Therefore, to understand evolutionary changes we need to engage in "population thinking."

The genetic constitution governing a heritable trait is called its **genotype** (see Chapter 10). A population evolves when individuals with different genotypes survive or reproduce at different rates. A single individual has only some of the alleles found in the population to which it belongs (Figure 20.1). The sum total of the alleles found in the population constitutes its **gene pool**.

The gene pool contains the variability on which agents of evolution act. Evolution may come about because heritable traits influence the ability of individuals to obtain mates or food or to avoid hazards. The agents of evolutionary change act on the physical expression of an organism's genotype, its **phenotype**. However, not all phenotypic variation is governed by genotypes. Some of the variation observed within populations is genetically determined, but some of it is not.

Some variation is environmentally induced

The shapes and sizes of many marine animals that live attached to a substratum (such as a rock) are affected by water temperature, concentration of nutrients (particularly calcium), competition with neighbors, and turbulence of the water. For example, limpets (*Acmaea*) growing high in the intertidal zone (the shallow water at the edge of land), where they experience heavy wave action, are more cone-shaped than are limpets of the same species growing in the subtidal zone, where they are protected from wave action. We know that this difference is not genetic, because individuals from high in the intertidal zone add new growth to their shells to produce a flatter, subtidal shape when they are transplanted to the subtidal zone (Figure 20.2a).

The cells of the leaves on a tree or shrub are normally genetically identical. Yet leaves on the same tree often differ in shape and size. Leaves closer to the top of an oak tree, for example, where they receive more wind and sunlight, may be more deeply lobed than leaves lower down on the same tree (Figure 20.2b). These within-plant variations, however, are not passed on to offspring. What *is* passed on is the ability to form various types of leaves from the same genotype in response to different environmental conditions.

Environmentally induced variation is often important in biology, but to investigate evolutionary questions, biologists must work with variation in traits that are heritable.

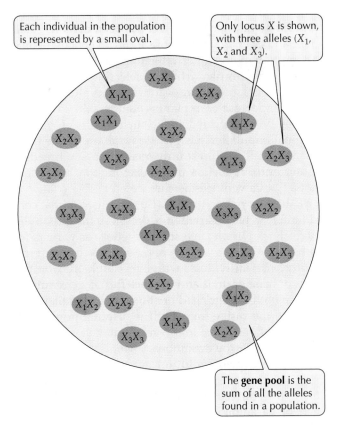

Each individual in the population is represented by a small oval.

Only locus X is shown, with three alleles (X_1, X_2 and X_3).

The **gene pool** is the sum of all the alleles found in a population.

20.1 A Gene Pool The allele frequencies in this drawing are 0.20 for X_1, 0.50 for X_2, and 0.30 for X_3.

(a)   Limpets (*Acmaea*)

In turbulent water, the form is more conical.

Moved from turbulent to nonturbulent site, new growth is less conical.

Protected from turbulent waves, the form is less conical.

High tide

Low tide

(b)   Leaves of white oak (*Quercus alba*)

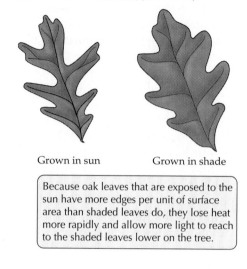

Grown in sun   Grown in shade

Because oak leaves that are exposed to the sun have more edges per unit of surface area than shaded leaves do, they lose heat more rapidly and allow more light to reach to the shaded leaves lower on the tree.

20.2 Environmentally Induced Variation   Individuals that are genetically identical may vary morphologically if they live in different environments.

Much variation is genetically based

High levels of genetic variation characterize nearly all natural populations. This fact has been demonstrated over and over again for thousands of years by people attempting to breed desirable traits in plants and animals. For example, artificial selection for different traits in a common European wild mustard produced many important crop plants (Figure 20.3). Plant and animal breeders can achieve such results only if the original population has genetic variation for the traits of interest. The almost universal success of breeders indicates that genetic variation is common, but it does not tell us how much there is.

Laboratory experiments also demonstrate that considerable genetic variation is present in most populations. For example, investigators chose as parents for subsequent generations of a species of fruit fly (*Drosophila*) individuals with either high or low numbers of bristles on their bodies. After 35 generations, flies in both lineages had bristle numbers that fell well outside the range found in the original population (Figure 20.4). Notice that considerable variation that can be acted on by evolutionary agents still remains.

To understand evolution, we need to know more precisely how much genetic variation populations have, the sources of that genetic variation, and how genetic variation is maintained and expressed in populations in space and over time. We also need to know the agents that change the genetic variation in popula-

tions, how they act, and their relative importance in affecting the direction of evolutionary changes.

How do we measure genetic variation?

A locally interbreeding group within a geographic population is called a **deme** or a Mendelian population. Demes are often the subjects of evolutionary studies. In this section we will describe how genetic variation within demes is measured and how it is influenced by evolutionary agents. To measure the gene pool of a deme we would need to count every allele at every locus in every organism in that deme. By measuring all the individuals in a deme, we could determine the relative proportions, or frequencies, of all alleles in the deme.

This would be an impossible task, but biologists can *estimate* **allele frequencies** by measuring numbers of alleles in a *sample* of individuals from a deme. Measures of frequency range from 0 to 1; the sum of all allele frequencies at a locus is equal to 1. The percentages of different alleles at each locus and the percentages of different genotypes in a deme describe its **genetic structure**. An allele's frequency is calculated using the following formula:

$$\frac{\text{number of copies of the allele in the population}}{\text{sum of alleles in the population}}$$

If only two alleles (*A* and *a*) for a given locus are found among the members of a diploid population, they may combine to form three different genotypes: *AA*, *Aa*, and *aa*. Using the formula above, we can calculate the relative frequencies of alleles *A* and *a* in a population of *N* individuals as follows:

Let N_{AA} be the number of individuals that are homozygous for the *A* allele (*AA*);

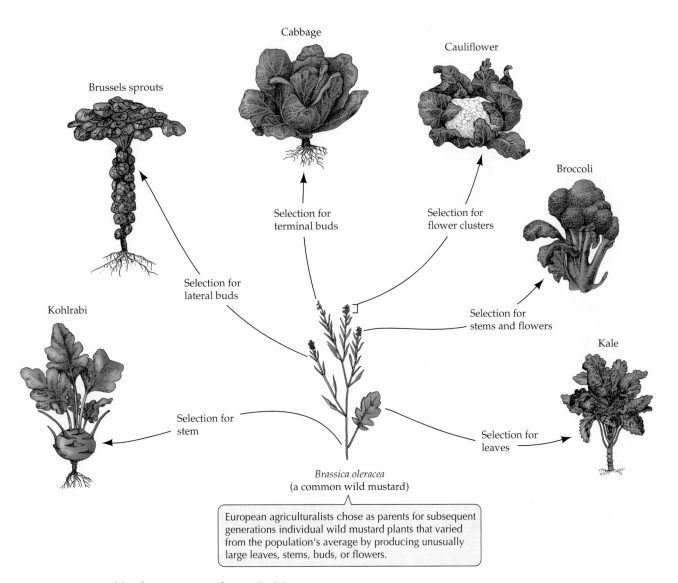

Brussels sprouts

Cabbage

Cauliflower

Broccoli

Selection for
terminal buds

Selection for
flower clusters

Selection for
lateral buds

Kohlrabi

Selection for
stems and flowers

Kale

Selection for
stem

Selection for
leaves

Brassica oleracea
(a common wild mustard)

European agriculturalists chose as parents for subsequent
generations individual wild mustard plants that varied
from the population's average by producing unusually
large leaves, stems, buds, or flowers.

20.3 Many Vegetables from One Species All of these
crop plants have been derived from one wild mustard
species. They illustrate the vast amount of variation that can
be present in a gene pool.

Let N_{Aa} be the number that are heterozygous (Aa);

and

Let N_{aa} be the number that are homozygous for the
a allele (aa).

Note that $N_{AA} + N_{Aa} + N_{aa} = N$, the total number of
individuals in the population, and that the total number of alleles present in the population is $2N$ because
each individual is diploid. Each AA individual has
two A alleles and each Aa individual has one A allele.
Therefore, the total number of A alleles in the population is $2N_{AA} + N_{Aa}$, and the total number of a alleles in
the population is $2N_{aa} + N_{Aa}$.

If p represents the frequency of A, and q represents
the frequency of a, then

$$p = \frac{2N_{AA} + N_{Aa}}{2N}$$

and

$$q = \frac{2N_{aa} + N_{Aa}}{2N}$$

To see how this works, let's calculate allele frequencies
in two populations, each consisting of 200 diploid individuals. Population 1 has mostly homozygotes (90
AA, 40 Aa, and 70 aa); population 2 has mostly heterozygotes (45 AA, 130 Aa, and 25 aa). In population 1,
where $N_{AA} = 90$, $N_{Aa} = 40$, and $N_{aa} = 70$,

$$p_1 = \frac{2N_{AA} + N_{Aa}}{2N} = \frac{180 + 40}{400} = 0.55$$

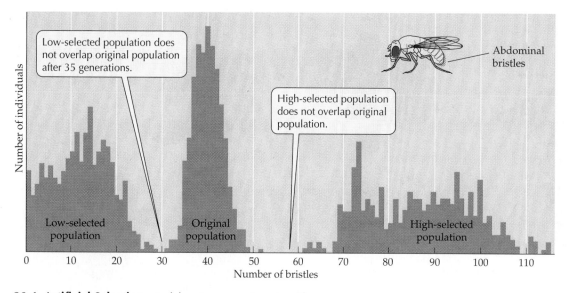

20.4 Artificial Selection In laboratory experiments with *Drosophila*, changes in bristle number evolved rapidly when artificially selected for.

and

$$q_1 = \frac{2N_{aa} + N_{Aa}}{2N} = \frac{140 + 40}{400} = 0.45$$

In population 2, where $N_{AA} = 45$, $N_{Aa} = 130$, and $N_{aa} = 25$,

$$p_2 = \frac{2N_{AA} + N_{Aa}}{2N} = \frac{90 + 130}{400} = 0.55$$

and

$$q_2 = \frac{2N_{aa} + N_{Aa}}{2N} = \frac{50 + 130}{400} = 0.45$$

These calculations demonstrate two important points. First, notice that for each population, $p + q = 1$. If there is only one allele in a population, its frequency is 1. If an allele is missing from a population, its frequency is 0, and the locus in that population is represented by one or more other alleles. Because $p + q = 1$, $q = 1 - p$, which means that when there are two alleles at a locus in a population, we can calculate the frequency of one allele and then easily obtain the second frequency by subtraction.

The second thing to notice from these calculations is that the two populations—one consisting mostly of homozygotes and the other mostly of heterozygotes—have the same allele frequencies for A and a. Therefore, they have the same gene pool for this locus. However, because the alleles in the gene pool are distributed differently among genotypes, the populations have *different genetic structures*.

Although we began our calculations with *numbers* of genotypes, for many purposes genotypes, like alleles, are best thought of as frequencies. In population 1 of our example, the genotype frequencies, which we calculate as the number of individuals that have the genotype divided by the total number of individuals in the population, are 0.45 *AA*, 0.20 *Aa*, and 0.35 *aa*. What are the genotype frequencies of population 2?

Preserving Genetic Variability: The Hardy–Weinberg Rule

A population that is not changing genetically, that has the same allele and genotype frequencies from generation to generation, is said to be at **equilibrium**. The conditions that result in an equilibrium population were discovered independently by the British mathematician Godfrey H. Hardy and the German physician Wilhelm Weinberg in 1908. Hardy wrote his equations in response to a question posed to him by the Mendelian geneticist Reginald C. Punnett at the Cambridge University faculty club. Punnett was puzzled by the fact that although the allele for short fingers was dominant and the allele for normal-length fingers was recessive, most people in Britain had normal-length fingers. Hardy's equations explain why dominant alleles do not replace recessive alleles in populations. They also explain other features of the genetic structure of populations.

The Hardy–Weinberg rule, as these equations are now collectively called, consists of a set of assumptions and two major results. The assumptions are that the population is very large, that mating is random, and that no agents of evolution are acting on the population. The first major result of the rule is that if these conditions hold, the frequencies of alleles at a locus will remain constant from generation to generation. The second result is that after one generation of random mating, the genotype frequencies will remain in the proportions

$$p^{2(AA)} + 2pq^{(Aa)} + q^{2(aa)} = 1$$

To see why these results are true, consider an example in which the frequency of *A* alleles (*p*) is 0.6. Because we assume that individuals select mates without regard to genotype, gametes carrying *A* or *a* combine at random—that is, as predicted by the frequencies *p* and *q*. The probability that any given sperm or egg in this example will bear an *A* allele rather than an *a* allele is 0.6. In other words, 6 out of 10 random selections of a sperm or an egg will bear an *A* allele. Because *q* = 1 − *p*, the probability of an *a* allele is 1 − 0.6 = 0.4.

To obtain the probability of two *A*-bearing gametes coming together at fertilization, we multiply the two independent probabilities of drawing them: $p \times p = p^2 = (0.6)^2 = 0.36$ (see the discussion of probability in Chapter 10). Therefore, 0.36, or 36 percent, of the offspring in the next generation will have the *AA* genotype. Similarly, the probability of bringing together two *a*-bearing gametes is $q \times q = q^2 = (0.4)^2 = 0.16$, so 16 percent of the next generation will have the *aa* genotype (Figure 20.5).

Figure 20.5 also shows that there are two ways of producing a heterozygote: an *A* sperm may combine with an *a* egg, the probability of which is $p \times q$; or an *a* sperm may combine with an *A* egg, the probability of which is also $p \times q$. Consequently, the overall probability of obtaining a heterozygote is $2pq$. What percentage of the next generation will be heterozygotes?

It is easy now to show that the allele frequencies *p* and *q* remain constant for each generation. Notice that the total of $p^2 + pq$ represents the total of the *A* alleles. The fraction that this frequency constitutes of all alleles is

$$\frac{p^2 + pq}{p^2 + 2pq + q^2} = \frac{p(p+q)}{(p+q)(p+q)} = \frac{p}{p+q} = \frac{p}{p+(1-p)} = p$$

Similarly, the frequency of *a* in the next generation will be

$$\frac{q^2 + pq}{p^2 + 2pq + q^2} = \frac{q(p+q)}{(p+q)(p+q)} = \frac{q}{p+q} = \frac{q}{(1-q)+q} = q$$

Thus the original allele frequencies are unchanged and the population is in Hardy–Weinberg equilibrium.

Why is the Hardy–Weinberg rule important?

The most important message of the Hardy–Weinberg rule is that allele frequencies remain the same from generation to generation unless a particular agent acts

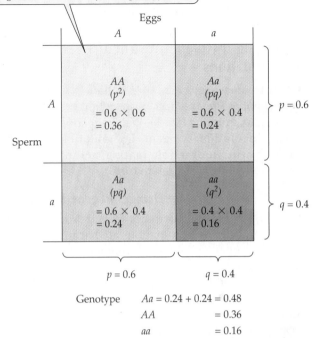

20.5 Calculating Hardy–Weinberg Genotype Frequencies
The probabilities of producing each genotype are calculated by assuming that mating is random. Because there are two ways of producing a heterozygote, the probability of this event occurring is the sum of the two *Aa* squares.

to change them. The rule also shows what distribution of genotypes to expect for a population at equilibrium at any value of *p* and *q*.

Given the stringency of the necessary conditions, you may already have recognized that populations in nature are rarely in Hardy–Weinberg equilibrium. Why, then, is the rule considered so important for the study of evolution? The answer is that without it, we cannot tell whether evolutionary agents are operating. If individuals in a population mate randomly and no other agents are operating to change allele frequencies, then the genotype frequencies we actually observe in the population will approximate those we calculate from the Hardy–Weinberg formula. However, if the frequencies of genotypes deviate significantly from the expected Hardy–Weinberg values, we have evidence that an agent of evolution is in action.

In practice it is not easy to measure allele frequencies precisely enough to determine if they are at Hardy–Weinberg proportions. Also, when we consider more than a single locus, the Hardy–Weinberg rule does not apply. Natural selection usually operates simultaneously on many traits or on traits governed jointly by more than one locus. Therefore to tell whether an evolutionary agent is influencing a population, biologists usually must measure the actual changes in allele frequencies in that population.

Changing the Genetic Structure of Populations

Evolutionary agents are forces that change the allele and genotype frequencies in a population. They cause deviations from the Hardy–Weinberg equilibrium. The known evolutionary agents are mutation, gene flow, genetic drift, nonrandom mating, and natural selection. We will discuss each in turn.

Mutations are random changes in genetic material

The origin of genetic variation is mutation (see Chapters 10 and 12). Most mutations are harmful or neutral to their bearers, but if the environment changes, previously neutral or harmful alleles may become advantageous. Mutation rates are very low for most loci that have been studied. Rates as high as one mutation in a thousand zygotes per generation are rare; one in a million is more typical. Nonetheless, these rates are sufficient to create considerable genetic variation over long time spans. In addition, mutations can restore to populations alleles that other evolutionary agents remove. Thus mutations both create and help maintain variation within populations.

Mutation rates are unusually high among some disease-causing microorganisms. This is especially true of viruses that are attacked by the immune systems of their hosts. For example, the human immunodeficiency virus type 1 (HIV-1; see Chapters 13 and 18), which is actively attacked by the human immune system, has such high mutation rates that it can overcome the host's defenses and evolve resistance faster than its host can respond. The viruses in a person with an advanced case of AIDS are genetically quite different from those that initially started the infection—they have evolved.

One condition for Hardy–Weinberg equilibrium is that there be no mutation. Although this condition is never strictly met, the rate at which mutations arise at single loci is usually so low that mutations result in only very small deviations from Hardy–Weinberg expectations. If large deviations are found, it is appropriate to dismiss mutation as the cause and to look for evidence of other evolutionary agents.

Migration of individuals followed by breeding produces gene flow

Because few populations are completely isolated from other populations of the same species, usually some migration between populations takes place. **Gene flow** happens when migration is *followed by breeding* in the new location. Gene flow ranges from extremely low to very high, depending on the number of migrating individuals and their genotypes.

Immigrants (individuals entering a population) may add new alleles to the pool of a population or may change the frequencies of alleles already present. Emigrants (individuals leaving a population) may completely remove alleles or may change the frequencies of alleles when they leave a population. For a population to be in Hardy–Weinberg equilibrium, there must be no net gene flow between it and other populations—in other words, the number of new alleles added by immigrants must equal the number of alleles removed by emigrants.

Genetic drift may cause large changes in small populations

In small populations, chance events can significantly alter allele frequencies. Such alteration is called **genetic drift**. Genetic drift is the reason that a population must be very large to be in Hardy–Weinberg equilibrium. If only a few individuals or a few gametes are drawn at random to form the next generation, the alleles they carry are not likely to be in the same proportions as alleles in the gene pool from which they were drawn. In very small populations, genetic drift may be strong enough to influence the direction of change of allele frequencies even when other evolutionary agents are pushing the frequencies in a different direction. Harmful alleles, for example, may increase because of genetic drift, and rare advantageous alleles may be lost. Two important causes of genetic drift are bottlenecks and founder effects.

BOTTLENECKS. Even organisms that normally have large populations may pass through occasional periods when only a small number of individuals survive. During these population **bottlenecks**, genetic variation can be lost by chance. This phenomenon is illustrated in Figure 20.6, which shows allele frequencies as proportions of red and yellow beans. Most of the beans that survive to germinate the next generation in this example are red, so the new population has a much higher frequency of red beans than the previous generation had.

Suppose we have performed a cross of $Aa \times Aa$ individuals of a species of *Drosophila* to produce an offspring population in which $p = q = 0.5$ and in which the genotype frequencies are 0.25 AA, 0.50 Aa, and 0.25 aa. If we randomly select four individuals from among the offspring to form the next generation, the allele frequencies in this small sample may differ markedly from $p = q = 0.5$. If, for example, we happen by chance to draw two AA homozygotes and two heterozygotes (Aa), the genotype frequencies in this "surviving population" are $p = 0.75$ and $q = 0.25$. If we replicate this sampling experiment 1,000 times, one of the two alleles will be missing entirely from about eight of the 1,000 "surviving populations."

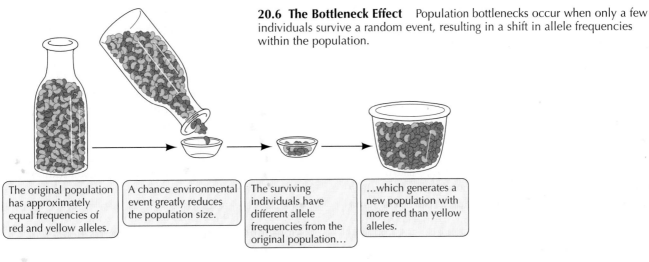

20.6 The Bottleneck Effect Population bottlenecks occur when only a few individuals survive a random event, resulting in a shift in allele frequencies within the population.

The original population has approximately equal frequencies of red and yellow alleles.

A chance environmental event greatly reduces the population size.

The surviving individuals have different allele frequencies from the original population…

…which generates a new population with more red than yellow alleles.

Populations in nature pass through bottlenecks for many different reasons. Predators may reduce populations of their prey to very small sizes. During the 1890s, hunting reduced the number of northern elephant seals to about 20 animals in a single population on the coast of Mexico. The actual breeding population may have been even smaller because only a few males mate with all the females and father the offspring in any generation (Figure 20.7).

By analyzing small samples of tissue collected from the current California population of northern elephant seals, which are descendants of the Mexican population that survived the bottleneck of the 1890s, scientists determined that these seals have less genetic variation than any other seal that has been studied. The investigators examined 24 proteins from each animal electrophoretically (see Chapter 16), looking for evidence of genetic variation among the seals at the loci encoding the proteins. They found no evidence of variation in any of the 24 proteins.

By contrast, the southern elephant seal, whose numbers were not severely reduced by hunting, has much more genetic variation. Currently, northern elephant seal populations are expanding rapidly, so their reduced genetic variability is not preventing high survival and reproductive rates. However, biologists worry that they may be vulnerable to a disease outbreak or other sudden environmental change.

FOUNDER EFFECTS. When a species expands into new regions, populations may be started by a small number of pioneering individuals. These pioneers are not likely to have all the alleles found in their source population. Even if they do, the allele frequencies are likely to differ from those in the source population. The situation is equivalent to that for a large population reduced by a bottleneck, but rather than a small surviving population, there is a small founding population. This type of genetic drift is called a **founder effect**. Because many plant species reproduce sexually by self-fertilization, a new plant population may be started by a single seed—an extreme example of a founder effect.

20.7 A Species with Low Genetic Variation
A few males sire most of the offspring in this northern elephant seal breeding colony, and the size of the breeding population is smaller than the whole population. This nonrandom mating, together with a bottleneck that occurred when the seals were overhunted in the late nineteenth century, resulted in a population with very little genetic variation.

Scientists were given an opportunity to study the genetic composition of a founding population when *Drosophila subobscura*, a well-studied European species of fruit fly, was discovered near Puerto Montt, Chile, in 1978 and at Port Townsend, Washington, in 1982. In both South and North America, populations of the flies grew rapidly and expanded their ranges. Today in North America, *D. subobscura* ranges from British Columbia, Canada, to central California. In Chile it has spread across 23° of latitude, nearly as wide a range as the species has in Europe (Figure 20.8).

The *D. subobscura* founders probably reached Chile and the United States from Europe aboard the same ship, because both populations are genetically very similar. For example, the North and South American populations have only 20 chromosomal inversions (see Chapter 12), 19 of which are the same on the two continents, whereas 80 inversions are known from European populations. New World populations also have fewer enzyme alleles than Old World populations. Only alleles that have a frequency higher than 0.1 in European populations are present in the Americas. Thus, as expected from a small founding population, only a small part of the total genetic variability found in Europe reached the Americas. Geneticists estimate that at least ten, but no more than a hundred, flies initially arrived in the New World.

EFFECTIVE POPULATION SIZE. Genetic drift can be important even in common species, many of which are divided into small, localized breeding populations. In addition, although there may be *N* individuals in a population, if some of them do not reproduce, the population is actually much smaller from a genetic point of view. For example, if only half of the individuals in a population of $N = 100$ breed, the population has an effective size, N_e, of 50, and N_e rather than N determines the rate of genetic drift. More generally, anything that causes some individuals to leave more offspring than others—such as unequal sex ratios, breeding systems in which only a few males in the population fertilize most of the females, and differential survival—reduces the effective population size.

Nonrandom mating increases the frequency of homozygotes

One Hardy–Weinberg assumption specifies that individuals do not choose mates on the basis of their genotypes; that is, mating must be random. In many cases, however, individuals with certain genotypes mate more often with individuals of either the same or different genotypes than would be expected on a random basis. Humans, for example, tend to choose mates that are similar in appearance to themselves. When such **assortative mating** takes place, in the next generation homozygous genotypes are overrepresented and heterozygous genotypes are underrepresented in comparison with Hardy–Weinberg expectations.

Self-fertilization (selfing), another form of nonrandom mating, is common in many groups of organisms, especially plants. Selfing tends to reduce the frequencies of heterozygous individuals in populations below Hardy–Weinberg expectations. Under assortative mating and self-fertilization, genotype frequencies change but allele frequencies remain the same. Nonrandom mating can alter allele frequencies, however, if some individuals are more successful in mating than others. We consider this situation, which is called **sexual selection**, in greater detail later in this chapter and in Chapter 50.

Natural selection leads to adaptation

Individuals vary in heritable traits that determine the success of their reproductive efforts. Not all individuals survive and reproduce equally well in a particular environment. Therefore, some individuals contribute more offspring to the next generation than do other individuals. As a result, allele frequencies in the population change.

The differential contribution of offspring resulting from variations in heritable traits was called **natural selection** by Charles Darwin. Natural selection is the only evolutionary agent that adapts organisms to their environments. Biologists investigate the action of natural selection by comparing genotype (or phenotype) frequencies between generations and attempting to

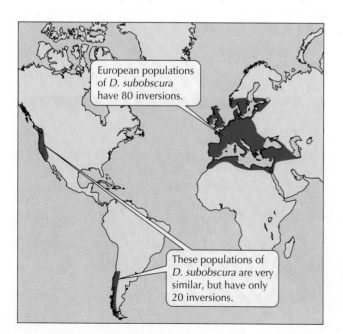

20.8 A Founder Effect Within two decades of arriving in the New World, populations of the fruit fly *Drosophila subobscura* have increased dramatically and spread widely in spite of having much-reduced genetic variation.

European populations of *D. subobscura* have 80 inversions.

These populations of *D. subobscura* are very similar, but have only 20 inversions.

determine the reasons that some genotypes (or phenotypes) survived and reproduced better than others.

Natural selection produces variable results

Depending on which traits are favored in a population, natural selection can produce any one of several quite different results. Selection may (1) preserve the characteristics of a population by favoring average individuals, (2) change the characteristics of a population by favoring individuals that vary in one direction from the mean of the population, or (3) change the characteristics of a population by favoring individuals that vary in both directions from the mean of the population.

In the example in Figure 20.9, the variable trait is size, a trait likely to be controlled by many loci. If many genetic and environmental factors contribute to size—and there is no selection—then the distribution of sizes in a population should approximate the bell-shaped curve shown in the top row of the figure.

If both the smallest and the largest individuals contribute relatively fewer offspring to the next generation than those closer to the center do, **stabilizing selection** is operating (Figure 20.9a). Stabilizing selection reduces variation but does not change the mean. Natural selection frequently acts in this way, countering increases in variation brought about by mutation or migration. We know from the fossil record that most populations evolve slowly most of the time. Rates of evolution are typically very slow because natural selection is usually stabilizing.

If individuals at one extreme of the size distribution—the larger ones, for example (as illustrated in Figure 20.9b)—contribute more offspring than other individuals do, then the mean size of individuals in the population will increase or decrease accordingly, and **directional selection** is operating. If directional selection operates over many generations, an evolutionary trend within the population results. Directional evolutionary trends often continue for many generations, but they may be reversed when the environment changes and different phenotypes are favored.

Individuals of plants growing on mine tailings may be able to tolerate high soil concentrations of heavy metals that are lethal to plants of the same species growing elsewhere. That ability resulted from directional selection. How tolerance might have evolved was demonstrated by an experimenter who planted seeds of

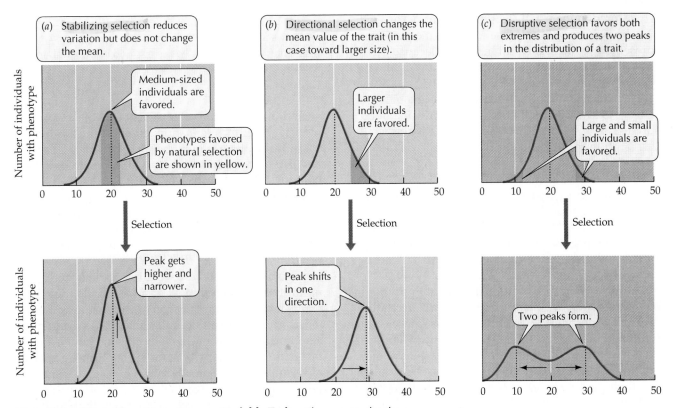

20.9 Natural Selection Operates on a Variable Trait The curves plot the distributions of phenotypes (body size) in a population before selection (top) and after selection (bottom). Natural selection may change the shape and position of the original curves.

individuals from a nontolerant population in soil contaminated with copper. He harvested seeds from the few individuals that survived and reproduced, and planted them in contaminated soil. He repeated this procedure over several generations and found that the proportion of seeds that produced viable adults increased each generation. Tolerance to high concentrations of copper evolved rapidly because only tolerant individuals survived long enough to reproduce.

Disruptive selection is selection that simultaneously favors individuals at both extremes of the distribution (Figure 20.9c). This type of selection apparently is rare. When disruptive selection operates, individuals at the extremes contribute more offspring than those in the center, producing two peaks in the distribution of a trait. The strikingly bimodal (two-peaked) distribution of bill sizes in the black-bellied seedcracker, *Pyrenestes ostrinus*, a West African finch (Figure 20.10), illustrates how disruptive selection can adapt populations in nature.

Seeds of two types of sedges (a marsh plant) are the most abundant food source for the finches during part of the year. Birds with large bills readily crack the hard seeds of the sedge *Scleria verrucosa*; birds with small bills crack *S. verrucosa* seeds only with difficulty, but small-billed birds feed more efficiently on the soft seeds of *S. goossensii* than do birds with larger bills.

Young finches whose bills deviate markedly from the two predominant bill sizes do not survive as well as finches whose bills are close to one of the two sizes represented by the distribution peaks. Because there are few abundant food sources in the environment and because the seeds of the two sedges do not overlap in hardness, birds with intermediate-sized bills are inefficient in utilizing either one of the principal food sources. Disruptive selection therefore maintains a bill size distribution with two peaks.

Sexual reproduction increases genetic variation

Recombination in sexually reproducing organisms multiplies the opportunities for natural selection to operate by increasing genetic variability. In asexually reproducing organisms, the daughter cells resulting from the mitotic division of a single cell normally contain identical genotypes at all loci.

In a population that reproduces asexually, there is no recombination of genetic material from different individuals, and every new individual is genetically identical to its parent unless there has been a mutation. However, when organisms exchange genetic material during sexual reproduction, the offspring differ from their parents because chromosomes assort randomly during meiosis and because fertilization brings together material from two different cells.

Sexual recombination generates an endless variety of genotypic combinations that increases the *evolutionary potential* of populations. Because it increases the

20.10 Natural Selection Alters Bill Sizes The distribution of bill sizes in the black-bellied seedcracker of West Africa is an example of disruptive selection, which favors individuals with larger and smaller bill sizes over individuals with intermediate-size bills.

variation among the offspring produced by an individual, sexual recombination improves the chance that some of the offspring will be successful in the varying and often unpredictable environments they will encounter. As the Hardy–Weinberg rule shows, sexual recombination does not influence the frequency of different alleles. Rather, it generates new combinations of genetic material on which natural selection can act.

Adaptation is studied in various ways

Biologists use several different methods to study how natural selection adapts organisms to their environments. One way is to measure the consequences of altering the form of a particular feature of an organism. Another, known as the comparative method, is to predict patterns of adaptation among species. We provide an example of each method.

ALTERING TAIL LENGTH IN BARN SWALLOWS. Both male and female barn swallows have long forked tails, but the tails of males are about 16 percent longer than the tails of females. Swallows molt their feathers every autumn, at which time they grow the new tails they

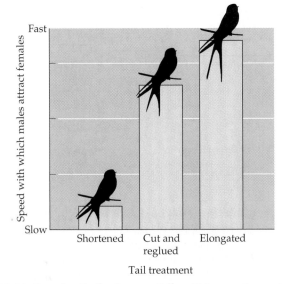

20.11 Females Prefer Longer Tails This experiment, in which the tails of male barn swallows were artificially shortened or lengthened, showed that female barn swallows prefer long-tailed males.

will have the next spring. By cutting off the ends of the tails of some males and elongating the tails of others by gluing additional feathers to them, an investigator showed that females prefer to mate with males that have longer than average tails. (Control males were handled and their tails were cut and reglued but not changed in length.)

The males with long tails attracted females faster than the males with short tails (Figure 20.11). However, flying around with a long tail is energetically costly. Males whose tails were artifically elongated one year grew tails that were shorter than normal the next year, and they took longer to attract mates than males whose tails had not been elongated. In barn swallows,

then, sexual selection by females results in males having longer tails than females.

Why do females preferentially mate with males that have longer tails? They evolved this preference because by exercising it they produce healthier offspring. Their offspring are healthier because males with naturally longer tails are more vigorous than are males with shorter tails. Males that are resistant to the mites that often infest swallow nests become heavier and they grow longer tails than less resistant males do. Offspring of heavily parasitized males are more likely to become infested, too, even if they are raised in foster nests away from their fathers.

MALE PLUMAGE AND PARASITE RESISTANCE. If resistance to parasites is heritable, females benefit by mating with resistant males. By choosing males on the basis of traits that signal their resistance, females may favor the evolution of a more striking expression of those traits. In many bird species, males with fewer parasites are healthier and thus more vividly colored than males who suffer with many parasites. Therefore, in species that typically are subject to heavy parasite infestation, females should pay particular attention to a male's bright plumage when choosing a mate; if, on the other hand, parasite infestation is not typically a problem for the species, females should pay little attention to the color of a male's feathers.

Many such evolutionary theories are tested by predicting and testing patterns among species. To test the prediction that colorful plumage would be more likely to occur in heavily parasitized than in lightly parasitized species, scientists compared the male plumages of many species. They did indeed find a positive correlation between brightly hued males and high normal levels of parasite infestation, giving the females of these species the opportunity to judge the health of prospective mates by their colors (Figure 20.12).

Piranga olivacea (scarlet tanager)   *Passerina ciris* (painted bunting)   *Icterus galbula* (northern oriole)

20.12 Brightly Colored Males Are Healthy Sexual selection for healthy males has produced the striking plumages of male birds in (from left) the scarlet tanager, painted bunting, and northern oriole.

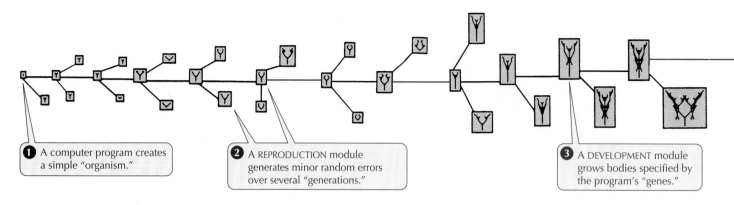

1 A computer program creates a simple "organism."

2 A REPRODUCTION module generates minor random errors over several "generations."

3 A DEVELOPMENT module grows bodies specified by the program's "genes."

Natural selection acts nonrandomly

Many people have erroneously concluded that Darwinian evolution is a random process because mutation, which provides the raw material on which natural selection acts, is random. Nothing could be further from the truth. Although mutations appear randomly, natural selection increases the frequency of only those mutant alleles that improve survival and reproductive success. Thus, evolutionary changes are the result of nonrandom processes continuing over many generations.

In addition to being nonrandom, adaptation by natural selection is a cumulative process. Natural selection acts on modifications to already existing organisms. As Richard Dawkins pointed out in his book *The Blind Watchmaker*, cumulative evolutionary change is like a computer program in which in every generation a module called REPRODUCTION takes the genes handed to it by the previous generation and hands them to the next generation, but with minor random errors.

After being reproduced, the genes are handed to another module, called DEVELOPMENT, which grows the body specified by the genes. The offspring carrying mutant genes have bodies slightly different from those of their parents. The modified bodies are acted on by another module, called EVOLUTION, that favors some body forms over others. The favored ones become the parents of the next generation. By this process, surprising changes can accumulate over a small number of generations, even though each change is small.

Figure 20.13 shows the results that Dawkins obtained during 29 generations of the computer program he created. The program started with a simple dot. Subsequent "mutations" caused the dot to branch, the branches to curve, and then the branches to branch again. This artificial example shows how complex forms can arise by nonrandom selection of slight modifications of previous body plans.

Fitness is the relative reproductive contribution of genotypes

A central evolutionary concept is **fitness**. The fitness of a genotype or phenotype is its reproductive contribution to subsequent generations *relative* to the contribution of other genotypes or phenotypes. The word "relative" is critical; the absolute number of offspring produced by an individual does not influence allele frequencies. Changes in *absolute* numbers of offspring are responsible for increases and decreases in the *size* of a population, but the *relative* success among genotypes within a population is what leads to changes in allele frequencies—that is, to evolution.

An individual may influence its own fitness in two ways. First, it may produce its own offspring, contributing to its **individual fitness**. Second, it may help the survival of relatives that have the same alleles by descent from a common ancestor—a phenomenon known as **kin selection**. Together, individual fitness and kin selection determine the inclusive fitness of the individual.

Among species that either are solitary or reproduce in groups no larger than a pair and its offspring, individual fitness strongly dominates inclusive fitness. However, among highly social species, such as social insects, some birds, and many primates, kin selection also may be very important. We will return to this topic in Chapter 50. Now we will consider the role of individual selection in the distribution and maintenance of genetic variation in natural poplations.

Maintaining Genetic Variation

Genetic drift, stabilizing selection, and directional selection all tend to reduce genetic variation within populations. Nevertheless, most populations show considerable genetic variation. Why isn't the genetic variation of a species lost over time and space? To answer this question we will distinguish between heritable and nonheritable variation and examine how genetic variation is distributed among and within populations of a species. By combining these components we will be able to understand why genetic variation is maintained.

Not all genetic variation affects fitness

Genetic variation that does not affect the fitness of an organism is called **neutral variation**. For example, some amino acid substitutions caused by mutations

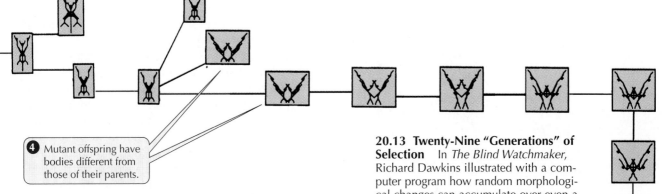

4 Mutant offspring have bodies different from those of their parents.

do not affect the functioning of the molecules of which they are a part. Such substitutions may be lost, or their frequencies may increase with time, untouched by natural selection. Therefore, much variation in neutral traits exists in most populations. Variation in the traits of organisms that we can observe with our unaided senses usually is not neutral, but much molecular variation apparently is neutral.

Modern molecular techniques enable us to measure neutral variation and provide the means by which to distinguish adaptive from neutral variation. How molecular techniques enable us to make those discriminations and how neutral variations can be used to estimate rates of evolution will be considered in Chapter 23.

Genetic variation is maintained spatially

Some of the genetic variation among populations of many species is correlated with the geographic distribution of the populations. For example, plant populations may vary geographically in the chemicals they synthesize to defend themselves against herbivores. Some individuals of the clover *Trifolium repens* produce the poisonous chemical cyanide. Poisonous individuals are less appealing to herbivores—particularly mice and slugs—than are nonpoisonous individuals. However, clover plants with cyanide are more likely to be killed by frost, because freezing damages cell membranes and releases the toxic cyanide into the plant's own tissues.

Thus in populations of *Trifolium repens*, the frequency of cyanide-producing individuals increases gradually from north to south and from east to west across Europe (Figure 20.14). Poisonous plants are abundant in clover populations only in areas where the winters are mild and the plants do not poison themselves. Cyanide-producing individuals are rare where winters are cold, even though herbivores graze them heavily in those areas because plants cannot use cyanide as a defense there.

Gradual geographic changes in phenotypes and genotypes, as illustrated by *Trifolium repens*, are called **clines**. Clines are widespread among most groups of organisms. In some regions, however, frequencies of

20.13 Twenty-Nine "Generations" of Selection In *The Blind Watchmaker*, Richard Dawkins illustrated with a computer program how random morphological changes can accumulate over even a small number of generations. Starting with a simple dot, the computer produced a complex design in only 29 generations of selection.

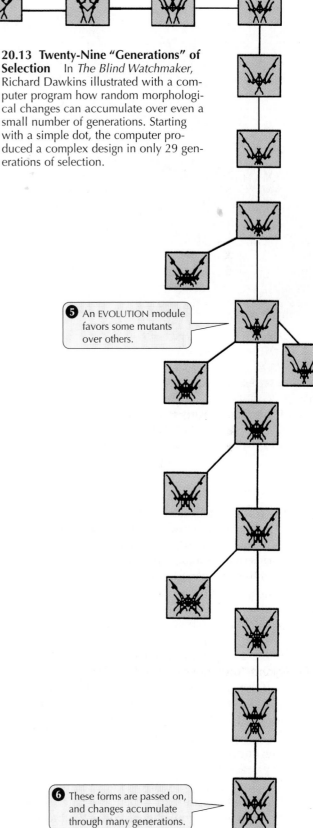

5 An EVOLUTION module favors some mutants over others.

6 These forms are passed on, and changes accumulate through many generations.

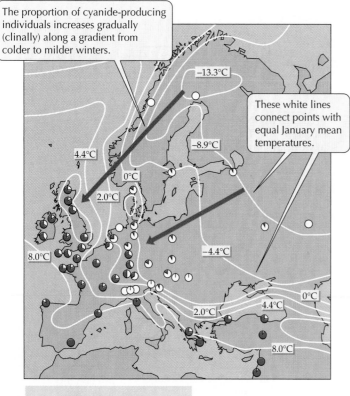

The proportion of cyanide-producing individuals increases gradually (clinally) along a gradient from colder to milder winters.

These white lines connect points with equal January mean temperatures.

−13.3°C
−8.9°C
4.4°C
0°C
2.0°C
8.0°C
−4.4°C
0°C
2.0°C
4.4°C
8.0°C

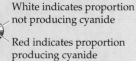

White indicates proportion not producing cyanide

Red indicates proportion producing cyanide

20.14 Geographic Variation in Poisonous Clovers The frequencies of cyanide-producing individuals in populations of white clover (*Trifolium repens*) are represented by the proportion of each circle that is red.

certain traits change abruptly, creating **step clines**. Color patterns in the rat snake, *Elaphe obsoleta*, are a good example (Figure 20.15). In this species the color differences are complex and striking. Single color patterns dominate extensive regions, and each change from one pattern to the next is abrupt. Because there are no obvious environmental changes in the regions where color patterns change, the causes of the abrupt changes are unknown.

Genetic variation exists within local populations

Natural selection often preserves variation by favoring different traits in different areas, as illustrated by clines and step clines. In such cases, much of the variation in a large population is preserved as differences between subpopulations. Natural selection also preserves variation as **polymorphisms**—genetic differences *within* a local population. For example, a polymorphism is maintained when the success of a genotype (or phenotype) depends on its frequency rel-

ative to other genotypes (or phenotypes), a process known as **frequency-dependent selection**.

A small fish that lives in Lake Tanganyika in east central Africa provides an example. The mouth of this scale-eating fish, *Perissodus microlepis*, opens either to the right or to the left as a result of an asymmetrical jaw joint (Figure 20.16). *P. microlepis* approaches its prey (another fish) from behind and dashes in to bite off several scales from its flank. "Right-mouthed" individuals always attack from the victim's left; "left-mouthed" individuals always attack from the victim's right. The distorted mouth enlarges the area of teeth in contact with the prey's flank, but only if the scale eater attacks from the appropriate side.

Prey fish are alert to approaching scale eaters, so attacks are more likely to be successful if the prey must watch both flanks. Guarding by the prey favors equal numbers of right-mouthed and left-mouthed *Perissodus*, because if one form were more common than the other, prey fish would pay more attention to potential attacks from the corresponding flank. Over an 11-year period in which the fish in Lake Tanganyika were studied, natural selection maintained the polymorphism, keeping the two forms of *P. microlepis* at about equal frequencies.

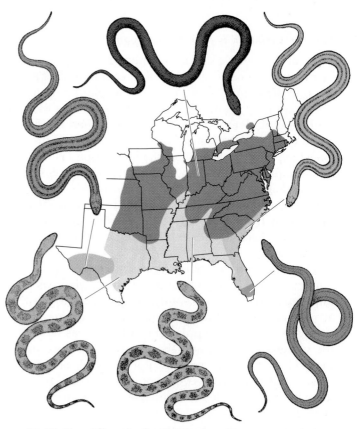

20.15 Step Clines in the Rat Snake There are no obvious environmental changes at the boundaries where the color patterns of these snakes change.

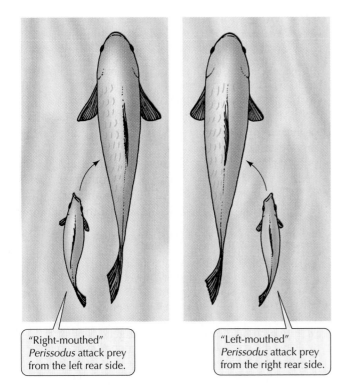

"Right-mouthed" *Perissodus* attack prey from the left rear side.

"Left-mouthed" *Perissodus* attack prey from the right rear side.

20.16 Balanced Polymorphism Natural selection maintains equal frequencies of left-mouthed and right-mouthed individuals of the scale-eating fish *Perissodus microlepis.*

Short-Term versus Long-Term Evolution

The short-term changes in populations—that is, changes in populations over years and decades—that we have been discussing in this chapter are microevolutionary changes. Studies of microevolution are an important part of evolutionary biology because short-term changes can be observed directly and can be manipulated experimentally.

Although studies of short-term changes reveal much about evolution, by themselves they cannot provide a complete explanation of the macroevolutionary changes we described in Chapter 19. We could measure the shapes of the legs of horses and correlate leg shape with the horses' running speed and their ability to escape from predators (microevolution). But such studies would not tell us why the small, four-toed, forest-dwelling horse *Hyracotherium* gave rise to large, plains-inhabiting descendants with a single toe (macroevolution).

Patterns of macroevolutionary changes can be strongly influenced by events that occur so infrequently or so slowly that they are unlikely to be observed during microevolutionary studies. The evolution of horses was influenced by long-term climatic changes that caused forests to shrink and grasslands to expand over large parts of Earth. In addition, the ways in which evolutionary agents act may change with time; even among the descendants of a single ancestral species, different lineages may be evolving in different directions.

Therefore, we cannot interpret the past simply by extending the results of short-term experiments and observations backward in time. Additional types of evidence, such as the occurrence of rare and unusual events and trends in the fossil record, must be gathered if we wish to understand the course of evolution over billions of years.

Summary of "The Mechanisms of Evolution"

- Biological evolution is a change over time in the genetic composition of members of a population.
- A population evolves when individuals having different genotypes survive or reproduce at different rates.
- Biological evolution results from the actions of evolutionary agents over millions of years.

Variation in Populations

- For a population to evolve, its members must possess genetic variation, which is the raw material on which agents of evolution act. High levels of genetic variation characterize nearly all natural populations. **Review Figures 20.3, 20.4**
- Allele frequencies measure the amount of genetic variation in a population. Genotype frequencies show how a population's genetic variation is distributed among its members.
- Biologists estimate allele frequencies by measuring a sample of individuals from a population. The sum of all allele frequencies at a locus is equal to 1.
- Populations that have the same allele frequencies may nonetheless have different genotype frequencies.

Preserving Genetic Variability: The Hardy–Weinberg Rule

- A population that is not changing genetically is said to be in equilibrium. Hardy–Weinberg equilibrium is possible only if a population is very large, mating is random, and no evolutionary agents are acting on the population.
- In a population at Hardy–Weinberg equilibrium, allele frequencies remain the same from generation to generation. In addition, genotype frequencies will remain in the proportions $p^2(AA) + 2pq(Aa) + q^2(aa) = 1$. **Review Figure 20.5**
- Biologists can determine if an agent of evolution is acting on a population by comparing the genotype frequencies of that population with Hardy–Weinberg equilibrium frequencies.

Changing the Genetic Structure of Populations

- Changes in allele frequencies and genotype frequencies within populations are caused by the actions of several different evolutionary agents: mutation, gene flow, genetic drift, nonrandom mating, and natural selection.
- The origin of genetic variation is mutation. Most mutations are harmful or neutral to their bearers, but some are advantageous, particularly if the environment changes.

- The migration of individuals from one population to another, followed by breeding in the new location, produces gene flow. Immigrants may add new alleles to a population or may change the frequencies of alleles already present. Emigrants may remove alleles from a population when they leave.
- Genetic drift alters allele frequencies primarily in small populations. Organisms that normally have large populations may pass through occasional periods (bottlenecks) when only a small number of individuals survive. New populations established by a few founding immigrants also have gene frequencies that differ from those in the parent population. **Review Figures 20.6, 20.8**
- When individuals mate more often with individuals that have the same or different genotypes than would be expected on a random basis—that is, when mating is not random—frequencies of homozygous and heterozygous genotypes differ from Hardy–Weinberg expectations.
- Self-fertilization, an extreme form of nonrandom mating, reduces the frequencies of heterozygous individuals below Hardy–Weinberg expectations.
- Natural selection is the only agent of evolution that adapts populations to their environments. Natural selection may preserve allele frequencies or cause them to change with time.
- Stabilizing selection, directional selection, and disruptive selection change the distributions of phenotypes governed by more than one locus. **Review Figures 20.9, 20.10**
- Sexual recombination generates an endless variety of genotypic combinations that increases the evolutionary potential of populations, but it does not influence the frequencies of alleles. Rather it generates new combinations of genetic material on which natural selection can act.
- Biologists study adaptation by experimentally altering organisms or their environments and by comparing traits among species. **Review Figures 20.11, 20.12**
- Natural selection acts nonrandomly, and adaptation by natural selection is a cumulative process extending over many generations. Cumulative evolutionary change results from directional selection acting on variation in populations over many generations. **Review Figure 20.13**
- The fitness of a genotype or phenotype is its contribution to subsequent generations relative to the contributions of other genotypes or phenotypes. An individual may influence its fitness by producing offspring, which contributes to its individual fitness, and by helping the survival of relatives. Individual fitness and kin selection in combination determine the inclusive fitness of the individual.

Maintaining Genetic Variation

- Natural selection maintains genetic variation within a species when different traits are favored in different places and when the direction of selection changes over time in a given place. Most species may vary geographically; the variation can be gradual or abrupt. **Review Figures 20.14, 20.15**
- Genetic variation within a population may be maintained by frequency-dependent selection. **Review Figure 20.16**

Short-Term versus Long-Term Evolution

- Patterns of macroevolutionary change can be strongly influenced by events that occur so infrequently or so slowly that they are unlikely to be observed during microevolutionary studies. Additional types of evidence must be gathered to understand macroevolution.

Self-Quiz

1. The phenotype of an organism is
   a. the type specimen of its species in a museum.
   b. its genetic constitution, which governs its traits.
   c. the chronological expression of its genes.
   d. the physical expression of its genotype.
   e. the form it achieves as an adult.

2. The appropriate unit for defining and measuring genetic variation is
   a. the cell.
   b. the individual.
   c. the population.
   d. the community.
   e. the ecosystem.

3. Which statement about allele frequences is *not* true?
   a. The sum of any set of allele frequencies is always 1.
   b. If there are two alleles at a locus and we know the frequency of one of them, we can obtain the frequency of the other by subtraction.
   c. If an allele is missing from a population, its frequency is 0.
   d. If two populations have the same gene pool for a locus, they will have the same proportion of homozygotes at that locus.
   e. If there is only one allele at a locus, its frequency is 1.

4. In a population at Hardy–Weinberg equilibrium in which the frequency of A alleles (p) is 0.3, the expected frequency of Aa individuals is
   a. 0.21.
   b. 0.42.
   c. 0.63.
   d. 0.18.
   e. 0.36.

5. Natural selection that preserves existing allele frequencies is called
   a. unidirectional selection.
   b. bidirectional selection.
   c. prevalent selection.
   d. stabilizing selection.
   e. preserving selection.

6. Laboratory selection experiments with fruit flies have demonstrated that
   a. bristle number is not genetically controlled.
   b. bristle number is not genetically controlled, but changes in bristle number are caused by the environment in which the fly is raised.
   c. bristle number is genetically controlled, but there is little variation on which natural selection can act.
   d. bristle number is genetically controlled, but selection cannot result in flies having more bristles than any individual in the original population had.
   e. bristle number is genetically controlled, and selection can result in flies having more bristles than any individual in the original population had.

7. Disruptive selection maintains bill size variability in the West African seedcracker because
   a. bills of intermediate shapes are difficult to form.
   b. the two major food sources of the finches differ markedly in size and hardness.

c. males use their large bills in displays.

d. migrants introduce different bill sizes into the population each year.

e. older birds need larger bills than younger birds.

8. A population is said to be polymorphic for a locus if it has at least

a. three different alleles at that locus.

b. two different alleles at that locus.

c. two genotypes for that locus.

d. three genotypes for that locus.

e. two genotypes for that locus, the rarest of which is more common than expected by mutation alone.

9. A cline is defined as

a. the distribution of an organism across a slope.

b. an abrupt change in frequencies of certain traits over time.

c. a gradual change in frequencies of certain traits over time.

d. an abrupt change in frequencies of certain traits over space.

e. a gradual change in frequencies of certain traits over space.

10. Adaptation is studied by

a . altering the form of an organism and observing the consequences.

b . testing predictions by comparing traits in many species.

c . developing theoretical models.

d . selecting for traits in the laboratory.

e. all of the above

Applying Concepts

1. During the past 50 years, more than 200 species of insects that attack crop plants have become highly resistant to DDT and other pesticides. Using your recently acquired knowledge of evolutionary processes, explain the rapid and widespread evolution of resistance. Propose ways of using pesticides that would slow down the rate of evolution of resistance. Now that DDT has been banned in the United States, what do you expect to happen to levels of resistance to DDT among insect populations? Justify your answer.

2. In nature, mating among individuals in a population is never truly random, and natural selection is seldom totally absent. Why, then, does it make sense to use the Hardy–Weinberg model, which is based on assumptions known generally to be false? Can you think of other models in science that are based on false assumptions? How are such models used?

3. An investigator is studying populations of house mice living in barns and sheds on a large farm. Each building has a population of between 25 and 50 mice. Populations in different buildings have strikingly different frequen-

cies of alleles determining coat color and tail length. By marking most individuals, the investigator determines that mice move only rarely between buildings. He interprets his study as providing evidence for random genetic drift. Could other agents of evolution plausibly account for the pattern? If so, how could they be distinguished?

4. As far as we know, natural selection cannot adapt organisms to future events. Yet many organisms exhibit responses in advance of natural events. For example, many mammals go into hibernation while it is still quite warm. Similarly, many birds leave the temperate zone for their southern wintering grounds long before winter has arrived. How can these "anticipatory" behaviors evolve?

5. Many people believe that species, like individual organisms, have life cycles. They believe that species are born by a process of speciation, grow and expand, and inevitably die out as a result of "species old age." Could any agent of evolution cause such a species life cycle? If not, how do you explain the high rates of extinction of species in nature?

Readings

Amábile-Cuevas, C. F., M. Cárdenas-García and M. Ludger. 1995. "Antibiotic Resistance." *American Scientist*, vol. 83, pages 320–329. An explanation of why bacteria are becoming increasingly resistant to antibiotics.

Dawkins, R. 1987. *The Blind Watchmaker*. W. W. Norton, New York. An engaging treatment showing how adaptation can happen in a world without design.

Dawkins, R. 1995. "God's Utility Function." *Scientific American*, November. An explanation of why patterns of seemingly intelligent design can be explained by basic evolutionary mechanisms.

Dennett, D. C. 1995. *Darwin's Dangerous Idea*. Simon and Schuster, New York. A compelling exposition of evolution and the meaning of life.

Endler, J. A. 1986. *Natural Selection in the Wild*. Princeton University Press, Princeton, NJ. A thorough review of the problems and successes in measuring natural selection in nature.

Futuyma, D. J. 1998. *Evolutionary Biology*, 3rd Edition. Sinauer Associates, Sunderland, MA. A comprehensive review of all aspects of evolutionary biology.

Hartl, D. L. and A. G. Clark. 1998. *Principles of Population Genetics*, 3rd Edition. Sinauer Associates, Sunderland, MA. An introduction to all aspects of modern population genetics.

Le Guenno, B. 1995. "Emerging Viruses." *Scientific American*, October. An explanation of the origins of the new viruses that increasingly cause outbreaks of disease.

Nowak, M. A. and A. J. McMichael. 1995. "How HIV Defeats the Immune System." *Scientific American*, August. A discussion of how proliferating, mutating viruses evolve faster than the immune system can change.

Williams, G. C. 1992. *Natural Selection: Domains, Levels, and Challenges*. Oxford University Press, New York. A thorough analysis of the mechanisms and meanings of natural selection as an evolutionary agent.

Chapter 21

Species and Their Formation

Trinidad Rain Forest
The mosquito that transmits malaria in Trinidad breeds in forests like this one.

During the 1940s officials in Trinidad launched an intensive campaign to control malaria. They spent much money on spraying and draining marshes, in the belief that malaria was transmitted by *Anopheles albimanus*, a swamp-breeding mosquito that is the principal vector of malaria in Latin America. The campaign failed because the principal vector of malaria in Trinidad was *Anopheles bellator*, a mosquito that breeds in water held within the leaves of bromeliads (relatives of pineapples) growing on the branches of palm trees.

Similarly, in Europe, malaria was believed to be transmitted by mosquitoes of a single species: *Anopheles maculipennis*. European efforts to control malaria sometimes succeeded and sometimes failed, because *A. maculipennis* is not a single species, but consists of at least 18 species that can be distinguished only by examination of their chromosomes. Some of the species breed in fresh water, others in brackish water. Some enter houses; others do not. Furthermore, which mosquito species is the vector of malaria changes regionally. Control efforts are successful only when directed against the species that actually transmits malaria in that area.

Millions of species inhabit Earth. Many of them are easy to identify, but some, such as the species in the *Anopheles maculipennis* complex, are not. All species, living and extinct, are believed to be descendants of a single ancestral species that lived several billion years ago. If speciation had been a rare event, the biological world would be very different than it is today. Speciation is an essential ingredient of evolutionary diversification, and species are the fundamental units of the biological classification systems we will discuss in Chapter 22.

But what are species and how did these millions of species form? How does one species become two? What factors stimulate such splitting? What conditions spur evolutionary radiations? Does speciation accelerate rates of evolutionary change? These and other related questions are the subject of this chapter.

What Are Species?

The word "Species" means, literally, "kinds;" but what do we mean by "kinds"? Someone who is knowledgeable about a group of organisms, such as orchids or lizards, usually can distinguish the different species of that group found in a particular area simply by examining them superficially. The patterns of morphological similarities that unite groups of organisms and separate them from other groups are familiar to all of us. The standard field guides to birds, mammals, insects, and flowers are possible only because most species are cohesive units that may change in appearance only gradually over large geographic distances. We can easily recognize red-winged blackbirds from New York and red-winged blackbirds from California as members of the same species (Figure 21.1).

But not all members of a species look that much alike. How do we decide whether similar but different individuals should be called different species or regarded as varieties within a species? The concept that has guided these decisions for a long time is genetic integration. If individuals within a population mate with one another but not with individuals of other populations, they constitute a reproductively isolated group within which genes recombine; that is, they are *independent evolutionary units*. These independent evolutionary groups are usually called species. Identifying

and naming species correctly is important because when different investigators report the results of studies of a particular species, we need to know that they are writing about the same species.

More than 200 years ago the Swedish biologist Carolus Linnaeus, who originated the system of naming organisms that we use today, described hundreds of species. Because he knew nothing about the mating patterns of the organisms he was naming, Linnaeus classified them on the basis of their appearance. Many species that were classified by their appearance, when nothing was known about their reproductive behavior, are actually independent evolutionary units. This is not surprising because the members of an evolutionary unit share genes inherited from common ancestors. These individuals look alike because they share many of the alleles that code for their structures.

The species definition that has been used by most biologists was proposed by Ernst Mayr in 1940. He stated, "Species are groups of actually or potentially interbreeding natural populations which are reproductively isolated from other such groups." The "groups" in this definition are collections of local populations. The words "actually or potentially" assert that, even if some members of a species are not in the same place and hence are unable to mate, they should not be placed in separate species if they would be likely to mate if they were together. The word "natural" is an important part of the definition because only *in nature* does the exchange of genes, which occurs within species, affect evolutionary processes; the interbreeding of two different species in captivity does not.

Gene exchange is the main reason why species are cohesive units. Individuals that mate with each other "recognize" one another as suitable mates. Many biologists study how individuals use visual, vocal, and chemical clues to recognize suitable mates and to avoid mating with individuals belonging to other species.

(a)   (b)

21.1 Redwings Are Redwings Everywhere   Both of these birds are obviously red-winged blackbirds, even though one (a) lives in Texas and the other (b) lives in Ontario.

Biologists attempting to reconstruct the evolutionary histories of organisms also try to identify independent evolutionary units. How biologists define species and identify evolutionary lineages will be the focus of Chapter 22.

How Do New Species Arise?

Not all evolutionary changes result in new species. Evolution creates two patterns across time and space: anagenesis and cladogenesis. **Anagenesis** is change in a single lineage through time. With sufficient time, the changes may be so great that the descendants are given another name, even though no "new" species has formed. Anagenetic changes are a common feature of the fossil record.

Cladogenesis (speciation) is the process by which one species splits into two species, which thereafter evolve as distinct lineages. Although Charles Darwin entitled his book *The Origin of Species*, he did not extensively discuss how a single species splits into two or more daughter species. Rather, he was concerned principally with demonstrating anagenesis, that species are altered by natural selection over time.

The critical process in the formation of two species from one ancestral species is the separation of the gene pool belonging to the ancestral species into two separate gene pools. Subsequently, within each isolated gene pool, allele and gene frequencies may change as a result of the action of evolutionary agents. If sufficient differences have accumulated during the period of isolation, the two populations may not exchange genes when they come together again.

Gene flow among populations may be interrupted in several ways, each of which characterizes a mode of speciation. The next three sections focus on these modes of speciation: sympatric speciation, allopatric speciation, and parapatric speciation.

Sympatric speciation occurs without physical separation

The subdividing of a gene pool even though members of the daughter species are not physically separated during the process is called **sympatric speciation** (*sym-*, "with"; *patris*, "country"). The most common means of sympatric speciation is **polyploidy**, a duplication of the number of chromosomes (see Chapter 9).

Polyploidy arises in two ways. One way is the accidental production during cell division of cells having four (tetraploid) rather than two (diploid) sets of chromosomes. This process produces an **autopolyploid** individual, one having more than two sets of chromosomes, all derived from a single species. This tetraploid individual cannot mate successfully with diploids, but if it self-fertilizes or mates with other tetraploids, a new evolutionary lineage may form.

A polyploid species can also be produced when individuals of two different species, whose chromosomes do not pair properly during meiosis, interbreed. The resulting individuals, called **allopolyploids**, are usually sterile, because the chromosomes from one species do not pair properly with those from the other species during meiosis, but they can reproduce asexually. After many generations, some of the individuals may become fertile as a result of further chromosome duplication.

Polyploidy can create a new species among plants much more easily than among animals because plants of many species can reproduce by self-fertilization as well as by crossing with a relatively unrelated individual. If the polyploidy arises in several offspring of a single parent, the siblings can fertilize one another. Speciation by polyploidy has been very important in the evolution of flowering plants. Botanists estimate that more than half of all species of flowering plants are polyploids. Most of these arose as a result of hybridization between two species, followed by self-fertilization.

The importance of allopolyploidy and the speed with which it can produce new species are illustrated by salsifies (*Tragopogon*), members of the sunflower family. Salsifies are weedy plants that thrive in disturbed areas around towns. People have inadvertently spread them around the world from their ancestral ranges in Eurasia. Three diploid species of salsify were introduced into North America early in this century: *T. porrifolius, T. pratensis,* and *T. dubius.* Two tetraploid hybrids—*T. mirus* and *T. miscellus*—between species of the original three were first reported in 1950. Both hybrids have spread since their discovery and today are more widespread than their diploid parents (Figure 21.2).

Studies of their cells have revealed that both types of hybrids have been formed more than once. Some populations of *T. miscellus,* for example, have the chloroplast genome of *T. pratensis,* whereas other populations have the chloroplast genome of *T. dubius.* Differences in the ribosomal genes of local populations of *T. miscellus* show that this allopolyploid has evolved independently at least three times. Scientists seldom know the dates and locations of species formation so well. The success of newly formed hybrid species of salsifies illustrates why so many species of flowering plants originated as polyploids.

Among animals, sympatric speciation by polyploidy is relatively rare, but a few cases are known. The tree frog *Hyla versicolor* is a tetraploid species with a broad range in eastern North America. *H. versicolor* arose recently as a result of at least three different hybridizations between individuals from eastern and western populations of its diploid relative *Hyla chrysocelis.* The eastern and western populations of *H. chrysocelis* were

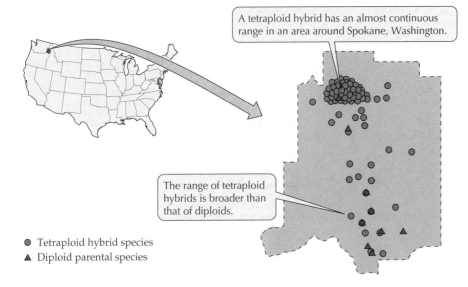

A tetraploid hybrid has an almost continuous range in an area around Spokane, Washington.

The range of tetraploid hybrids is broader than that of diploids.

● Tetraploid hybrid species
▲ Diploid parental species

21.2 Polyploids Can Outperform Their Parents *Tragopogon* species (salsifies) are members of the sunflower family. The map shows the distribution of the diploid parent species and the tetraploid hybrid species of *Tragopogon* in eastern Washington and adjacent Idaho.

isolated from one another for about 4 million years, enough time for substantial genetic differences to accumulate. Today the ranges of the two populations abut in a north-south line west of the Mississippi River (Figure 21.3). Eastern and western individuals rarely hybridize in nature, probably because the calls of the males are so different that females respond only to the calls of males of their own type when they have a choice.

Among animals, sympatric speciation as a result of habitat selection is more common than speciation by polyploidy. A good example is speciation in the picture-winged fruit fly, *Rhagoletis pomenella*, in New York. Until the mid 1800s, these fruit flies courted, mated, and deposited their eggs only on hawthorn berries. The larvae learned the odor of hawthorn as they fed on the berries, and when they emerged from their pupae they used this food-based memory to locate other hawthorn plants.

About 150 years ago, large commercial apple orchards were planted in the Hudson River Valley. A few

21.3 Gray Tree Frogs Speciated by Intraspecific Hybridization The tetraploid *Hyla versicolor* resulted from hybridization between eastern and western populations of the diploid *H. chrysocelis.*

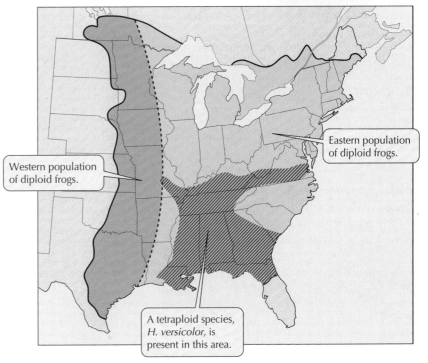

Eastern population of diploid frogs.

Western population of diploid frogs.

A tetraploid species, *H. versicolor,* is present in this area.

female *Rhagoletis* apparently laid their eggs on apples, perhaps by mistake. Their larvae did not grow as well as the larvae on hawthorn berries, but many did survive. These larvae had learned the odor of apples, so when they emerged as adults they sought out apple trees, where they mated with other flies reared on apples. Today there are two sympatric species of *Rhagoletis* in the Hudson River Valley. One feeds on hawthorn berries, the other on apples. The two species are reproductively isolated because they mate only with individuals raised on the same fruit, and because they emerge from their pupae at different times. In addition, apple-feeding flies have evolved so that they now grow more rapidly on apples than they originally did.

Allopatric speciation requires isolation by distance

Speciation that results when a population is divided by a barrier is known as **allopatric speciation** (*allo-*, "different"), or **geographic speciation** (Figure 21.4). Allopatric speciation is the dominant form of speciation among most groups of organisms. The range of a species may be divided by a physical barrier, such as water gaps for terrestrial organisms, dry land for aquatic organisms, and mountains. Barriers can form when continents drift, sea levels rise, or climates change. Populations separated in this way are often large initially. They evolve differences because the places in which they live are or become different.

Alternatively, allopatric speciation may result when some members of a population cross an existing barrier and found a new isolated population. Populations established in this way are typically small at first. They usually differ genetically from their parent populations because a small group of individuals has only an incomplete representation of the genes found in its parent population.

Allopatric speciation by this sampling effect is illustrated by the singing honeyeater, a common Australian bird that lives on the mainland and on coastal islands. The birds on Rottnest Island, 20 km off the coast of Western Australia, sing fewer song types than mainland birds do, and their songs have fewer syllables and notes. Evidently the island colonizers did not carry all of the alleles responsible for the full range and complexity of songs found in mainland individuals. Mainland singing honeyeaters do not respond to the songs of island birds and Rottnest birds do not respond to songs of mainland birds.

Dispersal across barriers often leads to species formation. For example, many of the hundreds of species of the fruit fly *Drosophila* in the Hawaiian Islands are restricted to a single island. They are almost

A single species is distributed over a broad range.

Time

Sea level rises and isolates species. Populations adapt to differing environments on opposite sides of the barrier.

Time

If the barrier to breeding is removed, the populations may recolonize the intervening area and mingle, but do not interbreed.

Range of overlap

21.4 Allopatric Speciation
Also known as geographic speciation, allopatric speciation may result when a population is divided by a physical barrier such as rising seas.

21.5 Founder Events Lead to Allopatric Speciation The extremely high level of speciation found among *Drosophila* in the Hawaiian Islands is almost certainly the result of founder events—new populations founded by individuals dispersing among the islands.

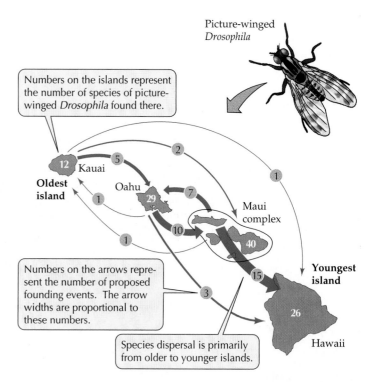

certainly the result of new populations founded by individuals dispersing among the islands, because the closest relative of a species on one island is often a species on a neighboring island rather than a species on the same island. On the basis of studies of their chromosomes, speciation among the picture-winged *Drosophila* is believed to have been caused by at least 45 founder events (Figure 21.5).

If the environments on the two sides of the physical barrier differ, evolutionary agents may cause the populations to diverge further genetically. Differences that accumulate while a barrier is in place may become so large that the populations will fail to establish gene flow if the barrier later breaks down; that is, they will have evolved to be different species.

The finches of the Galapagos Islands, 1,000 km off the coast of Ecuador, demonstrate the importance of geographic isolation for speciation. Darwin's finches (as they are usually called, because Darwin was the first scientist to study them) arose on the Galapagos Islands by speciation from a single South American species that colonized the islands. Today there are 14 species of Galapagos finches, all of which differ strikingly from the probable mainland ancestor (Figure 21.6).

The islands of the Galapagos Archipelago are sufficiently isolated from one another that the finches seldom migrate between them. Also, environmental conditions differ among the islands. Some are relatively flat and arid; others have forested mountain slopes. Populations of finches on different islands have differentiated enough that when occasional immigrants arrive from other islands, they either do not breed with the residents, or if they do, the resulting offspring usually do not survive as well as those produced by pairs composed of island residents. The genetic distinctness of different populations is thus maintained.

A barrier's effectiveness at preventing gene flow depends on the size and mobility of species. What is a firm barrier for a terrestrial snail may be no barrier at all to a butterfly or a bird. Populations of wind-pollinated plants are totally isolated at the maximum distance pollen is blown by the wind, but individual plants are effectively isolated at much shorter distances. Among animal-pollinated plants, the width of the barrier is the distance that pollinators travel while carrying pollen (see Chapter 36). Even animals with great powers of dispersal are often reluctant to cross narrow strips of unsuitable habitat. For animals that cannot swim or fly, narrow water-filled gaps may be effective barriers.

Parapatric speciation separates adjacent populations

Sometimes reproductive isolation develops among adjacent members of a population in the absence of a geographic barrier. Known as **parapatric speciation** (*para*, "beside"), this type of speciation is much less common than allopatric or sympatric speciation because gene flow usually prevents differentiation between populations in contact. Occasionally, however, a species boundary forms where there is a marked change in environment. That is, allopatric speciation becomes parapatric speciation when the geographical boundary separating species becomes extremely small.

Unusually abrupt changes in soil are created by mining activities that leave rubble (tailings) with high concentrations of heavy metals that are detrimental to plant growth. For example, the soils developing on the tailings at the Goginian lead mine near Aberystwyth, Wales, are highly contaminated with lead, but where the tailings end, they suddenly give way to normal rich pastureland (Figure 21.7).

The pasture grass *Agrostis tenuis* is common on both types of soils, but there is a sharp gradient in lead tolerance among plants less than 20 m apart. Plants on the mine tailings grow well in lead concentrations that would be lethal to plants growing just a few meters away. Nearly complete reproductive isolation exists between plants on contaminated and normal soil because they flower at different times. These two populations have not yet been designated as separate species, but reproductive isolation between them has already evolved, demonstrating that gene

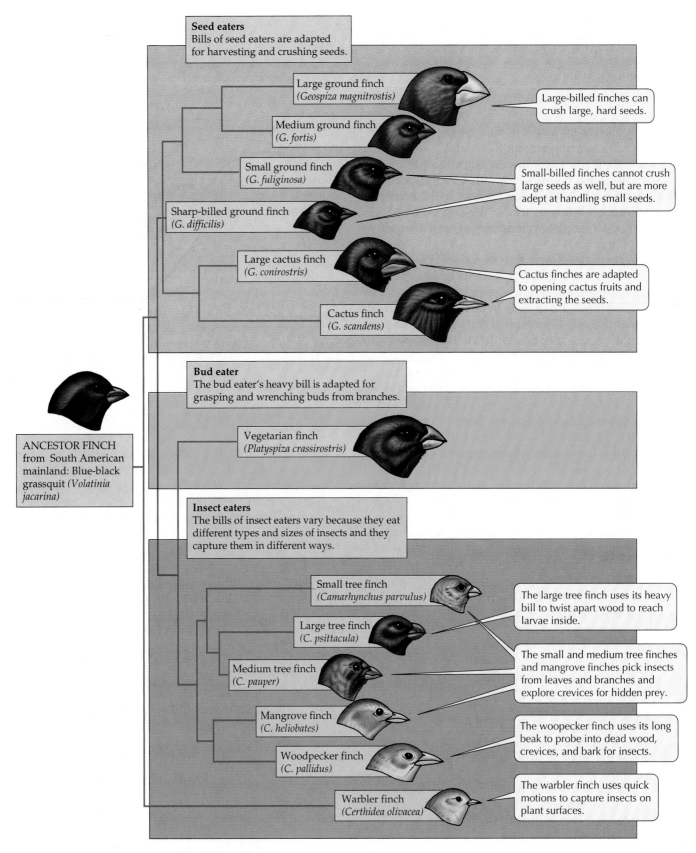

Seed eaters
Bills of seed eaters are adapted for harvesting and crushing seeds.

Large ground finch
(*Geospiza magnitrostis*)

Medium ground finch
(*G. fortis*)

Small ground finch
(*G. fuliginosa*)

Sharp-billed ground finch
(*G. difficilis*)

Large cactus finch
(*G. conirostris*)

Cactus finch
(*G. scandens*)

Large-billed finches can crush large, hard seeds.

Small-billed finches cannot crush large seeds as well, but are more adept at handling small seeds.

Cactus finches are adapted to opening cactus fruits and extracting the seeds.

Bud eater
The bud eater's heavy bill is adapted for grasping and wrenching buds from branches.

Vegetarian finch
(*Platyspiza crassirostris*)

ANCESTOR FINCH from South American mainland: Blue-black grassquit (*Volatinia jacarina*)

Insect eaters
The bills of insect eaters vary because they eat different types and sizes of insects and they capture them in different ways.

Small tree finch
(*Camarhynchus parvulus*)

Large tree finch
(*C. psittacula*)

Medium tree finch
(*C. pauper*)

Mangrove finch
(*C. heliobates*)

Woodpecker finch
(*C. pallidus*)

Warbler finch
(*Certhidea olivacea*)

The large tree finch uses its heavy bill to twist apart wood to reach larvae inside.

The small and medium tree finches and mangrove finches pick insects from leaves and branches and explore crevices for hidden prey.

The woopecker finch uses its long beak to probe into dead wood, crevices, and bark for insects.

The warbler finch uses quick motions to capture insects on plant surfaces.

21.6 Evolution among Galapagos Finches The descendants of the blue-black grassquits that colonized the Galapagos Islands several million years ago evolved into 14 species whose members are variously adapted to feed on seeds, buds, and insects. (The fourteenth species, not shown here, lives on Cocos Island, farther north in the Pacific Ocean.)

21.7 Parapatric Speciation *Agrostis tenuis* individuals growing on soil contaminated with heavy metals from this Welsh mine are reproductively isolated from nearby individuals growing on uncontaminated soil.

flow can slow or stop even in the absence of a distinct physical barrier.

Reproductive Isolating Mechanisms

Once a barrier to gene flow is established, by whatever means, the daughter populations may diverge genetically because of the action of evolutionary agents. Over many generations, differences that reduce the survivability of hybrid offspring may accumulate. In this way, reproductive isolation can evolve as an incidental by-product of other genetic changes in allopatric populations. For example, individuals in the two daughter populations may become so different that they are not recognized as suitable mates, as shown by the singing honeyeaters on Rottnest Island.

However, geographic isolation does not necessarily mean reproductive isolation. For example, American and European sycamores have been physically isolated from one another for at least 20 million years. Nevertheless, they are morphologically very similar (Figure 21.8), and they can form fertile hybrids. They lack traits that would prevent individuals of two different populations from producing fertile hybrids. In this section we examine the ways in which such traits—**reproductive isolating mechanisms**—arise. Then we explore what happens when reproductive isolation is incomplete.

Prezygotic barriers operate before mating

Individuals of different species may select different places in the environment in which to live. As a result, they may never come into contact during their respective mating seasons; that is, they are reproductively isolated by location (*spatial isolation*). Many organisms have mating periods that are as short as a few hours or days. If the mating periods of two species do not coincide, they will be reproductively isolated by time (*temporal isolation*). Differences in the sizes and shapes of reproductive organs may prevent the union of gametes

(a) *Plastanus occidentalis* (American sycamore)

(b) *Platanus hispanica* (European sycamore)

21.8 Geographically Separated, Morphologically Similar Although they have been separated on different continents for at least 20 million years, American and European sycamores are similar in appearance.

21.9 Prezygotic and Postzygotic Barriers to Gene Exchange Barriers to gene exchange exist both before and after mating.

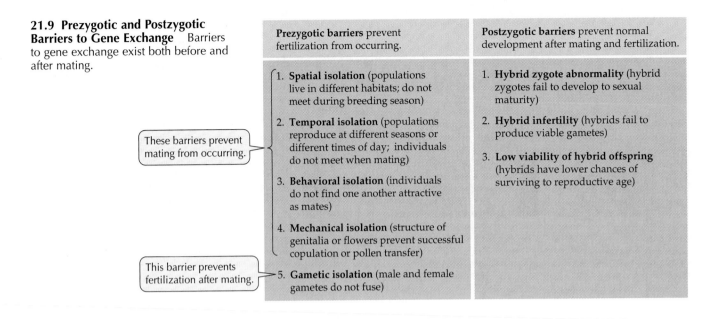

Prezygotic barriers prevent fertilization from occurring.	**Postzygotic barriers** prevent normal development after mating and fertilization.
These barriers prevent mating from occurring. { 1. **Spatial isolation** (populations live in different habitats; do not meet during breeding season) 2. **Temporal isolation** (populations reproduce at different seasons or different times of day; individuals do not meet when mating) 3. **Behavioral isolation** (individuals do not find one another attractive as mates) 4. **Mechanical isolation** (structure of genitalia or flowers prevent successful copulation or pollen transfer) This barrier prevents fertilization after mating. 5. **Gametic isolation** (male and female gametes do not fuse)	1. **Hybrid zygote abnormality** (hybrid zygotes fail to develop to sexual maturity) 2. **Hybrid infertility** (hybrids fail to produce viable gametes) 3. **Low viability of hybrid offspring** (hybrids have lower chances of surviving to reproductive age)

from different species (*mechanical isolation*). Sperm of one species may not be attracted to the eggs of another species because the eggs do not release the appropriate attractive chemicals, or the sperm may be unable to penetrate the egg because it is chemically incompatible (*gametic isolation*). These mechanisms, all of which operate before mating, are called **prezygotic reproductive barriers** (Figure 21.9).

Postzygotic barriers operate after mating

If individuals of two different populations still recognize one another and mate, **postzygotic reproductive barriers** may prevent gene exchange (see Figure 21.9). Genetic differences that accumulated while the daughter populations were allopatric are likely to reduce the survivability of offspring produced by matings between individuals from the two daughter populations.

The offspring of parents from genetically dissimilar populations are known as **hybrids** (see Chapter 10). Hybrid zygotes may be abnormal (*hybrid zygote abnormality*), or the hybrids may mature normally but be infertile when they attempt to reproduce (*hybrid infertility*). For example, the offspring of matings between horses and donkeys—mules—are vigorous, but mules are sterile; they produce no descendants (Figure 21.10).

If hybrid offspring are less viable than offspring resulting from matings within populations of the daughter species, postzygotic barriers may be reinforced by the evolution of more effective prezygotic barriers. More effective prezygotic barriers should evolve because individuals engaging in hybrid matings leave fewer surviving offspring than those that mate only within their group. Reinforcement of prezygotic barriers has been demonstrated in a few laboratory populations, but evidence for it in nature has been slow to accumulate.

Sometimes reproductive isolation is incomplete

Sometimes contact is reestablished between formerly geographically isolated populations before many genetic differences have accumulated. Then the individuals may interbreed freely with members of the

21.10 Sturdy but Sterile Mules are widely used as pack animals because of their stamina. For that purpose, their infertility is unimportant.

21.11 Formation of a New Hybrid Zone Blue and snow geese are forming mixed pairs because they now winter together in Louisiana and Texas rice fields where they choose mates.

other population and produce hybrid offspring that are as successful as those resulting from matings within each population. If successful hybrids spread through both populations and reproduce with other individuals, the gene pools combine quickly, and no new species result from the period of isolation.

Alternatively, rather than thoroughly combining their gene pools, the two populations may interbreed only where they come into contact, resulting in a **hybrid zone**. To determine what happens when formerly separate populations come together, studies ideally begin when contact is first established. Blue and snow geese provide an opportunity to observe the formation of a hybrid zone (Figure 21.11). These geese breed in Arctic North America and spend the winter in the southern United States. Birds with white plumage (snow geese) dominate breeding populations in the West; birds with dark plumage (blue geese) dominate in the East. Historical evidence shows that the two color forms were almost completely separated geographically until the 1930s.

The recent hybrid zone has resulted from a change in the winter feeding ranges of the birds. Birds of both types now winter in large flocks in the rice-growing regions of inland Texas and Louisiana. The geese select mates while on the wintering grounds, and pairs migrate to nest on the breeding grounds from which the female came. Interbreeding is now common between the two forms, and a hybrid zone is developing in a small region of the Canadian Arctic. Biologists are monitoring the spread of this hybrid zone to determine whether isolating mechanisms are developing or whether the zone will continue to spread.

We can measure genetic differences among species

If two species hybridize, we know that they are very similar genetically, but the absence of interbreeding tells us nothing about how *dis*similar two species are. Not until modern molecular tools were developed could biologists measure genetic differences among species.

Molecular studies are now demonstrating that many sympatric species differ from one another very little genetically. For example, flies of different species of Hawaiian *Drosophila* share nearly all of their alleles. Most morphological differences *among* the species are based on variability already present *within* each of the species. All of the hundreds of species of this genus that have evolved in Hawaii during the past 40 million years, even those that have diverged morphologically, are relatively similar genetically (Figure 21.12). Other research confirms that the differences among species generally are similar in type to the differences within species.

21.12 Morphologically Different, Genetically Similar Although these fruit flies, a small sample of the hundreds of species found only on the Hawaiian Islands, are extremely variable in appearance, they are genetically almost identical.

Drosophila silvestris

Drosophila conspicua

Drosophila balioptera

Variation in Speciation Rates

Some lineages of organisms have many species; others have few. The hundreds of species of *Drosophila* found in the Hawaiian Islands have evolved within the last 40 million years. In contrast, there is only one species of horseshoe crab, even though its lineage has survived more than 200 million years. Why do rates of speciation vary so widely among lineages? In the sections that follow we will examine several factors that influence speciation rates: species diversity and range size, life history traits, environment, and generation times.

Species richness may favor speciation

The larger the number of species, the larger the number of opportunities for new species to form. This is particularly true of speciation by polyploidy because more species are available to hybridize with one another. It is also partly true for geographic speciation, because the number of ranges bisected by a given barrier should be positively correlated with the number of species living in the area.

However, the relationship between range size and speciation rate is not simple, because ranges of individual species tend to be smaller where there are many species. The larger the range of a species, the more likely a physical barrier is to subdivide it. Conversely, the smaller the range size, the less likely a particular randomly placed barrier will subdivide it. Also, species with large ranges are more likely than species with small ranges to establish isolated peripheral populations that survive long enough to form new species.

Random variation in rates of events that create barriers is an important cause of variable speciation rates. Where and when geographic barriers arise, and where and when genetic accidents that result in polyploid individuals happen, are unpredictable and variable. Nonetheless, traits of species may influence how often their ranges are divided by barriers.

Life history traits influence speciation rates

Individuals of species with poor dispersal abilities are unlikely to establish new populations by dispersing across barriers, and even narrow barriers are effective among species whose individuals are highly sedentary. Populations of land snails, which have speciated profusely on Pacific islands, are separated by barriers as narrow as a city street (Figure 21.13).

Animals with complex behavior are likely to speciate at a high rate because they make sophisticated discriminations among potential mating partners. They distinguish members of their own species from members of other species, and they make subtle discriminations among members of their own species on the basis of size, shape, appearance, and behavior.

These discriminations may be based on the quality of the genes of the potential partner, the quality of parental care likely to be given, or both. Such behavioral discrimination can greatly influence which individuals are most successful in producing offspring. Therefore, mate selection is probably a major cause of rapid evolution of reproductive isolation between species.

21.13 Mobility Affects Speciation Rate Even a narrow street presents a geographic barrier to land snails. Because populations are readily isolated, genetic differences accumulate rapidly.

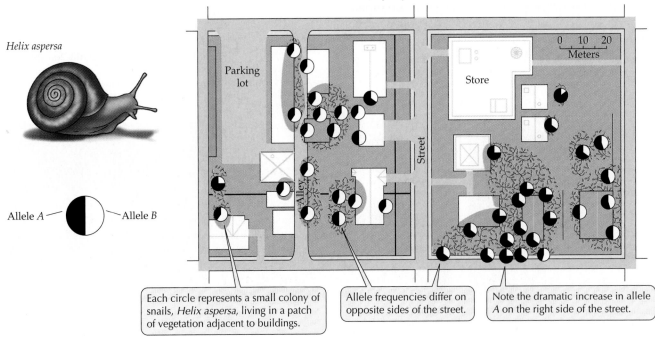

Helix aspersa

Allele *A* — Allele *B*

Each circle represents a small colony of snails, *Helix aspersa*, living in a patch of vegetation adjacent to buildings.

Allele frequencies differ on opposite sides of the street.

Note the dramatic increase in allele *A* on the right side of the street.

Heterogeneous environments favor high speciation rates

Speciation rates among different lineages of the large, hoofed mammals of Africa are correlated with diet: Grazers (which eat grass and other nonwoody plants) and browsers (which eat branches and leaves of woody plants) speciate faster than omnivores (which eat both plants and animals) and anteaters (Figure 21.14).

The grazers and browsers require large expanses of open grassland and woodland, respectively. In Africa, these resources disappeared from and reappeared in large areas during periods of climatic change, thus isolating populations and causing both high extinction rates and high rates of differentiation among populations between these isolated regions. Omnivores and anteaters, on the other hand, maintained more continuous populations during these climatic changes. Gene flow continued among their populations, and reproductive isolation was not established.

Short generation times enhance speciation

We have concentrated on factors that influence rates at which the ranges of species are subdivided by barriers. But the rate at which new species form also depends on how fast daughter populations diverge. The more rapidly they diverge, the sooner they are likely to evolve reproductive isolating mechanisms and the less likely they are to hybridize when they again become sympatric. Shorter generation times result in more generations per unit time and, as a result, generate the potential for more evolutionary changes per unit time.

Evolutionary Radiations: High Speciation and Low Extinction

As we learned in Chapter 19, the fossil record reveals that, at certain times in some lineages, speciation rates have been much higher than extinction rates. The result is an **evolutionary radiation** giving rise to a large number of daughter species. What conditions cause speciation rates to be much higher than extinction rates?

Evolutionary radiations are likely when a population colonizes an environment that has relatively few species. This condition typifies islands because many organisms disperse poorly across large water-filled gaps. Because islands lack many plant and animal groups found on the mainland, ecological opportunities exist that may stimulate rapid evolutionary changes. Water barriers also restrict gene flow among islands in an archipelago, so populations on different islands can evolve adaptations to their local environments. Together these two factors make it likely that speciation rates will exceed extinction rates.

Remarkable evolutionary radiations have occurred in the Hawaiian Archipelago, the most isolated islands in the world. The Hawaiian Islands lie 4,000 km from

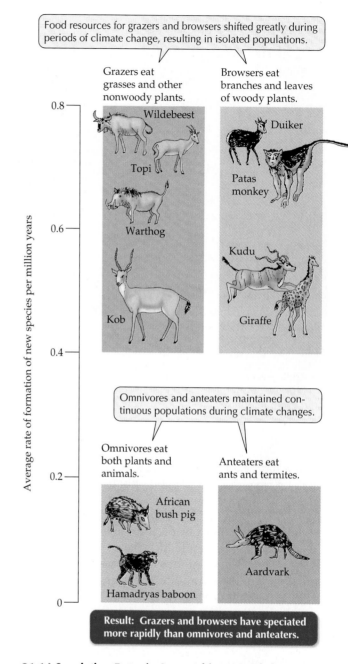

21.14 Speciation Rates in Some African Hoofed Mammals Speciation rates of these animals correlate with their diets.

the nearest major land mass and 1,600 km from the nearest group of islands. The islands are arranged in a line of decreasing age—the youngest islands to the southeast, the oldest to the northwest (see Figure 21.5).

The native biota of the Hawaiian Islands includes 1,000 species of flowering plants, 10,000 species of insects, 1,000 land snails, and more than 100 birds. However, there were no amphibians, no terrestrial reptiles, and only one native mammal—a bat—until humans introduced additional species. The 10,000

Argyoxiphium sandwichense

Wilkesia gymnoxiphium

Dubautia laxa

21.15 Rapid Evolution among Hawaiian Plants Three closely related genera of the sunflower family are believed to have descended from a single ancestor, a tarweed that colonized Hawaii from the Pacific coast of North America. Their rapid evolution makes them appear more distantly related than they actually are.

known native species of insects on Hawaii are believed to have evolved from only about 400 immigrant species; only seven immigrant species are believed to account for all the native Hawaiian land birds.

More than 90 percent of all plant species on the Hawaiian Islands are **endemic**—that is, they are found nowhere else. Several groups of flowering plants have more diverse forms and life histories on the islands and live in a wider variety of habitats than do their close relatives on the mainland. An outstanding example is the group of sunflowers called silverswords and tarweeds (the genera *Argyroxiphium, Dubautia,* and *Wilkesia*). Chloroplast DNA data show that these species share a relatively recent common ancestor, which is believed to be a species of tarweed from the Pacific coast of North America.

Whereas all mainland tarweeds are small, upright, nonwoody plants (herbs), Hawaiian silversword species include prostrate and upright herbs, shrubs, trees, and vines (Figure 21.15). They occupy nearly all the habitats of the islands, from sea level to above timberline in the mountains. Despite their extraordinary diversification, however, the silverswords have differentiated very little genetically. In other words, the rate of morphological evolution has been much more rapid than the rate of evolution of chloroplast DNA in these plants.

The island silverswords are more diverse in size and shape than the mainland tarweeds because the colonizers arrived on islands that had very few plant species. In particular, there were few trees and shrubs because such large-seeded plants only rarely disperse to oceanic islands; many island trees and shrubs have evolved from nonwoody ancestors. On the mainland, however, tarweeds have lived in ecological communities that contain tree and shrub lineages older than their own—that is, where opportunities to exploit the tree way of life were already preempted.

Evolutionary lineages may also radiate when they acquire a new adaptation that enables them to use the environment in new and varied ways. For example, ancestors of the 94 species of American blackbirds evolved powerful muscles for opening their bills. These muscles enable the birds to obtain food by opening their bills forcibly against objects they wish to move, exposing otherwise hidden prey (Figure 21.16). Such activity is called **gaping**. Birds lacking these powerful muscles can find prey only on exposed surfaces of objects. Blackbirds gape into wood, fruits, leaf clusters, and stems of nonwoody plants; under sticks, stones, and animal droppings; and into the soil. With this feeding method, they have come to occupy nearly all habitat types in North and South America, and they are among the most abundant birds throughout the region.

Speciation and Evolutionary Change

Does speciation stimulate evolutionary change? In 1972 Niles Eldredge and Stephen Jay Gould proposed that most evolutionary changes take place at the time

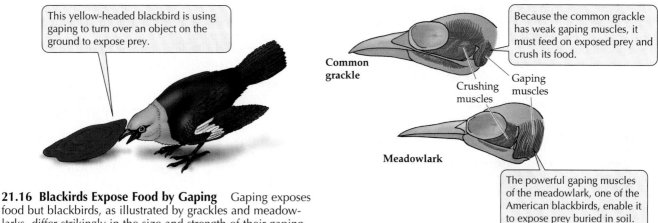

This yellow-headed blackbird is using gaping to turn over an object on the ground to expose prey.

Common grackle

Because the common grackle has weak gaping muscles, it must feed on exposed prey and crush its food.

Crushing muscles

Gaping muscles

Meadowlark

The powerful gaping muscles of the meadowlark, one of the American blackbirds, enable it to expose prey buried in soil.

21.16 Blackirds Expose Food by Gaping Gaping exposes food but blackbirds, as illustrated by grackles and meadowlarks, differ strikingly in the size and strength of their gaping muscles.

of speciation. They suggested that the isolation of small, peripheral populations was the most common event leading to rapid evolutionary changes. Their reasoning was that small founding populations lack some of the alleles found in the source populations and have different allele frequencies as a result of random genetic changes. These founding populations might change rapidly because they live in an environment that differs from the one from which they came. According to Eldredge and Gould, the speciation process stimulates evolutionary change, and once speciation has been completed, the better-integrated new genotypes resist change, leading to **stasis**—long periods of little change that are interrupted only by the next round of speciation. This pattern of evolution is called **punctuated equilibrium**.

The fossil record reveals examples of both stasis and long-term gradual evolution, but it does not tell us if periods of rapid change usually accompany times of speciation and are rare at other times. The fossil record is of limited help in testing this hypothesis because the ages of most fossils cannot be determined precisely enough. Molecular tools, on the other hand, enable evolutionary biologists to measure correlations between speciation and evolutionary change.

Lungless salamanders have been studied extensively at both molecular and morphological levels. The information from these studies allows us to compare the amount of genetic difference with the amount of morphological differences among these species to determine if speciation typically has been accompanied by morphological change. The results show that most speciation events are accompanied by almost no morphological changes. In these salamanders, speciation has proceeded at a much higher rate than morphological evolution. How often the rate of speciation exceeds the rate of morphological evolution is unknown.

The Significance of Speciation

The result of speciation processes operating over billions of years is a world in which life is organized into millions of species, each adapted to live in a particular place and to use environmental resources in a particular way. Earth would be very different if speciation had been a rare event in the history of life. How the millions of species are distributed over the surface of Earth and organized into ecological communities will be a major focus of Part Seven of this book, "Ecology and Biogeography," at which time we will also discuss how human activities are causing the extinction of many species and what we can do to reduce the rate of species loss.

Summary of "Species and Their Formation"

What Are Species?

• Species are independent evolutionary units. A generally accepted definition is that "species are groups of actually or potentially interbreeding natural populations which are reproductively isolated from other such groups."

How Do New Species Arise?

• Not all evolutionary changes result in new species.
• Evolution creates two patterns across time and space: anagenesis and cladogenesis. In anagenesis, a single lineage changes through time. In cladogenesis (speciation), one species splits into two separate species.
• Species may form quickly sympatrically by a multiplication of chromosome numbers because polyploid offspring are sterile in crosses with members of the parent species. Polyploidy has been a major factor in plant speciation but is rare among animals. **Review Figures 20.2, 20.3, 20.6**

• Allopatric (geographic) speciation is the most important means of speciation among animals and is common in other groups of organisms. **Review Figures 20.4, 20.5**
• Species may form parapatrically where marked environmental changes prevent gene flow among individual living in adjacent but different environments. **Review Figure 20.7**

Reproductive Isolating Mechanisms

• When previously allopatric species become sympatric, reproductive isolating mechanisms may prevent the exchange of genes.
• Barriers to gene exchange may operate before fertilization (prezygotic barriers) or after fertilization (postzygotic barriers). Hybrid zones may develop if barriers to gene exchange failed to develop during allopatry. **Review Figure 20.9**
• Hybrids may form if barriers break down before sufficient genetic differences have accumulated. Hybrids tell us that the two hybridizing species are very similar genetically, but species that do not hybridize may also differ from one another very little genetically.
• Genetic differences between species are similar in kind to those found within species, although differences between species usually are greater than differences within species.

Variation in Speciation Rates

• Rates of speciation differ greatly among lineages of organisms. Speciation rates are influenced by species diversity and range sizes, life history traits, environment, and generation times. **Review Figures 20.13, 20.14**

Evolutionary Radiations: High Speciation and Low Extinction

• Evolutionary radiations happen when speciation rates exceed extinction rates.
• High speciation rates often coincide with low extinction rates when species invade islands that have impoverished biotas or when a new way of exploiting the environment makes a different array of resources available to a species. **Review Figures 20.15, 20.16**

Speciation and Evolutionary Change

• Speciation may stimulate rapid evolutionary change, leading to a pattern known as punctuated equilibrium. Nonetheless, many speciation events are not accompanied by large evolutionary changes.

The Significance of Speciation

• As a result of speciation, Earth is populated with millions of species, each adapted to live in a particular place and to use environmental resources in a particular way.

Self-Quiz

1. A species is
   a. a group of actually interbreeding natural populations that is reproductively isolated from other such groups.
   b. a group of potentially interbreeding natural populations that is reproductively isolated from other such groups.
   c. a group of actually or potentially interbreeding natural populations that is reproductively isolated from other such groups.
   d. a group of actually or potentially interbreeding natural populations that is reproductively connected to other such groups.
   e. a group of actually interbreeding natural populations that is reproductively connected to other such groups.

2. Anagenesis is
   a. continuous change in a single lineage of organisms.
   b. the formation of two species by the splitting of one evolutionary lineage.
   c. the formation of a new species by the coming together of two evolutionary lineages.
   d. the reduction of two lineages by the extinction of one of them.
   e. the formation of new species by the reclassification of a group.

3. Allopatric speciation may happen when
   a. continents drift apart and separate previously connected lineages.
   b. a mountain range separates formerly connected populations.
   c. different environments on two sides of a barrier cause populations to diverge.
   d. the range of a species is separated by loss of intermediate habitat.
   e. all of the above

4. Finches speciated on the Galapagos Islands because
   a. the Galapagos Islands are a long way from the mainland.
   b. the Galapagos Islands are very arid.
   c. the Galapagos Islands are small.
   d. the islands in the Galapagos Archipelago are sufficiently isolated from one another that there is little migration among them.
   e. the islands in the Galapagos Archipelago are close enough to one another that there is considerable migration among them.

5. Which of the following is *not* a potential prezygotic isolating mechanism?
   a. Temporal segregation of breeding seasons
   b. Differences in chemicals that attract individuals
   c. Sterility of hybrids
   d. Spatial segregation of mating sites
   e. Inviability of sperm in female reproductive tracts

6. A common means of sympatric speciation is
   a. polyploidy.
   b. hybrid sterility.
   c. temporal segregation of breeding seasons.
   d. spatial segregation of mating sites.
   e. imposition of a geographic barrier.

7. Sympatric species are often very similar in appearance because
   a. appearances are often of little evolutionary significance.
   b. genetic changes accompanying speciation are often small.
   c. genetic changes accompanying speciation are usually large.
   d. speciation usually requires major reorganization of the genome.
   e. the traits that differ among species are not the same as the traits that differ among individuals within species.

8. Which statement about speciation is *not* true?
   a. It always takes thousands of years.
   b. It often takes thousands of years but may happen within a single generation.
   c. Among animals it usually requires a physical barrier.
   d. Among plants it often happens as a result of polyploidy.
   e. It has produced the millions of species living today.

9. Evolutionary radiations
   a. often happen on continents but rarely on island archipelagos.
   b. characterize birds and plants but not other taxonomic groups.
   c. have happened on all continents, as well as on islands.
   d. require major reorganizations of the genome.
   e. never happen in species-rich environments.

10. Speciation is often rapid within lineages in which species have complex behavior, because
    a. individuals of such species make very fine discriminations among potential mating partners.
    b. such species have short generation times.
    c. such species have high reproductive rates.
    d. such species have complex relationships with their environments.
    e. none of the above

Applying Concepts

1. Gene exchange between populations is prevented by geographic isolation, by behavioral responses before mating (for example, females rejecting courting males of other species), and by mechanisms that function after mating has occurred (for example, hybrid sterility). All of these are commonly called isolating mechanisms. In what ways are the three types very different? If you were to apply different names to them, which one would you call an isolating mechanism? Why?

2. The blue goose of North America has two distinct color forms: blue and white. As we have seen, matings between the two color types are common. On their breeding grounds in northern Canada, however, blue individuals pair with blue individuals much more frequently than would be expected by chance. Suppose that 75 percent of all mated pairs consisted of two individuals of the same color, what would you conclude about speciation processes in these geese? If 95 percent of pairs were of the same color? If 100 percent of pairs were of the same color?

3. Although many species of butterflies are divided into local populations among which there is little gene flow, these butterflies often show relatively little geographic variation. Describe the studies you would conduct to determine what maintains this morphological similarity.

4. Distinguish among the following: allopatric speciation, parapatric speciation, and sympatric speciation. For each of the three statements below, indicate which type of speciation is implied:
   a. This process in nature is most commonly a result of polyploidy.
   b. The present sizes of national parks and wildlife preserves may be too small to allow this type of speciation among organisms restricted to those areas.
   c. This process generally occurs in species that inhabit areas where sharp environmental contrasts exist.

5. Evolutionary radiations are common and easily studied on oceanic islands. In what types of mainland situations would you expect to find major evolutionary radiations? Why?

Readings

Endler, J. T. 1977. *Geographic Variation, Species, and Clines.* Princeton University Press, Princeton, NJ. A theoretical analysis of how sympatric and parapatric speciation might occur.

Futuyma, D. J. 1998. *Evolutionary Biology,* 3rd edition. Sinauer Associates, Sunderland, MA. Chapters 15 and 16 cover species and speciation in more detail.

Knowlton, N. 1994. "A Tale of Two Seas." *Natural History,* June. How allopatric speciation may have taken place when the Panamanian isthmus appeared, separating the Caribbean Sea from the Pacific Ocean.

Lambert, D. M. and H. G. Spencer. 1995. *Speciation and the Recognition Concept.* Johns Hopkins University Press, Baltimore. A collection of essays that examines the importance of mate recognition for processes of speciation.

Mayr, E. 1970. *Populations, Species, and Evolution.* Harvard University Press, Cambridge, MA. An abridged version of the most thorough work on speciation theory as applied to animals.

Otte, D. and J. A. Endler (Eds.). 1989. *Speciation and Its Consequences.* Sinauer Associates, Sunderland, MA. A comprehensive collection of essays on concepts, methods, and consequences of speciation. Includes general treatments and analyses of specific cases.

Ryan, M. J. 1990. "Signals, Species, and Sexual Selection." *American Scientist,* January/February. A study of the mate-recognition concept of species formation, using frog mating calls as the example.

Chapter 22

Constructing and Using Phylogenies

Asian Snails Can Transmit Schistosomiasis
Workers in the rice paddies of tropical Asia are at extreme risk of contracting schistosomiasis (known in some parts of the world as bilharzia). The disease is transmitted to humans via freshwater snails that thrive in the standing water of the paddies.

Schistosomiasis is a blood infection caused by a parasitic flatworm, *Schistosoma*. More than 200 million people in South America, Africa, China, Japan, and Southeast Asia have the disease. During part of its life cycle, *Schistosoma* inhabits a freshwater snail. People become infected when larval *Schistosoma* swim from a snail and penetrate their skin. The worm matures and lives in a person's abdominal blood vessels. The disease is progressively debilitating, causing a slow death.

For most of the twentieth century, only one species, *Schistosoma japonicum*, was known to infect humans, and it was thought to be transmitted by a single species of snail in the genus *Oncomelania*. Then in the 1970s researchers discovered that a different snail was transmitting *Schistosoma* to humans in the Mekong River in Laos. This discovery stimulated extensive field surveys and anatomical, genetic, and geographic research on the worms and snails in Southeast Asia.

Investigators found that *S. japonicum* was actually a cluster of at least six species. They also discovered that evolutionary relationships among snails determined which species could host *Schistosoma*. Evolutionary diversification from an ancestral stock of snails had produced a group of species of modern snails. Of these, only three retain the ability to host *Schistosoma*, and ten have a genetic trait that makes them unsuitable hosts for the disease.

This information is of great value in efforts to combat schistosomiasis. Most of the species of freshwater snails in Southeast Asia have not been described and named. By using information on evolutionary relationships among snails, scientists can quickly determine whether or not a newly discovered snail is a host for *Schistosoma*. Control efforts need to be directed toward only the snails that can transmit *Schistosoma* to humans, not to all freshwater snails in the region.

(a) *Campanula* sp.

(b) *Endymion nonscriptus*

(c) *Mertensia paniculata*

22.1 Many Different Plants Are Called Bluebells (a) These flowers from the plains of North Dakota are often called bluebells. (b) This English bluebell is a member of the lily family. (c) These Alaskan flowers are known as bluebells or chiming bells. None of these plants is closely related to the others.

How did investigators determine the evolutionary relationships among the snails that are hosts of *Schistosoma*? How could they determine that genes preventing snails from hosting *Schistosoma* arose three different times? How can evolutionary relationships be expressed in systems of classification that help guide further studies of organisms?

In this chapter, we present systematics, the science that provides answers to these questions. We consider the goals and methods of modern systematics, and we show how biologists determine evolutionary relationships among organisms—their **phylogeny**—and express those relationships in classification systems. Then we illustrate how knowledge of evolutionary relationships is used to solve other biological problems.

Classification systems are important. They improve our ability to explain relationships among things. Having classifications is especially important when biologists attempt to understand the evolutionary pathways that have produced the millions of species living today. Classification systems are also an aid to memory. It is impossible to remember the characteristics of many different things unless we can group them into categories based on shared characteristics.

Classification systems also provide unique names for organisms. If the names are changed, the systems provide means of tracing the changes. Common names, even if they exist (most organisms have no common names), are very unreliable and often confusing. For example, plants called bluebells are found in England, Scotland, Texas, and the Rocky Mountains—but none of the bluebells in any of those places is closely related evolutionarily to the bluebells in any of the other places (Figure 22.1).

Recognizing and interpreting similarities and differences among organisms is easier if the organisms are classified into groups that are ordered and ranked. Any group of organisms that is treated as a unit in a classification system is called a **taxon** (plural taxa). **Taxonomy** is the theory and practice of classifying organisms. **Systematics**, a larger field of which taxonomy is a part, is the scientific study of the diversity of organisms. Systematists study organisms to determine their evolutionary relationships and develop classification systems that reflect those relationships.

The Hierarchical Classification of Species

The biological classification system that is used today was developed by the great Swedish biologist Carolus Linnaeus in 1758. Linnaeus gave each species two names, one identifying the species itself and the other the genus to which the species belongs. This two-name system, referred to as **binomial nomenclature**,

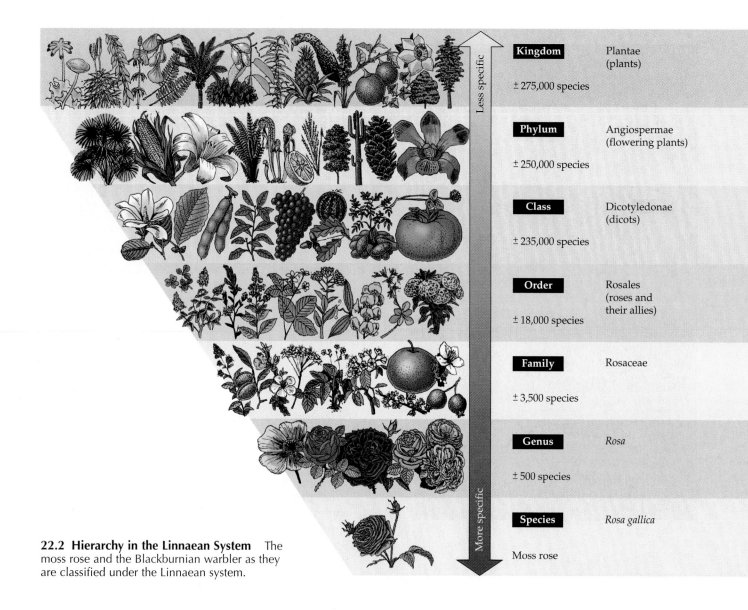

Kingdom	Plantae (plants)		
± 275,000 species			
Phylum	Angiospermae (flowering plants)		
± 250,000 species			
Class	Dicotyledonae (dicots)		
± 235,000 species			
Order	Rosales (roses and their allies)		
± 18,000 species			
Family	Rosaceae		
± 3,500 species			
Genus	*Rosa*		
± 500 species			
Species	*Rosa gallica*		
	Moss rose		

Less specific → More specific

22.2 Hierarchy in the Linnaean System The moss rose and the Blackburnian warbler as they are classified under the Linnaean system.

is universally employed in biology. Using this system, scientists throughout the world refer to the same organisms by the same names.

A **genus** (plural genera; adjectival form: generic) is a group of closely related species. In many cases the name of the taxonomist who first proposed the species name is added at the end. Thus, *Homo sapiens* Linnaeus is the name of the modern human species. *Homo* is the genus to which the species belongs, and *sapiens* identifies the species; Linnaeus proposed the species name *sapiens*. You can think of the generic name *Homo* as equivalent to your surname and the specific name *sapiens* as equivalent to your first name.

The generic name is always capitalized; the species name is not. Both names are always italicized, whereas common names are not. When referring to more than one species in a genus without naming each one, we use the abbreviation "spp." after the generic name (for example, "*Drosophila* spp." means more than one species in the genus *Drosophila*). The abbreviation "sp." is used after a generic name if the identity of the species is uncertain. Rather than repeating a generic name when it is used several times in the same discussion, biologists often spell it out only once and abbreviate it to the initial letter thereafter (for example, *E. coli* is the abbreviated form of *Escherichia coli*).

In the Linnaean system, species are grouped into higher taxonomic categories. The category (taxon) above genus in the Linnaean system is **family**. The names of animal families end in the suffix "-idae." Thus Formicidae is the family that contains all ant species, and the family Hominidae contains humans, a few of our fossil relatives, and chimpanzees and gorillas. Family names are based on the name of a member genus. Formicidae is based on *Formica*, and Hominidae is based on *Homo*.

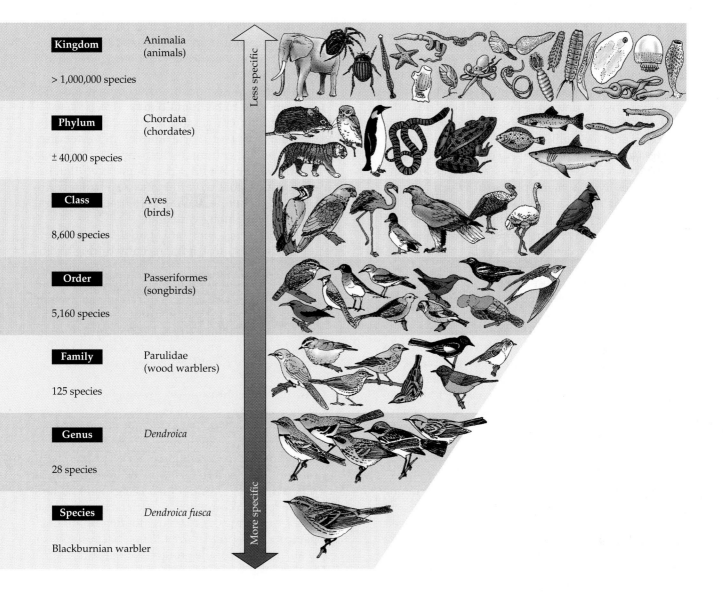

Kingdom	Animalia (animals)	
> 1,000,000 species		
Phylum	Chordata (chordates)	
± 40,000 species		
Class	Aves (birds)	
8,600 species		
Order	Passeriformes (songbirds)	
5,160 species		
Family	Parulidae (wood warblers)	
125 species		
Genus	*Dendroica*	
28 species		
Species	*Dendroica fusca*	
Blackburnian warbler		

Less specific → More specific

Plant classification follows the same procedures except that the suffix "-aceae" is used with family names instead of "-idae." Thus Rosaceae is the family that includes the genus of roses (*Rosa*) and its close relatives. Unlike generic and species names, family names are not italicized, but they are capitalized.

Families, in turn, are grouped into **orders**, and orders into **classes**. Classes are grouped into **phyla** (singular phylum), and phyla into **kingdoms**. The hierarchical units of this classification system, as applied to the Blackburnian warbler (*Dendroica fusca*) and the moss rose (*Rosa gallica*) are shown in Figure 22.2. For their detailed studies of particular groups, taxonomists use additional categories, such as subfamilies and subspecies, to indicate finer degrees of relationships, but we need not worry about them here.

Systematists attempt to solve two distinct but related problems. The first is how to determine evolutionary re-

lationships among organisms—that is, how to construct phylogenetic trees. The second is how to express evolutionary relationships in a classification system. We turn first to the methods used to deduce evolutionary relationships. There must be a single "true" phylogeny for a particular group of organisms, but identifying it is very difficult. The key events happened in the distant past, and evidence of those events is incomplete and sometimes contradictory.

Inferring Phylogenies

To infer phylogenies, systematists use information provided by both fossils and living organisms. And the more information, the better. In the pages that follow, we discuss what fossils tell us about relationships among organisms. Then we describe the set of methods, known as cladistics, used to construct phylogenies of organisms.

Fossils are the key to the past

Fossils tell us where and when organisms lived and what they looked like (see Chapter 19). When available, this information is valuable, but sometimes few or no fossils have been found for a taxon whose phylogeny we wish to determine. Even if there are many fossils, the first individuals of the group that lived in a particular area are unlikely to become fossilized and even less likely to be discovered. Fossils have been described from relatively few areas, which means that most species lived in many places for which we have no record. At best we can say that a species lived where we have found its remains and that it lived there at least as early as the earliest known fossil and as late as the most recent known fossil.

The incompleteness of the fossil record means that the chance of finding a fossil of an organism that was the direct ancestor of another fossil we happen to find is very small. Indeed, we can probably never determine accurately the ancestors of any group of organisms we are studying. But we can infer that two groups of organisms had a common ancestor at a particular time in the past even if we cannot describe that ancestor precisely. Fortunately, to construct phylogenetic trees, we do not need to identify ancestors; we need only know when they existed and be able to recognize some of their features.

Cladistics is a powerful evolutionary tool

Cladistics is a powerful new method for determining the evolutionary histories of organisms and expressing those relationships in treelike diagrams. Cladistics gets its name from **clade**, which is the entire portion of a phylogeny that is descended from a **common ancestor** (a single ancestral species). An evolutionary tree constructed using cladistic methods is called a **cladogram**. A cladogram does not attempt to describe ancestors; instead it shows points at which lineages diverged from a common ancestral form.

The goal of cladistics is to construct phylogenies by analyzing evolutionary changes in the traits of organisms. A **phylogenetic tree** is a cladogram to which additional information, such as evidence of the dates of separation of lineages, has been added. Phylogenetic trees are rather like pedigrees of lineages of organisms, except that they are usually constructed with the ancestor at the bottom rather than at the top. The base of the "trunk" of the tree represents the point in the past when the lineage consisted of only the ancestor.

What evidence do cladistic systematists use to infer how traits changed during evolution? How do they use that evidence to draw branches on a cladogram? In order to identify how traits have changed during evolution, cladists must determine the original state of the trait and then determine how it has been modified. That is, they must distinguish between ancestral and derived traits. A trait shared with a common ancestor is called an **ancestral trait**. A trait that differs from the form of the trait in the ancestor of a lineage is called a **derived trait**.

Any two traits descended from a common ancestral structure are said to be **homologous**. *General homologous traits* are shared by most or all organisms in the lineage being studied. For example, all vertebrates have a vertebral column, which appears to have evolved only once. We infer this single evolution because fossil ancestors of vertebrates also had a vertebral column. Therefore, having vertebrae is a general homologous trait among vertebrates.

Special homologous traits are shared by only a few species. Such traits arose more recently during evolution than general homologous traits did. Rats and mice, but not dogs or other mammals, have long, continuously growing incisor teeth, for example. Continuously growing incisors evidently developed in the common ancestor of rats and mice after their lineage separated from the one leading to dogs and other mammals, because no other mammals have that kind of incisor.

Thus, having continuously growing incisors is a special homologous trait among rats, mice, and their relatives. However, if we were trying to construct a phylogeny of a group of mice, continuously growing incisors would be a *general* homologous trait. Special homologous traits can be used to order the times of separation of lineages. A trait that is a general homology in one group of organisms, such as the vertebral column of vertebrates, is of no use for determining relationships among those organisms, because all, or nearly all, members of the group have the trait.

The first step in constructing a cladogram is to select the group of organisms whose phylogeny is to be determined. The next step is to choose the traits that will be used in the analysis and to identify the possible forms of those traits. A trait may be present or absent, or it may exist in more than one form. The last and usually the most difficult step is to determine the ancestral and derived forms of the traits.

IDENTIFYING ANCESTRAL TRAITS. Distinguishing derived traits from ancestral traits is difficult because traits may **diverge** (become dissimilar), making ancestral states unrecognizable. For example, the leaves of plants have diverged to form many different structures. Several lines of evidence, especially details of their structure and development, indicate that protective spines, tendrils, and brightly colored structures that attract pollinators (Figure 22.3) are all modified leaves; they are homologs of one another even though they do not resemble one another closely.

Not all resemblances are products of a common ancestry. If a trait evolves more than once and thus is pos-

The orange spines of the barrel cactus and the colorful bracts of *Heliconia* are both modified leaves.

Ferrocactus acanthodes (barrel cactus)

Heliconia rostrata

22.3 Homologous Structures Derived from Leaves The leaves of plants have diverged during their evolution to form many different structures, some of which bear very little resemblance to each other.

sessed by more than one species even though it was not found in their most recent common ancestor, it is said to exhibit **homoplasy**. Homoplasy can result when structures that were formerly very different come to resemble one another because they have been modified by natural selection to perform similar functions. We call this process **convergent evolution**.

Bats, birds, and insects fly by flapping their wings, but the structure of insect wings is totally different from that of bat and bird wings (Figure 22.4). The skeletons of bat and bird wings are homologous, but the supports for insect wings are not homologous with the wing bones of bats and birds. Similarly, the structures that aid plants in climbing over other plants have evolved from several different structures, including stipules, leaflets, leaves, and inflorescences. Structures that perform similar functions but have resulted from convergent evolution (homoplasy) are said to be **analogous** to one another.

The opening section of this chapter discussed the evolutionary diversification of the freshwater snails that act as hosts for the parasitic flatworm *Schistosoma* and transmit the parasite to humans. The ability of these snail species to host *Schistosoma* is a homologous, ancestral trait. However, Asian freshwater snails evolved resistance to the transmission of *Schistosoma* on three separate occasions; this resistance is an analogous trait (Figure 22.5).

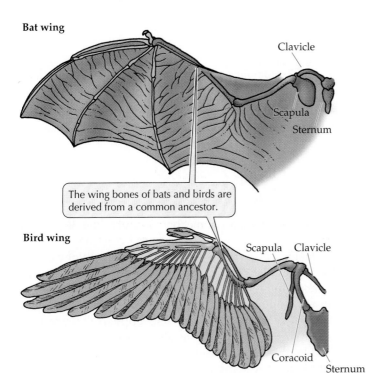

Bat wing

Clavicle

Scapula

Sternum

The wing bones of bats and birds are derived from a common ancestor.

Bird wing

Scapula Clavicle

Coracoid

Sternum

22.4 Wing Structures May Be Homologous or Analogous The supporting structures of bat and bird wings are derived from a common tetrapod (four-limbed) ancestor and are thus homologous. Although they are also used for flight (and thus are analogous), insect wings evolved independently and their supports are not homologous with the wing bones of bats and birds.

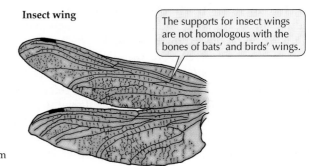

Insect wing

The supports for insect wings are not homologous with the bones of bats' and birds' wings.

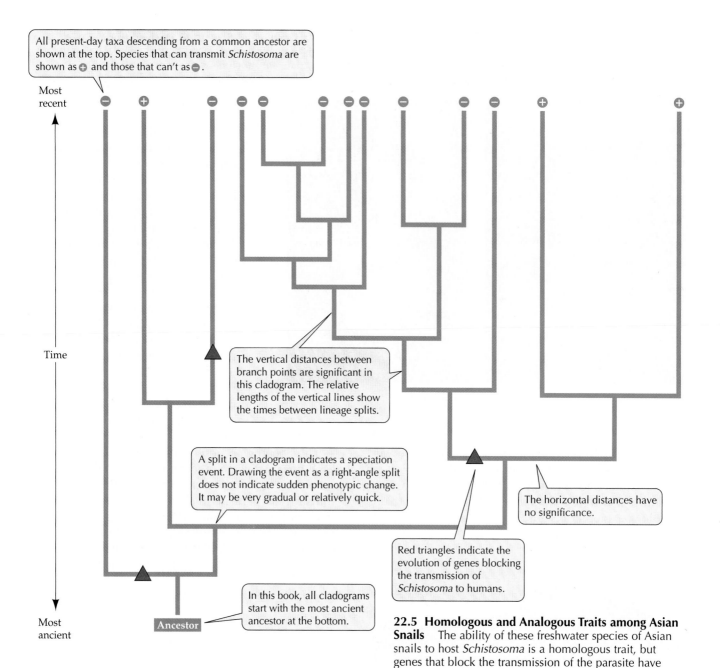

All present-day taxa descending from a common ancestor are shown at the top. Species that can transmit *Schistosoma* are shown as ⊕ and those that can't as ⊖.

Most recent

Time

The vertical distances between branch points are significant in this cladogram. The relative lengths of the vertical lines show the times between lineage splits.

A split in a cladogram indicates a speciation event. Drawing the event as a right-angle split does not indicate sudden phenotypic change. It may be very gradual or relatively quick.

The horizontal distances have no significance.

Red triangles indicate the evolution of genes blocking the transmission of *Schistosoma* to humans.

Most ancient

Ancestor

In this book, all cladograms start with the most ancient ancestor at the bottom.

22.5 Homologous and Analogous Traits among Asian Snails The ability of these freshwater species of Asian snails to host *Schistosoma* is a homologous trait, but genes that block the transmission of the parasite have arisen three distinct times during the snails' evolution; the trait of resistance is thus analogous.

Homoplasy is widespread in evolution. Its existence makes attempts to determine true phylogenies difficult because it is easy to assume that analogous traits are homologous. Taxonomists need a means by which to identify which forms of traits are ancestral and which are derived, and a means by which to distinguish homologies from analogies.

HENNIG'S METHOD FOR IDENTIFYING ANCESTRAL AND DERIVED TRAITS. A method for reconstructing phylogenies that serves both of these needs was developed by the German entomologist Willi Hennig in the 1950s. Hennig suggested that if two species possess the same trait, systematists should *provisionally* (that is, until

proven otherwise) assume that the trait is homologous in the two species.

Hennig also proposed that general homology could be distinguished from special homology as follows. A general homologous trait is one that is found not only in one or more species of a group whose phylogeny is being reconstructed, but also appears *outside* this group in what is known as an outgroup. An **outgroup** is a taxon that is closely related to the group whose phylogeny is being reconstructed but that branched off from the lineage of the group below its base on the evolutionary tree. Traits found only *within* the group, on the other hand, are special homologous traits.

In Hennig's system, the members of the group whose phylogeny is being reconstructed are ordered according to the number of special derived homologous traits they share. Species that share a recent common ancestor should share many general homologous traits and a few special derived homologous traits. However, they should share very few homoplastic traits, because little time has been available for those traits to arise.

Using this system, as more and more traits are measured, the data are increasingly likely to support a single phylogenetic pattern, and biologists can more readily distinguish between homologies and homoplasies. A few of the traits originally assumed to be homologies may turn out to be homoplasies, but the best way to determine the true status of shared traits is to assume that they are homologous until additional evidence suggests they aren't.

To see how a cladogram is constructed, consider seven vertebrate animals—hagfish, perch, pigeon, chimpanzee, salamander, lizard, and mouse. We will assume initially that a given derived trait evolved only once during the evolution of these animals and that no derived traits were lost from any of the descendant groups. For simplicity, we have selected traits that are either present (+) or absent (−). As will become evident in the survey of animals in Chapter 30, hagfishes are believed to be more distantly related to the other vertebrates than other vertebrates are to each other. Therefore, we will choose hagfishes as the outgroup for our analysis. Derived traits are those that have been acquired by other members of the lineage since they separated from hagfishes. The taxa and the traits we will consider are shown in Table 22.1.

Cladistic methods infer the branching points in evolutionary trees by minimizing the number of evolutionary changes that need to be assumed—that is, by minimizing homoplasy. For example, the chimpanzee and mouse share two unique traits, fur and mammary glands. Because these traits are absent in both the outgroup and the other species whose relationships we are attempting to determine, we infer that fur and mammary glands are derived traits that evolved in a common ancestor of chimpanzees and mice after that lineage separated from the ones leading to the other lineages. In other words, we assume that fur and mammary glands evolved only once among the animals we are classifying.

The pigeon has one unique trait: feathers. Similarly, we assume that feathers evolved only once, after the lineage leading to birds separated from that leading to the mouse and chimpanzee. By the same reasoning, we assume that claws or nails evolved only once, after the lineage leading to salamanders separated from the lineage leading to those animals that have claws or nails. We make the same assumption for lungs and jaws, continuing to minimize the number of evolutionary events needed to produce the patterns of shared traits among these seven animals. We can see the pattern clearly by ordering the animals in Table 22.1 according to the number of derived traits they share.

Using this information, we can construct a cladogram. The taxon with no derived traits, the hagfish, is the outgroup, and we assume, following Hennig's rule, that the animals that share unique derived traits have a common ancestor not shared with animals lacking those traits. That is, we assume that feathers, which are found only in the pigeon, evolved after the lineage leading to birds separated from the lineage leading to mice and chimpanzees. Otherwise we would need to assume that the ancestors of mice and chimpanzees also had feathers but that the trait was subsequently lost—an unnecessary additional assumption.

A cladogram for these taxa, based on the traits we used and the assumption that each derived trait evolved only once, is shown in Figure 22.6. Notice that

TABLE 22.1 Some Vertebrates Ordered According to Unique Shared Derived Traits

	DERIVED TRAIT[a]					
TAXON	**JAWS**	**LUNGS**	**CLAWS OR NAILS**	**FEATHERS**	**FUR**	**MAMMARY GLANDS**
Hagfish (outgroup)	−	−	−	−	−	−
Perch	+	−	−	−	−	−
Salamander	+	+	−	−	−	−
Lizard	+	+	+	−	−	−
Pigeon	+	+	+	+	−	−
Mouse	+	+	+	−	+	+
Chimpanzee	+	+	+	−	+	+

[a] A plus sign indicates the trait is present, a minus sign that it is absent.

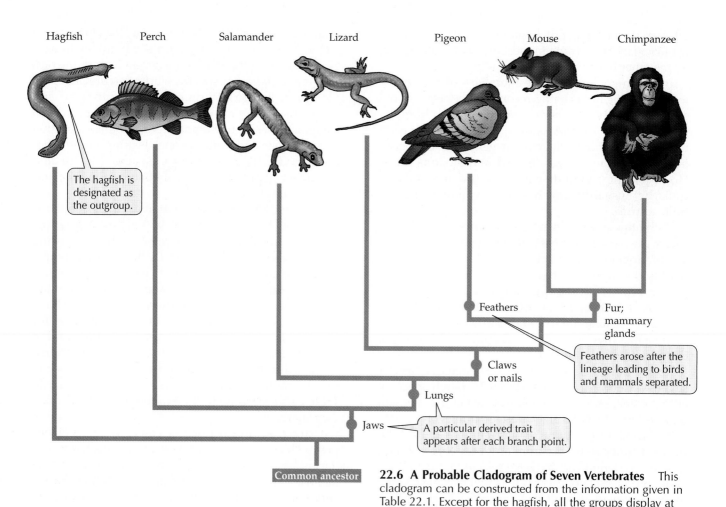

22.6 A Probable Cladogram of Seven Vertebrates This cladogram can be constructed from the information given in Table 22.1. Except for the hagfish, all the groups display at least one derived trait.

the cladogram does not describe ancestors or date the splits between lineages. A cladogram shows only the temporal order of splits between lineages; the oldest splits are at the bottom, and the more recent ones are near the top. Notice also that the *x* axis has no scale. Horizontal distances between taxa do not correlate with degree of similarity or difference between them, and we show lineage separations with right angles only for graphical convenience. In other words, do not interpret the cladograms in this book to mean that changes occur suddenly when lineages split.

The cladogram of these seven vertebrates was easy to construct because the traits we chose fulfilled the assumptions that derived traits appeared only once in the lineage and were never lost once they appeared. However, if we had included a snake (a reptile like the lizard) in the group, the second assumption would have been violated, because the ancestors of snakes had limbs, which were subsequently lost, along with their claws.

We would need to examine additional traits to determine that the lineage leading to snakes separated from the one leading to lizards long after the lineage leading to lizards separated from the others. In fact,

the analysis of many traits shows that snakes evolved from burrowing lizards that lost their limbs during a long period of subterranean existence.

Outgroups have also been used to identify ancestral and derived traits in lineages of butterflies. Species in two families—the brush-footed butterflies (Nymphalidae) and the monarchs (Danaidae)—have four functional and two very small legs, whereas the swallowtails (Papilionidae), the sulfurs (Pieridae), and all other butterflies have six functional legs. Biologists assume that having six legs is ancestral because moths and all other orders of insects have six functional legs. The four-legged trait in monarchs and brush-footed butterflies is thus inferred to be a derived trait of butterflies descended from six-legged ancestors (Figure 22.7). If other special shared traits also united brush-footed butterflies and monarchs, we would conclude that these two groups of butterflies share a more recent common ancestor than either group shares with any other group of butterflies. If not, we would conclude that the four-legged condition arose twice during insect evolution.

Pieris protodice

22.7 Six Legs Is the Ancestral Number Having six functional legs is the ancestral trait among butterflies; some species, however, have four legs—a derived trait.

Danaus plexippus

Many traits must be analyzed to reconstruct a phylogeny, and systematists use various methods to combine information from the different traits. Each method is based on specific operating rules—provisional assumptions about how evolution proceeds. A simple method—the one we used in our vertebrate example—makes two assumptions: (1) that the evolution of traits is irreversible (that an ancestral trait can change into a derived one, but the reverse change does not happen), and (2) that each trait can change only once within a lineage. However, we know from fossil and other evidence that the states of traits *can* reverse and that traits *can* change more than once. Therefore, systematists must often relax the rules and allow a derived trait to be lost or to evolve more than once.

A cladogram that postulates fewer reversals and changes in traits is more likely to be accurate than one that requires more changes, because reversals and multiple origins of traits are relatively rare events. Therefore, systematists typically employ the principle of **parsimony** in reconstructing a phylogeny. That is, they arrange the organisms such that the number of changes in traits that must be postulated to account for the inferred lineage is minimized. Parsimony is used as an operating rule in many types of biological investigations.

Using parsimony is helpful not because evolutionary changes are necessarily parsimonious, but because it is generally wiser not to adopt complicated explanations when simpler ones are capable of explaining the known facts. More complicated explanations are accepted only when evidence requires them. Thus, cladograms are hypotheses about evolutionary relationships that are repeatedly tested as additional traits are measured and as new fossil evidence becomes available.

Constructing Phylogenies

As we have seen, many different traits must be measured if we wish to distinguish homologies from homoplasies. Because organisms differ in many ways, systematists use many traits to reconstruct phylogenies. Some traits are readily preserved in fossils; others, such as behavior and molecular structure, rarely survive fossilization processes. By using cladistic techniques, systematists can take into consideration behavioral and molecular traits, which are not preserved in fossils, as well as structural traits in both living and fossil organisms. There is only one correct phylogeny, so the more traits that are measured, the more inferred phylogenies should converge on one another and on the true phylogeny. In this section, we will show how structural, developmental, and molecular traits of organisms are used to infer phylogenies.

Structure and development show ancient connections

An important source of information for taxonomists is **gross morphology**—that is, sizes and shapes of body parts. These traits are useful because they are under genetic control and are relatively stable, but they do change over evolutionary time. Because living organisms have been studied for centuries, we have a wealth of morphological data and extensive museum and herbarium collections of organisms whose traits, including molecular ones, can be measured. Also, this is the type of information most readily available from fossils. Sophisticated methods are now available for measuring and analyzing morphology and for estimating the amount of morphological variation among individuals, populations, and species.

The early developmental stages of many organisms reveal similarities with other organisms that are lost by the time of adulthood. For example, the larvae of the marine creatures called sea squirts have a supporting rod in their backs—the notochord—that disappears as they develop into adults. Many other animals—all the animals called vertebrates—also have such a structure at some time during their development. This shared structure is one of the reasons for believing that sea squirts are more closely related to vertebrates than would be suspected by examination of adults only (Figure 22.8).

Molecular traits: getting close to the genes

Like the sizes and shapes of their body parts, the molecules of organisms are heritable characteristics that may diverge among lineages over evolutionary time; this *molecular evolution* will be discussed in detail in Chapter 23. Among the most important molecular traits of organisms for constructing phylogenies are the structures of their proteins and nucleic acids (DNA and RNA).

PROTEIN STRUCTURE.   Relatively precise information about phylogenies can be obtained by comparison of the molecular structure of proteins. We estimate genetic differences between two taxa by obtaining homologous proteins from both and determining the number of amino acids that have changed since the lineages of the taxa diverged from a common ancestor. For example, determining the sequences of amino acids revealed a great deal about how natural selection influenced the evolution of cytochrome *c* (see Figure 23.1).

NUCLEIC ACID STRUCTURE.   The base composition of DNA provides excellent evidence of evolutionary relationships among organisms. Cells of eukaryotes have genes in their nuclei and mitochondria. Plant cells also have genes in their chloroplasts. The chloroplast genome (cpDNA), which is used extensively in phylogenetic studies of plants, consists of a circular, double-stranded DNA molecule. This molecule is evolutionarily highly stable. All land plants have nearly the same complement of 100 chloroplast genes that code transfer RNAs and some ribosomal RNA subunits and proteins, particularly those involved in photosynthesis. Mitochondrial DNA (mtDNA), which is very similar to cpDNA, has been used extensively for evolutionary studies of animals (see Chapter 23 for details).

DNAs can be compared, even if the precise sequences of their bases are not known, by a process called **DNA hybridization**, in which the double-stranded DNAs of two different species are combined, then separated into single strands, or *denatured* (see Chapter 14) and allowed to reassociate. The degree of mismatching of the resulting hybrid DNA is related to its thermal stability in a very consistent way. DNA hybridization reveals that the DNAs of humans and chimpanzees are much more similar than would be expected

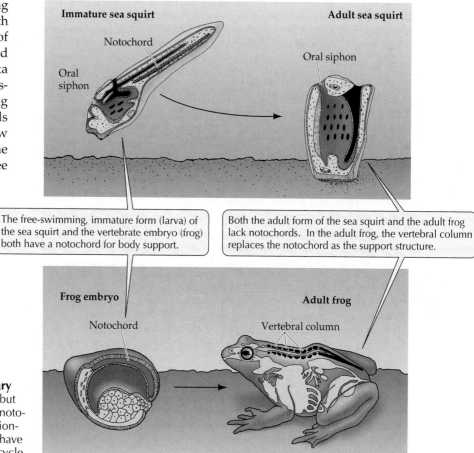

The free-swimming, immature form (larva) of the sea squirt and the vertebrate embryo (frog) both have a notochord for body support.

Both the adult form of the sea squirt and the adult frog lack notochords. In the adult frog, the vertebral column replaces the notochord as the support structure.

22.8 A Larva Reveals Evolutionary Relationships Sea squirt larvae, but not adults, have a well-developed notochord (blue) that reveals their relationship with vertebrates, all of which have one at some time during their life cycle.

given the considerable morphological differences between the two species (Table 22.2). This similarity indicates that humans and chimpanzees share a common ancestor that is more recent than previously thought.

A long-standing debate among biologists over whether the giant panda of China is more closely related to bears or to raccoons was resolved by DNA hybridization data, which clearly indicate that the giant panda is a bear (Figure 22.9). The non-bearlike features of the giant panda are recent adaptations to its specialized diet—bamboo.

Biological Classification and Evolutionary Relationships

Biological classification systems are designed to express relationships among organisms. The kind of relationships we wish to express determines which features we use to classify organisms. If, for instance, we were interested in a system that would help us decide what plants and animals were desirable as food, we might devise a classification based on tastiness, ease of capture, and the type of edible parts each organism

TABLE 22.2	Genetic Differences among Some Vertebrates as Estimated by DNA Hybridization	
TAXA COMPARED	**PERCENTAGE DIFFERENCE IN DNA SEQUENCES**
Human/chimpanzee	1.6
Human/gibbon[a]	3.5
Human/rhesus monkey	5.5
Human/galago[b]	28.0
House mouse/Norway rat	20.0
Cow/sheep	7.5
Cow/pig	20.0

[a] The gibbon is the smallest member of the ape family, which also includes chimpanzees and gorillas.

[b] Galagos are small, nocturnal primates found in Africa. They are more closely related to the lemurs of Madagascar than they are to monkeys.

possessed. Early Hindu classifications of plants were designed according to these criteria. Biologists do not use such systems today, but those systems served the needs of the people who developed them.

Ailurus fulgens (lesser panda)

Ailuropoda melanoleuca (giant panda)

22.9 The Giant Panda Is a Bear The DNA hybridization data used to construct the cladogram indicate that the giant panda of Asia is a bear. The lesser panda, also of Asia, was long thought to be closely related to the giant panda, but it is actually more closely related to raccoons and their allies of the Americas.

It is inappropriate to ask whether those classifications or any others, including contemporary ones, are right or wrong. Classification systems can be judged only in terms of their utility and consistency with their stated goals. To evaluate any classification system we must first ask the question, What relationships is it trying to express? Then, How well does it express those relationships? This section addresses these questions.

Early biological classifications were nonevolutionary

Many organisms were given species names and classified by Linnaeus and his followers before evolution became widely accepted as the central concept of biology. These workers described many features of organisms and grouped them according to which similarities seemed most important. They tried to develop "natural" systems of classification, but they had no basis for deciding what was natural or why some features of organisms were more important than others.

Because judging the "importance" of traits and identifying ancestors are difficult tasks, an approach to taxonomy that considered all traits to be of equal importance was developed in the 1950s. Numerical taxonomists, as practitioners of this approach came to be known, did not try to determine evolutionary relationships. Instead they measured as many traits as possible and used adding machines (at first) and computers to compute measures of differences between organisms. Today, many systematists use computer models developed by numerical taxonomists to assist them in developing cladograms and classifications based on evolutionary relationships.

Taxonomic systems should reflect evolutionary relationships

Most taxonomists today believe that classification systems should reflect the evolutionary relationships of organisms. However, they do not agree on the best criteria for doing so, even if they agree on the phylogenies. The reason for the lack of agreement is that lineages of organisms evolve at very different rates. Some "living fossils" have changed very little in the past 50 million years (Figure 22.10). Other lineages have undergone rapid evolutionary changes within the past few million years. Should a classification system be based only on the time since two lineages shared a common ancestor, or should it also reflect the rate of evolution of taxa after they separate from one another?

A clade contains all the descendants of a particular ancestor and no other organisms. For this reason, a clade is said to be **monophyletic**. A taxon consisting of members that do not share the same common ancestor is **polyphyletic**. A group that contains some but not all of the descendants of a particular ancestor is said to be **paraphyletic** (Figure 22.11). Taxonomists agree that polyphyletic groups are inappropriate as taxonomic

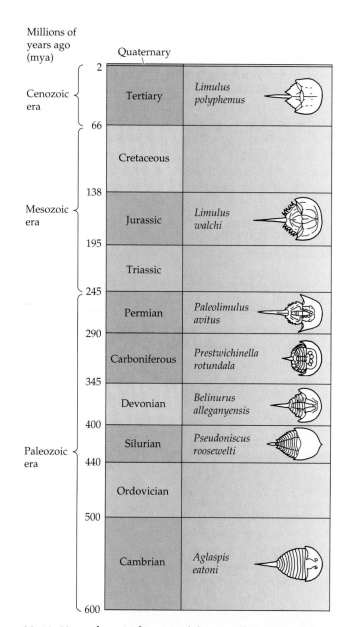

22.10 Horseshoe Crabs Are "Living Fossils" Horseshoe crabs that lived many millions of years ago are almost indistinguishable from modern horseshoe crabs. They have evolved very slowly.

units, but they disagree about the usefulness of paraphyletic groups.

Why they disagree can be illustrated by birds, crocodiles, and their relatives. We now know, primarily from fossil evidence, that birds and crocodilians (a group that includes crocodiles and alligators) share a more recent common ancestor than crocodilians share with snakes and lizards (Figure 22.12a). Until recently, crocodilians were grouped with snakes, lizards, and turtles as reptiles (class Reptilia). Birds were placed in a separate class, Aves (Figure 22.12b). This classification was used because crocodiles have evolved more slowly than birds since the two lineages separated. As

22.11 Mono-, Poly-, and Paraphyletic Taxa Polyphyletic groups are inappropriate as taxonomic units, but scientists disagree about the usefulness of paraphyletic groups.

a result, crocodilians are more similar in many features to snakes, lizards, and turtles than they are to birds. They look like very large lizards.

The phylogeny shown in Figure 22.12*a* indicates that the class Reptilia is paraphyletic because the class does not include all the descendants of its common ancestor; that is, birds are not included in the class. If only monophyletic taxa were permitted, birds would be included with crocodilians and their ancestors in a single taxon separate from snakes, lizards, and turtles (Figure 22.12*c*). Retaining birds as a separate class (that is, recognizing the reptiles as a paraphyletic group) emphasizes that birds have undergone rapid evolution since they separated from reptiles and have developed major unique derived traits.

The classifications used today still contain many polyphyletic groups because many organisms have not

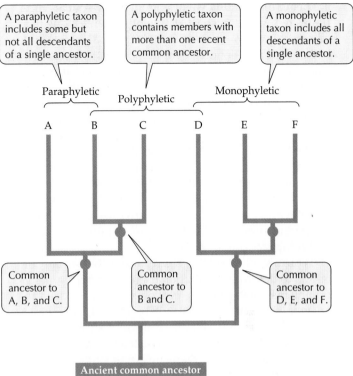

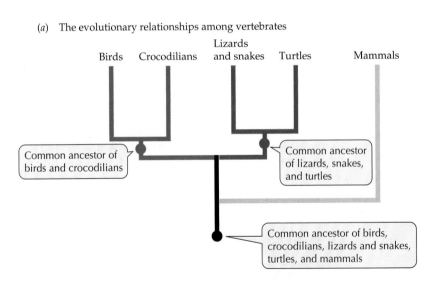

(a) The evolutionary relationships among vertebrates

been studied enough to distinguish between homologies and convergent evolution. However, as soon as they detect convergent evolution, systematists change their classifications to eliminate polyphyletic taxa. Nevertheless, many systematists favor retaining paraphyletic taxa because doing so highlights groups that have undergone especially rapid evolutionary change.

Another reason for retaining paraphyletic taxa is to maintain stability in the taxonomic system. Strict avoidance of paraphyletic taxa would require reclassifying many higher-level taxa every time new phylogenetic information became available. Some systematists believe that the resulting chaos would seriously reduce the value of the classification system.

The Future of Systematics

The development of molecular methods and powerful computers has ushered in a new era of taxonomy. Many phylogenies are being reconstructed and classifications

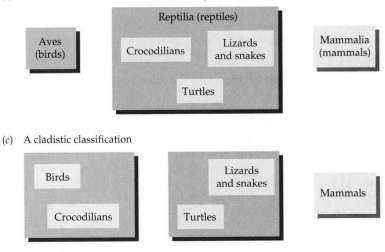

(b) The traditional classification of birds, reptiles, and mammals

(c) A cladistic classification

22.12 Phylogeny and Classification
A cladistic classification based on their evolutionary relationships includes the crocodilians in a subgroup with the birds.

are being revised. Information from many sources continues to be used in constructing phylogenies. The fossil record, which reveals when lineages diverged and began their independent evolutionary histories, is necessary to provide absolute dating for evolutionary events. Fossils provide important evidence that helps us distinguish ancestral from derived traits.

The range of data used in classification is likely to increase rather than decrease in the future because modern chemical, biochemical, and microscopic methods allow systematists to measure more traits of organisms than they could previously. Because systematics integrates activities from several different biological disciplines, a systematist needs to have a command of molecular techniques, natural history, and computer programming.

Typically, phylogenies are reconstructed as part of efforts to determine evolutionary relationships among organisms. Nevertheless, phylogenies can be used to answer many other types of biological questions. Indeed, it is difficult to think of any biological problem whose solution would not be assisted by having a reliable phylogeny of the organisms being studied.

Many biological statements are phylogenetic statements. Any statement claiming an association between a trait and a group of organisms is a claim about when during a lineage the trait first arose and about the fate of the trait since its first appearance. For example, the statement that DNA is the genetic material for all eukaryotes is a claim that organisms evolved DNA as their genetic material before eukaryotic cells appeared—that DNA is an ancestral and a general homologous trait among eukaryotes—and that DNA has been maintained as the genetic material in the subsequent evolution of all surviving eukaryote lineages.

The Three Domains and the Six Kingdoms

In Chapter 1 we introduced the three domains and six kingdoms into which organisms are generally classified. The domains are the most ancient divisions of the phylogenetic tree of life. They separated from one another so long ago that few fossils are available to help us determine when and in what order. The sequence of these ancient lineage splits that we use in this book is based on analyses of rRNAs and hundreds of amino acid sequences of dozens of enzymes from all the major groups of organisms.

These extensive molecular data suggest that the most ancient split in the phylogenetic tree of life, about 2,000 million years ago (mya), separated the ancestor of the domain Bacteria from the ancestor of all other life (Figure 22.13). The next split, about 1,800 mya, separated the ancestor of the domain Archaea from the ancestor of the domain Eukarya. Subsequent splits separated the ancestors of protists (1,230 mya),

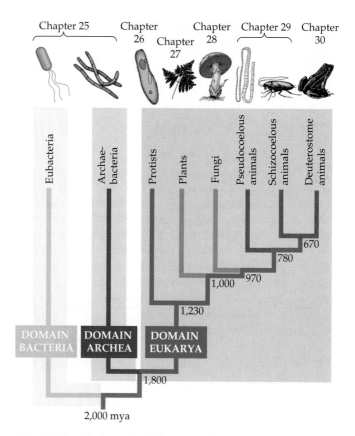

22.13 The Phylogeny of the Domains and Kingdoms
Approximate dates of lineage separations are shown at each node. mya, million years ago.

plants (1,000 mya), fungi (970 mya), and animals (780 mya). New information is likely to change the estimated dates of the splits, but the temporal ordering of the lineage separations is fairly certain.

Summary of "Constructing and Using Phylogenies"

• Classification systems improve our ability to explain relationships among things, aid our memory, and provide unique, universally used names for organisms.

The Hierarchical Classification of Species
• Biological nomenclature assigns to each organism a unique combination of a generic and a specific name. In the universally employed classification system, species are grouped into higher-level units called genera, families, orders, classes, phyla, and kingdoms. **Review Figure 22.2**

Inferring Phylogenies
• Systematists use data from fossils and the rich array of morphological and chemical data available from living organisms to determine evolutionary relationships.

- An ancestral trait is shared with a common ancestor. A derived trait differs from its form in the ancestors of a lineage.
- Homologous traits are descended from a common ancestor. Homoplastic traits evolved more than once. **Review Figures 22.3, 22.4**
- Structures that perform similar functions but have resulted from convergent evolution are said to be analogous to one another. **Review Figure 22.4**
- Cladistic methods were developed to help biologists distinguish between homologous and analogous traits. **Review Figure 22.5**
- To determine true evolutionary relationships, systematists must distinguish between ancestral and derived traits within a lineage, as well as between homologous and homoplastic traits. This task is often difficult because divergent evolution may make homologous traits appear dissimilar and convergent evolution may make nonhomologous traits appear similar. **Review Figures 22.3, 22.4, 22.5, 22.6, and Table 22.1**

Constructing Phylogenies

- Structures in early developmental stages sometimes show evolutionary relationships that are not evident in adults, and such early stages are often available in fossil material. **Review Figure 22.8**
- The structures of proteins and the base sequences of nucleic acids are important taxonomic data that can be obtained from living organisms.

Biological Classification and Evolutionary Relationships

- Taxonomists agree that taxa should share a common ancestor and that polyphyletic taxa are inappropriate taxonomic units. However, they disagree about the utility of paraphyletic taxa, which include some but not all of the descendants of a particular ancestor. **Review Figures 22.11, 22.12**
- Paraphyletic taxa may be retained to highlight the fact that members of some lineages evolved especially rapidly.

The Future of Systematics

- The development of molecular methods and powerful computers has ushered in a new era of systematics.
- Phylogenies are useful in solving many kinds of biological problems.

The Three Domains and the Six Kingdoms

- The domains are the most ancient divisions of the phylogenetic tree of life. They split more than 1,800 million years ago.
- Subsequent splits within the domain Eukarya separated the ancestors of protists (kingdom Protista), plants (kingdom Plantae), fungi (kingdom Fungi), and animals (kingdom Animalia). **Review Figure 22.13**

Self-Quiz

1. Which of the following is *not* a major role of a classification system?
   a. To aid memory
   b. To improve predictive powers
   c. To help explain relationships among things
   d. To provide relatively stable names for things
   e. To design identification keys

2. Any group of organisms treated as a unit in a classification system is
   a. a species.
   b. a genus.
   c. a taxon.
   d. a clade.
   e. a phylogen.

3. A genus is
   a. a group of closely related species.
   b. a group of genera.
   c. a group of similar genotypes.
   d. a taxonomic unit larger than a family.
   e. a taxonomic unit smaller than a species.

4. Outgroups are used in cladistic analyses to
   a. distinguish homoplasies from homologies.
   b. distinguish homoplasies from convergence.
   c. distinguish between general and special homologies.
   d. determine relationships between closely related taxa.
   e. distinguish between general and special homoplasies.

5. A clade contains
   a. all—and only—the descendants of a single ancestor.
   b. all the descendants of more than one ancestor.
   c. most but not all of the descendants of a single ancestor.
   d. members of two or more lineages.
   e. a few of the descendants of a single ancestor.

6. Which of the following is *not* a way of identifying ancestral traits?
   a. Determining which traits are found among fossil ancestors
   b. Using an outgroup in which the trait is also found
   c. Using more than one outgroup that has the trait
   d. Determining how many species in the lineage share the trait today
   e. Experimentally creating a known lineage

7. Traits that evolve very slowly are useful for determining relationships at the level of
   a. phyla.
   b. genera.
   c. orders.
   d. families.
   e. species.

8. Homologous traits are
   a. similar in function.
   b. similar in structure.
   c. similar in structure but derived from different ancestral structures.
   d. derived from a common ancestor whether or not they have the same function today.
   e. derived from different ancestral structures and have dissimilar structures.

9. The genes that are most extensively used to determine evolutionary relationships among plants are
   a. nuclear genes.
   b. chloroplast genes.
   c. mitochondrial genes.
   d. genes in flowers.
   e. genes in roots.

10. Which of the following is *not* a way in which phylogenies are used?
    a. To establish evolutionary relationships
    b. To determine how rapidly traits evolve
    c. To determine historical patterns of movement of organisms
    d. To help identify unknown organisms
    e. To infer evolutionary trends

Applying Concepts

1. The great blue heron (*Ardea herodias*) is found in most of North America. The very similar gray heron (*Ardea cinerea*) ranges over most of Europe and Asia. These two herons currently are treated as different species, but a colleague argues that they should be treated as a single species. What facts should you consider in evaluating your colleague's suggestion?

2. Why are systematists so concerned with identifying lineages descended from a single ancestor?

3. How are fossils used to identify ancestral and derived traits of organisms?

4. A student of the evolution of frogs has performed DNA hybridization experiments among about 25 percent of all frog species. As a result of these experiments, she proposes a new classification of frogs that differs strikingly from the traditionally accepted one. Should frog taxonomists immediately accept this new classification? Why or why not?

5. Linnaeus developed his system of classification before Darwin proposed his theory of evolution by natural selection, and most classifications of organisms were developed by people who were not evolutionists. Yet most of these classifications are still used today, with minor modifications, by evolutionists. Why?

Readings

Brooks, D. R. and D. H. McLennan. 1991. *Phylogeny, Ecology, and Behavior. A Research Program in Comparative Biology.* University of Chicago Press, Chicago. An excellent book that shows how rich insights can be derived by integrating phylogenetic analyses with studies of ecology and behavior. Includes a chapter describing the tools used to reconstruct phylogenies.

Eldredge, N. and J. Cracraft. 1980. *Phylogenetic Patterns and the Evolutionary Process.* Columbia University Press, New York. A good sampling of various perspectives on concepts and practices in systematics.

Harvey, P. H. and M. D. Pagel. 1991. *The Comparative Method in Evolutionary Biology.* Oxford University Press, Oxford. A good discussion of methods of reconstructing phylogenies and the use of comparative approaches in evolutionary studies.

Hillis, D. M., C. Moritz and B. K. Mable (Eds.). 1996. *Molecular Systematics,* 2nd Edition. Sinauer Associates, Sunderland, MA. A description of methods used currently in molecular systematics; designed to guide beginners through a molecular systematic study.

Mayr, E. 1982. *The Growth of Biological Thought.* Belknap Press, Cambridge, MA. A good treatment of the history of thinking about systematics, biological diversity, and evolution.

Ridley, M. 1986. *Evolution and Classification: The Reformation of Cladism.* Longman, London. A critical evaluation of current schools of thought in systematics that provides clear statements of the goals and methods of all approaches.

Wiley, E. O. 1981. *Phylogenetics: The Theory and Practice of Phylogenetic Systematics.* Wiley, New York. A clear overview of the phylogenetic approach to evolution and classification.

Chapter 23

Molecular Evolution

There is a specialized secretion system in many types of pathogenic eubacteria that was first noticed in 1991. These bacteria, which cause diseases such as bubonic plague in humans and fire blight in fruit trees, had acquired a set of apparently identical genes coding for an elaborate protein delivery system. The system, which is activated by contact with host cells, injects a variety of damaging proteins directly into the host's cell, thereby avoiding most of the host's defenses. The system was first discovered by scientists studying a species of a gram-negative bacterium in the genus *Yersinia*, which causes bubonic plague and intestinal infections in humans. Because these proteins seemed to be associated with the outer membranes of the bacteria, they were called **Yops** (*Yersinia* outer proteins). (Yops are not actually outer membrane proteins, but the name has persisted.)

Other researchers soon found Yops in other species of bacteria only distantly related to *Yersinia*. How did so many species of bacteria acquire the same set of genes? Could they all have inherited them from a common ancestor in the distant past? Or did they acquire them by transfer from other bacteria? Why do molecular biologists believe that the latter explanation is the correct one? To answer this question, we need to understand patterns and processes in molecular evolution.

For much of its history, evolutionary biology has consisted of the study of the obvious morphological features of organisms. Much of what Charles Darwin documented during his 5-year voyage aboard the *Beagle* were detailed observations about morphological

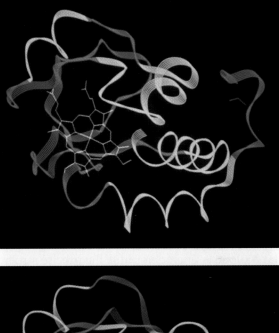

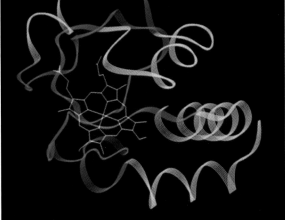

Protein Molecules Fold the Same Way in Different Species
Molecular sequencing techniques reveal that the three-dimensional structures of cytochrome *c* in rice (top) and tuna (bottom) are substantially similar. Such similarities indicate that all cytochrome *c* molecules have a common ancestor.

503

differences in the kinds of species found in different geographic areas. Darwin later synthesized these observations into descriptions of how species change over time.

His most famous example involved the finches of the Galapagos Islands. Darwin observed that insect-eating warblers and woodpeckers were absent from these isolated islands, but various species of finches, birds that usually eat seeds, had assumed the insect-eating habits of the missing species. In the process of adapting to their new diets, the finches had evolved beak morphologies that matched those of the missing species (see Figure 21.7).

Darwin was able to guess why many of these morphological changes had happened, but he could not determine the mechanisms by which those adaptations were expressed. Understanding of the mechanisms of morphological change had to await discoveries in biochemistry a century later. As you learned in Chapter 20, we now understand that genetic differences underlie the adaptive evolution of species. The goals of the study of *molecular evolution* are to determine the *patterns of evolutionary change* in the molecules of which organisms are composed and to determine the *processes that caused those changes*. If we understand the patterns and processes that underlie molecular evolution, we can use those insights to help us solve other biological problems.

To reveal patterns of molecular evolution, molecular biologists must determine the precise structure of molecules. Techniques for doing so were developed in the 1940s; the first complete sequence of a protein—insulin—was determined in 1952. Sequencing the insulins of different mammals revealed that amino acid substitutions were not randomly distributed across the molecules. Instead they were restricted to three positions on one chain of the molecule. In addition, the data revealed that most amino acid substitutions did not affect the biological activity of insulin. Insulin from one species was equally effective in other species.

As increasing numbers of proteins were sequenced, the same pattern was found: Most nucleotide substitutions were confined to particular regions of the molecules. This pattern could be explained by the assumption that most of the molecular changes being identified were substitutions that did not affect the functioning of the molecules; that is, they were neutral or nearly neutral substitutions.

The hypothesis of **neutral evolution** asserts that most of the variability in structures of molecules being measured by molecular biologists does not influence their functioning. In addition, the theory suggests that the rates at which neutral substitutions accumulate is not influenced by natural selection; it is determined simply by the mutation rate. As we will see, these discoveries and interpetations play a central role in the study of molecular evolution.

In this chapter, we briefly review how molecular biologists determine the structures of molecules. Then, by comparing the structures of molecules in living and fossil organisms, we show how biologists can infer both the patterns and the causal processes of molecular evolution. Finally, we will discuss how knowledge of the patterns of molecular evolution helps us solve other biological problems, including inferring phylogenetic relationships among organisms and determining how humans spread over Earth.

Determining the Structure of Molecules

As you learned in Chapter 11, biologists determined the structure of DNA by synthesizing evidence from several sources. The most important evidence came from X ray crystallography, which allows investigators to determine the positions of atoms in a crystalline substance by the diffraction pattern of X rays passed through the crystal. The fact that the total amount of purines in DNA equals the total amount of pyrimidines also provided important clues about the structure of DNA. By building three-dimensional models of possible molecular structures on the basis of this information, Francis Crick and James Watson proposed the structure of DNA that has been supported by all experimental evidence gathered since then.

The invention of the polymerase chain reaction (PCR) method (see Figure 16.12) to amplify DNA has allowed molecular biologists to determine the sequences of subunits of DNA from their fossilized remains, mummified tissues, or dried skins in museums, even though these objects contain only tiny amounts of DNA. DNA has been extracted and amplified from fossils up to 135 million years old (Table 23.1).

We infer past molecular structures by using the comparative method

Being able to measure directly the structure of molecules of fossil organisms is the most powerful way to determine patterns of molecular evolution. However, even when we cannot directly measure molecules from extinct organisms, we can nonetheless infer evolutionary patterns by comparing the molecules of living organisms. To make these comparisons, molecular evolutionists often use the cladistic methods described in Chapter 22. That is, they attempt to determine ancestral and derived states of molecules and develop cladograms based on shared derived traits.

A good example of the use of the comparative approach to determine patterns of molecular evolution is provided by studies of the enzyme cytochrome *c*. Cytochrome *c* is one component of the respiratory chain of mitochondria. Together with other proteins of the citric acid cycle and respiratory chain, cytochrome *c* is found in the cells of all eukaryotes.

TABLE 23.1 **Biological Tissues from which Ancient DNA Has Been Extracted**

TYPE OF MATERIAL	PROBABLE AGE IN YEARS
Human tissues	
Mummies	5,000
"Bog" bodies	7,500
Bones and teeth	10,000
Other animal material	
Feathers	130
Museum skins	140
Naturally preserved skins	13,000
Bones	>25,000
Amber specimens	135,000,000
Plant material	
Herbaria specimens	118
Charred seeds and cobs	4,500
Mummified seeds and embryos	44,600
Fossil magnolia leaf	20,000,000
Amber specimens	40,000,000

The amino acid sequences of cytochrome *c* are known for nearly 100 species of organisms, ranging from yeasts to humans (Figure 23.1). Among these cytochromes *c* are regions that accumulated changes relatively quickly; for example, positions 44, 89, and 100 differ among many of the organisms compared. There are also invariant positions, such as 14, 17, 18, and 80. This particular set of invariant residues is known to interact with the iron-containing heme group that is essential for enzyme functioning. Presumably, because any changes in these amino acids adversely affect the functioning of the heme group, they were removed by natural selection when they arose.

Molecular biologists now routinely use the comparative approach to identify regions of molecules that lack variation. They generally assume that changes in the amino acids in those positions would be likely to adversely affect the functioning of the molecule. In contrast, regions that change relatively quickly are believed to be functionally less significant. Consequently, amino acid substitutions in those regions are functionally neutral, and they are not influenced by natural selection.

The differences among the cytochromes *c* shown in Figure 23.1 are the result of accumulations of substitutions within the genomes of species over millions of years, but that is not the only means by which the genomes of organisms change. As you saw in Chapter 13, plasmids, episomes, and phage coats can transport genes from one bacterial cell to another. That is the mechanism by which Yops have been transferred among distantly related bacteria. Because Yops genes are absent in the species most closely related to the ones that have them, the genes were not inherited from a common ancestor, but rather were transferred horizontally directly between distantly related species.

Another type of gene transport mechanism that functions within individual cells is provided by transposable elements, segments of chromosomal or plasmid DNA that can be inserted at other points in the same or other DNA molecules. Transposable elements may have played a major role in the development of the genome of eukaryotes. When the nuclear envelope of eukaryotes evolved, it separated the transcription of DNA from its translation. As you learned in Chapter 14, the exons of a eukaryotic gene code a single protein. Exons are separated from one another by noncoding introns. Genomes that lack introns, as found in prokaryotes, are probably the ancestral state of the genetic material. Introns are probably the remains of transposable elements that invaded the genome after the evolution of a nuclear membrane and subsequently lost the ability to transpose.

Globin diversity evolved via gene duplication

Molecular evolution also proceeds by gene duplication, which has been well studied in the globin family of genes. Globins were among the first proteins to be sequenced and their amino acid sequences compared. Several globins have been crystallized, such that details of their three-dimensional structure are known. This knowledge has greatly facilitated the study of the relationship between the rate at which amino acid sequences diverge and how details of the three-dimensional structures of proteins influence their functional properties.

Humans have three families of globin genes: the *myoglobin family*, whose single member is located on chromosome 22; the *α-globin family* on chromosome 16; and the *β-globin family* on chromosome 11 (Figure 23.2). Two types of proteins are produced by these three families: myoglobin and hemoglobin. Comparisons of amino acid sequences strongly suggest that rather than arising from different genes that have independently converged on some similar functions, the different forms of globins arose as gene duplications (Figure 23.3). After a duplication event, each copy could evolve along its own evolutionary path. How long the genes have been evolving separately can be inferred by comparison of their amino acid sequences. The greater the number of amino acid differences between two globins, the farther back in time was their most recent common ancestor.

The earliest organisms known to have both α- and β-globin genes lived about 500 million years ago (mya). Thus, the initial duplication event at which myoglobins diverged from all other globins probably happened at least that long ago (Figure 23.4). Assuming that the rate of amino acid substitution has been relatively constant since then—about 100 substitutions per 500 million years—the two families, which differ at 77 sites, are estimated to have split about 450 mya.

Homologous gene families have similar effects on development

One way to detect genes that are homologous in distantly related organisms is to find identical or nearly identical families of genes that produce similar effects on developmental patterns. For example, some mutations in *Drosophila* cause appendages that are appro-

23.1 Amino Acid Sequences of Cytochrome *c* The protein sequence alignment for cytochrome *c* obtained from 33 species. Conservation of charge at a position is shown by the constancy of color in its column. Constancy of charge suggests that cytochrome *c* molecules in all these species fold in the same way and preserve the same three-dimensional structure; this structure is shown in the computer rendering below and at the beginning of this chapter.

Heme group

Acidic: D Aspartic acid   E Glutamic acid

Basic: H Histidine   K Lysine   R Arginine

Hydrophobic: F Phenylalanine   I Isoleucine   L Leucine   M Methionine
V Valine   Y Tyrosine   W Tryptophan   A Alanine (mildly hydrophobic)

Other: C Cysteine   P Proline   Q Glutamine   N Asparagine
S Serine   T Threonine   G Glycine (only amino acid lacking a side chain)

The number 1 indicates an invariant position in the cytochrome *c* molecule (i.e., all the organisms have the same amino acid in this position) and that the position is probably functionally very significant.

Side chains marked by red arrows interact with the heme group.

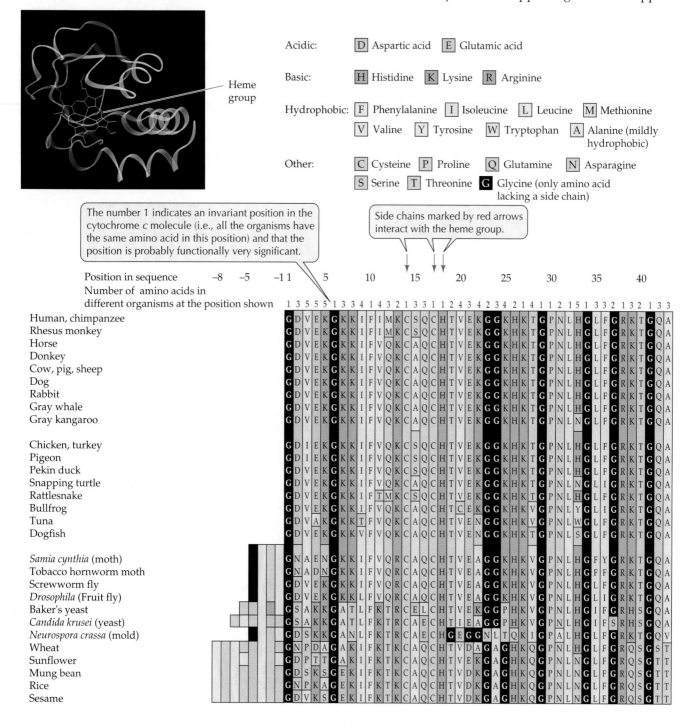

23.2 Chromosomal Arrangement of the Human Globin Gene Families The α-globin family of hemoglobin is found on chromosome 16, the β-globin family on chromosome 11, and the myoglobin gene on chromosome 22.

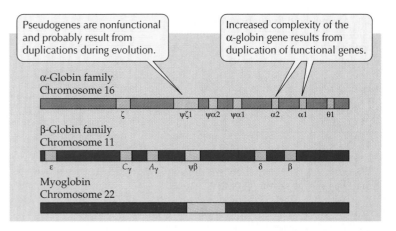

Pseudogenes are nonfunctional and probably result from duplications during evolution.

Increased complexity of the α-globin gene results from duplication of functional genes.

α-Globin family
Chromosome 16
ζ ψζ1 ψα2 ψα1 α2 α1 θ1

β-Globin family
Chromosome 11
ε Cγ Aγ ψβ δ β

Myoglobin
Chromosome 22

priate to one body segment to appear in another. Thus, leglike appendages may grow where there should be antennae, or a body segment may be duplicated (Figure 23.5). These unusual changes are caused by genes that occur in two tightly linked clusters that together constitute the **homeotic gene complex**. All homeotic genes contain a region, called the **homeobox**, which when transcribed specifies a sequence of 60 amino acids. Homeobox genes are active in specific body segments and, in the absence of mutations, are responsible for the appropriate development of those segments.

This pattern of developmental control is not unique to flies; more than 350 homeobox elements have been identified in fungi, plants, and animals. A similar set of genes, the *Hox* genes, has been found in house mice. The homeobox and *Hox* genes occur in the same order along the chromosome (see Figure 15.21). These struc-

Multiple amino acids at a position indicate a great deal of change and that the position is probably less significant.

Invariant Uncharged | Mostly charged | Uncharged | Rarely hydrophobic

45 50 55 60 65 70 75 80 85 90 95 100 104

23.3 Amino Acid Sequences of Residues 1 through 39 of the Globin Superfamily A comparison of partial myoglobin and partial representative α and β family hemoglobin amino acid sequences. The single-letter abbreviations for the amino acids are as given in Figure 23.1. Identical residues are boxed.

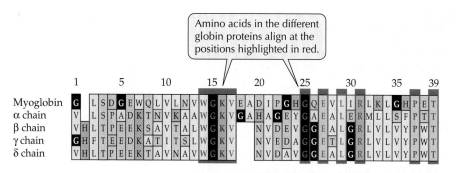

How Molecular Functions Change

Evolution as we know it would not have been possible if genes were unable to change their functional roles. As we have just seen with globin genes, gene duplication permits the genes, and the proteins they encode, to evolve different functions. Myoglobin, a monomer, has become the primary oxygen storage protein in muscle. Hemoglobin, a tetramer consisting of two α and two β chains, carries oxygen in blood. Myoglobin has evolved a specific affinity for oxygen that is much higher than that of hemoglobin.

In contrast to myoglobin, hemoglobin evolved to be much more refined and diversified in its role as the blood-oxygen carrier. Hemoglobin binds oxygen from the lungs or gills, where the partial pressure of oxygen is relatively high, transports it to regions of low oxygen partial pressure, and releases it in these areas. With its more complex, tetrameric structure, hemoglobin also is able to transfer hydrogen and carbon dioxide in the blood, bind four molecules of oxygen cooperatively, and respond to tissue acidity.

tural similarities strongly suggest that all homeobox genes have a common evolutionary origin, and that the mechanisms that determine the differentiation of the major morphological regions (body, head, trunk, and tail) may have arisen only once in animal evolution.

In humans, the α-hemoglobin family has four functional genes and three pseudogenes. The four functional genes diversified in function while the three pseudogenes lost all function. Thus, duplication events may result in increased genomic complexity (as seen with the alternate genes for α-hemoglobin) and produce nonfunctional DNA (pseudogenes). Pseudogenes may ultimately be removed via deletion.

Although point mutations can affect the function of a protein, duplication releases one copy of a gene from its original function. Duplication can result in the evolution of entirely novel functions, and, eventually, in the evolution of new gene families. The number of gene families in the human genome is not known, but much of the diversity within our genome arose as a result of ancient gene duplications. Ribosomal RNA and transfer RNA genes, lactate dehydrogenase genes, and pyruvate kinase genes are all examples of gene families that arose from gene duplication.

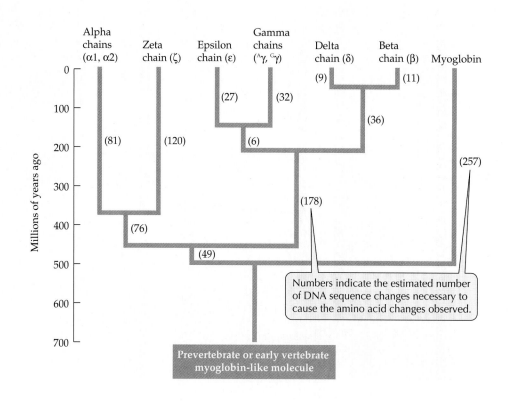

23.4 Globin Gene Tree The globin superfamily gene tree suggests that myoglobin diverged from modern hemoglobin precursors about 500 mya, at about the time of the origin of vertebrates.

Normal *Drosophila*

Second thoracic segment

Third thoracic segment

bithorax mutation

The third thoracic segment is mutated to produce an extra second thoracic segment.

23.5 The *bithorax* Mutation in *Drosophila* Mutations at one of the homeotic genes, *bithorax*, transform the third thoracic segment into a second copy of the second thoracic segment. The result is a fly with two sets of wings.

The globin genes illustrate that molecular functions may change after gene duplication, but how did these functional changes happen? We will explore this interesting component of molecular evolution by using lysozyme as an example.

Lysozyme evolved a novel function

Lysozyme is an enzyme found in almost all animals. It is produced in the tears, saliva, and milk of mammals and in the whites of bird eggs. Lysozyme digests the cell walls of bacteria, rupturing and killing them. As a result, lysozyme plays an important role as a first line of defense against invading bacteria. All animals digest bacteria, which is probably why all have lysozyme. However, in some animals lysozyme is used in the digestion of food.

Among mammals, a novel mode of digestion called *foregut fermentation* has evolved twice. The anterior part of the stomach (the foregut) becomes converted into a chamber for the bacterium-based digestion (fermentation) of ingested plant matter (see Chapter 47). The mammal uses bacteria to obtain nutrients from the otherwise indigestible cellulose of plant material. Foregut fermentation evolved independently in ruminants, such as cows, and certain leaf-eating monkeys, such as langurs.

We know these evolutionary events were independent because close relatives of langurs and ruminants do not ferment their food in their stomachs. In both foregut-fermenting lineages, lysozyme has been modified to play a new, nondefensive role in the stomach. Lysozyme ruptures some of the bacteria that normally live in the stomachs, releasing nutrients that the mammal absorbs through its stomach lining.

How many changes were incorporated in the lysozyme molecule to allow it to function amidst the harsh enzymes and low pH of the mammalian foregut? To answer this question, molecular evolutionists compared the amino acid sequences of lysozyme in foregut fermenters and several of their nonfermenting relatives. They then determined which amino acids differed and which were shared among the species (Table 23.2). Finally, they compared the patterns of these changes with the already known phylogenetic relationships among the species.

The most striking result is that amino acid changes have occurred about twice as rapidly in the lineage leading to langur lysozyme as in any other primate lineage. This increased rate of substitution shows that lysozyme went through a period of rapid adaptation in the stomach of langurs. The lysozymes of langurs and cows share five amino acid substitutions, all of which lie on the surface of the lysozyme molecule, well away from the active site. Several of the shared residues involve changes from arginine to lysine,

TABLE 23.2 **Comparisons of Lysozyme Amino Acid Sequences of Different Species**

SPECIES	LANGUR	BABOON	HUMAN	RAT	COW	HORSE
Langur*		14	18	38	32	65
Baboon	0		14	33	39	65
Human	0	1		37	41	64
Rat	0	1	0		55	64
Cow*	5	0	0	0		71
Horse	0	0	0	0	1	

Shown above the diagonal line is the number of amino acid sequence *differences* between the two species being compared; below the line are the number of sequences uniquely *shared* by the two species. Asterisks (*) indicate foregut-fermenting species.

which makes the proteins more resistant to attack by the pancreatic enzyme trypsin. By understanding the functional significance of amino acid substitutions, molecular evolutionists can explain the observed changes in amino acid sequences over time in terms of the changing function of the protein.

A large body of fossil, morphological, and physiological data shows that langurs and cows do not share a recent common ancestor. However, langur and ruminant lysozymes share many amino acid residues that neither animal shares with the lysozymes of their own closer relatives. In other words, the lysozymes have converged to a similar sequence despite having very distant ancestry. The amino acid residues they share give these lysozymes the ability to function in the unusual environment of the stomach. Langurs and ruminants both ferment their leafy food in their stomachs. If they are to profit from the metabolic activities of the bacteria that ferment the masticated food, they must have an enzyme capable of lysing bacteria that can function in that environment. Their nonfermenting relatives have no such need.

A group of birds uses lysozyme in digestion

An even more remarkable story emerges if we look at lysozyme in hoatzins, a group of Neotropical cuckoos that are the only known avian foregut fermenters. Hoatzins have an enlarged crop that contains resident bacteria and acts as a fermenting chamber, allowing hoatzins to survive on a diet of leaves. Lysozyme has also evolved to function in the crop of the hoatzin.

Many of the amino acid changes that occurred in the adaptation of hoatzin crop lysozyme are identical to the changes that evolved in ruminants and langurs. Thus, even though the three groups have evolved independently from one another for more than 300 million years, they have each evolved a similar molecule that enables them to recover increased amounts of nutrients from their fermenting bacteria in a highly acidic environment.

Genome Size: Surprising Variability

Organisms differ strikingly in genome size. Multicellular organisms have more DNA than do simpler organisms (Table 23.3). This is not surprising, since more complex instructions are needed for building and maintaining a large, complex organism than a small, simpler one. What is surprising is that lungfishes and lilies have about 40 times as much DNA as humans do. Clearly, a lungfish or a lily is not 40 times as complex as a human.

Some of the apparent difference disappears when we compare the percentage of DNA that actually codes functional RNA or proteins. The size of the *coding* genome of organisms increases in a way that makes sense. Eukaryotes have more coding DNA than prokaryotes have; vascular plants and invertebrate animals have more coding DNA than single-celled organisms have; invertebrates with wings, legs, and eyes have more coding DNA than a roundworm has; and vertebrates have more DNA than invertebrates have.

However, we still do not understand why the percentage of coding DNA varies so dramatically. Nor do we know how much genetic information is needed to program the development of a particular morphological structure.

Molecular Clocks

If we plot the time since the divergence of certain organisms, as determined by the fossil record, against the number of amino acids by which their cytochromes *c* differ, we find that differences in cytochrome *c* sequences have evolved at a relatively constant rate (Figure 23.6). Many proteins show this same constancy in the rate at which they have accumulated changes over time. If the structure of a molecule changes at a constant rate, we can use its **molecular clock** to determine dates of evolutionary events.

It would be convenient for biologists if the rates of change were the same for all molecules. Unfortunately,

TABLE 23.3 Genome Size and DNA Content of Some Organisms

ORGANISM	GENOME SIZE (BASE PAIRS $\times 10^9$)	PERCENT OF DNA THAT CODES FOR SOMETHING	TOTAL CODING DNA
Bacterium (*Escherichia coli*)	0.0004	100	0.00004
Yeast (*Saccharomyces*)	0.0009	70	0.00063
Roundworm (*Caenorhabditis*)	0.09	25	0.023
Fruit fly (*Drosophila*)	0.18	33	0.059
Newt (*Triturus*)	19.0	1.5–4.5	0.29–0.855
Human (*Homo*)	3.5	9–27	0.32–0.945
Lungfish (*Protopterus*)	140.0	0.4–1.2	0.56–1.18
Mustard (*Arabidopsis*)	0.2	31	0.062
Lily (*Fritillaria*)	130.0	0.02	0.026

however, different molecular clocks tick at different rates. These differences arise because different molecules have different functional constraints on their evolution. The molecular clock for rRNA ticks very slowly; the cytochrome *c* clock ticks at a slightly more rapid rate; the lysozyme clock ticks at an even faster rate that is not constant for all lysozymes.

Despite these differences, the rates at which many molecular clocks tick appear to be relatively constant. The reason is that the vast majority of molecular changes involve nucleotide or amino acid substitutions that do not affect the functioning of the molecule and, hence, the fitness of the organism. These *neutral changes* are not influenced by natural selection, and they accumulate at a rate roughly equal to the mutation rate. When changes occur that do affect fitness, such as the variants of lysozyme that became adapted to life in the stomach, the rate of evolution is quite different from the neutral rate. However, if *adaptive changes* are few relative to all the changes that distinguish the molecules of any pair of species—that is, if most substitutions are neutral—the molecular clock can still be used to date lineage separations in the distant past.

To use molecular clocks to date events in the remote past, we need to know what proportion of molecular clocks tick at a constant rate. Fairly accurate molecular clocks are known for many proteins, but detailed investigations sometimes indicate that the rates of these clocks occasionally change. We have seen such changes among lysozymes. Fortunately, if we have a good un-derstanding of the structure and function of a molecule, we can usually distinguish neutral from adaptive variation.

Even if the rate of ticking of a molecular clock changes slightly, the variations may not be great enough to seriously affect our estimates of the dates of lineage divergences. We can also check the constancy of molecular clocks by comparing the dates of lineage divergences calculated from molecular clocks with those estimated from well-dated fossil records. Typically these two estimates are close enough to one another to give us confidence in using molecular clocks to estimate dates of lineage separations among groups for which fossil data are rare or lacking.

Using Molecules to Infer Phylogenies

By comparing molecular structures from different species, we gain insights into how the molecules function and acquire a tool for inferring phylogenies. As we have seen, much evidence suggests that sequences of amino acids in proteins or base sequences in RNA and DNA that have changed very little during evolution probably have the same function in all species. Regions that have changed rapidly during evolution are likely to have fewer constraints on the functional

23.6 Cytochrome *c* Molecules Have Evolved at a Constant Rate Many proteins accumulate amino acid substitutions at a constant rate over time. This constancy provides scientists with a molecular clock that can sometimes be used to date events in the remote past.

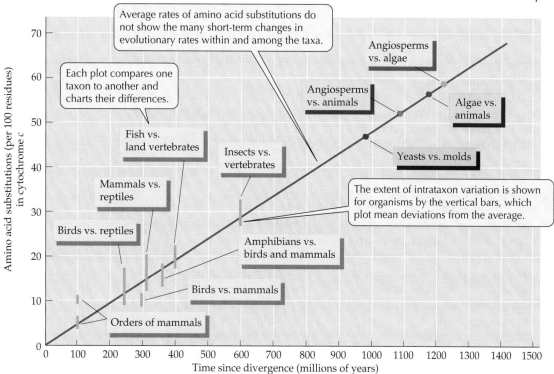

importance of specific sequences, or possibly have undergone major changes in function, as happened to lysozyme in foregut fermenters.

Because they differ in few amino acid or base sequences, molecules that have evolved slowly can be used to estimate relationships among organisms that diverged long ago. Molecules that have evolved rapidly are useful for studying organisms that share more recent common ancestors.

The first step in a molecular evolutionary analysis is to choose which molecule(s) to study. The best choice depends on the question at hand. If you are interested in determining the evolutionary relationships of all existing organisms, you must choose a molecule that all the organisms possess. Such a molecule might be a ribosomal RNA. Equally important is the fact that rRNAs experience very strong functional constraints (meaning that most changes in the RNA sequence prevent the ribosome from functioning properly). Thus, because rRNA evolves so slowly, comparisons of differences among the rRNAs of living organisms can be used to estimate when their lineages diverged, even if the split happened billions of years ago.

After the sequences of amino acids or bases have been determined, they must be compared—a task more difficult than might appear at a glance. A simple example illustrates. In Figure 23.7a two sequences (1 and 2) are being compared. The two sequences differ in number of residues and in identity. Our goal is to align these sequences so that we can compare homologous amino acids.

To do so, we first observe that, although the sequences appear quite different, they would become similar if we were to insert a space after the first amino acid in sequence 2 (after the leucine [leu] residue). In fact, these sequences then differ only by one amino acid at position 6 (serine or phenylalanine). A single insertion aligns the sequences in this case, but longer sequences and those that have diverged more extensively require more elaborate adjustments.

After we have aligned sequences, we can compare them in several ways. First, we can simply count the number of nucleotides or amino acids that differ between the sequences. Let's return to the previous example (see Figure 23.7a) and add additional sequences. We now have six amino acid sequences to align (Figure 23.7b). By adding up the number of similar and different amino acids, we can construct a **similarity matrix** (Figure 23.7c). The assumption is that the more recently two groups shared a common ancestor (for example, sequences 2 and 3 in Figure 23.7c), the more similar are their sequences. Having established similarities and differences between our sequences, we can then infer the evolutionary relationships among them.

A matrix of similarity (or difference) can be translated into a **molecular phylogeny**, or **gene tree**. A

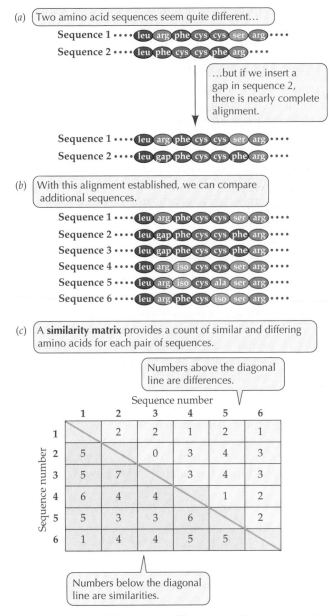

23.7 **Amino Acid Sequence Alignment** The insertion of gaps allows us to align these sequences so that we can compare homologous amino acids. The similarity matrix in (c) sums similarities and differences; the larger the number of similarities, the more recent the presumed common ancestor of the species.

gene tree is a visual representation of the evolutionary relationship inferred for the genes being compared. Several methods are used to translate a similarity matrix into a gene tree. Each differs in the assumptions it makes about how DNA or protein sequences evolve. In our previous example (see Figure 23.7c), we could use the number of amino acids shared to infer the ancestry of the six sequences. In this case, the branches of our tree represent the number of events, including the deletion, that distinguish each of the six sequences.

Although often the only data available with which to estimate ancient lineage divisions or the phylogenies of prokaryotes, molecular data are also regularly used in combination with morphological and fossil data. Why do we use molecules when morphology is available? The answer is simple: The more characters that are used (morphological, molecular, fossil, and so on) to deduce a phylogeny, the less likely we are to be misled by loss of traits or evolutionary convergence of characters (homoplasy), as discussed in Chapter 22.

Similarly, if we use data from more than one type of molecule, we can better detect convergent evolution. For example, if we were to infer a phylogeny from only lysozyme sequences, we might falsely conclude that langurs, cows, and hoatzins are more closely related to one another than morphological and fossil data indicate. However, if we also compared the DNA sequences of the genes that encode lysozyme, rather than just amino acid sequences, we would find that most of the nucleotide substitutions that have accumulated since the divergence of langurs, cows, and birds were not influenced by strong selection.

Molecules provide the best evidence for ancient lineage splits

No fossils exist to document the most ancient splits in the lineages of life on Earth. To infer the times of these lineage separations we must use molecules that are found in all organisms and that evolve very slowly. A good candidate is 16S rRNA molecules, which are found in all organisms and evolve very slowly. They have been used to identify the three major branches, or domains, of life: the Bacteria, the Archaea, and the Eukarya (see Figure 22.14).

The greatest advantage of employing molecules to infer ancient lineages follows from the fact that we cannot use morphological characters for comparisons between microbes and more advanced organisms—or, in many cases, between microbes themselves—because they lack comparable structures. Thus, only molecules can be used for an across-the-board comparison of all living organisms.

The structure of DNA extracted from extinct organisms is being used to determine the evolutionary relationships between those organisms and their surviving relatives. For example, DNA was obtained from the bones and mummified soft tissues of moas—large, flightless birds (weighing up to 200 kg) that lived in New Zealand until humans arrived a thousand years ago and hunted them to extinction. The amplified DNA was compared with members of other groups of flightless birds, such as kiwis and rheas.

These comparisons suggest that although kiwis and moas both lived in New Zealand, they are not each other's closest relatives (Figure 23.8). The closest relatives of the moas are unknown, extinct flightless birds that also gave rise to flightless birds on Australia. Kiwis came to New Zealand more recently; their closest relatives, emus and cassowaries, live in their ancestral home, Australia and New Guinea.

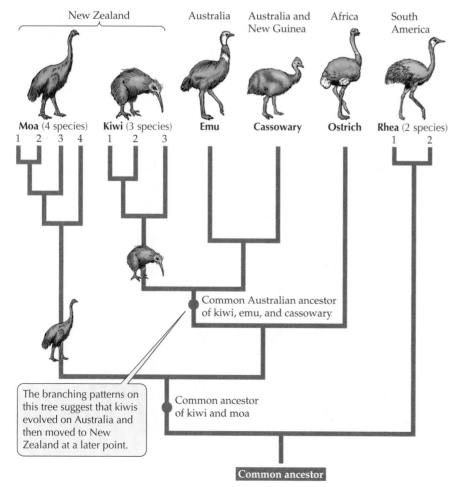

The branching patterns on this tree suggest that kiwis evolved on Australia and then moved to New Zealand at a later point.

23.8 Flightless Bird Phylogeny
Molecular phylogenetics of flightless birds, living and extinct, illustrates that although moas and kiwis recently lived together in New Zealand, they are not each other's closest relatives.

Cloning extinct organisms is impossible

It is tempting to imagine that we could use molecular methods to reconstruct extinct organisms, as envisioned in Michael Crichton's novel *Jurassic Park*. However, we are unlikely to be able to do so. We are generally able to amplify or clone only small fragments of ancient DNA, because DNA deteriorates with time. Further, we have no idea how to piece together the millions of small fragments of DNA extracted from fossilized remains, nor do we have any idea how to regulate their proper expression during the development of an embryo. What we can accomplish by studying ancient DNA is to determine the place of extinct organisms in the phylogeny of life.

Molecules and Human Evolution

Molecular evolution has influenced our understanding of our own evolution. In particular, it suggests that the human species is of relatively recent origin. Fossil evidence suggests that the hominoid lineage leading to modern humans diverged from a chimpanzee-like lineage about 6 million years ago (mya) in Africa. Shortly after it arose in Africa, about 1.7 mya, *Homo erectus* then spread to other continents. Fossil remains have been found in Africa, Indonesia, China, the Middle East, and Europe. The transition from *Homo erectus* to *Homo sapiens* occurred about 400,000 years ago, but there is considerable controversy about the place of origin of modern humans. The "out of Africa" hypothesis suggests a single origin in Africa followed by several dispersals. The "multiple regions" hypothesis, in contrast, proposes parallel origins of *Homo sapiens* in different regions of Europe, Africa, and Asia (Figure 23.9).

The limited number of human fossils and their patchy distribution do not allow us to choose between these views. However, DNA sequences of several mitochondrial genes from individuals from more than 100 ethnically distinct modern human populations provide the needed in-

formation. Mitochondrial DNA (mtDNA) is useful for studying the recent evolution of closely related species and populations within species because it accumulates mutations rapidly and because it is maternally inherited. The Y chromosome serves the same role for following male lineages. The mtDNA sequences for ethnically distinct modern humans imply or suggest a common ancestry of all mtDNAs about 200,000 years ago. The date of the shared ancestry was calculated using the number of nucleotide differences among existing humans, and the rate of mtDNA sequence divergence was calibrated using mammals with better fossil records.

In contrast, the multiple-origins hypothesis requires at least 1 million years of divergence since the last common ancestor. Thus, mtDNA lends support to the "out of Africa" hypothesis, suggesting that all modern human populations share a recent mitochondrial ancestor. Of importance, the mtDNA data support the idea of a common female ancestral lineage that ultimately gave rise to all contemporary mtDNAs. In principle, the history of genes passed through the male lineage could be different. Studies of 26 nuclear genes and a family of repeated sequences (called the *Alu* family) also support a recent African ancestry.

Molecular phylogenetic techniques have also helped us resolve some outstanding archaeological debates. One example concerns the origin of the early human inhabitants of Easter Island, a remote island in

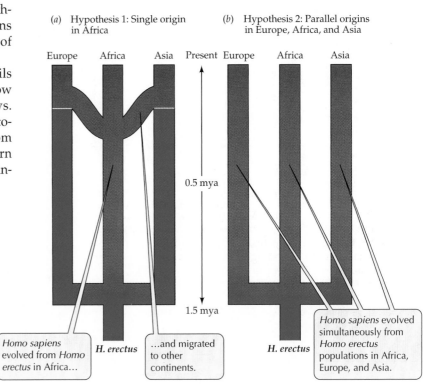

23.9 Two Models for the Origin of Modern Humans *Homo erectus* (blue lineage) arose in Africa and spread to other continents. However, there is considerable controversy among scientists as to whether the transition to *Homo sapiens* (red lineage) took place only in Africa (*a*), or the evolution of modern humans occurred simultaneously on three continents (*b*).

(*a*) Hypothesis 1: Single origin in Africa

(*b*) Hypothesis 2: Parallel origins in Europe, Africa, and Asia

Europe   Africa   Asia   Present   Europe   Africa   Asia

0.5 mya

1.5 mya

Homo sapiens evolved from *Homo erectus* in Africa…

H. erectus

…and migrated to other continents.

H. erectus

Homo sapiens evolved simultaneously from *Homo erectus* populations in Africa, Europe, and Asia.

the Pacific. Ancient DNA from human bones on Easter Island suggests that the inhabitants of this island were closely related to Polynesians from other Pacific islands, rather than to South American Indians, as was suggested by early Danish explorers.

Summary of "Molecular Evolution"

• The goals of the study of molecular evolution are to determine the patterns of evolutionary change in the molecules of which organisms are composed, to determine the processes that caused those changes, and to use those insights to help solve other biological problems. To achieve those goals, molecular evolutionists need to be able to determine the structures of molecules of living and fossil organisms.

Determining the Structure of Molecules

• The structure of DNA was deduced by combining data from X ray crystallography and base composition with three-dimensional models. The polymerase chain reaction method allows biologists to determine the sequence of DNA bases of organisms from their fossilized remains.
• Past molecular structures are inferred by comparison of molecules of existing organisms using cladistic techniques to identify ancestral and derived states.
• Changes evolve slowly in regions of molecules that are functionally significant, but more rapidly in regions where substitutions do not affect the functioning of the molecules. **Review Figure 23.1**
• Gene duplication, which frees one copy of a gene to evolve a novel function, has been responsible for much of the evolution of molecular diversity. **Review Figures 23.3, 23.4**
• Groups of genes that are aligned in the same order on chromosomes of distantly related species are likely to be homologs of one another. **Review Figures 23.5, 15.21**

How Molecular Functions Change

• Changes in the functions performed by molecules are stimulated by gene duplication, and by the traits that molecules must have to function in different situations, such as the acidic environment of the stomach.

Genome Size: Surprising Variability

• The genome sizes of organisms vary more than a hundredfold, but the amount of DNA that actually encodes protein varies much less. In general, eukaryotes have more coding DNA than do prokaryotes, vascular plants and invertebrate animals have more coding DNA than do single-celled organisms, and vertebrates have more coding DNA than do invertebrates. **Review Table 23.3**

Molecular Clocks

• Neutral molecular variation often accumulates at a constant rate determined by the mutation rate. Such a process is referred to as the ticking of a molecular clock.
• Molecular clocks tick more slowly for molecules, or parts of molecules, that experience strong constraints on their evolution than they do for molecules or parts of molecules influenced by strong directional selection. **Review Figure 23.6**
• We can assess the constancy of the ticking rates of molecular clocks by comparing the dates of lineage splits calculated assuming the operation of molecular clocks with those determined from accurately dated fossils. Typically these two estimates are reasonably similar, giving evolutionists confidence in using molecular clocks to date events for which there are no fossils.

Using Molecules to Infer Phylogenies

• Molecules are an important source of data that can be used to infer phylogenetic relationships among organisms. For ancient splits and phylogenies of prokaryotes, molecular data are the only source of information about phylogenetic relationships.
• The steps in a molecular evolutionary analysis are: choosing molecules to study; determining sequences of amino acids or bases; comparing the molecules; and constructing a gene tree. **Review Figure 23.7**
• Molecules that have evolved slowly are useful for determining ancient lineage splits. Molecules that evolve rapidly are useful for determining more recent lineage splits.

Molecules and Human Evolution

• Comparisons of mtDNA from more than 100 ethnically distinct modern human populations strongly suggest that all modern humans share a common African ancestor no more than 200,000 years old. **Review Figure 23.9**
• Molecules provide useful information about human evolution, helping to clarify the relationships between different peoples.

Self-Quiz

1. Questions about the *process* of molecular evolution address
   a. whether molecules evolve.
   b. how molecules change with time.
   c. how to reconstruct phylogenies of extinct organisms.
   d. the evolutionary relationships among molecules.
   e. the origin of organismal diversity.

2. Choosing the appropriate molecule for phylogenetic reconstruction does *not* require a consideration of
   a. the question at hand.
   b. the rate of evolution of the molecule.
   c. the phylogenetic distribution of the molecule.
   d. the function of the molecule.
   e. the completeness of the fossil record.

3. Ribosomal RNA sequences are useful for addressing the evolutionary relationships of highly divergent molecules because
   a. they evolve at a rapid rate.
   b. they have undergone convergent evolution in many lineages.
   c. they are molecules that all organisms have.
   d. they consist of mainly neutral characters.
   e. they are difficult to align.

4. Mitochondrial DNA sequences are useful in studying the recent evolution of closely related species because
   a. they accumulate mutations very rapidly.
   b. they are paternally inherited.
   c. they evolve only in a neutral fashion.

d. they are highly constrained in function.

e. they are easy to sequence.

5. Questions about the *pattern* of molecular evolution focus on
   a. the evolutionary relationships among molecules.
   b. the molecular clock.
   c. the rate of mutation for neutral characters.
   d. the importance of gene duplication in evolution.
   e. the neutral theory of evolution.

6. Molecules are used to reconstruct phylogenies, even if a fossil record is available, because
   a. the more characters the better.
   b. they are more accurate characters than are fossils.
   c. they undergo less homoplasy than do fossil characters.
   d. they are less subjective characters than are fossils.
   e. they give us the "right" phylogeny.

7. Neutral characters
   a. are not evolving under the influence of positive selection.
   b. have a neutral pH.
   c. are not useful in reconstructing phylogenies.
   d. are subject to strong functional constraints.
   e. are not likely to evolve.

8. *Jurassic Park* is unlikely ever to be a reality because
   a. we cannot obtain DNA from ancient organisms.
   b. we have no parks big enough for such large dinosaurs.
   c. although we can obtain DNA from dinosaurs, it is highly fragmented.
   d. the genetic code used by dinosaurs is different from all other organisms.
   e. we wouldn't know what to feed the dinosaurs.

9. The concept of a molecular clock implies
   a. that many proteins show a constancy in rate of change with time.
   b. that organisms evolve at a constant rate.
   c. that one can date evolutionary events with molecules alone.
   d. that all molecules change at the same rate in evolution.
   e. that we can predict how fast all genes will evolve.

10. The lysozyme story suggests that
    a. molecules cannot change function in evolution.
    b. that selection does not act at the molecular level.
    c. that molecules can help us understand the process of organismal evolution.
    d. that all organisms are capable of fermenting bacteria.
    e. that lysozyme has a very accurate molecular clock.

Applying Concepts

1. If you were interested in studying the molecular phylogeny of very closely related species of fruit flies, what kind of molecule(s) would you choose to examine? Why? If, on the other hand, you wanted to determine the phylogeny of all vertebrates, would you use the same molecule(s)? Why or why not?

2. How have our views about organismal evolution been affected by recent discoveries in molecular evolution?

3. Discuss the relative importance of molecular characters, versus morphological and fossil characters, in reconstructing the phylogeny of a group of organisms.

4. The existence of a "true" molecular clock is a contentious issue in molecular evolution. If it turns out that a molecular clock keeps very good time, of what use is it?

5. We are, by nature, interested in our own evolution. This chapter presented a brief introduction to the application of molecular methods to studying questions about human evolution. Make a short list of additional questions and develop a rough outline of the molecules and methods you might bring to bear in addressing these questions.

Readings

Avise, J. C. 1994. *Molecular Markers, Natural History and Evolution.* Chapman and Hall, New York. A thorough review of the use of molecules to infer evolutionary patterns.

De Robertis, E. M., O. Guillermo and C. V. E. Wright. 1990. "Homeobox Genes and the Vertebrate Body Plan." *Scientific American*, July. A well-illustrated review of current views regarding the evolution of the vertebrate body plan.

Gillespie, J. A. 1991. *The Causes of Molecular Evolution.* Oxford University Press, New York. A comprehensive and demanding review of the basic tenets of molecular evolution.

Kimura, M. 1983. *The Neutral Theory of Molecular Evolution.* Cambridge University Press, New York. An easy-to-read description of the process of neutral evolution by one of its original proponents.

Li, W.-H. 1997. *Molecular Evolution.* Sinauer Associates, Sunderland, MA. The best treatment of all aspects of molecular evolution.

Li, W.-H. and D. Graur. 1991. *Fundamentals of Molecular Evolution.* Sinauer Associates, Sunderland, MA. A brief, easy-to-read primer of the central topics in molecular evolution.

Olsen, G. J. and C. R. Woese. 1993. "Ribosomal RNA: A Key to Phylogeny." *Journal of the Federation of American Societies for Experimental Biology (FASEB)*, vol. 7, pages 113–123. A detailed and well-written summary of the use of rRNA in molecular phylogenetic reconstruction.

Chapter 24

The Origin of Life on Earth

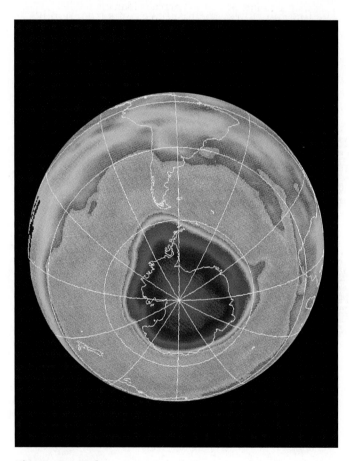

In 1985, scientists discovered that the high-altitude ozone layer, which shields organisms from harmful ultraviolet radiation, had thinned greatly over Antarctica. Ozone (O_3) is produced by the action of sunlight on atmospheric oxygen (O_2) in tropical regions and is transported to high latitudes, where it is destroyed. Chlorine compounds, produced mainly by humans, appear to be the main cause of the unusually high rates of ozone destruction. Ozone is now seriously depleted at very high latitudes, where conditions favor its destruction.

Because ultraviolet radiation damages DNA, it increases the incidence of mutation and cancer. A 1 percent decrease in atmospheric ozone is estimated to result in a 6 percent increase in the incidence of skin cancer. Average ozone concentrations have decreased about 3 percent, enough to have caused a 20 percent increase in skin cancers at midlatitudes. Skin cancer, which is sometimes lethal, is a serious human health problem.

The Ozone Hole
A satellite image showing ozone depletion over Antarctica in 1990. The areas of most severe loss appear in violet and pink; orange and yellow indicate approximately normal concentrations of ozone. The continents are outlined in white.

The modification of Earth's atmosphere by human activity is just the latest of a series of atmospheric changes caused by organisms. The most dramatic of these changes was the generation of O_2 by photosynthetic prokaryotes 2.5 billion years ago. Organisms have also changed Earth's waters and soils. Conditions on Earth, in turn, have influenced the evolution of life. In this chapter we examine how conditions on Earth 3.8 billion years ago influenced where and how life evolved and how life in turn, during the early phases of its evolution, changed Earth.

517

The phenomenon of reproduction distinguishes living organisms from nonliving and was central to discussions at many points in Parts One and Two of this book. How could something that is self-replicating have existed before there were cells? What was the earliest self-replicating unit, and how did it reproduce? Remarkably, scientific evidence is good enough that we can suggest answers to these questions.

Earth's oldest rocks contain no fossils, showing that for at least several hundreds of thousands years after its formation, Earth had no life. Today we are confident that all organisms come from other organisms. The first life, however, must have come from nonliving matter. How did this happen? Under what conditions did life originate on Earth?

Is Life Forming from Nonlife Today?

Until the nineteenth century, people believed in **spontaneous generation**—the regular formation of living organisms from nonliving matter. Flies and maggots were believed to arise from rotting meat and barnyard manure, eels and fish from sea mud, and frogs and mice from moist soil. For more than 2,000 years, belief in spontaneous generation was accepted by most scholars, clergy, and plain folk, although attempts to disprove it had been made.

In 1862, the great French scientist Louis Pasteur performed a series of meticulous experiments showing that microorganisms come only from other microorganisms and that a genuinely sterile solution remains lifeless indefinitely unless contaminated by living creatures. His most elegant experiment relied on swan-necked flasks that were open to the air. Pasteur filled the flasks with nutrient medium, heated them to kill any microorganisms present, then cooled them slowly. No new growth appeared in these flasks (Figure 24.1). The shape of the necks kept any new organisms or minute dust particles from falling into the medium. However, in flasks without the narrow swan necks, microorganisms entered and grew rapidly.

As a result of Pasteur's experiments, similar experiments by other scientists, and other discoveries of nineteenth-century science, most people accepted that all life comes from existing life. These experiments suggested that life was not forming from nonliving matter under the current conditions on Earth, but they shed no light on when or how life did form on Earth 3.8 billion years ago.

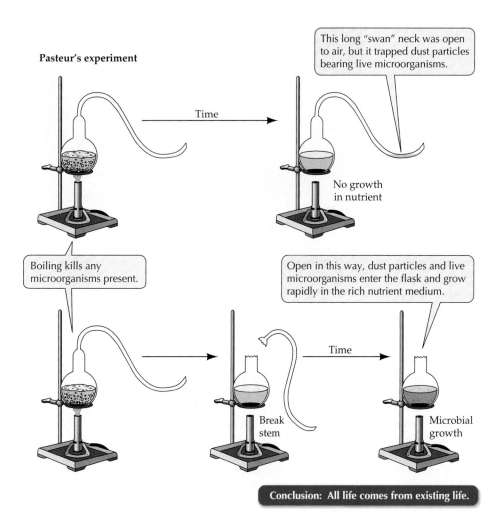

Pasteur's experiment

This long "swan" neck was open to air, but it trapped dust particles bearing live microorganisms.

Time

No growth in nutrient

Boiling kills any microorganisms present.

Open in this way, dust particles and live microorganisms enter the flask and grow rapidly in the rich nutrient medium.

Break stem

Time

Microbial growth

Conclusion: All life comes from existing life.

How Did Life Evolve from Nonlife?

To understand how life arose, we need to know the physical conditions that prevailed on Earth before there was life. And we need to know when life arose. The early conditions on Earth determined which types of chemical reactions could take place. If we know what types of reactions took place, we can suggest how those reactions could have led to the appearance of life.

For its first billion years, no life existed on Earth

How did the universe form? Scientists now believe that between 10 billion and 20 billion

24.1 Experiments Disproved the Spontaneous Generation of Life Louis Pasteur's classic experiments showed that a genuinely sterile solution remains lifeless indefinitely; only "contamination" by living organisms could cause life to appear in the flasks.

years ago there was a mighty explosion. The matter of the universe, which had been highly concentrated, began to spread apart rapidly. Eventually clouds of gases collapsed on themselves through gravitational attraction, forming the galaxies—great clusters of hundreds of billions of stars.

Somewhat less than 5 billion years ago, toward the outer edge of our galaxy (the Milky Way), our solar system (the sun, Earth, and our sister planets) took form. Earth probably formed by gravitational attraction of rocks of various sizes. As Earth grew by this process over millions of years, the weight of the outer layers compressed the interior of the planet. The resulting pressures, combined with the energy from radioactive decay, heated the interior until it melted. Within this viscous liquid, the heavier elements settled to produce a fluid iron and nickel core with a radius of approximately 3,700 km that persists to this day. Around the core lies a mantle of dense silicate materials that is 3,000 km thick. Over the mantle is a lighter crust, more than 40 km thick under the continents but as little as 5 km thick in some places under the oceans.

Conditions on Earth 3.8 billion years ago differed from those of today

Before the evolution of life, Earth's mantle and crust released carbon dioxide, nitrogen, and other heavier gases. These gases were held by Earth's gravitational field and gradually formed a new atmosphere. Earth accumulated an atmosphere consisting mostly of methane (CH_4), carbon dioxide (CO_2), ammonia (NH_3), hydrogen (H_2), nitrogen (N_2), and water vapor (H_2O). Eventually Earth cooled enough that the water vapor escaping from inside the planet condensed to liquid water and formed the seas. Violent electrical storms battered Earth's surface.

No free oxygen (O_2) was present in this early atmosphere, because oxygen reacted with hydrogen to form water and with components of Earth's crust and atmosphere to form iron oxides, silicates, carbon dioxide, and carbon monoxide. For more than 2 billion years, all oxygen was bound up with other elements, and Earth existed as a **re-**

ducing environment. (See Chapter 7 for a discussion of oxidation and reduction.) As a setting for chemical reactions, then, early Earth differed fundamentally from present-day Earth, which has an oxidizing atmosphere with large quantities of O_2.

The sections that follow describe the sequence of events leading to the appearance of the main stages in the evolution of life (Figure 24.2).

Conditions favored the synthesis of compounds containing carbon, nitrogen, and hydrogen

What sort of chemical reactions could have occurred in Earth's early reducing environment? Could such reactions have been the first step toward the origin of life? To investigate these questions, in the 1950s Stanley Miller studied chemical reactions proceeding under conditions that resembled those believed to exist on Earth 4 billion years ago.

Within a closed system of glass tubes, he established a reducing atmosphere of hydrogen, ammonia,

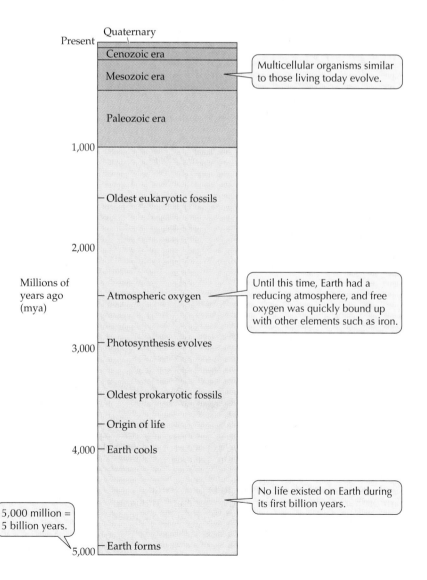

24.2 The Origin and Early Evolution of Life The sequence of events leading to the appearance of life and its earliest evolution, as described in this chapter and in Chapter 1.

Present — Quaternary
Cenozoic era
Mesozoic era — Multicellular organisms similar to those living today evolve.
Paleozoic era
1,000
— Oldest eukaryotic fossils
2,000
Millions of years ago (mya)
— Atmospheric oxygen — Until this time, Earth had a reducing atmosphere, and free oxygen was quickly bound up with other elements such as iron.
3,000 — Photosynthesis evolves
— Oldest prokaryotic fossils
— Origin of life
4,000 — Earth cools
— No life existed on Earth during its first billion years.

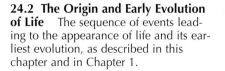

5,000 million = 5 billion years.

5,000 — Earth forms

methane gas, and water vapor. Through these gases, he passed a spark to simulate lightning (Figure 24.3). Within a few hours the system was found to contain numerous simple organic compounds, including hydrogen cyanide, formaldehyde, cyanogen, acetaldehyde, cyanoacetylene, and propionaldehyde. In water, these compounds dissolved and were rapidly converted into amino acids, simple acids, purines, and pyrimidines—the building blocks of life (Figure 24.4).

The same or similar compounds are produced under a variety of conditions, provided that free oxygen is absent—that is, if the environment is a reducing one. Thus, once Earth cooled enough for water to condense and form oceans, molecules of many kinds formed, and they probably accumulated until they reached relatively high concentrations. From such a prebiotic ("before life") molecular soup, life emerged from nonliving matter.

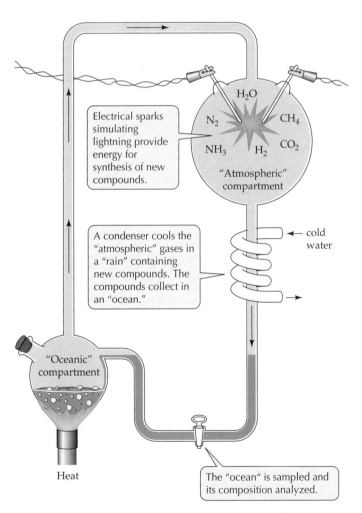

24.3 Synthesis of Molecules in an Experimental Atmosphere   Stanley Miller used an apparatus similar to this one to determine which molecules could be produced in a reducing atmosphere such as existed on early Earth.

Polymerization provided macromolecules with diverse properties

The next stage in the sequence leading to life was the generation of large molecules by **polymerization** of small molecules. The important large molecules of which organisms are composed (polysaccharides, proteins, and nucleic acids) are polymers formed by the combination of subunits called monomers (see Chapter 3). The polymerization reactions that generate these large molecules belong to a class of reactions called condensations, or dehydrations. During a condensation reaction, water forms (see Figure 3.4). Large molecules are assembled through repeated condensations of monomers, and each condensation reaction requires energy. In the prebiotic soup, polymers that formed faster or were more stable would have come to predominate.

Some polymers can direct the synthesis of molecules identical to themselves. Which of the molecules on prebiotic Earth were most likely to reproduce themselves? Nucleic acids, the basis of today's genetic code, are clearly capable of self-copying, and the purine and pyrimidine constituents of nucleotides are formed under conditions similar to those believed to have prevailed on early Earth.

Sugars, another basic component of living cells, are generated by the polymerization of formaldehyde, another molecule formed in Miller's experiments. Solutions of formaldehyde spontaneously polymerize to form several different five- and six-carbon sugars, but these sugars are unstable in aqueous solutions and break down to alcohols and carboxylic acids. However, three- and four-carbon sugars are very stable and can accumulate for hundreds of years in aqueous solutions.

Over millions of years, these organic molecules would have accumulated in the oceans. They would have reached even higher concentrations in drying ponds. High concentrations of polymers, in turn, would have stimulated further polymerization and chemical reactions leading to the synthesis of sugars, which were important as components of nucleotides and for energy storage in early cells.

Phosphate-based polymerizations occurred in the prebiotic soup

Monomers formed under prebiotic conditions can polymerize by phosphorylation (the addition of a phosphate group). Phosphorylated monomers are stable enough to accumulate in solution but reactive enough to polymerize further. The first biologically active polymers of nucleic acid bases may have been compounds formed from three- and four-carbon sugars. They could have self-replicated by forming complementary double-stranded molecules, just as pentose-based nucleic acid polymers do today. At some

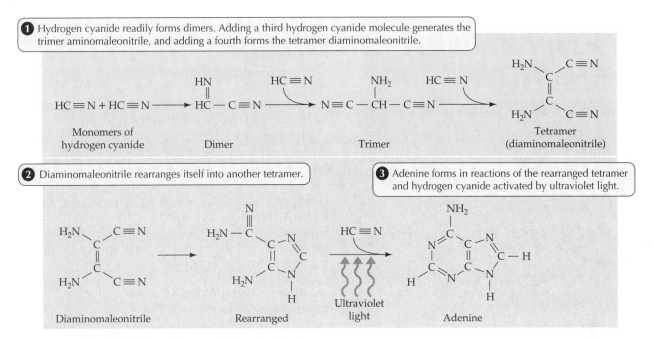

① Hydrogen cyanide readily forms dimers. Adding a third hydrogen cyanide molecule generates the trimer aminomaleonitrile, and adding a fourth forms the tetramer diaminomaleonitrile.

② Diaminomaleonitrile rearranges itself into another tetramer.

③ Adenine forms in reactions of the rearranged tetramer and hydrogen cyanide activated by ultraviolet light.

24.4 From Simple Carbon–Nitrogen Molecules to a Nucleic Acid Building Block In the oceans of prebiotic Earth, molecules of many kinds formed and accumulated. Like the chemicals in Miller's apparatus, these organic precursors were converted into the materials that become amino acids, simple acids, purines, and pyrimidines.

point, by as yet unidentified processes, polymers based on small sugars were replaced by those based on larger sugars, such as ribose, the sugar in the nucleotides of RNA.

RNA was probably the first biological catalyst

The enzymes that control the types and rates of reactions within organisms are proteins. As you learned in Chapter 12, proteins are synthesized by a process that begins with transcription of information from DNA to an RNA molecule that has a base sequence complementary to that of one strand of the DNA. This information is then translated into mRNA and is eventually used to synthesize a specific polypeptide from amino acids. Amino acids are brought to a ribosome by specific tRNA molecules and are attached sequentially to the growing polypeptide. But how could such a system have evolved if protein catalysts needed nucleic acids for their formation but nucleic acids needed proteins to catalyze their own replication? Which came first, if both are necessary?

The inability to solve this dilemma held up research on the origin of life for several decades. The discovery that provided a solution came in 1981 from researchers working with the unicellular protist *Tetrahymena thermophila*. These scientists were studying the excision of introns and the splicing together of exons. They found—entirely contrary to expectation—that excision

is catalyzed in the absence of enzymes. The intron itself—a 400-nucleotide sequence of RNA—carries out the excision and splicing. In addition, *Tetrahymena* ribosomes, which contain several molecules of RNA and a variety of proteins, have a catalytic RNA that operates in protein synthesis. RNAs that catalyze chemical reactions, called **ribozymes**, have now been found in many organisms (Figure 24.5).

The current system of macromolecular synthesis (DNA → RNA → protein) probably evolved gradually from much simpler processes. Biochemists believe that the first information-carrying molecules were short strands of RNA that replicated themselves without the help of enzymes. Evidence that RNAs can replicate themselves came first from experiments conducted by Manfred Eigen in the late 1970s. Eigen added RNA molecules to solutions containing monomers for making more RNA and found that sequences of five to ten nucleotides formed. If he added a simple inorganic molecule such as zinc, much longer sequences were copied.

Although these experiments showed that RNA could replicate itself, they did not demonstrate that RNA could catalyze the synthesis of other molecules as true enzymes do. Within two years, however, studies of a tRNA-processing enzyme that contains RNA and a protein (ribonuclease P) showed that the RNA alone can cut the pre-tRNA molecule at the correct spot, whereas the protein cannot.

Many scientists believe that the first genetic code was based on RNA that catalyzed both its own replication and other chemical reactions. In such conditions a high concentration of RNA would have been needed, so that it could participate in many different chemical

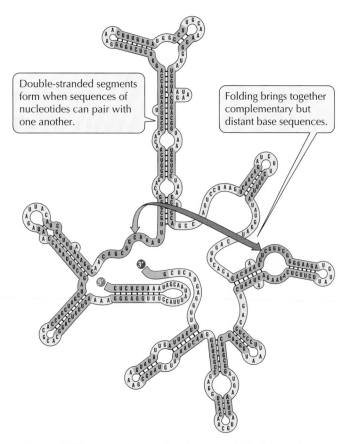

Double-stranded segments form when sequences of nucleotides can pair with one another.

Folding brings together complementary but distant base sequences.

24.5 A Ribozyme from a Protist The folded three-dimensional structure of this catalytic RNA, or ribozyme, enables it to catalyze chemical reactions during protein synthesis.

reactions. The accumulated products of RNA-catalyzed reactions could then participate in other reactions to form structures—for example, the production and accumulation of lipidlike molecules to form cell membranes, and the synthesis of proteins. However, after proteins evolved, they eventually took over most enzymatic functions because they are better catalysts than RNA and are capable of more diverse specificities.

In order to replicate, different RNAs would have competed with one another for monomers. Some RNA molecules would have been better at replicating in certain environments because their base sequences produced the most stable configurations (folded structures) under the particular conditions of temperature and salinity. With their higher rates of replication and greater stability, these molecules would have come to dominate the populations of RNA in the corresponding environments.

In the precellular world, RNA that contained both genetic information and catalytic capacity evolved. Molecules with these capacities have been produced in the laboratory by investigators who started with completely random-sequence RNA. These ribozymes

ligate (bind together) two RNA molecules in a reaction similar to that employed by the enzymes that synthesize RNA.

The investigators simulated the "evolution" of RNA molecules by selecting in test tubes for ribozymes with high ligation ability. By this method, they produced ribozymes with reaction rates 7 million times faster than the uncatalyzed reaction rate. These experiments suggest that ribozymes could have evolved rapidly when conditions on Earth became suitable for the formation of nucleic acids. But ribozymes in solution, even active ones, do not constitute life. Life required a barrier that permitted the homeostatic control of internal conditions; life could not appear until cells enclosed in membranes evolved. The prebiotic RNA world still lacked cells.

Membranes permitted a stable internal environment to form

The way in which ribozyme-based systems in solution might have evolved into cells was first suggested in the 1920s by the experiments of the Russian scientist Alexander Oparin, who spent much of his career studying complex solutions. Oparin observed that if he shook a mixture of a large protein and a polysaccharide, the drops that formed were divided into two "phases": an interior separated from but in contact with an exterior. These interiors, which were primarily protein and polysaccharide, with some water, were surrounded by an aqueous solution containing low concentrations of proteins and polysaccharides. These drops, known as **coacervates**, are quite stable and will form in solutions of many different types of polymers.

Coacervates have properties relevant to the origin of life. When added to a coacervate preparation, many substances are concentrated within the drops. Lipids coat the boundaries of drops with membranelike structures, which strengthen the drops and help contol the rates of passage of materials into and out of the drops. These properties simulate some characteristics of living cells. In addition, if enzyme molecules are added to them, coacervate drops exhibit a "metabolism": They can absorb substrates, catalyze reactions, and let the products diffuse back into the solution (Figure 24.6). Oparin even succeeded in making chlorophyll-containing coacervate drops that absorbed an oxidized dye from the solution, used light energy to reduce it, and released the reduced dye to the outside medium.

Coacervate drops are possible precursors to cells because they provide the physical structure by which internal conditions could be different from outside conditions. This structure permitted retention and concentration of some substances, new reactions, and release of products to the outside. Because drops in which chemical reactions were better controlled would have survived longer than drops with more poorly con-

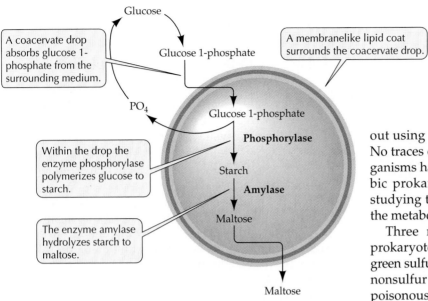

A coacervate drop absorbs glucose 1-phosphate from the surrounding medium.

Within the drop the enzyme phosphorylase polymerizes glucose to starch.

The enzyme amylase hydrolyzes starch to maltose.

A membranelike lipid coat surrounds the coacervate drop.

24.6 "Metabolism" of a Coacervate Drop
The properties of coacervate drops are similar to some of the properites of living cells. Held intact by a membranelike lipid coating, chemical reactions take place inside the drops in the presence of enzymes.

trolled reactions, refinements of metabolic processes by the use of catalysts could have evolved within them.

In the environment, billions of droplets would have competed with one another to acquire substrate materials. Droplets that evolved a better capacity to replicate internal reactions would have grown larger. But when coacervates become large, they are less stable, and physical agitation breaks them apart into smaller droplets. During this early form of reproduction, they would have passed on at least rough copies of their molecules to daughter droplets. At some point, such droplets may have accumulated enough RNA to display the key property we associate with living cells—the use of energy to maintain homeostasis.

DNA evolved from an RNA template

If the first cells used RNA as their hereditary molecule, then RNA must have provided the template for the synthesis of DNA. In solution, DNA is less stable than RNA. Therefore, DNA probably did not evolve until RNA-based life became contained in membrane-enclosed cells where water concentrations were lower than in the surrounding environment. Because DNA is a more stable storage location for genetic information than is RNA, once the appropriate environments were available, DNA probably evolved rapidly, replacing RNA as the genetic code for most organisms. But by then, RNAs had assumed their current roles as intermediaries in the translation of genetic information into proteins.

Early Anaerobic Cells

Earth's atmosphere lacked the gas O_2 for about a billion years after life evolved. Therefore, the earliest organisms must have processed energy chemically with-

out using O_2 as an electron acceptor (see Chapter 7). No traces of the metabolic pathways of those early organisms have been preserved, but some living anaerobic prokaryotes still employ similar pathways. By studying them, we can make reasoned guesses about the metabolism of ancient prokaryotes.

Three major types of anaerobic photosynthetic prokaryotes live today in sediments that lack oxygen: green sulfur bacteria, purple sulfur bacteria, and purple nonsulfur bacteria (see Chapter 25). In fact, oxygen is poisonous to them, as it must have been to organisms that evolved in a reducing atmosphere. These bacteria contain types of chlorophyll, called bacteriochlorophyll *a*, *c*, and *d*. Anaerobic photosynthetic bacteria also contain red and yellow carotenoids. These pigments absorb wavelengths of light that are not absorbed by chlorophyll and pass the absorbed energy along to chlorophyll for conversion into chemical energy.

The photosynthetic system of these bacteria is embedded in membrane complexes that also possess the electron transport chains and enzymes by which captured solar energy is converted to chemical energy and used to generate ATP (see Chapter 8). Photosynthetic prokaryotes similar in their metabolism to today's forms were so abundant about 3.4 billion years ago that their partly decomposed fossil remains formed extensive deposits of carbon, resembling the coal produced by vascular plant fossils 3 billion years later.

To reduce carbon dioxide (CO_2), a photosynthetic cell needs a source of electrons (hydrogen atoms). Many photosynthetic prokaryotes use light energy to generate ATP and NADPH + H^+. The particular waste product that they liberate depends on the source of hydrogen atoms they use. The green and purple sulfur photosynthetic bacteria obtain their hydrogen atoms from hydrogen sulfide (H_2S) and generate sulfur as a waste product (Figure 24.7). The purple nonsulfur bacteria typically obtain hydrogen atoms from organic compounds such as ethanol, lactic acid, or pyruvic acid, or directly from hydrogen gas (H_2).

In some environments today, the H_2 is produced by other prokaryotes as the end product of their fermentations. Under the anaerobic conditions of early Earth, hydrogen sulfide and other compounds containing hydrogen were more abundant than they are now. Today O_2 directly oxidizes H_2S, H_2, pyruvic acid, and similar compounds into water, carbon dioxide, and sulfur oxides.

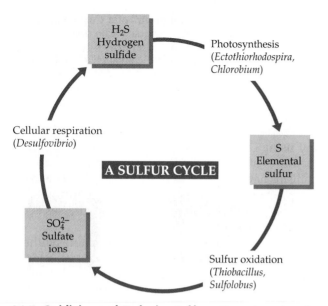

24.7 Oxidizing and Reducing Sulfur   Several species of bacteria and archaea may function together to carry out sulfur cycles, one of which is shown here. Representative genera able to conduct the processes are given in parentheses. Many such cycles are found in nature.

Aerobic photosynthesis is the source of atmospheric O_2

The evolution of aerobic photosynthesis, slightly more than 1 billion years ago, changed the course of evolution and changed Earth. The key change was the acquisition of the ability to use water as the source of hydrogen: $2 H_2O \rightarrow 4 H + O_2$. The chemical splitting of H_2O produced O_2 as a waste product and made available hydrogen atoms for reducing CO_2.

This ability appeared first in certain sulfur bacteria that evolved into cyanobacteria (see Chapter 25). Remains of these bacteria are abundantly preserved in fossil **stromatolites**. Cyanobacteria are still forming stromatolites in a few very salty places on Earth (Figure 24.8).

The ability to split water was doubtless the cause of the extraordinary success of the cyanobacteria. The O_2 they liberated opened the way for the evolution of oxidation reactions as the energy source for the synthesis of ATP. Their success made possible the evolution of the full respiratory chain of reactions now carried out by all aerobic cells. The evolution of life irrevocably changed the nature of our planet. Life created the O_2 of our atmosphere, and it removed most of the carbon dioxide from the atmosphere by transferring it to ocean sediments.

Oxygen was poisonous to the anaerobic organisms that lived on Earth, but the prokaryotes that evolved a tolerance to it were able to live successfully in environments empty of other organisms. Also, aerobic (oxygenated) metabolism could proceed much more rapidly and efficiently than the anaerobic metabolism that had dominated life until then.

Eukaryotic Cells: Increasing Complexity

The generation of atmospheric O_2 by cyanobacteria probably set the stage for the evolution of eukaryotes. Small, prokaryotic cells can obtain enough oxygen by simple diffusion even when oxygen concentrations are very low. Larger cells, however, have a lower surface area–to–volume ratio and need higher environmental O_2 concentrations to meet their needs.

About 1.5 billion years ago oxygen concentrations became high enough for larger cells to flourish and diversify. Some of these cells grew big enough to consume smaller cells; they were the first predators. Most of the prey died after they were engulfed, but some of them survived and eventually evolved into the mitochondria and chloroplasts that now maintain mutually beneficial relationships with their host.

Chromosomes organize large amounts of DNA

All early organisms were haploid. They reproduced by dividing, and each division was accompanied by a duplication of their circular chromosomes. The single circular DNA molecule of prokaryotes fits into a small cell, and it can be duplicated rapidly. Over time, the amount of DNA in the cells of early organisms increased, as it proved advantageous to encode information for the enzymatic catalysis of new reactions and to pass this information to progeny cells.

But replicating a long DNA molecule would take many hours if it could be started only at one spot. Replication can proceed much more rapidly if genes are separated into chromosomes, each of which has multiple initiation sites for duplication. Division of the genome into chromosomes also makes it less likely that the DNA molecules will become tangled when they divide. For these reasons, biologists believe that chromosomes evolved soon after the appearance of eukaryotes, about 2.5 billion years ago.

Variable offspring are advantageous in variable environments

All early organisms reproduced by simple fission (see Chapter 9), and each division was accompanied by a duplication of their chromosomes. Occasionally and accidentally, the duplication of DNA was not accompanied by a cell division, and the result was a diploid cell. A diploid cell has certain advantages. It can repair more kinds of chromosome damage than a haploid cell because the undamaged duplicate copy can guide the repair. But even if repair was not possible, a duplicate copy of the genome offered assurance that the life of the cell could continue.

Today, chromosome breakage can be repaired only by diploid cells. Because chromosome damage is espe-

(a)

(b)

24.8 Stromatolites Survive in a Few Places (a) A section through a fossil stromatolite. (b) Living stromatolites forming off the coast of Shark Bay in Western Australia.

cially likely in stressful environments, diploidy is advantageous in those conditions. Many present-day unicellular organisms that live in a haploid state most of the time produce diploid cells when they are environmentally stressed. Diploid cells also have protection from point mutations, most of which alter only one copy of a gene. The unaltered copy on the homologous chromosome continues to function normally.

A diploid cell can also result if two haploid cells fuse. The chromosomes of two haploid cells are unlikely to have the same damages. Thus, a diploid cell formed by acquiring chromosomes from another cell would be able to repair all its damaged chromosomes. This ability may have provided the original advantage of sexual recombination, but we can only speculate. Whatever its original advantages, sex appeared early during the evolution of life, and nearly all living organisms engage in sexual recombination at least occasionally during their life cycles. Those that do not are believed to have evolved recently from sexual ancestors; they are destined to have short evolutionary lives.

Sex is maintained in multicellular organisms today by a benefit different from the ones that initially favored it. Whereas the original benefits accrued directly to the individuals accepting genetic material, sex is maintained because of its effects on offspring. Whereas in unicellular organisms genes are not exchanged during reproduction (cell division), in multicellular organisms genes are nearly always exchanged at the time of reproduction, which is the only time during the life cycle when a single-cell stage exists.

The offspring of sexual multicellular organisms are genetically highly variable. In an environment that varies in space and time, the genetically uniform offspring of an asexual organism are less likely to find suitable conditions than are the variable offspring of a sexual organism.

Matter, Time, and Inevitable Life

Scientists have been able to gather information that provides many insights into the origin of life on Earth. In laboratory experiments they have studied chemical reactions under conditions similar to those believed to have prevailed on early Earth. Under laboratory conditions, they have witnessed the "evolution" of ribozymes with catalytic activity from random-sequence RNA.

Taken together, this information suggests that the evolution of life as we know it was highly probable under the conditions that prevailed on Earth 3.8 billion years ago. The molecules on which life is based form readily under such conditions, as does the organization of those molecules into larger units. Much remains to be learned about the early evolution of the complex metabolism of living organisms, but techniques are available to help us.

If the origin of life was almost inevitable, why is new life not still being assembled from nonliving matter on today's Earth? The reason is that simple biological molecules released into today's environment are quickly consumed by existing life. They cannot accumulate to the densities that characterized the prebiotic "soup," even in anaerobic environments. In aerobic environments these molecules are quickly oxidized to other forms. They would not accumulate even if they were not consumed. Life was generated from nonlife on Earth, but it was an event of the remote past. Once life had evolved, it prevented the formation of other life from nonlife.

Summary of "The Origin of Life on Earth"

Is Life Forming from Nonlife Today?
• For its first billion years, no life existed on Earth. Life originated from nonliving matter 3.8 billion years ago, but experiments by Louis Pasteur and others convinced scientists that life does not come from nonlife on Earth today. **Review Figure 24.1**

How Did Life Evolve from Nonlife?
• Conditions on Earth at the time of the origin of life differed from those of today. No free oxygen was present in Earth's early atmosphere.
• The reducing atmosphere of early Earth was the setting for chemical reactions that produced a molecular soup. Under conditions that resemble Earth's early reducing atmosphere, small molecules essential to living systems form and polymerize. **Review Figure 24.3**
• Before life appeared, polymerization reactions generated the carbohydrates, lipids, amino acids, and nucleic acids of which organisms are composed. These molecules accumulated in the seas because the rate of their formation was greater than the rate at which they were destroyed. **Review Figure 24.4**
• The first genetic material may have been RNA that had both a catalytic function and an information transfer function. Some RNAs—called ribozymes—still have catalytic functions today. **Review Figure 24.5**
• The earliest cells may have been similar to the coacervates that can be produced in the laboratory by the mixing of proteins and polysaccharides. Once cells evolved, they outcompeted all acellular proto-life. **Review Figure 24.6**
• DNA probably evolved after RNA-based proto-life became surrounded by membranes that provided an environment in which DNA is stable.

Early Anaerobic Cells
• For more than 2 billion years, the only organisms were anaerobic prokaryotes. Many prokaryotes that live today in environments lacking O_2 use metabolic pathways similar to those of early prokaryotes.
• Cyanobacteria, which evolved the ability to split water into hydrogen atoms (for energy and reduction) and O_2, proliferated and created atmospheric O_2. The accumulation of free O_2 in Earth's atmosphere made possible the evolution of aerobic metabolism, eukaryotes, sexual recombination, and multicellularity.

Eukaryotic Cells: Increasing Complexity
• Division of the genome into chromosomes increased the rate of DNA replication and decreased the likelihood that DNA molecules would become tangled when they divide.
• All early cells were haploid. The first diploid cells appeared probably when some cells failed to divide after their DNA was duplicated. These "accidental" diploids had important advantages over haploid cells. Diploid cells were better at repairing damaged chromosomes and were better protected from harmful point mutations.
• Sex evolved because it was directly advantageous to the individuals that accepted genes from others, but sex is main-

tained today because the variable offspring that sex produces are more likely than the genetically uniform offspring of asexual organisms to survive in the variable environments in which they must live.

Matter, Time, and Inevitable Life
• Because most of the chemical reactions that gave rise to life proceed readily under the conditions that prevailed on early Earth, the evolution of life was probably nearly inevitable.
• New life is not being assembled from nonliving matter today because simple biological molecules that form in the current environment are quickly consumed by existing life.

Self-Quiz

1. The atmosphere of early Earth consisted largely of
   a. water vapor.
   b. hydrogen.
   c. carbon dioxide.
   d. helium.
   e. nitrogen.

2. Pasteur's experiments and similar ones that followed convinced most people that spontaneous generation of life did not happen, because
   a. Pasteur was extremely meticulous.
   b. Pasteur used very fine mesh screens to cover his flasks.
   c. Pasteur did not boil his flasks for a long time.
   d. Pasteur's swan-necked flasks ruled out the objection that spoiled air could have contaminated his experiments.
   e. by the time Pasteur performed his experiments, many people no longer believed in spontaneous generation.

3. To determine which molecules might have formed spontaneously on early Earth, Stanley Miller used an apparatus with an atmosphere containing
   a. oxygen, hydrogen, and nitrogen.
   b. oxygen, hydrogen, ammonia, and water vapor.
   c. oxygen, hydrogen, and methane.
   d. hydrogen, oxygen, and carbon dioxide.
   e. hydrogen, ammonia, methane, and water vapor.

4. Most biologists think that RNA was the first genetic material because
   a. amino acids were produced in Stanley Miller's apparatus.
   b. DNA is the universal genetic material of eukaryotes.
   c. the existence of ribozymes suggests that early cells could have used RNA to catalyze chemical reactions and transfer information.
   d. RNA is simpler than DNA.
   e. DNA is not stable in hydrophobic environments.

5. Biologists believe that the current DNA → RNA → protein system is the result of a long period of evolution because
   a. the transcription of DNA to mRNA and translation of mRNA into proteins consists of many steps.
   b. DNA replication is complicated but relatively error-free.
   c. the current system is very complex and precise.
   d. evidence indicates that RNA preceded DNA as the genetic material.
   e. all of the above

6. The question whose answer enabled research on the origin of catalysis to proceed rapidly was,
   a. How could complex life have evolved on such a young Earth?
   b. How could the precise duplication of DNA have evolved?
   c. How could catalysis have evolved, given that RNA needs proteins for its synthesis and proteins need RNA for their synthesis?
   d. How could eukaryotes evolve from prokaryotes?
   e. How did the first cells form?

7. The key process in the formation of nucleic acid bases is
   a. the polymerization of hydrogen cyanide.
   b. the polymerization of formaldehyde.
   c. the spontaneous formation of monomers.
   d. the spontaneous formation of proteins.
   e. the polymerization of proteins.

8. The metabolism of living prokaryotes provides important insights into the chemical processes used by early organisms because
   a. many prokaryotes live in environments similar to those in which life first evolved.
   b. prokaryotes are simpler to study and hence are better known than are eukaryotes.
   c. many prokaryotes are obligate aerobes.
   d. many prokaryotes use oxygen as their oxidizing agent.
   e. fermentation evolved before aerobic respiration.

9. The most important advantage of diploidy is that
   a. it gives a cell more genes.
   b. more genes better fill the larger volume of eukaryotic cells.
   c. two copies of a gene are better than one.
   d. the duplicate chromosomes increase the ability of cells to repair chromosomal damage.
   e. diploid cells can reproduce faster than haploid cells in constant environments.

10. Sex is maintained in multicellular organisms because
   a. without it life would not be worth living.
   b. the offspring of sexual organisms survive better in variable environments than do the offspring of sexual organisms.
   c. sexual individuals are better than asexual individuals at repairing damaged chromosomes.
   d. sexual individuals can produce more offspring than asexual individuals can.
   e. sexual individuals can reproduce faster than asexual individuals.

Applying Concepts

1. Why is determining the composition of Earth's early atmosphere a key step in inferring how life arose?

2. Why is the ability of ribozymes to catalyze both their own synthesis and the synthesis of proteins so important for understanding the origin of life?

3. On the one hand, scientists are confident that life no longer arises from nonliving matter. On the other hand, most biologists believe that life did arise on this planet, billions of years ago, from nonliving matter. How can scientists hold both of these beliefs?

4. Why do biologists believe that the evolution of life was highly probable on early Earth?

5. How might each of the following have been involved in the evolution of coacervate drops?
   a. coating of drop boundaries with lipids
   b. wave action in bodies of water
   c. catalysts within the drops

Readings

Cech, T. R. 1986. "RNA as an Enzyme." *Scientific American,* November. A well-illustrated discussion of the discovery of the catalytic abilities of RNA and their implications for the origin and early evolution of life.

de Duve, C. 1995. "The Beginning of Life on Earth." *American Scientist,* vol. 83, pages 428–437. An argument that life was an inevitable outcome of prebiotic physics and chemistry.

Dyson, F. J. 1985. *The Origins of Life.* Cambridge University Press, New York. A concise argument in favor of multiple origins of life.

Fenchel, T. and B. J. Finlay. 1994. "The Evolution of Life without Oxygen." *American Scientist,* vol. 82, pages 22–29. A discussion of recent clues about the evolution of early eukaryotic cells.

Loomis, W. F. 1988. *Four Billion Years.* Sinauer Associates, Sunderland, MA. A very readable book on the evolution of genes and organisms, concentrating on the first billion years.

Margulis, L. 1984. *Early Life.* Jones and Bartlett, Boston. An engaging account of the earliest organisms and how they evolved. Good treatment of the endosymbiosis theory of the origin of eukaryotes by its principal proponent.

Maynard Smith, J. and E. Szathmáry. 1995. *The Major Transitions in Evolution.* W. H. Freeman, San Francisco. The first book to describe and synthesize the evolution of the major changes in the way genetic information is organized and transmitted from generation to generation.

Orgel, L. 1994. "The Origin of Life on Earth." *Scientific American,* October. A discussion of how life evolved after self-replicating molecules appeared.

Schopf, J. W. and C. Klein (Eds.). 1992. *The Proterozoic Biosphere.* Cambridge University Press, New York. An excellent source of information on interactions between life and early Earth.

Taylor, S. R. and S. M. McLennan. 1996. "The Evolution of Continental Crust." *Scientific American,* June. Earth appears to be the only planet in our solar system that has sustained enough geologic activity to create large, stable land masses.

Weinberg, S. 1993. *The First Three Minutes.* Basic Books, New York. A stimulating but demanding account of the origin of Earth for those interested enough to make the investment.

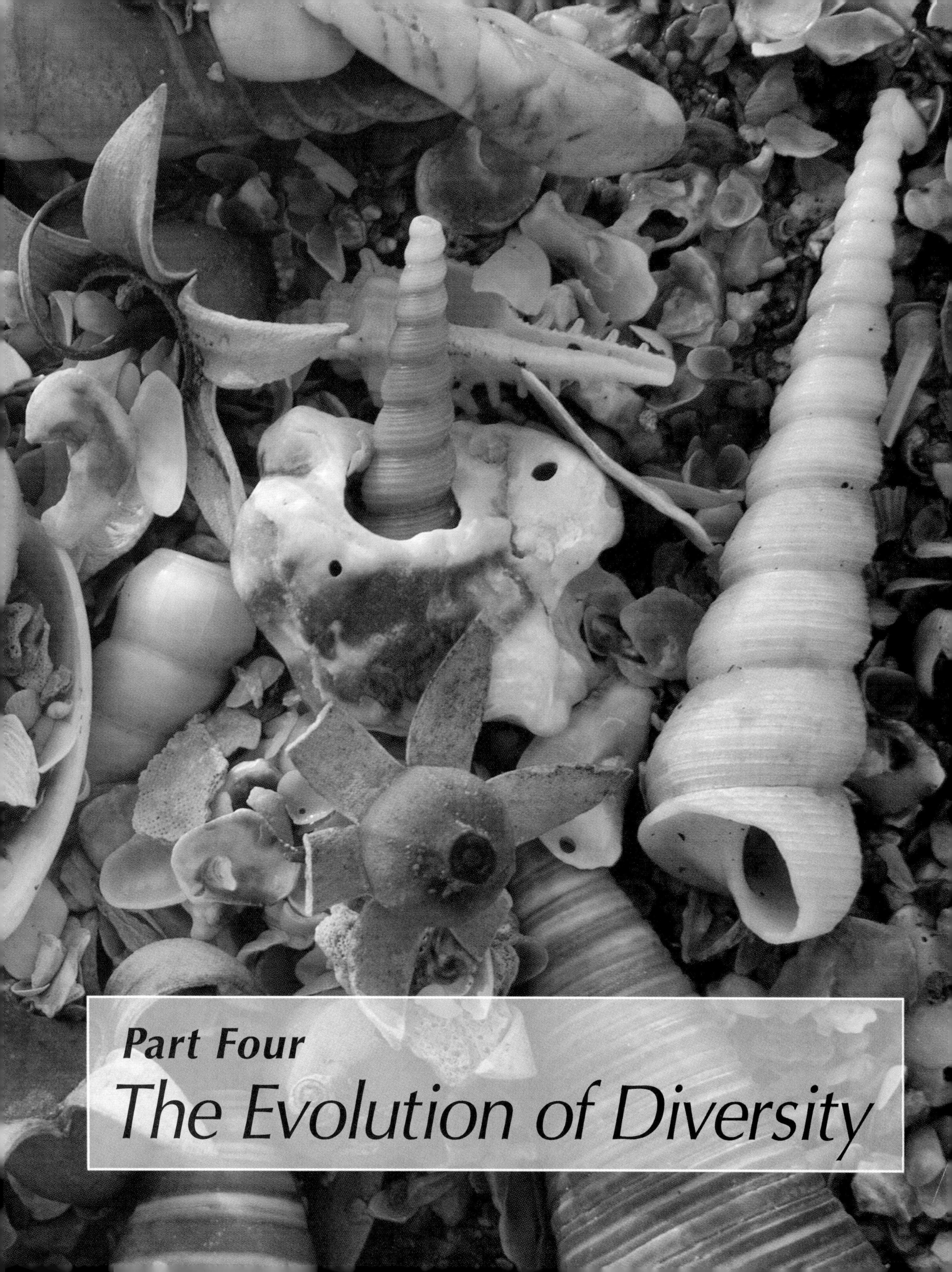

Part Four

The Evolution of Diversity

Chapter 25

Bacteria and Archaea: The Prokaryotic Domains

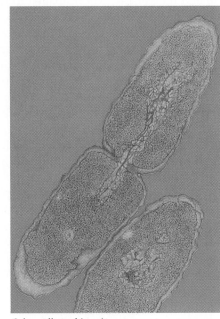
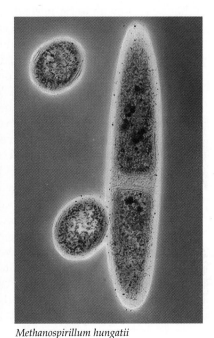

Salmonella typhimurium

Methanospirillum hungatii

Very Different Prokaryotes
In each image, one of the cells has nearly finished dividing. On the left are bacteria; on the right are archaea, which are more closely related to you than they are to the bacteria.

How can two kinds of microscopic creatures look so much alike and yet be so vastly different? After examining a micrograph of either a bacterium or an archaean, you might say, "Looks like a bacterium to me." However, the cells shown here on the right are more closely related to you and me than they are to the cell on the left!

Both of these simple-looking organisms are prokaryotes, but they differ in numerous biochemical and genetic ways. Not until the 1970s did biologists discover how radically bacteria and archaea differ from each other. And only in 1996, with the sequencing of an archaean genome, did we realize just how extensively archaea differ from bacteria and from eukaryotes, whose genomes had been sequenced a little earlier.

Scientists acknowledge the antiquity of the lineage separations and the importance of their differences by recognizing three domains of living things: Bacteria, Archaea, and Eukarya. The domain Bacteria comprises one kingdom, the kingdom Eubacteria ("true bacteria"); the domain Archaea (from the Greek *archaios*, "ancient") comprises one kingdom, the kingdom Archaebacteria ("ancient bacteria"). The domain Eukarya consists of four kingdoms that include all other living things on Earth. Dividing the living world in this way, with two prokaryotic domains and a single domain for all the eukaryotes, fits with current trends to reflect evolutionary relationships in classification systems (see Chapter 22).

In the six chapters of Part Four, we celebrate and describe the diversity of the living world—the products of evolution. This chapter focuses on the prokaryotic kingdoms. Chapters 26 to 30 deal with the kingdoms Protista, Plantae, Fungi, and Animalia.

In this chapter, we will pay close attention to the ways in which the two domains of prokaryotic organisms resemble one another, and how they differ. Then we will survey the surprising diversity of organisms within each of these domains, relating the characteristics of different prokaryotic groups to their roles in the biosphere and in our lives.

Why Three Domains?

What does it mean to be *different*? You and the person nearest you look very different—certainly more different than the two cells shown at the beginning of this chapter look. But the two of you are members of the same species, and those two tiny organisms are classified in entirely separate domains. You (in the domain Eukarya) and those two prokaryotes (in the domains Bacteria and Archaea) have a lot in common. Members of all three domains conduct glycolysis and replicate their DNA semiconservatively. In all three, the DNA encodes polypeptides that are produced by transcription and translation, and all cells have plasma membranes and ribosomes in abundance.

As a member of the domain Eukarya, *you* have cells with nuclei, membrane-enclosed organelles, and a cytoskeleton—things that no prokaryote has. However, a glance at Table 25.1 will show you that there are also major differences, most of which cannot be seen even under the microscope, between the two prokaryotic domains. In some ways the archaea are more like us; in other ways they are more like bacteria.

Comparisons of base sequences in ribosomal RNAs and of genomes themselves have led scientists to conclude that all three domains had a single common ancestor and that the present-day archaea share a more recent common ancestor with eukaryotes than they do with bacteria (Figure 25.1). In the same way, humans are more closely related to their sisters and brothers, with whom they share parents, than to their cousins, with whom they share grandparents.

Because of (1) the ancient time at which the three lineages diverged, (2) the major differences among these three kinds of organisms, and especially (3) the fact that the archaea are more closely related to the eukaryotes than are either of those groups to the bacteria, most biologists agree that it makes sense to treat the groups as domains—a higher taxonomic category than kingdom. To treat all the prokaryotes as a single kingdom within a five-kingdom classification of organisms would result in a kingdom that is paraphyletic. That is, the prokaryotic kingdom would not include all the descendants of their common ancestor (see Chapter 22, especially Figure 22.12, for a discussion of paraphyletic groups).

The Archaea, Bacteria, and Eukarya of today are all the products of billions of years of natural selection, and they are all highly adapted to present-day environments. None are "primitive." The common ancestor of the Archaea and the Eukarya lived 1.8 billion years ago by a conservative estimate, and the common

TABLE 25.1 The Three Domains of Life on Earth

| CHARACTERISTIC | DOMAIN | | |
	BACTERIA	ARCHAEA	EUKARYA
Membrane-enclosed nucleus	Absent	Absent	*Present*
Membrane-enclosed organelles	Absent	Absent	*Present*
Peptidoglycan in cell wall	*Present*	Absent	Absent
Membrane lipids	Ester-linked	*Ether-linked*	Ester-linked
	Unbranched	*Branched*	Unbranched
Ribosomes[a]	70S	70S	*80S*
Initiator tRNA	*Formylmethionine*	Methionine	Methionine
Operons	Yes	Yes	*No*
Plasmids	Yes	Yes	*Rare*
RNA polymerases	One	Several	Three
Sensitive to chloramphenicol and streptomycin	*Yes*	No	No
Ribosomes sensitive to diphtheria toxin	*No*	Yes	Yes
Some are methanogens	No	*Yes*	No
Some fix nitrogen	Yes	Yes	*No*
Some conduct chlorophyll-based photosynthesis	Yes	*No*	Yes

[a] 70S ribosomes are smaller than 80S ribosomes.

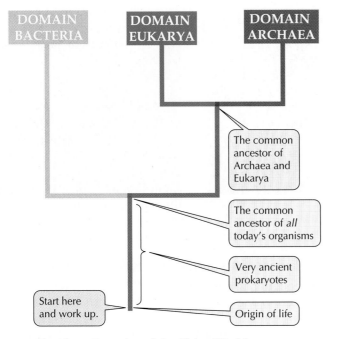

25.1 The Three Domains of the Living World

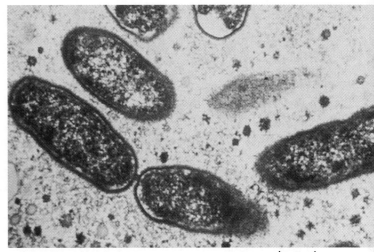

Rickettsia rickettsia   0.2 μm

25.2 Bacteria Living in Animal Cells These bacteria live within the mammalian body, where they can cause a lethal disease, Rocky Mountain spotted fever.

ancestor of the Archaea, the Eukarya, and the Bacteria lived more than 2 billion years ago.

The earliest prokaryotic fossils date back at least 3.5 billion years, and as we saw in Chapter 24, these ancient fossils indicate that there was considerable diversity among the prokaryotes even during the Archean eon. The prokaryotes reigned supreme on an otherwise sterile Earth for a very long time, adapting to new environments and to changes in existing environments.

General Biology of the Prokaryotes

There are many, many prokaryotes around us—everywhere. Although they are so small that we cannot see them with the naked eye, the prokaryotes are the most successful of all creatures on Earth if success is measured by numbers of individuals. The bacteria in one person's intestinal tract, for example, outnumber all the humans who have ever lived. Although small, prokaryotes play many critical roles in the biosphere, interacting in one way or another with every other living thing.

In this section on the general biology of the prokaryotes, we'll see that some perform key steps in the cycling of nitrogen, sulfur, and carbon. Other prokaryotes trap energy from the sun or from inorganic chemical sources, and some help animals digest their food. The members of the two prokaryotic domains outdo all other groups in metabolic diversity, and they occupy more—and more extreme—habitats than any other group.

They have spread to every conceivable habitat on the planet, from the coldest to the hottest, from the most acidic to the most alkaline, and to the saltiest. Some live where O_2 is abundant and others where there is no O_2 at all. They have established themselves at the bottom of the seas, in rocks more than 2 km into Earth's solid crust, and even inside other organisms, large and small (Figure 25.2). Their effects on our environment are diverse and profound.

Prokaryotes and their associations take a few characteristic forms

Three shapes are particularly common among the prokaryotes: spheres, rods, and curved or spiral forms (Figure 25.3). A spherical prokaryote is called a **coccus** (plural cocci). Cocci may live singly or may associate in two- or three-dimensional arrays as chains, plates, or blocks of cells. A rod-shaped prokaryote is called a **bacillus** (plural bacilli). Bacilli and spiral forms, the third main prokaryotic shape, may be single or may form chains.

Associations such as chains do not signify multicellularity, because each cell is fully viable and independent. Associations arise as cells adhere to one another after reproducing by fission, with each becoming two cells. Some bacteria associate in chains that become enclosed within delicate tubular sheaths. These associations are called filaments. All the cells of a filament divide simultaneously.

In addition to having distinguishing shapes, some prokaryotes are identified by other structural features. For example, some attach to their substrate by stalks that may be either an extension of the cell wall or a product secreted outside the cell.

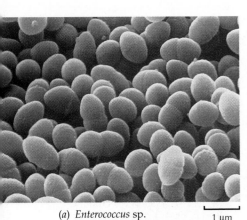

(a) *Enterococcus* sp.   1 µm

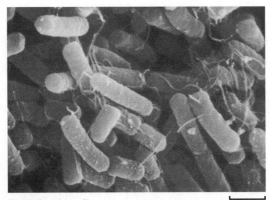

(b) *Escherichia coli*   1 µm

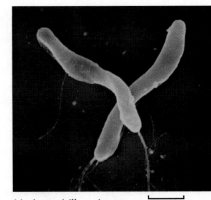

(c) *Aquaspirillum sinosum*   1 µm

25.3 Shapes of Prokaryotic Cells   (a) These spherical cocci of an acid-producing bacterium grow in the mammalian gut. (b) Rod-shaped *E. coli* are the most thoroughly studied of all bacteria—indeed, of almost any organism. (c) A freshwater spiral bacteria species. The cells move by means of the tufts of flagella at each pole.

Prokaryotes lack nuclei, organelles, and a cytoskeleton

The architectures of prokaryotic and eukaryotic cells were compared in Chapter 4 (you may wish to review Figures 4.3, 4.4, 4.5, 4.7, and 4.8). The basic unit of archaea and bacteria is the prokaryotic cell, which contains a full complement of genetic and protein-synthesizing systems, including DNA, RNA, and all the enzymes needed to transcribe and translate the genetic information into protein (see Chapter 12). The prokaryotic cell also contains at least one system for generating the ATP it needs (see Chapter 7).

The prokaryotic cell differs from the eukaryotic cell in three important ways. First, the organization and replication of the genetic material differs. The DNA of the prokaryotic cell is not organized within a membrane-enclosed nucleus. DNA molecules in prokaryotes are circular; typically there is a single chromosome, but there are often plasmids as well (see Chapter 13). The elaborate mechanism of mitosis is missing; prokaryotic cells divide by their own elaborate method, fission, after replicating their DNA.

Second, prokaryotes have none of the membrane-enclosed cytoplasmic organelles that modern eukaryotes have—mitochondria, chloroplasts, Golgi apparatus, endoplasmic reticulum—but the cytoplasm of a prokaryotic cell may contain a variety of infoldings of the plasma membrane (see Figure 4.4) and photosynthetic membrane systems not found in eukaryotes. Membranous infoldings frequently associate with new cell walls during cell division, and in electron micrographs the DNA of a bacterial cell is often seen attached to such an infolding, called a *mesosome* (Figure 25.4). Third, prokaryotic cells lack a cytoskeleton.

Prokaryotes have distinctive modes of locomotion

Although many prokaryotes are not motile, others can move by one of several means. Spirochetes use a rolling motion made possible by internal fibrils (Figure 25.5a). Many cyanobacteria and the "gliding bacteria" use various poorly understood gliding mechanisms, including rolling. Some aquatic prokaryotes, including some cyanobacteria, can move slowly up and down in the water by adjusting the amount of gas in gas vesicles (Figure 25.5b). By far the most common type of locomotion in prokaryotes is that driven by flagella.

Prokaryotic flagella are whiplike filaments that extend singly or in tufts from one or both ends of the cell (see Figure 25.3c and 25.8a), or all around it (Figure 25.6). A prokaryotic flagellum consists of a single fibril made of the protein *flagellin*, projecting from the cell surface. In contrast, the flagellum of eukaryotes is enclosed by the plasma membrane and usually contains a circle of nine pairs of microtubules surrounding two

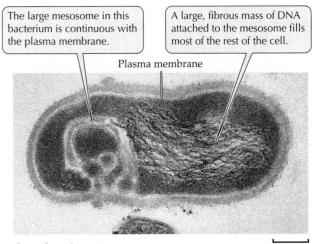

The large mesosome in this bacterium is continuous with the plasma membrane.

A large, fibrous mass of DNA attached to the mesosome fills most of the rest of the cell.

Plasma membrane

Corynebacterium parvum   0.3 µm

25.4 Some Prokaryotes Have Internal Membranes
Unlike eukaryotic organelles, the infolding in this bacterial cell is not a separate, membrane-enclosed compartment.

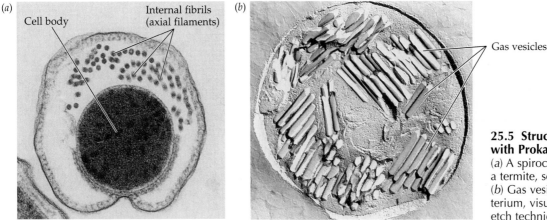

(a)

Cell body

Internal fibrils (axial filaments)

(b)

Gas vesicles

25.5 Structures Associated with Prokaryote Motility
(a) A spirochete from the gut of a termite, seen in cross section. (b) Gas vesicles in a cyanobacterium, visualized by the freeze-etch technique (see Chapter 5).

central microtubules, all made of the protein tubulin, along with other, associated proteins. The prokaryotic flagellum (see Figure 4.6) rotates about its base, rather than beating, as a eukaryotic flagellum or cilium does.

Prokaryotes have distinctive cell walls

Most prokaryotes have a thick and relatively stiff cell wall. This wall is quite different from the cell walls of plants and algae, which contain cellulose and other polysaccharides, and of fungi, which contain chitin. Almost all bacteria have cell walls containing peptidoglycan (a polymer of amino sugars). Archaean cell walls are of differing types, but most contain significant amounts of protein (often a glycoprotein). One group of archaea has pseudopeptidoglycan in its wall; as you have probably already guessed from the prefix

"pseudo-," pseudopeptidoglycan is similar to but distinct from the peptidoglycan of bacteria.

Peptidoglycan is a substance unique to bacteria; its absence from the walls of archaea indicates a key difference between the two prokaryotic domains, resulting from the separation of these two groups at the beginning of evolutionary history.

In 1884 Hans Christian Gram, a Danish physician, developed an uncomplicated staining process that has lasted into our high-technology era as the single most common tool in the identification of bacteria. The **Gram stain** separates most types of bacteria into two distinct groups: gram-positive and gram-negative, on the basis of their wall structure (Figure 25.7). A smear of cells on a microscope slide is soaked in a violet dye and treated with iodine; it is then washed with alcohol and counterstained with safranine (a red dye). **Gram-positive** bacteria retain the violet dye and appear blue to purple (Figure 25.7a). The alcohol washes the violet stain out of **gram-negative** cells; these cells then pick up the safranine counterstain and appear pink to red (Figure 25.7b). Gram-staining characteristics are a crucial consideration in classifying some kinds of bacteria and are important in determining the identity of bacteria in an unknown sample. Mycoplasmas, which lack walls, are not stained at all by the Gram stain.

The different staining reactions probably relate to differences in the physical structures of the cell walls of bacteria, but they correlate with the amount of peptidoglycan. The electron micrographs and associated sketches in Figure 25.7 show a thick layer of peptidoglycan outside the plasma membrane of gram-positive bacteria (see Figure 25.7a). The gram-negative cell wall usually has only one-fifth as much peptidoglycan, and outside the peptidoglycan layer the cell is surrounded by a second, outer membrane quite distinct in chemical makeup from the plasma membrane (see Figure 27.5b). The space between the inner (plasma) and outer membranes of gram-negative bacteria contains enzymes that are important in digesting some materi-

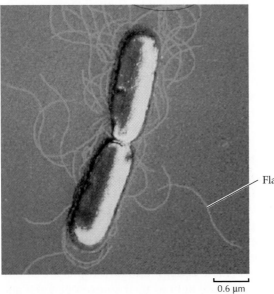

Flagellum

0.6 μm

25.6 Some Bacteria Use Flagella for Locomotion
Flagella surround the rod-shaped cells of this *Bacillus* species.

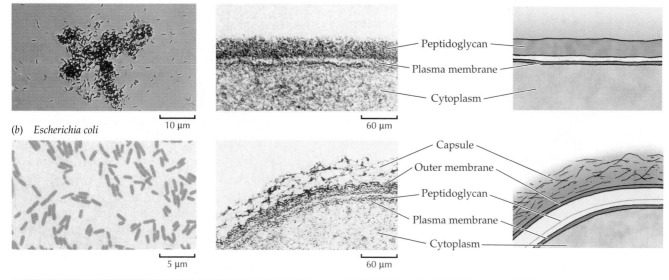

(a) *Bacillus subtilis*

Peptidoglycan
Plasma membrane
Cytoplasm

10 µm   60 µm

(b) *Escherichia coli*

Capsule
Outer membrane
Peptidoglycan
Plasma membrane
Cytoplasm

5 µm   60 µm

25.7 The Gram Stain and the Cell Wall When treated with a Gram stain, the cell wall components of different bacteria react in one of two ways. (a) Gram-positive bacteria retain the violet dye and appear deep blue or purple; the pink counterstain surrounds the cells in this micrograph. (b) Gram-negative bacteria internalize the counterstain and appear pink-to-red.

als, transporting others, and detecting chemical gradients in the environment.

The consequences of the different features of the cell walls are numerous and relate to the disease-causing characteristics of some prokaryotes. Indeed, the cell wall is a favorite target in medical combat against diseases that are caused by prokaryotes, because it has no counterpart in eukaryotic cells. Antibiotics and other agents that specifically attack peptidoglycan-containing cell walls tend to have little, if any, effect on the cells of humans and other eukaryotes.

Prokaryotes reproduce asexually, but genetic recombination does occur

Most prokaryotes reproduce by fission or by producing spores; both processes are asexual, or **vegetative**. Recall, however, that there are also processes—transformation, conjugation, and transduction—that allow the exchange of genetic information between some prokaryotes quite apart from either sex or reproduction.

Many prokaryotes multiply very rapidly. One of the fastest is the bacterium *Escherichia coli*, which under optimal conditions has a generation time of about 20 minutes. The shortest known prokaryote generation times are about 10 minutes; values of 1 to 3 hours are common; some extend to days. Bacteria living in rock deep in Earth's crust may suspend their growth for more than a century without dividing and then grow for a few days before suspending growth again.

Prokaryotes have exploited many metabolic possibilities

The long evolutionary history of bacteria and archaea, including their explorations of new environments, has led to extraordinary diversity of their metabolic "lifestyles"—their use or nonuse of O_2, their energy sources, the sources of their carbon atoms, and the materials they secrete.

ANAEROBIC VERSUS AEROBIC METABOLISM. Some prokaryotes can live only by anaerobic metabolism, because they are poisoned by oxygen gas. These oxygen-sensitive fermenters are called **obligate anaerobes**. Fermentation is not the only anaerobic way to obtain energy, as we'll see shortly when we discuss the denitrifiers.

Other organisms can shift their metabolism between anaerobic and aerobic modes (see Chapter 7) and thus are called **facultative anaerobes**. Some facultative anaerobes cannot conduct cellular respiration but are not damaged by oxygen when it is present. Many types of prokaryotes are facultative anaerobes that alternate between anaerobic metabolism (such as fermentation) and cellular respiration as conditions dictate.

At the other extreme from the obligate anaerobes, some prokaryotes are **obligate aerobes**, unable to survive for extended periods in the *absence* of oxygen.

NUTRITIONAL CATEGORIES. Biologists recognize four broad nutritional categories of organisms: photoautotrophs, photoheterotrophs, chemoautotrophs, and chemoheterotrophs. Prokaryotes are represented in all four groups (Table 25.2). **Photoautotrophs** are photosynthetic. They use light as their source of energy and carbon dioxide as their source of carbon. Like the photosynthetic eukaryotes, the cyanobacteria, one group of

TABLE 25.2 How Organisms Obtain Their Energy and Carbon

NUTRITIONAL CATEGORY	ENERGY SOURCE	CARBON SOURCE
Photoautotrophs (some Bacteria, some Eukarya)	Light	Carbon dioxide
Photoheterotrophs (some Bacteria)	Light	Organic compounds
Chemoautotrophs (some Bacteria)	Inorganic substances	Carbon dioxide
Chemoheterotrophs (found in all three domains)	Organic compounds	Organic compounds

photoautotrophs, photosynthesize with chlorophyll *a* as the key pigment and produce oxygen as a by-product of noncyclic photophosphorylation (see Chapter 8).

By contrast, the other photosynthetic bacteria use bacteriochlorophyll as their key photosynthetic pigment, and they do not release oxygen gas. Some of these photosynthesizers produce particles of pure sulfur instead because hydrogen sulfide (H_2S) rather than H_2O is the electron donor for photophosphorylation (Figure 25.8*a*). Bacteriochlorophyll absorbs light of longer wavelength than the chlorophyll used by all other photosynthesizing organisms does. As a result, bacteria using this pigment can grow in water beneath fairly dense layers of algae because light of the wavelengths they use is not appreciably absorbed by the algae (Figure 25.8*b*).

Photoheterotrophs use light as their source of energy but must obtain their carbon atoms from organic compounds made by other organisms. They use compounds such as carbohydrates, fatty acids, and alcohols as their organic "food." The purple nonsulfur bacteria, among others, are photoheterotrophs.

Chemoautotrophs obtain their energy by oxidizing inorganic substances, and they use some of that energy to fix carbon dioxide. Some chemoautotrophs use reactions identical to those of the photosynthetic carbon reduction cycle (see Chapter 8), but others use other pathways to fix carbon dioxide. The chemoautotrophs—some of them bacteria and others archaea—include the nitrifiers, which oxidize ammonia or nitrite ions to form nitrate ions that are taken up by plants, as well as other bacteria that oxidize hydrogen gas, hydrogen sulfide, sulfur, and other materials.

Scientists exploring the ocean bottom near the Galapagos Islands in 1977 discovered a spectacular example of chemoautotrophy. They found an entire ecosystem based on chemoautotrophic bacteria that are incorporated into a large community of crabs, mollusks, and giant worms (pogonophorans; see Figure 29)—all at a depth of 2,500 m, far below any hint of light from the sun but in the immediate neighborhood

25.8 Some Bacteria Photosynthesize (*a*) Cells of purple sulfur bacteria store granules of sulfur that they produce via anaerobic photosynthesis. (*b*) *Ulva*, a green alga, absorbs no light of wavelengths longer than 750 nm. Purple sulfur bacteria can conduct photosynthesis in the shade of the algae, using these longer wavelengths.

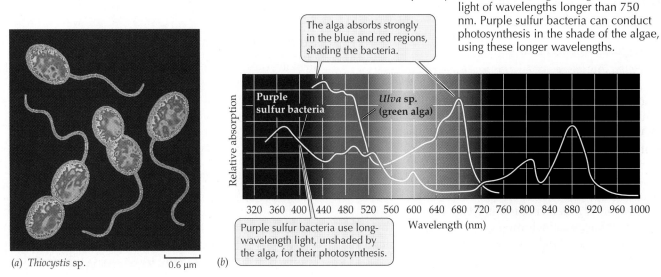

The alga absorbs strongly in the blue and red regions, shading the bacteria.

Purple sulfur bacteria use long-wavelength light, unshaded by the alga, for their photosynthesis.

Purple sulfur bacteria

Ulva sp. (green alga)

Relative absorption

320 360 400 440 480 520 560 600 640 680 720 760 800 840 880 920 960 1000
Wavelength (nm)

(*a*) *Thiocystis* sp. 0.6 μm (*b*)

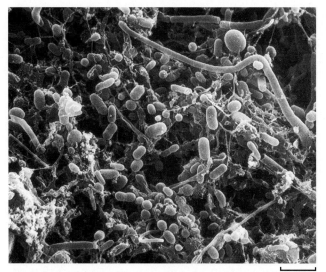

2 μm

25.9 Chemoautotrophs All of these archaea are chemoautotrophs that live near a hot-water vent along the Galapagos Rift in the eastern Pacific, where their fixing of carbon supports an entire community of organisms that thrives in total darkness.

of volcanic vents in the ocean floor (Figure 25.9). These bacteria obtain energy by oxidizing hydrogen sulfide and other substances released from the vents.

Finally, **chemoheterotrophs** typically obtain both energy and carbon atoms from one or more organic compounds. Most bacteria are chemoheterotrophs—as are all archaea, animals, and fungi, and many protists.

NITROGEN AND SULFUR METABOLISM. Some bacteria carry out respiratory electron transport without using oxygen as an electron acceptor. These forms use oxidized inorganic ions such as nitrate, nitrite, or sulfate as electron acceptors. Among these organisms are the **denitrifiers**, bacteria that return nitrogen to the atmosphere as nitrogen gas (N_2), completing the cycle of nitrogen in nature. These normally aerobic bacteria, mostly species of the genera *Bacillus* and *Pseudomonas*, use nitrate (NO_3^-) in place of oxygen if they are kept under anaerobic conditions:

$$2\ NO_3^- + 10\ e^- + 12\ H^+ \rightarrow N_2 + 6\ H_2O$$

Nitrogen fixers convert atmospheric nitrogen gas into chemical forms usable by the nitrogen fixers themselves and by other living things. For example,

$$N_2 + 6\ H \rightarrow 2\ NH_3 \text{ (ammonia)}$$

All organisms require nitrogen for their proteins, nucleic acids, and other important compounds. The vital process of nitrogen fixation is carried out by a wide variety of bacteria, including cyanobacteria, but by no other organisms. We'll discuss this process in detail in Chapter 34.

Ammonia is oxidized to nitrate by the process of **nitrification**. This process is carried out in the soil by bacteria called **nitrifiers**, mentioned earlier in our discussion of chemoautotrophs. Bacteria of two genera, *Nitrosomonas* and *Nitrosococcus*, convert ammonia to nitrite ions (NO_2^-), and *Nitrobacter* oxidizes nitrite to nitrate (NO_3^-). What do the nitrifiers get out of these reactions? These three genera are chemoautotrophs. Their chemosynthesis is powered by the energy released by oxidation of ammonia or nitrite.

For example, by passing the electrons from nitrite through an electron transport chain, *Nitrobacter* can make ATP, and using some of this ATP, it can also make NADH. With the ATP and NADH, the bacterium can convert CO_2 and H_2O to glucose and other foods. The nitrifiers base their entire biochemistry—their entire lives—on the oxidation of ammonia or NO_2^- ions. *Nitrobacter* can convert 6 molecules of CO_2 to 1 molecule of glucose for every 78 NO_2^- ions that they oxidize—not terribly efficient, but efficient enough to keep themselves living, growing, and reproducing.

What do plants get from the activities of nitrifiers? Under normal soil conditions, ammonium ions (NH_4^+) form readily from ammonia (NH_3) and bind to clay particles, which carry negative charges, in the soil. This binding makes NH_4^+ difficult for plants to take up. Nitrifiers have NH_4^+ transport proteins with a greater affinity than those of plants; they take up the NH_4^+ and oxidize it to NO_3^-, which, with its negative charge, doesn't bind to clay particles and hence is readily available to plants.

Numerous bacteria base their metabolism on the modification of sulfur-containing ions and compounds in their environment. As examples, we have already mentioned the photoautotrophic bacteria and chemoautotrophic archaea that use H_2S as an electron donor in place of H_2O. Such uses of nitrogen and sulfur have obvious environmental implications, as we'll see in the next section.

Prokaryotes in Their Environments

Prokaryotes live in and exploit all sorts of environments and are parts of many ecosystems. In the following pages, we'll examine prokaryotes in soils and water, and even in other living things, where they may exist in neutral, benevolent, or parasitic relationship with the host's tissues.

Prokaryotes are important players in soils and bodies of water

Animals depend on photosynthetic plants for their food, directly or indirectly. But plants depend on other organisms—prokaryotes—for their own nutrition. Without nitrogen fixers, no other life could exist, be-

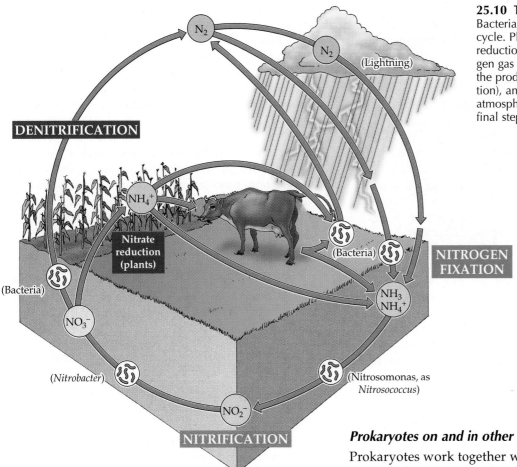

25.10 The Nitrogen Cycle
Bacteria carry out key steps in this cycle. Plants provide the nitrate reduction steps. Bacteria trap nitrogen gas (nitrogen fixation), convert the product to nitrate ions (nitrification), and return nitrogen gas to the atmosphere (denitrification) in the final step.

cause no other organisms can fix the nitrogen gas that surrounds us. Nitrifiers are also crucial to the biosphere, because plants often prefer nitrate ions (the end products of nitrification) over the products of nitrogen fixation for their own use. Nitrifiers produce the form of nitrogen most used by plants, which is the source of nitrogen compounds for animals and fungi.

If nitrogen fixers convert nitrogen gas into compounds that are used by the entire living world, why doesn't the atmosphere run out of N_2? Denitrifiers play a key role in keeping the nitrogen cycle going (Figure 25.10). Without denitrifiers, which convert nitrate ions back into nitrogen gas, all forms of nitrogen would leach from the soil and end up in lakes and oceans, making life on land impossible. Other prokaryotes crank a similar cycle for sulfur.

In the ancient past, the cyanobacteria had an equally dramatic impact: Their photosynthesis converted the global environment from anaerobic to aerobic. The result was the wholesale loss of species that couldn't tolerate the O_2 generated by the cyanobacteria, but this transformation made possible the evolution of cellular respiration and the subsequent explosion of eukaryotic life.

Prokaryotes on and in other organisms

Prokaryotes work together with eukaryotes in many ways. In fact, mitochondria and chloroplasts are descended from what were once free-living prokaryotes. Much later in evolutionary history, some plants learned to form associations with bacteria of the genus *Rhizobium* to form cooperative nitrogen-fixing nodules on their roots (see Chapter 34).

Many animals, including humans, harbor a variety of bacteria and archaea in their digestive tracts. Cows depend on prokaryotes in their complicated stomachs to perform important steps in digestion. Cows cannot produce cellulase, the enzyme needed to start the digestion of the cellulose that makes up the bulk of their food, but some of their stomach bacteria produce cellulase in sufficient quantity to process the cow's daily diet. We use some of the metabolic products—especially vitamins B_{12} and K—provided by the bacteria in our large intestine.

We are heavily populated, inside and out, by bacteria. Although very few of them are agents of disease, popular notions of bacteria as "germs" arouse our curiosity about those few, so we will briefly consider some bacterial pathogens.

A small minority of bacteria are pathogens

The late nineteenth century was a productive era in the history of medicine—a time during which bacteriologists, chemists, and physicians proved that many diseases are caused by microbial agents. During this time the German physician Robert Koch laid down a set of rules for testing the relationship between a disease and a microorganism.

According to Koch, the disease in question could be attributed to a particular microorganism if (1) the microorganism could always be found in diseased individuals; (2) the microorganism taken from the host could be grown in pure culture; (3) a sample of the culture produced the disease when injected into a new, healthy host; and (4) the newly infected host yielded a new, pure culture of microorganisms identical to that obtained in step 2. These rules—called **Koch's postulates**—were very important in a time when it was not widely accepted that microorganisms cause disease. Today, because of the availability of other diagnostic tools, medical science accepts less rigorous proofs in investigating new diseases.

Only a tiny percentage of all prokaryotes are pathogens (disease-producing organisms), and those that are are all bacteria. For an organism to be a successful pathogen, it must overcome several hurdles. It must arrive at the body surface of a potential host, enter the host, multiply inside the host, and, finally, prepare to infect the next host. Failure to overcome any of these hurdles ends the reproductive career of a pathogenic organism, and potential hosts have many defenses against pathogens. In Chapter 18 we considered the immune response and other modes of protection against diseases of microbial origin. But some bacteria *are* pathogenic, and they often succeed in infecting a host.

For the host, the consequences of an infection depend on several factors. One is the **invasiveness** of the pathogen—its ability to multiply within the body of the host. Another is its **toxigenicity**—its ability to produce chemical substances (toxins) harmful to the tissues of the host. *Corynebacterium diphtheriae*, the agent that causes diphtheria, has low invasiveness and multiplies only in the throat, but its toxigenicity is so great that the entire body is affected. By contrast, *Bacillus anthracis*, which causes anthrax (a disease primarily of cattle and sheep), has low toxigenicity but an invasiveness so great that the entire bloodstream ultimately teems with the bacteria.

There are two general types of bacterial toxins: exotoxins and endotoxins. **Endotoxins** are released when certain gram-negative bacteria lyse. They are lipopolysaccharides that form part of the outer membrane. Endotoxins are rarely fatal; they normally cause fever, vomiting, and diarrhea. Among the endotoxin producers are some species of *Salmonella* and *Escherichia*.

Exotoxins are proteins released by living, multiplying bacteria that may travel throughout the host's body. They are highly toxic—often fatal—but do not produce fevers. Many pathogenic bacteria produce exotoxins. Some examples of exotoxin-induced diseases are tetanus (from *Clostridium tetani*), botulism (from *Clostridium botulinum*), food poisoning (from *Bacillus cereus*), cholera (from *Vibrio cholerae*) and plague (from *Yersinia pestis*).

Remember that in spite of our frequent mention of human pathogens, only a small minority of the known prokaryotic species are pathogenic. Many more species play positive roles in our lives and in the biosphere. We make direct use of many bacteria and a few archaea in such diverse applications as cheese production, sewage treatment, and the industrial production of an amazing variety of antibiotics, vitamins, organic solvents, and other chemicals.

Prokaryote Phylogeny and Diversity

The prokaryotes comprise a diverse array of microscopic organisms. To explore this diversity, let's first consider how they are classified, and with what difficulty; then we'll look at some specific examples.

Nucleotide sequences of prokaryotes reveal evolutionary relationships

There are three primary motivations for classification schemes: to help identify unknown organisms, to reveal evolutionary relationships, and to provide universal names. Many scientists and medical technologists must be able to identify bacteria quickly and accurately. Lives depend on it.

Until recently taxonomists based their classification schemes for the prokaryotes on readily observable phenotypic characters such as color, motility, nutritional requirements, antibiotic sensitivity, and reaction to the Gram stain. Although such schemes have facilitated the identification of prokaryotes, they have failed miserably at giving insights into how these organisms evolved—a problem of great interest to microbiologists and to all students of evolution. The prokaryotes and protists (see Chapter 26) have long been major challenges to those who would classify them in a natural, evolution-based way. Only recently have systematists had the right tools for tackling this task.

Analyses of ribosomal RNAs now provide us with apparently reliable measures of evolutionary distance between taxonomic groups such as domains or genera. Ribosomal RNA (rRNA) is particularly useful for evolutionary studies of living organisms because (1) rRNA is evolutionarily ancient, (2) no living organism lacks rRNA, (3) rRNA plays the same role in translation in all organisms, and (4) rRNA itself has evolved slowly enough that sequence similarities between groups of organisms are easily found. (Most recently, biologists have been sequencing rRNA genes (DNA) rather than the rRNA itself, using the techniques described in Chapter 16.)

Comparisons of rRNAs from a great many sources showed that there are recognizable short base sequences characteristic of particular groups of organisms. These **signature sequences**, approximately 6 to 14 bases long, appear at the same approximate positions in rRNAs from related groups. For example, the signature sequence AAACUUAAAG occurs about 910 bases from one end of the RNA of the light subunit of ribosomes in 100 percent of the Archaea and Eukarya tested, but in *none* of the Bacteria tested. The signature sequence AAACUCAAA appears at the same position in all Bacteria but not in the Archaea or Eukarya. Several signature sequences distinguish each of the three domains. Similarly, each phylum of the kingdoms Archaebacteria and Eubacteria possesses a unique signature sequence.

These data sound promising, but as we will see, things aren't as easy as might seem possible. Although the evolutionary patterns revealed by signature sequences and other molecular tools are reliable, the groupings thus revealed are amazingly complex. A single phylum of bacteria or archaea may contain the most extraordinarily diverse species, and a species in one phylum may be phenotypically almost indistinguishable from one or many species in another phylum.

In the next section we'll explore some of the reasons for this spectacular diversity.

Mutations are the most important source of prokaryotic variation

Asexual reproduction, universal in prokaryotes, promotes genetic uniformity. Although prokaryotes can acquire different alleles by transformation, transduction, or conjugation (see Chapters 11 and 13), the most important source of genetic variation in populations of prokaryotes is mutation.

Mutations, especially recessive mutations, are slow to make their presence felt in populations of humans and other diploid organisms. In contrast, a mutation in a prokaryote, which is haploid, has immediate consequences for that organism, and if not lethal it will be transmitted to and expressed in the organism's daughter cells—and in their daughter cells, and so forth. A mutant allele spreads rapidly, if it is beneficial and favored by natural selection.

The rapid multiplication of many prokaryotes, coupled with mutation and selection, allows rapid changes in phenotype in a population. Important changes, such as loss of sensitivity to an antibiotic, can occur over broad geographic areas in just a few years. Think how many significant metabolic changes can occur over even modest time spans in relation to the history of life on Earth. When we introduce the largest phylum of bacteria, we will show how different groups within the phylum have easily and rapidly adopted and abandoned metabolic pathways under selective pressure from the environment.

The Bacteria

The great majority of prokaryotes are bacteria. They can at last be classified in a natural, phylogenetic way into 12 phyla, thanks to molecular tools such as rRNA sequencing that reveal evolutionary relationships. However, evolutionary relationships do not correspond closely with many of the important phenotypic traits of bacteria. The reason is that gene transfer has led to the abrupt appearance of certain traits (such as photosynthesis) multiple times during bacterial evolution, while other traits have been lost among the descendants of ancestors that had them. Therefore, the metabolic and ecological characteristics of bacteria, which are often the most important traits of the species, are shared among species that are only distantly related.

The rapid gain and loss of traits, although it makes classifying bacteria difficult, is in large part responsible for the success of these organisms. In the following discussion, rather than following evolutionary relationships, we will organize our analysis around some important functional traits of bacteria, such as their nutrition, locomotion, and ability to exploit habitats.

Metabolism in the Proteobacteria has evolved dramatically

By far the largest phylum of bacteria, in terms of number of species, is the phylum Proteobacteria, sometimes referred to as the Purple Bacteria. The classification problem begins right there—among the proteobacteria are many species of purple bacteria (gram-negative, bacteriochlorophyll-containing, sulfur-using photoautotrophs), but this phylum also includes a dramatically diverse group of bacteria that bear no resemblance to the purple bacteria in phenotype. It is this bacterial phylum to which the mitochondria of eukaryotes are most closely related.

No characteristic demonstrates the diversity of the proteobacteria more clearly than mode of nutrition (Figure 25.11). The common ancestor of all the proteobacteria was a photoautotroph. Early in evolution, two groups of proteobacteria lost their ability to photosynthesize and have been chemoheterotrophs ever since. The other three groups still have photoautotrophic members, but in *each* group, some evolutionary lines have abandoned photoautotrophy and taken up other modes of nutrition. There are chemoautotrophs and chemoheterotrophs in all of them. Why?

We can view each of the trends in Figure 25.11 as an evolutionary response to selective pressures encountered as these bacteria encountered new habitats that presented new challenges and opportunities. Much of the diversity of bacteria is metabolic, as illustrated here. But as we are about to see, there are other interesting differences.

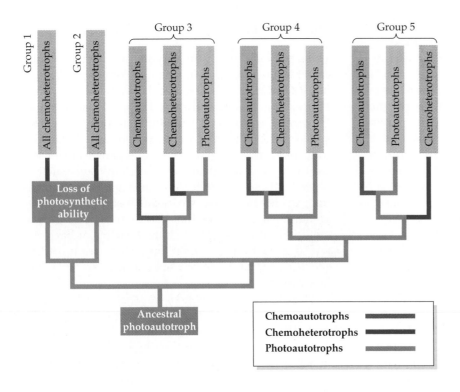

Group 1 — All chemoheterotrophs

Group 2 — All chemoheterotrophs

Group 3 — Chemoautotrophs / Chemoheterotrophs / Photoautotrophs

Group 4 — Chemoautotrophs / Chemoheterotrophs / Photoautotrophs

Group 5 — Chemoautotrophs / Photoautotrophs / Chemoheterotrophs

Loss of photosynthetic ability

Ancestral photoautotroph

Chemoautotrophs
Chemoheterotrophs
Photoautotrophs

25.11 The Evolution of Metabolism in the Proteobacteria Different groups of proteobacteria adopted different patterns of nutrition as they dealt with new environments.

Regardless of classification, the bacteria are highly diverse

In addition to the large phylum Proteobacteria, the domain Bacteria (kingdom Eubacteria) includes eleven other phyla. But the discussion that follows considers nine groups of bacteria defined by *shared phenotypes* rather than strictly by phylogenetic relationships. In most cases, these groups do not correspond neatly to phyla.

THERMOPHILES. The three phyla branching out earliest during bacterial evolution are all heat lovers, as are the most ancient of the archaea. This observation supports the hypothesis that the first organisms were thermophiles that appeared in an environment much hotter than the ones that predominate today.

CYANOBACTERIA. The cyanobacteria (blue-green bacteria) are very independent nutritionally. They appeared in a time of intense competition among heterotrophic prokaryotes for resources, and they prospered on the basis of their autotrophy. Cyanobacteria photosynthesize by using chlorophyll *a* and liberating oxygen gas, and many species also fix nitrogen. They require only water, nitrogen gas, oxygen, a few mineral elements, light, and carbon dioxide.

Cyanobacteria carry out the type of photosynthesis otherwise characteristic only of eukaryotic photosynthesizers. They contain elaborate and highly organized internal membrane systems: the photosynthetic lamellae, or *thylakoids* (Figure 25.12; see also Figure 4.4*a*).

The chloroplasts of photosynthetic eukaryotes are more closely related to the cyanobacteria than to any other group of bacteria.

Cyanobacteria associate in colonies or live free as single cells. Depending on the species and on growth conditions, colonies of cyanobacteria range from flat sheets one cell thick to spherical balls of cells. Some filamentous colonies differentiate into three cell types: vegetative cells, spores, and heterocysts.

Heterocysts are cells specialized for nitrogen fixation; the enzyme nitrogenase gives them this ability. All the known cyanobacteria with heterocysts fix nitrogen. Heterocysts also have a role in reproduction: When filaments break apart to reproduce, the heterocyst may serve as a breaking point. Figure 25.13*a* and *b* shows heterocysts within filaments.

GLIDING BACTERIA. The gliding bacteria are filaments or rods whose movement resembles gliding. Members of the genus *Beggiatoa* (Figure 25.14*a*) are an example. *Beggiatoa* move up and down in the soil, lying deeper during the day. They are aerobes that need O_2 to

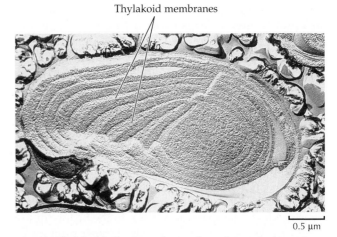

Thylakoid membranes

0.5 μm

25.12 Thylakoids in Cyanobacteria This cyanobacterial cell was prepared by freeze-etching (see Chapter 5) to emphasize the extensive system of internal membranes. These photosynthetic membranes are present through most of the cytoplasm.

Heterocyst   Resting spore

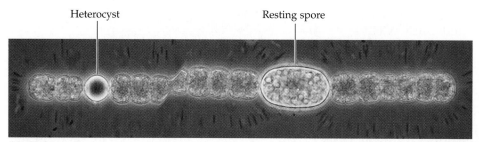

(a) *Anabaena* sp.

2 μm

A thick wall separates the cytoplasm of the nitrogen-fixing heterocyst from the surrounding environment.

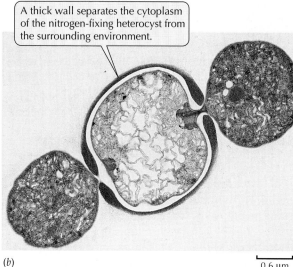

(b)   0.6 μm

(c)

25.13 Cyanobacteria (a) *Anabaena* is a genus of colonial, filamentous cyanobacteria; this colony displays both a heterocyst and a resting spore. (b) A thin neck attaches a heterocyst to each of two other cells in a colony. (c) Cyanobacteria appear in enormous numbers in some environments. This California pond has experienced eutrophication: Phosphorus and other nutrients generated by human activity have accumulated in the pond, feeding an immense green mat—commonly referred to as "pond scum"—made up of several species of unicellular cyanobacteria.

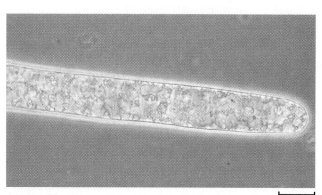

(a) *Beggiatoa* sp.   2 μm

25.14 Gliding Bacteria (a) This filamentous cell of a gliding bacterium was isolated from ocean mud. (b) Individual bacteria aggregated to make up the stalk and knobs of this fruiting body.

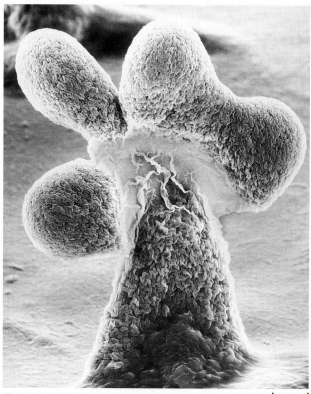

(b) *Stigmatella aurantiaca*   13 μm

metabolize, but they obtain their energy by oxidizing reduced sulfur-containing compounds such as H_2S. These sulfur compounds are produced deep in the soil, away from atmospheric O_2. Because O_2 itself oxidizes the sulfur compounds, *Beggiatoa* must grow in a narrow zone—a zone of compromise between being deep enough to have a supply of sulfur compounds but shallow enough to obtain enough O_2. Cyanobacteria also contribute to the O_2 supply, by photosynthesizing—which they do only during the day, allowing *Beggiatoa* to function deeper in the soil. At night, *Beggiatoa* must glide toward the surface to get enough O_2.

Some gliding bacteria form remarkable structures called fruiting bodies. A group of cells aggregates to make one of these structures (Figure 25.14*b*). The fruiting bodies of some species are simple globes more than 1 mm in diameter; those of other species are more complex, branched structures. Within the fruiting bodies of some species, single cells develop into thick-walled spores that can resist harsh environmental conditions; in other fruiting bodies, whole clusters of cells form a cyst that is resistant to drying. Both spores and cysts can germinate under favorable conditions to yield the next generation of typical gliding cells.

Beggiatoa and the gliding bacteria that form fruiting bodies belong to different groups of the phylum Proteobacteria; some other gliding bacteria belong to a different bacterial phylum. Gliding motion is not the signature of a single phylogenetic group.

SPIROCHETES. Spirochetes are characterized by unique structures called **axial filaments**, composed of flagella, running along the cell body between a thin, flexible cell wall and an outer envelope (see Figure 25.5*c*). In other respects the spirochete cell is similar to that of other gram-negative bacteria. The cell body is a long cylinder coiled into a spiral (Figure 25.15). The flagella constituting the axial filaments begin at either end of the cell and overlap in the middle. The axial filaments are responsible for the motility of these organisms,

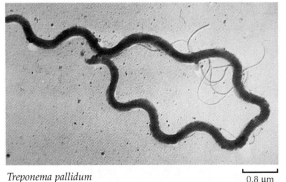

Treponema pallidum   0.8 μm

25.15 A Spirochete This corkscrew-shaped spirochete causes syphilis in humans.

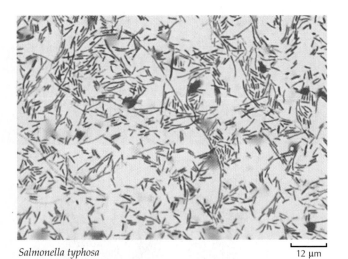

Salmonella typhosa   12 μm

25.16 Gram-Negative Rods The cause of typhoid fever is this gram-negative rod. Recall from Figure 25.7 that the pink color is the "negative" response to the Gram stain.

and there are typical basal rings where the flagella are attached to the cell wall. Many spirochetes live in humans as parasites. Others live free in mud or water.

GRAM-NEGATIVE RODS. Some gram-negative rods are aerobic, others are facultative anaerobes, and still others are obligate anaerobes. Some nitrogen-fixing genera such as *Rhizobium* (see Figure 34.10) are gram-negative rods, as are *Nitrobacter*, *Thiobacterium*, and other bacteria that use nitrogen or sulfur compounds instead of oxygen for respiration.

E. coli, the most studied organism, is a gram-negative rod. So, too, are many of the most famous human pathogens, such as *Yersinia pestis* (the cause of plague), *Shigella dysenteriae* (dysentery), *Vibrio cholerae* (cholera), and *Salmonella typhimurium* (a common agent of food poisoning in humans). A bacterium from this group is shown in Figure 25.16.

Certain gram-negative rods invade animal cells, where they survive and cause diseases. For example, *Yersinia pseudotuberculosis*, the agent of guinea pig plague, invades intestinal cells of guinea pigs and other mammals. Its ability to invade mammalian cells results from the possession of a single gene, called *inv*, that codes for a single large protein. Biologists using recombinant-DNA techniques have successfully transferred this gene from *Y. pseudotuberculosis* to *E. coli*, with the result that the recipient *E. coli* cells were able to invade mammalian cells, revealing the role of *inv*. Of course, these studies were carried out with great caution because of potential health hazards associated with such modified *E. coli* cells.

Most plant diseases are caused by fungi, and viruses cause others, but about 200 plant diseases are of bacterial origin. **Crown gall**, with its characteristic

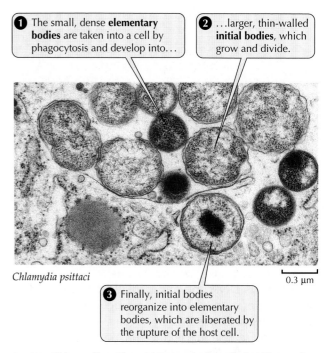

1 The small, dense **elementary bodies** are taken into a cell by phagocytosis and develop into...

2 ...larger, thin-walled **initial bodies**, which grow and divide.

Chlamydia psittaci

0.3 μm

3 Finally, initial bodies reorganize into elementary bodies, which are liberated by the rupture of the host cell.

25.18 Chlamydias Change Form during Their Life Cycle Elementary bodies and initial bodies are the two major phases of the life cycle of a chlamydia.

25.17 Crown Gall This massive growth on the trunk of a white oak tree is crown gall, a plant disease caused by the gram-negative rod *Agrobacterium tumefaciens*.

tumors (Figure 25.17), is one of the most striking. The causal agent of crown gall is *Agrobacterium tumefaciens*, a gram-negative rod. *A. tumefaciens* harbors a plasmid containing the genes responsible for the crown gall disease. The plasmid is used in recombinant-DNA studies as a vehicle for inserting genes into new plant hosts, where they multiply along with the crown gall tumor cells (see Chapter 16).

CHLAMYDIAS. Chlamydias are among the smallest bacteria (0.2 to 1.5 μm in diameter). They can live only as parasites within the cells of other organisms. These tiny spheres are unique prokaryotes because of their

complex reproductive cycle, which involves two different types of cells (Figure 25.18). In humans, various strains of chlamydias cause eye infections (especially trachoma), sexually transmitted disease, and some forms of pneumonia.

GRAM-POSITIVE BACTERIA. Some gram-positive bacteria produce **endospores** (Figure 25.19)—heat-resistant resting structures. When nutrients become scarce, the bacterium produces an endospore. The bacterium

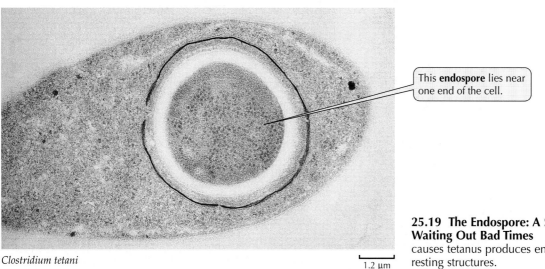

This **endospore** lies near one end of the cell.

Clostridium tetani

1.2 μm

25.19 The Endospore: A Structure for Waiting Out Bad Times The bacterium that causes tetanus produces endospores as resistant resting structures.

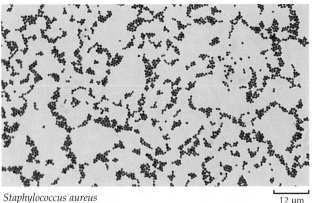

Staphylococcus aureus

12 µm

25.20 Gram-Positive Cocci   "Grape clusters" are the usual arrangement of gram-positive staphylococci.

replicates its DNA and encapsulates one copy, along with some of its cytoplasm, in a tough cell wall heavily thickened with peptidoglycan and surrounded by a spore coat. The parent cell then breaks down, releasing the endospore. *This is not a reproductive process*; the endospore merely replaces the parent cell. The endospore can survive harsh environmental conditions, such as high or low temperatures or dryness, because it is dormant—its normal activity is suspended. Later, if it encounters favorable conditions, the endospore germinates; that is, it becomes metabolically active and divides, forming new cells like the parent. Some endospores apparently can germinate even after more than a thousand years of dormancy.

Members of this endospore-forming group include the many species of *Bacillus* and *Clostridium*. The toxins produced by *C. botulinum* are among the most poisonous ever discovered; the lethal dose for humans is about one-millionth of a gram (1 µg).

There are many gram-positive cocci. Cells of the genus *Staphylococcus* (Figure 25.20), called staphylococci, are abundant on the human body surface and are responsible for boils and many other skin problems. *S. aureus* is the best-known human pathogen; it is found in 20 to 40 percent of normal adults (and in 50 to 70 percent of hospitalized adults) and can cause respiratory, intestinal, and wound infections, in addition to skin diseases. Staphylococci produce toxins that are a major cause of food poisoning and that cause toxic shock syndrome.

ACTINOMYCETES.   Actinomycetes develop an elaborately branched system of filaments (Figure 25.21). These bacteria closely resemble the filamentous bodies of fungi and, in fact, were once classified as fungi.

Some actinomycetes reproduce by forming chains of spores at the tips of the filaments. In the species that do not form spores, the branched, filamentous growth ceases and the structure breaks up into typical cocci or rods, which then reproduce by fission. The actino-

mycetes are part of the phylum called Gram-Positive Bacteria.

The actinomycetes include several medically important members: *Actinomyces israelii* causes infections in the oral cavity and elsewhere; *Mycobacterium tuberculosis* causes tuberculosis; and *Streptomyces* produces streptomycin, as well as hundreds of other antibiotics, including several dozen in general use. We derive most of our antibiotics from members of the actinomycetes. Why do these bacteria dedicate hundreds of genes to the production of antibiotics? We don't know for sure, but we do know that actinomycetes produce the antibiotics at the same time they produce their spores. Thus, the antibiotics may inhibit the growth of bacteria and fungi that would otherwise compete for nutrients with the germinating spores.

MYCOPLASMAS.   Mycoplasmas lack cell walls, although some have a stiffening material outside their plasma membrane. Some of them are the smallest cellular creatures ever discovered—they are even smaller than chlamydias (Figure 25.22). The smallest mycoplasmas capable of growth have a diameter of about 0.2 µm, and they are small in another crucial sense: They have less than half as much DNA as do most other prokaryotes. It has been speculated that the amount of DNA in a mycoplasma, which is about the same as that in *Thermoplasma* (an archaean discussed in the next section), may be the minimum amount required to code for the essential properties of a cell.

The mycoplasmas are classified as members of the Gram-Positive Bacteria—even though the mycoplasmas lack peptidoglycan-containing cell walls. This apparent irregularity of classification resulted from the fact that the mycoplasmas are most closely allied with *Clostridium* and its relatives, all of which are Gram-Positive Bacteria. Remove the cell wall of a *Clostridium*

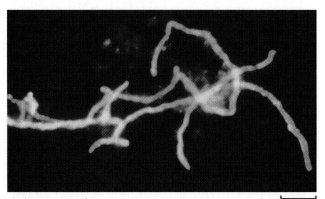

Actinomyces israelii

10 µm

25.21 Filaments of an Actinomycete   Branching filaments are visualized with a fluorescent stain. This species is part of the normal flora in the human tonsils, mouth, intestinal tract, and lungs but will invade body tissues and cause severe abscesses when afforded the opportunity.

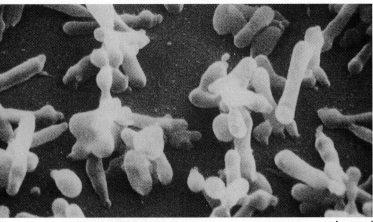

Mycoplasma gallisepticum

0.4 μm

25.22 The Tiniest Living Cells Containing only about one-fifth as much DNA as *E. coli*, mycoplasmas are the smallest known bacteria.

and you have a mycoplasma—but the mycoplasmas never form a wall.

In the bodies of animals, mycoplasmas live on the surfaces of mucous membranes. Lacking cell walls, they cannot be killed with penicillin, which kills other bacteria by interfering with wall synthesis. Diseases caused by mycoplasmas include urinary tract infections and some forms of pneumonia; they must be treated with antibiotics other than those that act on wall synthesis.

The Archaea

The domain Archaea consists of a few prokaryotic genera that live in habitats notable for characteristics such as extreme salinity (salt content), low oxygen concentration, high temperature, or high or low pH. On the face of it, the archaea do not seem to belong together as a group. However, they do share certain characteristics.

The Archaea share some unique characteristics

Two characteristics shared by all archaea are a definitive lack of peptidoglycan in their walls and the possession of lipids of distinctive composition (see Table 25.1). The base sequences of their ribosomal RNAs confirm the close relationship. Their separation from the Bacteria and Eukarya was clarified when biologists sequenced the first archaean genome; it consisted of 1,738 genes, *more than half of which* were unlike any genes ever found in the other two domains.

The unusual lipids in the membranes of archaea deserve some discussion. They are found in all archaea, and in no bacteria or eukaryotes. Most membrane lipids of bacteria and eukaryotes contain long-chain fatty acids connected to glycerol by **ester linkages**:

$$O$$
$$—C—O—C—$$

(see also Figure 3.4). The fatty acids are straight and unbranched, and the lipids form a bilayer in the membranes of bacteria and eukaryotes.

The most distinctive feature of archaean membrane lipids is that they contain long-chain hydrocarbons connected to glycerol by **ether linkages**:

$$—C—O—C—$$

In addition, their long-chain hydrocarbons are branched. One class of these lipids contains glycerol at *both* ends of the hydrocarbons. This structure still fits in a biological membrane, as shown in Figure 25.23. In spite of the striking difference in membrane lipids, all three domains have membranes with similar overall structure, dimensions, and functions.

Archaea live in amazing places

Among the homes of archaea are extremely hot environments (sometimes also very acidic), extremely salty environments, and the guts of animals. The domain Archaea (thus the kingdom Archaebacteria) can be divided into four phyla: Hyperthermophiles, Methanogens, Extreme Halophiles, and one phylum consisting entirely of the genus *Thermoplasma*.

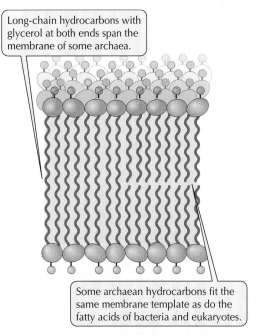

Long-chain hydrocarbons with glycerol at both ends span the membrane of some archaea.

Some archaean hydrocarbons fit the same membrane template as do the fatty acids of bacteria and eukaryotes.

25.23 Membrane Architecture in Archaea The long-chain hydrocarbons of archaean membranes are branched and may contain glycerol at both ends. This structure still fits into a biological membrane, however; in fact, all three domains have similar membranes.

25.24 Some Would Call It Hell; These Archaea Call It Home Masses of heat- and acid-loving archaea form a salmon-pink mat inside a volcanic vent on the island of Kyushu, Japan. Sulfurous residue is visible at the edges of the archaean mat.

HYPERTHERMOPHILES. Some archaea are both thermophilic (heat-loving) and acidophilic (acid-loving). *Sulfolobus* is one hyperthermophilic genus. Organisms of this genus live in hot sulfur springs at temperatures of 70 to 75°C. They die of "cold" at 55°C (131°F). Hot sulfur springs are also extremely acidic. *Sulfolobus* grows best in the pH range from 2 to 3, but it readily tolerates pH values as low as 0.9. Acidophilic hyperthermophiles that have been tested maintain an internal pH near 7 (neutral) in spite of the acidity of their environment (see Chapter 2 for a discussion of pH). These and other hyperthermophiles thus thrive where very few other organisms can even survive (Figure 25.24).

METHANOGENS. Some species of prokaryotes, previously assigned to unrelated bacterial groups, share the property of producing methane (CH_4) by reducing carbon dioxide. All these methanogens are obligate anaerobes, and methane production is the key step in their energy metabolism. Comparison of rRNA base sequences revealed the close evolutionary relationship among all methanogens.

Methanogens release approximately 2 billion tons of methane gas into Earth's atmosphere each year, accounting for all the methane in our air, including that associated with mammalian flatulence (the passing of gas). Approximately a third of the methane production comes from methanogens in the guts of grazing herbivores such as cows.

One methanogen, *Methanopyrus*, lives on the ocean bottom near blazing volcanic vents. *Methanopyrus* can survive and grow at 110°C; it is the current record holder for temperature tolerance. It grows best at 98°C and not at all at temperatures below 84°C.

EXTREME HALOPHILES. The extreme halophiles live exclusively in very salty environments. Because they contain pink carotenoids, they can be seen easily under some circumstances (Figure 25.25). Halophiles grow in the Dead Sea and in brines of all types: Pickled fish may sometimes show reddish pink spots that are colonies of halophilic archaea. Photographs of salt flats taken from orbiting satellites show a distinct pink tinge resulting from the presence of vast numbers of *Halobacterium* and its halophilic relatives.

Few other organisms can live in the saltiest of the homes that the strict halophiles occupy; most would "dry" to death, losing too much water by osmosis to the hyperosmotic (more concentrated) environment.

25.25 Extreme Halophiles Commercial seawater evaporating ponds such as these in San Francisco Bay are attractive homes for salt-loving archaea.

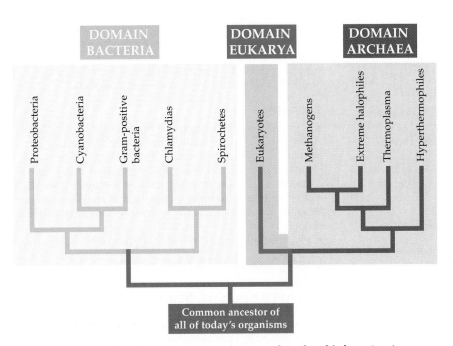

25.26 A Look Back at Two Domains and Forward to the Third A brief summary classification of the domains Bacteria and Archaea; the many groups of Eukarya will be discussed in the next five chapters.

Strict halophiles have been found in lakes with pH values as high as 11.5—the most alkaline environment inhabited by living organisms, almost as alkaline as household ammonia.

Some of the extreme halophiles have a unique system for trapping light energy and using the energy to form ATP—without using any form of chlorophyll—but only when oxygen is in short supply. They use the pigment retinal, also found in the vertebrate eye, combined with a protein to form **bacteriorhodopsin**, which is incorporated into the plasma membrane. Retinal is purple. Clusters of bacteriorhodopsin molecules form purple patches that cover as much as half of the cell surface. When one of these molecules absorbs light, protons are pumped through the membrane out of the cell. ATP forms by a chemiosmotic mechanism of the sort described in Chapters 7 and 8.

THERMOPLASMA. The fourth archaean phylum consists of a single genus, *Thermoplasma*. This prokaryote has no cell wall, it is thermophilic and acidophilic, its metabolism is aerobic, and it comes from coal deposits. It has the smallest genome among the archaea and perhaps the smallest genome (along with the mycoplasmas) of any free-living organisms—1,100 kilobase pairs.

The relationships of the three domains of the living world are summarized in Figure 25.26, which emphasizes the domains Bacteria and Archaea. For the rest of Part Four (Chapters 26 to 30) we'll consider the other domain: the Eukarya.

Summary of "Bacteria and Archaea: The Prokaryotic Domains"

Why Three Domains?

• Both the Archaea and the Bacteria are prokaryotic, but they differ from each other more radically than do the Archaea from the Eukarya, which constitute the rest of the living world.
• The evolutionary relationships of the three domains were first revealed by rRNA sequences. The common ancestor of all three domains lived more than 2 billion years ago, and the common ancestor of the Archaea and Eukarya at least 1.8 billion years ago. **Review Figure 25.1 and Table 25.1**

General Biology of the Prokaryotes

• The prokaryotes are the most numerous organisms on Earth, and they occupy an enormous variety of habitats.
• Most prokaryotes are cocci, bacilli, or spirals. Some link together to form associations, but none are truly multicellular. **Review Figure 25.3**
• Prokaryotes lack nuclei, membrane-enclosed organelles, and cytoskeletons. Their chromosomes are circular. They often contain plasmids. They reproduce asexually by fission.
• Many prokaryotes are motile by means of flagella, gas vesicles, or gliding mechanisms. Prokaryotic flagella rotate rather than beat.
• Prokaryotic cell walls differ from those of eukaryotes. Bacterial walls generally contain peptidoglycan. Differences in peptidoglycan content result in different reactions to the Gram stain. **Review Figure 25.7**
• Different prokaryotes have diverse metabolic pathways and nutritional modes. Prokaryotes are obligate anaerobes, facultative anaerobes, or obligate aerobes. The major nutri-

tional types are photoautotrophs, photoheterotrophs, chemoautotrophs, and chemoheterotrophs. Some prokaryotes base their energy metabolism on nitrogen- or sulfur-containing ions. **Review Figure 25.8 and Table 25.2**

Prokaryotes in Their Environments

• Some prokaryotes play key roles in global nitrogen and sulfur cycles. Important players in the nitrogen cycle are the nitrogen fixers, nitrifiers, and denitrifiers. **Review Figure 25.10**
• Photosynthesis by cyanobacteria generated the O_2 that resulted in the aerobic environment that permitted the evolution of aerobic respiration, enabling the appearance of present-day eukaryotes.
• Many prokaryotes live in or on other organisms, with neutral, beneficial, or harmful effects.
• A small minority of bacteria are pathogens. Pathogens vary with respect to their invasiveness and toxigenicity. Some produce endotoxins, which are rarely fatal; others produce exotoxins, which tend to be highly toxic.

Prokaryote Phylogeny and Diversity

• Some classification schemes are designed to help us identify unknown organisms; others reflect evolutionary relationships. No scheme simultaneously achieves both of these goals for the prokaryotic domains.
• Phylogenetic ("natural") classification of prokaryotes is now based on rRNA sequences, with particular attention to signature sequences that are characteristic of individual groups.
• Evolution, powered by mutation and natural selection, can proceed rapidly in prokaryotes because they are haploid and can multiply rapidly.

The Bacteria

• There are far more bacteria than archaea. One phylogenetic classification of the domain Bacteria (kingdom Eubacteria) groups them into 12 phyla.
• Among bacteria, phenotypic characters often correlate poorly with phylogeny.
• All four nutritional types are observed in the Bacteria—and all four occur in the largest bacterial phylum, Proteobacteria. Metabolism in different groups of proteobacteria has evolved along different lines. **Review Figure 25.11**
• The most ancient bacteria, like the most ancient archaea, are thermophiles, suggesting that life originated in a hot environment.
• Cyanobacteria photosynthesize using the same pathways as plants use, in contrast with other bacteria. Many cyanobacteria fix nitrogen.
• Some gliding bacteria move up and down in the soil on a daily cycle. And some form fruiting bodies in response to unfavorable environmental conditions.
• Spirochetes move by means of axial filaments.
• There are many kinds of gram-negative rods, including *E. coli*; they differ dramatically from one another in metabolism and habitat.
• Chlamydias are tiny parasites that live within the cells of other organisms.
• Gram-positive bacteria are diverse; some of them produce endospores as resting structures that resist harsh conditions.
• Actinomycetes, some of which produce important antibiotics, grow as branching filaments.

• Mycoplasmas, the tiniest living things, lack conventional cell walls. They, like the archaean *Thermoplasma*, have very small genomes.
• Most of these phenotypically defined bacterial groups bear little relationship to phylogenetic groupings.

The Archaea

• Archaea have cell walls that differ from those of bacteria and eukaryotes. Their walls lack peptidoglycan, and their membrane lipids differ radically from those of bacteria and eukaryotes, containing branched long-chain hydrocarbons connected to glycerol by ether linkages. **Review Figure 25.23**
• The domain Archaea (kingdom Archaebacteria) is divided into four phyla: Hyperthermophiles, Methanogens, Extreme Halophiles, and *Thermoplasma*.
• Hyperthermophiles are heat-loving and often acid-loving archaea.
• Methanogens produce methane by reducing carbon dioxide. Some methanogens live in the guts of herbivorous animals; others occupy high-temperature environments on the ocean floor.
• Extreme halophiles are salt lovers that often lend a pinkish color to salty environments; some halophiles also grow in extremely alkaline environments.
• Archaea of the genus *Thermoplasma* lack cell walls, are thermophilic and acidophilic, and have a tiny genome (1,100 kilobase pairs).
• The three domains of life descended from a common ancestor. **Review Figure 25.26**

Self-Quiz

1. Most prokaryotes
   a. are agents of disease.
   b. lack ribosomes.
   c. evolved from the most ancient protists.
   d. lack a cell wall.
   e. are chemoheterotrophs.

2. The division of the living world into three domains
   a. is strictly arbitrary.
   b. was inspired by the morphological differences between archaea and bacteria.
   c. emphasizes the greater importance of eukaryotes.
   d. was proposed by the early microscopists.
   e. is strongly supported by data on rRNA sequences.

3. Which statement about the archaean genome is true?
   a. It is much more similar to the bacterial genome than to eukaryotic genomes.
   b. More than half of its genes are genes that are never observed in bacteria or eukaryotes.
   c. It is much smaller than the bacterial genome.
   d. It is housed in the nucleus.
   e. No archaean genome has yet been sequenced.

4. Which statement about nitrogen metabolism is *not* true?
   a. Certain prokaryotes reduce atmospheric N_2 to ammonia.
   b. Nitrifiers are soil bacteria.
   c. Denitrifiers are strict anaerobes.
   d. Nitrifiers obtain energy by oxidizing ammonia and nitrite.
   e. Without the denitrifiers, terrestrial organisms would lack a nitrogen supply.

5. All photosynthetic bacteria
   a. use chlorophyll *a* as their photosynthetic pigment.
   b. use bacteriochlorophyll as their photosynthetic pigment.
   c. release oxygen gas.
   d. produce particles of sulfur.
   e. are photoautotrophs.

6. Gram-negative bacteria
   a. appear blue to purple following Gram staining.
   b. are the most abundant of the bacterial groups.
   c. are all either rods or cocci.
   d. contain no peptidoglycan in their walls.
   e. are all photosynthetic.

7. Endospores
   a. are produced by viruses.
   b. are reproductive structures.
   c. are very delicate and easily killed.
   d. are resting structures.
   e. lack cell walls.

8. Actinomycetes
   a. are important producers of antibiotics.
   b. belong to the kingdom Fungi.
   c. are never pathogenic to humans.
   d. are gram-negative.
   e. are the smallest known bacteria.

9. Which statement about mycoplasmas is *not* true?
   a. They lack cell walls.
   b. They are the smallest known cellular organisms.
   c. They contain the same amount of DNA as do other prokaryotes.
   d. They cannot be killed with penicillin.
   e. Some are pathogens.

10. Archaea
   a. have cytoskeletons.
   b. have distinctive lipids in their plasma membranes.
   c. survive only at moderate temperatures and near neutrality.
   d. all produce methane.
   e. have substantial amounts of peptidoglycan in their cell walls.

Applying Concepts

1. Why do systematic biologists find rRNA sequence data more useful than data on metabolism or cell structure for classifying prokaryotes?

2. Differentiate among the members of the following sets of related terms:
   a. prokaryotic/eukaryotic
   b. obligate anaerobe/facultative anaerobe/obligate aerobe
   c. photoautotroph/photoheterotroph/chemoautotroph/chemoheterotroph
   d. gram-positive/gram-negative

3. For each type of organism listed below, give a single characteristic that may be used to differentiate it from the related organism(s) in parentheses.
   a. spirochetes (spiral bacteria)
   b. *Bacillus* (*Lactobacillus*)
   c. mycoplasmas (other bacteria)
   d. cyanobacteria (other photoautotrophic bacteria)

4. Until fairly recently, the cyanobacteria were called blue-green algae and were not grouped with the bacteria. Suggest several reasons for this (abandoned) tendency to separate the bacteria and cyanobacteria. Why are the cyanobacteria now grouped with the other bacteria?

5. Hyperthermophiles are of great interest to molecular biologists and biochemists. Why? What practical concerns might motivate that interest?

Readings

Balows, A., H. G. Trüper, M. Dworkin, W. Harder and K.-H. Schleifer (Eds.). 1992. *The Prokaryotes*, 2nd Edition. Four volumes. Springer-Verlag, New York. The ultimate reference on the prokaryotes, describing ecophysiology, isolation, identification, and applications.

Fischetti, V. A. 1991. "Streptococcal M Protein." *Scientific American*, June. A discussion of how rheumatic fever and strep throat bacteria evade the body's defenses.

Fredrickson, J. K. and T. C. Onstott. 1996. "Microbes Deep inside the Earth." *Scientific American*, October. An account of the discovery of bacteria and archaea by drilling. Probes the question, Are there microbes under the surface of Mars or one or more moons in the solar system?

Koch, A. L. 1990. "Growth and Form of the Bacterial Cell Wall." *American Scientist*, vol. 78, pages 327–341. An exploration into the question of how this cell wall, a single peptidoglycan molecule, allows the cell to grow but keeps it from bursting.

Losick, R. and D. Kaiser. 1997. "Why and How Bacteria Communicate." *Scientific American*, February. An examination of the important chemical means by which prokaryotes communicate with each other and with plants and animals.

Madigan, M. T. and B. L. Marrs. 1997. "Extremophiles." *Scientific American*, April. An interesting account of the archaea and their hardy enzymes.

Madigan, M. T., J. M. Martinko and J. Parker. 1997. *Brock Biology of Microorganisms*, 8th Edition. Prentice Hall, Upper Saddle River, NJ. An excellent general textbook, covering diversity, industrial microbiology, microbial ecology, and clinical and epidemiological topics.

McEvedy, C. 1988. "The Bubonic Plague." *Scientific American*, February. An account of how bubonic plague, which still exists, has shaped world history.

Shapiro, J. A. 1988. "Bacteria as Multicellular Organisms." *Scientific American*, June. A description of the behavior of highly regular bacterial colonies.

Woese, C. 1981. "Archaebacteria." *Scientific American*, June. An early treatment of the subject by its leading scholar.

Chapter 26

Protists and the Dawn of the Eukarya

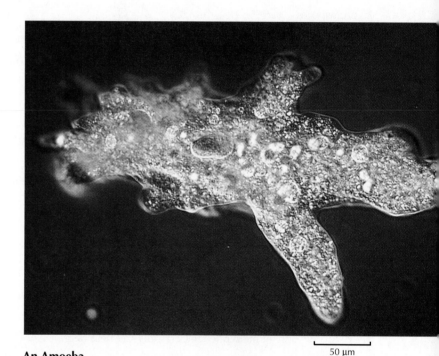

An Amoeba

Amoebas have a nucleus and several kinds of organelles; they are members of the domain Eukarya. Their flowing pseudopods are constantly changing shape as the amoeba moves and feeds.

50 μm

*A*fter their origin, prokaryotes had the living world to themselves for more than 1.5 billion years. Prokaryotes constitute two of the three domains of today's biosphere. As we saw in Chapter 25, the bacteria and the archaea differ sharply in several important ways—but neither looks much like the single-celled organism shown here. What strikes you the most about this amoeba? Probably the most obvious visible difference between it and the prokaryotes is that the amoeba has numerous compartments—membrane-enclosed organelles.

Amoebas are eukaryotic organisms: They have a nucleus enclosed by a nuclear envelope, they have several kinds of organelles, and they differ from members of the two prokaryotic domains in other important ways. They are members of the domain Eukarya. When eukaryotes appeared in the course of evolution, various members of the group experimented in many ways, resulting in a profusion of body forms. The evolution of the eukaryotes has produced great diversity, but also many cases of convergent evolution; for example, amoebalike organisms arose several times.

Look at the three eukaryotic organisms shown in Figure 26.1. Figures 26.1*a* and *b* show tiny unicellular organisms, both of which have a nucleus. The dinoflagellate is a swimming photosynthesizer. *Giardia* is a parasite that causes diarrhea and other symptoms in humans. Figure 26.1*c* shows a giant kelp—multicellular, photosynthetic, and very big, sometimes achieving lengths greater than that of a football field.

All three of these organisms belong to the Eukarya, but what kingdoms should we put them in? Is *Giardia* simply an unusually tiny animal? Are the photosynthetic organisms plants? Some biologists assign all three to a single

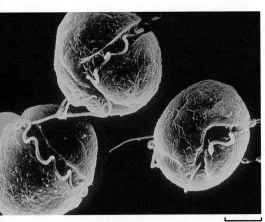

(a) *Gonyaulax* sp.   15 µm

26.1 Three Eukaryote Protists
(a) Dinoflagellates are photosynthetic unicellular algae. (b) *Giardia* is a unicellular parasite of humans. (c) Giant kelps are some of the world's longest organisms.

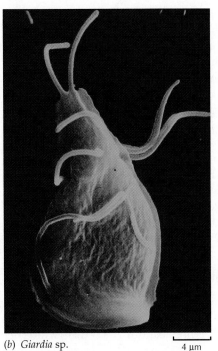

(b) *Giardia* sp.   4 µm   (c) *Macrocystis* sp.

kingdom, the Protista. Some others put them in three different kingdoms (Protista, Archezoa, and Chromista, respectively). Some consider the giant kelp (but not the microscopic dinoflagellate) a plant.

To keep the number of new terms to a minimum, and to simplify the presentation of the diversity in this fascinating group of "other eukaryotes," we will treat them as a single, paraphyletic kingdom, Protista. We hope this will be a convenient way of looking at these organisms. If you prefer to view them as representing more than one kingdom, it won't bother them or us.

The origin of the eukaryotic cell was one of the pivotal events in evolutionary history. In this chapter on the Protista we describe and celebrate the origin and early diversification of the eukaryotes and the complexity achieved in some single cells. Most of the organisms we call protists here are unicellular, but many are multicellular. This kingdom is a great evolutionary grab bag defined, for the purposes of this book, largely by exclusion: *the protists are all the eukaryotes that are not plants, fungi, or animals.* All protists are eukaryotic, and all evolved from prokaryotes.

We'll consider the origin of the eukaryotic cell, the shared characteristics of the protists, some of the diversity of protist body forms, and the relationships of certain protist groups to the other eukaryotic kingdoms.

Protista and the Other Eukaryotic Kingdoms

Part of the difficulty in placing certain eukaryotic organisms in the appropriate kingdom is a natural consequence of the fact that the other eukaryotic kingdoms have their evolutionary origin in the kingdom Protista. The other eukaryotic kingdoms—Plantae, Fungi, and Animalia—arose from protists in various ways.

Deciding just where to draw the lines between the Protista and the other eukaryotic kingdoms is difficult. Some protists, formerly classified as animals, are sometimes referred to as **protozoans**, although many biologists regard this term as inappropriate because it lumps together protist groups that are phylogenetically distant from one another. We will use the term "protozoan" for convenience in reference to certain protists, most of which ingest food by endocytosis. There are several kinds of photosynthetic protists that some biologists still refer to as **algae**. There are also some protists—the slime molds and water molds—that were once classified as fungi, and there are many others that look like nothing else on Earth.

In this book, we assign most unicellular and colonial eukaryotic organisms to the kingdom Protista. We base the separation of protists from fungi on the composition of the cell wall. Only fungi encase their absorptive (feeding) cells in a chitin-containing cell wall, and protists lack the dikaryotic phase that many fungal life cycles have. Development is what sets photosynthetic protists (algae) apart from plants as defined in this book: Whereas plants develop from embryos protected by tissues of the parent plant, algae develop from a single cell that has no such protection. The separation between protists and animals is relatively easy: An organism is an animal if it is a multicellular heterotroph with ingestive metabolism, passes through an embryonic stage called a blastula (see Chapter 40), and has an extracellular matrix containing collagen.

Tracing the natural phylogeny of the Eukarya presents some problems, and the phylogeny of protists is an area of intense, exciting research. The marvelous diversity of protist body forms and metabolic lifestyles seems reason enough for a fascination with these organisms, but questions about whether and how the multicellular eukaryotic kingdoms originated from the Protista stimulate further interest. Fortunately, the tools of molecular biology—rRNA sequencing in particular—make it possible to explore evolutionary relationships in greater detail and with greater confidence than previously (see Chapters 23 and 25).

The Origin of the Eukaryotic Cell

The eukaryotic cell differs in many ways from prokaryotic cells. How did it originate? Given the nature of evolutionary processes, the differences cannot all have arisen simultaneously. We think we can make some reasonable guesses about the sequence of events, bearing in mind that the global environment underwent an enormous change—from anaerobic to aerobic—during the course of these events. As you read this chapter, keep in mind that the steps we suggest are just that: guesses. This version of the story is one of a few under current consideration. We present it as a framework for thinking about this challenging problem, *not* as a set of facts.

The modern eukaryotic cell arose in several steps

The essential steps in the origin of the eukaryotic cell include

1. The origin of a flexible cell surface
2. The origin of a nuclear envelope
3. The appearance of digestive vesicles
4. The origin of a cytoskeleton
5. The endosymbiotic acquisition of certain organelles

WHAT A FLEXIBLE CELL SURFACE ALLOWS. Most present-day prokaryotic cells have firm cell walls. The first step toward the eukaryotic condition may have been the loss of the cell wall by an ancestral cell. This may not seem like an obvious first step, but consider the possibilities open to a flexible cell without a wall.

First, think of cell size. As a cell grows, its surface area-to-volume ratio decreases (see Chapter 4). Unless the surface is flexible and can fold inward and elaborate itself, creating more surface area for gas and nutrient exchange (Figure 26.2), the cell volume will reach an upper limit. With a surface flexible enough to allow infolding, the cell can exchange materials with its environment rapidly enough to sustain a larger volume and more rapid metabolism. Further, a flexible surface may pinch off bits of the environment, bringing them into the cell by endocytosis as compartments in which digestion may occur (Figure 26.3).

Also recall that the chromosome of a prokaryotic cell is attached to a site on its plasma membrane (see

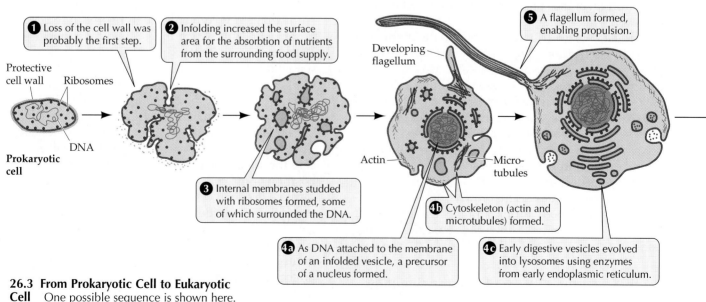

26.3 From Prokaryotic Cell to Eukaryotic Cell One possible sequence is shown here. The steps labeled 4a, 4b, and 4c all took place at the same time.

1 Loss of the cell wall was probably the first step.

2 Infolding increased the surface area for the absorbtion of nutrients from the surrounding food supply.

3 Internal membranes studded with ribosomes formed, some of which surrounded the DNA.

4a As DNA attached to the membrane of an infolded vesicle, a precursor of a nucleus formed.

4b Cytoskeleton (actin and microtubules) formed.

4c Early digestive vesicles evolved into lysosomes using enzymes from early endoplasmic reticulum.

5 A flagellum formed, enabling propulsion.

Protective cell wall

Ribosomes

DNA

Prokaryotic cell

Developing flagellum

Actin

Microtubules

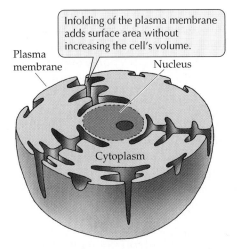

Infolding of the plasma membrane adds surface area without increasing the cell's volume.

Plasma membrane

Nucleus

Cytoplasm

26.2 Membrane Infolding The loss of the rigid prokaryotic cell wall meant the newly flexible cell could elaborate inward and create more surface area.

Figure 25.4). If that region of the plasma membrane were to fold into the cell, the first step would be taken toward the evolution of a nucleus, the key feature of the eukaryotic cell.

CHANGES IN CELL STRUCTURE AND FUNCTION. Early steps in the evolution of the eukaryotic cell are likely to have included three advances: the formation of ribosome-studded internal membranes, some of which surrounded the DNA (steps 3 and 4a in Figure 26.3); the appearance of actin fibers and microtubules—a cytoskeleton to manage changes in shape and to move materials from one part of the now much larger cell to

other parts (step 4b); and the evolution of the early digestive vesicles into lysosomes (step 4c).

From this intermediate kind of cell, the next advance was probably to a truly eukaryotic cell that we could call a phagocyte—a motile cell that could prey on other cells by engulfing and digesting them. The first true eukaryote possessed a nuclear envelope and an associated endoplasmic reticulum and Golgi apparatus, and perhaps one or more flagella of the eukaryotic type. Notice how much of the progress to this point was made possible by the loss of the wall and the elaboration of what was originally the plasma membrane.

ENDOSYMBIOSIS AND ORGANELLES. During the processes already outlined, the cyanobacteria were very busy, generating oxygen gas as a product of photosynthesis. The increasing O_2 levels in the atmosphere had disastrous consequences for most other living things, because most living things of the time (archaea and bacteria) were unable to tolerate the newly aerobic, oxidizing environment. But some prokaryotes managed to cope, and—fortunately for us—so did some of the ancient phagocytes.

According to one hypothesis, the key to the survival of early eukaryotes was the ingestion and incorporation of a prokaryote that became symbiotic within the phagocytes and evolved into the peroxisomes of today (step 6 in Figure 26.3). These organelles were able to disarm the toxic products of oxygen action, such as hydrogen peroxide. This association may have been the first important endosymbiosis in the evolution of the eukaryotic cell.

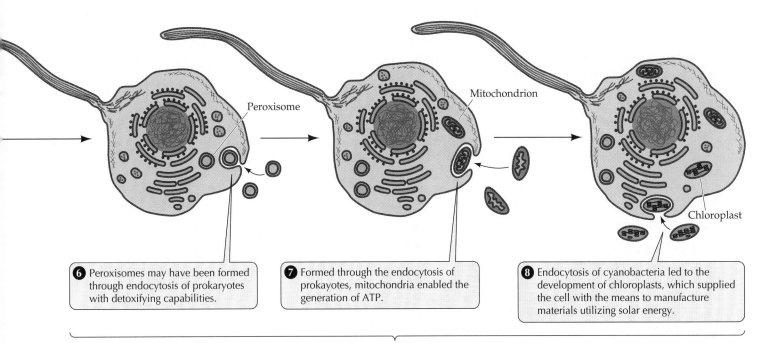

Peroxisome

Mitochondrion

Chloroplast

6 Peroxisomes may have been formed through endocytosis of prokaryotes with detoxifying capabilities.

7 Formed through the endocytosis of prokayotes, mitochondria enabled the generation of ATP.

8 Endocytosis of cyanobacteria led to the development of chloroplasts, which supplied the cell with the means to manufacture materials utilizing solar energy.

Endosymbiosis

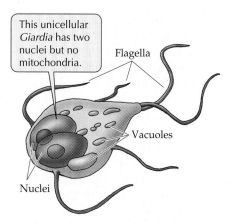

This unicellular *Giardia* has two nuclei but no mitochondria.

Flagella

Vacuoles

Nuclei

26.4 *Giardia*, a Protist without Mitochondria   Present-day organisms such as this lend support to the hypothesis that the nucleus appeared before mitochondria in the evolution of eukaryotes.

Another key endosymbiotic event was the incorporation of a prokaryote, related to the proteobacteria, that was the precursor of mitochondria (step 7 in Figure 26.3). On completion of this step, the basic modern eukaryotic cell was complete. Some very important eukaryotes are the result of yet another endosymbiotic step, the incorporation of prokaryotes, related to today's cyanobacteria, that are now chloroplasts (step 8).

Reviewing the likely origin of eukaryotic cells, endosymbiosis played multiple roles, and elaboration of the plasma membrane resulted in various advances.

"Archezoans" resemble a proposed intermediate stage

The hypothesis that the eukaryotic nucleus evolved before the mitochondrion gains support from the existence of a polyphyletic group of protists sometimes called archezoans, consisting of a few organisms such as *Giardia*. *Giardia lamblia* is a familiar parasite that contaminates water supplies and causes the intestinal disease giardiasis (Figure 26.4). This tiny organism has no mitochondria, chloroplasts, or other membrane-enclosed organelles, but it contains a nucleus bounded by a nuclear envelope, and it has a cytoskeleton. We view *Giardia* as a modern descendant of a transitional stage in the evolution of the eukaryotes.

General Biology of the Protists

Most protists are aquatic. Some live in marine environments, others in fresh water, and still others in the body fluids of other organisms. The slime molds inhabit damp soil and the moist, decaying bark of rotting trees. Many other protists also live in soil water, some of them contributing to the global nitrogen cycle by preying on soil bacteria and recycling their nitrogen compounds to nitrates.

Protists are strikingly diverse in their metabolism, perhaps second only to the prokaryotes. Nutritionally, some are autotrophs, some absorptive heterotrophs, and others ingestive heterotrophs. Some switch with ease between the autotrophic and heterotrophic modes of nutrition.

Three protist phyla consist entirely of nonmotile organisms, but all the other phyla include cells that move by amoeboid motion, by ciliary action, or by means of flagella. Most unicellular protists are tiny, but the multicellular plantlike protists include the giant kelps (see Figure 26.1c), which live in the oceans and are among the longest organisms in existence.

Vesicles perform a variety of functions

Unicellular organisms tend to be of microscopic size. An important reason that cells are small is that they need enough membrane surface area in relation to their volume to support the exchange of materials required for them to live (see Figures 4.2 and 26.2). The size that unicellular protists can achieve is limited by their surface area–to–volume ratio. Many relatively large unicellular protists minimize this problem by having membrane-enclosed vesicles of various types that increase their effective surface area.

For example, many freshwater protists address their osmotic problems by using vesicles that contract to excrete excess water. Members of several of the protist phyla have such **contractile vacuoles**, which help them cope with their hypoosmotic environments.

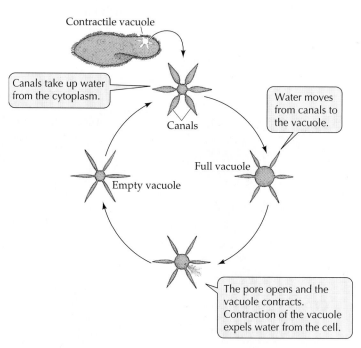

Contractile vacuole

Canals take up water from the cytoplasm.

Water moves from canals to the vacuole.

Canals

Full vacuole

Empty vacuole

The pore opens and the vacuole contracts. Contraction of the vacuole expels water from the cell.

26.5 Contractile Vacuoles Bail Out Excess Water   Water constantly enters the cell by osmosis. A pore in the cell surface allows the contractile vacuole to expel the water it accumulates.

Because these organisms have a more negative water potential than their freshwater environment does, they constantly take in water by osmosis. Excess water collects in the contractile vacuole and is then pushed out (Figure 26.5).

A beautifully simple experiment confirms that bailing out water is the principal function of the contractile vacuole. First we observe cells under the microscope and note the rate at which the vacuoles are contracting—they look like little eyes winking. Then we place other cells of the same type in solutions of differing osmotic potential. The less negative the osmotic potential of the surrounding solution, the more hyperosmotic the cells are and the faster the water rushes into them, causing the contractile vacuoles to pump more rapidly. Conversely, the contractile vacuoles will stop pumping if the solute concentration of the medium is increased so that it is isosmotic with the cells.

A second important type of vesicle found in many protists is the **food vacuole**. Protists such as *Paramecium* engulf solid food, forming food vacuoles within which the food is digested (Figure 26.6). Smaller vesicles containing digested food pinch away from the food vesicle and enter the organism's cytoplasm. These tiny vesicles provide a large surface area across which the products of digestion may be absorbed by the rest of the cell.

The cell surfaces of protists are diverse

A few protists, such as some amoebas, are surrounded by only a plasma membrane, but most have stiffer surfaces that maintain the structural integrity of the cell. Many algae and other protists have cell walls, which are often complex in structure.

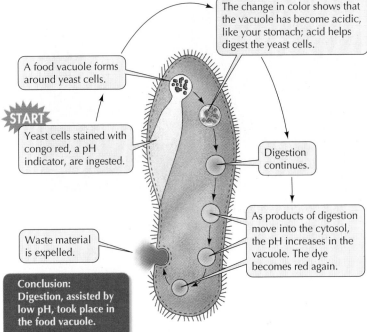

26.6 Food Vacuoles Handle Digestion and Excretion
An experiment with *Paramecium* demonstrates the function of food vacuoles.

The protozoans, lacking cell walls, have a variety of ways of strengthening their surface. Some have internal "shells," which either the organism itself produces, as foraminiferans do, or is made of bits of sand and thickenings immediately beneath the plasma membrane, as in some amoebas (Figure 26.7).

26.7 Diversity in Protozoan Cell Surfaces (a) Foraminiferan shells are made of protein hardened with calcium carbonate. (b) This shelled amoeba constructed its shell by cementing sand grains together. (c) Spirals of protein make this *Paramecium*'s surface—known as its pellicle—flexible but resilient.

(a)

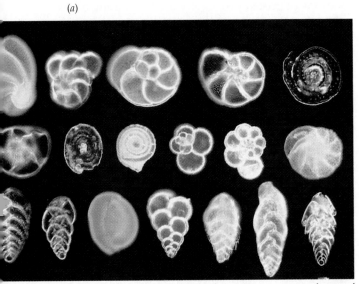

1 mm

(b) *Difflugia* sp.

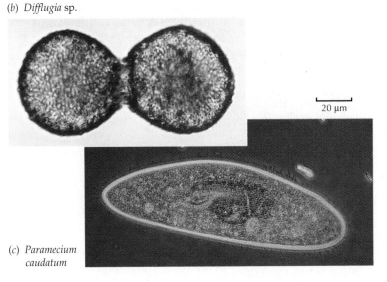

20 µm

(c) *Paramecium caudatum*

26.8 Protists within Protists Photosynthetic algae living as endosymbionts within these radiolarians provide food for the radiolarians, as well as part of the pigmentation seen through the glassy skeletons. Both the algae and the radiolarians are protists.

Many protists contain endosymbionts

In Chapter 4 we introduced the concept of *endosymbiosis* (organisms living together, one inside the other). Endosymbiosis is very common among the protists, and in some instances both the host and the endosymbiont are protists. Many radiolarians, for example, harbor photosynthetic protists as endosymbionts (Figure 26.8). As a result, these radiolarians appear greenish or yellowish, depending on the type of endosymbiont they contain.

This arrangement is beneficial to the radiolarian, for it can make use of the food produced by its photosynthesizing guest. The guest, in turn, may make use of metabolites made by the host, or it may simply receive physical protection. Alternatively, the guest may be a victim, exploited for its photosynthetic products while receiving no benefit itself.

Endosymbiosis is important in the lives of many protists. This and other phenomena have contributed to the great success of the kingdom Protista, a kingdom that flourished for hundreds of millions of years before the first multicellular species evolved. Another source of this extraordinary success of the protists is the remarkable diversity of their sexual and asexual reproductive strategies.

Both asexual and sexual reproduction occur in the kingdom Protista

Although most protists indulge in both asexual and sexual reproduction, some groups lack sexual reproduction. As we will see, some asexually reproducing protists do engage in genetic recombination, even though it does not relate directly to reproduction.

Asexual reproductive processes in the kingdom Protista include binary fission (simple splitting of the cell), multiple fission, budding (the outgrowth of a new cell from the surface of an old one), and the formation of spores. Sexual reproduction also takes various forms. In some protists, as in animals, the gametes are the only haploid cells. In many algae, by contrast, both diploid and haploid cells undergo mitosis, giving rise to an *alternation of generations* (which will be described later in the chapter; see Figure 26.26).

The diversity of form, habitat, metabolism, locomotion, reproduction, and life cycles of protists reflects the diversity of avenues pursued in the early evolution of eukaryotes. Many of these avenues led to great success, judging from the abundance and diversity of today's protists and other eukaryotes.

Protozoan Diversity

All protozoans are unicellular. Most ingest their food by endocytosis. Their diversity, which includes many of the most abundant or commonly observed protist types, is summarized in Table 26.1. In this and subsequent sections of this chapter, please be aware that our goal is to give you a clear sense of the *diversity* of protists, more than it is to reflect taxonomic groupings. Explorations of protist phylogeny are fascinating and rewarding, but they are not our primary concern here.

We will identify several groups of organisms as phyla, but other groupings, less natural, will not be

TABLE 26.1   Some Groups of Protozoans

COMMON NAME	PHYLUM	FORM	LOCOMOTION	EXAMPLES
Flagellates	(Several)	Unicellular, some colonial	One or more flagella	*Trypanosoma, Euglena,* Choanoflagellida
Amoebas	(Several)	Unicellular, no definite shape	Pseudopods	*Amoeba, Entamoeba, Chaos*
Actinopods	Actinopoda	Unicellular	Pseudopods	Radiolarians, heliozoans
Foraminiferans	Foraminifera	Unicellular	Pseudopods	Foraminiferans
Apicomplexans	Apicomplexa	Unicellular	None	*Plasmodium*
Ciliates	Ciliophora	Unicellular	Cilia	*Paramecium, Blepharisma, Vorticella*

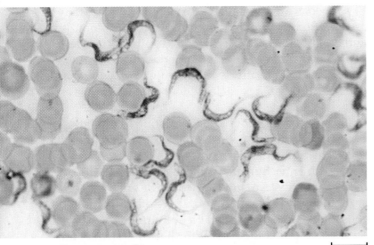

Trypanosoma gambiense

25 µm

26.9 A Parasitic Flagellate Trypanosomes cause sleeping sickness in mammals. A flagellum runs along one edge of the cell as part of a structure called the undulating membrane. The other cells in the micrograph are mammalian red blood cells.

given the title of phylum. For example, certain body plans, such as those of amoebas and those of protists with flagella, arose again and again during evolution, in groups only distantly related to one another. We will emphasize diversity of body plans over assignment of species to phyla.

In this section on protozoan diversity, we'll describe euglenoids and other flagellates, amoebas, actinopods, foraminiferans, apicomplexans, and ciliates.

Many protists have flagella

Many protists—thousands of species—possess one or more flagella and hence are called flagellates. Flagellates do not constitute a single phylum. There are many flagellate groups, and they appear in different parts of the kingdom Protista. Flagellates reproduce vegetatively by binary fission (mitosis and cytokinesis)—the simplest and most direct way.

Some free-living flagellate species survive by preying on other protists. An impressive variety of other flagellates live as internal parasites on animals, including humans (Figure 26.9). Within the guts of certain wood-eating roaches and termites live an array of huge flagellates that possess some of the most bizarre and complicated body forms found anywhere among the protists.

One group of flagellates, the Choanoflagellida, is thought to comprise the closest relatives of the animals. Members of this group are especially closely related to the sponges, the most ancient of the surviving phyla of animals (see Chapter 29). Sponges are colonial, rather than truly multicellular. That is, sponges lack organized tissues, and their cells can be separated

and recombined. The Choanoflagellida bear a striking resemblance to the most characteristic type of cell found in the sponges (see Figure 29.5).

Some flagellates are human pathogens. Sleeping sickness, one of the most dreaded diseases of Africa, is caused by the parasitic flagellate *Trypanosoma* (see Figure 26.9). The vector (intermediate host) for sleeping sickness is an insect, the tsetse fly. Carrying its deadly cargo, the tsetse fly bites livestock, wild animals, and even humans, infecting all of them with *Trypanosoma*. *Trypanosoma* then multiplies in the mammalian bloodstream and produces toxic substances. When these parasites invade the nervous system, the neurological symptoms of sleeping sickness appear— and are followed by death. Another disease-causing flagellate is *Trichomonas vaginalis*, which causes a common but usually mild sexually transmitted disease.

Some euglenoids are photosynthetic

The 800 species of euglenoids* used to be claimed by the zoologists as animals and by the botanists as plants. They are unicellular flagellates, but many members of the group photosynthesize, as do the algae.

Figure 26.10 depicts a cell of the genus *Euglena*. Like most other members of the group, this common freshwater organism has a complex cell plan. It propels itself through the water with one of its two flagella, which may also serve as an anchor to hold the organism in place. The flagellum provides power by means of a wavy motion that spreads from base to tip. The second flagellum is often rudimentary. *Euglena* reproduces vegetatively by mitosis and cytokinesis.

---

* You may be wondering why we capitalized "Choanoflagellida" in the previous section but not "euglenoids" here. The official names of phyla and other taxonomic groups are capitalized; less formal names are not. Euglenoids belong to more than one phylum.

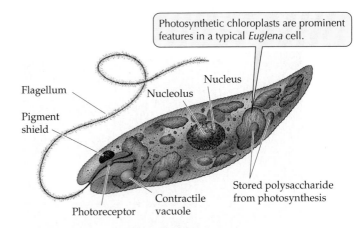

Photosynthetic chloroplasts are prominent features in a typical *Euglena* cell.

Flagellum

Pigment shield

Photoreceptor

Nucleolus

Nucleus

Contractile vacuole

Stored polysaccharide from photosynthesis

26.10 A Photosynthetic Flagellate Several species of *Euglena* are among the best-known flagellates.

Euglena has very flexible nutritional requirements. In sunlight it is fully autotrophic, using its chloroplasts to synthesize organic compounds through photosynthesis. When kept in the dark, the organism loses its photosynthetic pigment and begins to feed exclusively on dead organic material floating in the water around it. Such a "bleached" cell of *Euglena* resynthesizes its photosynthetic pigment when it is returned to the light and hence becomes autotrophic again.

Euglena cells treated with certain antibiotics or mutagens lose their photosynthetic pigment completely; neither they nor their descendants are ever autotrophs again. Those descendants, however, function well as heterotrophs. We believe that a photosynthetic prokaryote became an endosymbiont inside an ancestor of the photosynthetic euglenoids, endowing its host with the ability to photosynthesize, in much the same way that chloroplasts became integral parts of the cells of green algae and plants.

Amoebas form pseudopods

Amoebas are protists that form **pseudopods**, extensions of their constantly changing body mass (see Figure 4.22*b* and the photo at the beginning of this chapter). As we have mentioned, the amoebas do not constitute a single, coherent group of organisms; rather, this body plan has appeared by convergent evolution in various groups.

Amoebas have often been portrayed in popular writing as simple blobs—the simplest form of "animal" life imaginable. Superficial examination of a typical amoeba shows how such an impression might have been obtained. An amoeba consists of a single cell. It feeds on small organisms and particles of organic matter by phagocytosis, engulfing them with its pseudopods. Particles of food are sealed off in food vacuoles within the cytoplasm of the amoeba. The material is then slowly digested and assimilated into the main body of the organism. Pseudopods are also the organs of locomotion. (The mechanism of amoeboid motion will be discussed in Chapter 44.)

Amoebas are specialized forms of protists. Many are adapted for life on the bottoms of lakes, ponds, and other bodies of water. Their creeping locomotion and their manner of engulfing food particles fit them for life close to a relatively rich supply of sedentary organisms or organic particles. Other amoebas are even more specialized.

All amoebas are animal-like, existing as predators, parasites, or scavengers. None are photosynthetic. Amoebas of the free-living genus *Naegleria*, some of which can enter humans and cause a fatal disease of the nervous system, have a two-stage life cycle, one stage having amoeboid cells and the other flagellated cells. Some amoebas are shelled, living in casings of sand grains glued together or in shells secreted by the organism itself (see Figure 26.7*b*).

Actinopods have thin, stiff pseudopods

The actinopods are recognizable by their thin, stiff pseudopods, which are reinforced by microtubules. The pseudopods play at least four roles: (1) They greatly increase the surface area of the cell for exchange of materials with the environment; (2) they help the cell float in its marine or freshwater environment; (3) they provide locomotion in heliozoans, a group of actinopods that roll along the substrate by shortening and elongating their pseudopods; and (4) they are the cell's feeding organs; the pseudopods trap smaller organisms, often taking up prey by endocytosis and transporting it to the main cell body.

Radiolarians, actinopods that are exclusively marine, are perhaps the most beautiful of all microorganisms (Figure 26.11*a*). Almost all radiolarian species secrete glassy endoskeletons (internal skeletons) from which needlelike pseudopods project. Part of the skeleton is a central capsule within the cytoplasm. The skeletons of the different species are as varied as

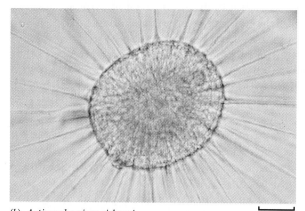

(*b*) *Actinosphaerium eichorni*   150 μm

26.11 Actinopods   (*a*) A radiolarian displays its glassy skeleton of delicate intricacy. (*b*) A heliozoan with long pseudopods.

(*a*) Radiolarian (species not identified)

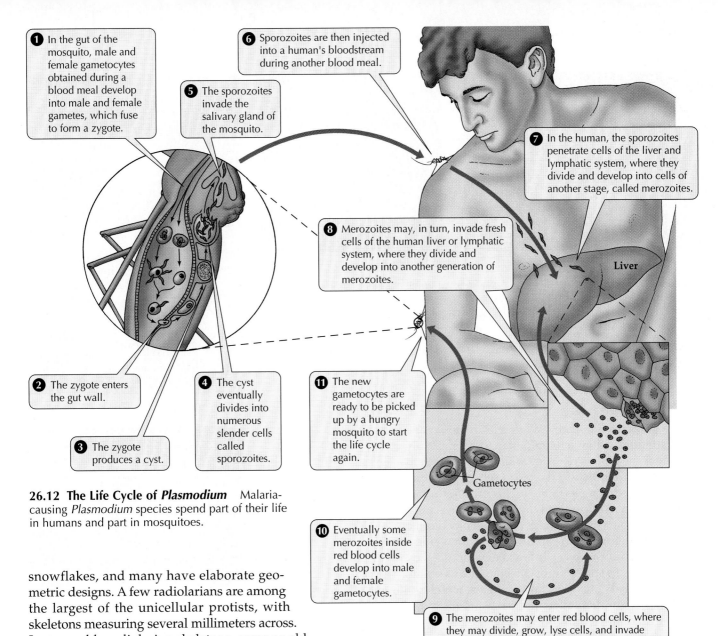

1 In the gut of the mosquito, male and female gametocytes obtained during a blood meal develop into male and female gametes, which fuse to form a zygote.

6 Sporozoites are then injected into a human's bloodstream during another blood meal.

5 The sporozoites invade the salivary gland of the mosquito.

7 In the human, the sporozoites penetrate cells of the liver and lymphatic system, where they divide and develop into cells of another stage, called merozoites.

8 Merozoites may, in turn, invade fresh cells of the human liver or lymphatic system, where they divide and develop into another generation of merozoites.

Liver

2 The zygote enters the gut wall.

4 The cyst eventually divides into numerous slender cells called sporozoites.

11 The new gametocytes are ready to be picked up by a hungry mosquito to start the life cycle again.

3 The zygote produces a cyst.

Gametocytes

10 Eventually some merozoites inside red blood cells develop into male and female gametocytes.

9 The merozoites may enter red blood cells, where they may divide, grow, lyse cells, and invade fresh red blood cells on a 48-hour cycle.

26.12 The Life Cycle of *Plasmodium* Malaria-causing *Plasmodium* species spend part of their life in humans and part in mosquitoes.

snowflakes, and many have elaborate geometric designs. A few radiolarians are among the largest of the unicellular protists, with skeletons measuring several millimeters across. Innumerable radiolarian skeletons, some as old as 700 million years, form the sediment under some tropical seas.

Heliozoans are actinopods that lack an endoskeleton (Figure 26.11*b*). Most heliozoans live in fresh water.

Apicomplexans are parasites with unusual spores

Exclusively parasitic protozoans, the apicomplexans are thus named because the apical end of their spore contains a mass of organelles. These organelles help the apicomplexan spore invade its host tissue. Unlike many other protists, apicomplexans lack contractile vacuoles. Because their rigid cell walls limit expansion, they do not take in excess water.

Apicomplexans generally have an indefinite body form like that of an amoeba. This body form has evolved over and over again in parasitic protists. The form has appeared, for example, even in parasitic dinoflagellates, a group of algae whose nonparasitic relatives have highly distinctive, regular body forms. Like many animal obligate parasites, apicomplexans

have elaborate life cycles featuring asexual and sexual reproduction by a series of very dissimilar life stages. Often these stages are associated with two types of host organisms.

The best-known apicomplexans are the malaria parasites of the genus *Plasmodium*, a highly specialized group of organisms that spend part of their life cycle within human red blood cells (Figure 26.12). Malaria continues to be a major problem in some tropical countries, although it has been almost eliminated from the United States; indeed, in terms of number of people infected, malaria is one of the world's most serious diseases.

Female mosquitoes of the genus *Anopheles* transmit *Plasmodium* to humans. *Plasmodium* enters the human circulatory system when an infected *Anopheles* mosquito penetrates the human skin in search of blood.

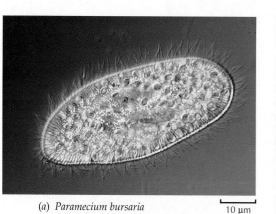

(a) *Paramecium bursaria*   10 µm

(b) *Epistylis* sp.   60 µm

26.13 Diversity in the Ciliates (a) A free-swimming organism, this paramecium belongs to the ciliate subgroup called holotrichs, which have many cilia of uniform length. (b) Members of the peritrich subgroup of Ciliophora have cilia on their mouthparts. (c) Tentacles replace cilia as suctorians (another subgroup) develop. (d) This ciliate "walks" on fused cilia called cirri that project from its body. Other cilia are fused into flat sheets that sweep in food particles; this individual has just fed on green algae.

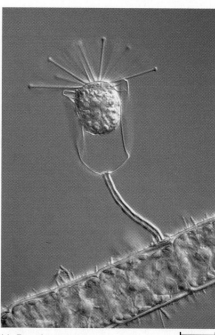

(c) *Paracineta* sp.   20 µm

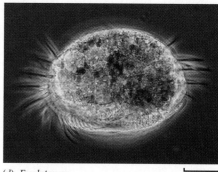

(d) *Euplotes* sp.   25 µm

The parasites find their way to cells in the liver and the lymphatic system, change their form, multiply, and reenter the bloodstream, attacking red blood cells. The attackers multiply in a red blood cell for approximately 2 days, producing as many as 36 new *Plasmodium* cells each. The victimized cell then bursts, releasing a new swarm of parasites to attack other red blood cells.

If another *Anopheles* bites the victim, some of the parasitic *Plasmodium* cells are taken into the mosquito along with the blood, thus infecting the mosquito. The infecting cells develop into gametes, which unite to form zygotes that lodge in the mosquito's gut, reproduce, and move into the salivary glands, from which they can be passed to another human host.

Plasmodium is an intracellular parasite in the human host and an extracellular parasite in the mosquito. The *Plasmodium* life cycle that spreads malaria is best broken by the removal of stagnant water, in which mosquitoes breed. The use of insecticides to reduce the *Anopheles* population can be effective, but one must weigh possible ecological, economic, and health risks posed by the insecticides themselves.

Malaria kills more than a million people each year, and *Plasmodium* has proved to be a singularly difficult pathogen to attack. However, there is now new hope in the form of a genome-sequencing project that targets *Plasmodium falciparum*. Scheduled to be completed by the year 2002, that project may provide the information needed to end this epidemic.

Foraminiferans have created vast limestone deposits

Foraminiferans are marine creatures—some floating as plankton and many others living at the bottom of the sea—that secrete shells of calcium carbonate (see Figure 26.7a). Their long, threadlike, branched pseudopods reach out through numerous microscopic pores in the shell and interconnect to create a sticky net, which the foraminiferan uses to catch smaller plankton (free-floating microscopic organisms).

After foraminiferans reproduce—by mitosis and cytokinesis—the daughter cells abandon the parent shell and make new shells of their own. The discarded skeletons of ancient foraminiferans make up extensive limestone deposits in various parts of the world, forming a covering hundreds to thousands of meters deep over millions of square kilometers of ocean bottom. Foraminiferan skeletons also make up the sand of some beaches. A single gram of such sand may contain as many as 50,000 foraminiferan shells.

The shells of individual foraminiferan species have distinctive shapes and are easily preserved as fossils in marine sediments. Each geologic period has distinctive foraminiferan species. For this reason, and be cause they are so abundant, the remains of foraminiferans are especially valuable as indicators in the classification and dating of sedimentary rocks, as well as in oil prospecting.

Ciliates have two types of nuclei

Because members of the phylum Ciliophora characteristically have hairlike cilia, they have the common name ciliates. This protozoan group ranks with the flagellates in diversity and ecological importance (Figure 26.13). Almost all ciliates are heterotrophic (a few contain photosynthetic endosymbionts), and they are much more specialized in body form than are most flagellates and other protists. Ciliates are also characterized by the possession of two types of nuclei, a large **macronucleus** and, within the same cell, from 1 to as many as 80 **micronuclei**.

The micronuclei, which are typical eukaryotic nuclei, are essential for genetic recombination. The macronucleus contains many copies of the genetic information, packaged in units containing very few genes each; the macronuclear DNA is transcribed and translated to control the life of the cell. Although we do not know how this system of macro- and micronuclei came into being, we know something about the behavior of these nuclei, which we will discuss after describing the body plan of one important ciliate, *Paramecium*.

A CLOSER LOOK AT ONE CILIATE. *Paramecium*, a frequently studied ciliate genus, exemplifies the complex structure and behavior of ciliates (Figure 26.14*a*). The slipper-shaped cell is covered by an elaborate **pellicle**, a structure composed principally of an outer membrane and an inner layer of closely packed, membrane-enclosed sacs (called alveoli) that surround the bases of the cilia. Defensive organelles called trichocysts are also present in the pellicle. A microscopic explosion expels the trichocysts in a few milliseconds, and they emerge as sharp darts, driven forward at the tip of a long, expanding shaft (Figure 26.14*b*).

The cilia provide a form of locomotion that is generally more precise than that made possible by flagella or pseudopods. A paramecium can direct the beat of its cilia to propel itself either forward or backward in a spiraling manner (Figure 26.15). A paramecium can

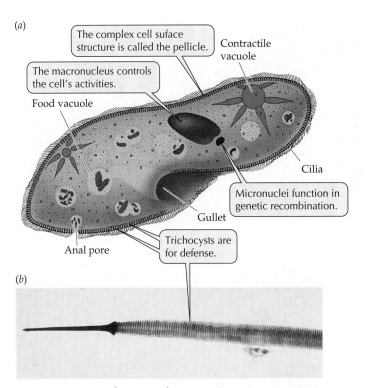

(*a*)

The complex cell surface structure is called the pellicle.

The macronucleus controls the cell's activities.

Contractile vacuole

Food vacuole

Cilia

Micronuclei function in genetic recombination.

Gullet

Anal pore

Trichocysts are for defense.

(*b*)

26.14 Anatomy of *Paramecium* (*a*) The major structures of a typical paramecium. (*b*) A trichocyst discharged from beneath the pellicle of a paramecium has a sharp point and a straight filament.

back off swiftly when it encounters a barrier or a negative stimulus. Some of these large ciliates hold the speed record for the kingdom Protista—faster than 2 mm/s. The coordination of ciliary beating is probably the result of a differential distribution of calcium and other ion channels near the two ends of the cell.

REPRODUCTION WITHOUT SEX, AND SEX WITHOUT REPRODUCTION. Paramecia reproduce by cell division. The micronuclei divide mitotically; the macronucleus simply pinches apart to yield two daughter macronuclei. Paramecia also have an elaborate sexual behavior called **conjugation** (Figure 26.16). Two paramecia line

26.15 "Swimming" with Cilia Beating its cilia in coordinated waves that progress from one end of the cell to the other, a paramecium can move in either direction with respect to the long axis of the cell; this one is moving from left to right. The cell also rotates in a spiral as it travels.

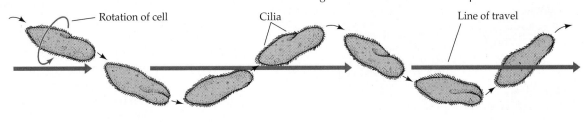

Rotation of cell

Cilia

Line of travel

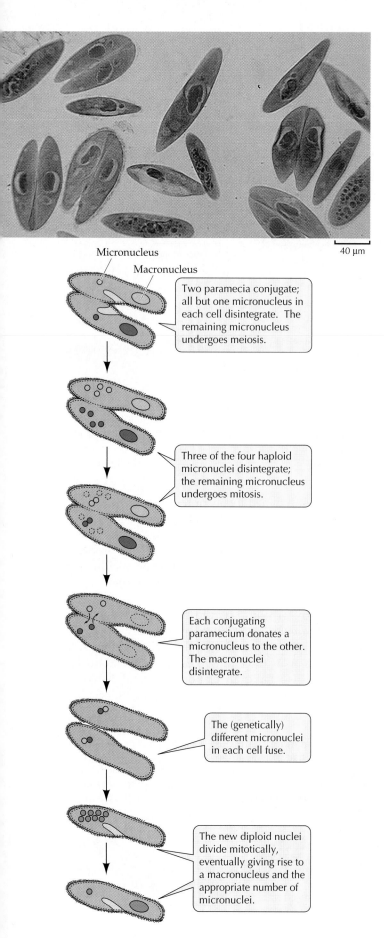

40 μm

26.16 Paramecia Achieve Genetic Recombination by Conjugating Conjugating *Paramecium* individuals exchange micronuclei, thereby permitting genetic recombination. After conjugation, the cells separate and continue their lives as two individuals.

Micronucleus

Macronucleus

Two paramecia conjugate; all but one micronucleus in each cell disintegrate. The remaining micronucleus undergoes meiosis.

Three of the four haploid micronuclei disintegrate; the remaining micronucleus undergoes mitosis.

Each conjugating paramecium donates a micronucleus to the other. The macronuclei disintegrate.

The (genetically) different micronuclei in each cell fuse.

The new diploid nuclei divide mitotically, eventually giving rise to a macronucleus and the appropriate number of micronuclei.

up tightly against each other and fuse in the oral region of the body. Nuclear material is extensively reorganized and exchanged during the next several hours. The reorganization includes both meiosis and the fusion of gametic nuclei from the two conjugating partners. The exchange of gametic nuclei is fully reciprocal—each of the two paramecia gives and receives an equal amount of DNA. Afterward the two organisms separate and go their own ways, each equipped with new combinations of alleles by the genetic recombination that occurred during conjugation.

Conjugation in *Paramecium* is a *sexual* process of genetic recombination, but it is not a *reproductive* process. The same two cells that begin the process are there at the end, and no new cells are created. As a rule, each clone of paramecia must periodically conjugate. Laborious experimentation has shown that if some species are not permitted to conjugate, the asexual clones can live through no more than approximately 350 cell divisions before they die out.

CYTOPLASMIC ORGANIZATION IN THE CILIATES. Most ciliates possess all the traits of *Paramecium*. Some, however, are notable for the exceptional degree of development of their individual organelle systems. Certain ciliates, for example, have the equivalent of legs. Fused cilia called cirri move in an independent, but coordinated, fashion and enable the organism to "walk" over surfaces (see Figure 26.13d). Cytoskeletal elements leading to individual cirri assist this locomotion. The coordination is lost if these structures are experimentally cut.

Many types of ciliates possess **myonemes**, muscle-like fibers within the cytoplasm. The contraction of myonemes causes a rapid retraction of the stalk in ciliates such as *Vorticella* (see Figure 26.13b) when the organism is disturbed. What may be the ultimate cytoplasmic organization is displayed by the highly specialized ciliates that live in the digestive tracts of cows and many other hoofed mammals. These ciliates possess not only myonemes and elaborately fused cilia, but also a cytoplasmic "skeleton" and a "cellular gut" complete with a "mouth," an "esophagus," and an "anus" (Figure 26.17).

When examining the intricate structures of these ciliates and many other protists, we must pause and remember that we are looking at only one cell. Structural complexity in multicellular organisms—fungi, animals, and plants—is based on the diversity

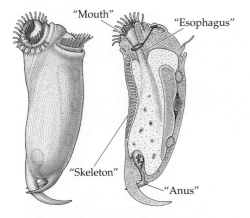

"Mouth"

"Esophagus"

"Skeleton"

"Anus"

26.17 An Exceptional Ciliate Surface and cutaway views of *Diplodinium dentatum*, an amazingly complex single cell.

and coordination of cell types. These protists, on the other hand, owe their complexity to the diversity and coordination of organelles within a single cell.

Protists Once Considered To Be Fungi

Unlike fungi, the protists lack chitinous cell walls in their feeding phase and have no dikaryotic stage (in which cells contain two haploid nuclei each) in their life cycle (see Chapter 28). Table 26.2 summarizes some groups of protists that have been considered by some biologists to be fungi. In this section we'll describe four phyla: the Myxomycota, the Dictyostelida, the Acrasida, and the Oomycota.

The members of three of these groups seem so similar at first glance that they were once grouped in a single phylum. However, these so-called slime molds are actually so different that some biologists classify them in different *kingdoms.*

The three groups of slime molds share some characteristics. All are motile, all ingest particulate food by endocytosis, and all form spores on erect fruiting bodies. They undergo striking changes in organization during their life cycles, and one stage consists of isolated cells that engage in absorptive nutrition. Some

slime molds may attain areas of 1 meter or more in diameter while in their less-aggregated stage. Such a large slime mold may weigh more than 50 grams. Slime molds of all three types favor cool, moist habitats, primarily in forests. They range from colorless to brilliantly yellow and orange.

Acellular slime molds form multinucleate masses

If the nucleus of an amoeba began rapid mitotic division, accompanied by a tremendous increase in cytoplasm and organelles, the resulting organism might resemble the vegetative phase of the *acellular* slime molds (phylum Myxomycota). During most of its life history, an acellular slime mold is a wall-less mass of cytoplasm with numerous diploid nuclei. This mass streams very slowly over its substrate in a remarkable network of strands called a **plasmodium** (Figure 26.18*a*). A plasmodium of a myxomycete is an example of a **coenocyte,** a body in which many nuclei are enclosed in a single plasma membrane. The outer cytoplasm of the plasmodium (closest to the environment) is normally less fluid than the interior cytoplasm and thus provides some structural rigidity.

Myxomycetes such as *Physarum* (a popular research subject) provide a dramatic example of **cytoplasmic streaming**. The outer cytoplasmic region becomes more fluid in places, and cytoplasm rushes into those areas, stretching the plasmodium. This streaming somehow reverses its direction every few minutes as cytoplasm rushes into a new area and drains away from an older one, moving the plasmodium over its substrate in search of food.

The plasmodium engulfs food particles—predominantly bacteria, yeasts, spores of fungi, and other small organisms, as well as decaying animal and plant remains. Sometimes an entire wave of plasmodium moves across the substrate, leaving strands behind. Actin filaments and a contractile protein called myxomyosin interact to produce the streaming movement.

An acellular slime mold can grow almost indefinitely in its plasmodial stage, as long as the food supply is adequate and other conditions, such as moisture and pH, are favorable. However, one of two things can

TABLE 26.2 Classification of Protists with Absorptive Nutrition

PHYLUM	COMMON NAME	FORM	LOCOMOTION	EXAMPLES
Myxomycota	Acellular slime molds	Single cells and coenocytes	Amoeboid	*Physarum*
Dictyostelida	Dictyostelid cellular slime molds	Single cells and aggregates	Amoeboid	*Dictyostelium*
Acrasida	Acrasid cellular slime molds	Single cells and aggregates	Amoeboid	Acrasids
Oomycota	Water molds and downy mildews	Coenocytic mycelium	None	*Saprolegnia, Achlya, Phytophthora*

(a) *Physarum polycephalum*

(b) *Physarum* sp.

26.18 Acellular Slime Mold (a) Plasmodia of yellow slime mold cover a rock in Nova Scotia. (b) The fruiting structures—sporangiophore (yellow) and sporangia (black)—of *Physarum*.

happen if conditions become unfavorable. The plasmodium can form a resistant structure, an irregular mass of hardened cell-like components called a **sclerotium**, which rapidly becomes a plasmodium again when favorable conditions are restored; or the plasmodium can transform itself into spore-bearing fruiting structures (Figure 26.18*b*). Rising from heaped masses of plasmodium, these stalked or branched fruiting structures—called **sporangiophores**—derive their rigidity from the thickening of the walls of their component cells.

The nuclei of the plasmodium are diploid, and they divide by meiosis during the development of the sporangiophore. One or more knobs, called **sporangia**, develop on the end of the stalk. Within a sporangium, haploid nuclei become surrounded by walls and form spores. Eventually, as the sporangiophore dries, it sheds its spores.

The spores germinate into wall-less, flagellated, haploid cells called **swarm cells**, which either can divide mitotically to produce more haploid swarm cells or can function as gametes. Swarm cells can live on their own and can become walled and resistant cysts when conditions are unfavorable. When conditions improve again, the cysts release flagellated swarm cells. Two swarm cells can fuse to form a diploid zygote, which divides by mitosis (but without a wall forming between the nuclei) and thus forms a new, coenocytic plasmodium.

Cells retain their identity in the dictyostelid slime molds

The phylum Dictyostelida consists of *cellular* slime molds. Whereas the plasmodium is the basic vegeta-

tive (feeding) unit of acellular slime molds, an amoeboid cell is the vegetative unit of the dictyostelid slime molds. Large numbers of cells called **myxamoebas**, which have single haploid nuclei, engulf bacteria and other food particles by endocytosis and reproduce by mitosis and fission. This simple developmental stage, consisting of swarms of independent, isolated cells, can persist indefinitely (as long as food and moisture are available).

When conditions become unfavorable, however, dictyostelids aggregate and form fruiting structures, as do their acellular counterparts. The apparently independent myxamoebas aggregate into a mass called a **pseudoplasmodium** (Figure 26.19*a*). Unlike the true plasmodium of the acellular slime molds (see Figure 26.18*a*), this structure is not simply a giant sheet of cytoplasm with many nuclei; the individual myxamoebas retain their plasma membranes and, therefore, their identity.

The chemical signal that causes the myxamoebas of dictyostelid slime molds to aggregate into a pseudoplasmodium is 3′,5′-cyclic adenosine monophosphate (cAMP), a compound that plays many important roles in chemical signaling in animals (see Chapter 38). A pseudoplasmodium may migrate over its substrate for several hours before becoming motionless and reorganizing to construct a delicate, stalked fruiting structure (Figure 26.19*b*). Cells at the top of the fruiting structure develop into thick-walled spores. The spores are released; later, under favorable conditions, they germinate, releasing myxamoebas.

The cycle from myxamoebas through a pseudoplasmodium and spores to new myxamoebas is asexual. There is also a sexual cycle, in which two myxamoebas (possibly of different mating types; see Chapter 28) fuse. The product of this fusion develops into a spherical structure that ultimately germinates, releasing new haploid myxamoebas.

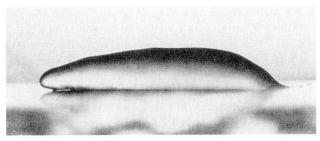

(a) Dictyostelium discoideum

26.19 A Cellular Slime Mold (*a*) A pseudoplasmodium migrates over its substrate. (*b*) Fruiting structures in various stages of development.

(b) Dictyostelium discoideum

Acrasids are also cellular slime molds

Members of the third group of slime molds, the phylum Acrasida, are also cellular slime molds, but they are only distantly related to the Dictyostelida and Myxomycota. Among their other differences from the dictyostelids, acrasids do not appear to use cAMP as an aggregation signal.

The oomycetes include water molds and their relatives

The three phyla of slime molds consist of motile organisms that feed by endocytosis. Members of the fourth phylum of protists that sometimes have been classified as fungi, the Oomycota, are filamentous and stationary, and they feed by absorption.

The phylum Oomycota consists in large part of the water molds and their funguslike terrestrial relatives, such as the downy mildews. If you have seen a whitish, cottony mold growing on dead fish or dead insects in water, it was probably a water mold of the common genus *Saprolegnia* (Figure 26.20). *Saprolegnia*

Saprolegnia sp.

26.20 Water Mold The filaments of a water mold radiate from the carcass of an insect.

itself is a common target of parasitism by the fungus *Rhizidiomyces*, described in Chapter 28.

The oomycetes are coenocytic: Their filaments have no cross-walls to separate the many nuclei into discrete cells. Their cytoplasm is continuous throughout the body of the mold, and there is no single structural unit with a single nucleus, except in certain reproductive stages. A distinguishing feature of the oomycetes is their flagellated reproductive cells. Oomycetes are diploid throughout most of their life cycle and have cellulose in their cell walls.

The water molds, such as *Saprolegnia*, are all aquatic and saprobic (feeding on dead organic matter). Some other ooymcetes are terrestrial. Although most terrestrial oomycetes are harmless or helpful decomposers of dead matter, a few are serious plant parasites that attack crops such as avocados, grapes, and potatoes. The mold *Phytophthora infestans*, for example, is the causal agent of late blight of potatoes, which brought about the great Irish potato famine of 1845 to 1847. *P. infestans* destroyed the entire Irish potato crop in a matter of days in 1846. Among the consequences of the famine were a million deaths from starvation and the emigration of about 2 million people, mostly to the United States.

Albugo, another oomycete, is a well-known parasitic genus that causes a mealy blight on sweet-potato leaves, morning glories, and numerous other plants. An obligate parasite, *Albugo* has never been grown on any medium other than its plant host.

The Algae

Algae (singular alga) are photosynthetic protists, carrying out probably 50 to 60 percent of all the photosynthesis on Earth (plants account for most of the rest). The overall contribution of cyanobacteria and other photosynthetic prokaryotes is smaller, although it is locally important in some aquatic ecosystems. Algae differ from plants in that the zygote of an alga is on its own; the parent gives the zygote no protection. A plant zygote, on the other hand, grows into a multicellular embryo that is protected by parental tissue.

Algae exhibit a remarkable range of growth forms. Some are unicellular; others are filaments composed either of distinct cells or of coenocytes (multinucleate structures that lack cross-walls). Still others—including the algae commonly known as seaweeds—are multicellular and intricately branched or arranged in leaflike extensions. The bodies of a few types of algae are even subdivided into tissues and organs. Certain algal phyla—for example, the phylum Chlorophyta (the green algae)—include representatives of almost all these growth forms.

Algal life cycles show extreme variation, but all algae except members of the phylum Rhodophyta (red algae) have forms with flagellated motile cells in at least one stage of their life cycle. Some algae—for example, the dinoflagellates in the phylum Pyrrophyta—are unicellular and motile throughout most of their existence.

Table 26.3 summarizes the classification of algae. Note that the algae do not constitute a natural group within the Protista—they are polyphyletic. For convenience, we describe them together here to emphasize the diversity of protists.

Some algae are unicellular flagellates

All members of the phylum Pyrrophyta are unicellular. A distinctive mixture of photosynthetic and accessory pigments gives their chloroplasts a golden-brown color. The dinoflagellates, the major group within this phylum, are of great ecological, evolutionary, and morphological interest. Dinoflagellates are probably second in importance only to the diatoms (see the next section) as primary photosynthetic producers of organic matter in the oceans.

Many dinoflagellates are endosymbionts, living within the cells of other organisms, including various invertebrates and even other marine protists. Dinoflagellates are particularly common endosymbionts in corals, to whose growth they contribute mightily by photosynthesis. Some dinoflagellates are nonphotosynthetic and live as parasites within other marine organisms.

Dinoflagellates are distinctive cells (see Figure 26.1a). They have two flagella, one in an equatorial groove around the cell, the other starting at the same point as the first and passing down a longitudinal groove before extending free into the surrounding medium. Most dinoflagellates are marine organisms. Some dinoflagellates reproduce in enormous numbers in warm and somewhat stagnant waters. The result can be a "red tide," so called because of the reddish color of the sea that results from pigments in the dinoflagellates (Figure 26.21). During a red tide, the concentration of dinoflagellates may reach 60 million cells per liter of ocean water. Certain red-tide species produce a potent nerve toxin that can kill tons of fish. The genus *Gonyaulax* produces a potent toxin that can accumulate in shellfish in amounts that, although not fatal to the shellfish, may kill a person who eats the shellfish.

Many dinoflagellates are bioluminescent. In complete darkness, cultures of these organisms emit a faint glow. If you suddenly stir or bubble air through a culture containing these dinoflagellates, the organisms each emit numerous bright flashes, producing a light that is perhaps a thousandfold brighter than the dim glow of an undisturbed culture. The flashing then rapidly subsides. A ship passing through a tropical ocean that contains a rich growth of these species produces a bow wave and a wake that glow eerily as billions of these dinoflagellates discharge their light systems.

TABLE 26.3 Classification of Photosynthetic Protists

PHYLUM	COMMON NAME	FORM	LOCOMOTION	EXAMPLES
Pyrrophyta	Dinoflagellates (and others)	Unicellular	Two flagella	*Gonyaulax, Ceratium, Noctiluca*
Chrysophyta	Diatoms	Usually unicellular	Usually none	*Diatoma, Fragilaria, Ochromonas*
Phaeophyta	Brown algae	Multicellular	Two flagella on reproductive cells	*Macrocystis, Fucus*
Rhodophyta	Red algae	Multicellular or unicellular	None	*Chondrus,* coralline algae
Chlorophyta	Green algae	Unicellular, colonial, or multicellular	Most have flagella at some stage	*Chlorella, Ulva, Acetabularia*

Architectural magnificence on a microscopic scale is the hallmark of the diatoms (Figure 26.22*a*). Despite their remarkable morphological diversity, however, all diatoms are symmetrical—either bilaterally (division along only one plane results in identical halves) or radially (division along any plane that passes through the center results in identical halves).

Many diatoms deposit silicon in their cell walls. The cell wall of some species is constructed in two pieces, with the wall of the top overlapping the wall of the bottom like the top and bottom of a petri plate (Figure 26.22*b*). The silicon-impregnated walls have intricate, unique patterns; in fact, the taxonomy of these marine or freshwater organisms is based entirely on their wall patterns.

Diatoms reproduce both sexually and asexually. Asexual reproduction is by cell division and is somewhat constrained by the stiff, silica-containing cell wall. Both the top and the bottom of the "petri plate" become tops of new "plates" without changing appreciably in size; as a result, the new cells made from former bottoms are smaller than the parent cells (Figure 26.23). If the process continued indefinitely, one cell line would simply vanish, but sexual reproduction largely solves this potential problem. Gametes form, shed their cell walls, and fuse. The resulting zygote then increases substantially in size before a new wall is laid down.

Diatoms are everywhere in the marine environment and are frequently present in great numbers, making them the leading photosynthetic producers in the oceans. Diatoms are also common in fresh water. Because the silicon-containing walls of dead diatom cells resist decomposition, certain sedimentary rocks are composed almost entirely of these silica-containing skeletons that sank to the seafloor. Diatomaceous earth, which is obtained from such rocks, has many industrial uses—from insulation and filtration to metal polishing. It has also been used as an "Earth-friendly" insecticide that clogs the tracheae (breathing structures) of insects.

26.21 A Red Tide of Dinoflagellates In astronomical numbers, the dinoflagellate *Gonyaulax tamarensis* causes a toxic red tide, as seen here off the coast of Baja California.

Diatoms in their glass "houses" account for most oceanic photosynthesis

Diatoms and their relatives constitute the phylum Chrysophyta. Some species are single-celled; others are filamentous. Many have sufficient carotenoids in their chloroplasts to give them a yellow or brownish color. All make chrysolaminarin (a carbohydrate) and oils as photosynthetic storage products.

(*a*)   30 µm

(*b*)   7 µm

26.22 Diatom Diversity (*a*) Diatoms exhibit a splendid variety of species-specific forms. (*b*) The dark and light areas of this scanning electron micrograph of a diatom emphasize the distinct two-piece construction of the cell wall.

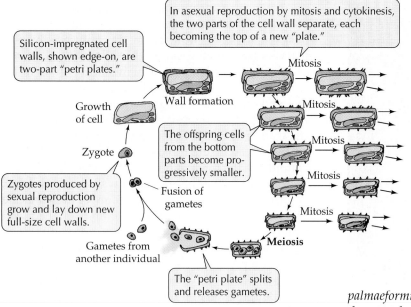

Silicon-impregnated cell walls, shown edge-on, are two-part "petri plates."

In asexual reproduction by mitosis and cytokinesis, the two parts of the cell wall separate, each becoming the top of a new "plate."

Growth of cell

Wall formation

Mitosis

Mitosis

Zygote

The offspring cells from the bottom parts become progressively smaller.

Mitosis

Zygotes produced by sexual reproduction grow and lay down new full-size cell walls.

Fusion of gametes

Mitosis

Mitosis

Gametes from another individual

Meiosis

The "petri plate" splits and releases gametes.

26.23 Diatom Reproduction Half of the cells created by asexual reproduction are always smaller than the parent cells; the sexual reproduction phase creates new parent cells with full-size cell walls.

may be up to 60 meters long (see Figure 26.1*c*). The brown algae are almost exclusively marine. Some float in the open ocean; the most famous example is the genus *Sargassum*, which forms dense mats of vegetation in the Sargasso Sea in the mid-Atlantic. Most brown algae, however, are attached to rocks near the shore. A few thrive only where they are regularly exposed to heavy surf; a notable example is the sea palm *Postelsia palmaeformis* of the Pacific coast (Figure 26.25*a*). All of the attached forms develop a specialized structure, called a **holdfast**, that literally glues them to the rocks (Figure 26.25*b*).

Some brown algae may differentiate extensively into stemlike stalks and leaflike blades, and some develop gas-filled cavities or bladders. For biochemical reasons that are only poorly understood, these gas cavities often contain as much as 5 percent carbon monoxide—a concentration high enough to kill a human.

In addition to organ differentiation, the larger brown algae also exhibit considerable tissue differentiation. Most of the giant kelps have photosynthetic filaments only in the outermost regions of the stalks and

Some brown algae are the largest protists

All members of the phylum Phaeophyta, commonly called brown algae, are multicellular, composed either of branched filaments or of leaflike growths called **thalli** (singular thallus) (Figure 26.24). The brown algae obtain their namesake color from the carotenoid fucoxanthin, which is abundant in the plastids. The combination of this yellow-orange pigment with the green of chlorophylls *a* and *c* yields a dirty brown color.

The Phaeophyta include the largest of the protists. Giant kelps, such as those of the genus *Macrocystis*,

(*a*) *Hormosira* sp.

26.24 Brown Algae (*a*) An intertidal brown alga growing in Australia. (*b*) A filamentous brown alga seen through a light microscope.

(*b*) *Ectocarpus* sp.

(b)

(a) *Postelsia palmaeformis*

26.25 Brown Algae in a Turbulent Environment Algae growing in the intertidal zone on an exposed rocky shore take a tremendous pounding by the surf. (a) Sea palm growing along the California coast. (b) The tough, branched holdfast that anchors the sea palm.

metes fuse (syngamy), a diploid organism forms (Figure 26.26). The haploid organism, the diploid organism, or both may also reproduce asexually.

The two organisms (spore-producing and gamete-producing) differ genetically, in that one has haploid cells and the other has diploid cells, but they may or may not differ morphologically. In *heteromorphic* alternation of generations, the two organisms differ morphologically; in *isomorphic* alternation of generations, they do not, despite their genetic difference. We will see examples of both heteromorphic and isomorphic alternation of generations as we consider some representative algae.

Gametes are not generally produced directly by meiosis in plants or multicellular algae. Instead, specialized cells of the diploid organism, called **sporocytes**, divide meiotically to produce four spores. The spores may eventually germinate and divide mitotically to produce multicellular haploid organisms, the

blades. Within the photosynthetic region lie filaments of long cells that closely resemble the food-conducting tissue of plants (see Chapter 32). Called trumpet cells because they have flaring ends, these tubes rapidly conduct the products of photosynthesis through the body of the alga.

The cell walls of brown algae may contain as much as 25 percent alginic acid, a gummy polymer of sugar acids. Alginic acid cements cells and filaments together and provides good holdfast glue. It is used commercially as an emulsifier in ice cream, cosmetics, and other products.

Many algal life cycles feature an alternation of generations

In **alternation of generations**, a multicellular, diploid, spore-producing organism called the **sporophyte** gives rise to a multicellular, haploid, gamete-producing organism called the **gametophyte**. When two ga-

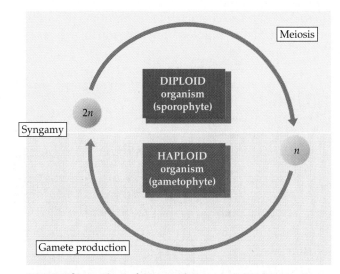

26.26 Alternation of Generations A diploid generation that produces spores alternates with a haploid generation that produces gametes.

gametophyte generation, which produces gametes—by *mitosis* and cytokinesis.

Unlike spores, gametes can produce new organisms only by fusing with other gametes. The fusion of two gametes produces a diploid zygote, which then undergoes mitotic divisions to produce a diploid organism: the sporophyte generation. The sporocytes of the sporophyte generation undergo meiosis at some point and produce haploid spores, starting the cycle anew.

BROWN ALGAE HAVE ALTERNATING GAMETOPHYTE AND SPORO- PHYTE GENERATIONS. The brown algae exemplify the extraordinary diversity found among the algae. One genus of simple brown algae is *Ectocarpus* (see Figure 26.24*b*). Its branched filaments, a few centimeters long, commonly grow on shells and stones. The gametophyte and sporophyte phases of the alternation of generations of *Ectocarpus* can be distinguished only by chromosome number or reproductive products (zoospores or gametes). Thus the generations are isomorphic.

By contrast, some kelps of the genus *Laminaria* and some other brown algae show a more complex heteromorphic alternation of generations. The larger and more obvious generation of these species is the sporophyte. Meiosis in special fertile regions of the leaflike fronds produces haploid zoospores. These germinate to form a tiny, filamentous gametophyte that produces either eggs or sperm.

The genus *Fucus* carries reduction in gametophyte size still further: It has no multicellular haploid phase—only a multi*nucleate* haploid phase. The gametes themselves are formed directly by meiosis.

Red algae may have donated organelles to other protist phyla

Almost all Rhodophyta (red algae) are multicellular (Figure 26.27). The characteristic color of the red algae is a result of the pigment phycoerythrin, which is found in relatively large amounts in the chloroplasts of many species. In addition to phycoerythrin, red algae contain phycocyanin, carotenoids, and chlorophyll. The red algae include species that grow in the shallowest tide pools, as well as the algae found deepest in the ocean (as deep as 260 meters if nutrient conditions are right and the water is clear enough to permit the penetration of light). Very few red algae inhabit fresh water. Most grow attached to a substrate by a holdfast.

In a sense the red algae, along with several other groups of algae, are misnamed. They have the capacity to change the relative amounts of their various photosynthetic pigments depending on the light conditions where they are growing. Thus the leaflike *Chondrus crispus*, a common North Atlantic red alga, may appear bright green when it is growing at or near the surface of the water and deep red when growing at greater depths.

The pigmentation—the ratio of pigments present—depends to a remarkable degree on the intensity of the light that reaches the alga. In deep water, where the light is dimmest, the algae accumulate large amounts of phycoerythrin, an accessory photosynthetic pigment (see Figure 8.7). The algae in deeper water have as much chlorophyll as the green ones near the surface, but the accumulated phycoerythrin makes them look red.

In addition to being the only algae with phycoerythrin and phycocyanin among their pigments, the red

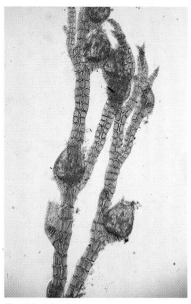

(*b*) *Polysiphonia* sp.

26.27 Red Algae (*a*) Dulse is a large, edible red alga growing on rocks in New Brunswick. (*b*) Both vegetative and reproductive structures of this alga can be seen under the light microscope.

(*a*) *Palmaria palmata*

algae have two other unique characteristics: They store the products of photosynthesis in the form of floridean starch, which is composed of very small, branched chains of approximately 15 glucose units. And they produce no motile, flagellated cells at any stage in their life cycle. The male gametes are naked and slightly amoeboid, and the female gametes are completely immobile.

Some red algal species enhance the formation of coral reefs. They share with the coral animals the biochemical machinery for depositing calcium carbonate both in and around their cell walls. After the death of the algal cells, the calcium carbonate persists, sometimes forming substantial rocky masses.

Some red algae also produce large amounts of mucilaginous polysaccharide substances, which contain the sugar galactose with a sulfate group attached. This material readily forms solid gels and is the source of agar, a substance widely used in the laboratory for making a solid aqueous medium on which tissue cultures and many microorganisms may be grown.

Some marine red algae are parasitic on other red algae. The hosts are photosynthetic, but the parasites are often colorless and nonphotosynthetic, deriving their nutrition from the host. The parasitic red alga *Choreocolax* inserts its nuclei into cells of its host red alga, *Polysiphonia*. This is the first example found of regular introduction of parasite nuclei into living host cells. Apparently some parasite genes are expressed in the host cytoplasm, diverting the host's metabolism.

Other red algae may at one time have been endosymbionts within the cells of certain other, nonphotosynthetic protists, eventually being reduced to chloroplasts. This seems to have been the evolutionary origin of the chloroplasts of a group of algae called cryptomonads. Endosymbiotic red algae may also be the ancestors of chloroplasts of the brown algae and the diatoms. Endosymbiosis has played several roles in the origins of different kinds of eukaryotic cells, as we have seen by now. Endosymbiotic cyanobacteria became chloroplasts, including those of the red algae, and red algal cells gave rise to the chloroplasts of certain other algae.

The green algae and plants form a monophyletic lineage

Many systematists place the green algae in the kingdom Plantae, and evidence from rRNA sequencing shows that the green algae and the plants form a monophyletic lineage. We will treat the green algae as the protist phylum Chlorophyta instead, as do some other systematists. Our primary motivation here is to group material for your convenience in learning; the topics in Part Five deal entirely with plants that have embryos and not at all with the green algae.

The kingdom Plantae evolved from one or more representatives of the phylum Chlorophyta. The green algae have uniform pigmentation and a characteristic storage product, starch. The Chlorophyta and the euglenoids are the only protist groups that contain the full complement of photosynthetic pigments that are also characteristic of the kingdom Plantae.

Chlorophyll *a* predominates, and a major pigment is chlorophyll *b*, which none of the other algae have. The carotenoids found in these groups, predominantly β-carotene and certain xanthophylls (carotenoids with one or more hydroxyl groups), are likewise those characteristic of plants. The principal photosynthetic storage product, like that of the plant kingdom, is long, straight or branched chains of glucose that together make up starch.

BODY SHAPE AND CELLULARITY IN THE CHLOROPHYTA. We find in the green algae an incredible variety in shape and construction of the algal body. *Chlorella* is an example of the simplest type: unicellular and flagellated. Surprisingly large and well-formed colonies of cells are found in such freshwater groups as the genus *Volvox* (Figure 26.28*a*). The cells are not differentiated into tissues and organs as in plants and animals, but the colonies show vividly how the preliminary step of this great evolutionary development might have been taken.

The intermediate stages between the one-celled state and the extreme colonial state of *Volvox* are preserved in loosely colonial forms, such as *Gonium* and *Pandorina*. By contrast, *Oedogonium* is multicellular and filamentous, and each of its cells has only one nucleus. *Cladophora* is multicellular, but each cell is multinucleate. *Bryopsis* is tubular and coenocytic, forming cross-walls only when reproductive structures form. *Acetabularia* is a single, giant uninucleate cell a few centimeters long and with remarkable morphology, becoming multinucleate only at the end of the reproductive stage. *Ulva lactuca* is a membranous sheet two cells thick; its unusual appearance justifies its common name: sea lettuce (Figure 26.28*b*). Finally, the remarkable unicellular desmids have elaborately sculptured cell walls (Figure 26.28*c*).

LIFE CYCLES IN THE CHLOROPHYTA. The life cycles of green algae are diverse. We will examine two algal life cycles in detail, beginning with that of the sea lettuce *Ulva lactuca* (Figure 26.29). The diploid sporophyte of this common seashore alga is a "leaf" a few centimeters in diameter. Specialized cells (sporocytes) differentiate and undergo meiosis and cytokinesis, producing motile haploid spores (zoospores). These swim away, each propelled by four flagella, and some eventually find a suitable place to settle. The spores then lose their flagella and begin to divide mitotically, producing a thin filament that develops into a broad sheet only two cells thick. The gametophyte thus produced looks just like the sporophyte.

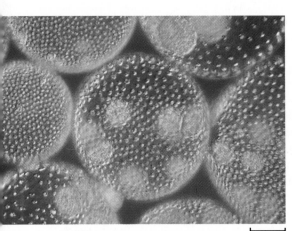

(a) *Volvox* sp.

18 μm

(b) *Ulva lactuca*

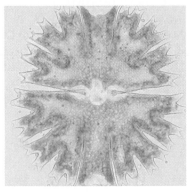

(c) *Micrasterias* sp.

26.28 Green Algae (a) Colonies showing the precise spacing of cells and containing numerous daughter colonies. (b) A stand of sea lettuce, submerged in a tidal pool. (c) A microscopic desmid. A narrow, nucleus-housing isthmus joins two elaborate semicells—halves of the unicellular organism. A single large, ornate chloroplast fills much of the volume of each semicell.

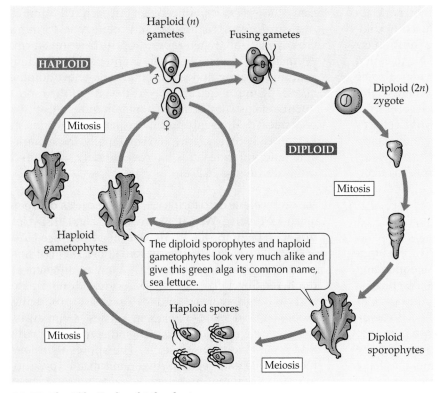

26.29 The Life Cycle of *Ulva lactuca*

Each spore contains genetic information for just one mating type (see Chapter 28), and a given gametophyte can produce only male or female gametes—never both. The gametes arise mitotically within single cells (gametangia), rather than within a specialized multicellular structure, as in plants (see Chapter 27). Both types of gametes bear two flagella (in contrast to the four flagella of a haploid spore) and hence are motile.

In most species of *Ulva* the female and male gametes are indistinguishable structurally, making those species **isogamous**—having gametes of identical appearance. Some other algae and absorptive protists are also isogamous. Yet other algae, including some species of *Ulva*, are **anisogamous**—having female gametes that are distinctly larger than the male gametes. Female and male gametes come to-

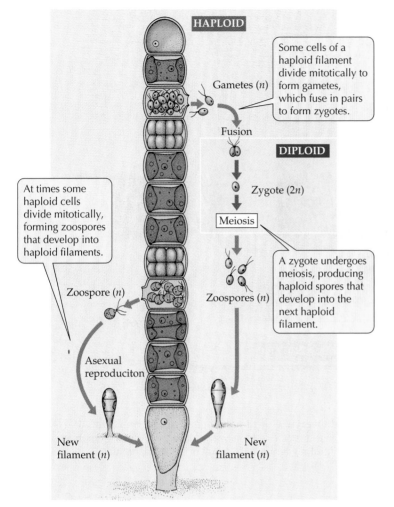

HAPLOID

Gametes (n)

Some cells of a haploid filament divide mitotically to form gametes, which fuse in pairs to form zygotes.

Fusion

DIPLOID

Zygote (2n)

Meiosis

At times some haploid cells divide mitotically, forming zoospores that develop into haploid filaments.

Zoospore (n)

Zoospores (n)

A zygote undergoes meiosis, producing haploid spores that develop into the next haploid filament.

Asexual reproduciton

New filament (n)

New filament (n)

26.30 A Haplontic Life Cycle In the life cycle of *Ulothrix,* a filamentous, multicellular gametophyte generation alternates with a sporophyte generation consisting of a single cell.

gether and unite, losing their flagella as the zygote forms and settles. After resting briefly, the zygote begins mitotic division, producing a multicellular sporophyte. Any gametes that fail to find partners can settle down on a favorable substrate, lose their flagella, undergo mitosis, and produce a new gametophyte directly; in other words, the gametes can also function as zoospores. Few algae other than *Ulva* have motile gametes that can also function as zoospores.

A life cycle such as that of *Ulva* is isomorphic: Sporophyte and gametophyte generations are identical in structure. By contrast, in many other algae, the generations are heteromorphic: Sporophyte and gametophyte generations differ in structure. In one variation of the heteromorphic cycle—the **haplontic** life cycle (Figure 26.30)—a multicellular haploid individual produces gametes that fuse to form a zygote. The zygote functions directly as a sporocyte, undergoing

meiosis to produce spores, which in turn produce a new haploid individual. In the entire haplontic life cycle, only one cell—the zygote—is diploid. The filamentous green algae of the genus *Ulothrix* are examples of haplontic algae.

Other algae have a **diplontic** life cycle like that of many animals. In the diplontic life cycle, meiosis of sporocytes produces gametes directly; the gametes fuse, and the resulting zygote divides mitotically to form a new multicellular sporophyte. In such organisms, every cell except the gametes is diploid. Between these two extremes are algae whose gametophyte and sporophyte generations are both multicellular, but that have one phase (usually the sporophyte) that is much larger and more prominent than the other.

EVOLUTIONARY TRENDS IN THE CHLOROPHYTA. Among the green algae, some evolutionary trends have been identified. For example, some groups show a trend toward multicellularity. Another trend leads from the mass release of large numbers of tiny, seemingly undifferentiated gametes (isogamy) toward increased protection of a single, large egg (anisogamy). For example, the female gamete of *Oedogonium* is large and immobile, and the male gamete is free-swimming and flagellated—an extreme case of anisogamy known as oogamy.

Phylogeny of the Protists

Can we make sense of protist diversity? Molecular systematists still wrestle with the relationships among the groups we've considered here—some of which are themselves not monophyletic, natural groups. Just as in the kingdom Eubacteria, the systematics of which is complicated by the seemingly easy gain and loss by bacteria of metabolic pathways, it appears that chloroplasts have been gained and lost by protists.

Still, progress is being made, particularly but not exclusively in the application of molecular tools such as rRNA sequencing. Figure 26.31 summarizes the most convincing phylogeny currently available, focusing primarily on the protists. Evidence from molecular systematics supports each feature of the figure, but not all the evidence is consistent. We show only a few of the numerous protist groups, and some of the ones we omit present problems.

Biologists disagree about the number of kingdoms of living things. Figure 26.31 is not a lot of help in resolving the problem, and adding groups that we've omitted from the figure would not make it any easier. Biologists continue to pursue this vexing problem—how many kingdoms are there?—because we want to understand the full and true course of evolutionary history. Perhaps the kingdom concept will fade away and be replaced by an emphasis on the numerous protist lineages.

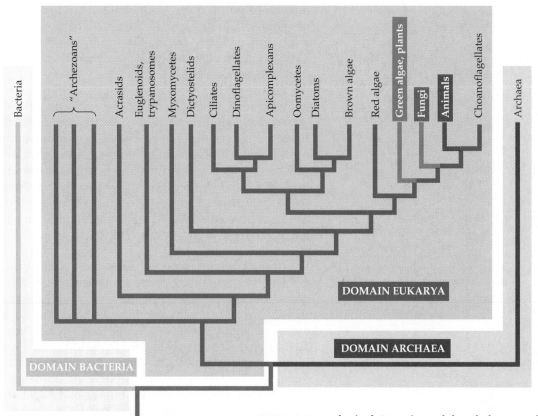

26.31 A Hypothetical Overview of the Phylogeny of Life on Earth We don't know for certain that it happened this way, but this phylogeny (emphasizing the protists, shown in blue) agrees with many of the available data.

Summary of "Protists and the Dawn of the Eukarya"

Protista and the Other Eukaryotic Kingdoms

• In this book we define the kingdom Protista as all eukaryotes that are not plants, fungi, or animals.

• The Protista arose from prokaryotic origins. The other eukaryotic kingdoms (Plantae, Fungi, Animalia) arose from protists.

• We distinguish protists from animals on the bases of how food enters the body (animals are ingestive heterotrophs), how development proceeds (animals have a blastula stage), and the presence (animals) or absence (protists) of an extracellular matrix.

• Protists that were once considered to be fungi differ from fungi in that they lack chitinous walls in their absorptive phase and their life cycles have no dikaryotic phase.

• Photosynthetic protists (algae) differ from plants in that they lack a protected embryo.

The Origin of the Eukaryotic Cell

• The modern eukaryotic cell arose in several steps. One possible scenario is as follows: An ancestral prokaryotic cell lost its cell wall, becoming more flexible. The plasma membrane folded into the cytoplasm, increasing the cell's surface area, thus allowing an increase in cell size and leading to the formation of vesicles that gained digestive and other capabilities. **Review Figure 26.2**

• In subsequent steps in the origin of the eukaryotic cell, infolded plasma membrane attached to the chromosome may have led to the formation of a nuclear envelope. Ribosomes became associated with internal membranes, some vesicles developed into lysosomes, and a primitive cytoskeleton evolved. **Review Figure 26.3**

• The first truly eukaryotic cell was much larger than the ancestral prokaryote, and it probably possessed one or more flagella of the eukaryotic type.

• The incorporation of prokaryotic cells as endosymbionts gave rise to eukaryotic organelles. Peroxisomes, protecting the host cell from an O_2-rich atmosphere, may have been the first of the organelles of endosymbiotic origin. Mitochondria evolved from once free-living proteobacteria. Then chloroplasts evolved from once free-living cyanobacteria. According to this hypothetical scenario, cells with nuclei appeared before the first cells with mitochondria. **Review Figure 26.3**

• Archezoans resemble a proposed intermediate evolutionary stage that lends support to the hypothesis that the eukaryotic nucleus evolved before the mitochondrion. **Review Figure 26.4**

General Biology of the Protists

- Most protists are aquatic; some live within other organisms. The great majority are unicellular and microscopic, but many are multicellular and a few are enormous.
- Protists vary widely in their nutrition, metabolism, and locomotion. Some protist cells contain contractile vacuoles, and some digest their food in food vacuoles. **Review Figures 26.5, 26.6**
- Protists have a variety of cell surfaces, some of them protective. **Review Figure 26.7**
- Many protists contain endosymbiotic prokaryotes. Some protists are endosymbionts in other cells, often other protists. Some endosymbiotic protists perform photosynthesis, to the advantage of their hosts.
- Most protists reproduce both asexually and sexually.

Protozoan Diversity

- As defined here, all protozoans are unicellular.
- Flagellates—protists with flagella—are the largest group of protozoans, but they are not a single, natural group. Some are parasitic, and some of these parasitic flagellates are important human pathogens. One flagellate group, the Choanoflagellida, is closely related to both the animal and fungal kingdoms. The animals and Choanoflagellida share an ancestor that no other organisms have. Some euglenoids, although flagellates, are photosynthetic. **Review Figure 26.10 and Table 26.1**
- Amoebas, which appear in many protist phyla, move by means of pseudopods. **Review Table 26.1**
- Actinopods have thin, stiff pseudopods that serve various functions, including food capture. **Review Table 26.1**
- Foraminiferans (phylum Foraminifera) also use pseudopods for feeding, and they secrete shells of calcium carbonate. **Review Table 26.1**
- Apicomplexans (phylum Apicomplexa) are parasites that have rigid cell walls and whose spores are adapted to the invasion of host tissue. The apicomplexan *Plasmodium*, which causes malaria, uses two alternate hosts (humans and *Anopheles* mosquitoes). **Review Figure 26.12 and Table 26.1**
- Ciliates (phylum Ciliophora) move rapidly by means of cilia and have two kinds of nuclei: a macronucleus and one or more micronuclei. The macronucleus controls the cell by way of transcription and translation. The micronuclei are responsible for genetic recombination, accomplished by conjugation that is sexual but not reproductive. Some ciliates have a remarkably complex internal structure. **Review Figures 26.14, 26.15, 26.16, 26.17**

Protists Once Considered To Be Fungi

- Acellular slime molds (phylum Myxomycota) and cellular slime molds (phyla Dictyostelida and Acrasida) are superficially very similar, moving as slimy masses and producing stalked fruiting structures. However, they differ at the cellular level. Acellular slime molds are coenocytes with diploid nuclei. Cellular slime molds consist of individual haploid cells that aggregate into masses consisting of distinct cells. **Review Table 26.2**
- Oomycetes (phylum Oomycota) include water molds, downy mildews, and some other protists. Like the acellular slime molds, the oomycetes are coenocytic. They are diploid for most of their life cycle. A few oomycetes are serious plant pathogens. **Review Table 26.2**

The Algae

- Algae are photosynthetic protists responsible for about half of the photosynthesis on Earth. Unlike plants, algae have no multicellular protection for their zygotes. As a group, the algae are polyphyletic.
- Dinoflagellates (phylum Pyrrophyta) are unicellular; they are major contributors to world photosynthesis. Many are endosymbionts; in that role they are important contributors to coral reef growth. Dinoflagellates are responsible for toxic "red tides." **Review Table 26.3**
- Diatoms (phylum Chrysophyta) are unicellular and have complex, two-part, glassy cell walls. They contribute extensively to world photosynthesis. **Review Figure 26.23 and Table 26.3**
- The brown algae (phylum Phaeophyta) and red algae (phylum Rhodophyta) are predominantly multicellular. The two differ in their pigmentation and in other ways. The brown algae include the largest of all protists, and some show considerable tissue differentiation. **Review Table 26.3**
- In many algae, both haploid and diploid cells undergo mitosis, leading to an alternation of generations. The diploid sporophyte generation forms spores by meiosis, and the spores develop into haploid organisms. This haploid gametophyte generation forms gametes by mitosis, and their fusion yields zygotes that develop into the next generation of sporophytes. **Review Figure 26.26**
- Red algae have a characteristic storage product (floridean starch) and differ from the other algal groups in lacking flagellated reproductive cells. Some red algae are parasitic; others gave rise to chloroplasts in some other algae.
- The green algae (phylum Chlorophyta) are often multicellular. They have the same chloroplast pigments and storage product as do plants, which descended from an ancestral green alga. Several life cycles are used by different green algae; among these are the isomorphic alternation of generations of *Ulva* and the haplontic cycle of *Ulothrix*. **Review Figures 26.29, 26.30**

Phylogeny of the Protists

- The evolutionary relationships among the various protist groups, and even their exact composition, remain controversial. However, some proposals for a natural phylogeny have been made, based in part on rRNA sequencing and other molecular techniques. Biologists have different views on the number of kingdoms into which the living world should be divided. **Review Figure 26.31**

Self-Quiz

1. Flagellates
   - *a.* appear in several protist phyla.
   - *b.* are all algae.
   - *c.* all have pseudopods.
   - *d.* are all colonial.
   - *e.* are never pathogenic.

2. Which statement about amoebas is *not* true?
   - *a.* They are specialized.
   - *b.* They use amoeboid movement.
   - *c.* They include both naked and shelled forms.
   - *d.* They possess pseudopods.
   - *e.* They appeared only once in evolutionary history.

3. The Apicomplexa
    a. possess flagella.
    b. possess chloroplasts.
    c. are all parasitic.
    d. are algae.
    e. include the trypanosomes that cause sleeping sickness.

4. The Ciliophora
    a. move by means of flagella.
    b. use amoeboid movement.
    c. include *Plasmodium*, the agent of malaria.
    d. possess both a macronucleus and micronuclei.
    e. are autotrophic.

5. The Myxomycota
    a. are also called the acellular slime molds.
    b. lack fruiting bodies.
    c. consist of large numbers of myxamoebas.
    d. consist at times of a mass called a pseudoplasmodium.
    e. possess flagella.

6. The Dictyostelida
    a. are also called the acellular slime molds.
    b. lack fruiting bodies.
    c. form a plasmodium that is a coenocyte.
    d. use cAMP as a "messenger" to signal aggregation.
    e. possess flagella.

7. Which statement about algae is *not* true?
    a. They differ from plants in lacking protected embryos.
    b. They are photosynthetic autotrophs.
    c. They have chitin in their cell walls.
    d. They include both unicellular and multicellular forms.
    e. Their life cycles show extreme variation.

8. Which statement about the Phaeophyta is *not* true?
    a. They are all multicellular.
    b. They use the same photosynthetic pigments as do plants.
    c. They are almost exclusively marine.
    d. A few are among the largest organisms on Earth.
    e. Some have extensive tissue differentiation.

9. The Rhodophyta
    a. are mostly unicellular.
    b. are mostly marine.
    c. owe their red color to a special form of chlorophyll.
    d. have flagella on their gametes.
    e. are all heterotrophic.

10. Which statement about the Chlorophyta is *not* true?
    a. They use the same photosynthetic pigments as do plants.
    b. Some are unicellular.
    c. Some are multicellular.
    d. All are microscopic in size.
    e. They display a great diversity of life cycles.

Applying Concepts

1. For each type of organism below, give a single characteristic that may be used to differentiate it from the other, related organism(s) in parentheses.
    a. foraminiferans (radiolarians)
    b. *Vorticella* (*Paramecium*)
    c. *Euglena* (*Volvox*)
    d. *Trypanosoma* (*Giardia*)
    e. amoeba (flagellate)
    f. *Physarum* (*Dictyostelium*)

2. For each of the following groups, give at least two characteristics used to distinguish the group from other groups. (*a*) Ciliophora; (*b*) Apicomplexa; (*c*) Phaeophyta; (*d*) Rhodophyta.

3. Identify one major role in the world that is played by the photosynthetic protists. Identify one major role in the world that is played by the absorptive protists.

4. Giant seaweed (mostly brown algae) have "floats" that aid in keeping their fronds suspended at or near the surface of the water. Why is it important that the fronds be suspended?

5. Justify the placement of *Euglena* among the flagellate protozoans. Why might it be considered part of the phylum Chlorophyta?

6. Why are algal pigments so much more diverse than those of plants?

Readings

Anderson, D. M. 1994. "Red Tides." *Scientific American*, August. A discussion of the origin and consequences of the dinoflagellate blooms.

Brusca, R. C. and G. J. Brusca. 1990. *Invertebrates*. Sinauer Associates, Sunderland, MA. Chapter 5, on the protozoans, also covers many algae—showing once again that different authorities follow different taxonomic schemes for the kingdom Protista.

de Duve, C. 1996. "The Birth of Complex Cells." *Scientific American*, April. One hypothesis, offered by a leading scholar of cell structure and function, concerning the evolution of the eukaryotic cell by loss, elaboration, and gain of parts.

Donelson, J. E. and M. J. Turner. 1985. "How the Trypanosome Changes Its Coat." *Scientific American*, February. A description of how trypanosomes evade the host's immune system by constantly switching on new genes that code for different surface antigens.

Farmer, J. N. 1980. *The Protozoa*. Mosby, St. Louis. A full treatment of the animal-like protists.

Fenchel, T. and B. J. Finlay. 1994. "The Evolution of Life without Oxygen." *American Scientist*, vol. 82, pages 22–29. Studies on eukaryote origins.

Godon, G. N. 1985. "Molecular Approaches to Malaria Vaccines." *Scientific American*, May. The life cycle and molecular biology of a problem pathogen.

Harrison, F. W. and J. O. Corliss (Eds.). 1991. *Microscopic Anatomy of Invertebrates*. Volume 1: *Protozoa*. Wiley-Liss, New York. Includes marvelous illustrations.

Jacobs, W. P. 1994. "Caulerpa." *Scientific American*, December. An exploration into the biology of the largest unicellular organism.

Kabnick, K. S. and D. A. Peattie. 1991. "*Giardia*: A Missing Link between Procaryotes and Eucaryotes." *American Scientist*, vol. 79, pages 34–43. A description of the intestinal parasite that appears to be an intermediate step in the origin of the eukaryotic cell.

Margulis, L. 1993. *Symbiosis in Cell Evolution*, 2nd Edition. W. H. Freeman, New York. It is interesting to review the theory as you study the protists, which may illustrate some of the stages proposed by Margulis.

Vidal, G. 1984. "The Oldest Eukaryotic Cells." *Scientific American*, February. A look at ancient marine protists.

Chapter 27

Plants: Pioneers of the Terrestrial Environment

What's missing?

*A*utotrophs are the vital base of every present-day food web—they feed the entire living world. Photosynthetic protists and bacteria feed the great bulk of oceanic life, and chemoautotrophic bacteria support thermal vent communities in the ocean deeps. But what about the terrestrial environment, the part of Earth that *we* inhabit? Terrestrial Earth is supported by the photosynthesis of plants. But at one time there were no plants.

What a thrill it would be to see the face of Earth at different stages of its evolution. Successive photographs from space would show the drifting of the continents. On Earth's surface we would see scenes scarcely imaginable to us today. For more than 4 billion years the terrestrial environment was basically a mass of rock. Its appearance changed relatively little until plants colonized the land and slowly spread over its surface.

Earth did not take on a green tint until less than half a billion years ago, long after the ancestors of today's plants had invaded the land sometime during the Paleozoic era (see Table 19.1). The earliest land plants were tiny, but their metabolic activities

Here they are!

helped convert native rock to soil that could support the needs of their successors. Evolution led rapidly (in geologic terms) to larger and larger plants, and in the Carboniferous period (345 to 290 million years ago) great forests were widespread. Those forests would look unfamiliar to us, for few of their trees were like those we know today.

579

During the tens of millions of years since, these trees were replaced by others more familiar to us—pinelike conifers and their relatives and, in the last 100 million years, the broad-leaved forms of modern vegetation. Today Earth is a patchwork of widely differing environments, supporting differing communities of plants, animals, fungi, protists, archaea, and bacteria. What do we notice when we look at any of these environments? Consciously or not, we see its *plants*—or, in a very harsh environment, its relative *lack* of plants.

In this chapter we will see how the plant kingdom differs from the other kingdoms and how it conquered the land and then evolved. We will survey the diverse products of plant evolution, from evolutionarily ancient groups to the one that appeared most recently, the flowering plants.

The Plant Kingdom

Broadly defined, a plant is a *multicellular photosynthetic eukaryote.* The definition is not precise, because a few organisms that can be regarded only as plants are not photosynthetic. These parasitic species, such as Indian pipe (Figure 27.1), are clearly related to photosynthetic plants (they have leaves, roots, flowers, and so on) but possess adaptations that provide them with alternative modes of nutrition. The definition of plants casts an enormous net over a wide range of organisms that, while all multicellular and photosynthetic, differ in size, cellular organization, photosynthetic and associated pigments, and cell wall chemistry—and do not form a monophyletic group. At present we prefer to narrow the definition of a plant to exclude algae, which we discussed in Chapter 26.

According to this narrower definition, plants are multicellular, photosynthetic eukaryotes that have the following additional properties: *They develop from embryos protected by tissues of the parent plant.* (This characteristic is definitive; an older classification scheme used the term Embryophyta to refer to precisely those organisms that we assign here to the kingdom Plantae.) Their cells have walls that contain cellulose as the major strengthening polysaccharide. Their chloroplasts contain chlorophylls *a* and *b* and a limited array of specific carotenoids (see Chapter 8). Their major storage carbohydrate is starch. Their life cycles have important features in common, notably an alternation of haploid and diploid generations.

Life cycles of plants feature alternation of generations

A universal feature of the life cycles of plants is the alternation of generations (see Chapters 9 and 26). If we consider the plant life cycle to begin with a single cell, the diploid zygote, then the first phase of the cycle features the formation, by mitosis and cytokinesis, of a

Monotropa uniflora

27.1 Stretching the Definition Indian pipe is not photosynthetic, but it is a plant in all other respects.

multicellular embryo and eventually the mature diploid plant (see Figure 26.26).

This multicellular, diploid plant is the **sporophyte** ("sporophyte" means "spore plant"). Cells contained in **sporangia** (singular sporangium, "spore reservoir") on the sporophyte undergo meiosis to produce haploid, unicellular spores. By mitosis and cytokinesis a spore forms a haploid plant. This multicellular, haploid plant, the **gametophyte** ("gamete plant"), produces haploid gametes. The fusion of two gametes (syngamy, or fertilization) results in the formation of a diploid cell, the zygote, and the cycle begins again.

The sporophyte generation extends from the zygote through the adult, multicellular, diploid plant; the gametophyte generation extends from the spore through the adult, multicellular, haploid plant to the gamete. The transitions between the phases are accomplished by fertilization and meiosis.

The gametophyte and sporophyte of any plant look totally different, unlike the generations of some green algae, such as *Ulva*, which are indistinguishable to the eye (see Figure 26.29). In all plants and green algae, however, the sporophyte and gametophyte differ genetically: One has diploid cells, the other haploid cells. This alternation of diploid and haploid generations is found in the life cycles of all phyla of plants.

There are twelve surviving phyla of plants

The surviving members of the kingdom Plantae fall naturally into 12 phyla (Table 27.1). We will group some of these phyla to emphasize the evolutionary trends you'll be learning, but bear in mind that these groups are for convenience and that they are not natural groups in the sense that the phyla are.

All members of nine of the phyla possess well-developed "plumbing systems" that transport materials throughout the plant body. We call these nine phyla,

TABLE 27.1   Classification of Plants[a]

PHYLUM	COMMON NAME	CHARACTERISTICS
Nontracheophytes		
Hepatophyta	Liverworts	No filamentous stage
Anthocerophyta	Hornworts	Embedded archegonia
Bryophyta	Mosses	Filamentous stage
Tracheophytes		
Nonseed tracheophytes		
Lycophyta	Club mosses	Simple leaves in spirals
Sphenophyta	Horsetails	Simple leaves in whorls
Psilophyta	Whisk ferns	No true leaves
Pterophyta	Ferns	Complex leaves
Seed plants		
Gymnosperms		
Cycadophyta	Cycads	Fernlike leaves; swimming sperm; seeds in strobili
Ginkgophyta	Ginkgo	Deciduous; fan-shaped leaves
Gnetophyta	Gnetophytes	Vessels in vascular tissue
Coniferophyta	Conifers	Many have seeds in cones; needlelike leaves
Angiosperms		
Angiospermae	Flowering plants	Double fertilization; seeds in fruit

[a] No extinct groups are included in this classification.

collectively, the **tracheophytes**. The remaining three phyla (liverworts, hornworts, and mosses) were once considered classes of a single larger phylum, of which the most familiar examples are mosses. Now we use the term **nontracheophytes** to refer collectively to these three phyla. Although the older term "nonvascular plants" is a time-honored name, it is misleading in that some mosses, unlike liverworts and hornworts, have a limited amount of vascularlike tissue. The nontracheophytes are sometimes collectively called bryophytes, but we will reserve that term for the mosses.

Plants arose from green algae

The plant kingdom arose from the green algae (the protist phylum Chlorophyta). The characteristics of green algae that originally identified them as the most likely ancestors of the plants include the possession of photosynthetic and accessory pigments in their plastids that are similar to those of plants, the use of starch as their principal storage carbohydrate, and the presence of cellulose as the principal component of their cell walls.

In addition, some green algae (not all) show the same oogamous type of reproduction—which features a large, stationary egg—that is characteristic of sexual reproduction in plants. Like plants, a few green algae have a life cycle that includes both a multicellular gametophyte and a multicellular sporophyte generation. Furthermore, some green algae have bulky, three-dimensional bodies like plants, although most are unicellular, filamentous, or two-dimensional. And the chloroplasts of plants and many green algae have thylakoids that are arranged into grana. In addition, some green algae produce new cell walls after cell division using a mechanism similar to that in plants. No other phylum of algal protists shares so many traits with the Plantae. Molecular biological techniques such as rRNA sequencing have confirmed that plants evolved from green algal ancestors.

Much evidence indicates that the closest relatives of the plants are two groups of living green algae (stoneworts and *Coleochaete*-like algae), but we don't yet know which is the true sister group. The stoneworts are characterized by multicellular sex organs and complex body patterns, and they resemble the plants in terms of their rRNA and DNA sequences, peroxisome contents, mechanics of mitosis and cytokinesis, and chloroplast structure (Figure 27.2). Other strong evidence, from morphology-based cladistic analysis, suggests that the sister group of the plants is a group of green algae that includes the genus *Coleochaete*. Those algae have features, such as plas-

Chara sp. (stonewort)

27.2 The Closest Relatives of Plants
The plant kingdom evolved from an ancient organism that was also ancestral to this protist.

modesmata and embryolike structures, found in plants. The ancestral algae, whether more similar to stoneworts or to *Coleochaete*, lived at the margins of ponds or marshes, ringing them with a green mat. From the margins of ponds, plants moved onto the land.

Plants colonized the land

Plants or their immediate ancestors in the green mat pioneered and modified the terrestrial environment. That environment differs dramatically from the aquatic environment in several ways. The density of the plant body is much closer to that of water than of air, so in water aquatic plants are buoyant and are supported against gravity.

A multicellular plant on land must either have a support system against gravity or else sprawl unsupported on the ground. It also must have the means to transport water and minerals from the soil to its aerial (raised) parts or else live where water is abundant enough to bathe all of its parts regularly. It must be able to resist desiccation (drying). And it must use different mechanisms for dispersing its gametes and progeny than do its aquatic relatives. How did such organisms arise from aquatic ancestors to thrive in such a challenging environment?

Most present-day plants have vascular tissue

An ancestral organism gave rise to the first plants, which were, in turn, ancestral to all present-day plants. The first plants were truly nonvascular, lacking both water-conducting and food-conducting cells. Later, the first tracheophytes arose from a nontracheophyte ancestor (Figure 27.3).

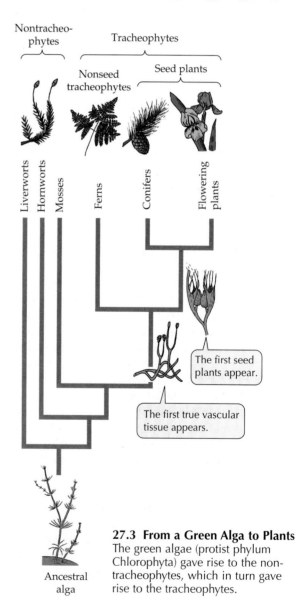

27.3 From a Green Alga to Plants
The green algae (protist phylum Chlorophyta) gave rise to the nontracheophytes, which in turn gave rise to the tracheophytes.

27.4 Some Nontracheophytes Form Mats   Dense moss forms hummocks in a valley on New Zealand's South Island.

Liverworts, hornworts, and mosses never have been large plants. Except for some of the mosses, they have no water-transporting tissue, yet some are found in dry environments. Many grow in dense masses (Figure 27.4), through which water can move by capillary action. The "leafy" structure of nontracheophytes readily catches and holds water that splashes onto them. These plants are small enough that minerals can be distributed internally by diffusion. They lack the leaves, stems, and roots that characterize the tracheophytes, although they have structures analogous to each.

The tracheophytes include the ferns, conifers, and flowering plants. Tracheophytes differ from liverworts, hornworts, and mosses in crucial ways, one of which is the possession of a well-developed **vascular system** consisting of specialized tissues for the transport of materials from one part of the plant to another. One tissue, the **xylem**, conducts water and minerals from the soil to aerial parts of the plant; the other, the **phloem**, conducts the products of photosynthesis from sites of production or release to sites of utilization or storage (see Chapters 31 and 32).

The tracheophytes appear earlier than the nontracheophytes in the fossil record. The oldest tracheophyte fossils date back more than 410 million years, whereas the oldest nontracheophyte fossils are about 350 million years old, dating from a time when tracheophytes were already widely distributed. This does not mean that the tracheophytes are evolutionarily more ancient than the nontracheophytes; they are simply more likely to form fossils, given their structure and the chemical makeup of their cell walls.

Adaptations to life on land distinguish plants from algae

Most of the characteristics that distinguish plants from algae are evolutionary adaptations to life on land. Both nontracheophytes and tracheophytes have protective coverings that prevent drying, and both have means of taking up water from the soil (most tracheophytes have roots; nontracheophytes have rhizoids). Plants derive support against gravity from the internal fluid pressure (turgor pressure; see Chapter 32) of plant cells, from a woody stem in some tracheophytes, or from thickened cell walls in nontracheophytes. Unlike the algae, plants form embryos, young sporophytes contained within a protective structure.

We will examine the adaptations of the tracheophytes later in the chapter, but let's concentrate first on the nontracheophytes.

Nontracheophytes: Liverworts, Hornworts, and Mosses

Most liverworts, hornworts, and mosses are small and grow in dense mats in moist habitats (see Figure 27.4). The largest of these plants are only about 1 meter tall, but most are only a few centimeters tall or long. Why have larger nontracheophytes never evolved? The probable answer is that they lack an efficient system for conducting water and minerals from the soil to distant parts of the plant body.

Most nontracheophytes live on the soil or on other plants, but some grow on bare rock, dead and fallen tree trunks, and even on the buildings in which we live and work. These small plants have been here much longer than we have, and perhaps a quarter of a billion years longer than the tracheophytes have. Nontracheophytes are widely distributed over six continents and exist very locally on the coast of the seventh (Antarctica). They are very successful plants, well adapted to their environments. Most are terrestrial. Some live in wetlands. Although a few nontracheophytes live in fresh water, these aquatic forms are descended from terrestrial ones. There are no marine nontracheophytes. All nontracheophytes have a waxy covering that retards water loss, and their embryos are protected within layers of maternal tissue.

Nontracheophyte sporophytes are dependent on gametophytes

The life cycles of the nontracheophytes differ sharply from those of other plants in that the conspicuous green nontracheophyte that we recognize is the gametophyte, whereas the familiar forms of ferns and seed plants are sporophytes. The nontracheophyte gametophyte is photosynthetic and therefore nutritionally independent, whereas the sporophyte may or may not be photosynthetic but is *always* dependent on the gametophyte and remains permanently attached to it.

A sporophyte produces unicellular, haploid spores as products of meiosis. A spore germinates, giving rise to a multicellular, haploid gametophyte whose cells contain chloroplasts and are thus photosynthetic. Eventually, gametes form within specialized sex organs. The **archegonium** is a multicellular, flask-shaped female sex organ with a long neck and a swollen base (Figure 27.5a). The base contains a single egg. The **antheridium** is a male sex organ in which sperm, each bearing two flagella, are produced in large numbers (Figure 27.5b).

Once released, the sperm must swim or be splashed by raindrops to a nearby archegonium on the same or a neighboring plant. The sperm are aided in this task by chemical attractants released by the egg or the archegonium. Before sperm can enter the archegonium, certain cells in the neck of the archegonium must break down, leaving a water-filled canal through which the sperm swim to complete their journey. Note the dependence on liquid water for all these events.

On arrival at the egg, one of the sperm nuclei fuses with the egg nucleus to form the zygote. Mitotic divisions of the zygote produce a multicellular, diploid sporophyte embryo. The base of the archegonium grows to protect the embryo during its early growth. Eventually the developing sporophyte elongates sufficiently to break out of the archegonium, but it remains connected to the gametophyte by a "foot" that is embedded in the parent tissue and absorbs water and nutrients from it (see Figure 27.8). The sporophyte remains attached to the gametophyte throughout its life. The sporophyte produces a sporangium, or **capsule**, within which meiotic divisions produce spores and thus the next gametophyte generation.

The structure and pattern of elongation of the sporophyte differ among the three phyla of nontracheophytes—the liverworts (Hepatophyta), hornworts (Anthocerophyta), and mosses (Bryophyta). The evolutionary relationships of the three phyla and the tracheophytes can be seen in Figure 27.3.

Archegonia develop at the tip of a gametophyte. In the archegonium the egg will be fertilized and begin development into a sporophyte.

(a)

The large egg cell in the center of the archegonium looks like an eye.

Antheridia are also located at the tip of the gametophyte.

(b)

These male organs (antheridia) contain a large number of sperm. When released, the sperm must locate an archegonium and swim down its neck to the egg.

27.5 Nontracheophyte Sex Organs Archegonia (*a*) and antheridia (*b*) of the moss *Mnium* (phylum Bryophyta).

Liverwort sporophytes have no specific growing zone

The gametophytes of some liverworts (phylum Hepatophyta) are "leafy" and prostrate. The simplest liverwort gametophytes, however, are flat plates of cells, a centimeter or so long, that produce antheridia or archegonia on their upper surfaces and rhizoids on the lower. Liverwort sporophytes are shorter than those of mosses and hornworts, rarely exceeding a few millimeters.

The sporophyte has a stalk that connects capsule and foot. The stalk elongates and thus raises the cap-

The finger-headed structures bear archegonia.

The disc-headed structures bear antheridia.

These cups are filled with gemmae—small, lens-shaped outgrowths of the body, each capable of developing into a new plant.

(a) *Marchantia* sp.   (b) *Marchantia* sp.   (c) *Lunularia* sp.

27.6 Liverwort Structures   Members of the phylum Hepatophyta display various characteristic structures. (*a*) Gametophytes. (*b*) Antheridia and archegonia. (*c*) Gemmae.

sule above ground level, favoring dispersal of spores when they are released. The sporophyte elongates over the entire length of the stalk, which has no specific growing zone. The liverwort sporophyte has no stomata (pores allowing gas exchange between the atmosphere and the plant's interior).

The capsules of liverworts are simple: a globular capsule wall surrounding a mass of spores. In some species of liverworts spores are not released by the sporophyte until the surrounding capsule wall rots. In other liverworts, however, the spores are disseminated by structures called **elaters** located within the capsule. Elaters are long cells that have a helical thickening of the cell wall. As an elater loses water, the whole cell shrinks longitudinally to a fraction of its former length, thus compressing the helical thickening like a spring. When the stress becomes sufficient, the compressed "spring" snaps back to its resting position, throwing spores in all directions.

Among the most familiar liverworts are species of the genus *Marchantia* (Figure 27.6*a*). *Marchantia* is easily recognized by the characteristic structures on which its male and female gametophytes bear their antheridia and archegonia (Figure 27.6*b*). Like most liverworts, *Marchantia* also reproduces vegetatively by simple fragmentation of the gametophyte. *Marchantia* and some other liverworts and mosses also reproduce vegetatively by means of **gemmae** (singular gemma), lens-shaped clumps of cells loosely held in structures called gemma cups (Figure 27.6*c*).

Hornwort sporophytes grow at their basal end

The phylum Anthocerophyta, comprising the hornworts—so named because their sporophytes look like little horns—appear at first glance to be liverworts with very simple gametophytes (Figure 27.7*a*). These gametophytes consist of flat plates of cells, a few cells thick. However, hornworts—unlike liverworts—have stomata. And hornworts have two characteristics that distinguish them from both liverworts and mosses. First, the archegonia are embedded in the gametophytic tissue instead of being borne on stalks. Second, of all the nontracheophyte sporophytes, those of the hornworts come closest to being capable of indefinite growth (without a set limit).

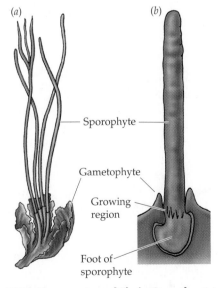

(a)   (b)

Sporophyte

Gametophyte

Growing region

Foot of sporophyte

27.7 Hornworts and Their Growth   (*a*) The hornwort *Anthoceros*, drawn slightly larger than actual size. (*b*) The sporophyte of a hornwort grows from its basal end.

The stalk of either the liverwort or moss sporophyte stops growing as the capsule matures, so elongation of the sporophyte is strictly limited. In a hornwort such as *Anthoceros,* however, there is no stalk, but a basal region of the capsule remains capable of indefinite cell division, continuously producing new spore-bearing tissue above (Figure 27.7*b*). Sporophytes of some hornworts growing in mild and continuously moist conditions can become as tall as 20 cm, making them some of the tallest known nontracheophytes.

Whereas the photosynthetic cells in other plants have many small chloroplasts, each cell of a hornwort has only a single, large chloroplast. Hornworts have internal cavities filled with a mucilage, often populated by cyanobacteria that convert atmospheric nitrogen gas into a nutrient form usable by the host plant (see Chapter 34).

Moss sporophytes grow at their apical end

The most familiar nontracheophytes are the mosses (phylum Bryophyta). There are more species of mosses than of liverworts and hornworts combined, and these hardy little plants are found in almost every terrestrial environment. The mosses are sister to the tracheophytes (see Figure 27.3). The moss gametophyte that develops following spore germination is a branched, filamentous plant, or **protonema** (plural protonemata), that looks much like a filamentous green alga and is unique to this phylum. Some of the filaments contain chloroplasts and are photosynthetic; others are nonphotosynthetic and anchor the protonema to the substrate. The nonphotosynthetic filaments, called **rhizoids**, are the nontracheophyte counterpart of the root hairs (single-celled outgrowths) of tracheophytes, but they have no structural resemblance to true roots. After a period of growth, cells close to the tips of the photosynthetic filaments divide rapidly in three dimensions to form buds. The buds eventually differentiate a distinct apex and produce the familiar leafy moss plant with the "leaves" spirally arranged.

These leafy shoots produce antheridia or archegonia at their tips (Figure 27.8*a*). The antheridia release sperm that travel through liquid water to the archegonia, where they fertilize the eggs. Sporophyte development in most mosses follows a precise pattern, resulting ultimately in the formation of an absorptive foot, a stalk, and, at the tip, a swollen capsule. In contrast to hornworts, which grow from the base (see Figure 27.7*b*), the moss sporo-

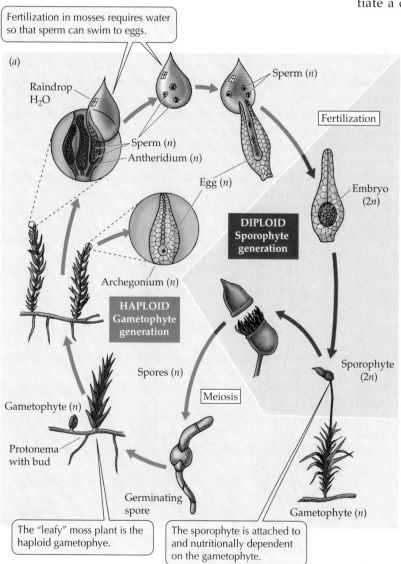

Fertilization in mosses requires water so that sperm can swim to eggs.

(a)

Raindrop H₂O

Sperm (*n*)

Sperm (*n*)
Antheridium (*n*)

Egg (*n*)

Fertilization

Archegonium (*n*)

DIPLOID Sporophyte generation

Embryo (2*n*)

HAPLOID Gametophyte generation

Spores (*n*)

Meiosis

Sporophyte (2*n*)

Gametophyte (*n*)

Protonema with bud

Germinating spore

Gametophyte (*n*)

The "leafy" moss plant is the haploid gametophye.

The sporophyte is attached to and nutritionally dependent on the gametophyte.

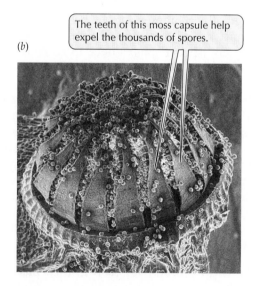

The teeth of this moss capsule help expel the thousands of spores.

(b)

27.8 The Moss Life Cycle

27.9 Peat Bogs   The moss *Sphagnum* forms peat, which is used for fuel. This peat bog in Ireland is being harvested for commercial use.

phyte grows at its apical end, as do the tracheophytes. Cells at the tip of the stalk divide, supporting elongation of the structure and giving rise to the capsule. Unlike liverworts, moss sporophytes often possess stomata, allowing gas exchange with the atmosphere—the uptake of CO_2 for photosynthesis and the release of O_2.

The archegonial tissue grows rapidly as the stalk elongates, for a time keeping pace with the rapidly expanding sporophyte. Eventually, however, the archegonium is outgrown and is torn apart around its middle. The top portion of the archegonium frequently persists on the top of the rapidly elevating capsule as a little pointed cap, called the **calyptra**.

The top of the capsule is shed after meiosis and development of numerous mature spores within. Groups of cells just below the lid form a series of "teeth" surrounding the opening. Highly responsive to humidity, the teeth arch into the mass of spores when the atmosphere is dry and then out again, shoveling out spores, when the atmosphere becomes moist (Figure 27.8*b*). The spores are thus dispersed when the surrounding air is moist—that is, when conditions favor their subsequent germination.

Only a few mosses depart from this pattern of capsule development. A familiar exception is the genus *Sphagnum*, which occurs in tremendous quantities in northern bogs and tundra extending well into the Arctic (Figure 27.9) and whose global biomass probably exceeds that of all other mosses combined. Species

in this genus have a very simple capsule with an air chamber in it. Air pressure builds up in this chamber, eventually causing the capsule lid to pop open, dispersing the spores with an audible explosion.

Many mosses contain a type of cell, called a **hydroid**, that dies and leaves a tiny channel through which water may travel. The hydroid likely is a progenitor of the characteristic water-conducting cell of the tracheophytes, but it lacks lignin (a waterproofing substance) and the wall structure found in tracheophyte water-conducting cells. The possession of hydroids and of a limited system for transport of foods by some mosses shows that the old term "nonvascular plant" is somewhat misleading when applied to mosses.

Introducing the Tracheophytes

Although an extraordinarily large and diverse group, the tracheophytes can be said to have been launched by a single evolutionary event. Sometime during the Paleozoic era, probably well before the Silurian period, the sporophyte generation of a now long-extinct organism produced a new cell type, the **tracheid**. The tracheid is the principal water-conducting element of the xylem in all tracheophytes except the angiosperms (flowering plants); even in angiosperms the tracheid persists along with a more specialized and efficient system of vessels and fibers that evolved from tracheids.

The evolutionary appearance of a tissue composed of tracheids had two important consequences. First, it provided a pathway for long-distance transport of water and mineral nutrients from a source of supply to regions of need. Second, it provided something almost completely lacking—and unnecessary—in the largely aquatic algae: rigid structural support. Support is important in a terrestrial environment because land plants tend to grow upward as they compete for sunlight to power photosynthesis. Thus the tracheid set the stage for the complete and permanent invasion of land by plants.

The present-day evolutionary descendants of the early tracheophytes belong to nine distinct phyla (Figure 27.10). We can sort these phyla into two groups: those that produce seeds and those that do not. In the nonseed tracheophytes the haploid and diploid generations are independent at maturity. The sporophyte is the large and obvious plant that one normally notices in nature (in contrast to the nontracheophyte sporophyte, which is attached to and dependent on the gametophyte). Gametophytes of the nonseed tracheophytes are rarely more than 1 or 2 centimeters long and are short-lived, whereas the sporophyte of a tree fern, for example, may be 15 or 20 meters tall and may live for years.

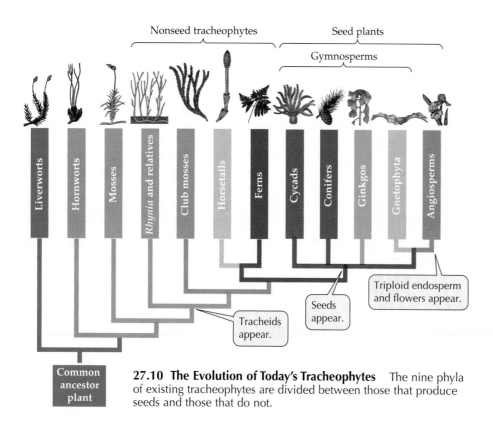

27.10 The Evolution of Today's Tracheophytes The nine phyla of existing tracheophytes are divided between those that produce seeds and those that do not.

The most prominent resting stage in the life cycle of a nonseed tracheophyte is the single-celled spore. This feature makes this life cycle similar to those of the fungi, the algae, and the nontracheophytes but not, as we will see, to that of the seed plants. Nonseed tracheophytes must have an aqueous environment for at least one stage of their life cycle because fertilization is accomplished by a motile, flagellated sperm.

We will discuss the life cycle of seed plants later in this chapter, but now we turn to a more detailed account of the evolution of the tracheophytes.

The tracheophytes have been evolving for almost half a billion years

The green algal ancestors of the plant kingdom successfully invaded the terrestrial environment between 400 and 500 million years ago. The evolution of a water-impermeable cuticle (waxy outer covering) and of protective layers for the gamete-bearing structures helped make the invasion successful, as did the initial absence of herbivores (plant-eating animals).

By the late Silurian period (see Table 19.1) tracheophytes were being preserved as fossils that we can study today. Several remarkable developments arose during the Devonian period, 400 to 345 million years ago. Three groups of nonseed tracheophytes that still exist made their first appearances during that period: the lycopods (club mosses), horsetails, and ferns. The proliferation of these plants made the terrestrial envi-

ronment ever more hospitable to animals; amphibians and insects arrived soon after the plants became established. Fossil remains about 360 million years old provide the first evidence of seed plants.

Trees of various kinds appeared in the Devonian, and came into their own in the Carboniferous period (345 to 290 million years ago). Mighty forests of lycopods up to 40 meters tall, horsetails, and tree ferns flourished in the tropical swamps of what would become North America and Europe (Figure 27.11). In the subsequent Permian period the continents came ponderously together to form a single, gigantic land mass, called Pangaea. The continental interior become warmer and drier, but late in the period glaciation was extensive. The 200-million-year reign of the lycopod–fern forests came to an end, to be replaced by forests of nonflowering seed plants (gymnosperms) that ruled throughout the Triassic and Jurassic periods. The gymnosperm forests changed with time as the gymnosperm groups evolved. Gymnosperm forests dominated during the era in which the continents drifted apart and dinosaurs strode the Earth.

The oldest evidence of angiosperms (flowering plants) dates to the Cretaceous period, about 120 million years ago. The angiosperms radiated almost explosively and, over a period of about 55 million years, became the dominant plant life of the planet.

The earliest tracheophytes lacked roots and leaves

The first tracheophytes belonged to the now-extinct phylum Rhyniophyta. The rhyniophytes appear to have been the only tracheophytes in the Silurian period of the Paleozoic era. The landscape at that time probably consisted of bare ground, with stands of rhyniophytes in low-lying moist areas. Early versions of the structural features of all the other tracheophyte phyla appeared in the plants of that time. These shared features strengthen the case for the origin of all tracheophytes from a common nontracheophyte ancestor.

In 1917 the British paleobotanists Robert Kidston and William H. Lang reported well-preserved fossils of tracheophytes embedded in Devonian rocks near Rhynie, Scotland. The preservation of these plants was remark-

27.11 An Ancient Forest A little more than 300 million years ago, a forest grew in a setting similar to tropical river delta habitats of today. Most of the plants depicted here were nonseed tracheophytes 10 to 20 m tall. Far in the distance, early seed plants—giants up to 40 m tall—towered over the forest.

able, considering that the rocks were more than 395 million years old. These fossil plants had a simple vascular system consisting of phloem and xylem (tracheids or hydroids). Flattened scales on the stems of some of the plants lacked vascular tissue and thus were not comparable with the true leaves of any other tracheophytes.

Lacking roots, the plants were apparently anchored in the soil by horizontal portions of stem (**rhizomes**) that bore rhizoids. These rhizomes also bore aerial branches. Sporangia—homologous with the nontracheophyte capsule—were found at the tips of the stems. Branching was dichotomous; that is, the shoot apex divided to produce two equivalent new branches, each pair diverging at approximately the same angle from the original stem (Figure 27.12). Scattered fragments of such plants had been found earlier, but never in such profusion or so well preserved as those discovered near Rhynie by Kidston and Lang.

The presence of xylem in these plants indicated that they, named *Rhynia* after the site of their discovery, were tracheophytes. But were they sporophytes or gametophytes? Close inspection of thin sections of fossil sporangia revealed that the spores were in groups of four. In almost all living nonseed tracheophytes (with no evidence to the contrary from fossil forms), the four

products of a meiotic division and cytokinesis remain attached to one another during their development into spores. The spores separate only when they are mature, and even after separation their walls reveal the exact geometry of how they were attached. Thus a group of four closely packed spores is found only immediately after meiosis, and a plant that produces

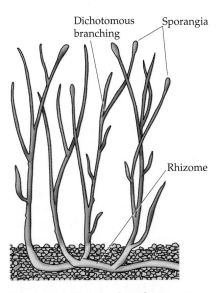

27.12 The First Tracheophytes This extinct plant in the genus *Rhynia* (phylum Rhyniophyta) lacked roots and leaves. The rhizome is a horizontal underground stem, not a root. The aerial shoots were less than 50 cm tall, and some were topped by sporangia.

such a group of four must be a diploid sporophyte. The gametophytes of the Rhyniophyta were branched. Depressions at the apices of the branches contained archegonia and antheridia.

Although apparently ancestral to the other tracheophyte phyla, the rhyniophytes themselves are long gone. None of their fossils appear anywhere after the Devonian period.

Early tracheophytes added new features

Within a few tens of millions of years, during the Devonian period, three new phyla—Lycophyta, Sphenophyta, and Pterophyta—of tracheophytes appeared on the scene, arising from rhyniophyte ancestors. These new groups featured specializations over the rhyniophytes, including one or more of the following: true roots, true leaves, and a differentiation between two types of spores.

THE ORIGIN OF ROOTS. *Rhynia* and its close relatives lacked true roots. They had only rhizoids, arising from a prostrate rhizome (see Figure 27.12), with which to gather water and minerals. How, then, did subsequent groups of tracheophytes come to have the complex roots we see today?

A French botanist, E. A. O. Lignier, proposed an attractive hypothesis in 1903 that is still widely accepted today. Lignier argued that the ancestors of the first tracheophytes branched dichotomously. This explanation accounts for the dichotomous branching observed in the rhyniophytes themselves. Lignier suggested that such a branch could bend and penetrate the soil, branching there (Figure 27.13). The underground portion could anchor the plant firmly, and even in this primitive condition it could absorb water and minerals. The subsequent discovery of fossil plants from the Devonian period, all having horizontal stems (rhizomes) with both underground and aerial branches, supported Lignier's hypothesis.

The underground branches, in an environment sharply different from that above the ground, were subjected to very different selection during the succeeding millions of years. Thus the two parts of the plant axis (the shoot and root systems) diverged in structure and came to have distinct internal and external anatomies (see Chapter 31). In spite of these differences, we believe that the root and shoot systems of tracheophytes are homologous—that that they were once part of the same organ.

THE ORIGIN OF TRUE LEAVES. Thus far, we have used the term "leaf" rather loosely. We spoke of "leafy" mosses; we also commented on the absence of "true leaves" in rhyniophytes. In the strictest sense, a **leaf** is a flattened photosynthetic structure emerging laterally from a main axis or stem and possessing true vascular tissue.

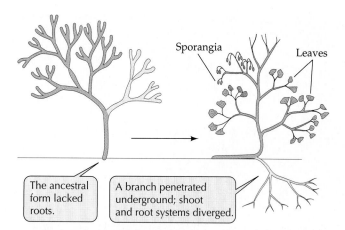

27.13 Is This How Roots Evolved? According to Lignier's hypothesis, branches from ancestral rootless plants could have penetrated the soil, where they gradually evolved into a root system.

This precise definition allows a closer look at true leaves in the tracheophytes, which shows us that there are two different types of leaves, very likely of different evolutionary origins.

The first type of leaf is usually small and only rarely has more than a single vascular strand, at least in plants alive today. Plants in the phylum Lycophyta

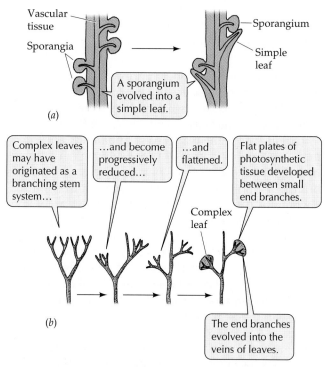

27.14 The Evolution of Leaves (a) Simple leaves are thought to have evolved from sterile sporangia. (b) The complex leaves of ferns and seed plants may have arisen as photosynthetic tissue developed between complex branching patterns.

(club mosses), of which only a few genera survive, have such *simple* leaves. The evolutionary origin of this type of leaf is thought to be sterile sporangia (Figure 27.14a). The principal characteristic of this type of leaf is that its vascular strand departs from the vascular system of the stem in such a way that the structure of the stem's vascular system is scarcely disturbed. This was true even in the fossil lycopod trees of the Carboniferous period, many of which had leaves several centimeters long.

The other type of leaf is encountered in ferns and seed plants. This larger, more *complex* type of leaf is thought to have arisen from the flattening of a dichotomously branching stem system, with the development of extensive photosynthetic tissue between the branch members (Figure 27.14b). The complex leaf may have evolved several times, in different phyla of tracheophytes.

HOMOSPORY AND HETEROSPORY. In the most ancient of the present-day tracheophytes, both the gametophyte and the sporophyte are independent and usually photosynthetic. Spores produced by the sporophytes are of a single type, and they develop into a single type of gametophyte, bearing both female and male reproductive organs. Such plants, which bear a single type of spore, are said to be **homosporous** (Figure 27.15a). The sex organs on the gametophytes of homosporous plants are of two types. The female organ is a multicellular archegonium, typically containing a single egg. The male organ is an antheridium, containing many sperm.

A different system, with two distinct types of spores, evolved somewhat later. Plants of this type are said to be **heterosporous** (Figure 27.15b). One type of spore, the **megaspore**, develops into a larger, specifically female gametophyte (megagametophyte) that produces only eggs. The other type, the **microspore**, develops into a smaller, male gametophyte (microgametophyte) that produces only sperm. Megaspores are produced in small numbers in megasporangia on the sporophyte, and microspores in large numbers in microsporangia.

The most ancient tracheophytes were all homosporous. Heterospory evidently evolved several times, independently, in the early evolution of the tracheophytes descended from the rhyniophytes. The fact that heterospory evolved re-

peatedly suggests that it affords selective advantages. As we will see, subsequent evolution in the plant kingdom featured ever greater specialization of the heterosporous condition.

The Surviving Nonseed Tracheophytes

Ferns are now the most abundant and diverse phylum of nonseed tracheophytes, although the club mosses and horsetails were once dominant elements of Earth's

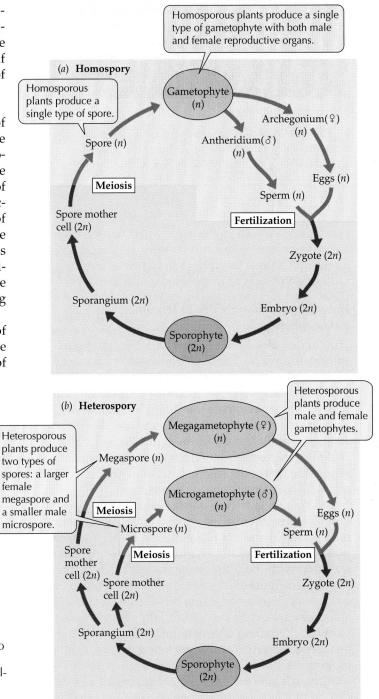

27.15 Homospory and Heterospory (a) Homosporous plants bear a single type of spore. Each gametophyte has two types of sex organs, antheridia (male) and archegonia (female). (b) Heterospory, with two types of spores that develop into distinctly male and female gametophytes, evolved later.

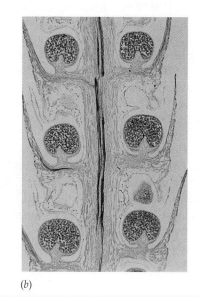

(a) *Lycopodium flabellitorme*

(b)

27.16 Club Mosses (a) Conelike strobili are visible at the tips of this club moss. Club mosses have simple leaves arranged spirally on their stems. (b) Thin section through a strobilus of another club moss.

vegetation. A fourth phylum, the whisk ferns, contains only two genera. In this section we'll look at the characteristics of these four phyla and at some of the evolutionary advances that appeared in them.

The club mosses are sister to the other tracheophytes

The club mosses (lycopods, phylum Lycophyta) have ancient origins; the remaining tracheophytes share an ancestor that was not ancestral to the Lycophyta. There are relatively few surviving species of club mosses. They have roots that branch dichotomously. They bear only simple leaves, and the leaves are arranged spirally on the stem (Figure 27.16a). Growth in club mosses comes entirely from groups of dividing cells at the tips of the stems. Thus stem growth is apical, as it is in many flowering plants.

The sporangia in club mosses are contained within conelike structures called **strobili** (singular strobilus; Figure 27.16b) and are tucked in the upper angle between a specialized leaf and the stem. This placement contrasts with the terminal sporangia of the rhyniophytes (see Figure 27.12). There are both homosporous species and heterosporous species of club mosses. Like all other nonseed tracheophytes, they have a heteromorphic alternation of generations with a large, independent sporophyte and a small, independent gametophyte.

Although only a minor element of present-day vegetation, the Lycophyta are one of two phyla that ap-

pear to have been the dominant vegetation during the Carboniferous period. One abundant type of coal, Cannel coal, is formed almost entirely from fossilized spores of a tree lycopod named *Lepidodendron;* this finding is an indication of the abundance of this genus in the forests of that time. The other major element of the Carboniferous vegetation was the phylum Sphenophyta, the horsetails.

Horsetails grow at the bases of their segments

Like the club mosses, the horsetails (phylum Sphenophyta) are represented by only a few present-day species. They are sometimes called scouring rushes because silica deposits found in the cell walls made them once useful for cleaning. They have true roots that branch irregularly, as do the roots of all tracheophytes except the club mosses. Their sporangia curve back toward the stem on the ends of short stalks (sporangiophores) (Figure 27.17a). Horsetails have a large sporophyte and a small gametophyte, both independent.

The leaves of horsetails are simple and form distinct whorls (circles) (Figure 27.17b). Growth in horsetails originates to a large extent from discs of dividing cells just above each whorl of leaves, so each segment of the stem grows from its base. Such basal growth is uncommon in plants, but it is also found in the grasses, a major group of flowering plants.

Present-day whisk ferns resemble the most ancient tracheophytes

There once was some disagreement about whether rhyniophytes are entirely extinct. The confusion arose because of the existence today of two genera of rootless, spore-bearing plants, *Psilotum* and *Tmesipteris.*

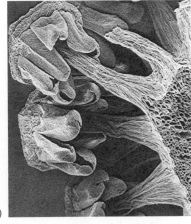

27.17 Horsetails (a) Sporangia and sporangiophores of the horsetail *Equisetum arvense.* (b) Vegetative and fertile shoots of the marsh horsetail. Leaves form in spaced whorls at nodes on the stems of the vegetative shoot on the right; the fertile shoot on the left is ready to disperse its spores.

(b) *Equisetum palustre*

Psilotum nudum (Figure 27.18) has only minute scales instead of true leaves, but plants of the genus *Tmesipteris* have flattened photosynthetic organs with well-developed vascular tissue. Are these two genera the living relics of the rhyniophytes, or do they have more recent origins?

Psilotum and *Tmesipteris* once were thought to be evolutionarily ancient descendants of anatomically simple ancestors. That hypothesis was weakened by an enormous hole in the geologic record—between the rhyniophytes, which apparently became extinct more than 300 million years ago, and *Psilotum* and *Tmesipteris,* which are modern plants. DNA sequence data finally settled the question in favor of a more modern origin. Most botanists now treat these two genera as their own phylum, the Psilophyta, or whisk ferns.

The whisk ferns are highly specialized plants that evolved fairly recently from anatomically more complex ancestors. Today *Psilotum* is widely distributed in the Tropics and Subtropics.

Ferns evolved large, complex leaves

The sporophytes of the ferns and seed plants have roots, stems, and leaves. Their leaves are typically large and have branching vascular strands. Some species have small leaves as a result of evolutionary reduction, but even the small leaves have more than one vascular strand. The sporangia of these plants are on the lower surfaces of leaves or leaflike structures or, more rarely, on their margins.

The true ferns constitute the phylum Pterophyta, which first appeared during the Devonian period and today consists of about 12,000 species. Ferns are characterized by fronds (large leaves with complex vasculature) (Figure 27.19), by the absence of seeds, and by a requirement for water for the transport of the male gametes to the female gametes. Most ferns inhabit shaded, moist woodlands and swamps. Some, the tree ferns, reach heights of up to 20 m. Tree ferns lack the rigidity of woody plants, and thus do not grow in sites exposed directly to strong winds but rather in ravines or beneath trees in forests.

During its development, the fern frond unfurls from a tightly coiled "fiddlehead" (Figure 27.19c). Some fern leaves become climbing organs and may grow to be as much as 30 m long. The sporangia are found on the undersurfaces of the leaves, sometimes covering the whole undersurface and sometimes only at the edges; in some species the sporangia are clustered in groups called **sori** (singular sorus) (Figure 27.20).

Devonian fossil beds have yielded ferns with some characteristics that are rhyniophyte and some that resemble those of other tracheophyte phyla. For example, like a modern fern, the fossil genus *Protopteridium* had flattened branch systems with extensive photosynthetic tissue between the branches; but like a rhyniophyte, it bore terminal sporangia and lacked

Psilotum nudum

27.18 A Modern Whisk Fern Aerial branches of a whisk fern, a plant once considered by some to be a surviving rhyniophyte and by others to be a fern. It is now included in the phylum Psilophyta.

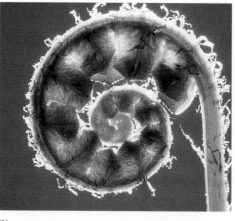

(a)

(b)

(c) *Marsilea mutica*

27.19 Fern Fronds Take Many Forms (a) Fern fronds blanket a forest floor in the Florida Everglades. (b) "Fiddleheads" (developing fronds) of a common forest fern; this structure will unfurl and expand to give rise to a complex adult frond such as those in (a). (c) The tiny fronds of a water fern.

true roots (Figure 27.21). During late Paleozoic times, the ferns underwent considerable evolutionary experimentation in the structure of their leaves and particularly in the arrangement of their vascular tissue.

The sporophyte generation dominates the fern life cycle

The undersides or edges of fern fronds carry sporangia in which cells undergo meiosis to form haploid spores (Figure 27.22). Once shed, spores often travel great distances and eventually germinate to form small, independent gametophytes. These gameto-

phytes produce antheridia and archegonia, although not necessarily at the same time or on the same gametophyte.

Sperm swim through water to archegonia, often on other gametophytes, where they unite with an egg. The resulting zygote develops into a new sporophyte embryo. The young sporophyte sprouts a root and can thus grow independently of the gametophyte. In the alternating generations of a fern, the gametophyte is small, delicate, and short-lived, but the sporophytes can be very large and can sometimes survive for hundreds of years.

Most ferns are homosporous. However, two groups of aquatic ferns, the Marsileales and Salviniales, have evolved heterospory. Megaspores and microspores of these plants (which germinate to produce female and male gametophytes, respectively) are produced in different sporangia, and the male spores are always

Dryopteris intermedia

27.20 Fern Sori Contain Sporangia Sori, each with many spore-producing sporangia, form on the underside of a frond of the midwestern fancy fern.

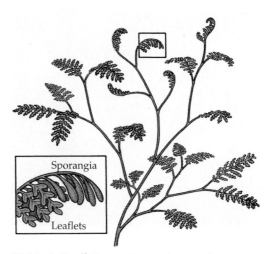

27.21 A Fossil Fern During their evolutionary history, ferns exhibited combinations of structures. This fossil fern (*Protopteridium*, from the Devonian period) lacked true roots and had branches with both leaflets and terminal sporangia.

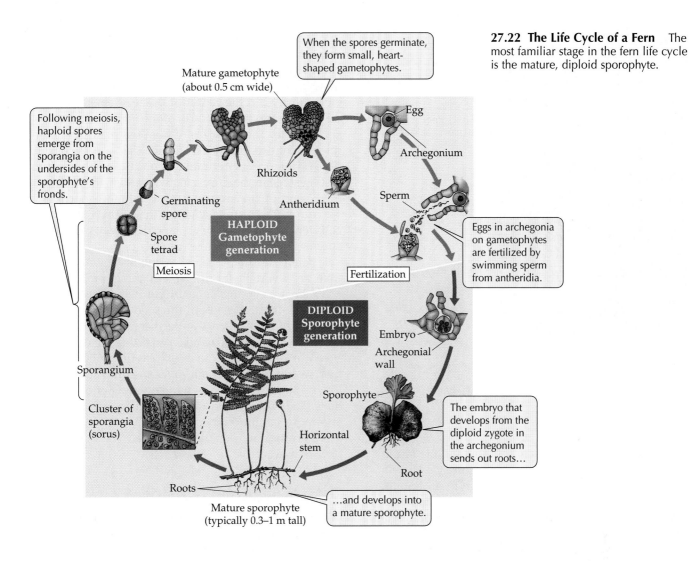

27.22 The Life Cycle of a Fern   The most familiar stage in the fern life cycle is the mature, diploid sporophyte.

When the spores germinate, they form small, heart-shaped gametophytes.

Mature gametophyte (about 0.5 cm wide)

Following meiosis, haploid spores emerge from sporangia on the undersides of the sporophyte's fronds.

Egg

Archegonium

Rhizoids

Germinating spore

Antheridium

Sperm

Spore tetrad

HAPLOID Gametophyte generation

Meiosis

Fertilization

Eggs in archegonia on gametophytes are fertilized by swimming sperm from antheridia.

DIPLOID Sporophyte generation

Embryo

Archegonial wall

Sporangium

Sporophyte

Cluster of sporangia (sorus)

Horizontal stem

The embryo that develops from the diploid zygote in the archegonium sends out roots...

Root

Roots

Mature sporophyte (typically 0.3–1 m tall)

...and develops into a mature sporophyte.

much smaller and greater in number than the female spores.

A few genera of ferns produce a tuberous, fleshy gametophyte instead of the characteristic flattened, photosynthetic structure described earlier. These tuberous gametophytes depend on a mutualistic fungus for nutrition*; in some genera, even the sporophyte embryo must become associated with the fungus before extensive development can proceed. In Chapter 28 we will see that there are many important plant–fungus associations.

The Seeds of Success

The most recent group to appear in the evolution of the tracheophytes is the seed plants: the paraphyletic **gymnosperms** (such as pines and cycads) and the monophyletic **angiosperms** (flowering plants). In seed plants the gametophyte generation is reduced even

further than it is in ferns. The haploid gametophyte develops partly or entirely while attached to and nutritionally dependent on the diploid sporophyte. Among the seed plants, only the earliest types of gymnosperms and their few survivors (cycads, for example) had swimming sperm. All other seed plants have evolved other means of bringing gametes together. The culmination of this striking evolutionary trend in plants was independence from liquid water for the purposes of reproduction.

Seed plants are heterosporous, forming separate megasporangia and microsporangia—female and male sporangia, respectively—on modified leaves or leaflike structures that are grouped on short axes to form strobili (see Figure 27.16), such as the cones of conifers and the flowers of angiosperms.

As in other plants, spores of seed plants are produced by meiosis within the sporangia, but in this case they are not shed. Instead, the gametophytes develop within the sporangia and depend on them for food and water. In most species only one of the meiotic products in a megasporangium survives. The surviv-

*In a mutualistic association, both partners—here, the gametophyte and the fungus—profit.

27.23 Gymnosperms Are Wind-Pollinated   Pollen grains are the male gametophytes of gymnosperms.

ing haploid nucleus divides mitotically, and the resulting cells (without walls) divide again to produce a small multicellular female gametophyte. In the angiosperms, female gametophytes do not normally include more than eight nuclei. The female gametophyte of a seed plant is retained within the megasporangium, where it matures, is fertilized, and undergoes the early development of the next sporophyte generation.

Within the microsporangium the meiotic products develop into microspores that divide one or a few times to form male gametophytes, the familiar **pollen grains** (Figure 27.23; see also Figure 36.2). Distributed by wind, by an insect, by a bird, or by a plant breeder, a pollen grain that reaches the appropriate surface of a sporophyte, near the female gametophyte, develops further. It produces a slender **pollen tube** that grows and digests its way through the sporophyte tissue toward the female gametophyte.

When the tip of the male gametophyte's pollen tube reaches the female gametophyte, one or two sperm cells are released from the tube and fertilization occurs. The resulting diploid zygote divides repeatedly, forming a young sporophyte that develops to an embryonic stage at which growth becomes temporarily suspended (often referred to as a dormant stage). The end product at this stage is a **seed**.

A seed may contain tissues from three generations. The seed coat develops from tissue of the diploid parent sporophyte. Within the seed coat is a layer of haploid female gametophyte tissue from the next generation (this tissue is fairly extensive in most gymnosperm seeds, but in angiosperm seeds its place is taken by a tissue called endosperm, which we will discuss shortly). In the center of the seed package is the third generation, in the form of the embryo of the new sporophyte. The embryos of nonseed plants develop directly into sporophytes, which either survive or die, depending on environmental conditions; there is no resting stage in the life cycle. By contrast, the multicellular seed of a gymnosperm or an angiosperm is a well-protected resting stage for the embryo. Layers of cells enclose the embryo, and the seeds of some species may remain viable (capable of growth and development) for many years, germinating when conditions are favorable for the growth of the sporophyte.

When the young sporophyte begins to grow, it draws on food reserves in the seed. During the dormant stage the seed coat protects the embryo from excessive drying and may also protect against potential predators that would otherwise eat the embryo or the food reserves. Many seeds have structural adaptations that promote dispersal by wind or, more often, by animals. The possession of seeds is a major reason for the enormous evolutionary success of seed plants, which are the prominent elements of Earth's modern land flora in most areas.

The Gymnosperms: Naked Seeds

The gymnosperms are a paraphyletic group of seed plants that never form flowers. Although there are probably fewer than 750 species of living gymnosperms, these plants are second only to the angiosperms (flowering plants) in their dominance of the land masses.

There are four phyla of gymnosperms living today. The cycads (phylum Cycadophyta) are palmlike plants of the Tropics, growing as tall as 20 meters (Figure 27.24a). Ginkgos (phylum Ginkgophyta), which were common during the Mesozoic era, are represented today by a single genus and species, *Ginkgo biloba*, the maidenhair tree (Figure 27.24b). The phylum Gnetophyta consists of three very different genera that share a characteristic type of cell (the vessel element) in their xylem tissue that is found in no other group of plants except the angiosperms, which shared an ancestor with the Gnetophyta (see Figure 27.10). One member of this phylum is *Welwitschia* (Figure 27.24c), a long-lived desert plant with just two straplike leaves that sprawl on the sand and can become as long as 3 m. Far and away the most abundant of the gymnosperms are the conifers (phylum Coniferophyta), cone-bearing plants such as pines and redwoods (Figure 27.24d).

All living gymnosperms have active secondary growth (growth in the diameter of stems and roots; see Chapter 31), and all but the Gnetophyta have only tra-

(a) *Cycas revoluta*

(c) *Welwitschia mirabilis*

(b) *Ginkgo biloba*

27.24 Gymnosperms (a) This sago palm belongs to the cycads, the most ancient of the present-day gymnosperms. Many cycads have growth forms that resemble both ferns and palms. (b) Characteristic fruit and broad leaves of the maidenhair tree. (c) A gnetophyte growing in the Namib Desert of Africa. Two huge, straplike leaves grow throughout the life of the plant, breaking and splitting as they grow. (d) A dramatic conifer, this giant sequoia grows in Yosemite National Park, California.

(d) *Sequoiadendron giganteum*

cheids as water-conducting and support cells in the xylem. Despite this water transport and support system (which might seem suboptimal compared with that of angiosperms), the gymnosperms include some of the tallest trees known. The coastal redwoods of California are the tallest gymnosperms; the largest are well over 100 m tall. Xylem produced by gymnosperms is the principal resource of the lumber industry.

In this section on gymnosperms, we'll take a brief look at their fossil history and then examine the life cycle of the conifers in some detail.

We know the early gymnosperms only as fossils

The earliest fossil evidence of gymnosperms is found in Devonian rocks. The early gymnosperms combined characteristics of rhyniophytes and heterosporous ferns, but they had tracheids of the same type found in modern gymnosperms. They also differed from the plants around them by their extensively thickened stems.

By the Carboniferous period, several new lines of gymnosperms had evolved, including various seed ferns that possessed fernlike foliage but had characteristic gymnosperm seeds attached to the leaf margins.

The first true conifers appeared at approximately the same time. Either they were not dominant trees or they did not grow where conditions were right for fossilization, so we have few preserved examples. During the Permian period, however, the conifers and cycads came into their own. Gymnosperms dominated *all* forests until less than 80 million years ago, and they still dominate some present-day forests.

Conifers have cones but no motile cells

The great Douglas fir and cedar forests of the northwestern United States and the massive forests of pine, fir, and spruce that clothe the northern continental regions and upper slopes of mountain ranges rank among the great vegetation formations of the world (the "boreal forest"; see Chapter 54). All these trees belong to one phylum of gymnosperms, Coniferophyta—the conifers, or cone-bearers. Male and female spores are produced in separate male and female cones. Female cones are larger than male cones.

We will use the life cycle of a pine to illustrate reproduction in gymnosperms (Figure 27.25). The production of male gametophytes as pollen grains frees the plant once and for all from its dependence on external liquid water for fertilization. Instead of water, wind assists conifer pollen grains in their first stage of travel to the female gametophyte. The pollen tube provides the means for the last stage of travel by growing and digesting its way through maternal sporophytic tissue and eventually releasing a sperm cell near the egg.

The megasporangium, which will form the female gametophyte containing eggs within archegonia, is enclosed in a special layer of sporophytic tissue, called the **integument**, that will eventually develop into the seed coat. The integument, the megasporangium inside it, and the tissue attaching it to the maternal sporophyte form the **ovule**. The pollen grain travels through the small opening in the integument at the apex of the ovule, the **micropyle**.

Gymnosperms (meaning literally "naked-seeded") derive their name from the fact that their seeds are not protected by fruit tissue. Most conifer ovules (which on fertilization develop into seeds) are borne exposed on the upper surfaces of cone scales. The only protection from the environment is that the scales are tightly pressed against each other within the cone. Some pines have such tightly closed female cones that only fire suffices to split them open and release the seeds. One example is the lodgepole pine, which covers vast fire-prone areas in the Rocky Mountains and elsewhere.

About half of the conifer species have fruitlike, fleshy tissues associated with their seeds; examples are the "berries" of juniper and yew. Animals may eat these tissues and then disperse the seeds via the waste they deposit, often carrying them considerable distances—which may spread the conifer population. True fruits are one of the characteristics of the plant phylum that is dominant today: Angiospermae.

The Angiosperms: Flowering Plants

The phylum Angiospermae* consists of the flowering plants, also known as the angiosperms (literally "enclosed-seeded"). This highly diverse phylum includes about 275,000 species. In other chapters, when we mention plants in discussing processes such as long-distance transport in the xylem and phloem, or the chemical regulation of development, generally we are referring to the angiosperms.

The angiosperms represent the current extreme of an evolutionary trend that runs throughout the tracheophytes, in which the *sporophyte* generation becomes *larger* and *more independent* of the gametophyte, while the *gametophyte* generation becomes *smaller* and *more dependent* on the sporophyte. Angiosperms differ from other plants in several ways, although exceptions exist.

Double fertilization was long considered the single most reliable distinguishing characteristic of the angiosperms. *Two* male gametes, contained within a single microgametophyte (pollen grain), participate in fertilization events within the megagametophyte of an angiosperm. One sperm combines with the egg to produce a diploid zygote, the first cell of the sporophyte generation. The other sperm nucleus usually combines with two other haploid nuclei of the female gametophyte to form a triploid ($3n$) nucleus. This nucleus, in turn, divides to form a triploid tissue, the **endosperm**, that nourishes the embryonic sporophyte during its early development.

Double fertilization occurs in all present-day angiosperms and probably in all three existing genera of Gnetophyta: *Ephedra*, *Gnetum*, and *Welwitschia*. William Friedman, at the University of Georgia, confirmed in 1990 that *Ephedra* has double fertilization but that both fertilizations produce diploid products. Thus, we are left with the formation of an extensive triploid endosperm as the most definitively angiosperm trait. We are not sure when and how double fertilization evolved because there is no fossil evidence on this point.

A second consistent characteristic of angiosperms is the possession of specialized water-transporting cells called **vessel elements** in the xylem, but these are also found, in different form, in a few gymnosperms and

---

* The flowering plants have also been referred to as the phylum Anthophyta. However, that term is now typically used to denote the monophyletic group consisting of the Gnetophyta and the flowering plants.

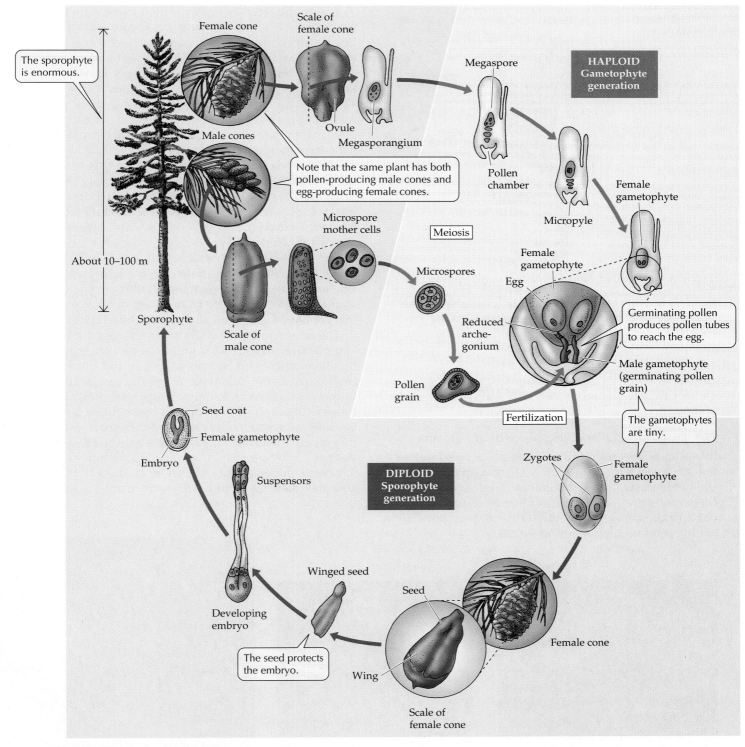

27.25 The Life Cycle of a Pine Tree

ferns, and even in a club moss and a horsetail. Another distinctive cell in angiosperm xylem is the **fiber**, which plays an important role in supporting the plant body. The name "angiosperm" refers to the diagnostic character that the seeds of these plants are enclosed in a modified leaflike structure called a *carpel.* Of course, the most evident diagnostic feature of angiosperms is that they have flowers.

In the following sections we'll examine the structure and function of flowers, evolutionary trends in flower structure, the functions of pollen and fruits, the

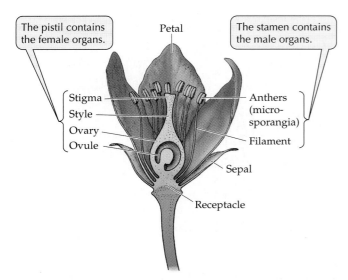

The pistil contains the female organs.

The stamen contains the male organs.

Petal

Stigma
Style
Ovary
Ovule

Anthers (micro-sporangia)

Filament

Sepal

Receptacle

27.26 A Generalized Flower   Not all flowers possess all the structures shown here, but they must possess stamens, pistils, or both in order to play their role in sexual reproduction. Flowers that have both, such as this one does, are referred to as perfect.

angiosperm life cycle, the two major groups of angiosperms, and the origin and evolution of flowering plants.

The sexual structures of angiosperms are flowers

If you examine any familiar flower, you will notice that each part has one or more veins and, in the case of the outer parts, looks somewhat like a leaf. In fact, these parts are modified from leaves to function as parts of the flower. Thus the terms that we have already used for other plants apply here.

A generalized flower (for which there is no exact counterpart in nature) is shown in Figure 27.26. In the flower, structures bearing microsporangia are called **stamens**, each composed of a filament and two **anthers** that contain pollen-producing sporangia; structures bearing megasporangia are called **carpels**. A structure composed of one carpel or two or more fused carpels is called a **pistil**. The swollen base of the pistil, containing one or more ovules (each containing a megasporangium), is an **ovary**; the apical stalk of the pistil is a **style**; and the terminal surface that receives the pollen is called a **stigma**. In addition, a flower often has several specialized sterile (non-spore-bearing) leaves: The inner ones are called **petals** (collectively, the **corolla**), and the outer, **sepals** (collectively, the **calyx**). The corolla and calyx, which can be quite showy, often play roles in attracting animal pollinators to the flower. The calyx more commonly protects the flower in bud. From base to apex, the sepals, petals, stamens, and carpels are usually arranged in whorls and attached to a central stalk called the receptacle.

The flower in Figure 27.26 has both megasporangia and microsporangia and is said to be **perfect**, meaning that it contains both functional female and functional male parts. Many angiosperms produce two types of flowers, one type with only megasporangia and the other with only microsporangia; consequently, either the stamens or the carpels are nonfunctional or absent in a given flower, and the flower is referred to as **imperfect**.

Species such as corn or birch in which both female and male flowers occur on the same plant are said to

(a) *Daucus carota*

(b) *Echinacea purpurea*

(c) *Pennisetum setaceum*

27.27 Inflorescences   (a) The inflorescence of Queen Anne's lace is an umbel. Each umbel bears flowers on stalks that arise from a common center. (b) Cornflowers are members of the aster family; their inflorescence is a head. In a head, each of the long, petal-like structures is a ray flower; the central portion of the head consists of dozens to hundreds of disc flowers. (c) Grasses such as this fountain grass have inflorescences called spikes.

(a)

(b)

27.28 Flower Form and Evolution (*a*) A magnolia flower shows the major features of early flowers: radial symmetry with the individual tepals, carpels, and stamens separate, numerous, and attached at their bases in a spiral arrangement. (*b*) Orchids have a bilaterally symmetrical structure that evolved much later than the form of the magnolia flower in (*a*). One of the three petals evolved into the complex lower "lip." Inside, the stamen and pistil are fused, and there is a single anther.

be **monoecious** (meaning "one-housed"—but, it must be added, one house with separate rooms). The sexes are completely separated in some other species of angiosperms, such as willows and date palms; in these species, a given plant produces either male or female flowers, but never both. Such species are said to be **dioecious** ("two-housed"). In other words, there are truly female plants and truly male plants.

In the generalized flower of Figure 27.26, we illustrated distinct petals and sepals arranged in distinct whorls. In nature, however, the petals and sepals sometimes are arranged in a continuous spiral and are indistinguishable. Such appendages are called **tepals**. In other cases petals, sepals, or tepals are completely absent.

Flowers may be single, or grouped together to form an **inflorescence**. Different families of flowering plants have their own, characteristic types of inflorescences, such as the umbels of the carrot family, the heads of the aster family, and the spikes of many grasses, among others (Figure 27.27).

There still are questions about evolutionary trends in flower structure

Botanists disagree about which type of flower is evolutionarily the most ancient. One of the earliest types has many tepals (or sepals and petals), carpels, and stamens, all spirally arranged (Figure 27.28*a*). Evolutionary change within the angiosperms included some striking modifications from this early condition: reduction in the number of each type of organ, differen-

tiation of petals from sepals, stabilization of each type of organ to a fixed number, arrangement in whorls, and finally, change in symmetry from radial (as in a lily) to bilateral (as in a sweet pea or orchid), often accompanied by an extensive fusion of parts (Figure 27.28*b*). A great variety of corolla types have emerged in the course of evolution, as you will realize if you think of some of the flowers you recognize.

According to one theory, the first carpels to evolve were modified leaflike structures, folded but incompletely closed and thus differing from the scales of the gymnosperms and the true carpels of the angiosperms. In the groups of angiosperms that evolved later, the carpels fused and became progressively more buried in receptacle tissue (Figure 27.29*a*); in the flowers of the latest groups to evolve, the other flower parts are attached at the very top of the ovary rather than at the bottom. The stamens of the most-ancient flowers may have been leaflike (Figure 27.29*b*), little resembling those of the generalized flower in Figure 27.26. Botanists disagree as to whether carpels, stamens, and even petals evolved from leaves or from another structure.

Why do so many flowers have pistils with long styles and anthers with long filaments? Natural selection has favored length in both of these structures probably because length increases the likelihood of successful pollination. Long filaments may bring the anthers in contact with insect bodies, or they may put the anthers where they catch the wind better. Similar arguments apply to long styles.

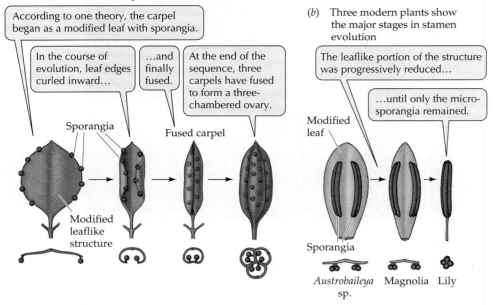

27.29 Evolution of Carpels and Stamens Carpels and stamens may have evolved from leaflike structures. The small picture below each illustration is a cross section.

A long style may serve another purpose as well. If several pollen grains land on one stigma, a pollen tube will start growing from each grain toward the ovary. If there are more pollen grains than ovules, there is a "race" for the ovules. The race down the style can be viewed as "mate selection" by the plant holding that style. Pollen has played another major role in evolution, as we are about to see.

Pollen played a crucial role in the coevolution of angiosperms and animals

Whereas many gymnosperms are wind-pollinated, most angiosperms are animal-pollinated. Animals visit flowers to obtain nectar or pollen and in the process often carry pollen from one flower to another, or from one plant to another. Thus in its quest for food, the animal contributes to the genetic diversity of the plant population. Insects—especially bees—are among the most important pollinators; birds and some species of bats also play major roles.

For more than 100 million years, angiosperms and their animal pollinators have coevolved in the terrestrial environment. The animals have affected the evolution of the plants, and the plants have affected the evolution of the animals. Pollination by just one or a very few committed animal species provides a plant species with a reliable mechanism for transferring pollen from one to another of its members. Flower structure has become incredibly diverse under selection.

Some of the products of coevolution are highly specific; for example, some yucca species are pollinated by one and only one species of moth. Most plant–pollinator interactions are much less specific; that is, many different animal species pollinate the same plant species, and the same animal species pollinate many plant species. However, even these less specific interactions have developed some specialization. Bird-pollinated flowers are often red and odorless; insect-pollinated flowers often have characteristic odors and may have conspicuous markings ("nectar guides") that are evident only in the ultraviolet region of the spectrum, where insects have better vision than in the red region.

We treat coevolution and other aspects of plant–animal interactions in more detail in Chapter 52. Here we'll consider what happens to the flower after fertilization.

Angiosperms produce fruits

The ovary of a flowering plant (together with its seeds) develops into a fruit after fertilization. Fruit production is thus another diagnostic character of angiosperms. A fruit may consist only of the mature ovary and its seeds, or it may include other parts of the flower or structures closely related to it.

A **simple fruit**, such as a cherry (Figure 27.30a), is one that develops from a single carpel or several united carpels. A raspberry is an example of an **aggregate fruit** (Figure 27.30b)—one that develops from several separate carpels of a single flower. Pineapples and figs are examples of **multiple fruits** (Figure 27.30c),

(a)

(b)

(c)

(d)

27.30 Fruits Come in Many Forms and Flavors (a) A simple fruit: sour cherry. (b) An aggregate fruit: raspberries. (c) A multiple fruit: pineapple. (d) An accessory fruit: pear.

formed from a cluster of flowers (an inflorescence). Fruits derived from parts in addition to the carpel and seeds are called **accessory fruits** (Figure 27.30d); examples are apples, pears, and bananas.

The development, ripening, and dispersal of fruits will be considered in Chapters 35 and 36.

The angiosperm life cycle features double fertilization

The life cycle of the angiosperms will be considered in detail in Chapter 36. The summarized life cycle in Figure 27.31 shows that the angiosperm life cycle has similarities with and differences from the conifer life cycle (see Figure 27.25). Like all other seed plants, angiosperms are heterosporous. The female gametophyte is even more reduced than in the gymnosperms. The ovules are contained within carpels, rather than being exposed on the surfaces of scales as in most gymnosperms. The male gametophytes are, again, pollen grains.

The ovule develops into a seed containing the products of the double fertilization that characterizes angiosperms. The triploid endosperm serves as storage tissue for starch or lipid reserves, storage proteins, and other reserve substances. The diploid zygote develops into an embryo consisting of an embryonic axis and one or two **cotyledons**. Also called seed leaves, cotyledons have different fates in different plants. In many, they serve as absorptive organs that take up and digest the endosperm. In others, they enlarge and become photosynthetic when the seed germinates. Often they play both of these roles.

There are two major groups of flowering plants

The angiosperms may be divided into two classes: Monocotyledones (the monocots; Figure 27.32) and Dicotyledones (the dicots; Figure 27.33). The monocots are so called because they have a single embryonic cotyledon; the dicots have two. And there are other major differences between the two classes (see Figure 31.5). Included are differences in leaf vein patterns, in the arrangement of vascular tissue in the stem and root, in the number of flower parts, and in the presence or absence of secondary growth (produced by a cambium; see Chapter 31). The cotyledons of some, but not all, dicots store the reserves originally present in the endosperm.

The monocots include grasses, cattails, lilies, orchids, and palm trees. The dicots include the vast number of familiar seed plants: most of the herbs, vines, trees, and shrubs. Among them are oaks, willows, violets, snapdragons, and sunflowers.

The origin and evolution of the flowering plants

How did the angiosperms arise? Modern cladistic analyses (see Chapter 22) have settled this once vexing question. It is widely agreed that the angiosperms and two groups of gymnosperms, the Gnetophyta and the long-extinct cycadeoids (so named because their leaves resembled cycad leaves), arose from a single common ancestor. A close relationship between the angiosperms and the Gnetophyta was long suspected, primarily on the grounds that some gnetophytes have vessel elements, which characterize the angiosperms. (Ironically, this was a poor argument, because vessel elements probably evolved independently in the two groups.)

In 1990 this theory was strengthened by the confirmation via the light microscope of double fertilization in *Ephedra*, a member of the Gnetophyta. The cycadeoids, which became extinct about the same time as the dinosaurs did, shared several important characteristics with the Gnetophyta and the angiosperms. The reproductive organ of one of the cycadeoids, although clearly a gymnosperm structure with naked seeds, was suggestively similar to the flower of *Magnolia*.

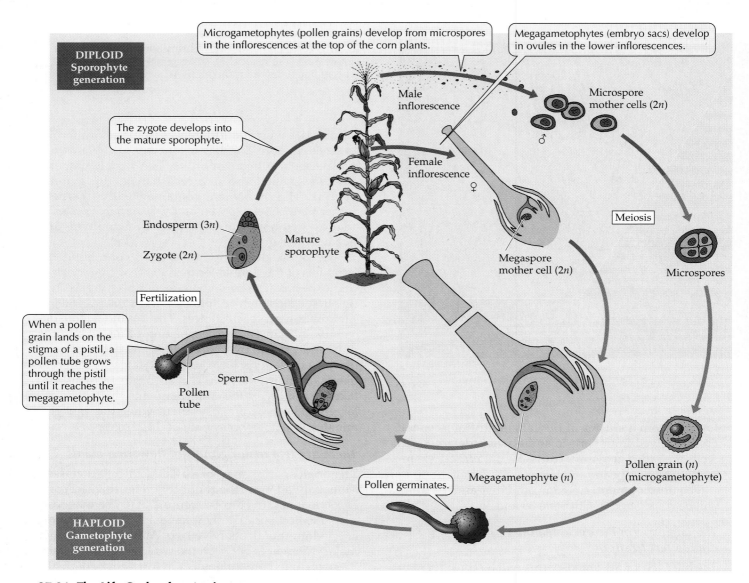

27.31 The Life Cycle of an Angiosperm *Zea mays*, popularly known as corn, is a thoroughly studied angiosperm with great economic value to humans.

(a)   (c)

27.32 Monocots (a) Palms are among the few monocot trees. Date palms are a major food source in some areas of the world. (b) Grasses such as this cultivated durham wheat and the fountain grass in Figure 27.27c are monocots. (c) Monocots include popular garden flowers such as these daylilies. Orchids (Figure 27.29b) are another highly prized monocot flower.

(b)

27.33 Dicots (a) The cactus family is a large group of dicots, with about 1,500 species in the Americas. This cactus bears scarlet flowers for a brief period of the year. (b) The flowering dogwood is a small dicot tree. (c) These climbing Cape Cod roses are members of the family Rosaceae, as are the familiar roses from your local florist.

(a)   (b)   (c)

27.34 Herbs and Tundra The landscape of this Alaskan tundra is dominated by herbs.

through good timing—growing, photosynthesizing, and reproducing early in the season, before the foliage develops fully on the trees above. When the leafy canopy of the forest becomes dense and blocks sunlight from reaching the forest floor, the shoots of these herbs die back, leaving underground organs (rhizomes or tubers) with stored food for the beginning of the following year's growth.

Herbs predominate in at least one extreme environment: tundra (Figure 27.34). Here the ability of some herbs to store photosynthate in underground organs stands them in particularly good stead, because they leave no aboveground parts to be damaged during the long cold season. The cold is so intense that shrubs and trees are unable to survive, but the dormant underground parts of herbs succeed.

Determining which angiosperms were the first flowering plants is a question of great controversy. Two candidates are the magnolia family (see Figure 27.28*a*) and another family, the Chloranthaceae, whose flowers are much simpler than those of the magnolias. Other candidates have also been suggested. One complication in the search for the earliest fossil angiosperms is that we are unlikely to be able to tell whether an ancient fossil plant practiced double fertilization. There continues to be disagreement as to whether the first angiosperms were trees or herbs. However that disagreement is being resolved, today the predominant ground cover consists of herbaceous angiosperms.

Herbaceous plants predominate in many situations

When a piece of ground becomes available for new colonization—perhaps as a result of fire, or the appearance of a new sandbar, or clearing by humans—the first colonizers are generally herbs. Their seeds may have been present in a dormant state in the soil, or they may have been brought in by wind or animals. The rapidly growing herbs produce seeds that germinate and contribute to the growth of the herb population.

Before the land is cleared, the foliage of shrubs and trees may shade the ground, making little light available for photosynthesis by small herbs. When the taller plants are removed, the herbs have their day in the sun. Later, as the land is modified by these herbaceous plants, other, larger forms appear (or reappear) and generally take over as succession proceeds (see Chapter 52).

Although not dominating the scene, some herbaceous angiosperms do well in forests. They succeed

Summary of "Plants: Pioneers of the Terrestrial Environment"

The Plant Kingdom

- Plants are multicellular, photosynthetic eukaryotes that develop from embryos protected by parental tissue.
- Plant life cycles feature alternation of generations: gametophyte (haploid) and sporophyte (diploid).
- There are twelve surviving phyla of plants grouped into two main categories: tracheophytes and nontracheophytes. **Review Table 27.1**
- The plant kingdom arose from the green algae—specifically, a common ancestor of the stoneworts or of *Coleochaete*. Descendants of this ancestor colonized the land.
- Tracheophytes, characterized by possession of a vascular system, consisting of water- and mineral-conducting xylem and food-conducting phloem, evolved from nontracheophytes. **Review Figure 27.3**

Nontracheophytes: Liverworts, Hornworts, and Mosses

- The nontracheophytes include the liverworts (phylum Hepatophyta), hornworts (phylum Anthocerophyta), and mosses (phylum Bryophyta). **Review Table 27.1**
- Nontracheophytes either lack vascular tissues completely or, in the case of certain mosses, have only a rudimentary system of water- and food-conducting cells.
- The nontracheophyte sporophyte generation is smaller than the gametophyte generation and depends on the gametophyte for water and nutrition. **Review Figures 27.5, 27.8**
- Liverwort sporophytes have no specific growing zone. Hornwort sporophytes grow at their basal end, and moss sporophytes grow at their apical end. **Review Figure 27.7**
- The hydroids of mosses, through which water may travel, are probably ancestral to the water-conducting cells of the tracheophytes.

Introducing the Tracheophytes

• The tracheophytes have vascular tissue with tracheids and other specialized cells designed to conduct water, minerals, and foods.

• Present-day tracheophytes are grouped into nine phyla that form two major groups: nonseed tracheophytes and seed plants.

• In tracheophytes the sporophyte generation is larger than the gametophyte and independent of the gametophyte generation.

• The earliest tracheophytes, known to us only in fossil form, lacked roots and leaves. Roots may have evolved from branches that penetrated the ground. Simple leaves are thought to have evolved from sporangia, and complex leaves may have resulted from the flattening and reduction of a branching stem system. **Review Figures 27.12, 27.13, 27.14**

• Heterospory, the production of distinct female megaspores and male microspores, evolved on several occasions from homosporous ancestors. **Review Figure 27.15**

The Surviving Nonseed Tracheophytes

• Whisk ferns (phylum Psilophyta) look very much like the ancient (and extinct) first tracheophytes. Club mosses (phylum Lycophyta) have simple leaves arranged spirally. Horsetails (phylum Sphenophyta) have simple leaves in whorls. Leaves with more complex vasculature are characteristic of all other phyla of tracheophytes. **Review Table 27.1**

• Ferns (phylum Pterophyta) have complex leaves and a dominant sporophyte generation. **Review Figure 27.22 and Table 27.1**

The Seeds of Success

• The seed plants (gymnosperms and angiosperms) are heterosporous and have much reduced gametophytes.

• Most modern seed plants have no swimming gametes and are independent of liquid water for fertilization. The male gametophyte, the pollen grain, is dispersed by wind or by animals.

• The seed is a well protected resting stage that often contains food that supports the growth of the embryo.

The Gymnosperms: Naked Seeds

• The gymnosperms, once the dominant vegetation of all of Earth, still dominate forests in the northern parts of the Northern Hemisphere and at high elevations.

• The four surviving gymnosperm phyla are the Cycadophyta (the most ancient), Ginkgophyta (consisting of a single genus, the maidenhair tree), Gnetophyta (the sister group to the angiosperms), and Coniferophyta (the familiar cone-bearing trees). **Review Table 27.1**

• Modern gymnosperms all have abundant xylem and extensive secondary growth.

• Conifers have a life cycle in which naked seeds are produced on the scales of female cones. Male cones are smaller than female cones. Pollen is transferred from male to female cones by wind. **Review Figure 27.25**

The Angiosperms: Flowering Plants

• Angiosperms (phylum Angiospermae) are distinguished by the production of flowers and fruits. They demonstrate double fertilization resulting in a triploid nutritive tissue, the endosperm. Double fertilization is also characteristic of the Gnetophyta. **Review Figure 27.31**

• The vascular tissues of angiosperms contain cell types (such as vessel elements and fibers) rarely found elsewhere in the plant kingdom.

• Flowers are made up of various combinations of carpels, stamens, petals, and sepals. Perfect flowers have both carpels (female parts) and stamens (male parts). **Review Figure 27.26**

• Monoecious plant species have both female and male flowers on the same plant. Dioecious species have separate female and male plants.

• Carpels and stamens may have evolved from leaflike structures. **Review Figure 27.29**

• Angiosperms and the animals that pollinate them have coevolved.

• The two classes of flowering plants, Monocotyledones and Dicotyledones, differ in vein patterns in the leaves, in the number of cotyledons, and in the number of flower parts. Grasses are common monocots, and broad-leaved plants such as oaks and roses are common dicots.

Self-Quiz

1. Plants differ from algae in that only plants
   a. are photosynthetic.
   b. are multicellular.
   c. possess chlorophyll.
   d. have multicellular embryos protected by the parent.
   e. are eukaryotic.

2. Which statement about the alternation of generations in plants is *not* true?
   a. It is heteromorphic.
   b. Meiosis occurs in sporangia.
   c. Gametes are always produced by meiosis.
   d. The zygote is the first cell of the sporophyte generation.
   e. The gametophyte and sporophyte differ genetically.

3. Which statement is *not* evidence for the origin of plants from the green algae?
   a. Some green algae have multicellular sporophytes and multicellular gametophytes.
   b. Both plants and green algae have cellulose in their cell walls.
   c. The two groups have the same photosynthetic and accessory pigments.
   d. Both plants and green algae produce starch as their principal storage carbohydrate.
   e. All green algae produce large, stationary eggs.

4. The nontracheophytes
   a. lack a sporophyte generation.
   b. grow in dense masses, allowing capillary movement of water.
   c. possess xylem and phloem.
   d. possess true leaves.
   e. possess true roots.

5. The rhyniophytes
   a. possessed vessel elements.
   b. possessed true roots.
   c. possessed sporangia at the tips of stems.
   d. possessed leaves.
   e. lacked branching stems.

6. Club mosses and horsetails
   a. have larger gametophytes than sporophytes.
   b. possess small leaves.
   c. are represented today primarily by trees.
   d. have never been a dominant part of the vegetation.
   e. produce only simple fruits.

7. Which statement about ferns is *not* true?
   a. The sporophyte is larger than the gametophyte.
   b. Most are heterosporous.
   c. The young sporophyte can grow independently of the gametophyte.
   d. The frond is a large leaf.
   e. The gametophytes produce archegonia and antheridia.

8. The gymnosperms
   a. dominate all land masses today.
   b. have never dominated land masses.
   c. have active secondary growth.
   d. all have vessel elements.
   e. lack sporangia.

9. Which statement about flowers is *not* true?
   a. Pollen is produced in the anthers.
   b. Pollen is received on the stigma.
   c. An inflorescence is a cluster of flowers.
   d. A species having female and male flowers on the same plant is dioecious.
   e. A flower with both mega- and microsporangia is said to be perfect.

10. Which statement about fruits is *not* true?
    a. They develop from ovaries.
    b. They may include other parts of the flower.
    c. A multiple fruit develops from several carpels of a single flower.
    d. They are produced only by angiosperms.
    e. A cherry is a simple fruit.

Applying Concepts

1. Mosses and ferns share a common trait that makes water droplets a necessity for sexual reproduction. What is this trait?

2. Ferns display a dominant sporophyte stage (with large fronds). Describe the major advance in anatomy that enables most ferns to grow much larger than mosses.

3. What features distinguish club mosses from horsetails? What features distinguish these groups from rhynio-phytes and psilophytes? from ferns?

4. Suggest an explanation for the great success of the angiosperms in occupying terrestrial habitats.

5. Contrast simple leaves with complex leaves in terms of structure, evolutionary origin, and occurrence among plants.

6. In many locales, large gymnosperms predominate over large angiosperms. Under what conditions might gymnosperms have the advantage, and why?

Readings

Burnham, C. R. 1988. "The Restoration of the American Chestnut." *American Scientist*, vol. 76, pages 478–487. An account of the comeback of a species almost completely wiped out by disease. If you read the article on chestnut blight recommended in the next chapter, you may be encouraged to see that there are methods, based on Mendelian genetics, to rebuild the native population of this important tree species.

Crosson, P. R. and N. J. Rosenberg. 1989. "Strategies for Agriculture." *Scientific American,* September. A discussion of approaches to increasing yields for an expanding population, emphasizing that social and economic changes will also be required.

Friedman, W. E. 1990. "Double Fertilization in *Ephedra,* a Nonflowering Seed Plant: Its Bearing on the Origin of Angiosperms." *Science,* vol. 247, pages 951–954. A description of a breakthrough in understanding the origin of the flowering plants.

Graham, L. E. 1985. "The Origin of the Life Cycle of Land Plants." *American Scientist,* vol. 73, pages 178–186. A discussion of how plants made it to the terrestrial environment.

Heyler, D. and C. M. Poplin. 1988. "The Fossils of Montceau-les-Mines." *Scientific American,* September. A description of plants and animals of the Carboniferous period, discovered in a rich fossil lode in central France. Presents a detailed picture of life in a time long past.

Hinman, C. W. 1986. "Potential New Crops." *Scientific American,* July. A look at plants such as jojoba, buffalo gourd, and others as prospects for food and other materials.

Niklas, K. J. 1986. "Computer-Simulated Plant Evolution." *Scientific American,* March. An examination of some hypotheses concerning plant evolution. The hypotheses were modeled on computers, generating testable suggestions. The article presumes no knowledge of computers.

Niklas, K. J. 1987. "Aerodynamics of Wind Pollination." *Scientific American,* July. A discussion of the many adaptations of wind-pollinated plants that favor successful pollination.

Raven, P. H., R. F. Evert and S. Eichhorn. 1992. *Biology of Plants,* 5th Edition. Worth, New York. An excellent general botany textbook.

Chapter 28

Fungi: A Kingdom of Recyclers

Fairy-Tale Fungi
Forests full of elves might be one of the images evoked by members of the mushroom family of the kingdom Fungi. In fact fungi have many growth forms and play an indispensable role in the biosphere.

Surely you are familiar with mushrooms—in the market, on your plate, or in whimsical fairy-tale portraits, pleasant and picturesque. But what about the black spots that appear on your home-made bread, or the green fuzz that forms on your orange? These are molds, and they are neither pleasant nor picturesque. Molds and mushrooms are both fungi, and Earth would be a messy place without them. Because they are superbly adapted to absorptive nutrition, fungi are at work in forests, fields, and garbage dumps, breaking down the remains of dead organisms and even manufactured substances such as some plastics. For almost a billion years, the ability of fungi to decompose substances has been important for life on Earth, chiefly because by breaking down carbon compounds, fungi return carbon and other elements to the environment, where they can be used again by other organisms.

Already crucial for the continuation of life as we know it, the fungi may soon play enhanced roles in decomposing toxic environmental pollutants. This task is necessary because our species has been loading toxic substances into the environment for some time. Might the removal of these substances be a job for the fungi, naturally occurring or genetically engineered? Indeed it might, and biologists and chemists are exploring the ability of different fungi, as well as bacteria, to break down pesticides and other toxic materials into harmless substances. This overall process of degradation and decomposition of toxic materials is known as **bioremediation**.

28.1 Parasitic Fungi Attack Other Living Organisms (*a*) The gray masses on this ear of corn are the parasitic fungus *Ustilago maydis*, commonly called corn smut. (*b*) The tropical fungus whose fruiting body is growing out of the carcass of this ant has developed from a spore ingested by the ant. The spores of this fungus must be ingested by insects before they will germinate and develop. The growing fungus absorbs organic and inorganic nutrients from the ant's body, eventually killing it, after which the fruiting body produces a new crop of spores. (*c*) An amoeba (below) being parasitized by a fungus (above) of the genus *Amoebophilus* (which means "amoeba lover").

(*a*)

(*b*)

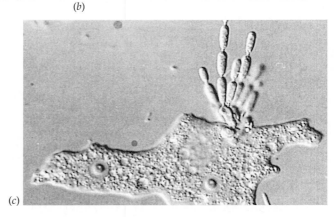

(*c*)

In this chapter we will examine the general biology of this very important group of organisms—the kingdom Fungi—which differs in interesting ways from the other kingdoms. We will also explore the diversity of body forms, reproductive structures, and life cycles of the four phyla of fungi, as well as the mutually beneficial associations of certain fungi with other organisms. As we begin our study, recall that the fungi and the animals are descended from a common ancestor—we are more closely related to molds and mushrooms than we are to the roses we admired in the last chapter (see Figure 26.31).

General Biology of the Fungi

We define the kingdom Fungi as encompassing *heterotrophic organisms* with *absorptive nutrition.* Fungi are **saprobes** (organisms that live on dead matter), **parasites** (organisms that absorb nutrients from living hosts; Figure 28.1), or **mutualists** (organisms that live in mutually beneficial symbiosis with other organisms). All fungi form spores, but only in one phylum (Chytridiomycota) do single-celled stages possess flagella. Fungi reproduce sexually in a variety of ways, ranging from fusion of filaments of different mating types to fusion of distinct gametes. The walls that enclose the filaments of all fungi contain at least some **chitin**, a polysaccharide that is also found in the skeletons of arthropods. Most fungi have complex forms.

These criteria enable us to distinguish between the fungi and some protists. The slime molds consist of two protist phyla (Dictyostelida and Acrasiomycota), whose members take up food by phagocytosis rather than by absorption, and a third protist phylum (Myxomycota), whose members have cells with flagella. The other funguslike protists (Oomycota) also have flagellated cells, and they have cellulose, rather than chitin, in their cell walls.

The kingdom Fungi consists of four phyla: Chytridiomycota, Zygomycota, Ascomycota, and Basidiomycota (Table 28.1). We distinguish the fungal phyla on the basis of the methods and structures they use for sexual reproduction and, to a lesser extent, on criteria such as the presence or absence of cross-walls separating their cell-like compartments.

Some fungi, called imperfect fungi, do not form sexual structures by which they might be easily identified as members of one of the four phyla. However, techniques of molecular taxonomy, such as DNA sequencing, now allow us to identify many imperfect fungi as asexual zygomycetes, ascomycetes, or basidiomycetes.

In the sections that follow we'll consider some aspects of the general biology of the fungi, including their body structure and its intimate relationship with the environment, their nutrition, and some special aspects of their sexual reproductive cycles, especially the occurrence of mating types, of a dikaryotic ($n + n$) nuclear condition, and of dual hosts.

The body of a fungus is composed of hyphae

The vegetative, feeding body of a fungus is called a **mycelium** (plural mycelia). It is composed of rapidly growing individual filaments called **hyphae** (singular

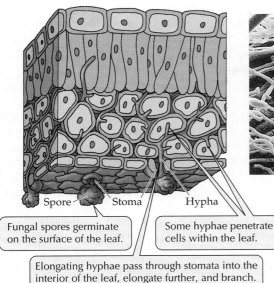

Fungal spores germinate on the surface of the leaf.

Some hyphae penetrate cells within the leaf.

Elongating hyphae pass through stomata into the interior of the leaf, elongate further, and branch.

Spore   Stoma   Hypha

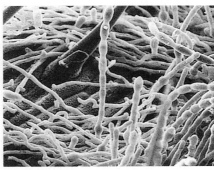

28.2 A Fungus Attacks a Leaf The white structures in the micrograph are the hyphae of the fungus *Blumeria graminis*, which is growing on the dark surface of the leaf of a grass.

hypha), which are usually subdivided into cell-like compartments by *incomplete* partitions called **septa** (singular septum). In most hyphae, organelles (even nuclei) can move around, and there is no division into separate cells. Certain modified hyphae, the **rhizoids**, anchor saprobic fungi to their substrate (the dead organism or other matter upon which they feed). Parasitic fungi may have modified hyphae that take up nutrients from the host. The total hyphal growth of a mycelium (not the growth of an individual hypha) may exceed 1 km per day. The hyphae may be highly dispersed or may clump together in a cottony mass. Sometimes, when sexual spores are produced, the mycelium is organized into elaborate fruiting bodies such as mushrooms.

The way in which a parasitic fungus attacks a plant illustrates the roles of some fungal structures (Figure 28.2). The hyphae of a fungus invade a leaf through pores called stomata (see Chapter 8), through wounds, or, in some cases, by direct penetration of epidermal cells. Once inside, the hyphae form a mycelium. Some hyphae grow into living plant cells, absorbing the nutrients within the cells. Eventually fruiting bodies form, either within the plant body or on its surface.

Fungi are in intimate contact with their environment

The tubular hyphae of a multicellular fungus give it a unique relationship with its physical environment: The fungal mycelium has an enormous surface area–to–volume ratio compared with that of most large multicellular organisms. This large ratio of surface area to volume is a marvelous adaptation for absorptive nutrition, in which nutrients are absorbed across the hyphal surfaces. Throughout the mycelium, except in fruiting bodies, all the hyphae are very close to their environmental food source.

Another characteristic of some fungi is their tolerance for highly hyperosmotic environments (those with very negative osmotic potential; see Chapter 5). Many fungi are more resistant than are bacteria to damage in hyperosmotic surroundings. For example, jelly in the refrigerator will not become a growth medium for bacteria, because the jelly is too hyperosmotic to the bacteria, but it may eventually harbor mold colonies. You have probably seen the green mold *Penicillium* growing on oranges in the refrigerator. The refrigerator example illustrates another trait of many fungi: tolerance of temperature extremes. Many fungi

TABLE 28.1	Classification of Fungi		
PHYLUM	**COMMON NAME**	**FEATURES**	**EXAMPLES**
Chytridiomycota	Chytrids	Aquatic; gametes have flagella	*Allomyces*
Zygomycota	Zygomycetes	No regularly occurring hyphal cross-walls; usually no fleshy fruiting body	*Rhizopus*
Ascomycota	Ascomycetes	Ascus; perforated cross-walls	*Neurospora*, baker's yeast
Basidiomycota	Basidiomycetes	Basidium; perforated cross-walls	*Puccinia*, mushrooms

tolerate temperatures as low as 5 or 6°C below freezing, and some tolerate temperatures as high as 50°C or more.

Fungi are absorptive heterotrophs

All fungi are heterotrophs that obtain food by direct absorption from the immediate environment. The vast majority are saprobes, obtaining their energy, carbon, and nitrogen directly from dead organic matter by the action of enzymes. However, as we've learned already, some are parasites, and still others form mutualistic associations with other organisms.

Saprobic fungi, along with bacteria, are the major decomposers of the biosphere, contributing to decay and thus to the recycling of the elements used by living things. In the forest, for example, the invisible mycelia of fungi obtain nutrients from fallen trees, thus decomposing the trees. Fungi are the principal decomposers of cellulose and lignin, the main components of plant cell walls (most bacteria cannot break down plant cell wall material).

Because saprobic fungi are able to grow on artificial media, we can determine their exact nutritional requirements. Sugars are the favored source of carbon, and most fungi obtain nitrogen from proteins or the products of protein breakdown. Many fungi can use nitrate (NO_3^-) or ammonium (NH_4^+) ions as their sole source of nitrogen. No known fungus gets its nitrogen directly from nitrogen gas, as can some bacteria and plant–bacteria associations (see Chapter 34). Vitamins may play a role in fungal nutrition. Most fungi are unable to synthesize their own thiamin (vitamin B_1) or biotin (another B vitamin) and hence must absorb these vitamins from their environment. Other compounds that are vitamins for animals can be made by fungi. Like all other organisms, fungi also require some mineral elements.

Nutrition in the parasitic fungi is particularly interesting. **Facultative** parasites can grow parasitically or by themselves on defined artificial media. Biologists can work out the exact nutritional requirements by varying the composition of the growth medium. **Obligate** parasites cannot be grown on any available defined medium; they can grow only on

their specific hosts, usually plants. Obligate parasites include various mildews. Because their growth is so limited, they must have unusual nutritional requirements. Biologists are thus very interested in learning more about them.

Some fungi have adaptations that enable them to function as active predators, trapping nearby microscopic protists or animals, from which they obtain nitrogen and energy. The most common approach is to secrete sticky substances from the hyphae so that passing organisms stick tightly to them. Fungal hyphae then quickly invade the prey, growing and branching within it, spreading through its body, absorbing nutrients, and eventually killing it.

A more dramatic adaptation for predation is the **constricting ring** formed by some species of *Arthrobotrys, Dactylaria,* and *Dactylella* (Figure 28.3). All of these fungi grow in soil; when nematodes (tiny roundworms) are present, these fungi form three-celled rings that have a diameter that just fits a nematode. A nematode crawling through one of these rings stimulates the ring, causing the cells of the ring to swell and trap the worm. Fungal hyphae quickly invade and digest the unlucky victim. Some species of wood-decaying mushrooms have the ability to supplement their nitrogen supply by feeding on nematodes.

Certain highly specific associations between fungi and other organisms have nutritional consequences for the fungal partner. **Lichens** are associations of a fungus with either a unicellular alga or a cyanobacterium; in this union the fungus draws nutrition from the photosynthetic bacterium or alga. **Mycorrhizae** (singular mycorrhiza) are associations between spe-

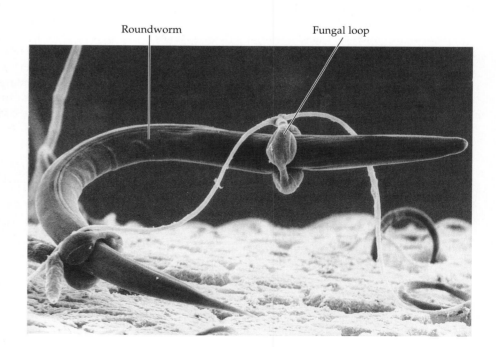

Roundworm   Fungal loop

28.3 Some Fungi Are Predators
A nematode (roundworm) is trapped in sticky loops of the soil-dwelling fungus *Arthrobotrys anchonia.*

Workers of the Costa Rican ant species *Atta cephalotes* add a cut piece of leaf to their fungal garden.

The fungus is the white material. In response to this care, the fungus grows and serves as food for the ants.

28.4 A Fungus Garden Central American leaf cutter ants cultivate fungi. These species of fungi are found only in association with the ant "farms."

cific fungi and the roots of plants. In such associations the fungus is fed by the plant but provides minerals (primarily phosphorus) to the plant root, so the plant's nutrition is also promoted. Seed germination in most orchid species depends on the presence of a specific mutualistic fungus, which itself derives nutrients from the seed and seedling of the orchid. We will discuss lichens and mycorrhizae more thoroughly later in this chapter.

Perhaps the most striking fungal associations are with insects. Some leaf-cutting ants "farm" fungi, feeding the fungi and later harvesting and eating them (Figure 28.4). The ants collect leaves and flower petals and chew them into small bits, on which they "plant" fungal mycelium. The ants even "weed" these gardens by removing other fungal species. The species of fungus cultivated by the ants are found nowhere other than in these tiny gardens. Scale insects live in association with the fungus *Septobasidium*. The fungus spreads over a colony of the insects, infecting some of them and thus parasitizing them—without killing them. As new insects hatch from the eggs within the colony, some of them become infected with the fungus. They take the fungus along as they establish new colonies. The fungus protects the colony against drying and against some predators but also draws its nutrition from the insects.

Most fungi reproduce asexually and sexually, and they have mating types

Both asexual and sexual reproduction are common in the fungi. Asexual reproduction takes several forms.

One form is the production of haploid spores within structures called sporangia. The second form is the production of naked spores at the tips of hyphae; these spores, called **conidia** (from the Greek *konis*, "dust"), are not produced in sporangia. The third form of asexual reproduction, performed by unicellular fungi, is cell division—either a relatively equal division or an asymmetric division in which a tiny bud is produced. The fourth form of asexual reproduction, seen in some hyphal fungi, is simple breakage of the mycelium.

Sexual reproduction in many fungi features an interesting twist. There is no distinction between female and male structures, or between female and male organisms. Rather, there is a genetically determined distinction between two or more **mating types**. Individuals of the same mating type cannot mate with one another, but they can mate with individuals of another mating type. This distinction prevents self-fertilization. Individuals of different mating types differ genetically from one another but are often visually and behaviorally indistinguishable.

The nuclei of most fungi are haploid, except in the zygotes formed in sexual reproduction, which are diploid. The zygotes undergo meiosis, producing haploid nuclei that become incorporated into spores. Haploid fungal spores, produced sexually or asexually, germinate, and their nuclei divide mitotically to produce hyphae. Fungal hyphae may have a nuclear configuration—a dikaryon stage, as described in the next section—other than the familiar haploid and diploid.

Many fungal life cycles include a dikaryon stage

Sexual reproduction begins in some fungi in an unusual way: The cytoplasms of two individuals of opposite mating type fuse long before their nuclei fuse, so *two genetically different haploid nuclei exist within the same hypha*. This hypha is called a **dikaryon** (having *two* nuclei). Because the two nuclei differ genetically, the hypha is also called a **heterokaryon** (having *different* nuclei).

Eventually specialized fruiting structures form, within which pairs of dissimilar nuclei—one from each parent—fuse, giving rise to zygotes long after the original "mating." The zygote nucleus—which may be the only diploid nucleus in the entire life cycle—undergoes meiosis, producing four haploid nuclei. The mitotic descendants of those nuclei become the nuclei of the next generation of hyphae.

The reproduction of such fungi displays several unusual features. First, there are no gamete cells, only gamete nuclei. Second, there is never any true diploid tissue, although for a long period during development the genes of both parents are present in the dikaryon and can be expressed. In effect, the cells are neither diploid (2*n*) nor haploid (*n*); rather they are dikaryotic

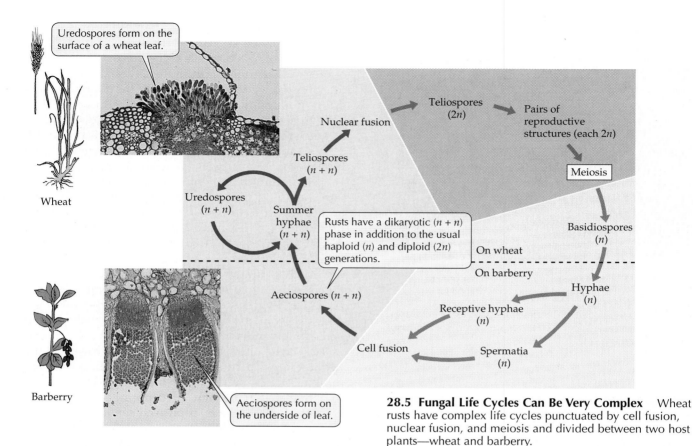

Uredospores form on the surface of a wheat leaf.

Wheat

Barberry

Aeciospores form on the underside of leaf.

Nuclear fusion

Teliospores (2n)

Pairs of reproductive structures (each 2n)

Meiosis

Teliospores (n + n)

Uredospores (n + n)

Summer hyphae (n + n)

Rusts have a dikaryotic (n + n) phase in addition to the usual haploid (n) and diploid (2n) generations.

Basidiospores (n)

On wheat

On barberry

Aeciospores (n + n)

Hyphae (n)

Receptive hyphae (n)

Cell fusion

Spermatia (n)

28.5 Fungal Life Cycles Can Be Very Complex Wheat rusts have complex life cycles punctuated by cell fusion, nuclear fusion, and meiosis and divided between two host plants—wheat and barberry.

(n + n). A harmful recessive mutation in one nucleus may be compensated for by a normal allele on the same chromosome in the other nucleus. Dikaryosis is perhaps the most significant of the genetic peculiarities of the fungi.

Finally, although these organisms grow in moist places, the gamete nuclei are not motile and are not released into the environment; therefore, liquid water is not required for fertilization. (Some members of the plant kingdom require liquid water for the meeting of gametes, as do some aquatic animals.)

The life cycles of some parasitic fungi involve two hosts

Some parasitic fungi are very specific about what host organism they use as a source of nutrition, and some even use different hosts for different stages of their life cycle. (Remember that a host is an organism that harbors another organism—generally a parasite—and provides it with nutrition.) One of the most striking examples of a fungal life cycle that involves two different hosts is the complicated life cycle of *Puccinia graminis*, the agent of a major agricultural disease, black stem rust of wheat. In the epidemic of 1935, *P. graminis* was responsible for the loss of about one-fourth of the wheat crop in Canada and the United States.

Let's examine a year in the life cycle of *P. graminis* (Figure 28.5). During the summer, dikaryotic (n + n) hyphae of *P. graminis* proliferate in the stem and leaf tissues of wheat plants, drawing their nutrition from the wheat and damaging it severely. These dikaryotic hyphae produce extensive amounts of summer spores called **uredospores**. The one-celled, orange uredospores are scattered by the wind and infect other wheat plants, on which the summer hyphae then proliferate. Like the hyphae, the uredospores are dikaryotic.

Dark brown winter spores—**teliospores**—begin to appear on certain special hyphae in late summer. Each teliospore consists initially of two dikaryotic cells. The two haploid nuclei in each cell fuse to form a single, diploid (2n) nucleus. These are the only truly diploid cells in the entire life cycle. Both cells have thick walls and usually survive freezing. The teliospores remain dormant until spring, when they germinate. Each of the two cells develops into a reproductive structure, within which the nucleus divides meiotically to produce four haploid **basidiospores** (discussed later in this chapter, under the phylum Basidiomycota).

The basidiospores are of two different mating types: plus (+) and minus (–). They are carried by the wind, and if they land on a leaf of the common barberry plant, they germinate and produce haploid hyphae (+ or –, depending on the mating type of the germinating basidiospore). These hyphae invade the barberry leaf.

The hyphae form flask-shaped structures on the upper surfaces of the leaf, within which some of the hyphae pinch off tiny, colorless, haploid **spermatia** from their tips. The hyphae also form another type of structure, called an **aecial primordium**, near the lower surface of the barberry leaf. Thus the leaf contains "flasks" on the upper surface and aecial primordia near the lower one, and these are connected by the hyphae. Insects attracted to a sweetish liquid produced in the flasks carry spermatia from one flask to another, initiating the next step in the cycle.

A spermatium of one mating type fuses with a receptive hypha of the other type within a flask. The nucleus of the spermatium repeatedly divides mitotically, and the products move through the hyphae into immature aecial primordia, where they produce dikaryotic cells with two haploid nuclei, one of each mating type. These cells develop into **aeciospores**, each containing two unlike nuclei. The aeciospores are scattered by the wind, and some germinate on wheat plants, continuing the life cycle. When the dikaryotic aeciospores germinate, they produce the summer hyphae with two nuclei in each cell.

To summarize, in the stages from the aeciospore to the teliospore, *P. graminis* is dikaryotic; that is, the individual cells contain nuclei from both "parents." Teliospores, at first dikaryotic, become diploid. Basidiospores are produced by meiosis and cytokinesis, so they are haploid, each having a mating type of + or –.

The different types of spores produced during the life cycle of *P. graminis* play very different roles. Windborne uredospores are the primary agents for spreading the rust from wheat plant to wheat plant and to other fields. Resistant teliospores allow the rust to survive the harsh winter but contribute little to the spreading of the rust. Basidiospores spread the rust from wheat to barberry plants. Spermatia and receptive hyphae initiate the sexual cycle of the rust. Finally, aeciospores spread the rust from barberry to wheat plants.

The traditional means of combating *P. graminis* in wheat country was to remove barberry from the area, because without this obligate (necessary) alternate host, the fungus cannot complete its life cycle. However, this approach had limited success because some uredospores survived the winter and could infect a new generation of wheat. Modern control of the disease focuses on the breeding of resistant strains of wheat—a difficult task, given the rapid evolution of the rust. A new wheat variety carrying new resistance genes has to be released every year to keep up with the genetic changes in the rust.

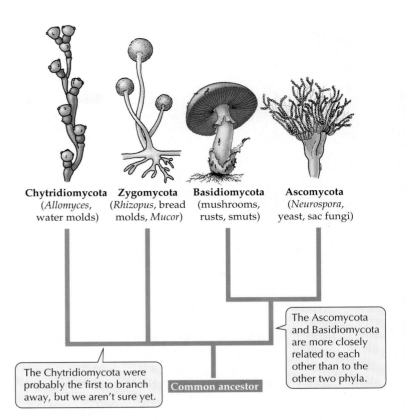

Chytridiomycota
(*Allomyces*, water molds)

Zygomycota
(*Rhizopus*, bread molds, *Mucor*)

Basidiomycota
(mushrooms, rusts, smuts)

Ascomycota
(*Neurospora*, yeast, sac fungi)

The Chytridiomycota were probably the first to branch away, but we aren't sure yet.

Common ancestor

The Ascomycota and Basidiomycota are more closely related to each other than to the other two phyla.

28.6 Phylogeny of the Fungi In addition to these four phyla, the imperfect fungi, or Deuteromycetes, is a "holding group" for fungal species whose status is yet to be determined.

Diversity in the Kingdom Fungi

Each phylum of the kingdom Fungi appears to be monophyletic, so our consideration of fungal diversity probably corresponds with a natural phylogeny (Figure 28.6). Because the imperfect fungi (deuteromycetes) almost certainly are polyphyletic, we will not give them phylum status. In this section on fungal diversity, we will consider the four phyla—Chytridiomycota, Zygomycota, Ascomycota, and Basidiomycota—and we'll discuss the status of the deuteromycetes.

Chytrids probably resemble the ancestral fungi

The most ancient fungal phylum is the Chytridiomycota—the chytrids, a group of aquatic microorganisms sometimes classified as protists. We place chytrids among the fungi because their cell walls consist primarily of chitin and because molecular evidence indicates that they are monophyletic with the other fungi.

Chytrids are either parasitic (on organisms such as algae, mosquito larvae, and nematodes) or saprobic, obtaining nutrients by breaking down dead organic

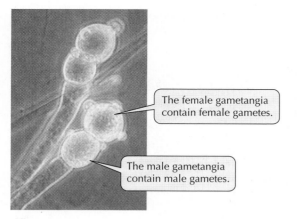

The female gametangia contain female gametes.

The male gametangia contain male gametes.

Allomyces sp.

28.7 Reproductive Structures of a Chytrid The haploid gametes produced in these gamete cases will fuse and form diploid zoospores.

matter. (Chytrids in the compound stomachs of cud-chewing animals may be an exception, living in a mutualistic association with their hosts.) Most chytrids live in freshwater habitats or in moist soil, but some are marine. Some chytrids are unicellular; others have mycelia made up of branching chains of cells. Chytrids reproduce both sexually and asexually.

Allomyces, a well-studied genus of chytrids, displays an alternation of haploid and diploid generations. A haploid **zoospore** (spore with flagella) comes to rest on dead plant or animal material in water and germinates to form a small haploid organism. This organism produces female and male **gametangia** (gamete cases) (Figure 28.7). The male gametangia are smaller than the female gametangia and possess a light orange pigment. Mitosis in the gametangia results in the formation of haploid gametes, each with a single nucleus.

Both female and male gametes have flagella. The motile female gamete produces a chemical attractant (pheromone; see Chapter 42) that attracts the swimming male gamete. The gametes fuse in pairs, and then their nuclei fuse to form a diploid zygote. Cell divisions give rise to a small diploid organism that produces numerous diploid flagellate zoospores by mitosis and cytokinesis. These diploid zoospores germinate to form more of the diploid organisms. Eventually, the diploid organism produces thick-walled resting sporangia that can survive unfavorable conditions such as dry weather or freezing. Nuclei in the resting sporangia eventually undergo meiosis, giving rise to haploid zoospores that are released into the water and begin the cycle anew.

Another chytrid, *Coelomomyces*, has a life cycle similar to that of *Allomyces*, but it is a parasite. *Coelomomyces* also requires two animal hosts: a mos-quito larva and a copepod (a type of crustacean discussed in the next chapter).

Chytrids are the only fungi that have flagella. We speculate that the protist ancestor of the fungi possessed flagella, because the phylum Chytridiomycota was the first fungal group to diverge from the others (see Figure 28.6). As we learned in Chapter 26, the same protist ancestor gave rise to the protist phylum Choanoflagellida and to the animal kingdom. A key event when the chytrids branched off was the loss of flagella in the other branch of the fungal kingdom.

Zygomycetes reproduce sexually by fusion of two gametangia

Most zygomycetes (phylum Zygomycota) have hyphae without regularly occurring septa (cross-walls). They produce no motile cells, and only one diploid cell, the zygote, appears in the entire life cycle. Most zygomycetes form no fleshy fruiting body; rather, the hyphae spread in an apparently random fashion, with occasional stalked sporangia reaching up into the air (Figure 28.8). A very important group of zygomycetes serves as the fungal partners in the most common type of mycorrhizal association with plant roots (we discuss mycorrhizae in depth later in the chapter). Almost 900 species of zygomycetes have been described. A zygomycete that you may have seen at one time or another is *Rhizopus stolonifer*, the black bread mold. The mycelium of a zygomycete spreads over its substrate, growing forward by means of specialized hyphae. In vegetative reproduction of *Rhizopus*, many

Phycomyces sp.

28.8 A Zygomycete This small forest of filamentous structures is made up of sporangiophores. The stalks end in tiny, rounded sporangia.

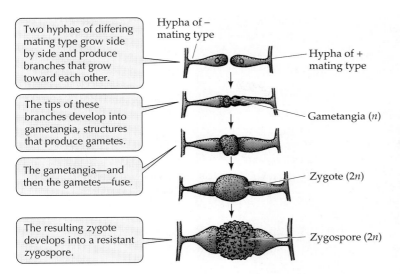

Two hyphae of differing mating type grow side by side and produce branches that grow toward each other.

Hypha of – mating type

Hypha of + mating type

The tips of these branches develop into gametangia, structures that produce gametes.

Gametangia (*n*)

The gametangia—and then the gametes—fuse.

Zygote (2*n*)

The resulting zygote develops into a resistant zygospore.

Zygospore (2*n*)

28.9 Conjugation in a Zygomycete The micrograph shows a zygospore produced by conjugation in the black bread mold.

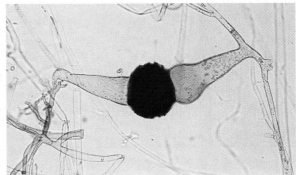

Rhizopus stolonifer

stalked sporangiophores are produced, each bearing a single sporangium containing hundreds of minute spores. Other zygomycetes have sporangiophores with many sporangia. As in other filamentous fungi, the spore-forming structure is separated from the rest of the hypha by a wall.

Zygomycetes reproduce sexually when adjacent hyphae of two different mating types grow together and produce gametangia that fuse to form zygotes (Figure 28.9). Zygotes develop into thick-walled, highly resistant **zygospores** contained within the thickened walls of the old gametangia. Zygospores may remain dormant for months before their nuclei undergo meiosis.

A sporangium sprouts from the zygospore. The sporangium contains the products of meiosis: haploid nuclei that are incorporated into spores that germinate to form a new generation of haploid hyphae. What causes the hyphae to grow together and fuse? The two mating types release pheromones that direct this conjugation process. (Recall that a pheromone directs gamete attraction in *Allomyces* as well.)

The sexual reproductive structure of ascomycetes is a sac

The ascomycetes (phylum Ascomycota) are a large and diverse group of fungi distinguished by the production of sacs called **asci** (singular ascus) (Figure 28.10). The ascus is the characteristic sexual reproductive structure of the ascomycetes. Ascomycete hyphae are segmented by more or less regularly spaced septa. A pore in each septum permits extensive movement of cytoplasm and organelles (including the nuclei) from one "cell" to the next.

The approximately 30,000 known species of ascomycetes can be divided into two broad groups, depending on whether the asci are contained within a specialized fruiting structure. Species that have a fruit-

ing structure, the **ascocarp**, are collectively called euascomycetes ("true ascomycetes"); those without ascocarps are called hemiascomycetes ("half ascomycetes").

HEMIASCOMYCETES. In general, the hemiascomycetes are microscopic, and many species are unicellular. Perhaps the best known are the yeasts, especially baker's or brewer's yeast (*Saccharomyces cerevisiae*; Figure 28.11*a*). The yeasts are among the most important domesticated fungi. *S. cerevisiae* metabolizes glu-

Ascus

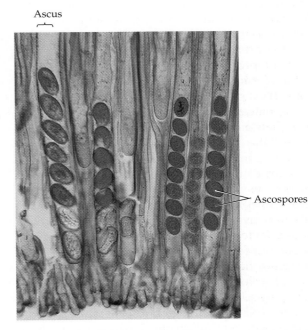

Ascospores

28.10 Asci and Ascospores Each ascus contains eight ascospores—the products of meiosis followed by a single mitotic division and cytokinesis. The ascospores are stained red for this micrograph.

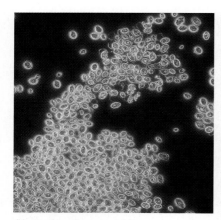

(a) *Saccharomyces cerevisiae*

(c) *Morchella esculenta*

(b) *Aleuria* sp.

28.11 Some Ascomycetes
(a) Some of these cells of baker's yeast are budding. These hemiascomycetes are facultative anaerobes—they can grow in either the presence or the absence of free oxygen. The brilliant red cups in (b) are cup fungi, as are the two yellow morels in (c). Morels, which have spongelike caps and a subtle flavor, are considered a gourmet delicacy.

cose obtained from its environment to ethanol and carbon dioxide. Carbon dioxide bubbles form in bread dough and give baked bread its light texture. Although baked away in bread making, the ethanol and carbon dioxide are both retained in beer. Other yeasts live on fruits such as figs and grapes and play an important role in the making of wine.

Yeasts reproduce asexually either by fission or, in the better-known genera, by **budding** (the outgrowth of a new cell from the surface of an old one). Sexual reproduction takes place only occasionally between two adjacent haploid cells of opposite mating type. (We discussed the genetics of yeast mating types in Chapter 14.) In some species, the resulting zygote buds to form a diploid cell population; in others, the zygote nucleus undergoes meiosis immediately. When diploid nuclei undergo meiosis, the entire cell becomes an ascus. Depending on whether the products of meiosis undergo mitosis, a yeast ascus usually has either eight or four **ascospores** (see Figure 28.10). The ascospores germinate to become haploid cells. Yeasts have no dikaryon stage.

Yeasts, especially *Saccharomyces cerevisiae*, are heavily used in molecular biological research. Just as *E. coli* is the most-studied prokaryote, *S. cerevisiae* is perhaps the most-studied eukaryote. Recall, for example, the creation and use of yeast artificial chromosomes, described in Chapter 16.

EUASCOMYCETES. Among the euascomycetes are several common molds, including *Neurospora*, the pink molds. *Neurospora* are found in a wide variety of habitats, and they are a serious contaminant in many laboratories. Recall that Beadle and Tatum used *Neurospora crassa* in their pioneering work on biochemical genetics (see Figure 12.1). Many euascomycetes are serious parasites on higher plants. Chestnut blight and Dutch elm disease are caused by euascomycetes. The powdery mildews are euascomycetes that infect cereal grains, lilacs, and roses, among many other plants. They can be a serious problem to grape growers, and a great deal of research has focused on ways to control these agricultural pests.

The euascomycetes also include the cup fungi (Figure 28.11b and c). In most of these organisms the fruiting structures are cup-shaped and can be as large as several centimeters across. The inner surfaces of the cups are covered with a mixture of both sterile filaments and asci, and they produce huge numbers of spores. Although these fleshy structures appear to be composed of distinct tissue layers, microscopic examination shows that their basic organization is still filamentous—a tightly woven mycelium. Such fruiting structures are formed only by the dikaryotic mycelium.

Two particularly delicious cup fungus fruiting structures are morels (Figure 28.11c) and truffles. Truffles grow in a mutualistic association with the roots of some species of oaks. Europeans traditionally used pigs to find truffles because some truffles secrete a substance that has an odor similar to a pig's sex attractant. Unfortunately, pigs also eat truffles, so dogs are now the usual truffle hunters.

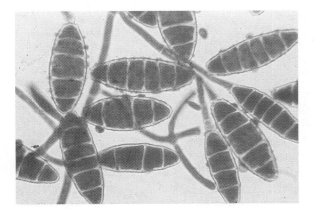

28.12 Conidia The large oval shapes are 4- to 6-celled macroconidia ("large conidia") at the tips of hyphae of an ascomycete. Each cell of a macroconidium can develop into a multicellular haploid individual, completing the asexual reproductive cycle.

The euascomycetes reproduce asexually by means of conidia that form at the tips of specialized hyphae (Figure 28.12). These small chains of conidia are produced by the millions and are sufficiently resistant to survive for weeks in nature. The conidia are what give molds their characteristic colors.

The sexual cycle of euascomycetes includes the formation of a dikaryon stage (Figure 28.13). Most euascomycetes form mating structures, some "female" and some "male." Nuclei from a male structure on one hypha enter a female mating structure on a hypha of compatible mating type. Dikaryotic ascogenous (ascus-forming) hyphae develop from the now dikaryotic female mating structure. The introduced nuclei divide simultaneously with the host nuclei. Eventually asci form at the tips of the ascogenous hyphae. Only with the formation of asci do the nuclei finally fuse. Both nuclear fusion and the subsequent meiosis of the resulting diploid nucleus take place within individual asci. The meiotic products are incorporated into ascospores that are ultimately shed by the ascus to begin the new haploid generation.

The sexual reproductive structure of basidiomycetes is a specialized cell bearing spores

About 25,000 species of basidiomycetes (phylum Basidiomycota) have been described. Basidiomycetes produce some of the most spectacular fruiting structures found anywhere among the fungi. These amazing fruiting structures are the puffballs (which may be more than half a meter in diameter), mushrooms of all kinds (more than 3,250 species, including the familiar *Agaricus campestris* you may enjoy on your pizza, as well as various poisonous mushrooms, such as members of the genus *Amanita*), and the giant bracket fungi often encountered on trees and fallen logs in a damp

forest (Figure 28.14). Bracket fungi do great damage to cut lumber and to stands of timber. Some basidiomycetes are among the most damaging plant pathogens, including wheat rust (*Puccinia graminis;* see Figure 28.5) and the smut fungi (see Figure 28.1a) that parasitize cereal grains and some noncereal crops. In sharp contrast, other basidiomycetes contribute to the well-being of plants as fungal partners in mycorrhizae (which we'll discuss shortly).

Basidiomycete hyphae characteristically have septa with small, distinctive pores. (Recall that hyphae of the zygomycetes lack regularly occurring septa.) The **basidium** (plural basidia), a swollen cell at the tip of a hypha, is the characteristic sexual reproductive structure of the basidiomycetes. It is the site of nuclear fusion and meiosis (Figure 28.15). Thus the basidium plays the same role in the basidiomycetes as the ascus does in the ascomycetes.

After the nuclei fuse in the basidium, the resulting diploid nucleus undergoes meiosis, and the four haploid nuclei migrate from the basidium into haploid **basidiospores**, which form on tiny stalks. These basidiospores typically are forcibly discharged from their basidia and then germinate, giving rise to haploid hyphae. As these hyphae grow, haploid hyphae of different mating types meet and fuse, forming dikaryotic hyphae, each cell of which contains two nuclei, one from each parent hypha. The dikaryotic mycelium grows and eventually produces fruiting structures. The dikaryotic phase may persist for years—some basidiomycetes live for decades or even centuries.

The elaborate fruiting structure of some fleshy basidiomycetes, such as the gill mushroom in Figure 28.15, is topped by a cap, or pileus, which has gills on its underside. Basidia develop in enormous numbers between the gills. The basidia discharge their spores into the air spaces between adjacent gills, and the spores sift down into air currents for dispersal and germination as new haploid mycelia.

The exact pattern of the gills and the spore color are criteria for distinguishing mushroom species. If a mature cap is placed gill side down on a piece of paper for a few hours in a quiet place, the ejected basidiospores settle from between the gills, leaving on the paper an elegant replica of the gill pattern and visible evidence of spore color.

Imperfect fungi lack a sexual stage

Mechanisms of sexual reproduction readily distinguish the zygomycetes, ascomycetes, and basidiomycetes. But many fungi, both saprobes and parasites, lack sexual stages entirely; presumably these stages have been lost during evolution. Classifying these fungi as belonging to any of the three major phyla was at one time difficult, but biologists now can classify most such fungi on the basis of DNA sequences.

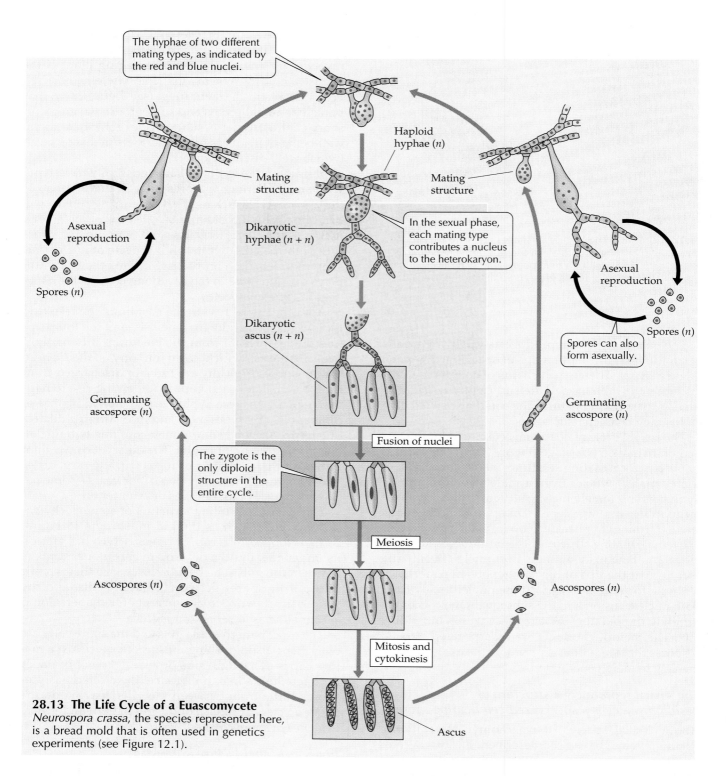

The hyphae of two different mating types, as indicated by the red and blue nuclei.

Haploid hyphae (n)

Mating structure

Mating structure

In the sexual phase, each mating type contributes a nucleus to the heterokaryon.

Asexual reproduction

Asexual reproduction

Dikaryotic hyphae (n + n)

Spores (n)

Spores (n)

Spores can also form asexually.

Dikaryotic ascus (n + n)

Germinating ascospore (n)

Germinating ascospore (n)

Fusion of nuclei

The zygote is the only diploid structure in the entire cycle.

Meiosis

Ascospores (n)

Ascospores (n)

Mitosis and cytokinesis

28.13 The Life Cycle of a Euascomycete
Neurospora crassa, the species represented here, is a bread mold that is often used in genetics experiments (see Figure 12.1).

Ascus

Fungi that lack sexual stages are found in the three phyla already discussed. Fungi that have not yet been placed in the existing phyla are grouped together in the "orphanage" known as the imperfect fungi, or deuteromycetes. Thus, the deuteromycete group is a holding area for species whose status is yet to be resolved. At present, about 25,000 species are classified as imperfect fungi.

If sexual structures are found on a fungus classified as a deuteromycete, the fungus is reassigned to the ap-

(a) *Lycoperdon perlatum*

(b) *Amanita muscaria*

(c) *Laetiporus sulphureus*

28.14 Basidiomycete Fruiting Structures   Although some persist only a few days or weeks, the fruiting structures of the basidiomycetes are probably the most familiar structures produced by fungi. (a) When raindrops hit them, these puffballs will release clouds of spores for dispersal. (b) A member of a highly poisonous mushroom genus, *Amanita*. (c) This bracket fungus is parasitizing a tree.

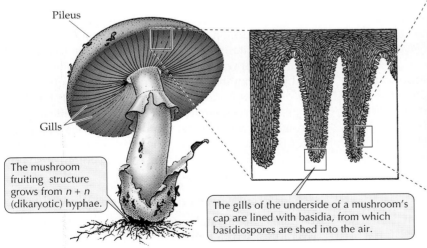

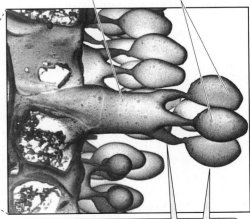

Basidium   Basidiospores

Pileus

Gills

The mushroom fruiting structure grows from *n* + *n* (dikaryotic) hyphae.

The gills of the underside of a mushroom's cap are lined with basidia, from which basidiospores are shed into the air.

Meiosis and further development lead to the production of four or eight narrow projections through which nuclei and other organelles squeeze, forming basidiospores.

28.15 A Mushroom's Fruiting Structures   The basidium is the characteristic sexual reproductive structure of basidiomycetes and is the site of nuclear fusion. Basidiospores form on tiny stalks and are then forcibly dispersed to germinate into haploid hyphae, from which the familiar fruiting structure eventually grows.

propriate phylum. That happened, for example, with a fungus that produces plant growth hormones called gibberellins (see Chapter 35). Originally classified as the deuteromycete *Fusarium moniliforme*, this fungus was later found to produce asci, whereupon it was renamed *Gibberella fujikuroi* and transferred to the phylum Ascomycota.

Penicillium is a deuteromycete genus of green molds, of which some species produce the antibiotic penicillin, presumably for defense against competing bacteria. Two species, *P. camembertii* and *P. roquefortii*, are the organisms responsible for the characteristic flavors of Camembert and Roquefort cheeses, respectively.

Another deuteromycete genus of importance in some diets is *Aspergillus*, the genus of brown molds. *A. tamarii* acts on soybeans in the production of soy sauce, and *A. oryzae* is used in brewing the Japanese alcoholic beverage sake. Some species of *Aspergillus* that grow on nuts such as peanuts and pecans produce extremely carcinogenic (cancer-inducing) compounds called aflatoxins.

Fungal Associations

Earlier in this chapter we spoke briefly about mycorrhizae and lichens; in both, fungi live in intimate association with other organisms. Now that we have learned a bit about fungal diversity, it is appropriate to consider mycorrhizae and lichens in greater detail.

Mycorrhizae are essential to many plants

Many plants, including almost all tree species, depend on a mutually beneficial symbiotic association with fungi for an adequate supply of water and mineral elements. Unassisted, the root hairs of such plants do not absorb enough of these materials to sustain maximum growth. However, the roots become infected with fungi, forming the association called a *mycorrhiza* (Figure 28.16).

In ectomycorrhizae, the fungus wraps around the root; in endomycorrhizae, the infection is internal to the root, with no hyphae visible on the root surface. Infected roots characteristically branch extensively and become swollen and club-shaped. The hyphae of the fungi attached to the root increase the surface area for the absorption of water and minerals, and the mass of the mycorrhiza, like a sponge, holds water efficiently in the neighborhood of the root.

Most families of flowering plants contain some species that form mycorrhizae, as do liverworts, ferns, club mosses, and gymnosperms (see Chapter 27). Fossils of mycorrhizal structures more than 300 million years old have been found. Certain plants that live in nitrogen-poor habitats, such as cranberry bushes and orchids, invariably have mycorrhizae.

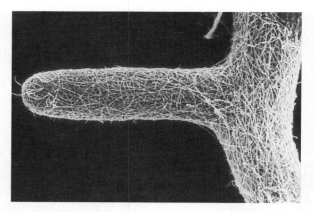

28.16 A Mycorrhizal Association Hyphae of the fungus *Pisolithus tinctorius* cover this eucalyptus root, forming a mycorrhiza.

Orchid seeds will not germinate in nature unless they are already infected by the fungus that will form their mycorrhizae. Plants that lack chlorophyll always have mycorrhizae, which are often shared with the roots of green, photosynthetic plants.

The symbiotic fungus–plant association of a mycorrhiza is important to both partners. The fungus obtains important organic compounds, such as sugars and amino acids, from the plant. In return, the fungus greatly increases the absorption of water and minerals (especially phosphorus) by the plant. The fungus also often provides certain growth hormones (see Chapter 35), and it protects the plant against attack by microorganisms. Plants that have active mycorrhizae typically are a deeper green and may resist drought and temperature extremes better than plants of the same species that have little mycorrhizal development. Attempts to introduce some plant species to new areas have failed until a bit of soil from the native area (presumably containing the fungus necessary to establish mycorrhizae) was provided.

The partnership between plant and fungus results in a plant better adapted for life on land. It has been suggested that the evolution of this symbiotic association was the single most important step leading to the colonization of the terrestrial environment by living things. A hardy group of present-day colonizers is the lichens, which we consider next.

Lichens grow where no eukaryote has succeeded

A lichen is not a single organism, but is a meshwork of two radically different organisms: a fungus and a photosynthetic microorganism. Together the organisms constituting a lichen survive some of the harshest environments on Earth (Figure 28.17). In spite of this hardiness, lichens are very sensitive to air pollution because they are unable to excrete toxic substances that

28.17 Lichens in Frigid Environments The many types of lichens shown here are growing on an Alaskan tundra.

they absorb. Hence they are not common in industrialized cities. Lichens are good biological indicators of air pollution because of their sensitivity.

The fungal components of most lichens are ascomycetes, but some are basidiomycetes or imperfect fungi (only one zygomycete serving as the fungal component of a lichen has been reported). The photosynthetic component may be either a cyanobacterium or a green alga. Relatively little experimental work has focused on lichens, perhaps because they grow so slowly—typically less than 1 cm per year. Thus only recently have workers been able to culture the fungal

and photosynthetic partners separately and then reconstruct a lichen from the two.

There are about 13,500 "species" of lichens. They are found in all sorts of exposed habitats: tree bark, open soil, or bare rock. Reindeer "moss" (actually not a moss at all, but the lichen *Cladonia subtenuis*) covers vast areas in arctic, subarctic, and boreal regions, where it is an important part of the diets of reindeer and other large mammals. Lichens come in various forms and colors (Figure 28.18). Crustose (crustlike) lichens look like colored powder dusted over their substrate; foliose (leafy) and fruticose (shrubby) lichens may appear quite complex.

The most widely held interpretation of the lichen relationship is that it is a type of mutually beneficial symbiosis. Hyphae of the fungal mycelium are tightly pressed against the photosynthetic cells of the alga or cyanobacterium and sometimes even invade them. The bacterial or algal cells not only survive these indignities but continue their growth and photosynthesis. In fact, algal cells in a lichen "leak" photosynthetic products at a greater rate than do similar cells growing on their own. On the other hand, photosynthetic cells

(a)

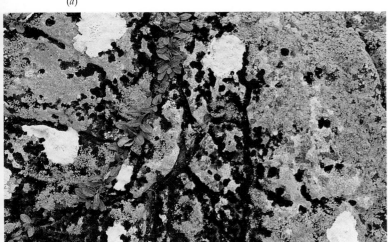

(b)

(c)

28.18 Lichen Body Forms Lichens fall into three principal classes based on their form. (a) Crustose lichens such as the orange, white, and black species in this photograph often grow on otherwise bare rock, as shown here, or on tree bark. (b) A wet foliose lichen growing close to the ground. (c) A fruticose lichen grows among fallen leaves.

28.19 Lichen Anatomy (*a*) Soredia of a fruticose lichen. (*b*) Cross section showing layers of a lichen.

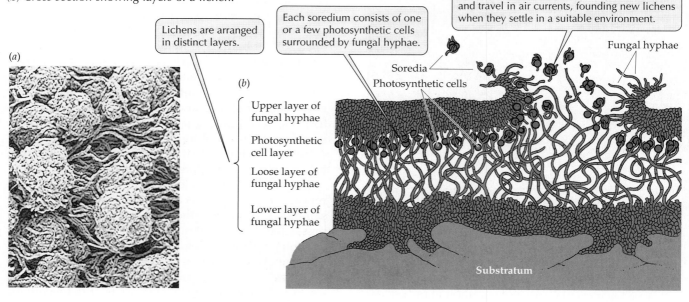

Lichens are arranged in distinct layers.

Each soredium consists of one or a few photosynthetic cells surrounded by fungal hyphae.

Soredia detach readily from the parent lichen and travel in air currents, founding new lichens when they settle in a suitable environment.

Fungal hyphae

Soredia

Photosynthetic cells

(*a*)

(*b*)

Upper layer of fungal hyphae

Photosynthetic cell layer

Loose layer of fungal hyphae

Lower layer of fungal hyphae

Substratum

from lichens grow more rapidly on their own than when combined with a fungus. On this ground, we may consider lichen fungi as parasitic on their photosynthetic partners.

Lichens can reproduce simply by fragmentation of the vegetative body, which is called the **thallus**, or by specialized structures called **soredia** (singular soredium). Soredia consist of one or a few photosynthetic cells surrounded by fungal hyphae (Figure 28.19*a*). The soredia become detached, move in air currents, and upon arriving at a favorable location, develop into a new lichen. If the fungal partner is an ascomycete or a basidiomycete, it may go through its sexual cycle, producing either ascospores or basidiospores. When these are discharged, however, they disperse alone, unaccompanied by the photosynthetic partner, and thus may not be capable of reestablishing the lichen association. Nevertheless, many lichens produce characteristic fruiting structures in which the asci or basidia are located.

Visible in a cross section of a typical foliose lichen are a tight upper region of fungal hyphae alone, a layer of cyanobacteria or algae, a looser hyphal layer, and finally hyphal rhizoids that attach the whole structure to its substrate (Figure 28.19*b*). The meshwork has properties that enable it to hold water fairly tenaciously. Some nutrients for the photosynthetic cells arrive through the fungal hyphae, the meshwork provides a suitably moist environment for the photosynthetic cells, and the fungi derive fixed carbon from the photosynthesis of the algal or cyanobacterial cells.

Lichens are often the first colonists on new areas of bare rock. They satisfy most of their needs from the air

and from rainwater, augmented by minerals absorbed from their rocky substrate. A lichen begins to grow shortly after a rain, as it begins to dry. As it grows, the lichen acidifies its environment slightly, and this acid contributes to the slow breakdown of rocks, an early step in soil formation. After further drying, the lichen's photosynthesis ceases. The water content of the lichen may drop to less than 10 percent of its dry weight, at which point the lichen becomes highly insensitive to extremes of temperature. The flora of Antarctica features more than 100 times as many species of lichens as of plants.

Whether living on their own or in lichen associations, fungi have spread successfully over much of Earth since their origin from a protist ancestor. That ancestor also gave rise to the animal kingdom, the group we'll consider in the next two chapters.

Summary of "Fungi: A Kingdom of Recyclers"

• Fungi are the principal degraders of dead organic matter in the biosphere.

General Biology of the Fungi

• Fungi are heterotrophic eukaryotes with absorptive nutrition. They are saprobes, parasites, or mutualists. The body of a fungus is composed of chitinous-walled hyphae, often massed to form a mycelium. The filamentous hyphae give fungi a large surface area–to–volume ratio, enhancing their ability to absorb food.

- Fungi reproduce asexually by means of spores formed within sporangia, by conidia formed at the tips of hyphae, by budding, or by fragmentation. **Review Figures 28.5, 28.13**
- Fungi reproduce sexually when hyphae or motile cells of different mating types meet and fuse.
- The gametes and zoospores of chytrids are the only cells that have flagella in the kingdom Fungi. **Review Figure 28.9**
- In addition to the haploid and diploid states, many fungi demonstrate a third nuclear condition: the dikaryotic, or $n + n$, state. Some fungal life cycles involve two hosts. **Review Figures 28.5, 28.13**

Diversity in the Kingdom Fungi

- The kingdom Fungi consists of four phyla: Chytridiomycota, Zygomycota, Ascomycota, and Basidiomycota. These differ in their reproductive structures, spore formation, and less importantly, the cross-walls (if any) of their hyphae.
- Chytrids, with their flagellated zoospores and gametes, probably resemble the ancestral fungi.
- Zygomycetes reproduce sexually by fusion of gametangia. **Review Figure 28.9**
- The sexual reproductive structure of ascomycetes is an ascus containing ascospores. The ascomycetes are divided into two groups, euascomycetes and hemiascomycetes, on the basis of whether they have an ascocarp. **Review Figures 28.10, 28.12, 28.13**
- The sexual reproductive structure of basidiomycetes is a basidium, a swollen cell bearing basidiospores. **Review Figure 28.15**
- Imperfect fungi (deuteromycetes) lack sexual structures, but DNA sequencing helps identify the phyla to which they belong. **Review Figure 28.6 and Table 28.1**

Fungal Associations

- Mycorrhizae, associations of fungi with plant roots, enhance the ability of the roots to absorb water and nutrients.
- Lichens, mutualistic combinations of a fungus with a green alga or a cyanobacterium, are found in some of the most inhospitable environments on the planet. **Review Figure 28.19**

Self-Quiz

1. Which statement about fungi is *not* true?
   a. A hyphal fungus has a body called a mycelium.
   b. Hyphae are composed of individual mycelia.
   c. Many fungi tolerate highly hyperosmotic environments.
   d. Many fungi tolerate low temperatures.
   e. Most fungi are anchored to their substrate by rhizoids.

2. The absorptive nutrition of fungi is aided by
   a. dikaryon formation.
   b. spore formation.
   c. the fact that they are all parasites.
   d. their large surface area–to–volume ratio.
   e. their possession of chloroplasts.

3. Which statement about fungal nutrition is *not* true?
   a. Some fungi are active predators.
   b. Some fungi form mutualistic associations with other organisms.
   c. All fungi require mineral nutrients.
   d. Fungi can make some of the compounds that are vitamins for animals.
   e. Facultative parasites can grow only on their specific hosts.

4. Which statement about heterokaryosis is *not* true?
   a. The cytoplasm of two cells fuses before their nuclei fuse.
   b. The two haploid nuclei are genetically different.
   c. The two nuclei are of the same mating type.
   d. The heterokaryotic stage ends when the two nuclei fuse
   e. Not all fungi have a heterokaryotic stage.

5. Reproductive structures consisting of a photosynthetic cell surrounded by fungal hyphae are called
   a. ascospores.
   b. basidiospores.
   c. conidia.
   d. soredia.
   e. gametes.

6. The zygomycetes
   a. have hyphae without regularly occurring cross-walls.
   b. produce motile gametes.
   c. form fleshy fruiting bodies.
   d. are haploid throughout their life cycle.
   e. have structures homologous to those of the ascomycetes.

7. Which statement about ascomycetes is *not* true?
   a. They include the yeasts.
   b. They form reproductive structures called asci.
   c. Their hyphae are segmented by cross-walls.
   d. Many of their species have a dikaryotic stage.
   e. All have fruiting structures called ascocarps.

8. The basidiomycetes
   a. often produce fleshy fruiting structures.
   b. have hyphae without cross-walls.
   c. have no sexual stage.
   d. never produce large fruiting structures.
   e. form diploid basidiospores.

9. The deuteromycetes
   a. have distinctive sexual stages.
   b. are all parasitic.
   c. include some commercially important species.
   d. include the ascomycetes.
   e. are never components of lichens.

10. Which statement about lichens is *not* true?
    a. They can reproduce by fragmentation of their vegetative body.
    b. They are often the first colonists in a new area.
    c. They render their environment more basic (alkaline).
    d. They contribute to soil formation.
    e. They may contain less than 10 percent water by weight.

Applying Concepts

1. You are shown an object that looks superficially like a pale green mushroom. Describe at least three criteria (including anatomical and chemical traits) that would enable you to tell whether the object is a piece of a plant or a piece of a fungus.

2. Differentiate between the members of each of the following pairs of related terms.
   a. hypha/mycelium
   b. euascomycete/hemiascomycete
   c. ascus/basidium
   d. ectomycorrhiza/endomycorrhiza

3. For each type of organism listed below, give a single characteristic that may be used to differentiate it from the other, related organism(s) in parentheses.
   a. Zygomycota (Ascomycota)
   b. Basidiomycota (deuteromycetes)
   c. Ascomycota (Basidiomycota)
   d. baker's yeast (*Neurospora crassa*)

4. Many fungi are dikaryotic during part of their life cycle. Why are dikaryons described as $n + n$ instead of $2n$?

5. Review Figure 28.6. Why are the Ascomycota and Basidiomycota grouped together, apart from the other two phyla?

6. If all the fungi on Earth were suddenly to die, how would the surviving organisms be affected? Be thorough and specific in your answer.

Readings

Alexopoulos, C. J., C. W. Mims and M. Blackwell. 1996. *Introductory Mycology*, 4th Edition. John Wiley & Sons, New York. A new edition of a leading textbook on the biology of fungi.

Kendrick, B. 1992. *The Fifth Kingdom*, 2nd Edition. Focus Information Group, Newburyport, MA. An introduction to the fungi, with useful emphasis on taxonomy.

Kosikowski, E. V. 1985. "Cheese." *Scientific American*, May. A description of how fungi are involved in the production of many cheeses.

Moore-Landecker, E. 1996. *Fundamentals of Fungi*, 4th Edition. Prentice-Hall, Englewood Cliffs, NJ. Another introduction to the kingdom.

Newhouse, J. R. 1990. "Chestnut Blight." *Scientific American,* July. A description of a new biological method for controlling the fungus responsible for a disease that almost eliminated an important tree species.

Raven, P. H., R. F. Evert and S. Eichhorn. 1992. *Biology of Plants*, 5th Edition. Worth, New York. Includes an excellent discussion of the fungi.

Sternberg, S. 1994. "The Emerging Fungal Threat," *Science*, December 9, pages 1632–1634. Enumeration of the dangers posed to AIDS patients by fungi.

Chapter 29

Protostomate Animals: The Evolution of Bilateral Symmetry and Body Cavities

A Sea of Animals
The coral reef is formed by the skeletons of myriad animals, and a host of other animal species call the reef habitat home.

Tropical coral reefs abound with life. The corals themselves dominate the landscape, and their skeletons form its structures. Growing among the corals are many kinds of organisms. Some of them are algae; but many of them, although they may *look* like plants, are actually animals that do not at all resemble the terrestrial animals with which we are most familiar.

The members of the kingdom Animalia are among the most conspicuous living things in the world around us and, as members of this kingdom ourselves, we have a special interest in its other members. Humans tend to be most aware of the large animals that share our terrestrial environment, but these are only a few of Earth's myriad animal species. Even a brief excursion into a coral reef brings us into contact with individuals from many animal phyla—some of which we might not even recognize as animals.

The identification of Earth's animal species has barely begun. Millions of animal species live on Earth today, and describing them is an enormous task. Many are so inconspicuous they evade all but the most alert and trained eye. Many live in remote or harsh areas that are difficult for humans to visit. Some animals appear vastly different at different stages of their lives (the caterpillar and the butterfly are a familiar example) and without study and observation we might think they were two different species. Other animals that look very much alike might in fact belong to different species. Parasitic species, which live at least part of their life cycles inside other animals, are especially poorly known.

627

If many kinds of animals seem so unlike the animals with which we are familiar, how do we decide that they are indeed animals? We class them all as animals because they share both a common ancestry and certain traits not found in organisms belonging to the other kingdoms. In this chapter we will first discuss how biologists infer the evolutionary relationships among animals and some of the defining characteristics of the animal way of life. Then we'll describe the diversity of animals in one of the two great animal evolutionary lineages defined by their mode of embryonic development: the protostomes. Chapter 30 treats the other major animal lineage, the deuterostomes.

Descendants of a Common Ancestor

Biologists have long debated whether the animals arose once or several times from protist ancestors, but recent evidence strongly indicates that all animals are descendants of a single lineage. First, similarities in their 5S and 18S ribosomal RNAs suggest that all animals have a common ancestor. In addition, all animals share the same complex set of extracellular matrix (ECM) molecules, first discussed in Chapter 4.

These molecules, which make up the connective tissues of animals, form the basal laminae that underlie the sheets of the epithelial cells that all animals have. The ECM is the fabric that connects the cells of both embryos and adult animals (Figure 29.1). The molecular composition, structure, and functioning of the extracellular matrices of all animals are almost identical, and the ECM develops in similar ways in the embryos of all animals. The likelihood that such a complex system evolved more than once is remote.

Animals probably arose evolutionarily from ancestral colonial flagellated protists as a result of division of labor among their aggregated cells. Division of labor probably evolved because a mass of identical cells exchanges materials with its environment relatively slowly, and because some functions are performed better by specialized cells.

Within the ancestral colonies of cells—perhaps similar to those still existing in the chlorophyte *Volvox* or some colonial choanoflagellates (see Figure 26.28a)—some cells became specialized for movement, others became specialized for nutrition, and still others differentiated into gametes. Once the division of labor had begun, the units continued to differentiate, while improving their coordination with other working groups of cells. These coordinated groups of cells evolved into the larger and more complex organisms that we now call animals.

The development of a multicellular organism with differentiated cells requires the integration of several processes (see Chapter 15). First, genes must be regulated so that even though all cells contain all of the

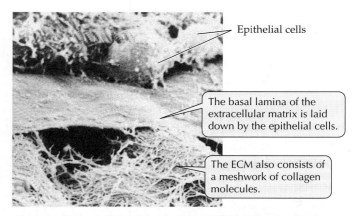

Epithelial cells

The basal lamina of the extracellular matrix is laid down by the epithelial cells.

The ECM also consists of a meshwork of collagen molecules.

29.1 An Extracellular Matrix is Found in All Animals
Extracellular matrix molecules secreted by animal cells form connective tissue that holds the cells together. The similarity of this structure in all animals indicates a common ancestral origin for the kingdom.

animal's genes, particular genes are active only in particular cells at certain times during development. Second, the differentiation status of a cell must be transmitted during cell division. That is, muscle cells must give rise to muscle cells and epithelial cells must give rise to epithelial cells. Third, differentiated cells must become arranged in a specific and reliable spatial pattern. These processes require primarily changes in controls over developmental processes rather than new cellular structures or functions not present in unicellular organisms.

Food Procurement:
The Key to the Animal Way of Life

Animals require a variety of complex organic molecules as sources of energy, and they obtain these molecules by active expenditure of energy. This energy is used to move the animals through their environment, to cause the environment and the food it contains to move to the animals, or to position the animals where food will pass by them. The foods animals eat include most other members of the animal kingdom, as well as members of all other kingdoms. Much of the diversity of animal sizes and shapes evolved as animals acquired the ability to capture and eat many different kinds of food.

The need to move in search for food has favored structures that provide animals with detailed information about their environment, and structures able to receive and coordinate this information. Consequently, most animals are behaviorally much more complex than plants. Because animals ingest chemically complex foods, they expend considerable energy to maintain a relatively constant internal composition while they take in foods that vary chemically.

A real appreciation of animal structure and functioning is best achieved through firsthand experience in the field and laboratory. The accounts in this chapter and the next serve as an orientation to the major groups of animals, their similarities and differences, and the evolutionary pathways that resulted in the current number and variety of animal species.

Clues to Evolutionary Relationships among Animals

Biologists try to classify animals in ways that reflect their evolutionary relationships. Determining the early evolutionary relationships of animals is difficult because fossils of most animal phyla appear simultaneously near the beginning of the Cambrian period, and because animal evolution is replete with examples of the convergence of traits. Therefore, biologists use a variety of traits in their attempts to infer animal phylogenies. Clues to these relationships are found in the fossil record, in the patterns of embryological development of animals, in the comparative morphology and physiology of living and fossil animals, and in their molecular biology (see Chapters 22 and 23).

Body plans are basic structural designs

The entire animal, its organ systems, and the integrated functioning of its parts are known as its **body plan**. Body plans are basic structural designs that reflect, and provide clues to, the evolutionary history of animal lineages. Consequently, we use them as one way to organize our treatment of animal groups. Animals in many, but not all, lineages evolved increasing body complexity. As you will see, animals with complex bodies can perform and control complex movements.

A fundamental aspect of an animal's body plan is its overall shape, described as its **symmetry**. A symmetrical animal can be divided along at least one plane into similar halves. Animals that have no plane of symmetry are said to be **asymmetrical** (Figure 29.2*a*). Many sponges are asymmetrical, but most animals have some kind of symmetry.

The simplest form is **spherical symmetry**, in which body parts radiate out from a central point. An infinite number of planes passing through the central point can divide a spherically symmetrical organism into similar halves. Spherical symmetry is widespread among protists (for example, radiolarians; see Figure 26.11*a*), but most animals possess other forms of symmetry.

An organism with **radial symmetry** has one main axis around which its body parts are arranged (Figure 29.2*b*). A perfectly radially symmetrical animal can be divided into similar halves by any plane that contains the main axis. Some simple sponges and a few other animals have such symmetry, but most radially symmetrical animals are modified such that only two

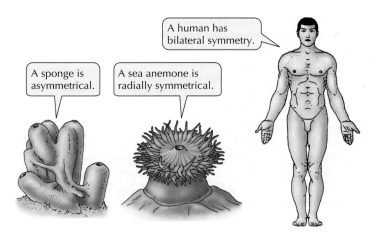

29.2 Body Symmetry Most animals above the level of sponges display some form of body symmetry. Any plane passing through the main axis of a radially symmetrical animal will divide it into equal halves, whereas there is only one plane that will divide a bilaterally symmetrical animal in half.

planes can divide them into similar halves. These animals are said to have **biradial symmetry**. Three animal phyla—Cnidaria, Ctenophora, and Echinodermata—are composed primarily of radially or biradially symmetrical animals. As we'll see, these animals move only slowly, if at all.

A **bilaterally symmetrical** animal can be divided into mirror images only by a single plane that passes through the midline of its body from the front (anterior) to the back (posterior) end (Figure 29.2*c*). The other (left–right) axis divides the body into two dissimilar sides; the side of a bilaterally symmetrical animal without a mouth is its **dorsal surface**; the side with a mouth is its **ventral surface**. Bilateral symmetry is a common characteristic of animals that move through their environments. Such symmetry is strongly correlated with the development of sense organs and central nervous tissues at the anterior end of the animal, a process known as **cephalization**. Cephalization may have been selected for among motile animals because the anterior end typically encounters new environments first.

Symmetry is an important way of comparing animal body plans and relating the plans to how the animals live. However, it is not very useful for determining evolutionary relationships among animals, because animals in closely related phyla with different lifestyles may have quite different symmetries.

Developmental patterns are often evolutionarily conservative

The early development of embryos has sometimes been **evolutionarily conservative**; that is, changes in early development have evolved very slowly in some lineages. For this reason, developmental patterns reveal a great deal about evolutionary relationships among some animals.

During development from a single-celled zygote to a multicellular adult, animals form layers of cells. These layers behave as units during early embryonic development and give rise to different tissues and organs in the adult animal. The embryos of **diploblastic** animals have only two cell layers: an outer *ectoderm* and an inner *endoderm.* The embryos of **triploblastic** animals have, in addition to ectoderm and endoderm, a third layer, the *mesoderm,* which lies between the ectoderm and the endoderm.

On the basis of differences in their early developmental patterns, combined with other shared, derived traits, zoologists divide animals other than sponges, cnidarians, and ctenophores into two major lineages. In the **protostomate lineage**, the pattern of early cleavage is spiral. The plane of division is oblique to the long axis of the egg, causing the cells to be arranged in a spiral pattern. Cleavage of the fertilized egg of protostomes is **determinate**; that is, if the egg is allowed to divide a few times and the cells are then separated, each cell develops into only a partial embryo (see Chapter 40).

In the **deuterostomate lineage**, the fertilized egg cleaves radially. Cells divide along a plane either parallel to or at right angles to the long axis of the fertilized egg. In addition, cleavage in deuterostomes typically is **indeterminate**; that is, cells separated after several cell divisions can still develop into complete embryos.

Finally, among deuterostomes, the mouth of the embryo originates some distance away from the embryonic structure called the *blastopore,* which becomes the anus. Among protostomes, the mouth arises from or near the blastopore. Table 29.1 summarizes the

TABLE 29.1 Developmental Differences between Protostomes and Deuterostomes.	
PROTOSTOMES	**DEUTEROSTOMES**
Spiral cleavage	Radial cleavage
Determinate cleavage	Indeterminate cleavage
Blastopore becomes mouth	Blastopore becomes anus
Mesoderm derives from cells on lip of blastopore	Mesoderm derives from walls of developing gut
Mesoderm splits to form coelom	Mesoderm usually outpockets to form coelom

early embryological differences between protostomes and deuterostomes.

Fluid-filled spaces, called **body cavities**, lie between the cell layers of the bodies of many kinds of animals. These body cavities are of great functional significance to animals. The type of body cavity that animals have, in combination with other traits, is useful for comparing grades (levels of structural complexity) in animals. The type of body cavity an animal has strongly influences how it can move. Therefore, knowledge of a group's body cavities tells us a great deal about the way of life of its members. However, body cavities are not completely reliable traits for determining evolutionary relationships, because as animals adapt to different ways of life, their body cavities may be altered or even lost.

Some animals, such as flatworms, are called **acoelomates** because they lack an enclosed body cavity (Figure 29.3a). Their only internal cavity is the digestive cavity; the space between the gut and the body wall is filled with masses of cells that are collectively

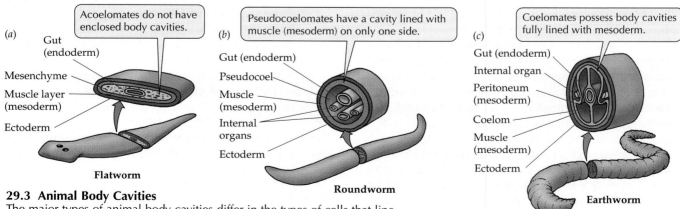

(a)

Acoelomates do not have enclosed body cavities.

Gut (endoderm)
Mesenchyme
Muscle layer (mesoderm)
Ectoderm

Flatworm

(b)

Pseudocoelomates have a cavity lined with muscle (mesoderm) on only one side.

Gut (endoderm)
Pseudocoel
Muscle (mesoderm)
Internal organs
Ectoderm

Roundworm

(c)

Coelomates possess body cavities fully lined with mesoderm.

Gut (endoderm)
Internal organ
Peritoneum (mesoderm)
Coelom
Muscle (mesoderm)
Ectoderm

Earthworm

29.3 Animal Body Cavities
The major types of animal body cavities differ in the types of cells that line them. Tissues derived from ectoderm are colored blue, those from mesoderm are pink, and those from endoderm are yellow; this color convention will be continued in Chapter 40, where these three cell layers will be discussed in detail.

called *mesenchyme.* Another group of animals, the **pseudocoelomates,** have a body cavity, called the **pseudocoel,** derived from the first cavity formed inside the proliferating ball of embryonic cells (Figure 29.3*b*). The pseudocoel provides a liquid-filled space in which many of the body organs are suspended, but control over body shape is crude because a pseudocoel has muscles on only one side.

Coelomate animals have a **coelom** (Figure 29.3*c*), a body cavity that develops within the embryonic mesoderm. It has muscles on both sides and is lined with a special structure derived from mesoderm, called the **peritoneum.** The internal organs of coelomates are slung in pouches of the peritoneum rather than being suspended within the cavity. Coelomate animals include peanut worms, echiuran worms, pogonophorans, annelids, mollusks, arthropods, echinoderms, and chordates. Among arthropods, the coelom evolved into a *hemocoel*—a cavity that is filled with blood.

The phylogeny of animals we adopt in this book is based on a combination of many developmental, structural, and molecular traits. Figure 29.4 shows the postulated order of splitting of the major lineages in animal evolution. Later in the chapter we will give details of phylogenies of protostomes and deuterostomes. New information continues to modify and refine our understanding of the details of phylogenetic relationships, and different types of traits often suggest different relationships. Nonetheless, some of the proposed lineages, such as the protostomes and deuterostomes, are supported by many types of data; they are unlikely to be altered by new information.

Early Animal Diets: Prokaryotes and Protists

When the first animals evolved, the primary food items available to them were algae floating in the water (called **phytoplankton**), extensive algal mats covering the shallow sea bottoms, protists, and prokaryotes.

The earliest animals were probably floating colonies of flagellated cells that fed on prokaryotes and protists. Some of these animals developed stinging tentacles housed in special cells. With such tentacles they could capture larger prey. Others evolved physical and behavioral adaptations for grazing in the algal mats. Both of these changes favored the evolution of larger animals that could move about.

Not surprisingly, the presence of these larger animals created opportunities for carnivores that fed on them and parasites that lived on or inside them. As defenses against predators, natural selection favored the evolution of shells and other protections, the ability to burrow, the use of safe refuges such as caves and crevices, and faster movement. These evolutionary themes of protection, defense, refuge, and movement continue today.

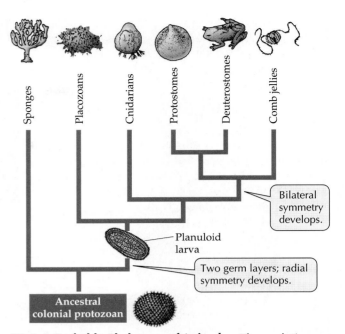

29.4 A Probable Phylogeny of Animals The evolutionary tree of animals we will use in this chapter and the next one postulates that animals are monophyletic. The two major animal lineages—protostomes and deuterostomes—are ancient divisions in animal evolution.

For the moment, however, let's examine how these selection pressures operated during the early evolution of animals, which began about 2.5 billion years ago. Remember that all life was confined to the seas before and during much of animal evolution.

Simple Aggregations: Sponges and Placozoans

Animals probably arose from protists whose cells remained together after division, forming a multicellular mass called a colony. It is difficult to distinguish between a protist colony and some simple multicellular animals that have little differentiation or coordination among their cells. The lineage leading to modern sponges separated from the lineage leading to all other animals very early in the evolution of animals (see Figure 29.4). Some living sponges are still very similar to the probable ancestral colonial protists.

Sponges are loosely organized

The sponges (phylum Porifera, Latin for "pore bearers") are *sessile*: They live attached to the substratum and do not move about. All sponges, even large ones, which may reach a meter in length, have a very simple body plan. The body of a sponge is a loose aggregation of cells built around a water canal system. A sponge has no mouth or digestive cavity, no muscles, and no nervous system. In fact, there are no organs in the usual sense of the word.

29.5 The Body Plan of a Simple Sponge The flow of water through the sponge is shown by blue arrows.

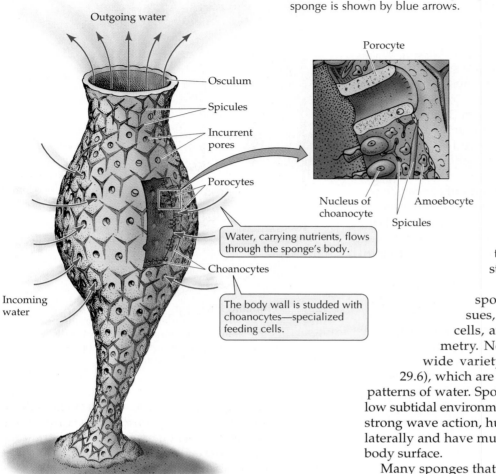

Outgoing water

Osculum

Spicules

Incurrent pores

Porocytes

Water, carrying nutrients, flows through the sponge's body.

Choanocytes

The body wall is studded with choanocytes—specialized feeding cells.

Incoming water

Porocyte

Nucleus of choanocyte

Amoebocyte

Spicules

through one or more larger openings (Figure 29.5).

Between the thin epidermis and the choanocytes lies a layer of cells, some of which are similar to amoebas and move about in the body. A supporting skeleton is also present, either in the form of simple or branching spines or as an elastic network of spongin fibers. The spicular skeletons of some sponges may be very complex, as is evident in the sponge skeletons you can find in a hardware store.

Unlike most other animals, sponges have no distinct body tissues, no cavity between the layers of cells, and no recognizable body symmetry. Nonetheless, sponges come in a wide variety of sizes and shapes (Figure 29.6), which are adapted to different movement patterns of water. Sponges living in intertidal or shallow subtidal environments, where they are subjected to strong wave action, hug the substratum. They spread laterally and have multiple pores scattered over their body surface.

Many sponges that live in calm waters are simple, with a single large opening on top of the body (see Figure 29.5). Water is taken in through pores on the sides of the body and expelled upward through the large opening on top. Sponges that live in flowing water do not need to exert much energy to move water through their bodies. Most of them are flattened and are oriented at right angles to the direction of current flow; they intercept water as it flows past them. Water is drawn out of the top pores of such tall, thin sponges just as air is drawn out of a tall chimney by the wind.

Sponges reproduce both sexually and asexually. In most species, a single individual produces both eggs and sperm. Water currents carry sperm from one individual to another. Asexual reproduction is by budding and other processes that produce fragments able to develop into new sponges. Most of the 10,000 species of sponges are marine animals, but about 50 species are found in fresh water.

Placozoans are the simplest multicellular animals

Molecular and morphological evidence suggests that the next evolutionary split separated placozoans from other animal lineages (see Figure 29.4). Placozoans

A sponge is so loosely organized that even if it is completely disassociated by being strained through a filter, its cells can reassociate into a new sponge. Nevertheless, sponges have at least 34 different cell types, most of which are specialized mobile cells. Thus sponges are more complex than their colonial flagellate ancestors.

Throughout most of its life, an individual sponge is attached to a substratum. It feeds by drawing water into itself and filtering out the small organisms and nutrient particles that flow past the walls of its inner cavity. Unique feeding cells called **choanocytes** line the inside of the water canals of sponges; these cells have a collar consisting of cytoplasmic extensions that surround a flagellum. The flagellated choanocytes set up the water currents. Water flows into the animal either by way of small pores that perforate special epidermal cells (in simple sponges) or through intercellular pores (in complex sponges). The water passes into small chambers within the body where food particles are captured by the choanocytes. Water then exits

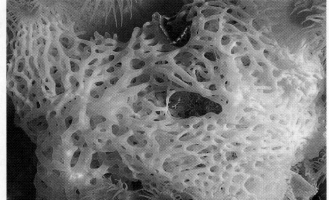

29.6 Sponges Differ in Size and Shape Two of the many different growth forms of sponges are illustrated.

Aplysina fistularia

Clathrina coriacea

(phylum Placozoa) are much smaller than sponges, less than 3 mm in diameter. Their bodies consist of no more than a few thousand cells and only four cell types, in contrast to the 34 cell types found in sponges. Placozoans lack any kind of symmetry, and they have no body cavity or distinct tissues or organs (Figure 29.7). Their body plan is a flat plate consisting of two layers of flagellated cells that enclose a fluid-filled area that contains fibrous cells.

For a long time biologists thought that placozoans were larvae of some type of sponge or cnidarian, but in the late 1960s placozoans were observed to achieve sexual maturity and produce gametes. Placozoans are widely distributed in shallow tropical ocean waters. They feed by endocytosis, taking prokaryotes or small protists into their own cells, or by secreting enzymes onto their protist prey, which then digest the prey out-

side the placozoan's body. The cells of the placozoan then absorb the prey's digested remains. Only two species have been described in this unusual phylum.

The Evolution of Cell Layers: The Cnidarians

Animals in all phyla other than Porifera and Placozoa have distinct cell layers and symmetrical bodies. Although the members of most animal phyla have three cell layers, the next lineage to split off from the main line of animal evolution resulted in a phylum of animals—the cnidarians—having only two cell layers (see Figure 29.4).

Cnidarians are simple but specialized carnivores

Cnidarians (phylum Cnidaria) appeared early in evolutionary history and radiated into many different species; they may have constituted more than half of the late Precambrian animal species. About 10,000 modern cnidarian species—jellyfish, sea anemones, corals, and hydrozoans—live today. All but a few species are marine.

The smallest cnidarians can hardly be seen without a microscope; the largest jellyfish is 2.5 meters wide and has tentacles 25 meters long. All cnidarians have only two cell layers and are radially symmetrical. A layer called the mesoglea between the two cell layers eventually develops from the ectoderm, but it never produces complex internal organs like those found in animals that have three true cell layers.

A key feature of cnidarians is their **cnidocytes**, specialized cells that contain stinging structures called **nematocysts** that can discharge toxins into their prey (Figure 29.8). Cnidocytes, which are borne on tentacles, allow cnidarians to capture large prey. The nematocysts paralyze and help hold prey; they are responsi-

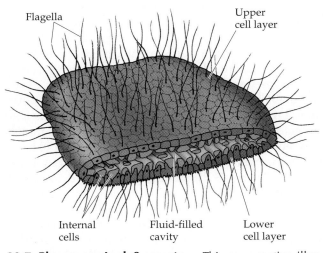

Flagella

Upper cell layer

Internal cells

Fluid-filled cavity

Lower cell layer

29.7 Placozoans Lack Symmetry This cross section illustrates the simple structure of a placozoan.

Portuguese man-of-war

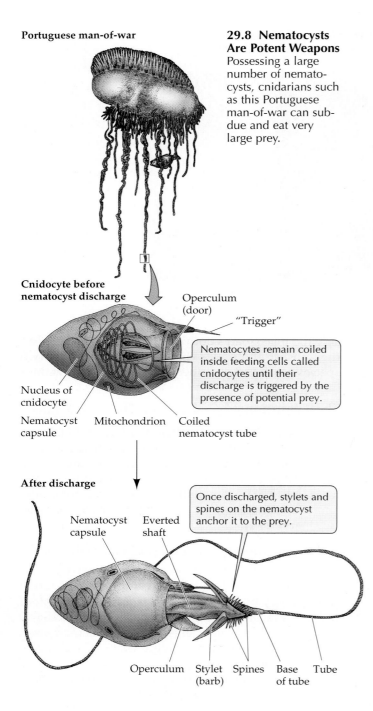

29.8 Nematocysts Are Potent Weapons Possessing a large number of nematocysts, cnidarians such as this Portuguese man-of-war can subdue and eat very large prey.

Cnidocyte before nematocyst discharge

Operculum (door)

"Trigger"

Nematocytes remain coiled inside feeding cells called cnidocytes until their discharge is triggered by the presence of potential prey.

Nucleus of cnidocyte

Nematocyst capsule   Mitochondrion   Coiled nematocyst tube

After discharge

Once discharged, stylets and spines on the nematocyst anchor it to the prey.

Nematocyst capsule   Everted shaft

Operculum   Stylet (barb)   Spines   Base of tube   Tube

ble for the sting that some jellyfish and other cnidarians can inflict on human swimmers. At the extreme, the tropical Pacific sea wasp (genus *Chironex*) can cause fatal injuries to humans.

A cnidarian transports its captured prey to its mouth by specialized structures called *tentacles.* The mouth is connected to a dead-end sac called the **gastrovascular cavity,** which functions in digestion, circulation, and gas exchange. The same opening, derived from the blastopore, serves as both mouth and anus in

cnidarians. Cnidarians also have epithelial cells with muscle fibers whose contractions enable the animals to move, as well as nerve nets that integrate their body activities.

CNIDARIAN LIFE CYCLES. The generalized cnidarian life cycle has two distinct stages (Figure 29.9)—polyp and medusa—but one stage is lacking in many species. The **polyp** stage is a cylindrical stalk with tentacles surrounding a mouth that is at the opposite end from the site of attachment to the substratum. This stage is usually asexual, but individual polyps may reproduce by budding, thereby forming a colony.

The **medusa** (plural medusae) is a familiar, free-swimming, sexual stage shaped like a bell or an umbrella. It typically floats with its mouth and tentacles facing downward. Medusae produce eggs and sperm and release them into the water. When an egg is fertilized, it develops into a free-swimming, ciliated larva called a **planula** (plural planulae) that eventually settles to the bottom and transforms into a polyp.

Although the polyp and medusa stages appear very different, they share a similar body plan. A medusa is essentially a polyp without a stalk. Most of the outward differences between polyps and medusae are due to the development of the **mesoglea,** a middle body layer, composed of jellylike material, that is largely devoid of cells and has a very low metabolic rate. In polyps, the mesoglea is usually thin; in medusae it is very thick, constituting the bulk of the animal.

HYDROZOANS. Among members of the class Hydrozoa—the group containing the only freshwater cnidarians—life cycles are diverse. The polyp commonly dominates the life cycle, but some species have only medusae and others only polyps. A few species have solitary polyps, but most hydrozoans are colonial. A single planula eventually gives rise to a colony of many polyps, all interconnected and sharing a continuous gastrovascular cavity (Figure 29.10). Within such a colony, the polyps often differentiate. Some polyps have tentacles with many nematocysts; they capture prey for the colony. Others lack tentacles and are unable to feed, but are specialized for the production of medusae. Still others are fingerlike and defend the colony. However, all of these types are ultimately derived from a single, sexually produced planula.

The siphonophores, such as the Portugese man-of-war (see Figure 29.8), are free-floating hydrozoans in which medusae and polyps combine to form complex colonies. Individual units are modified for specific functions: to act as gas-filled floats, to move the colony through the water by jet propulsion, or to defend it. And of course there are also feeding and reproductive medusae.

29.9 The Life Cycle of a Cnidarian
Cnidarians typically have two body forms, one asexual and the other sexual.

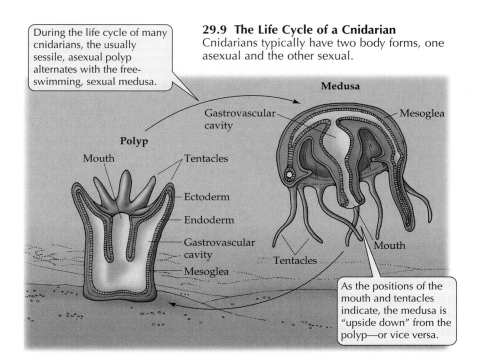

During the life cycle of many cnidarians, the usually sessile, asexual polyp alternates with the free-swimming, sexual medusa.

Medusa

Gastrovascular cavity

Mesoglea

Polyp

Mouth

Tentacles

Ectoderm

Endoderm

Gastrovascular cavity

Mesoglea

Mouth

Tentacles

As the positions of the mouth and tentacles indicate, the medusa is "upside down" from the polyp—or vice versa.

SCYPHOZOANS. The several hundred species of the class Scyphozoa are all marine. Some are as large as 2.5 meters in diameter. The mesoglea of their medusae is very thick and firm, giving rise to their common name, jellyfish. The medusa typically has the form of an inverted cup, and the tentacles with nematocysts extend downward from the margin of the cup.

Contraction of muscles ringing the margin expels water from the cup. When the muscles relax, the mesoglea expands the cup, which again fills with water. This muscular contraction cycle allows scyphozoan medusae to swim through the water. Food captured by the tentacles is passed to the mouth and distributed to one of four gastric pouches, where enzymes begin digesting the food.

The medusa, rather than the polyp, dominates the life cycle of scyphozoans. Gonads (sex organs) develop in tissues close to the gastric pouches. An individual medusa is male or female, releasing eggs or sperm into the open sea. The fertilized egg de-

velops into a small, heavily ciliated planula that quickly settles on a substratum and changes into a small polyp. This polyp feeds and grows and may produce additional polyps by budding.

After a period of growth, the polyp begins to bud off small medusae by transverse division of its body column (Figure 29.11). These small medusae feed, grow, and transform themselves into adult medusae, which are commonly seen during summer in harbors and bays. Thus a polyp that grows from a single fertilized egg is capable of producing many genetically identical medusae that will eventually reproduce sexually.

ANTHOZOANS. The roughly 6,000 species of sea anemones and corals that constitute the class Anthozoa are all marine. Unlike other cnidarians, anthozoans entirely lack the medusa stage of the life cycle. The polyp produces eggs and sperm, and the fertilized egg develops into a planula that metamorphoses directly into another polyp. Many species can also reproduce asexually, by budding or fission.

Like all other cnidarians, anthozoans are carnivores that capture prey with nematocyst-studded tentacles. However, the digestive cavity of anthozoans is more complex than that of other cnidarians. It is partitioned

29.10 Many Hydrozoans Have Colonial Polyps
The polyps within a colony may differentiate to perform specialized tasks.

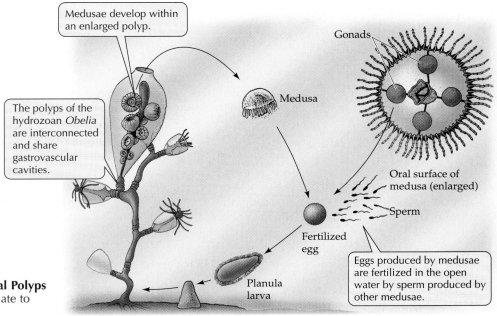

Medusae develop within an enlarged polyp.

The polyps of the hydrozoan *Obelia* are interconnected and share gastrovascular cavities.

Medusa

Gonads

Oral surface of medusa (enlarged)

Sperm

Fertilized egg

Planula larva

Eggs produced by medusae are fertilized in the open water by sperm produced by other medusae.

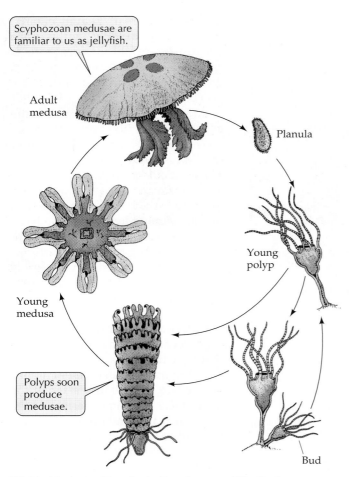

Scyphozoan medusae are familiar to us as jellyfish.

Adult medusa

Planula

Young medusa

Young polyp

Polyps soon produce medusae.

Bud

29.11 Medusae Dominate Scyphozoan Life Cycles
Scyphozoan medusae are the familiar jellyfish of coastal waters. The small polyps quickly produce medusae.

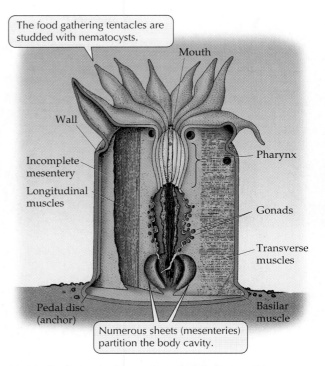

The food gathering tentacles are studded with nematocysts.

Mouth

Wall

Incomplete mesentery

Longitudinal muscles

Pharynx

Gonads

Transverse muscles

Pedal disc (anchor)

Basilar muscle

Numerous sheets (mesenteries) partition the body cavity.

29.12 Anthozoans Have Complex Polyps   This sea anemone has the typical muscular structure of an anthozoan.

by numerous sheets, called *mesenteries*, that increase the surface area available for secreting digestive enzymes and absorbing nutrients (Figure 29.12). Gonads also develop on these mesenteries.

Sea anemones such as the one shown in the chapter-opening photograph are solitary anthozoans that lack specialized protective coverings. They are widespread in both warm and cold ocean waters. Many sea anemones are able to crawl slowly on the discs with which they attach to the substratum; a few species can swim.

Corals, by contrast, are usually sessile and colonial. The polyps of most corals secrete a matrix of organic molecules upon which calcium carbonate, the eventual skeleton of the colony, is deposited. The forms of coral skeletons are species-specific and highly diverse (Figure 29.13*a*). The common names of coral groups—horn corals, brain corals, staghorn corals, organ pipe corals, sea fans, and sea whips, among others—describe their appearance.

As a coral colony grows, old polyps die and leave their calcareous skeletons intact. The living members form a layer on top of a growing reef of skeletal remains. Reef-building corals are restricted to clear, warm waters. They are especially abundant in the Indo-Pacific region, where they form chains of islands and reefs. The Great Barrier Reef along the northeast coast of Australia is a system of coral formations more than 2,000 km long and as wide as 150 km. A continuous coral reef hundreds of kilometers long in the Red Sea has been calculated to contain more material than all the buildings in the major cities of North America combined.

Corals flourish in nutrient-poor, clear, tropical waters. For a long time scientists wondered how corals obtain enough nutrients to grow as rapidly as they do. The answer is that highly modified dinoflagellates live symbiotically within the cells of the corals. By their photosynthesis, the dinoflagellates provide carbohydrates to their hosts and help with calcium deposition. In turn, the dinoflagellates within the coral's tissues are protected from predators. This symbiotic relationship explains why reef-forming corals are restricted to surface waters, where light levels are high enough to allow photosynthesis (Figure 29.13*b*).

KEYS TO CNIDARIAN DIVERSITY.   Cnidarians were the dominant marine organisms late in the Precambrian period, more than 600 million years ago, and they remain

(a)

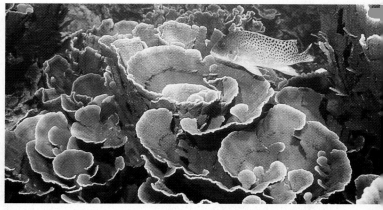

(b) *Montipora foliosa*

29.13 Corals (a) Many different species of corals grow together on this reef in Indonesia. (b) The ends of the branches of a cabbage coral spread outward, maximizing the amount of sunlight received by the photosynthetic dinoflagellates that live symbiotically inside their cells.

important components of marine ecological communities today (Figure 29.14). The cnidarian body plan combines a low metabolic rate with the ability to capture relatively large prey. The bulk of the body of medusae and many polyps is made up of the largely inert mesoglea. As a result, even a large cnidarian such as a sea anemone requires relatively little food and can fast for weeks or months.

Nematocysts allow cnidarians to subdue prey, such as fishes, that are more active and structurally com-

plex than the cnidarians themselves. Nonetheless, many cnidarians eat microscopic prey, and ancestral cnidarians (as well as some living species) probably consumed bacteria and protists. Many corals and other cnidarians have symbiotic associations with photosynthetic dinoflagellates that enable them to grow where food is scarce.

These traits allow cnidarians to survive in environments where encounter rates with prey are much lower than required to sustain animals with higher

(a) *Gonothyraea loveni*

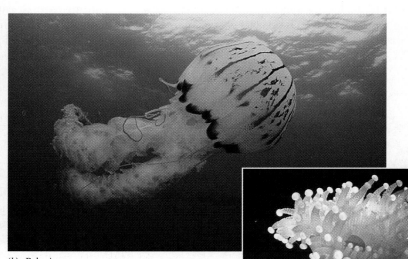

(b) *Pelagia panopyra*

(c) *Corynactis californica*

29.14 Diversity among Cnidarians (a) The structure of the polyps of a North Atlantic coastal hydrozoan is visible here. (b) This purple jellyfish illustrates the complexity of some scyphozoan medusae. (c) The nematocyst-studded tentacles of the strawberry anemone are poised to capture large prey carried to the animal by water movement.

metabolic needs. Cnidarians have persisted with few modifications for hundreds of millions of years.

The Evolution of Bilateral Symmetry

Although some cnidarians can move about slowly, they lack a front end that typically encounters new environments first. Bilateral symmetry probably arose first in simple organisms consisting of flattened masses of cells. These animals crawled over a substratum, feeding as they went, just as placozoans do today. Placozoans are no more complex than the earliest of these simple animals probably were, but the lineages that arose from these simple ancestors evolved into a great diversity of animals, many of which are highly complex structurally.

As inferred from traits of extant species, the first split in the lineage of bilaterally symmetrical animals separated the ctenophores (comb jellies) from the lineage leading to all other animals (see Figure 29.4).

Ctenophores capture prey with mucus and tentacles

Ctenophores, also known as comb jellies, constitute the phylum Ctenophora. Because the body plans of ctenophores and cnidarians are superficially similar, they were long considered to be closely related. Both have two cell layers separated by a thick, gelatinous mesoglea, and both have radial symmetry and feeding tentacles. Unlike cnidarians, however, ctenophores have a complete gut, with two anal pores through which wastes are voided.

Ctenophores have eight comblike rows of fused plates of cilia, called *ctenes*, and they move by beating these fused cilia rather than by muscular contractions. Ctenophoran tentacles are solid and lack nematocysts; instead, the tentacles are covered with sticky filaments to which prey adhere (Figure 29.15). After capturing

its prey, the ctenophore retracts its tentacles to bring the food to its mouth. In some species, the entire surface of the body is coated with a sticky mucus that captures prey. All of the 100 known species of ctenophores are carnivorous marine animals.

Most ctenophores cannot capture large prey, but one group lacks tentacles and feeds on other ctenophores by ingesting them whole, using a muscular pharynx. The typical ctenophore feeding method is to dangle sticky tentacles in the water, where planktonic prey adheres to them. Like cnidarians, ctenophores have low metabolic rates because they are composed primarily of inert mesoglea. They are common in open seas, where prey are often scarce.

Ctenophore life cycles are simple. Gametes from gonads located on the walls of the gastrovascular cavity are liberated into the cavity and then discharged through the mouth or through pores. Fertilization takes place in the open seawater. In nearly all species the fertilized egg develops directly into a miniature ctenophore that gradually grows into an adult.

Protostomes and Deuterostomes: An Early Lineage Split

The next major split in the evolution of animal lineages separated the two groups that dominate today's biota—the protostomes and deuterostomes (see Figure 29.4). Because of the many differences in the structures and embryological development of protostomes and deuterostomes (see Table 29.1), evolutionary biologists agree that these are two monophyletic groups that have been evolving separately since the Cambrian period.

The major shared, derived traits that unite protostomes are a central nervous system consisting of an anterior brain that surrounds the entrance to the digestive tract; paired or fused longitudinal nerve cords, and a free-floating larva with a food-collecting system consisting of compound cilia on multiciliate cells. The major shared, derived traits that unite deuterostomes are a dorsal nervous system and larvae with a food-collecting system consisting of cells with single cilia.

Protostomes: A Diversity of Body Plans

Most of the world's living animal species are protostomes, among which are rotifers, roundworms, segmented worms, arthropods, and mollusks. The diversity of protostome body plans and lifestyles has posed many challenges to zoologists attempting to infer the evolutionary relationships among these animals. Developmental, structural, and molecular evidence often disagree, and the biology of some marine groups is poorly known.

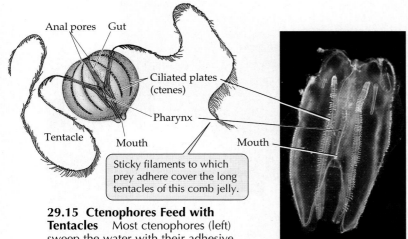

Anal pores   Gut
Ciliated plates (ctenes)
Pharynx
Tentacle
Mouth
Mouth

Sticky filaments to which prey adhere cover the long tentacles of this comb jelly.

29.15 Ctenophores Feed with Tentacles Most ctenophores (left) sweep the water with their adhesive tentacles, capturing plankton. *Leucothea* (right) has much shorter tentacles.

Leucothea

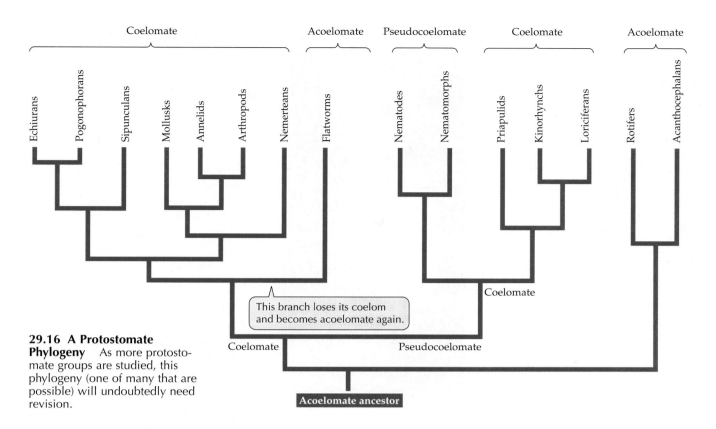

29.16 A Protostomate Phylogeny As more protostomate groups are studied, this phylogeny (one of many that are possible) will undoubtedly need revision.

The phylogeny we have adopted (Figure 29.16) is not accepted by all investigators, and changes are certain to be made in the near future. Our chosen phylogeny is based on the assumption that some animal groups with relatively simple body plans had more complex ancestors. In particular, we assume that possession of a pseudocoel arose independently several times during animal evolution in lineages that had coelomate ancestors.

Movement is a key feature of protostome evolution

Speed is often advantageous for both prey and the predators that pursue them. Fast-moving prey and predators evolved in the early Cambrian period (550 mya). Animals probably evolved fluid-filled body cavities because such cavities facilitate swift movement.

Hydrostatic skeletons, as fluid-filled cavities are called, facilitate movement because they are incompressible. When muscles around part of a fluid-filled body cavity contract, the fluid must move to another part of the cavity. If the body tissues around the cavity are flexible, fluids moving from one region cause the other region to expand. Fluids can thus move specific body parts or even the whole animal, provided that temporary attachments can be made to the substratum.

The types and numbers of body cavities provide a major key to the lives of these animals and the degree to which they can control and change their shapes and the complexity of the movements they can perform.

Rotifers are small but structurally complex

An abundant and widespread group of pseudocoelomate animals is the phylum Rotifera. Rotifers are bilateral, unsegmented animals that have three cell layers. Most rotifers are tiny (50 to 500 μm long)—smaller than some ciliate protists—but they have highly developed organs (Figure 29.17). A complete gut passes from an anterior mouth to a posterior anus, and the pseudocoel functions as a hydrostatic skeleton. Most rotifers are active, propelling themselves through the water by means of rapidly beating cilia rather than by muscular contraction. This type of movement is effective because rotifers are so small.

The most distinctive organs of rotifers are those used to collect and process food. A conspicuous, ciliated organ (the *corona*) surmounts the head of many species. Coordinated beating of cilia provides the force for locomotion and also sweeps particles of organic matter from the water into the mouth and down to a complicated structure (the *mastax*), which has teeth that grind the food. By contracting the muscles that surround the pseudocoel, a few rotifer species that prey on protists and small animals can protrude the mastax through the mouth and seize small objects with it.

Some rotifers are marine, but most of the 1,800 known living species live in fresh water. Members of a few species rest on the surface of mosses and lichens in a desiccated, inactive state until it rains. When rain

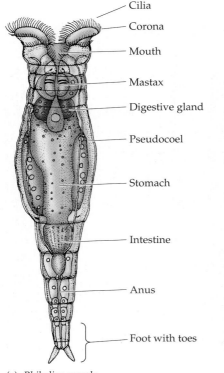

Cilia
Corona
Mouth
Mastax
Digestive gland
Pseudocoel
Stomach
Intestine
Anus
Foot with toes

(a) *Philodina roseola*

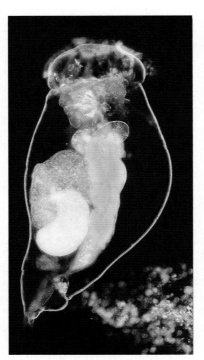

(b) *Epiphanes senta*

29.17 Rotifers (a) This rotifer reflects the general structure of many of the free-living species in this phylum. (b) The internal anatomy of a rotifer is clear in this micrograph.

Most of them are believed to be descendants of a lineage that has evolved independently since early animal evolution. This body form enables animals to move efficiently through muddy and sandy marine sediments.

Roundworms are simple but abundant

Roundworms (phylum Nematoda) have a thick, multilayered cuticle secreted by the underlying epidermis that gives their body its shape (Figure 29.19a). As a roundworm grows, it sheds and resecretes its cuticle four times. Because the cross-sectional shape of its body is round, a roundworm has a relatively small outer body surface for exchanging oxygen and other materials with the environment.

falls, they absorb water and swim about and feed in the films of water that temporarily cover the plants. Most rotifers live no longer than 1 or 2 weeks.

All spiny-headed worms are parasites

The spiny-headed worms, members of the phylum Acanthocephala (from the Greek *akantha*, "spine," and *kephale*, "head"), are obligate intestinal parasites of vertebrates, primarily freshwater fishes. Most species are less than 20 cm long, but a few extend nearly a meter. The anterior end has a hook-bearing structure by which the animal attaches to its host.

Acanthocephalans have no gut; they absorb their food directly through their body wall. Larvae develop in the body cavity of the female and are released with the host's feces. The larvae are eaten by an intermediate host (an insect or crustacean), which, in turn, is eaten by a fish or other vertebrate, thereby completing the life cycle (Figure 29.18). Many acanthocephalans change the behavior of their hosts in ways that enhance their transmission to the next host in their life cycle.

Wormlike Bodies: Movement through Sediments

Animals in 16 protostomate phyla are wormlike; that is, they are bilaterally symmetrical, legless, soft-bodied, and at least several times longer than they are wide.

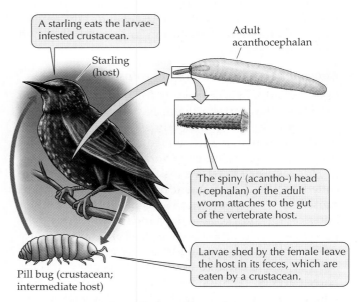

A starling eats the larvae-infested crustacean.

Starling (host)

Adult acanthocephalan

The spiny (acantho-) head (-cephalan) of the adult worm attaches to the gut of the vertebrate host.

Larvae shed by the female leave the host in its feces, which are eaten by a crustacean.

Pill bug (crustacean; intermediate host)

29.18 An Acanthocephalan Life Cycle The life cycle of a typical spiny-headed worm includes an arthropod and a vertebrate host. The parasites alter the behavior of the arthropods to increase their chances of being eaten by the vertebrate host.

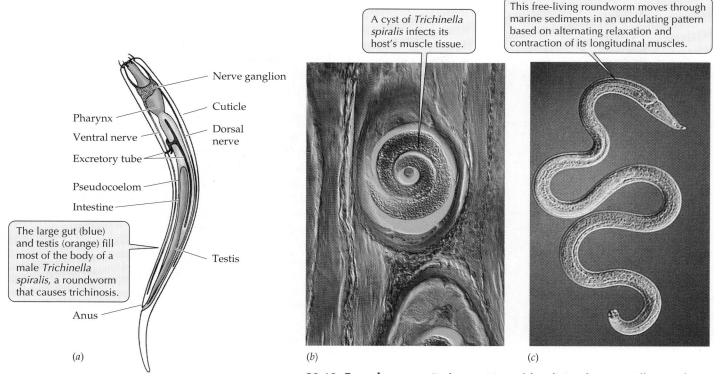

A cyst of *Trichinella spiralis* infects its host's muscle tissue.

This free-living roundworm moves through marine sediments in an undulating pattern based on alternating relaxation and contraction of its longitudinal muscles.

Nerve ganglion

Cuticle

Pharynx

Dorsal nerve

Ventral nerve

Excretory tube

Pseudocoelom

Intestine

The large gut (blue) and testis (orange) fill most of the body of a male *Trichinella spiralis*, a roundworm that causes trichinosis.

Testis

Anus

(a)

(b)

(c)

29.19 Roundworms   Both parasitic and free-living forms are illustrated.

Oxygen and nutrients are exchanged not only through the cuticle, but also through the intestine, which is only one cell layer thick. Materials are moved through the gut by rhythmic contraction of a highly muscular organ at the anterior end. The pseudocoelom of roundworms is small; the body organs fill most of the internal space.

Roundworms are one of the most abundant and universally distributed of all animal groups. We unintentionally eat and drink enormous numbers of roundworms in our lifetimes. For example, examination of a single rotting apple from the ground of an orchard revealed 90,000 roundworms contained in the flesh of the fruit. One square meter of mud off the coast of The Netherlands yielded 4,420,000 individuals. The topsoil of rich farmland has up to 3 billion nematodes per acre.

Countless numbers of roundworms live as scavengers in the upper layers of the soil, on the bottoms of lakes and streams, and as parasites in the bodies of most kinds of plants and animals. The largest known roundworm, which reaches a length of 9 m, is found in the placenta of female sperm whales. About 20,000 species have been described, but the actual number of living species may exceed 1 million.

The diets of roundworms are as varied as their habits. Many roundworms live parasitically within their hosts. Many are predators, preying on protists and other small animals (including other round-worms). The roundworms that are parasites of people, cats, dogs, cows, sheep, and economically important plants have been studied intensively in an effort to find ways of controlling them.

The structure of parasitic roundworms (see Figure 29.19a and b) does not differ much from that of free-living species (Figure 29.19c), but the life cycles of many parasitic species have special stages that facilitate the transfer of individuals among hosts. *Trichinella spiralis*, the parasite that causes the disease trichinosis, has a relatively simple life cycle. The larvae of *Trichinella* form cysts in the muscles of their mammalian hosts (see Figure 29.19b). If present in great numbers, these cysts cause severe pain or death.

Trichinella is transmitted to a mammal that eats the flesh of an infected individual. The activated larvae leave the cysts and attach to the intestinal wall, where they feed. Later, they bore through the host's intestinal wall and are carried in the bloodstream to muscles, where they form cysts. The alternate host is likely to be another mammal (usually a pig in the case of human infections). No special stage in the *Trichinella* life cycle lives in an alternate host. However, other roundworm life cycles are more complex, involving one or more alternate hosts.

Horsehair worms are parasites

The larvae of horsehair worms (phylum Nematomorpha) live inside terrestrial and aquatic insects and

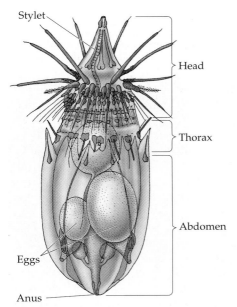

Stylet

Head

Thorax

Abdomen

Eggs

Anus

29.20 A Loriciferan
Even though they are abundant, the tiny (~0.3 μm) members of this phylum went unnoticed until 1983.

crabs. The gut is much reduced and is probably non-functional. Horsehair worms may feed only as larvae, absorbing nutrients from their hosts across their body wall, but many continue to grow after they have left their hosts, suggesting that adults may also absorb nutrients from their environment.

Several phyla of small worms live in ocean sediments

Members of several other small phyla of wormlike animals live in muddy and sandy ocean bottoms. Most

are small and poorly known. Among these are members of a phylum, the Loricifera (Figure 29.20), that was first discovered in 1983. Loriciferans live at great depths and cling so tightly to ocean sediments that it is difficult to separate them from sand. As a result, even though abundant, these tiny worms escaped human detection for centuries.

Flatworms move by beating cilia

Flatworms (phylum Platyhelminthes) are bilaterally symmetrical animals whose internal organs, though simple, are more complex than those of cnidarians and ctenophores. Flatworms have no enclosed body cavity, they lack organs for transporting oxygen to internal tissues, and they have only simple organs for excreting metabolic wastes. This body plan dictates that each cell must be near a body surface in order to respire, a requirement aided by the flattened body form.

Flatworms have traditionally been thought to represent a protostomate lineage that was ancestral to nearly all other animal linages, but recent analyses suggest that many features of these worms are derived rather than ancestral traits. The digestive tract of a flatworm, if there is one, consists of a mouth opening into a dead-end sac. However, the sac is often highly branched, forming intricate patterns that increase the surface area available for absorption of nutrients. All living flatworms feed on animal tissues—some as carnivores and parasites, others as scavengers. Motile flatworms glide over surfaces, powered by broad layers of cilia. This form of movement is very slow but it is sufficient for small, scavenging animals.

The flatworms probably most similar to the ancestral forms are the turbellarians (class Turbellaria): small, free-living marine and freshwater animals (a

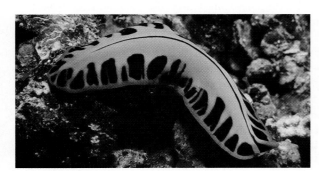

(a) *Pseudobiceros* sp.

29.21 Flatworms (a) Some flatworm species are free-living, like this marine flatworm of the South Pacific. (b) This flatworm lives parasitically in the gut of sea urchins. (c) The sheep liver fluke, also a parasite, is filled with highly branched large gonads.

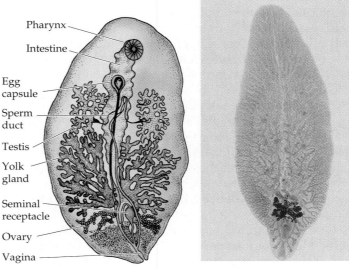

Pharynx

Intestine

Egg capsule

Sperm duct

Testis

Yolk gland

Seminal receptacle

Ovary

Vagina

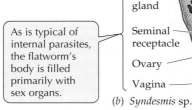

As is typical of internal parasites, the flatworm's body is filled primarily with sex organs.

(b) *Syndesmis* sp.

(c) *Fasciola hepatica*

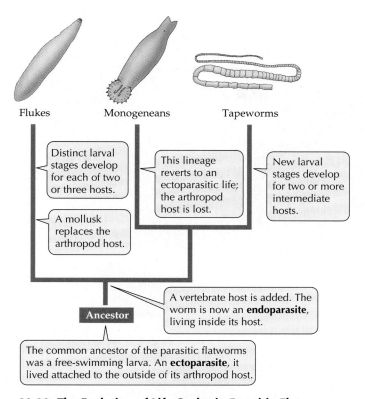

29.22 The Evolution of Life Cycles in Parasitic Flatworms
Hosts have been added and subtracted during the evolution of parasitic flatworms.

few live in moist terrestrial habitats). Freshwater turbellarians of the genus *Dugesia*, better known as planarians, are the most familiar species of flatworms. At one end they have a head with chemoreceptor organs, two simple eyes, and a tiny brain composed of anterior thickenings of the longitudinal nerve cords.

Although the earliest flatworms were free-living (Figure 29.21*a*), the flatworm body plan readily adapted to a parasitic existence. A likely evolutionary transition was from feeding on dead organisms, to feeding on the body surfaces of dying hosts, to invading and consuming parts of living, healthy hosts. Most of the 25,000 species of living flatworms—the tapeworms (class Cestoda) and flukes (class Trematoda; Figure 29.21*b* and *c*)—are parasitic. These worms inhabit the bodies of many vertebrates and cause serious human diseases, such as trichinosis, filariasis, and elephantiasis. Monogeneans (class Monogenea) are external parasites of fishes and other aquatic vertebrates.

Cladistic analyses have been used to suggest the sequence in which lineages of parasitic flatworms added or subtracted hosts from their life cycle (Figure 29.22). For example, monogeneans, all of which are ectoparasites with simple life cycles, evolved from ancestors that had more than one host in their life cycle.

Parasites live in nutrient-rich environments in which food is delivered to them, but they face other challenges. To complete their life cycle, parasites must overcome the defenses of their host. Because they die when their host dies, they must disperse their offspring to new hosts. The eggs of some parasitic flatworms are voided with the host's feces. Later, these eggs are ingested directly by other host individuals. However, most parasitic species have complex life cycles involving two or more hosts and several larval stages (Figure 29.23).

The complex life cycles of internal parasites often evolve because intermediate hosts increase the probability that individuals of the primary host will be infected. An intermediate host may provide a good feeding environment for the parasite, allowing it to develop and reproduce within an organism that is likely to be eaten by the primary host. However, these very advantages provide scientists with opportunities to reduce infections by interrupting a transmission stage in the parasite's life cycle.

Coelomate Protostomes

Although the evolution of body cavities provided many new movement capabilities, control over body shape is crude if the cavity has muscles on only one side, as a pseudocoel does (see Figure 29.3*b*). A coelom, which is surrounded by muscles (see Figure 29.3*c*), allows better control over movement of the fluids it contains. Even with a coelom, however, control over movement is limited if an animal has only a single, large body cavity.

This limitation is overcome if a large cavity is separated into compartments or segments so that localized changes in shape are produced by circular and longitudinal muscles in individual segments. Thus the animal can change the shape of each segment independently of the others. Segmentation of the coelom evolved several different times among both protostomes and deuterostomes. Among the coelomate protostomes, segmentation is found in annelids, arthropods, and mollusks, but there are several lineages of unsegmented, coelomate worms. They are believed to be descendants of a lineage that long ago separated from the lineage in which body segmentation evolved (see Figure 29.16).

Ribbon worms are dorsoventrally flattened

Ribbon worms (phylum Nemertea) are dorsoventrally flattened and have nervous and excretory systems similar to those of flatworms, but unlike flatworms, they have a complete digestive tract with a mouth at one end and an anus at the other. Food items move in one direction through the digestive tract of a ribbon worm and are acted on by a series of digestive enzymes.

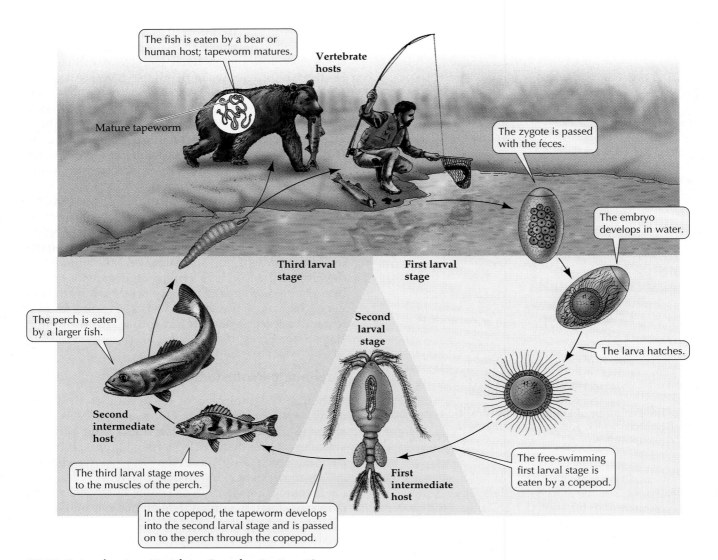

The fish is eaten by a bear or human host; tapeworm matures.

Vertebrate hosts

Mature tapeworm

The zygote is passed with the feces.

The embryo develops in water.

Third larval stage

First larval stage

Second larval stage

The perch is eaten by a larger fish.

The larva hatches.

Second intermediate host

First intermediate host

The third larval stage moves to the muscles of the perch.

The free-swimming first larval stage is eaten by a copepod.

In the copepod, the tapeworm develops into the second larval stage and is passed on to the perch through the copepod.

29.23 Returning to a Host by a Complex Route The broad fish tapeworm *Diphyllobothrium latum* must pass through the bodies of a copepod (a type of crustacean) and a fish before it can reinfect its primary host, a mammal. Such complex life cycles assist the flatworm's recolonization of hosts, but they also offer opportunities for humans to break the cycle with hygienic measures.

In the body of almost all ribbon worms is a fluid-filled cavity called the **rhynchocoel**, within which floats a hollow, muscular proboscis. The **proboscis** is the feeding organ; it may extend much of the length of the worm. Contraction of the muscles surrounding the rhynchocoel causes the proboscis to be ejected explosively through an anterior opening to obtain food (Figure 29.24). Thus the rhynchocoel transmits forces that move the proboscis rapidly, although it does not move the rest of the animal.

The proboscis of most ribbon worms is armed with a sharp hook (stylet) that pierces the prey. Paralysis-causing toxins produced by the proboscis are dis-

charged into the wound made by the stylet. Some species lack a hook and capture prey by wrapping the muscular proboscis around it. The ribbon worm then withdraws its proboscis into the rhynchocoel by means of a retractor muscle and takes the prey into its mouth.

Small ribbon worms move by beating their cilia. Larger ones employ waves of contraction of body muscles to move on the surface of sediments or to burrow. Movement by both of these methods is slow. In addition to capturing prey, the rhynchocoel helps the ribbon worm to burrow. For burrowing, the worm pushes its proboscis into the substratum for attachment; when the proboscis contracts and fattens, the worm is pulled forward. Similar burrowing mechanisms are found in other phyla.

Nearly all of the approximately 900 species of ribbon worms are marine, but some are found in fresh water and a few live in moist tropical terrestrial environments. All ribbon worms are carnivores, feeding mostly on arthropods and small worms in several different phyla.

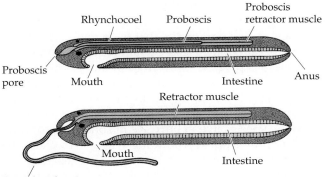

29.24 A Ribbon Worm The proboscis is the nemertean feeding organ. Floating in a cavity called the rhynchocoel, the proboscis can be moved rapidly. The worm, however, moves slowly.

Coelomate worms have varied body plans

Some phyla of coelomate worms are dismissed as "minor" groups because combined they have fewer than 600 species. However, because their body plans are quite different from those of other animals, they are worthy of mention.

The 250 species of peanut worms (phylum Sipuncula) are marine animals that burrow into sediments, live under rocks, or attach to the holdfasts of algae. They are common inhabitants of coral reefs, where they often burrow into calcareous coral skeletons. The body of a peanut worm is sausage-shaped and has a large, unsegmented coelom and feeding tentacles at the anterior end (Figure 29.25a). The coelom is filled with water and functions as a hydrostatic skeleton.

The echiuran worms (phylum Echiura) are similar in structure to peanut worms, and they often live in similar environments (Figure 29.25b). They are exclusively marine and most of them burrow in sand or mud. They gather their food with a proboscis that they extend over the surface of the substratum, where it gathers organic detritus.

Many early protostomes were small animals with thin body coverings. Gases and waste products were routinely exchanged across the body wall. Marine waters and sediments contain abundant food particles in the form of bacteria and dissolved organic matter that also can be taken in directly through the body wall.

Members of one lineage of protostomes, the phylum Pogonophora, evolved into burrowing forms with a crown of tentacles through which gases are exchanged, and they entirely lost their digestive tracts. Pogonophorans were not discovered until the twentieth century, when deep ocean exploration revealed them living many thousands of meters below the surface (Figure 29.25c). In these deep oceanic sediments, pogonophorans are abundant, reaching densities of many thousands per square meter. About 145 species have been described.

The coelom of a pogonophoran consists of an anterior compartment, into which the tentacles can be withdrawn, and a long, subdivided cavity that extends much of the length of its body. Experiments using radioactively labeled molecules have shown that pogonophorans take up dissolved organic matter at high rates from either the sediments in which they live or the surrounding water.

The largest and most remarkable pogonophorans, which grow to 2 m in length, live near deep-ocean hy-

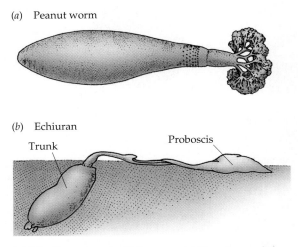

The red bodies of these deep-sea-vent pogonophorans project from the long white tubes in which they live.

29.25 Unsegmented Worms (a) The sipunculids, or peanut worms, are burrowing marine animals. (b) Echiurans are also burrowing worms; they gather food by extending their proboscis. (c) Pogonophorans live in tubes from which they extend their large feeding tentacles.

drothermal vents—openings in the sea floor through which hot, sulfur-rich water pours. The tissues of these species are filled with prokaryotes that fix carbon using energy obtained from oxidation of hydrogen sulfide (H_2S). The pogonophorans either consume these prokaryotes directly or live on their metabolic by-products, while keeping the prokaryotes in optimal environments near the vents.

Hydrothermal vent ecosystems are not based on light as their energy source. They have attracted considerable interest because they may be the type of environment in which life on Earth originated (see Chapter 24).

Segmented Bodies: Improved Locomotion

A body cavity segmented into compartments allows an animal to alter the shape of its body in complex ways and to control its movements precisely. Fossils of segmented worms are known from the middle Cambrian; the earliest forms are thought to have been burrowing marine animals. Segmentation evolved several times among protostomes.

Annelids are the dominant segmented worms

The annelids (phylum Annelida) are a very diverse group of worms that have a segmented body cavity (Figure 29.26). The approximately 15,000 described annelid species live in marine, freshwater, and terrestrial environments. The bulging waves that undulate up and down the length of the body to move the worm are made possible by its segmentation. A separate nerve center controls each segment, but the centers are connected by nerve cords that coordinate their functioning. The coelom in each segment is isolated from those in other segments.

Most annelids lack a rigid, external protective covering. The thin body wall serves as a general surface for gas exchange in most species, but this thin, permeable body surface restricts annelids to moist environments; they lose body water rapidly in dry air.

POLYCHAETES. More than half of all annelid species are placed in the class Polychaeta. Nearly all polychaetes are marine animals. Most have one or more pairs of eyes and one or more pairs of tentacles

at the anterior end of their body. The body wall in most segments extends laterally as a series of thin outgrowths, called **parapodia**, that have many blood vessels and function in gas exchange and locomotion. Stiff bristles protruding from each parapodium form temporary attachments to the substratum and prevent the animal from slipping backward when its muscles contract.

Many species live in burrows in soft sediments and capture prey from surrounding water with elaborate feathery tentacles (Figure 29.27a). The sexes are separate in most polychaetes. Typically, polychaetes release gametes into the water, where they become fertilized. A fertilized egg develops into a ciliated larva known as a **trochophore** (Figure 29.27b). As a trochophore develops, it forms body segments at its posterior end; eventually it metamorphoses into a small adult worm.

OLIGOCHAETES. More than 90 percent of the approximately 3,000 described species of oligochaetes (class Oligochaeta) live in freshwater or terrestrial habitats. Oligochaetes have no parapodia, eyes, or anterior tentacles, and they have relatively few bristles (setae). Earthworms—the most familiar oligochaetes—are scavengers and ingesters of soil, from which they extract food particles.

Unlike polychaetes, all oligochaetes are **hermaphroditic**: Each individual is both male and female. Sperm are exchanged simultaneously between two copulating individuals (Figure 29.27c). Eggs are laid in a cocoon outside the adult's body. The cocoon is shed, and

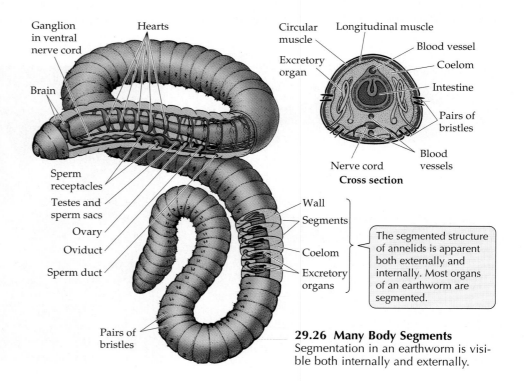

Ganglion in ventral nerve cord

Hearts

Brain

Sperm receptacles

Testes and sperm sacs

Ovary

Oviduct

Sperm duct

Pairs of bristles

Circular muscle

Longitudinal muscle

Excretory organ

Blood vessel

Coelom

Intestine

Pairs of bristles

Blood vessels

Nerve cord

Cross section

Wall

Segments

Coelom

Excretory organs

The segmented structure of annelids is apparent both externally and internally. Most organs of an earthworm are segmented.

29.26 Many Body Segments
Segmentation in an earthworm is visible both internally and externally.

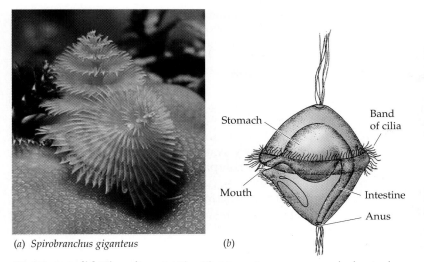

(a) *Spirobranchus giganteus*

Stomach

Band
of cilia

Mouth

Intestine

Anus

(b)

During copulation, each individual both donates and receives sperm.

(c) *Lumbricus* sp.

(d) Leech (species not identified)

29.27 Annelid Diversity (a) The Christmas tree worm, a polychaete, has striking feeding tentacles. (b) The trochophore is the polychaete larval form. (c) Individual earthworms are hermaphroditic (simultaneously male and female). (d) This freshwater leech has conspicuous anterior and posterior suckers.

when development is complete, miniature worms emerge and begin independent life.

LEECHES. Leeches (class Hirudinea) probably evolved from oligochaete ancestors. Most species live in freshwater or terrestrial habitats and, like oligochaetes, they lack parapodia and tentacles. Leeches are hermaphroditic; each individual serves as a sperm donor and a sperm recipient during copulation. The coelom of leeches is not divided into compartments, and the coelomic space is largely filled with mesenchyme tissue. Therefore, the movement of leeches differs radically from that of other annelids.

Groups of segments at each end of a leech are modified to form suckers, which serve as temporary anchors (Figure 29.27d). With its posterior anchor attached, the leech extends its body by contracting its circular muscles. The anterior sucker is then attached, the posterior one detached, and the leech shortens itself by contracting its longitudinal muscles.

Most leeches are external parasites of other animals, but some species also eat snails and other invertebrates. The mouth has three toothed jaws, with which the leech makes an incision in its host. It feeds on the host's blood. A leech feeds infrequently, but it can ingest so much blood in a single feeding that its body may enlarge several times. A substance secreted by the leech into the wound keeps the host's blood flowing.

For hundreds of years leeches were widely employed in medicine for bloodletting. Even today they are used to reduce fluid pressure in tissues damaged by, for example, a snake bite, and to eliminate pools of coagulated blood.

Mollusks lost segmentation but evolved shells

Mollusks (phylum Mollusca) underwent one of the most remarkable of animal evolutionary radiations. This dramatic radiation, which began with a segmented ancestor, was based on a body plan with three major structural components: the foot, the mantle, and the visceral mass, plus a rasping feeding structure known as the **radula** (plural radulae) at the anterior end.

Animals that appear very different, including snails, clams, and squids, are all built from these components, although individual components have been lost in some lineages (Figure 29.28). These three unique shared derived characteristics are the reason that zoologists place all 100,000 species of mollusks in one phylum.

The molluscan **foot** is a large, muscular structure that originally was the molluscan organ of locomotion, as well as the support for internal organs. In the lineage leading to squids and octopuses, the foot was modified to form arms and tentacles borne on a head with complicated sense organs. In other groups, such as clams, the foot was transformed into a burrowing organ. In some lineages the foot is greatly reduced.

The **mantle** is a fold of body wall that secretes the shell at its lip. The mantle partly encloses an external space called the mantle cavity. The internal organs lie in this cavity, covered by the outer layer of the **visceral mass**—the major internal organs. The gills of mollusks, which are used for gas exchange and feeding, lie

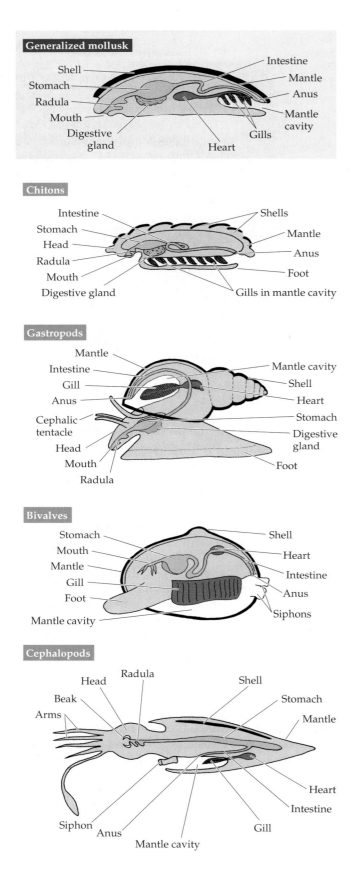

Generalized mollusk

Shell
Stomach
Radula
Mouth
Digestive gland
Intestine
Mantle
Anus
Mantle cavity
Gills
Heart

Chitons

Intestine
Stomach
Head
Radula
Mouth
Digestive gland
Shells
Mantle
Anus
Foot
Gills in mantle cavity

Gastropods

Mantle
Intestine
Gill
Anus
Cephalic tentacle
Head
Mouth
Radula
Mantle cavity
Shell
Heart
Stomach
Digestive gland
Foot

Bivalves

Stomach
Mouth
Mantle
Gill
Foot
Mantle cavity
Shell
Heart
Intestine
Anus
Siphons

Cephalopods

Head
Radula
Beak
Arms
Siphon
Anus
Mantle cavity
Shell
Stomach
Mantle
Heart
Intestine
Gill

29.28 Molluscan Body Plans The diverse modern mollusks all display variations of a general body plan that includes the foot, the mantle, and a visceral mass of internal organs.

in the mantle cavity. When the cilia on the gills beat, they create a flow of oxygenated water over the gills.

The coelom of mollusks is much reduced, but the open circulatory system has large fluid-filled cavities that are major components of the hydraulic skeleton. The radula was originally an organ for scraping algae from rocks, a function it retains in many living mollusks. However, in some mollusks the radula has been modified into a drill or a poison dart. In others, such as clams, it is absent.

Mollusks range in size from snails a mere 1 mm high to giant squids more than 18 meters long—the largest known protostomes. No fossils of the most ancient mollusks have yet been found, but they were probably segmented, as some mollusks are today.

MONOPLACOPHORANS. Unlike all other living mollusks, monoplacophorans (class Monoplacophora) have multiple gills, muscles, and excretory structures that are repeated along the length of the body. However, the shell is united into a single structure that covers the gills and other body organs. The gills are located in a large cavity under the shell, through which oxygen-bearing water circulates.

Monoplacophorans are believed to be the most ancient molluscan group. They were the most abundant mollusks during the Cambrian period, 550 million years ago. For a long time, it was thought that they had become extinct many millions of years ago, but in 1952, off the Pacific coast of Costa Rica, an oceanographic vessel dredged up ten specimens of an unusual little mollusk with a cap-shaped shell. Finding these animals, which were placed in the genus *Neopilina*, created a sensation because they turned out to be monoplacophorans.

CHITONS. Chitons (class Polyplacophora) have multiple gills and segmented shells, but other body parts are not segmented (Figure 29.29*a*). The chiton body is symmetrical, and its internal organs, particularly the digestive and nervous systems, are relatively simple. Chitons have trochophore larvae that are almost indistinguishable from those of annelids (see Figure 29.27*b*). Most chitons are marine herbivores that scrape algae from rocks with their sharp radulae. An adult chiton spends most of its life glued tightly to rock surfaces by its large, muscular, mucus-covered foot. A chiton can move slowly by means of rippling waves of muscular contractions in its foot.

(a) *Tonicella lineata*

(b) *Tridacna* sp.

(c) *Tridachiella diomedea*

(d) *Helix aspera*

(e) *Octopus horridus*

(f) *Nautilus belavensis*

29.29 Mollusk Diversity (a) This chiton is common in the intertidal zone of the Pacific coast of North America. (b) Giant clams are common inhabitants of tropical coral reefs. (c) Terrestrial and marine slugs are gastropods that have lost their shells through evolution; this "Spanish dancer" nudibranch is very conspicuous. (d) These garden snails are shelled gastropods. (e) Octopuses are active predators. This one is a banded octopus found in the Indian Ocean. (f) The boundaries of the chambers are clearly visible on the outer surface of the shell of this nautilus.

BIVALVES AND GASTROPODS. One lineage of early mollusks developed a two-part shell that extended over the sides of the body as well as the top, giving rise to the bivalves (class Bivalvia)—the familiar clams, oysters, scallops, mussels, and other important edible shellfish, together with many similar, less familiar forms (Figure 29.29b). The name "bivalve" derives from the structure of the shell of these animals, which has two major pieces connected by a flexible hinge. Bivalves are largely sedentary and have greatly reduced heads. The foot is compressed and, in many clams, is used for burrowing into mud and sand. Bivalves capture food from the water using their large gills, which are also the main sites of gas exchange.

Another lineage of early mollusks evolved into the gastropods (class Gastropoda), which includes the snails. Most gastropods are motile, using their large foot to move slowly across the substratum or to burrow through it. The shell and visceral mass of a gastropod larva undergo a 180° counterclockwise torsion relative to the foot during development. The result is that the digestive tract and nervous system are twisted so that the anus, the opening of the mantle, and the gills move to the front of the body, just behind and over the head. All gastropods undergo torsion, but in some the torsion is reversed later during development and they are not twisted as adults. The gills of gastropods are the primary sites of gas exchange. In some species, they are also feeding devices.

Gastropods are the most diverse and widely distributed of the molluscan classes (Figure 29.29c,d). Some species can crawl, including a rich variety of snails, whelks, limpets, slugs, abalones, and the often brilliantly ornamented nudibranchs. Still others—the sea butterflies and heteropods—have a modified foot that functions as a swimming organ with which they move through open ocean waters. The only mollusks that live in terrestrial environments—many snails and slugs—are gastropods. In terrestrial species the mantle cavity is modified into a highly vascularized lung.

CEPHALOPODS. One lineage of mollusks evolved an exit tube for currents leaving the mantle cavity. At first, this tube probably simply improved the flow of water over the gills; subsequently it became modified to allow the early cephalopods (class Cephalopoda) to control the direction in which water leaves the shell cavity. By closing off shell chambers and then pumping out the water, the animals could also control their buoyancy. Together, these adaptations allow cephalopods to live in open water.

Cephalopods were the first large, shelled animals able to move vertically in the ocean. The modification of the mantle into a device for forcibly ejecting water from the cavity enabled them to move rapidly through the water. With this greatly enhanced mobility, some cephalopods, such as squids, became the major predators in open ocean waters (Figure 29.29e). They are still important marine predators today. As is typical of active predators, cephalopods have complicated sense organs, most notably eyes that are comparable to those of vertebrates in their ability to resolve images (see Chapter 42).

Cephalopods appeared near the beginning of the Cambrian period, about 600 million years ago, and by the Ordovician period a wide variety of types were present. Increases in size and reductions in external hard parts characterize the subsequent evolution of many lineages. The cephalopod foot is closely associated with the large, branched head that bears tentacles and a siphon. The large, muscular mantle is a solid external supporting structure. The gills hang within the mantle cavity. Cephalopods capture and subdue their prey with a sharp beak.

The earliest cephalopod shells were divided by partitions penetrated by tubes through which liquids could be removed. As fluid moves out of a chamber, gas diffuses into it, changing the buoyancy of the animal. Nautiloids (genus *Nautilus*) are the only cephalopods with external chambered shells that survive today (Figure 29.29f).

Arthropods: Segmented External Skeletons

Fluid-filled body cavities acted on by muscles surrounding them provide the skeletal support for most of the animals we have discussed so far. The body plans of these animals vary according to the size and extent of the cavities, how they are lined, and whether they are subdivided. Most of these animals have external coverings that are relatively thin and flexible and allow gas exchange (although some, such as roundworms, have tough cuticles).

In Precambrian times, the body covering in some wormlike lineages became thickened by the incorporation of layers of protein and a strong, flexible, waterproof polysaccharide called **chitin**. After this change, which was initially probably protective in function, the body covering acquired support and locomotor functions—it became an **exoskeleton**, a hard covering on the outside of the body.

To locomote, an animal with a rigid exoskeleton and no cilia needs appendages that are moved by muscles. Such appendages evolved several times in late Precambrian times, leading to the phyla collectively called **arthropods**. The appendages of these proto-arthropods were initially unjointed, but jointed appendages evolved in most arthropod lineages. The species with unjointed limbs that survive today show that effective locomotion is possible without joints.

The muscles of animals that have exoskeletons attach to the inside of the skeleton, and each segment

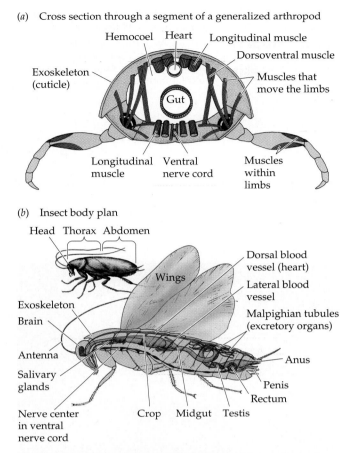

(a) Cross section through a segment of a generalized arthropod

Hemocoel   Heart   Longitudinal muscle

Dorsoventral muscle

Exoskeleton (cuticle)

Muscles that move the limbs

Gut

Longitudinal muscle   Ventral nerve cord   Muscles within limbs

(b) Insect body plan

Head Thorax Abdomen

Wings

Dorsal blood vessel (heart)

Lateral blood vessel

Malpighian tubules (excretory organs)

Exoskeleton

Brain

Antenna

Salivary glands

Anus

Penis

Rectum

Nerve center in ventral nerve cord

Crop   Midgut   Testis

29.30 External and Jointed Skeletons *(a)* A cross section shows the arthropod exoskeleton with jointed appendages. *(b)* The body plan of this insect differs in many details from that of other arthropods, but the basic theme of a segmented body with modified appendages is general to most arthropod lineages.

has muscles that operate a particular segment and the appendages attached to it (Figure 29.30*a*). Arthropod appendages serve many functions, including walking and swimming, food capture and manipulation, copulation, and sensory perception. The division of labor among body regions afforded a versatility to each individual arthropod species not found in its less complex ancestors.

An exoskeleton affected its possessors in other ways. First, the coelom lost its function as a hydrostatic skeleton and became much reduced. The pseudocoel became a hemocoel, filled with blood that directly bathes the animal's organs and allows hydraulic extension of limbs. Second, because the exoskeleton slows gas exchange, arthropods evolved new means of taking up oxygen and releasing carbon dioxide: Many arthropods evolved gills for gas exchange; others evolved tubes, called *tracheae*, that carry oxygen deep into their bodies.

A rigid, nonliving exoskeleton prevents an animal from gradually increasing in size. Arthropods grow by **molting**: a periodic shedding of the exoskeleton followed by the rapid hardening of a new and larger exoskeleton formed from cells under the old one. Soon after the old exoskeleton is shed, the new one is pumped up, expands, and hardens. While this process is under way, movement is difficult or impossible and the animals are highly vulnerable to predators. Soft-shelled crabs, for example, are individuals captured just after they shed their old exoskeletons.

Arthropods diversified dramatically

The evolution of an exoskeleton had another profound influence on arthropod evolution. Encasement within armor does more than protect an animal from predators. It also provides support for walking on dry land, and, if it has special waterproofing, it keeps the animal from drying out quickly in air. Arthropods were, in short, excellent candidates to invade the land, and they did so—repeatedly.

Several lineages of arthropods colonized the land, but all other groups of arthropods are overshadowed in number and diversity by the insects. The great majority not only of arthropods, but of all animal species, are insects.

Ancient arthropod lineages are a mystery

The earliest known arthropods date from Precambrian times. The roots of the lineage are so ancient, and the rate of evolutionary diversification was so rapid, that the ancestors of arthropods are as yet unknown. The divisions among arthropod lineages are so ancient that it is reasonable to divide the species into four phyla: Trilobita, Chelicerata, Crustacea, and Uniramia, the latter three of which survive today. Because the sequence of splitting of the lineages is unknown, the phylogeny we have adopted shows the three lineages separating at the same time (Figure 29.31).

No trilobites survive

A once-dominant line of arthropods, the trilobites (phylum Trilobita) flourished in Cambrian and Ordovician seas, but were extinct by the close of the Paleozoic era, 245 million years ago. Trilobites were heavily armored, and their body segmentation and appendages followed a relatively simple, repetitive plan (Figure 29.32). Why trilobites declined in abundance and eventually became extinct is unknown.

Chelicerates invaded the land

The bodies of all chelicerates (phylum Chelicerata) are divided into two major regions, the anterior of which bears two pairs of appendages modified to form mouthparts and four pairs of walking legs. The 63,000 described species are usually placed in three classes,

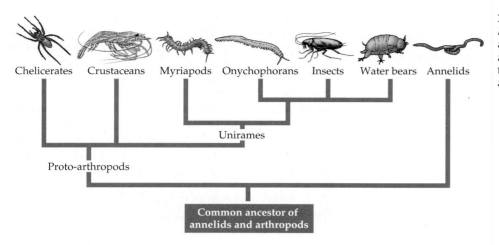

29.31 A Probable Phylogeny of Arthropods Lineage separations among surviving arthropods are so ancient that we divide them into three phyla: Chelicerata, Crustacea, and Uniramia.

only one of which contains many species.

The pycnogonids (class Pycnogonida), or sea spiders, are a small group of marine species that are seldom seen except by marine biologists (Figure 29.33*a*). The class Merostomata contains a single order, the Xiphosura, or horseshoe crabs. These marine animals, which have changed very little during their long fossil history, have a large horseshoe-shaped covering over most of the body. They are common in shallow waters along the eastern coasts of North America and Southeast Asia, where they scavenge and prey on bottom-dwelling invertebrates. Periodically they crawl into the intertidal zone to mate and lay eggs (Figure 29.33*b*).

The arachnids (class Arachnida) are abundant in terrestrial environments. Most arachnids have a simple life cycle in which miniature adults hatch from eggs and begin independent lives almost immediately. But some species have a more complex life cycle. Still others retain their eggs during development and give

(*a*) *Decolopoda* sp.

(*b*) *Limulus* sp.

Dalmanites limulurus

29.32 Trilobites The relatively simple, repetitive segments of the now-extinct trilobites are illustrated here by fossils of a species that lived during the Silurian period.

29.33 Minor Chelicerates (*a*) Although they are not spiders, it is easy to see why sea spiders were given their common name. (*b*) This spawning aggregation of horseshoe crabs was photographed on a sandy beach in Delaware.

(a) Uroctonus mondax

(c) Harvestman (species not identified)

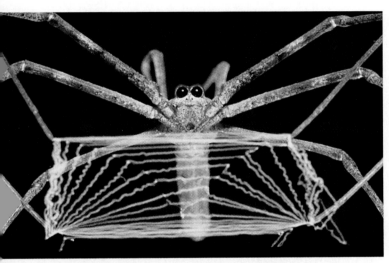

(b) Salticid (jumping) spider (species not identified)

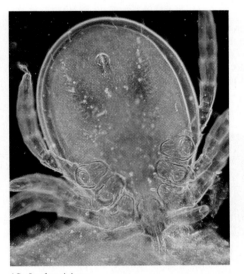

(d) Ixodes ricinus

29.34 Arachnid Diversity *(a)* Scorpions are nocturnal predators. *(b)* The evolution of webs is characteristic of spiders, which use them in many ways. This netcasting spider uses the web to envelop prey. *(c)* Harvestmen, often called daddy longlegs, are scavengers. *(d)* Ticks are blood-sucking, external parasites on vertebrates. This wood tick is piercing the skin of its human host.

birth to live young. The most diverse and ecologically important arachnids are the scorpions, harvestmen, spiders, mites, and ticks (Figure 29.34).

Spiders are important terrestrial predators. Some have excellent vision that enables them to chase and seize their prey. Others spin elaborate webs to snare prey. The webs of different groups of spiders are strikingly varied and enable spiders to position their snares in many different environments. Spiders also use protein threads to construct safety lines during climbing, and as homes, mating structures, protection for developing young, and dispersal. The threads are produced by modified abdominal appendages that are connected to internal glands that secrete the proteins of which the threads are constructed.

Crustaceans are diverse and abundant

Crustaceans (phylum Crustacea) are the dominant arthropods of the oceans. The individuals of one group alone, the copepods, are so numerous that they may be the most abundant of all animals. Most of the 40,000 species of crustaceans have a body that is divided into three regions: head, thorax, and abdomen. The segments of the head are fused together, and the head bears five pairs of appendages. Each of the multiple thoracic and abdominal segments usually bears one pair of appendages. In many species, a fold of the exoskeleton, the **carapace**, extends dorsally and laterally back from the head to cover and protect some of the other segments (Figure 29.35*a*).

The sexes are separate in nearly all crustaceans; males and females come together to copulate. The fertilized eggs of most crustacean species attach to the

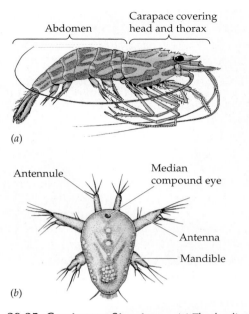

Abdomen | Carapace covering head and thorax

(a)

Antennule | Median compound eye | Antenna | Mandible

(b)

29.35 Crustacean Structure (a) The bodies of crustaceans are divided into three regions, each of which bears appendages. Two regions, the head and thorax, are covered by the protective carapace in many species. (b) A nauplius has one compound eye and three pairs of appendages.

(a) Decapod crustacean (species not identified)

(b) *Ligia occidentalis*

outside of the female's body, where they are held during their early development. At hatching, the young of some species are released as larvae; those of other species are released as juveniles similar in form to the adults. But some species release fertilized eggs into the water or attach them to an object in the environment. The typical crustacean larva, called a **nauplius**, has three pairs of appendages and one compound eye (Figure 29.35b).

The most familiar crustaceans are shrimp, lobsters, crayfish, and crabs (all decapods; Figure 29.36a); sow bugs (isopods; Figure 29.36b); and sand fleas (amphipods). Also included in the crustaceans are a wide variety of other small species, many of which superficially resemble shrimps. The abundant copepods (class Copepoda) already mentioned form one of these groups (Figure 29.36c). Barnacles (class Cirripedia) are unusual crustaceans that are sessile as adults (Figure 29.36d). With their calcareous shells, they superficially

(c) *Eucheta* sp.

(d) *Lepas anatifera*

29.36 Crustacean Diversity (a) This blue mountain crayfish is a terrestrial decapod crustacean from Australia. (b) This sow bug is an intertidal isopod found on the California coast. (c) A typical planktonic copepod. (d) Gooseneck barnacles attach to a substratum and feed by protruding and retracting a feeding appendage from their shells.

(a) *Scolopendra heros*

(b) Millipede (species not identified)

29.37 Myriapods (a) Centipedes have powerful jaws for capturing active prey. (b) Millipedes, which are scavengers and plant eaters, have smaller jaws and legs.

resemble mollusks, but as the zoologist Louis Agassiz remarked more than a century ago, a barnacle is "nothing more than a little shrimp-like animal, standing on its head in a limestone house and kicking food into its mouth."

Unirames are primarily terrestrial

The body of a unirame (phylum Uniramia) is divided into two or three regions. The anterior regions have few segments, but the posterior region, the abdomen, has many segments. Unirames are primarily terrestrial animals; most have elaborate systems of channels that bring oxygen to the cells of their internal organs.

MYRIAPODS. Centipedes, millipedes, and two other groups of animals (subphylum Myriapoda) have two body regions: a head and a trunk. Centipedes and millipedes have a well-formed head and a long, flexible, segmented trunk that bears many pairs of legs (Figure 29.37). Centipedes prey on insects and other small animals. Millipedes scavenge and eat plants. More than 3,000 species of centipedes and 10,000 species of millipedes have been described; many species remain unknown. Although most myriapods are less than a few centimeters long, some tropical species are ten times that size.

ONYCHOPHORANS. The 75 species of onychophorans (subphylum Onychophora) are similar to myriapods (Figure 29.38a). The soft bodies of onychophorans are covered by a thin, flexible cuticle that contains chitin; they use their body cavities as hydrostatic skeletons. Their soft, fleshy, unjointed, claw-bearing legs are formed by outgrowths of the body.

WATER BEARS. Like the onychophorans, water bears (subphylum Tardigrada) have fleshy, unjointed legs and use their fluid-filled body cavities as hydrostatic skeletons (Figure 29.38b). Unlike onychophorans, water bears are all extremely small, and they lack a circulatory system and gas exchange organs. The 550 extant species of water bears live in marine sands and gravels and on temporary water films on plants. When these films dry out, water bears also lose water and shrink to small, barrel-shaped objects that can survive at least a decade in a dehydrated state. Onychophorans and water bears are probably not ancestors of other arthropod lineages, but they may be similar in appearance to the early arthropods.

(a) *Peripatus* sp.

(b) *Echiniscoides sigismundi*   50 µm

29.38 Arthropods with Unjointed Legs
(a) Onychophorans have unjointed legs and use the body cavity as a hydrostatic skeleton. (b) The appendages and general anatomy of water bears superficially resemble those of onychophorans.

Insects are the dominant unirames

The 1.5 million species of insects (subphylum Insecta) that have been described are believed to be only a small fraction of the total number living on Earth today. They are the most numerous of all animal species, and the vast biological subdiscipline of entomology—the study of insects—attests to the importance of their interactions with humans.

Insects are found in nearly all terrestrial and freshwater habitats, and they utilize as food nearly all species of plants and many species of animals. Some are internal parasites of plants and animals; others suck blood externally or consume surface body tissues. Insects transmit many viral, bacterial, and protist diseases among plants and animals. Very few insect species live in the oceans. In freshwater environments, on the other hand, they are sometimes the dominant animals, burrowing through substrata, extracting suspended prey from the water, and actively pursuing other animals.

Insects have three basic body parts (head, thorax, abdomen), a single pair of antennae on the head, and three pairs of legs attached to the thorax (see Figure 29.30b). Insects exchange gases by means of air sacs and tubular channels called **tracheae** (singular trachea) that extend from external openings inward to tissues throughout the body. The adults of most flying insects have two pairs of stiff, membranous wings attached to the thorax—except for flies, which have only one pair, and beetles, in which the forewings form heavy, hardened wing covers.

Most insects have the full number of adult segments when they hatch from their eggs, but species differ strikingly in their state of maturity at hatching and in the processes by which they achieve adulthood. Wingless insects (class Apterygota), of modern insects probably the most similar to insect ancestors, have **simple development**. They hatch from the egg looking like small adults.

Development in the winged insects (class Pterygota) is more complex. The hatchlings are less similar in form to adults, and they undergo substantial changes at each molt in the process of becoming an adult. The immature stages of insects between molts are called **instars**. If changes between its instars are gradual, an insect is said to have **incomplete metamorphosis**. If dramatic changes occur between one instar and the next, an insect is said to have **complete metamorphosis** (see Figures 38.4).

Entomologists divide the winged insects into about 28 different orders; Figure 29.39 shows some of them. We can make sense out of this bewildering variety by recognizing three major types.

Members of one lineage cannot fold their wings back against the body. Although they are often excel-

29.39 Insect Diversity (a) The firebrat is a typical member of the apterygote (wingless) order Thysanura. (b) Unlike most insects, this adult mayfly (order Ephemeroptera) cannot fold its wings over its back. Representatives of some of the largest insect orders are (c) a broad-winged katydid (Orthoptera), (d) harlequin bugs (Hemiptera), (e) a predaceous diving beetle (Coleoptera), (f) a Great Mormon butterfly (Lepidoptera), (g) a giant robberfly (Diptera), and (h) a red-tailed bumblebee (Hymenoptera). ▶

lent flyers, they require a great deal of open space in which to maneuver. The only surviving members of these insects are the orders Odonata (dragonflies and damselflies) and Ephemeroptera (mayflies). All members of these two orders have complete metamorphosis. Their aquatic larvae change into flying adults after they crawl out of the water. Dragonflies and damselflies are active predators as adults, but adult mayflies lack functional digestive tracts and live only long enough to mate and lay eggs.

A second major evolutionary lineage includes the orders Orthoptera (grasshoppers, crickets, roaches, mantids, and walking sticks), Isoptera (termites), Plecoptera (stone flies), Dermaptera (earwigs), Thysanoptera (thrips), Hemiptera (true bugs), and Homoptera (aphids, cicadas, and leafhoppers). These insects have incomplete metamorphosis and are able to fold their wings back. Hatchlings are sufficiently similar in form to adults to be recognizable. They acquire adult organ systems, such as wings and compound eyes, gradually through several juvenile instars.

Insects belonging to the third lineage also are able to fold their wings back, but like the first lineage, they undergo complete metamorphosis, with different life stages specialized for living in different environments and using different food sources. In many species the larvae are adapted for feeding and growing, and the adults are specialized for reproduction and dispersal. The adults of some species do not feed at all, living only long enough to mate, disperse, and lay eggs. In many species whose adults do feed, adults and larvae use different food resources.

The most familiar example of complete metamorphosis is a caterpillar that changes into a butterfly, but other insects, including beetles and flies, undergo similar transformations. Their wormlike larvae transform into the adult form during a specialized "inactive" phase, the **pupa**, in which many larval tissues are broken down and the adult form develops. About 85 percent of all winged insects have complete metamorphosis. Familiar examples are the orders Neuroptera (lacewings and their relatives), Coleoptera (beetles), Trichoptera (caddisflies), Lepidoptera (butterflies and moths), Diptera (flies), and Hymenoptera (sawflies, bees, wasps, and ants).

(*a*) *Thermobia domestica*

(*b*) Mayfly (species not identified)

(*c*) *Microcentrum rhombifolium*

(*d*) *Murgantia histrionica*

(*e*) *Dytiscus marginalis*

(*f*) *Papilio memnon*

(*g*) *Blepharotes asilidae*

(*h*) *Bombus lapidarius*

Because they can fold their wings over their backs, insects belonging to the latter two adaptive types are able to fly from one place to another and then, upon landing, tuck their wings out of the way and crawl into crevices and other tight places. Several orders, including the Phthiraptera (lice) and Siphonaptera (fleas) are parasitic. Although descended from flying ancestors, these insects have lost the ability to fly.

Why have the insects undergone such incredible evolutionary diversification? Insects may have originated from a centipedelike ancestor at least as far back as the Devonian period, more than 350 million years ago. With this early start, they were able to exploit the newly formed forests and other ancient forms of land vegetation. The terrestrial environments penetrated by insects were like a new planet, an ecological world with more complexity than the surrounding seas, but one with relatively few species of terrestrial animals.

By Carboniferous times, a great diversity of insect types already swarmed over the land, and winged insects—the first animals to fly—had appeared. Because they can fly, insects are able to move efficiently through complex vegetation, and they eat nearly all types of plant tissues.

Themes in Protostome Evolution

Most protostome evolution took place in the oceans. Because water is an incompressible liquid, early animals used body fluids as the basis for their support. When acted on by surrounding circular and longitudinal muscles, these fluid-filled spaces function as skeletons that enable large animals to move.

Subdivisions of the body cavity allow better control of movement and permit different parts of the body to be moved independently of one another. Thus some protostome lineages gradually evolved the ability to change their shape in complex ways and to move with great speed on and through sediments or in the water.

Predation may have been the major selective pressure for the development of hard, external body coverings. Such coverings evolved independently in many protostomate phyla. Originally protective in function, they became key elements in the development of new systems of locomotion. Locomotory abilities permitted prey to escape more readily from predators but also allowed predators to pursue their prey more effectively. Thus, the evolution of protostomes has been and continues to be a complex saga of arms races among predators and prey.

During much of protostome evolution, the only food in the water consisted of dissolved organic matter and very small organisms. Consequently, many different lineages of animals evolved feeding structures designed to extract small prey from water, as well as structures for moving water through or over their prey-collecting devices.

Because water flows readily, bringing food with it, sessile lifestyles also evolved repeatedly during protostome evolution. Most protostomate phyla today have at least some sessile members. A sessile animal gains access to local resources but forfeits access to more distant resources. A sessile animal is exposed to physical agents and predators from which it cannot escape by movement.

Sessile animals cannot come together to mate; instead most sessile animals rely on the fertilization of gametes that they have ejected into the water. Some species eject both eggs and sperm into the water; others retain their eggs within their bodies and extrude only their sperm, which are carried by the water to other individuals. Species whose adults are sessile often have motile larvae, many of which have complicated mechanisms for locating suitable sites on which to settle. Many colonial sessile protostomes are able to grow in the direction of better resources or into sites offering better protection.

For both animals and plants, a frequent consequence of a sessile existence is intense competition for space that provides access to light and other resources. Such competition is intense among plants in most terrestrial environments. In the sea, especially in shallow waters, animals also compete directly for space. They have evolved mechanisms for overgrowing one another and for engaging in toxic warfare where they come into contact.

Individual members of colonies, if they are directly connected, can share resources. The ability to share resources enables some individuals to specialize for particular functions, such as reproduction, defense, or feeding. The nonfeeding individuals can derive their nutrition from their feeding associates.

Although we have concentrated on the evolution of greater complexity in protostomate lineages, many lineages that remained simple survive today. Cnidarians are common in the oceans; roundworms are abundant in most aquatic and terrestrial environments. Parasites have retained simple body plans but have evolved complex life cycles.

All the phyla of protostomate animals had evolved by the Cambrian period, 600 million years ago, but extinction and diversification within those lineages continue. The characteristics of the existing protostomate phyla are summarized in Table 29.2. Many of the evolutionary trends demonstrated by protostomes also dominated the evolution of deuterostomes, the lineage that includes the chordates, the group to which humans belong. Hard external body coverings evolved and were later abandoned by many lineages. In the next chapter, after considering the evolution of deuterostomes, we will return to the major themes in animal evolution and describe patterns that characterize all groups of animals.

TABLE 29.2   General Characteristics of Protostomate Phyla

PHYLUM	SYMMETRY	BODY CAVITY	DIGESTIVE TRACT	CIRCULATORY SYSTEM
Rotifera	Bilateral	Pseudocoelom	Complete	None
Acanthocephala	Bilateral	Pseudocoelom	None	None
Nematoda	Bilateral	Pseudocoelom	Complete	None
Nematomorpha	Bilateral	Pseudocoelom	Greatly reduced	None
Loricifera	Bilateral	Pseudocoelom	Complete	None
Platyhelminthes	Bilateral	None	Dead-end sac	None
Nemertea	Bilateral	Coelom	Complete	Closed
Sipunculida	Bilateral	Coelom	Complete	None
Echiura	Bilateral	Coelom	Complete	None
Pogonophora	Bilateral	Coelom	None	None
Annelida	Bilateral	Coelom	Complete	Closed or open
Mollusca	Bilateral	Reduced coelom	Complete	Open except in cephalopods
Chelicerata	Bilateral	Hemocoel	Complete	Closed or open
Crustacea	Bilateral	Hemocoel	Complete	Closed or open
Uniramia	Bilateral	Hemocoel	Complete	Closed or open

Summary of "Protostomate Animals: The Evolution of Bilateral Symmetry and Body Cavities"

Descendants of a Common Ancestor

• All members of the kingdom Animalia are believed to have a common flagellated protist ancestor. **Review Figure 29.4**

• The specialization of cells by function made possible the complex, multicellular body plan of animals.

Food Procurement: The Key to the Animal Way of Life

• Animals obtain their food—complex organic molecules—by active expenditure of energy.

Clues to Evolutionary Relationships among Animals

• Most animals have either radial or bilateral symmetry. Radially symmetrical animals move slowly or not at all. Bilateral symmetry is strongly correlated with more rapid movement and the development of sense organs at the anterior end of the animal. **Review Figure 29.2**

• The two major animal lineages—protostomes and deuterostomes—are believed to have separated early in animal evolution, because they differ in several components of their early embryological development. **Review Figure 29.4 and Table 29.1**

The body cavity of an animal is strongly related to its ability to move. On the basis of body cavity, animals are classified as acoelomates, pseudocoelomates, or coelomates. **Review Figure 29.3 and Table 29.2**

Early Animal Diets: Prokaryotes and Protists

• The earliest animals probably ate prokaryotes and protists, but some evolved specialized cells with stinging tentacles that allowed them to capture larger prey.

Simple Aggregations: Sponges and Placozoans

• All sponges (phylum Porifera) are simple animals that lack cell layers and body symmetry, but they have many different cell types.

• Sponges feed via choanocytes, feeding cells that draw water through the sponge body and filter out small organisms and nutrient particles. **Review Figure 29.5**

• Sponges come in a wide variety of sizes and shapes that are adapted to different movement patterns of water.

• Placozoans (phylum Placozoa) are very small and simple and lack symmetry. **Review Figure 29.7**

The Evolution of Cell Layers: The Cnidarians

• Cnidarians (phylum Cnidaria) are radially symmetrical and have only two cell layers, but with their nematocyst-studded tentacles, they can take prey larger and more complex than themselves. **Review Figure 29.8**

• Most cnidarian life cycles have a sessile polyp and a free-swimming, sexual, medusa stage, but some species lack one of the stages. **Review Figures 29.9, 29.10, 29.11, 29.12**

• Cnidarians have persisted through evolution because of their low metabolic needs.

The Evolution of Bilateral Symmetry

• Bilateral symmetry probably arose first in simple animals consisting of flattened masses of cells.

• Ctenophores (phylum Ctenophora), descendants of the first split in the lineage of bilaterally symmetrical animals, are marine carnivores that have simple life cycles. **Review Figure 29.15**

Protostomes and Deuterostomes: An Early Lineage Split

• Protostomes and deuterostomes are monophyletic groups that have been evolving separately since the Cambrian period. Protostomes have a central nervous system, paired nerve cords, and larvae with compound cilia. Deuterostomes have a dorsal nervous system and larvae with single cilia. **Review Figure 29.4 and Table 29.1**

Protostomes: A Diversity of Body Plans

• Most of Earth's living animal species are protostomes. The diversity of protostomate body plans and lifestyles makes it a challenge to determine their evolutionary relationships. **Review Figure 29.16**

• Improved control of movement, which is determined by body cavity, is a key feature of protostome evolution.

• Although no larger than many ciliated protists, rotifers (phylum Rotifera) have highly developed internal organs. **Review Figure 29.17**

• Spiny-headed worms (phylum Acanthocephala) are parasites of vertebrates, primarily freshwater fishes. They have complex life cycles. **Review Figure 29.18**

Wormlike Bodies: Movement through Sediments

• Members of 16 protostomate phyla are wormlike.

• Roundworms (phylum Nematoda) are one of the most abundant and universally distributed of animal groups. Many are parasitic. **Review Figure 29.19**

• Flatworms (phylum Platyhelminthes) have no body cavity, lack organs for oxygen transport, have only one entrance to their gut, and move by beating their cilia. Many species are parasitic. **Review Figures 29.21, 29.23**

• Flatworms have added and subtracted hosts from their life cycle throughout evolution. **Review Figure 29.22**

Coelomate Protostomes

• A coelom, which is surrounded by muscles, allows better control over movement than a pseudocoel, which has muscles on only one side. **Review Figure 29.3**

• Ribbon worms (phylum Nemertea) have a complete digestive tract and capture prey with an ejectable proboscis. **Review Figure 29.24**

• Several small phyla (Sipuncula, Echiura, and Pogonophora) of unsegmented worms with simple coeloms live in marine sediments.

Segmented Bodies: Improved Locomotion

• Segmentation of the coelom, which evolved several times among protostomate lineages, allows precise control of movement. **Review Figure 29.26**

• Annelids (phylum Annelida) are a diverse group of segmented worms that live in marine, freshwater, and terrestrial environments.

• Mollusks (phylum Mollusca) evolved from segmented ancestors but subsequently became unsegmented. The molluscan body plan has three basic components: foot, mantle, and visceral mass. **Review Figure 29.28**

• The molluscan body plan has been modified to yield a diverse array of animals that superficially appear very different from one another.

Arthropods: Segmented External Skeletons

• The arthropod lineages evolved external skeletons and jointed appendages, with which they move. **Review Figure 29.30**

• The arthropod skeleton provides support for walking on dry land and, if waterproofed, keeps the animal from drying out quickly in air. Several lineages of arthropods colonized terrestrial environments.

• The arthropods contain more species than do any other animal groups, and most arthropods are insects.

• Arthropods are divided into three extant phyla: Chelicerata, Crustacea, and Uniramia. The roots of each of these lineages are very ancient and thus unknown. **Review Figure 29.31**

• Insects, of which there is staggering variety, live in nearly all terrestrial and freshwater environments. Some insects evolved wings, becoming the first animals to fly.

Themes in Protostome Evolution

• Most of protosome evolution took place in the oceans; early protostomes used their body fluids as hydrostatic skeletons.

• Protostomes in several lineages evolved improved abilities to change their shapes and control their movements.

• Protostomes evolved the ability to capture and eat a great variety of sizes and shapes of prey.

• Most protostomate phyla have some sessile members, many of which are colonial.

Self-Quiz

1. The body plan of an animal is
   a. its general structure.
   b. the functional interrelationship of its parts.
   c. its general form and the functional interrelationship of its parts.
   d. its general form and its evolutionary history.
   e. the functional interrelationship of its parts and its evolutionary history.

2. A bilaterally symmetrical animal can be divided into mirror images by
   a. any cut through the midline of its body.
   b. any cut from its anterior to its posterior end.
   c. any cut from its dorsal to its ventral surface.
   d. only a cut through the midline of its body from its anterior to its posterior end.
   e. only a cut through the midline of its body from its dorsal to its ventral surface.

3. Among protostomes, cleavage of the fertilized egg is
   a. delayed while the egg continues to mature.
   b. determinate; cells separated after a few divisions develop into only partial embryos.
   c. indeterminate; cells separated after a few divisions develop into complete embryos.
   d. triploblastic.
   e. diploblastic.

4. The sponge body plan is characterized by
   a. a mouth and digestive cavity but no muscles or nerves.
   b. muscles and nerves but no mouth or digestive cavity.
   c. a mouth, digestive cavity, and spicules.
   d. muscles and spicules but no digestive cavity or nerves.
   e. no mouth, digestive cavity, muscles, or nerves.

5. The phyla of diploblastic animals are
   a. Porifera and Cnidaria.
   b. Cnidaria and Ctenophora.
   c. Cnidaria and Platyhelminthes.
   d. Ctenophora and Platyhelminthes.
   e. Porifera and Ctenophora.

6. Cnidarians are abundant, perhaps because of their ability
   a. to live in both salt and fresh water.
   b. to move rapidly in the water column.
   c. to capture and consume large numbers of small prey.
   d. to capture large prey, and their low metabolic rate.
   e. ability to capture large prey and to move rapidly.

7. Many parasites evolved complex life cycles because
   a. they are too simple to disperse readily.
   b. they are poor at recognizing new hosts.
   c. they were driven to it by host defenses.
   d. having an intermediate host usually increases the probability of transfer to a new individual of the primary host.
   e. their ancestors had complex life cycles and they simply retained them.

8. Which of the following is *not* part of the molluscan body plan?
   a. Mantle
   b. Foot
   c. Radula
   d. Visceral mass
   e. Jointed skeleton

9. Insects that hatch from eggs into juveniles that resemble the adults are said to have
   a. instars.
   b. neopterous development.
   c. simple development.
   d. incomplete metamorphosis.
   e. complete metamorphosis.

10. Many lineages of protostomes evolved feeding structures designed to extract small prey from the water because
   a. during much of protostome evolution, the only food available was dissolved organic matter and very small organisms.
   b. during much of protostome evolution, small animals were more abundant than large animals.
   c. large animals were available as food but they were difficult to capture.
   d. to be successful in competition for space, protostomes had to feed on small prey.
   e. water flowed naturally over their feeding structures, so early protostomes did not have to work to get food.

Applying Concepts

1. Differentiate among the members of each of the following sets of related terms:
   a. radial symmetry/bilateral symmetry
   b. protostome/deuterostome
   c. indeterminate cleavage/determinate cleavage
   d. spiral cleavage/radial cleavage
   e. incomplete metamorphosis/complete metamorphosis
   f. coelomate/pseudocoelomate/acoelomate

2. For each of the types of organisms listed below, give a single trait that may be used to distinguish them from the organisms in parentheses:
   a. cnidarians (sponges)
   b. gastropods (all other mollusks)
   c. polychaetes (other annelids)
   d. ribbon worms (roundworms)

3. Segmentation has arisen various times during protostome evolution. What advantages does segmentation provide? Given these advantages, why do so many unsegmented animals survive?

4. Many animals extract food from the surrounding medium. What protostomate phyla contain animals that extract suspended food from the water column? What structures do these animals use to capture prey?

5. Which protostomate phyla have some species that form large colonies of attached individuals? What advantages does coloniality provide to animals?

6. Discuss the structures that different lineages of animals evolved that enable them to capture large prey.

7. A major factor influencing the evolution of animals was predation. What major animal features appear to have evolved in response to predation? How do they help their bearers avoid becoming a meal for another animal?

Readings

Barrington, E. J. W. 1979. *Invertebrate Structure and Function*, 2nd Edition. Halsted-Wiley, New York. Engaging coverage of the invertebrates, with an emphasis on morphology.

Brusca, R. C. and G. J. Brusca. 1990. *Invertebrates.* Sinauer Associates, Sunderland, MA. A thorough account of the invertebrates that provides detailed treatments of body plans and includes excellent discussions of phylogenies.

Chapman, R. F. 1982. *The Insects: Structure and Function*, 3rd Edition. Harvard University Press, Cambridge, MA. A good overview of the rich diversity of insect life.

Cloudsley-Thompson, J. L. 1976. *Insects and History.* St. Martin's Press, New York. A vivid and readable account of instances in which insects have affected human society; special emphasis is placed on insects as carriers of disease organisms.

Kozloff, E. N. 1990. *Invertebrates.* Saunders, Philadelphia. An excellent reference book on all invertebrate taxa.

Morris, S. C. and H. B. Whittington. 1979. "The Animals of the Burgess Shale." *Scientific American*, July. A brief overview of the rich collection of fossils from this important site.

Nielsen, C. 1995. *Animal Evolution: Interrelationships among the Living Phyla.* Oxford University Press, Oxford. A book that emphasizes evolutionary relationships among animal phyla and patterns of animal evolution.

Noble, E. R. and G. A. Noble. 1982. *Parasitology: The Biology of Animal Parasites*, 5th Edition. Lea & Febiger, Philadelphia. A general text covering the major parasitic protists and worms that infect animals, including people.

Ruppert, E. E. and R. D. Barnes. 1994. *The Invertebrates*, 6th edition,. Saunders, Philadelphia. A thorough general textbook with a strong functional approach to the invertebrates.

Chapter 30

Deuterostomate Animals: Evolution of Larger Brains and Complex Behaviors

Predators as Pets
All these animals belong to the same species, *Canis familiaris*, the domestic dog. The great variety in their appearance is the result of artificial selection by humans.

P redators are typically behaviorally more complex than their prey because capturing an animal with well-developed escape behavior is more difficult than finding suitable leaves to chew. As a result, predators make more interesting pets than herbivores do. Dogs were probably the first animals to be domesticated by people, and a remarkable variety of dog breeds has been produced by artificial selection. Some breeds were developed for purely utilitarian purposes, such as herding of sheep, assistance in hunting, transportation in snow, and protection of camps against predators. But many breeds have been developed simply for aesthetic or fanciful purposes, as a visit to any dog show will demonstrate.

Increasing behavioral complexity has been a common theme during evolution within many deuterostomate lineages, but early deuterostomes were structurally and behaviorally simple animals, as are many of their living descendants. The ancestral traits shared by all members of the deuterostomate lineage include indeterminate cleavage in the early embryo, formation of the mesoderm from outpocketing of the embryonic gut, a blastopore that becomes the anus, three body layers, and a well-developed coelom (see Table 29.1). Living deuterostomes vary in their type of body symmetry and circulatory systems (Table 30.1). No fossils of ancestral deuterostomes that lived before the lineage split into several lineages have been found.

There are fewer lineages and many fewer species of deuterostomes than of protostomes (Figure 30.1), but we have a special interest in deuterostomes because humans are members of that lineage. In this chapter, we first describe and

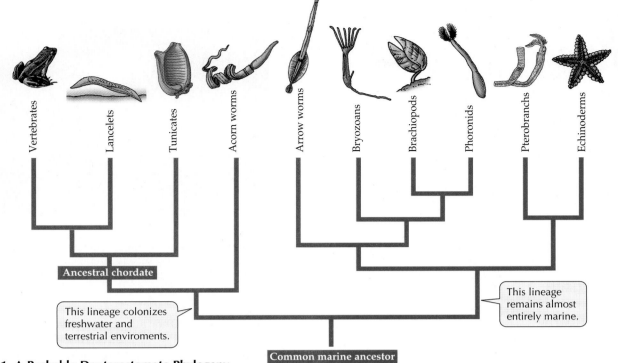

30.1 A Probable Deuterostomate Phylogeny
There are fewer lineages and many fewer species of deutero-stomes than of protostomes.

Labels on figure: Vertebrates, Lancelets, Tunicates, Acorn worms, Arrow worms, Bryozoans, Brachiopods, Phoronids, Pterobranchs, Echinoderms. Ancestral chordate. This lineage colonizes freshwater and terrestrial enviroments. This lineage remains almost entirely marine. Common marine ancestor

discuss the major deuterostomate phyla. Then we direct special attention to the primate lineage that gave rise to our own species. Finally, we provide an overview of the major themes in the evolution of animals on Earth.

Three Body Parts: An Ancient Body Plan

The most ancient division of the deuterostomes resulted in two lineages that followed very different evolutionary pathways (see Figure 30.1). One lineage remained almost entirely marine and gave rise to the lophophorate animals (phoronids, brachiopods, bry-

ozoans, and pterobranchs), arrow worms, and echinoderms. The other lineage invaded freshwater and terrestrial environments, where it underwent a rich evolutionary radiation that gave rise to the chordates.

Early deuterostomes probably obtained food by filtering it from ocean waters, a trait they shared with many protostomes. Members of one deuterostomate lineage evolved a new structure—the **lophophore**—with which they filtered their food. The most conspicuous feature of these animals, the lophophore is a circular or U-shaped ridge around the mouth that bears one or two rows of ciliated, hollow tentacles (Figure 30.2). This large and complex structure is an organ for both food collection and gas exchange. It is held in position and moved by contraction of muscles surrounding the middle body part, the mesocoel. All adult lophophorate animals are sessile, and they use the tentacles and cilia of their lophophores to capture phytoplankton and zooplankton.

The body of a lophophorate is divided into three parts: the *prosome* (anterior), *mesosome* (middle), and *metasome* (posterior). In most species, each region has a separate coelomic compartment: the protocoel, mesocoel, and metacoel, respectively. Typically, these animals secrete a tough outer body covering. All lophophorates also have a U-shaped gut; the anus is located close to the mouth but outside the tentacles.

Four phyla of lophophorate animals survive today: Phoronida, Brachiopoda, Bryozoa, and Pterobranchia. Nearly all members of these phyla are marine; only a

TABLE 30.1	General Characteristics of Deuterostomate Animal Phyla[a]	
PHYLUM	**SYMMETRY**	**CIRCULATORY SYSTEM**
Lophophorate phyla	Bilateral	None in most
Chaetognatha	Bilateral	None
Echinodermata	Biradial	Open or none
Hemichordata	Bilateral	Closed
Chordata	Bilateral	Closed

[a]Members of all phyla have a coelom and a complete digestive tract.

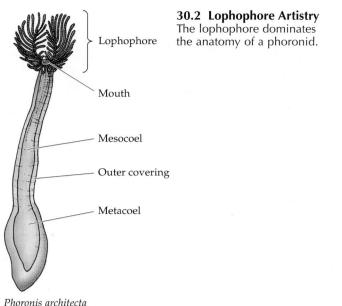

30.2 Lophophore Artistry
The lophophore dominates the anatomy of a phoronid.

Lophophore

Mouth

Mesocoel

Outer covering

Metacoel

Phoronis architecta

The lophophores of these phoronids are extended in the feeding position.

Phoronis sp.

30.3 Phoronids Phoronids live attached to the sediments of their marine habitat. They secrete a tube made mainly of chitin and live within it, extending their lophophores to feed.

few species live in fresh water. About 4,500 living species are known, but many times that number of species existed during the Paleozoic and Mesozoic eras.

Phoronids live in chitinous tubes

The 15 known species of phoronids (phylum Phoronida) are sedentary worms that live in muddy or sandy sediments or attached to a rocky substratum. Phoronids are found in waters ranging from intertidal zones to about 400 m deep. They range from 5 to 25 cm in length, and they secrete chitinous tubes in which they live.

The lophophore is the most conspicuous external feature of phoronids (Figure 30.3). Cilia drive water into the top of the lophophore. Water exits through the narrow spaces between the tentacles. Suspended food particles are caught and transported by ciliary action to the food groove and into the mouth.

Brachiopods superficially resemble bivalve mollusks

Brachiopods, or lampshells (phylum Brachiopoda), are solitary, marine, lophophorate animals that superficially resemble bivalve mollusks (Figure 30.4). Most brachiopods are between 4 and 6 cm long, but some are as large as 9 cm. Brachiopods have a mantle and a two-valved shell that can be pulled shut to protect the soft body. The shell differs from that of bivalves in that the two halves are dorsal and ventral rather than lateral. The two-armed lophophore of a brachiopod is located within the shell. The beating of cilia on the lophophore draws water into the slightly opened shell. Food is trapped in the lophophore and directed to a ridge along which it is transferred to the mouth.

Brachiopods are either attached to a solid substratum or are embedded in soft sediments. Most species are attached by means of a long, flexible stalk that holds the animal above the substratum. Gases are exchanged across nonspecialized body surfaces, especially the tentacles of the lophophore. Most brachiopods release their gametes into the water, where they are fertilized. The larvae, which resemble the adults, remain in the plankton for only a few days before they settle and metamorphose into adults.

Brachiopods reached their peak abundance and diversity in Paleozoic and Mesozoic times. More than 26,000 fossil species have been described. Only about 350 species survive, but they are common in some marine environments.

Lophophore

Laqueus sp.

30.4 Brachiopods You can see the lophophore of this North Pacific brachiopod between the valves of the shell.

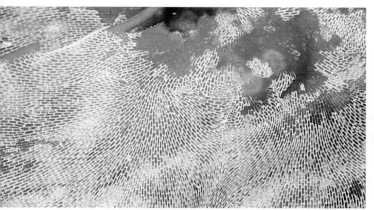

(a) *Membranipora membranacea*

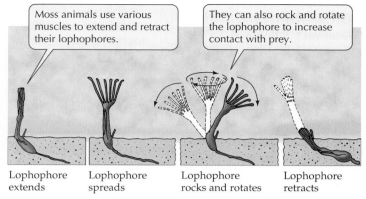

Moss animals use various muscles to extend and retract their lophophores.

They can also rock and rotate the lophophore to increase contact with prey.

Lophophore extends

Lophophore spreads

Lophophore rocks and rotates

Lophophore retracts

(b)

30.5 Moss Animals (a) Branching colonies of moss animals appear plantlike. Some members of the colony are specialized for certain functions. (b) Moss animals have greater control over their lophophores than members of other lophophorate species.

Bryozoans are colonial lophophorates

Moss animals (phylum Bryozoa) are colonial lophophorates with a "house" secreted by the body wall. A colony consists of many small individuals connected by strands of tissues along which materials can be moved. Bryozoans are called moss animals because their colonies appear plantlike (Figure 30.5a). Most moss animals are marine, but a few live in fresh water. They are the only lophophorates able to completely retract their lophophores, which they can also rock and rotate to increase contact with prey (Figure 30.5b).

A colony of moss animals is created by the asexual reproduction of its founding members. One colony may contain as many as 2 million individuals. In some species, individual colony members are specialized for feeding, reproduction, defense, or support. Moss animals reproduce sexually by releasing sperm into the water, where they are collected by other individuals. Eggs are fertilized internally, and developing embryos are brooded before they exit as larvae that seek suitable sites for attachment.

Pterobranchs capture prey with their tentacles

The fourth lophophorate group of animals consists of the pterobranchs (phylum Pterobranchia), which has only ten living species. They appear to have changed relatively little from the ancestors of their lineage. Pterobranchs are sedentary animals up to 12 mm long that live in tubes secreted by a proboscis, which is homologous to the prosome of phoronids and brachiopods. Some species are solitary; others form colonies of individuals joined together (Figure 30.6). Behind the proboscis is a collar with one to nine pairs of arms bearing long tentacles that capture prey and

permit gas exchange. The digestive tract is U-shaped, with the anus situated next to the tentacles. The proboscis encloses a coelomic cavity that has a pair of openings to the exterior, through which excretory wastes leave the body.

Arrow worms are active marine predators

Arrow worms (phylum Chaetognatha) also have three-part, streamlined bodies, but they do not have lophophores. Most of them swim in the open sea, but a few live on the seafloor. Their abundance as fossils indicates that they were already common more than 500 million years ago (mya). The 100 or so living species of arrow worms are small marine carnivores, all less than 12 cm long (Figure 30.7). They are so small that their gas exchange and excretion requirements are met by diffusion through the body surface.

Arrow worms lack a circulatory system. Wastes and nutrients are moved around the body in the coelomic fluid, which is propelled by cilia that line the coelom. The body plan is based on a coelom that is divided into head, trunk, and tail compartments. There is no

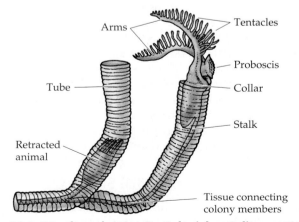

30.6 Pterobranchs May Be Colonial or Solitary This drawing of *Rhabdopleura* depicts two members of a colony.

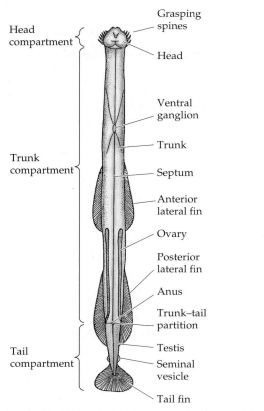

Head compartment {
Trunk compartment {
Tail compartment {

Grasping spines
Head
Ventral ganglion
Trunk
Septum
Anterior lateral fin
Ovary
Posterior lateral fin
Anus
Trunk–tail partition
Testis
Seminal vesicle
Tail fin

30.7 Arrow Worms Arrow worms have a tripartite body plan: head, trunk, and tail. The fins and grasping spines are adaptations for a predatory life.

distinct larval stage. Miniature adults hatch directly from eggs that are released into the water.

Arrow worms are among the dominant predators of small organisms in the open oceans. They typically lie motionless in the water until movement of the water signals the approach of a prey item, which range from small protists to young fish as large as an arrow worm. At that time the arrow worm darts forward and grasps the prey with the stiff spines adjacent to its mouth. Arrow worms are stabilized in the water by means of one or two pairs of lateral fins and a caudal fin.

Arrow worms are not powerful enough to swim against strong currents, but several species undertake daily vertical migrations of up to several hundred meters, moving to deeper water during the day and back to the surface at night. (Many other small oceanic animals make such vertical treks, probably because surface waters are dangerous during the day, when visually hunting predators are active.)

Echinoderms: Complex Radial Symmetry

Members of one lineage whose ancestors had lophophores evolved calcified internal plates covered by thin layers of skin and some muscles. The calcified plates of the early ancestors became enlarged and thickened until they fused inside the entire body, giving rise to an internal skeleton. This skeleton gave its bearers protection against predators and enabled them to live above bottom sediments.

A second major change in this lineage was the evolution of a **water vascular system**, a series of seawater channels and spaces derived by the enlargement and extension of the mesocoel. The water vascular system is a network of hydraulic canals leading to extensions called **tube feet** that function in gas exchange, locomotion, and feeding (Figure 30.8a). Seawater enters the water vascular system through a perforated *sieve plate* from which a calcified canal leads to another canal that rings the esophagus. From this *ring canal*, other canals radiate, extending through the arms (in species that have arms) and connecting with the tube feet (Figure 30.8a and b). The development of these two structural innovations—calcified internal skeleton and water vascular system—resulted in one of the most striking evolutionary radiations, the one that gave rise to the echinoderms (phylum Echinodermata).

Echinoderms have an extensive fossil record. About 23 classes have been described, of which only six survive. About 7,000 species of echinoderms exist today, but an additional 13,000 species, probably only a small fraction of those that actually lived, have been described from their fossil remains. Nearly all living species have a bilaterally symmetrical, ciliated larva that feeds for a while as a planktonic organism before settling and transforming into a radially symmetrical adult (Figure 30.8c).

Living echinoderms are members of two lineages: subphyla Pelmatozoa and Eleutherozoa. The two groups differ in the form of their water vascular systems and in their number of arms.

PELMATOZOANS. Sea lilies and feather stars (class Crinoidea) are the only surviving pelmatozoans. Sea lilies were abundant 300 to 500 mya, but only about 80 species survive today. Sea lilies attach to a substratum by means of a flexible stalk consisting of a stack of calcareous discs. The main body of the animal is a cup-shaped structure that contains a tubular digestive system. Five to several hundred arms, usually in multiples of five, extend outward from the cup. Jointed calcareous plates that cover the arms enable them to bend. A ciliated groove runs down the center of each arm. On both sides of the groove are tube feet covered with mucus-secreting glands.

A sea lily feeds by orienting its arms in passing water currents. Food particles strike and stick to the tube feet, which transfer the particles to the groove, where the action of the cilia carries the food to the mouth. The tube feet of sea lilies are also used for gas exchange and elimination of nitrogenous wastes.

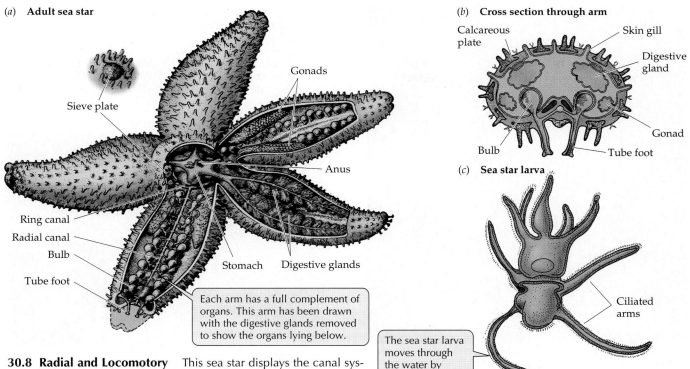

(a) **Adult sea star**

Sieve plate

Gonads

Anus

Ring canal

Radial canal

Bulb

Tube foot

Stomach

Digestive glands

Each arm has a full complement of organs. This arm has been drawn with the digestive glands removed to show the organs lying below.

(b) **Cross section through arm**

Calcareous plate

Skin gill

Digestive gland

Gonad

Bulb

Tube foot

(c) **Sea star larva**

Ciliated arms

The sea star larva moves through the water by beating its cilia.

30.8 Radial and Locomotory This sea star displays the canal system and tube feet of the echinoderm water vascular system, as well as a calcified internal skeleton.

Feather stars are similar to sea lilies, but they have flexible appendages with which they grasp the substratum while they are feeding and resting (Figure 30.9a). Feather stars feed in much the same manner as sea lilies. Feather stars can walk on the tips of their arms or swim by rhythmically beating their arms. About 600 living species of feather stars have been described.

ELEUTHEROZOANS. Most surviving echinoderms are members of the eleutherozoan lineage. The most familiar echinoderms are the sea stars (class Asteroidea; Figure 30.9b). Tube feet serve as organs of locomotion, and because their walls are thin, they are important sites for gas exchange. Each tube foot of a sea star is also a little adhesive organ consisting of an internal bulb connected by a muscular tube to an external sucker. The tube foot is moved by hydraulic expansion and contraction. When the bulb and the circular muscles of the tube contract, the tube foot elongates. The tube foot adheres to a surface by secreting a sticky substance around the sucker.

Many sea stars prey on polychaetes, gastropods, bivalves, and fishes. With hundreds of tube feet acting simultaneously, a sea star can exert an enormous and continuous force. It can grasp a clam in its arms, anchor the arms with tube feet, and, by steady contraction of the muscles in its arms, gradually exhaust the clam's muscles.

Sea stars that feed on bivalves are able to push their stomach out through their mouth and through the narrow space between the two halves of the clam's shell. The sea star's stomach then secretes enzymes into the soft parts of the bivalve, which digest the clam. Other sea star species feed on smaller prey or suspended particles and do not extrude their stomachs. Sea stars are important predators in many marine environments, such as coral reefs and rocky intertidal zones.

Brittle stars (class Ophiuroidea) are similar in structure to sea stars, but their flexible arms are composed of jointed hard plates (Figure 30.9c). Brittle stars can move by thrashing their arms, but burrowing forms and young individuals of most species also use their tube feet to move. Brittle stars generally have five arms, but each arm may branch several times. Most of the 2,000 species of brittle stars ingest particles from the surfaces of sediments and assimilate the organic material from them. They eject indigestible particles through their mouths, because, unlike most other echinoderms, brittle stars have only one opening to their digestive tract. Some brittle star species remove suspended food particles from the water; others capture small animals.

The remaining three classes of echinoderms lack arms. The sea daisies (class Concentricycloidea) were not discovered until 1986, and little is known about

(a) Feather star (species not identified)

(b) *Pentaceraster cumingi*

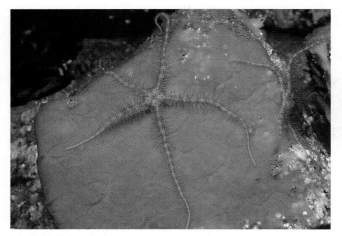

(c) *Ophiothrix spiculata*

30.9 A Diversity of Echinoderms (*a*) The flexible arms of these golden feather stars are clearly visible. (*b*) The cushion star is a typical five-armed sea star. Some species have many more arms. (*c*) This brittle star is resting on a sponge. (*d*) Purple sea urchins are important grazers of algae in the intertidal zone of the Pacific coast of North America. (*e*) This sea cucumber extends its feeding tentacles so food particles can adhere.

(d) *Strongylocentrotus purpuratus*

(e) *Parastichopus californicus*

them. They have tiny disc-shaped bodies with a ring of marginal spines but no arms. Sea daisies are the only echinoderms in which the water vascular system has two ring canals and in which the tube feet are arranged in a circle around the edge of the disc rather than along grooves radiating from the center. Sea daisies are found on rotting wood in deep ocean waters. They apparently feed on prokaryotes, which they digest and absorb either through a membrane that covers the oral surface or via a shallow, saclike stomach.

Sea urchins (class Echinoidea; Figure 30.9*d*) are armless, hemispherical animals that are covered with spines attached to the underlying skeleton via ball-

and-socket joints. Grooves run up the sides of the body toward the apex, instead of along the undersurfaces of arms as in sea stars. The spines of sea urchins come in varied sizes and shapes; a few produce highly toxic substances. Many sea urchins consume algae, scraping them from the rocks with a complex rasping structure. Others feed on small organic debris that they collect with their tube feet or spines.

Sea cucumbers (class Holothuroidea; Figure 30.9*e*) resemble stretched, flexible sea urchins that lack spines and have greatly reduced skeletal plates. Tube feet located on either side of five grooves along the body of the animal are used primarily for attaching to the substratum rather than for moving. Some species have tube feet only at the anterior end. The anterior tube feet are modified into large, feathery tentacles that can be protruded around the mouth. The tentacles are coated with a sticky substance to which prey or the surrounding substratum adhere. Periodically, a sea cucumber sticks its tentacles into its mouth, wipes them off, and then digests the adhered material.

Modified Lophophores: New Ways of Feeding

Evolution in the other major lineage of deuterostomes resulted in several different modifications of the lophophore and the coelomic cavity. These modifications, which include the enlargement of the proboscis and the development of pharyngeal gill slits, provided new ways of capturing and handling food. Some living representatives of this lineage, such as acorn worms, are wormlike animals that live buried in marine sands or muds, under rocks, or attached to algae. They are probably similar to the ancestors of this group of animals. Another lineage evolved the strikingly different body plan of the chordates.

Acorn worms capture prey with a proboscis

The acorn worms (phylum Hemichordata) have not changed much from the ancestral condition of animals in this lineage (Figure 30.10). They have a three-part body plan. The three regions of the body—proboscis, collar, and trunk—appear to be homologous to the prosome, mesosome, and metasome of lophophorates. In the hemichordate lineage, the lophophore was apparently lost and the proboscis grew larger and became a digging organ. The survivors of this lineage are the 70 species of acorn worms. These animals live in burrows in muddy and sandy sediments.

The enlarged proboscis is coated with a sticky mucus that traps prey items. The mucus and its attached prey are conveyed by ciliary action to the mouth. In the esophagus, the food-laden mucus is compacted into a ropelike mass that is moved through the digestive tract by ciliary action. Behind the mouth is a pharynx that opens to the outside through several

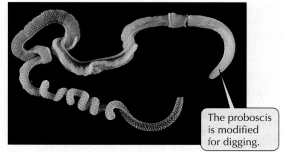

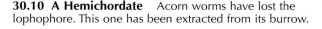

Saccoglossus kowalevskii

30.10 A Hemichordate Acorn worms have lost the lophophore. This one has been extracted from its burrow.

slits that allow water to exit. Highly vascularized tissue surrounding the slits serves as a gas exchange apparatus. An acorn worm breathes with the anterior end of its gut by pumping water into its mouth and out through its **pharyngeal slits**.

In chordates, the pharynx became a feeding device

The requirement for effective gas exchange—a large surface area—also serves well for capturing prey. In many lophophorates, the lophophore serves this dual function. The pharyngeal slits, which originally functioned as sites for gas exchange and eliminating water, as they do in modern acorn worms, were further enlarged in a sister lineage. This enlargement of the pharyngeal slits eventually led to remarkable evolutionary developments that gave rise to the chordates.

Members of the chordate lineage lost the lophophore and proboscis, replacing them with enlarged pharyngeal slits as a feeding device. The chordates (phylum Chordata) are bilaterally symmetrical animals that have pharyngeal slits at some stage in their development. The main shared ancestral features of their body plan are a dorsal, hollow nervous system; a ventral heart; and a tail that extends beyond the anus. All species have a dorsal supporting rod, the **notochord**, at some stage during their development (see Figure 30.12). In some species, the notochord is lost during metamorphosis to the adult stage. In other species, it is replaced by other skeletal structures that have the same function.

The tunicates (subphylum Urochordata) may be similar to the ancestors of all chordates. All 2,500 species of tunicates are marine animals, most of which are attached to a substratum as adults. The swimming, tadpolelike larvae reveal the close evolutionary relationships between tunicates and other chordates (see Figure 22.9).

In addition to its pharyngeal slits, a tunicate larva has a dorsal, hollow nerve cord and a notochord. Muscles attach to the notochord, which provides relatively rigid support. After a short time as a member of

30.11 Tunicates In a photograph these transparent "sea squirts" take on the blue of their environment. Pharyngeal baskets occupy most of the tunicate body cavity.

the plankton, the larva settles on the seafloor and transforms into a sessile adult that feeds by extracting plankton from the water with its pharynx, which is enlarged into a pharyngeal basket.

More than 90 percent of known species of tunicates are sea squirts (class Ascidiacea). Some sea squirts are solitary, but others produce colonies by asexual budding from a single founder. Individual sea squirts range in size from less than 1 mm to 60 cm, but colonies may measure several meters across. The baglike bodies of the adults are surrounded by a tough tunic, composed of protein and a complex polysaccharide, secreted by the epidermal cells. Much of the body is occupied by the large pharyngeal basket lined with cilia, whose beating moves water through the animal (Figure 30.11). The cilia also move the thin layer of mucus that lines the basket and to which the food particles adhere. Water enters the body through an anterior opening, passes through the pharyngeal basket into a chamber that is enclosed by the tunic, and out through another opening well removed from the site where the water entered.

The larvaceans (class Appendicularia) become reproductively mature and complete their life cycle as small planktonic organisms. They swim, filtering prey through screens made of mucopolysaccharides that collect small food particles. The larval form that functioned as a dispersal stage in their ancestors became a new lifestyle in these animals. There are only a few species of larvaceans, but they are widespread in the world's oceans.

The 25 species of lancelets (subphylum Cephalochordata) are small, fishlike animals that rarely exceed 5 cm in length. With their notochord, extending the entire length of the body throughout their lives, and their pharyngeal baskets, they resemble small fishes. Lancelets live partly buried in soft marine sediments worldwide, and they extract small prey from the water with their pharyngeal baskets (Figure 30.12). They may be living descendants of the ancestors of the vertebrates, another group of chordates.

The Origin of the Vertebrates: Sucking Mud

In one chordate lineage, which gave rise to the **vertebrates** (subphylum Vertebrata), the pharyngeal basket became enlarged. With its many exit openings, the enlarged basket was effective in extracting prey from mud, where many inedible particles are ingested along with the food. In the late Cambrian period, more than 500 mya, these early vertebrates evolved improved structures for extracting food from mud and sand and for swimming on the surface of the substratum.

A jointed, dorsal **vertebral column** replaced the notochord as the primary support in these chordates. Attached to the vertebral column, which gives the vertebrates their name, are two pairs of appendages. The faster locomotion made possible by these appendages favored the evolution of an anterior skull with a large brain (Figure 30.13). Vertebrates have a large coelom in which the body organs are slung, but the coelom serves neither as a hydrostatic skeleton nor as a gas exchange structure. Instead, a well-developed circulatory system, driven by contractions of a ventral heart, delivers oxygen to internal organs.

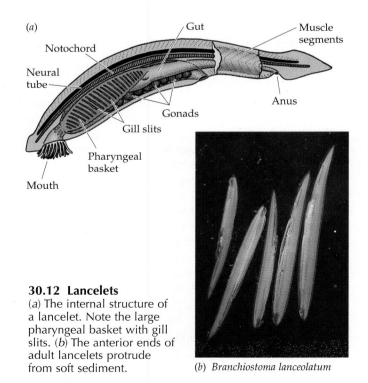

30.12 Lancelets
(a) The internal structure of a lancelet. Note the large pharyngeal basket with gill slits. (b) The anterior ends of adult lancelets protrude from soft sediment.

(b) *Branchiostoma lanceolatum*

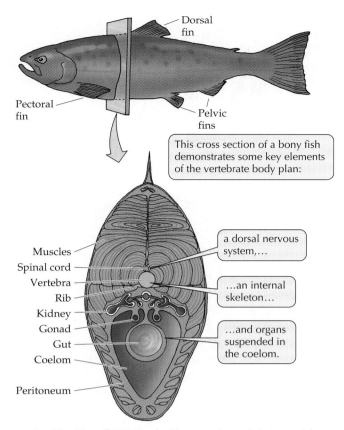

This cross section of a bony fish demonstrates some key elements of the vertebrate body plan:

Dorsal fin

Pectoral fin

Pelvic fins

Muscles
Spinal cord
Vertebra
Rib
Kidney
Gonad
Gut
Coelom
Peritoneum

a dorsal nervous system,…

…an internal skeleton…

…and organs suspended in the coelom.

30.13 The Vertebrate Body Plan A bony fish is used here to illustrate the elements common to all vertebrates.

The unique ability of living fishes to maintain nearly constant internal salt concentrations when environmental salinities change (see Chapter 37) leads many biologists to believe that vertebrates evolved in estuaries. Salinities change frequently in estuaries, and animals able to maintain relatively constant internal salt concentrations under those conditions would have been able to exploit environments from which other animals were excluded.

According to this view, filter-feeding ancestors of vertebrates evolved the ability to live in es-

30.14 A Probable Vertebrate Phylogeny This phylogeny incorporates the view that vertebrates evolved in estuaries, where their ability to handle varying salinities allowed them to exploit habitats not available to marine animals.

tuaries and mouths of rivers, where they escaped from their predators and competitors, most of which could not handle varying salinities. The early vertebrates fed on microscopic organisms, and after growing for a while in estuaries, they could have returned to the oceans larger and better able to compete and defend themselves. They probably did not venture into rivers, because at that time rivers and terrestrial environments lacked multicellular plants or animals on which the early vertebrates could feed.

The first vertebrates lacked jaws. These jawless fishes (class Agnatha) probably swam over the bottom, sucking mud and extracting microscopic food from it as they moved. The lineage leading to modern hagfishes separated first from other groups and returned to live entirely in the sea (Figure 30.14). Hagfishes lack the osmoregulatory mechanisms found in all other fishes, suggesting that they have had a long period of life in the sea.

Other early jawless fishes probably continued to live primarily in estuaries. One group, called ostracoderms, meaning "shell-skinned," evolved a bony external armor. With their heavy armor, these small fishes could swim only slowly, but swimming above the substratum was easier than having to burrow through it, as all previous sediment feeders had done.

This new mobility enabled vertebrates to exploit their environments in new ways. One of those ways was to attach to dead, rotting flesh and use the pharynx to create a suction to pull fluids and partly decomposed tissues into the mouth. Hagfishes and lampreys, the only jawless fishes to survive beyond the Devonian period, feed in this way (Figure 30.15). These fishes have tough scaly skins instead of external armor. Hagfishes ingest the tissues of dead animals; most

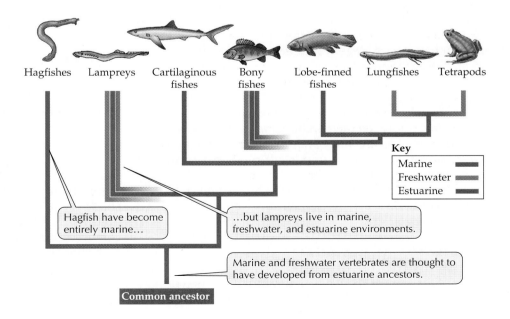

Hagfishes Lampreys Cartilaginous fishes Bony fishes Lobe-finned fishes Lungfishes Tetrapods

Key
Marine
Freshwater
Estuarine

Hagfish have become entirely marine…

…but lampreys live in marine, freshwater, and estuarine environments.

Marine and freshwater vertebrates are thought to have developed from estuarine ancestors.

Common ancestor

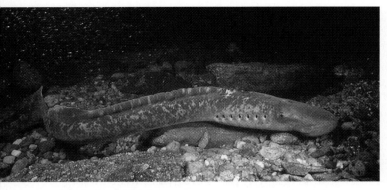

Petromyzon marinus

30.15 An Agnathan This sea lamprey has a large, jaw-less mouth, which it uses to suck blood and flesh from other fishes.

adult lampreys suck the blood of living fishes or eat the flesh of dying fishes. The round mouth is a sucking organ with which the animals attach to their prey and rasp at the flesh. Lampreys live in both fresh and salt water; many species move between the two environments, spawning in rivers and maturing in the sea.

Jaws were a key evolutionary novelty

During the Devonian period, about 400 mya, many new kinds of fishes evolved in the seas, estuaries, and fresh waters. Although most of these were jawless, members of one lineage evolved jaws from some of the skeletal arches that supported the gill region (Figure 30.16). A jaw allows a fish to grasp and subdue relatively large, living prey. Further development of the jaws and teeth among fishes led to the ability to chew both soft and hard body parts of prey. Chewing aided chemical digestion and improved nutrition obtained from prey. Although many intermediates must have existed between jawless fishes and the fully jawed ancestors of modern fishes, it is not difficult to imagine how each stage would have functioned better than those that preceded it.

The most important early jawed fishes were the heavily armored placoderms (class Placodermi). Some of these fishes evolved elaborate fins and sleek body forms that improved their ability to maneuver in open water. A few became huge and, together with squids, were probably the most important predators in the Devonian oceans. Despite their early abundance, however, most placoderms disappeared by the end of the Devonian period, 345 mya; none survived to the end of the Paleozoic era.

Cartilaginous and bony fishes evolved fins to control motion

Two other groups of fishes that survive today—the cartilaginous fishes and the bony fishes—became nu-

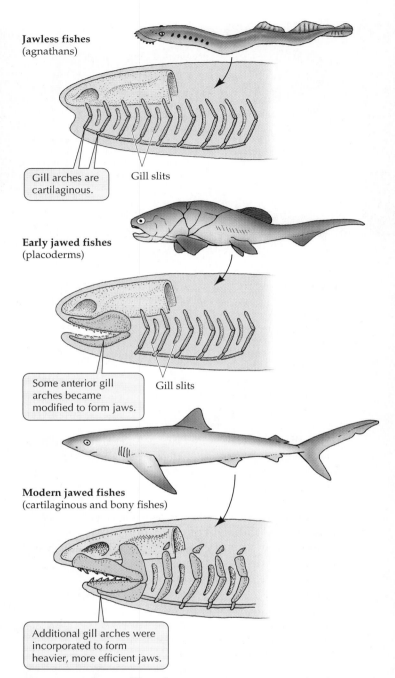

30.16 Jaws from Gill Arches A probable scenario for the evolution of jaws from the anterior gill arches of fishes.

merically important during the Devonian period. The sharks, skates and rays, and chimaeras—all cartilaginous fishes (class Chondrichthyes)—have a skeleton composed entirely of a firm but pliable material called **cartilage** (Figure 30.17). Their skin is flexible and leathery, sometimes bearing bristly projections that give it the consistency of sandpaper. The loss of external armor, which increased mobility and ability to es-

(a) *Triaenodon obesus*

(b) *Trygon pastinaca*

30.17 Cartilaginous Fishes (a) Most sharks, such as this Pacific white tip reef shark, are active predators living in open waters. (b) Skates and rays, represented here by a stingray, have their mouths on the ventral surface of their body and feed on the ocean bottom.

cape from predators, was accompanied by the evolution of rapid swimming.

Control of swimming is provided by pairs of unjointed fins: a pair of pectoral fins just behind the gill slits and a pair of pelvic fins just in front of the anal region. A dorsal median fin stabilizes the fish as it moves (see Figure 30.13). Sharks move forward by means of their tail and pelvic fins. Skates and rays propel themselves by means of the undulating movements of their greatly enlarged pectoral fins.

Most sharks are predators, but some feed by straining plankton. The world's largest fish, the whale shark (*Rhincodon typhus*), which may grow to more than 15 m long and weigh more than 9,000 kg, feeds on plankton. Most skates and rays live on the ocean floor, where they feed on mollusks and other invertebrates buried in the sediments. Chimaeras feed on mollusks, whose shells they crack with their hard, flat teeth. Nearly all cartilaginous fishes live in the oceans.

Bony fishes evolved buoyancy

Fishes in another lineage that evolved in estuaries have internal skeletons of bone rather than cartilage, giving them their common name, bony fishes (class Osteichthyes) (see Figure 30.14). The pharyngeal slits in bony fishes open into a single chamber covered by a hard flap. Movement of the flap improves the flow of water over the gills and brings more oxygen in contact with the gas exchange surfaces.

Early bony fishes also evolved lunglike sacs that supplemented the gills in respiration. These features enabled these fishes to live where oxygen was periodically in short supply, as it often is in estuarine and freshwater environments. In most fishes, the lunglike sacs evolved into larger **swim bladders**, which serve as organs of buoyancy that help keep the fish suspended in water. By adjusting the amount of gas in its swim bladder, a fish can control the depth in the water column at which it is stable. Lungfishes are the only fishes that still use their lunglike sacs for respiration.

The external armor of bony fishes is greatly reduced, but most species are covered with flat, smooth, thin, lightweight scales that provide some protection. With their light skeletons and their swim bladders, some bony fishes recolonized the seas to become major components of marine ecological communities.

The more than 20,000 species of bony fishes have a remarkable diversity of sizes, shapes, and lifestyles (Figure 30.18). The smallest bony fish is a goby that is only 1 cm long as an adult. The largest are ocean sunfishes that weigh as much as 900 kg. Bony fishes exploit nearly all types of aquatic food sources. In the oceans they filter plankton from the water, rasp algae from rocks, eat corals and other colonial invertebrates, dig invertebrates from sediments, and prey on all other vertebrates except large whales and dolphins. In fresh water they eat plankton, devour insects of all aquatic orders, consume fruits that fall into the water in flooded forests, and prey on other aquatic vertebrates.

Many bony fishes live buried in soft sediments, where they capture passing prey or from which they emerge at night to feed in the water column above. Many are solitary, but in open water others form large aggregations called schools. Many fishes perform complicated behaviors by means of which they maintain schools, build nests, court and choose mates, and

(a) *Acipenser* sp.

(c) *Salmo gairdneri*

30.18 A Diversity of Bony Fishes (a) Sturgeon are survivors of an ancient lineage of bony fishes. (b) Coral reef fishes such as this Australian sweetlips are usually colorful and
small. (c) These rainbow trout are members of a commercially important group, the salmonids. (d) This wolf eel lacks
pelvic and dorsal fins but has well developed pectoral fins.

(b) Sweetlips (species not identified)

(d) *Anarrhichthys ocellatus*

care for their young. They are a group in which behavior has stimulated many evolutionary changes.

With their fins and swim bladders, fishes can readily control their position in open water, but their eggs
tend to sink. Therefore, most fishes attach their eggs to
a substratum. A few species, however, discharge their
very small eggs directly into surface waters, where
they are buoyant enough to complete their development before they sink very far. Most marine fishes
move to food-rich shallow waters to lay their eggs,
which is why coastal waters and estuaries are so important in the life cycles of many species of fishes.
Some fishes, such as salmon, ascend rivers to spawn in
freshwater streams and lakes. Conversely, some
species, such as eels, that live most of their lives in
fresh water, migrate to the sea to lay their eggs there.

Colonizing the Land

Although the evolution of lunglike sacs in early bony
fishes was a response to the inadequacy of gill respiration in oxygen-poor waters, it also set the stage for the
invasion of land by some of their descendants. Early
bony fishes probably used their lungs to supplement
their gills when oxygen levels in the water were low.
This ability would also have allowed them to leave the
water temporarily when pursued by predators unable
to breathe air. But with their unjointed fins, bony
fishes could only flop around on land as most fishes
do today if placed out of water.

The evolution of joints in their fins enabled fishes to
move over land to find new bodies of water when
those in which they had been living dried up or became overpopulated. Descendants of these fishes
began to use terrestrial food sources and become more
fully adapted to life on land.

A group of bony fishes thought to be similar to the
fishes that invaded the land are the lobe-finned fishes
(subclass Crossopterygii). Lobe-fins flourished from
the Devonian period (400 mya) until about 25 mya,
when they were thought to have become extinct.
However, in 1939 a lobe-fin was caught by a commercial fisherman off the east coast of Africa. Since that
time, several dozen specimens of this extraordinary
fish, *Latimeria chalumnae*, have been collected.

30.19 A Modern Lobe-Fin *Latimeria chalumnae* is the sole survivor of a lineage that was thought to be extinct.

Latimeria, a predator on other fishes, reaches a length of about 1.5 m and weighs up to 82 kg (Figure 30.19). The skeleton of *Latimeria* is composed mostly of cartilage rather than bone, and because it lives in very deep water, its swim bladder contains fat, which is less compressible than the gases found in the swim bladders of the early crossopterygians.

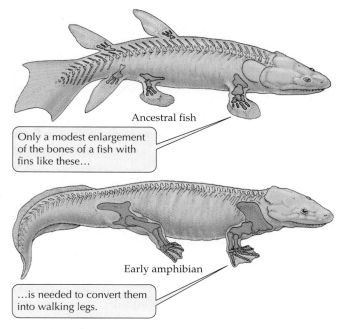

Only a modest enlargement of the bones of a fish with fins like these…

Ancestral fish

Early amphibian

…is needed to convert them into walking legs.

30.20 Legs from Fins The evolutionary step from lobe-fin fish to amphibian is a fairly small one.

Amphibians were the first invaders of the land

During the Devonian period (400 to 345 mya), amphibians (class Amphibia) arose from an ancestor they shared with lungfishes (see Figure 30.14). These ancestral fishes had thin skins through which they exchanged gases, and stubby, jointed fins that evolved into walking legs (Figure 30.20). The design of these legs has remained largely unchanged throughout the evolution of terrestrial vertebrates.

Devonian predecessors of amphibians were the first *tetrapods*—animals with two pairs of limbs. They gradually evolved to be able to live on swampy land and, eventually, on drier land. They were probably able to crawl from one pond or stream to another by pulling themselves along on their finlike legs, as do some modern species of catfishes and other fishes. Living amphibians have relatively small lungs, and most species still exchange gases through their skin.

About 4,500 species of amphibians live on Earth today, many fewer than those known only from fossils. The living amphibians belong to three orders (Figure 30.21): the worm-

(a) *Dermophus mexicanus*

(b) *Scaphiophryne gottlebei*

(c) *Salamandra salamandra*

30.21 Amphibians (a) Burrowing caecilians superficially look more like worms than amphibians. (b) A rare frog species discovered in a national park on the island of Madagascar. (c) A European fire salamander.

like, tropical, burrowing caecilians (order Gymnophiona); frogs and toads (order Anura, "tail-less"); and salamanders (order Urodela or Caudata, "tailed"). Most species of frogs and toads live in tropical and warm temperate regions, although a few are found at very high latitudes. Salamanders are more diverse in temperate regions, but many species are found in cool, moist environments in the mountains of Central America.

Most amphibians live in water at some time in their lives. In the typical life cycle, an amphibian spends part or all of its adult life on land, usually in a moist habitat, but adults return to fresh water to lay their eggs (Figure 30.22). An amphibian egg must remain moist because it is surrounded by a delicate envelope through which it loses water readily if its surroundings are dry. The egg of most species gives rise to a larva that lives in water until it metamorphoses into a terrestrial adult.

There are interesting variations on this life cycle. Some amphibians are entirely aquatic, never leaving the water. Others are entirely terrestrial, laying their eggs in moist places on land. Many lungless salamanders live in rotting logs or moist soil and exchange gases entirely through their skin and mouth lining. Terrestrial species are confined to moist environments because most amphibians rapidly lose water through their skins when exposed to dry air, but some toads have tough skins that enable them to live for long periods of time in dry places.

Amphibians are the focus of much attention today because populations of many species are declining rapidly. For example, the golden toad has disappeared from the Monteverde Cloud Forest Reserve in Costa Rica, which was established primarily to protect this rare species. The reasons for the declines are not known, but biologists are monitoring amphibian populations closely to learn more about the causes of their difficulties and to determine the implications of amphibian declines for other organisms, including humans.

30.22 In and Out of the Water   Most stages in the life cycle of many temperate-zone frogs live in the water. (Many tropical frogs, by contrast, are fully terrestrial and lay their eggs in wet places on land.)

Reptiles colonized dry environments

Two morphological changes allowed one lineage of vertebrates to control water loss and therefore to exploit the full range of terrestrial habitats. The first was an egg with a shell that is relatively impermeable to water and can be laid in dry places. The second was a combination of traits that included a tough skin impermeable to water and kidneys that could excrete concentrated urine.

The vertebrates that evolved both of these morphological changes are called *amniotes*. They were the first vertebrates to become common over much of the terrestrial surface of Earth. The amniote egg has a leathery or brittle calcium-impregnated shell that retards evaporation of the fluids inside but permits O_2 and CO_2 to pass. Within the shell and surrounding the embryo are membranes that protect the embryo from desiccation and assist its excretion and respiration. The egg also supplies the embryo with large quantities of food—yolk—that permit it to attain a relatively advanced state of development before it hatches and must feed itself (Figure 30.23).

An early amniote lineage, the reptiles (class Reptilia) arose from early tetrapods in the Carboniferous period, some 300 mya. About 6,000 species of reptiles live today. Most reptiles do not care for their eggs after laying them. In some species eggs do not develop shells, but are retained inside the female's body until they hatch. Still other species evolved placentas that nourish the developing embryos (see Chapter 39).

The skin of a reptile, which is covered with horny scales that greatly reduce loss of water from the body surface, is unavailable as an organ of gas exchange. In reptiles gases are exchanged almost entirely by the lungs, which are proportionally much larger in surface area than those of amphibians. A reptile forces air into and out of its lungs by bellowslike movements of its ribs. The reptilian heart is divided into chambers that separate oxygenated from unoxygenated blood. With this heart, reptiles can generate higher blood pressures than amphibians, and can sustain higher levels of muscular activity, although they tire much more rapidly than do birds or mammals.

MODERN REPTILES ARE MEMBERS OF THREE LINEAGES. Only three of the many reptilian lineages survive today. Turtles and tortoises (subclass Chelonia) have an armor of dorsal and ventral bony plates that form a shell into which the head and limbs can be withdrawn (Figure 30.24*a*). Most turtles live in lakes and ponds, but tortoises are terrestrial, and sea turtles spend their entire lives at sea except when they come ashore to lay eggs. Most turtles and tortoises are herbivores that eat a variety of aquatic and terrestrial plants, but some species are strongly carnivorous.

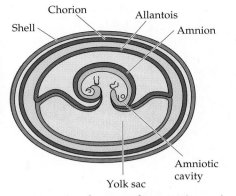

30.23 An Egg for Dry Places The evolution of the amniote egg, with its shell, extraembryonic membranes (amnion, chorion, and allantois), and embryo-nourishing yolk sac, was a major step in colonization of the terrestrial environment.

The only survivors of the reptilian subclass Sphenodontida are the two species of tuataras restricted to a few islands off the coast of New Zealand (Figure 30.24*b*). Tuataras superficially resemble lizards, but they differ in several internal anatomical features.

The third reptilian subclass, Squamata, includes the lizards, amphisbaenians—a group of legless, wormlike, burrowing animals with greatly reduced eyes—and snakes (Figure 30.24*c* and *d*). Most lizards are insectivores, but some are herbivores, and still others prey on other vertebrates. The largest lizards, growing as long as 3 m, are some species of monitors that live in the East Indies. Most lizards walk on four limbs, but some are limbless, as are all snakes.

Recent phylogenetic analyses show that snakes are monophyletic and that they evolved from burrowing lizards. All snakes are carnivores that can swallow objects much larger than their own diameter. Three groups of snakes evolved poison glands and inject venom into their prey with their teeth. The largest snakes—the pythons—are more than 10 m long.

Surviving Dinosaur Lineages

During the Mesozoic era (245 to 66 mya) one amniote lineage, the thecodonts (class Archosauria) split from other reptiles and underwent an extraordinary evolutionary diversification. One thecodont lineage gave rise to crocodilians (subclass Crocodylia)—crocodiles, caimans, gharials, and alligators—which are confined to tropical and warm temperate environments (Figure 30.25).

Crocodilians spend much of their time in water, but they build nests on land or on floating piles of vegetation. The eggs, which are warmed by heat generated by the decaying organic matter of the nest, typically are tended by the female until they hatch. All crocodil-

(a) *Chelonia mydas*

(b) *Sphenodon punctatus*

(c) *Chamaeleo* sp.

(d) *Tropidolaemus* sp.

30.24 Reptilian Diversity (a) The green sea turtle is widely distributed in tropical oceans. (b) The tuatara looks like a typical lizard, but it is one of only two survivors of a lineage that separated from lizards long ago. (c) The carpet chameleon from Madagascar has a long tail with which it can grasp branches and large eyes that move independently in their sockets. (d) The McGregor's viper is a tree snake found in the Philippines.

(a) *Alligator mississippiensis*

(b) *Crocodylus niloticus*

30.25 Crocodilians (a) Most crocodilians are tropical, but alligators live in warm temperate environments in Asia and, like this one, in the southeastern United States. (b) A Nile crocodile photographed in the Serengeti of Tanzania.

ians are carnivorous; they prey on vertebrates of all classes, including large mammals.

Another thecodont lineage led to the dinosaurs (subclass Dinosauria), the prevalent large terrestrial reptiles for millions of years and the largest terrestrial animals ever to inhabit Earth. Some of the largest dinosaurs weighed up to 100 tons, and many were agile and fast-moving (see Figure 19.15). The ability to move actively on land was not achieved easily. The first terrestrial vertebrates probably moved only very slowly, much more slowly than their aquatic relatives. The reason is that they apparently could not walk and breathe at the same time.

Not until the evolution of the lineages leading to the mammals, dinosaurs, and birds did the legs assume more vertical positions, which reduced the lateral forces on the body during locomotion. Special ventilatory muscles that enabled the lungs to be filled and emptied while the limbs moved also evolved. These muscles are visible in living birds and mammals. We can infer their existence in dinosaurs from the structure of the vertebral column and the capacity of many dinosaurs for bounding, bipedal (two-legged) locomotion.

The ability to breathe and run simultaneously, which we take for granted, was a major innovation in the evolution of terrestrial vertebrates. This capacity enabled its bearers to maintain steady, high levels of activity, which generated enough heat to result in high body temperatures.

Ancestral birds evolved feathers

During the Mesozoic era, about 175 mya, another thecodont lineage gave rise to the birds (subclass Aves). Fossils of the earliest birds have not yet been found. The oldest known avian fossil, *Archaeopteryx*, was covered with feathers that are almost identical to those of modern birds, and had well-developed wings and a long tail (Figure 30.26). It also had a wishbone, which in modern birds is an anchoring site for flight muscles. *Archaeopteryx* had typical perching bird claws, suggesting that it lived in trees and shrubs and used the clawed fingers on its forearms to assist it in clambering over branches. Thus, *Archaeopteryx* was already a highly evolved bird 150 mya.

During the Cretaceous period (138 to 66 mya), birds underwent an extensive evolutionary radiation. The dominant Cretaceous lineage was the "opposite birds," so named because the tarsal bones of their legs fused in the opposite direction from the way in which fusion happens in all modern birds. All lineages of opposite birds died out at the end of the Cretaceous, but paleontologists disagree over how many other avian lineages survived the mass extinction. Some believe, on the basis of the fossil record, that members of only one lineage, collectively known as the transitional

30.26 A Mesozoic Bird An artist's recreation of *Archaeopteryx* shows its modern feathers and arboreal habits.

shorebirds, survived (Figure 30.27*a*). Others believe, on the basis of molecular data, that at least some representatives of many Cretaceous avian lineages were not exterminated (Figure 30.27*b*).

The single most characteristic feature of birds is their feathers, which are highly modified scales. The flying surface of the wing is created by large quills that arise from the forearm and the reduced, stubby fingers. Other strong feathers sprout like a fan from the shortened tail and serve as stabilizers during flight. The contour feathers and down feathers, which arise from well-defined tracts, cover the body like a garment and provide insulation to control loss of body heat.

The light but strong bones of birds are hollow and have internal struts for strength. The sternum (breastbone) forms a large, vertical keel to which the breast muscles are attached. These muscles pull the wings downward during the main propulsive movement in flight. Flight is metabolically expensive, and a flying bird consumes energy at a very high rate.

Because birds have high metabolic rates, they generate large amounts of heat. They control the rate of heat loss using their feathers, which may be held close

30.27 The Evolution of Birds
(*a*) Some paleontologists believe that modern birds are descendants of a small group that survived the mass extinction at the end of the Cretaceous. (*b*) Others believe that many lineages of birds persisted during the transition from the Cretaceous to the Tertiary.

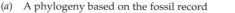
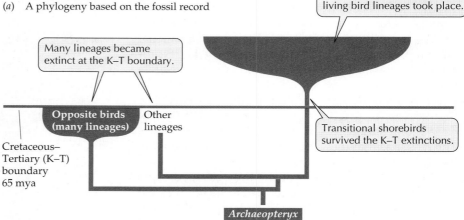

(*a*) A phylogeny based on the fossil record

Explosive radiation of all other living bird lineages took place.

Many lineages became extinct at the K–T boundary.

Opposite birds (many lineages)

Other lineages

Cretaceous–Tertiary (K–T) boundary 65 mya

Transitional shorebirds survived the K–T extinctions.

Archaeopteryx

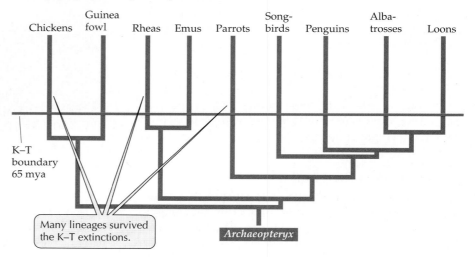

(*b*) A phylogeny based primarily on molecular data

Chickens Guinea fowl Rheas Emus Parrots Song-birds Penguins Alba-trosses Loons

K–T boundary 65 mya

Many lineages survived the K–T extinctions.

Archaeopteryx

to the body or elevated to alter the amount of air trapped as insulation. A bird's metabolic rate is so high that it consumes about eight times the amount of energy per day as a lizard of the same weight. The brain of a bird is relatively large in proportion to its body size, primarily because the cerebellum, the center of sight and muscular coordination, is enlarged. The beaks of modern birds lack teeth.

Most birds lay their eggs in a nest, where they are incubated by the body heat of an adult. Because birds have high body temperatures, the eggs of most species hatch in about 2 weeks. Nestlings of many species hatch at a relatively helpless stage and are fed for some time by their parents. Young of other species can feed themselves shortly after hatching. Adults of all species attend their offspring for some period of time, warning them of and protecting them from predators, guiding them to good foraging places, protecting them from bad weather, and feeding them.

As a group, birds eat almost all types of animal and plant material. A few aquatic species have bills modified for filtering small food particles from the water. In terrestrial environments, insects are the most important dietary item for birds. In addition, birds eat fruits and seeds, nectar and pollen, leaves and buds, carrion, and other vertebrates. Birds are major predators of flying insects during the day, and some species exploit that food source at night. By eating the fruits and seeds of vascular plants, birds serve as major agents of seed dispersal.

As adults, birds range in size from the 2-g bee hummingbird of the West Indies to 150-kg ostriches. Some flightless birds of Madagascar and New Zealand were even larger, but they were exterminated by early hu-mans when the humans first reached those islands. Although there are about 8,600 species of living birds, more species than in any other vertebrate group except fishes, birds are less diverse structurally than are other vertebrates, probably because of the constraints imposed by flying (Figure 30.28).

The Tertiary: Mammals Diversify

Mammals (the chordate class Mammalia) appeared in the early part of the Mesozoic era (about 225 mya), branching from the now extinct, mammal-like reptiles known as therapsids. Small mammals coexisted with reptiles and dinosaurs for at least 150 million years, but when the large reptiles and dinosaurs disappeared during the mass extinction at the close of the Mesozoic era, mammals increased dramatically in number, diversity, and size.

(a) *Aptenodytes patagonicus*

(b) *Eclectus roratus*

(c) *Cardinalis cardinalis*

30.28 A Diversity of Birds (a) Penguins such as these king penguins are adapted to the harsh Antarctic environment. They are expert swimmers, although they do not fly. (b) Parrots are a diverse group of birds, especially in the tropics of Asia and the Pacific islands. This eclectus parrot is native to Australia and New Guinea. (c) The perching birds are found worldwide. This female northern cardinal is a common example from eastern North America. (d) The turkey vulture, the most widespread New World vulture, glides on updrafts in search of carrion.

(d) *Cathartes aura*

Mammals acquired simplified skeletons

Skeletal simplification accompanied the evolution of mammals from their therapsid ancestors. During mammalian evolution, most lower-jaw bones became incorporated into the middle ear, leaving a single bone in the lower jaw, and the number of bones in the skull decreased. The bulk of both the limbs and the bony girdles from which they are suspended was reduced, and the limbs became oriented beneath the body, as they are in dinosaurs and birds, rather than poking out to the side and then down, as in reptiles. Fossils of later therapsids suggest that their legs also were positioned underneath the body. Thus the early mammals represent a continuation of changes that were already under way (Figure 30.29).

The skeletal features we have been discussing are readily fossilized, but the important soft parts of mammals are seldom preserved in fossils. Mammalian features such as mammary glands, sweat glands, hair, and a four-chambered heart may have evolved among the later therapsids, but the existing record does not tell us when this happened.

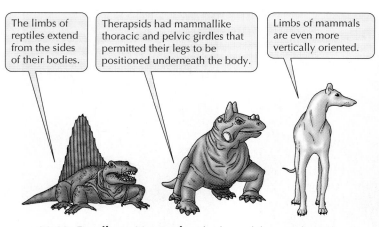

> The limbs of reptiles extend from the sides of their bodies.

> Therapsids had mammallike thoracic and pelvic girdles that permitted their legs to be positioned underneath the body.

> Limbs of mammals are even more vertically oriented.

30.29 Reptile to Mammal The legs of therapsid reptiles gradually became positioned under the body; early mammals represent a continuation of this evolutionary trend.

Ornithorhyncus anatinus

30.30 A Monotreme The Australian duck-billed platypus is one of three surviving species of monotremes.

Mammals are unique among animals in that they suckle their young with a nutritive fluid (milk) secreted by mammary glands. Mammalian eggs are fertilized within the female's body, and the embryos develop somewhat within the uterus before being born. In addition, mammals have a protective and insulating covering of hair, which is luxuriant in some species but has been largely lost in adult whales and dolphins in favor of thick layers of insulating fat (blubber).

Mammals have far fewer, but more highly differentiated, teeth than those of reptiles. Differences in the number, type, and arrangement of teeth in mammals reflect their varied diets. Mammals range in size from tiny shrews weighing only about 2 g to the blue whale, which measures up to 31 m long and weighs up to 160,000 kg—the largest animal ever to live on Earth.

Mammals have varied reproductive patterns

The approximately 4,000 species of living mammals are placed into three major groups: prototherians, marsupials, and eutherians. The subclass Prototheria contains a single order, the Monotremata, represented by two families and a total of three species, which are found only in Australia and New Guinea. These mammals, the duck-billed platypus and spiny anteaters, or echidnas, differ from other mammals in that they lay eggs and have some reptilelike anatomical features (Figure 30.30). Monotremes nurse their young on milk, but they have no nipples on their mammary glands; rather the milk simply oozes out and is lapped off the fur by the offspring.

The other two groups of mammals are members of the subclass Theria. Females of one group, the order Marsupialia, which contains about 240 species, have a ventral pouch in which they carry and feed their young (Figure 30.31). Gestation (pregnancy) in marsupials is short; the young are born tiny but with well-developed forelimbs, with which they climb to the pouch. Once her offspring has left the uterus, a female marsupial may become sexually receptive again. She can then

(a) Macropus rufus *(b) Caluromys phicander*

30.31 Marsupials (a) Australia's kangaroos are perhaps the most familiar marsupials. This female red kangaroo is carrying her offspring in her distinctive pouch. (b) The wooly opossum is a South American marsupial from Guyana.

(a) Citellus parryi

(c) Felis onca

(b) Eptesicus fuscus

30.32 Eutherian Diversity (*a*) The Arctic ground squirrel is one of many species of small, diurnal rodents of western North America. (*b*) With their powers of echolocation (see Chapter 42), many bats can locate and capture prey even in complete darkness. This big brown bat is about to capture a large moth. (*c*) Cats ambush their prey and capture them after short chases. The jaguar, whose coloration camouflages it in the foliage, is the largest cat in the Americas. (*d*) Large hoofed mammals are important herbivores over much of Earth. This caribou bull is grazing by himself, although caribou are often seen in huge herds (see Figure 51.15*b*).

(d) Rangifer tardanus

carry a fertilized egg capable of initiating development and replacing the offspring in the pouch should something happen to it. The marsupial mode of reproduction functions well in varying arid environments, where adults must travel long distances to find food and where droughts may cause young offspring to die.

At one time marsupials were widely distributed on Earth, but today most species are restricted to the Australian region, with a modest representation in South America (Figure 30.31*a*). One species, the Virginia opossum, is widely distributed in the United States. Marsupials radiated into terrestrial herbivores, insectivores, and carnivores, but no species are marine or can fly. The largest living marsupial is the red kangaroo of Australia (Figure 30.31*b*), which weighs up to 90 kg, but much larger marsupials existed in Australia until quite recent times. These large marsupials were probably exterminated by humans soon after they reached Australia about 50,000 years ago.

Most living mammals are eutherians (sometimes called placentals, but this is not a good name because some marsupials also have placentas). Eutherians are more highly developed at birth than are marsupials, and no external pouch houses them after birth. The nearly 4,000 species of eutherians are placed into 16 major groups (Figure 30.32), the largest of which is the rodents, with about 1,700 species. The next-largest group, the bats, has about 850 species, followed by the insectivores (moles and shrews), with slightly more than 400 species.

Several lineages of terrestrial mammals subsequently colonized marine environments to become whales, dolphins, seals, and sea lions. Today the largest mammals are marine, but some terrestrial mammals, such as elephants and rhinoceroses, weigh more than several thousand kilograms.

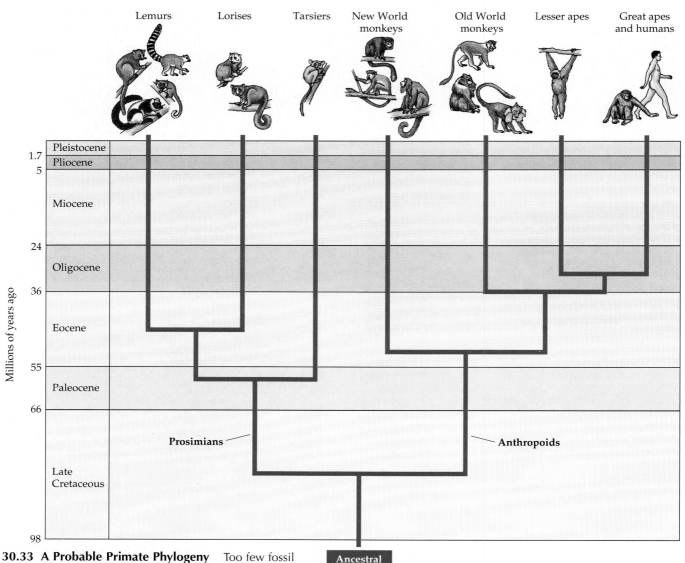

30.33 A Probable Primate Phylogeny Too few fossil primates have been discovered to reveal with certainty their evolutionary relationships, but this phylogenetic tree is consistent with existing evidence.

Eutherians are extremely varied in form and ecology. They are, or were until recently, the most important grazers and browsers in most terrestrial ecosystems. Grazing and browsing by eutherian mammals has been an evolutionary force intense enough to select for the spines, tough leaves, and difficult-to-eat growth forms found on many plants.

Humans Evolve: Earth's Face Is Changed

The primate lineage, to which humans belong, has undergone extensive recent evolutionary radiation. Primates probably descended from small, arboreal (tree-inhabiting) insectivores sometime during the Cretaceous period. The major traits that distinguish primates from other mammals are all adaptations to arboreal life. They include dexterous hands with opposable thumbs that can grasp branches and manipulate food, nails rather than claws, eyes on the front of the face that provide good depth perception, and very small litters (usually one) of offspring that receive extended parental care.

The primate lineage split into two main branches—prosimians and anthropoids—early in its evolutionary history (Figure 30.33). The prosimians—lemurs, tarsiers, pottos, and lorises (Figure 30.34)—once lived on all continents, but today they are restricted to Africa, tropical Asia, and Madagascar. All mainland species are arboreal and nocturnal, but on Madagascar, the site of a remarkable prosimian radiation, there are also diurnal and terrestrial species. Until the recent arrival of *Homo sapiens*, there were no other primates on Madagascar.

(a) *Propithecus verreauxi*

(b) *Tarsius syrichta*

(c) *Nycticebus coucang*

30.34 Prosimians (a) The sifaka lemur is one of many lemur species of Madagascar, where it is part of a unique assemblage of plants and animals. (b) An inhabitant of the rainforests of the Philippines, the tarsier seems other-worldly to our eyes. (c) The slow loris lives in southeast Asia.

The anthropoids—monkeys, apes, and humans (see Figure 30.33)—evolved from an early primate stock about 55 mya in Africa or Asia. New World monkeys have been evolving separately from Old World monkeys long enough that they could have reached South America from Africa when those two continents were still close to one another. Perhaps because tropical America has been heavily forested for a long time, all New World monkeys are arboreal (Figure 30.35a). Many of them have long, prehensile tails with which

(a) *Leontopithecus rosalia*

30.35 Monkeys (a) Golden lion tamarins are endangered New World monkeys living in coastal Brazilian rainforests. (b) Many Old World species, such as these Barbary macaques, live and travel in social groups. Here two members of the group groom each other.

(b) *Macaca sylvanus*

(a) *Hylobates lar*

(b) *Pan troglodytes*

(c) *Pongo pygmaeus*

(d) *Gorilla gorilla*

30.36 Apes (a) The gibbons are the smallest of the apes. This common gibbon is found in Asia, from India to Borneo. (b) A mother chimpanzee carries her offspring on her back. Her "knuckle walk" position is characteristic of apes when they move on the ground. (c) Intelligent and endangered, orangutans face massive habitat destruction in their native Indonesia. (d) This mother also demonstrates the knuckle walk of lowland gorillas in Africa.

they can grasp branches. Many Old World primates are arboreal, but a few species are terrestrial, some of which, such as baboons and macaques, live and travel in large groups (Figure 30.35b). No Old World primates have prehensile tails.

About 35 mya the lineage leading to modern apes separated from other Old World primates. The first apes were arboreal, but some species came to live in drier habitats with scattered trees, where they obtained most of their food from the ground. Jawbones of apes that lived between 15 and 8 mya have been found in Africa, the Near East, and Asia. These extinct apes, genus *Ramapithecus*, have features suggesting that they were the beginning of the lineage leading to humans.

Like us, ramapithecines had short muzzles and small canines. Their chewing teeth were worn down flat, indicating that they chewed from side to side as we do, rather than up and down as chimpanzees and gorillas do. The four living genera of apes—gorillas (*Gorilla*), chimpanzees (*Pan*), orangutans (*Pongo*), and gibbons (*Hylobates*)—are restricted to tropical Africa and Asia (Figure 30.36).

Human ancestors descended to the ground

Some experts believe that *Ramapethicus* was the direct ancestor of both modern apes and humans; others believe that *Ramapithecus* is a member of only the hominid (human) lineage. But there is no disagreement that

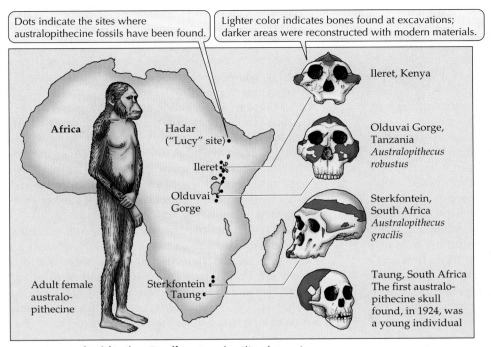

Dots indicate the sites where australopithecine fossils have been found.

Lighter color indicates bones found at excavations; darker areas were reconstructed with modern materials.

Ileret, Kenya

Olduvai Gorge, Tanzania
Australopithecus robustus

Sterkfontein, South Africa
Australopithecus gracilis

Taung, South Africa
The first australopithecine skull found, in 1924, was a young individual

Africa

Hadar ("Lucy" site)

Ileret

Olduvai Gorge

Sterkfontein

Taung

Adult female australopithecine

30.37 Australopithecine Fossils Few fossilized remains are complete, but skull shapes can be reconstructed accurately.

members of another lineage, the australopithecines, are direct human ancestors. The australopithecines had distinct morphological adaptations for **bipedalism**, locomotion in which the body is held erect and moved exclusively by movements of the hind legs. Bipedal locomotion frees the hands to manipulate objects and to carry them while walking. It also elevates the eyes, enabling the animal to see over tall vegetation to spot predators and prey. Both advantages were probably important for early australopithecines.

The first australopithecine skull was found in South Africa in 1924; since then, other fragments have been found at various sites in Africa (Figure 30.37). The oldest and most complete fossil skeleton of an australopithecine, approximately 3.5 million years old, was discovered in Ethiopia in 1974. That individual, a young female known to the world as Lucy, attracted a great deal of attention because her remains were so complete and well preserved. Lucy has been assigned to the species *Australopithecus afarensis*, the most likely ancestor of later hominids. All the evidence from different parts of her skeleton suggest that Lucy was only about 1 m tall and walked upright.

From *Australopithecus afarensis* ancestors, several species of australopithecines evolved. Several million years ago, two distinct types of australopithecines lived together over much of eastern Africa. The more robust type (about 40 kg) is represented by at least two species, both of which died out suddenly about 1.5

mya. The smaller (25–30 kg), more slender *A. africanus* is much rarer as a fossil, suggesting that it was less common than the other species.

Because they were less agile, members of the robust species probably stayed relatively close to trees, to which they retreated at night and when predators were near. Members of the smaller *A. africanus* were able to run faster and, because they were smaller, needed less food per day to survive. They probably lived in more open, drier savannas where food was less abundant than in the moister areas inhabited by the more robust species.

Humans arose from australopithecine ancestors

Many experts believe that a population of *Australopithecus africanus* or a similar species gave rise to the genus *Homo* about 2.5 mya. Early members of *Homo* lived contemporaneously with australopithecines for perhaps half a million years. Two major changes accompanied the evolution of *Homo* from *Australopithecus*: an increase in body size and a doubling of brain size.

The oldest fossil remains of members of the genus *Homo*, named *Homo habilis*, were discovered in the Olduvai Gorge, Tanzania, and are estimated to be 2 million years old. Other fossils of *Homo habilis* have been found in Kenya, Ethiopia, and South Africa, indicating that the species had a wide range in Africa. Tools used by these early hominids to obtain food were found with the fossils.

Homo habilis lived in relatively dry areas where, for much of the year, the main food reserves are subterranean roots, bulbs, and tubers. To exploit these food resources an animal must dig into hard, dry soils, something that cannot be done with an unaided hand. However, roots can be harvested in large quantities in a relatively short time by an individual with a simple digging tool. *Homo habilis* females carrying infants could have done so, freeing males to hunt animal prey to provide the proteins that roots lack.

The only other known extinct species of our genus, *Homo erectus*, evolved in Africa about 1.6 mya. Soon thereafter it had spread as far as eastern Asia. As it expanded its range and increased in abundance, *Homo erectus* may have exterminated *Homo habilis*.

Members of *Homo erectus* were as large as modern people, but their bones were somewhat heavier. *Homo erectus* used fire for cooking and for hunting large animals, and made characteristic stone tools that have been found in many parts of the Old World. These tools were probably used for digging, capturing animals, cleaning and cutting meat, scraping hides, and cutting wood. Although *Homo erectus* survived in Eurasia until about 250,000 years ago, it was replaced in tropical regions by our species, *Homo sapiens*, about half a million years ago.

Brains steadily become larger

The trends that accompanied the transition from *Australopithecus* to *Homo erectus* continued during the evolution of our own species. The earliest humans (*Homo sapiens*) had larger brains than did members of the earlier species of *Homo*, a change that was probably favored by an increasingly complex social life. The ability of group members to communicate with one another was valuable for cooperative hunting and gathering, for sharing information about the location and use of food sources, and for improving one's status in the complex social interactions that must have characterized those societies, just as they do ours today.

Several types of *Homo sapiens* existed during the mid-Pleistocene epoch, from about 1.5 million to about 300,000 years ago. All were skilled hunters of large mammals, but plants continued to be an important component of their diet. During this period two other distinctly human traits emerged: rituals and a concept of life after death. Deceased individuals were buried with tools and clothing in their graves, presumably for their existence in the next world.

One type of *Homo sapiens*, generally known as Neanderthal because it was first discovered in the Neander Valley in Germany, was widespread in Europe and Asia between about 75,000 and 30,000 years ago. Neanderthals were short, stocky, and powerfully built humans whose massive skulls housed brains somewhat larger than our own. They manufactured a variety of tools and hunted large mammals, which they probably ambushed and subdued in close

combat. For a short time, their range overlapped that of Cro-Magnon people, a more modern form of *Homo sapiens*, but then the Neanderthals abruptly disappeared. Many scientists believe that they were exterminated by the Cro-Magnons, just as *Homo habilis* may have been exterminated by *Homo erectus*.

Cro-Magnon people made and used a variety of sophisticated tools. They created the remarkable paintings of large mammals, many of them showing scenes of hunting, that have been discovered in caves in various parts of Europe (Figure 30.38a). The animals depicted were characteristic of the cold steppes and grasslands that occupied much of Europe during periods of glacial expansion.

Cro-Magnon people spread across Asia, reaching North America perhaps as early as 20,000 years ago, although the date of their arrival in the New World is still uncertain. As they rapidly spread southward through North and South America, they may have exterminated, by overhunting, populations of many species of large mammals that had lived on those continents.

Humans evolve language and culture

As our ancestors evolved larger brains, they also increased their behavioral capabilities, especially the capacity for language. Most nonhuman animal communication consists of a limited number of signals, which pertain mostly to immediate circumstances and are associated with changed emotional states induced by those circumstances. (The language of honeybees is unusual in that it contains a symbolic component referring to events distant in both space and time; see Chapter 49.)

Human language is far richer in its symbolic character than are any other animal vocalizations. Our words can refer to past and future times and to distant places. We are capable of learning thousands of words, many of them referring to abstract concepts. We can rearrange words to form sentences with complex meanings.

The expanded mental abilities of humans are largely reponsible for the development of **culture**, the process by which knowledge and traditions are passed along from one generation to another by teaching and observation. Culture can change rapidly because genetic changes are not necessary for a cultural trait to spread through a population. The primary disadvantage of culture is that its norms must be taught to each generation.

The tools and other implements associated with human fossils, as well as the cave paintings that early humans created, reveal cultural traditions. Cultural learning greatly facilitated the spread of domestic plants and animals and the resulting conversion from societies in which food was obtained by hunting and

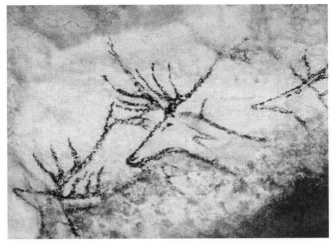

(a)

(b)

(c)

30.38 Hunting, Pastoralism, and Agriculture (a) Lascaux Cave in France provides striking examples of the artistry of Cro-Magnons. Cave walls often depict animals of the glacial steppes; these paintings may have served as part of rituals to increase success during hunts. (b) Pastoralism displaced hunting in many societies. This hut is the temporary dry-season quarters of a herder in Burkina Faso, Africa. (c) Agricultural development has totally transformed the landscape in these hills above Port-au-Prince, Haiti.

gathering to societies in which **pastoralism** (the raising of domestic animals to provide food and other products) and **agriculture** dominated (Figure 30.38*b* and *c*). The agricultural revolution, in turn, led to an increasingly sedentary life, the development of cities, greatly expanded food supplies, rapid growth of the human population, and the appearance of occupational specializations, such as artisans and shamans.

Agriculture developed in the Middle East approximately 11,000 years ago. From there it spread rapidly northwestward across Europe, finally reaching the British Isles about 4,000 years ago. The plants that these early agriculturalists domesticated were cereal grains such as wheat and barley; legumes—beans, lentils, and peas; and woody plant crops such as grapes and olives. Other plants, such as rye, cabbage, celery, and carrots, were domesticated as agriculturalists spread across Europe. Cattle, sheep, goats, horses, dogs, and cats were their most important domesticated animals.

Agriculture developed independently in eastern Asia, contributing to our modern diets soybeans, rice, citrus fruits, and mangoes, and pigs and chickens. There was some exchange, even in early times, among agricultural centers in the Old World, but when people crossed the cold and barren Bering land bridge into the New World, they apparently brought no domesticated plants with them. These people subsequently developed rich and varied agricultural systems based on corn, tomatoes, kidney and lima beans, peanuts, potatoes, chili peppers, and squashes.

The human population has had three major growth increases. The first, stimulated by toolmaking, lasted about a million years. During that time, human numbers increased to about 5 million. During the second surge, which followed the domestication of plants and animals and the invention of agriculture, the human population increased to about 500 million people within 8,000 years. We are currently in the middle of the third great surge, which was triggered by the Industrial Revolution (Figure 30.39).

The current human population is nearly 6 billion and is projected to increase to more than 11 billion within 50 years. The first two population surges were followed by periods of relative stability. Whether the current one will follow the same path, at what size the population might level off, and the environmental consequences of continued growth are questions that are fiercely debated (see Chapter 55).

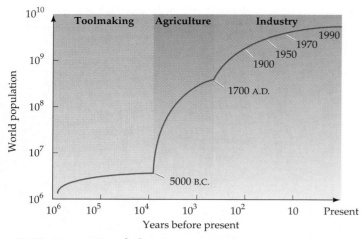

30.39 Human Population Surges The human population surged after the invention of tools, the domestication of plants and animals, and the Industrial Revolution. Note that the axes are scaled logarithmically.

which the animal is stable (Figure 30.40*b*). Similar planktonic larval stages evolved in marine members of many protostomate and deuterostomate phyla; all of these fed on tiny planktonic organisms in the open water.

Both protostomes and deuterostomes colonized the land—the former via beaches, the latter via fresh water—but the consequences were very different. The jointed external skeletons of arthropods, although they provide excellent support and protection in air, are not suitable for large animals. In addition, an arthropod must shed its skin and become temporarily vulnerable in order to grow. But the internal, jointed skeletons of vertebrates permit growth to large size without temporary vulnerable stages. Consequently, although arthropods are abundant and diverse on land, they never evolved into large animals.

Terrestrial lineages of vertebrates recolonized aquatic environments several times. Suspension feed-

Deuterostomes and Protostomes: Shared Evolutionary Themes

Table 30.2 summarizes the traits of all animal phyla. Deuterostome evolution paralleled protostome evolution in several important ways. Both lineages exploited the abundant food supplies buried in soft marine substrata, attached to rocks, or suspended in water columns. Because of the ease with which water can be moved, many groups of both lineages developed elaborate structures for moving water and extracting prey from it (Figure 30.40*a*).

In both groups, a coelomic cavity evolved and subsequently became divided into compartments that allowed better control of body shape and movement. Both groups evolved locomotor abilities. Some members of both groups evolved mechanisms for controlling their buoyancy in water, using gas-filled internal spaces whose contents can be adjusted to control the depth at

30.40 Parallel Evolution Devices for filtering food from the water (*a*) and maintaining buoyancy in the water (*b*) evolved in both protostomate and deuterostomate lineages.

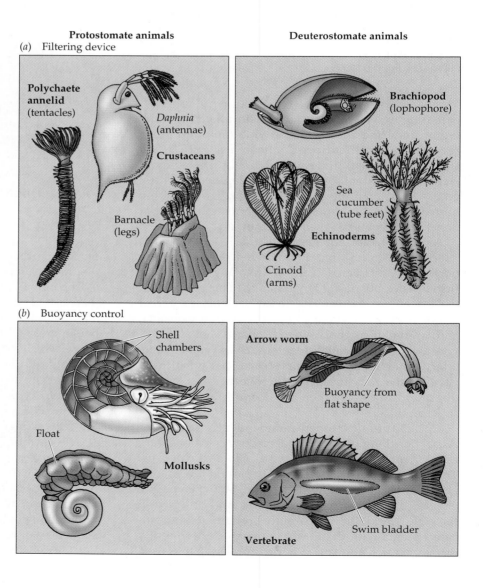

TABLE 30.2   Summary of Living Members of the Kingdom Animalia[a]

PHYLUM	NUMBER OF LIVING SPECIES DESCRIBED	SUBGROUPS
Porifera: Sponges	10,000	
Cnidaria: Cnidarians	10,000	Hydrozoa: Hydrozoans Scyphozoa: Jellyfish Anthozoa: Corals, sea anemones
Ctenophora: Comb jellies	100	
Protostomes		
Rotifera: Rotifers	1,800	
Nematoda: Roundworms	20,000	
Platyhelminthes: Flatworms	25,000	Turbellaria: Free-living flatworms Trematoda: Flukes (all parasitic) Cestoda: Tapeworms (all parasitic) Monogenea: Ectoparasites of fish
Nemertea: Ribbon worms	900	
Pogonophora: Pogonophorans	145	
Annelida: Segmented worms	15,000	Polychaeta: Polychaetes (all marine) Oligochaeta: Earthworms, freshwater worms Hirudinea: Leeches
Mollusca: Mollusks	100,000	Monoplacophora: Monoplacophorans Polyplacophora: Chitons Bivalvia: Clams, oysters, mussels Gastropoda: Snails, slugs, limpets Cephalopoda: Squids, octopuses, nautiloids
Chelicerata: Chelicerates	63,000	Merostomata: Horseshoe crabs Arachnida: Scorpions, harvestmen, spiders, mites, ticks
Crustacea: Crustaceans	40,000	Crabs, shrimp, lobsters, barnacles, copepods
Uniramia: Unirames	1,500,000	Myriapoda: Centipedes, millipedes Onychophora: Onychophorans Tardigrada: Water bears Insecta: Insects
Deuterostomes		
Phoronida: Phoronids	15	
Brachiopoda: Lampshells	350	More than 26,000 fossil species described
Bryozoa: Moss animals	4,000	
Chaetognatha: Arrow worms	100	
Echinodermata: Echinoderms	7,000	Crinoidea: Sea lilies, feather stars
Pterobranchia: Pterobranchs		Asteroidea: Sea stars Ophiuroidea: Brittle stars Concentricycloidea: Sea daisies Echinoidea: Sea urchins Holothuroidea: Sea cucumbers
Hemichordata: Acorn worms	100	
Chordata: Chordates	40,000	Urochordata: Tunicates Cephalochordata: Lancelets Agnatha: Hagfishes, lampreys Chondrichthyes: Cartilaginous fishes Osteichthyes: Bony fishes Amphibia: Amphibians Reptilia: Reptiles Archosauria: Dinosaurs, crocodilians, birds Mammalia: Mammals

[a]A few small phyla are not included.

ing re-evolved in several of these lineages. The largest living mammals, the baleen whales (the toothless whales, including blue whales, humpback whales, and right whales), feed on relatively small prey that they extract from the water with large plates in their mouth.

Unlike the oceans, where the dominant photosynthetic plants are unicellular algae, most photosynthesis on land is carried out by vascular plants. Dominant plants in most terrestrial environments are large, complex organisms whose tissues are difficult to digest. Herbivores must ingest large quantities of fibers and defensive chemicals along with the energy-rich molecules they need. Because larger animals can exist on food of poorer quality than small animals can, a steady increase in body size is a common pattern in herbivore evolution.

This pattern is strikingly illustrated by the evolution of reptiles and the later evolution of mammalian herbivores. The evolution of large herbivores, in turn, favored the evolution of larger carnivores able to attack and overpower them. This evolutionary trend may have come to a temporary halt because of the invention of weapons by a moderately sized, omnivorous primate—*Homo sapiens*.

Summary of "Deuterostomate Animals: Evolution of Larger Brains and Complex Behaviors"

• There are fewer lineages and fewer species of deuterostomes than of protostomes, but as members of the lineage we have a special interest in its other members. **Review Figure 30.1**

Three Body Parts: An Ancient Body Plan

• Members of one deuterostomate lineage have bodies divided into three segments. They evolved a new structure—the lophophore—for extracting prey from the water. **Review Figure 30.2**
• Four lophophorate phyla survive today: Phoronida, Brachiopoda, Bryozoa, and Pterobranchia. Nearly all the members of these phyla are marine, but a few live in fresh water.
• Arrow worms, though divided into three body parts—head, trunk, and tail—lack a lophophore.

Echinoderms: Complex Radial Symmetry

• Echinoderms are deuterostomes that have a radially symmetrical body plan, a unique water vascular system, and a calcified internal skeleton. **Review Figure 30.8a**
• Nearly all living species of echinoderms have a bilaterally symmetrical, ciliated larva that feeds for a while as a planktonic organism. **Review Figure 30.8c**
• Of the 23 echinoderm lineages that have been described, only 6 groups survive. Some groups of echinoderms have arms, but others do not.

Modified Lophophores: New Ways of Feeding

• Evolution among acorn worms (phylum Hemichordata) and chordates (phylum Chordata) led to new ways of capturing and handling food.
• In the lineage leading to acorn worms, the lophophore was lost, but a proboscis grew larger and became a digging organ.
• Members of another lineage, the chordates (phylum Chordata), lost both the lophophore and the proboscis, replacing them with enlarged pharyngeal slits as a feeding device. They also evolved a dorsal supporting rod, the notochord.
• Tunicates (the chordate subphylum Urochordata) are sessile as adults and filter prey from seawater with large pharyngeal baskets. Only their larvae have notochords.
• Lancelets (the chordate subphylum Cephalochordata) are small, fishlike animals that live partly buried in soft marine sediments and extract small prey from the water with pharyngeal baskets.

The Origin of the Vertebrates: Sucking Mud

• Vertebrates (the chordate subphylum Vertebrata) evolved jointed internal skeletons that enabled them to swim rapidly. Early vertebrates lacked jaws and fed by filtering small animals from mud. **Review Figures 30.13, 30.14**
• Jaws, which evolved from anterior gill arches, enabled their possessors to grasp and chew their prey, expanding food sources and nutrition. Jawed fishes rapidly became dominant animals in both marine and fresh waters. **Review Figure 30.16**
• Cartilaginous and bony fishes evolved two pairs of unjointed fins, with which they control their swimming movements and stabilize themselves in the water.
• Bony fishes evolved lunglike sacs that help keep them suspended in open water. Bony fishes come in a wide variety of sizes and shapes, and many species have complex social systems.

Colonizing the Land

• One lineage of fishes evolved jointed fins that enabled them to move more effectively over land to find new bodies of water. One genus of these lobe-finned fishes—*Latimeria*—survives today.
• One group of lobe-fins gave rise to the amphibians, the first terrestrial vertebrates. **Review Figure 30.20**
• About 4,500 species of amphibians live today. They belong to three orders: caecilians, frogs and toads, and salamanders.
• Most amphibians live in water at some time in their lives, and their eggs must remain moist. **Review Figure 30.22**
• Amniotes, the common ancestors of reptiles and mammals, evolved eggs with shells impermeable to water and thus became the first vertebrates to be independent of water for breeding. **Review Figure 30.23**
• Modern reptiles are members of three lineages: turtles and tortoises, tuataras, and snakes and lizards.

Surviving Dinosaur Lineages

• One amniote lineage, the thecodonts, gave rise to the crocodilians.
• Another group of thecodonts led to dinosaurs, the dominant terrestrial reptiles for millions of years. The nearly 9,000 species of birds are the descendants of a third thecodont lineage. Birds are characterized by their feathers, high metabolic rates, and parental care.

The Tertiary: Mammals Diversify

• Mammals evolved during the Mesozoic era from now extinct therapsids. **Review Figure 30.29**

• The eggs of mammals are fertilized within the bodies of females, and mammalian embryos develop for some time before being born. Mammals are unique in that they suckle their young with milk secreted by mammary glands.

• The few species of monotremes lay eggs, but all other mammals give birth to live young.

• Marsupials give birth to tiny young that are housed and nursed in a pouch. Most marsupial species live in Australia.

• Eutherian mammals, which give birth to relatively well developed offspring, are placed in 16 major groups.

Humans Evolve: Earth's Face Is Changed

• The primates, to which humans belong, split into two major lineages, one leading to the prosimians (lemurs, lorises, and tarsiers), the other leading to the anthropoids (monkeys, apes, and humans). **Review Figure 30.33**

• Humans evolved in Africa from terrestrial, bipedal, australopithecine ancestors. **Review Figure 30.37**

• Early humans evolved large brains, language, and culture. They manufactured and used tools, developed rituals, and domesticated plants and animals. In combination, these traits enabled humans to increase greatly in number and to transform the face of Earth.

• The human population has increased greatly three times. We are currently in the middle of the third population surge. When and how it will end is hotly debated. **Review Figure 30.39**

Deuterostomes and Protostomes: Shared Evolutionary Themes

• Devices for extracting prey from water, for controlling buoyancy, and for moving rapidly evolved many times during the evolution of both protostomes and deuterostomes. Members of both lineages colonized terrestrial environments. **Review Figure 30.40**

Self-Quiz

1. Which of the following are deuterostomate phyla with a three-part body plan?
   a. Rotifera, Phoronida, Bryozoa, and Brachiopoda
   b. Phoronida, Bryozoa, Brachiopoda, and Hemichordata
   c. Phoronida, Bryozoa, Hemichordata, and Chordata
   d. Echinodermata, Bryozoa, Brachiopoda, and Chordata
   e. Phoronida, Bryozoa, Hemichordata, and Echinodermata

2. The structure used by brachiopods to capture food is a
   a. pharyngeal gill basket.
   b. proboscis.
   c. lophophore.
   d. mucous net.
   e. radula.

3. The water vascular system of echinoderms is a
   a. series of seawater channels derived by enlargement and extension of a coelomic cavity.
   b. series of seawater channels derived by enlargement and extension of the pharyngeal cavity.
   c. series of channels derived by enlargement and extension of a coelomic cavity, filled with coelomic fluid.
   d. series of channels derived by enlargement and extension of a coelomic cavity and filled with fresh water.
   e. series of channels that can be filled to different levels with water, enabling the animal to control its buoyancy.

4. The pharyngeal gill slits of chordates originally functioned as sites for
   a. uptake of oxygen only.
   b. release of carbon dioxide only.
   c. both uptake of oxygen and release of carbon dioxide.
   d. removal of small prey from the water.
   e. forcible expulsion of water to move the animal.

5. The key to the vertebrate body plan is
   a. a pharyngeal gill basket.
   b. a vertebral column to which internal organs are attached.
   c. a vertebral column to which two pairs of appendages are attached.
   d. a vertebral column to which a pharyngeal gill basket is attached.
   e. a pharyngeal gill basket and two pairs of appendages.

6. Which of the following fishes do *not* have a cartilaginous skeleton?
   a. Chimaeras
   b. Lungfishes
   c. Sharks
   d. Skates
   e. Rays

7. In most fishes, lunglike sacs evolved into
   a. pharyngeal gill slits.
   b. true lungs.
   c. coelomic cavities.
   d. swim bladders.
   e. none of the above

8. Most amphibians return to water to lay their eggs because
   a. water is isotonic to egg fluids.
   b. adults must be in water while they guard their eggs.
   c. there are fewer predators in water than on land.
   d. amphibians need water to produce their eggs.
   e. amphibian eggs quickly lose water and desiccate if their surroundings are dry.

9. The horny scales that cover the skin of reptiles prevent them from
   a. using their skin as an organ of gas exchange.
   b. sustaining high levels of metabolic activity.
   c. laying their eggs in water.
   d. flying.
   e. crawling into small spaces.

10. Which statement about bird feathers is *not* true?
    a. They are highly modified reptilian scales.
    b. They provide insulation for the body.
    c. They arise from well-defined tracts.
    d. They help birds fly.
    e. They are important sites of gas exchange.

11. Monotremes differ from other mammals in that they
    a. do not produce milk.
    b. lack body hairs.
    c. lay eggs.
    d. live in Australia.
    e. have a pouch in which the young are raised.

12. Bipedalism is believed to have evolved in the human lineage because
    a. bipedal locomotion is more efficient than quadrupedal locomotion.
    b. bipedal locomotion is more efficient than quadrupedal locomotion, and it frees the hands to manipulate objects.
    c. bipedal locomotion is less efficient than quadrupedal locomotion, but it frees the hands to manipulate objects.
    d. bipedal locomotion is less efficient than quadrupedal locomotion, but bipedal animals can run faster.
    e. bipedal locomotion is less efficient than quadrupedal locomotion, but natural selection does not act to improve efficiency.

Applying Concepts

1. In what animal phyla has the ability to fly evolved? How do structures used for flying differ among the animals in these phyla?

2. Extracting suspended food from the water column is a common mode of foraging among animals. Which groups contain species that extract prey from the air? Why is this mode of obtaining food so much less common than extracting prey from water?

3. Compare the buoyancy systems of cephalopods and fishes.

4. Why does possession of an external skeleton limit the size of a terrestrial animal more than possession of an internal skeleton does?

5. Large size both confers benefits and poses certain risks. What are these risks and benefits?

Readings

Alexander, R. M. 1975. *The Chordates*. Cambridge University Press, New York. A comprehensive and readable account of the biology of members of the phylum Chordata.

Bakker, R. T. 1975. "Dinosaur Renaissance," *Scientific American*, April. A discussion of the relationships between birds and dinosaurs, presenting evidence that dinosaurs were warm-blooded.

Bond, C. E. 1979. *Biology of Fishes*. Saunders, Philadelphia. A leading text for courses on ichthyology.

Carroll, R. C. 1987. *Vertebrate Paleontology and Evolution*. W. H. Freeman, San Francisco. A thorough account of the fascinating evolutionary history of the vertebrates.

Colbert, E. H. 1980. *Evolution of the Vertebrates: A History of the Backboned Animals through Time*, 3rd Edition. Wiley-Interscience, New York. A thoughtful discussion of the origins and evolutionary radiations of the vertebrate groups.

Diamond, J. 1992. *The Third Chimpanzee*. Harper Collins, New York. An engaging account of human evolution and how we acquired the unique traits that characterize our species.

Gill, F. B. 1990. *Ornithology*. W. H. Freeman, San Francisco. A technically accurate and readable introduction to bird biology for students at any level.

Langston, W., Jr. 1981. "Pterosaurs," *Scientific American*, February. An account of the largest animals ever to fly, including notes on evolutionary relationships among birds and reptiles.

Pough, F. H., J. B. Heiser, and W. N. McFarland. 1989. *Vertebrate Life*. Macmillan, New York. An excellent treatment of the evolution and ecology of the vertebrates.

Vaughan, T. A. 1978. *Mammalogy*, 2nd Edition. Saunders, Philadelphia. The leading textbook on mammals. Offers good coverage of both the orders of mammals and general aspects of mammalian biology.

Willson, M. F. 1984. *Vertebrate Natural History*. Saunders, Philadelphia. A thorough treatment of all aspects of the lives of vertebrates.

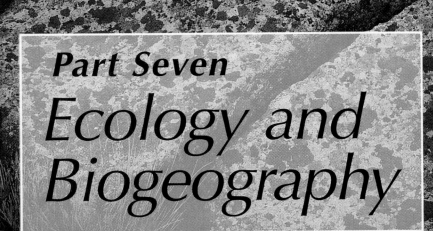

Part Seven
Ecology and Biogeography

Chapter 50

Behavioral Ecology

Choices Are the Foundation of Ecology
Choices made locally may influence distant events, and vice versa. Our choices in a supermarket may influence interactions among species that live far from where we shop. Do we prefer locally grown food or food imported from remote places? Do we buy fresh or processed foods? Do we choose foods produced by environmentally friendly methods?

*I*ndividuals of all species interact in various ways with individuals of their own and other species and with the physical environment. The task of ecology is to understand the nature and consequences of these interactions. Ecologists study patterns of distribution and abundance of organisms to determine how patterns are established and maintained, and how they change over short and long time periods. From its roots in descriptive natural history, ecology has developed into a complex field of inquiry dealing with levels of organization ranging from relationships of individual organisms to their physical and biological environments to the structure of communities and ecosystems.

In this first chapter on ecology, we will discuss how animals behave, how they make the decisions that influence their survival and reproductive success, how ecologists study these decisions, and what they have learned from their studies. Animals decide where to carry out their activities and how to select the resources they need—food, water, shelter, nest sites. Animals respond to predators and competitors, and decide how to interact with other members of their own species. (The use of the words "decision" and "decide" here does not imply that the choices animals make are conscious, but rather that these choices influence the survival and reproductive success of the animals and thus are molded by natural selection.) Individual choices are the foundation of much of ecology. Changes in densities and distributions of populations are the cumulative results of the decisions of myriad individuals.

The term *environment*, as used by ecologists, includes physical and chemical factors such as water, nutrients, light, temperature, and wind. It also includes all other organisms that influence the lives of individuals. Because species are adapted for life in many different environments, their interactions with their physical and biological environments are also varied. An environmental factor that exerts a strong influence on individuals of one species may have no influence on individuals of another species.

Interactions between organisms and their environments are two-way processes. Organisms both influence and are influenced by their environments. Indeed, managing environmental changes caused by our own species is one of the major problems of the modern world. For this reason, ecologists are often asked to help analyze causes of environmental problems and to assist in finding solutions for them. However, it is important not to confuse the science of ecology with the term "ecology" as it is often used in popular writing, to describe nature as some kind of superorganism.

Choosing Environments

Where an individual chooses to live, and how it uses that environment, strongly influence its survival and reproductive success. We will consider how animals choose places that provide adequate food and shelter, and how they select their food.

Features of an environment may indicate its suitability

Selecting a place to live is one of the most important decisions an individual makes. The environment in which an organism normally lives is called its **habitat**. Habitat selection requires a sequence of decisions, each one of which limits the options available for successive choices. Once a habitat is chosen, an animal must seek its food, resting places, nest sites, and escape routes within that habitat.

The cues organisms use to select suitable habitats are as varied as the organisms themselves, but all habitat selection cues have a common feature: they are good predictors of general conditions suitable for future survival and reproduction. For example, a young red abalone, a kind of gastropod mollusk, begins its life as an egg that is fertilized in the open ocean. About 14 hours after fertilization the egg hatches, but the motile larva emerges with enough yolk to continue developing for another 7 days without eating. At the end of 7 days the larva stops developing, swims to the seafloor, chooses a place in which to settle, and metamorphoses.

Red abalone larvae settle only on coralline algae, upon which they feed (Figure 50.1). They recognize coralline algae by a chemical signal—a water-soluble

Haliotis rufescens

50.1 Chemistry Provides the Cue   Red abalone larvae (dark ovals) settle only where they recognize the chemical composition of their food source, coralline algae.

peptide containing about 10 amino acids—that all these algae produce. In the laboratory, abalone larvae will settle on any surface on which this molecule has been placed, but in nature *only* coralline algae produce it. By using this simple cue, the larvae always settle on a surface that is suitable for their future development. Because red abalone larvae grow to adulthood by feeding on a single patch of coralline algae, their habitat is their food. However, most animals make many choices about where and how to seek and select food after they have settled in a habitat.

How do animals choose what to eat?

After choosing a habitat, individuals use the local resources, such as shelters from the physical environment, nest sites, and food. Because food is so important, we consider it here in some detail. When an animal is looking for food, how much time should it spend searching before moving to another site? When many different types of prey are available, which ones should a predator take, and which ones should it ignore? **Foraging theories** were developed to help answer these questions. To construct a theory to predict how a foraging animal should behave, a scientist first specifies the objective of the behavior and then attempts to determine the behavioral choices that would best achieve that objective. This approach is known as

optimality modeling, and its underlying assumption is that natural selection has molded the behavior of animals so that they solve problems by making the best choices available to them. There are many foraging theories, because a forager may be attempting to maximize the rate at which it obtains energy, to get enough vitamins or minerals, or to minimize its risk while foraging.

As an example, consider how a predator should choose among available prey in order to maximize the amount of energy it obtains. This is a reasonable objective because the more rapidly a predator captures food, the more time it will have for other activities, such as reproduction. Therefore, a more efficient predator should produce more offspring than a less efficient one, and animals should evolve to regularly make prey choice decisions that maximize their rate of energy intake.

To build a model of a predator that chooses prey in a way that maximizes its energy intake rate, we characterize each type of prey by the amount of time it takes the predator to pursue, capture, and consume one of them, and by the amount of energy an individual prey contains. We then rank the prey according to the amount of energy the predator gets relative to the amount of time the predator spends capturing and handling the prey. The most valuable prey type is the one that yields the most energy per unit of time invested.

With this information, we can build a model to calculate the rate at which a predator would obtain energy given a particular prey selection strategy. We can then compare alternative foraging strategies and determine the one that yields the highest rate of energy intake. The interesting result of such calculations is that, if the most valuable prey type is abundant enough, a predator gains the most energy per unit of time spent foraging by taking only the most valuable prey type and ignoring all others. The reason is that while the predator is capturing and consuming a less valuable prey individual, it could have found one of the most valuable ones. However, as the abundance of the most valuable prey type decreases, an energy-maximizing predator adds less valuable prey items to its diet in order of the energy per unit of time that those prey yield. Whether or not a certain prey type is included in the predator's diet does not depend on the abundance of that prey type. It depends only on the abundances of more valuable prey types.

Ecologists tested this theoretical model using bluegill sunfish (Figure 50.2a). They performed laboratory experiments with bluegills to measure the energy content of different prey types (water fleas of different sizes), the time needed to capture and eat different prey types, the energy spent searching for and capturing prey, and actual encounter rates with prey under different prey densities. Using these measurements,

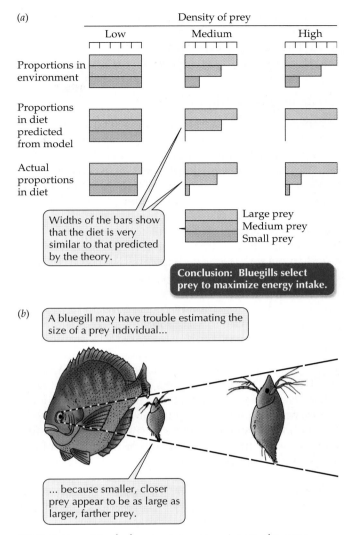

50.2 Energy Maximizers In an experiment, the prey choices of bluegill sunfish were very similar to those predicted by an energy-maximizing model.

the investigators predicted the proportions of large, medium, and small water fleas that bluegills would capture in environments stocked with different densities and proportions of water fleas of those sizes. They predicted that in an environment stocked with low densities of all three sizes of prey, the fish would take every water flea that they encountered, but that in an environment with abundant large water fleas, the bluegills would ignore smaller water fleas.

To test their predictions, the investigators put the bluegills in three different environments and observed the proportions of the water fleas of different sizes they actually captured (Figure 50.2b). The proportions of large, medium, and small water fleas taken by the fish were very close to those predicted by the model. Such tests of foraging theories using many different kinds of animals have provided ecologists with a set of rules showing how animals find and choose their prey. They have also provided estimates of the costs and benefits of foraging behavior.

Costs and Benefits of Behaviors

As shown by the example of prey selection behavior, ecologists who use optimality modeling to understand the evolution of behavior often analyze their observations in terms of costs and benefits. A cost–benefit analysis is based on the principle that an animal has only a limited amount of time, energy, and materials to devote to its activities. Animals generally do not perform behaviors whose total costs are greater than the sum of their benefits—the improvements in survival and reproductive success that the animal achieves by performing the behavior.

Of course, animals do not consciously calculate costs and benefits, but over many generations, natural selection molds behavior in accordance with costs and benefits. A cost–benefit approach provides a framework within which behavioral ecologists can design experiments and make observations that enable them to understand why behavior patterns evolve as they do. Even when costs and benefits cannot be measured directly, experiments can reveal which ones are important.

To understand what we mean by costs and benefits, consider the reproductive behavior of male elephant seals. A male elephant seal defends a small area of beach by threatening other males and fighting with challengers (Figure 50.3). Adult male elephant seals, which are several times larger than females, have large

Mirounga angustirostris

50.3 Loud Threats These mature male elephant seals are engaged in a mock battle for territory. This time, no blood will be drawn.

canine teeth with which they fight and thick skins that serve as shields. Their odd elephantlike snouts are resonating chambers that help amplify the roars that accompany their threat displays. Females congregate on a small number of beaches during the breeding season, where they give birth to their pups and mate again. Males that control good pupping beaches achieve most of the matings. Less than 10 percent of male elephant seals ever mate with a female. Of those that do, some mate with more than 100 females during their lifetimes. By contrast, nearly all females that survive to adulthood breed, but a female rarely weans as many as 10 pups during her lifetime.

The **energetic cost** of a behavior is the difference between the energy the animal would have expended had it rested and the energy expended in performing the behavior. A male elephant seal that fights with rivals expends more energy than a resting male. He therefore exhausts his fat reserves faster than if he had not attempted to defend a parcel of beach.

The **risk cost** of a behavior is the increased chance of being injured or killed as a result of performing it, compared with resting. A displaying male elephant seal is more likely to be injured by a rival than a male that avoids fights. The **opportunity cost** of a behavior is the sum of the benefits the animal forfeits by not being able to perform other behaviors during the same time interval. A male elephant seal cannot search for food while he defends a section of beach, and the longer he does so, the more time he needs to regain his energy reserves once he returns to the sea. The **benefit** a male elephant seal receives from incurring all those costs is access to reproductive females. A male elephant seal that does not defend a parcel of beach cannot mate with females and sire offspring.

Investigators used a cost–benefit approach to determine why female red-winged blackbirds regularly sing while they are on their nests. Most birds are silent while on their nests, presumably to reduce the chance that predators will find the nest. A behavioral ecologist guessed that these birds gained a benefit from singing that was greater than the risk of nest predation. To test this hypothesis, he placed redwing nests containing artificial eggs in breeding areas and placed loudspeakers near them. He broadcast recordings of female songs from some of the loudspeakers; the speakers at the control nests were silent. Ten of 15 nests near a noisy loudspeaker were destroyed by predators within 2 days, but none of the control nests were taken by predators. This experiment confirmed that predators find nests more readily if females are singing near them—a measure of the cost of singing.

To measure the benefits of singing, the ecologist compared the behavior of females that successfully raised their young with the behavior of those whose nests were lost to predators. He found that successful females sang significantly more often at their nests than unsuccessful females did. He then placed a model of a crow, an important nest predator in the area, near his mock nests. Male redwings scolded and attacked the crow more vigorously at nests located near loudspeakers that broadcast female songs than at silent control nests. Thus, both the costs and benefits of singing at the nest relate to predators: singing attracts both predators and males that defend the nest against them. The fact that most females sing regularly at their nests suggests that, on average, the benefits of singing more than compensates for the costs. Mock nests were often destroyed because males typically paid little attention to them.

Mating Tactics and Roles

Individuals choose their associates, how to interact with them, and when to leave them. The most important choice of associates an animal makes is mate selection. Mating behavior involves only a small set of choices. The most basic mating decision is choosing a partner of the correct species. Once the correct species has been determined, additional decisions can be based on the qualities of a potential mate, on the resources—food, nest sites, escape places—it controls, or on a combination of the two. Among those species in which individuals do not control any resources, traits of the partner are the only criteria for mate selection. Here we will discuss how individuals choose their mating partners and show why males and females approach courtship so differently.

Abundant small sperm and scarce large eggs drive mating behavior

The reproductive behavior of males and females is often very different. Males usually initiate courtship, and they often fight for opportunities to mate with females. Females seldom fight over males, and they often reject courting males. Why do males and females approach courtship and copulation so differently?

The answer lies in the costs of producing sperm and eggs. Because sperm are small and cheap to produce, one male produces enough to sire a very large number of offspring—usually many more than the number of eggs a female can produce or the number of young she can nourish. Therefore, males of most species can increase their reproductive success by mating with many females. Eggs, on the other hand, are typically much larger than sperm and are expensive to produce. Consequently, a female is unlikely to increase her reproductive output very much by increasing the number of males with which she mates. The reproductive success of a female depends primarily upon the resources her mate controls, the amount of assistance he provides in the care of her offspring, and the quality of the genes she receives from him. By their choices among males, females cause the evolution of exaggerated traits that signal male quality.

Sexual selection leads to exaggerated traits

Traits may evolve among individuals of one sex as a result of **sexual selection**, the spread of traits that confer advantages to their bearers during courtship or when they compete for mates or resources. Successful competitors gain exclusive access to mates that are attracted to the resources they control. Traits that improve success in courtship evolve as a result of mating preferences by individuals of the opposite sex.

Sexual selection is responsible for the evolution of the remarkable tails of African long-tailed widowbirds, which are longer than their heads and bodies combined. To examine the role of the tail in sexual selection, an ecologist shortened the tails of some males by cutting them, and lengthened the tails of others by gluing on additional feathers. Both short-tailed and long-tailed males successfully defended their display sites, indicating that the long tail does not confer an advantage in male–male competition. However, males with artificially elongated tails attracted about four times more females than males with shortened tails (Figure 50.4). Given these experimental results, why don't male widowbirds have even longer tails than they do? A likely answer is that there are costs to producing and maintaining long tails that were not measured during these short-term experiments.

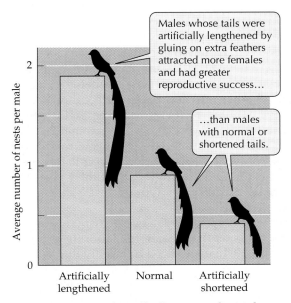

50.4 The Longer the Tail, the Better the Male   Males with shortened tails defended their display sites successfully but attracted few females. Experiments showed how sexual selection favored long tails.

50.5 A Male Wins His Mate The male hanging fly on the left has just presented a moth to his mate, thus demonstrating his foraging skills. She feeds on the moth while they copulate.

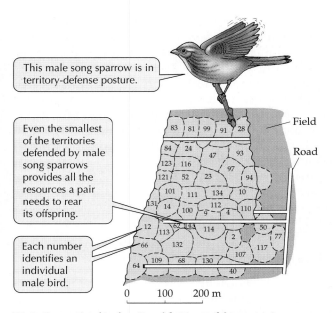

This male song sparrow is in territory-defense posture.

Even the smallest of the territories defended by male song sparrows provides all the resources a pair needs to rear its offspring.

Each number identifies an individual male bird.

Field

Road

0   100   200 m

50.6 Some Territories Provide Everything Male song sparrows defend territories that contain food, nesting sites, and protective cover.

Males attract mates in varied ways

Males employ a variety of tactics to induce females to copulate with them. If a male controls no resources, he uses courtship behavior that signals in some way that he is in good health, that he is a good provider of parental care, or that he has a good genotype. For example, males of some species of hanging flies court females by offering them dead insects. A female hanging fly will mate with a male only if he provides her with a morsel of food. The bigger the food item, the longer she copulates with him, and the more of her eggs he fertilizes (Figure 50.5). A female gains from this behavior because she obtains a better supply of energy for egg production. Also, because males fight for possession of dead insects, a male with a large insect is likely to be a good fighter and in good health.

Whether a male fertilizes the eggs of a female with whom he has copulated depends on when they copulate and whether she copulates with other males. Males have evolved behavior patterns that increase the probability that it will be *their* sperm that fertilize the eggs. The simplest method is to remain with the female and prevent other males from copulating with her, but this method has high opportunity costs because a male cannot do anything else while he is guarding a female.

Males of many species have evolved sperm competition mechanisms that are more elaborate but take less time. A male black-winged damselfly that grabs a receptive female uses his penis to scrub out sperm deposited by other males in her sperm storage chamber. The copulating male removes between 90 and 100 percent of competing sperm before he inserts his own sperm into the chamber. Males of many insects deposit a plug that effectively seals the opening to the

female's genital chamber and prevents other sperm from entering.

Males of some species defend territories that contain food, nesting sites, or other resources. Some territories are all-purpose; they provide mating sites, nesting sites, and the food necessary to rear offspring (Figure 50.6). Other territories include a large breeding and nesting area, but do not supply all of the food necessary to rear young. Male red-winged blackbirds defend this type of territory in emergent vegetation in marshes (Figure 50.7). These territories provide nesting sites over water that are protected from some terrestrial mammalian predators, but both males and females get much of their food from upland areas near the marshes. The territories of many aquatic birds, such as gannets, penguins, and cormorants, are very small areas that individuals can defend while sitting on their nests (Figure 50.8).

Males of some species gather at display grounds, called **leks**, that may be used for many years. Within a lek each male defends a small display territory. Females come to the lek to choose a mate, but then leave the area and raise their young with no help from the males. Males battle intensely for possession of central display territories, and females usually mate with males holding central territories (Figure 50.9).

Females are the choosier sex

Females can improve their reproductive success if they can assess the genetic quality and health of potential mates, the quantity of parental care they may provide,

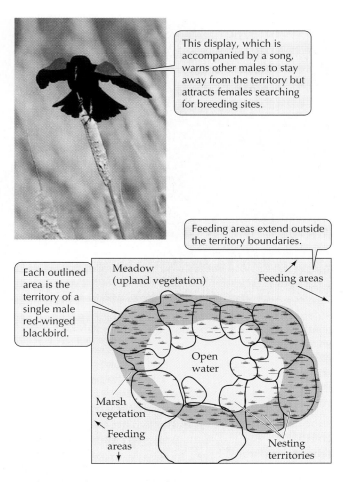

This display, which is accompanied by a song, warns other males to stay away from the territory but attracts females searching for breeding sites.

Feeding areas extend outside the territory boundaries.

Each outlined area is the territory of a single male red-winged blackbird.

Meadow (upland vegetation)

Feeding areas

Open water

Marsh vegetation

Feeding areas

Nesting territories

50.7 A Territory with Nesting Sites and Some Food
Red-winged blackbird nesting territories are located in emergent marsh vegetation, which provides some protection from predators. The territories are not large enough to provide all the food the birds need to raise their offspring.

Sula bassana

50.8 Some Territories Are Small The spacing of individual breeding territories of cape gannets is determined by how far an incubating bird can reach to peck its neighbors without getting off its eggs.

and the quality of the resources they control. But how can females make such assessments when all males attempt to signal that they are good in all three of these traits? The answer is that by paying particular attention to those signals at which males cannot cheat, females have favored the evolution of "reliable" signals. Possession of a large dead insect signals good fighting ability in a male hanging fly. A male elephant seal that controls a section of good pupping beach is certainly a high-quality mate.

Social and genetic partners may differ

Behavioral ecologists have known for many years that animals always copulate with their mates—the individuals with which they have established pair bonds—but that they sometimes also copulate with other individuals. However, until the recent development of DNA fingerprinting methods, investigators

had to assume that mated individuals were the parents of the offspring they raised. Recently, by using these new methods to compare the genetic material of offspring with that of their supposed parents and other individuals, ecologists have found that nestling

50.9 Lek Territories Are Display Grounds Fallow deer gather at leks, within which each antlered male establishes a small display ground to which he attempts to attract females. Dominant males command the territories closest to the center of the lek, and males may fight with each other over a specific display ground.

birds are nearly always the offspring of the female attending the nest—that is, females rarely lay eggs in other females' nests. However, the nestlings often have different fathers. For example, 34 percent of nestlings in a population of red-winged blackbirds in Washington State were fathered by a male other than the owner of the territory in which the nest was located. All these other fathers were males holding nearby territories; fertile females went to those territories and solicited copulations from the males.

Females that copulated with more than one male raised more offspring than females that remained faithful to their mates. Their reproductive success improved because neighboring males that had copulated with a female were more likely to defend her nest against predators than males that had not copulated with her. These males also let females with whom they had copulated look for food on their territories. Also, there were fewer infertile eggs in nests with multiple fathers than in nests with single fathers. Males try to prevent their mates from copulating with other males, but they must leave them unguarded at times, both to feed and to seek copulations with other females. Because all fathers hold territories, on average males gain and lose equally from extra-pair copulations.

Social Behavior: Costs and Benefits

When animals interact with individuals of the same species, we refer to their behavior as **social**. Social behavior evolves when individuals that cooperate with others of the same species have, on average, higher rates of survival and reproduction than those achieved by solitary individuals. Associations for reproduction may consist of little more than a coming together of eggs and sperm, but individuals of many species associate for longer times to provide care for their offspring. Associating with conspecific individuals may also improve survival for reasons unrelated to reproduction, such as reduced risk of predation.

We will describe only a few animal social systems, but these examples demonstrate two important concepts. First, social systems are best understood not by asking why they benefit the species as a whole, but by asking how the individuals that join together benefit by the association. Second, social systems are dynamic; individuals constantly communicate with one another and adjust their relationships.

Group living confers benefits

Living in groups may confer many types of benefits. It may improve hunting success or expand the range of prey that can be captured. For example, by hunting together, social carnivores improve their efficiency in bringing down prey (Figure 50.10). Such cooperative hunting was a key component of the evolution of

50.10 Group Hunting Improves Foraging Efficiency By hunting together lionesses can kill larger animals than a single female could subdue.

human sociality. By hunting in groups, our ancestors were able to kill large mammals they could not have subdued as individual hunters. These social humans could also defend their prey and themselves from other carnivores (Figure 50.11).

Many small birds form tight flocks when they are attacked by a hawk (Figure 50.12a). Clumping deters the predator because it risks injury if it hits one of the prey with its wing while dashing into a compact group. Also, predators may have greater difficulty in approaching groups of prey individuals without being detected. When a trained goshawk was released near wood pigeons in England, the hawk was most successful when it attacked solitary pigeons. Success in capturing a pigeon decreased as the number of pigeons in the flock increased (Figure 50.12b). To be successful, a goshawk must get close to foraging pigeons without being spotted so that it can launch a surprise attack. The larger the flock of pigeons, the sooner some individual in the flock spotted the hawk.

Group living imposes costs

Living in a group typically imposes costs as well as benefits because individuals in groups may compete for food, interfere with one another's foraging, injure one another's offspring, inhibit one another's reproduction, or transmit diseases to their associates. The effects of group living on the survival and reproductive success of an individual also depend on its age, sex, size, and physical condition. Individuals may be larger or smaller than the average for their age and sex. Variation in skills, competitive abilities, and at-

50.11 Early People Defended Their Prey Having killed their prey, a band of *Homo habilis* drives rival predators—spotted hyenas and sabertooth cats—from a fallen dinothere (an extinct relative of modern elephants).

50.12 Flocking Provides Defense against Hawks
Birds in flocks are less likely to be captured by a predator because (a) a falcon risks injury if it dashes into a compact flock and (b) the more birds in a flock, the sooner they spot the predator.

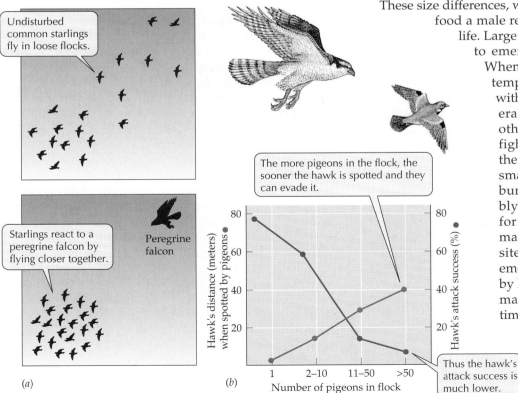

tractiveness to potential mates is often associated with these size differences.

The relative sizes of individuals may determine their success when performing different types of behavior. For example, the largest males of *Centris pallida*, a solitary bee of the American Southwest, are three times the size of the smallest males (Figure 50.13). These size differences, which depend on the amount of food a male receives as a larva, are fixed for life. Large males search for females about to emerge from their buried pupae. When they detect a female, they attempt to dig her up and copulate with her, but the digging takes several minutes and often attracts other males. These new arrivals fight with the original male, and the largest male usually wins. If a small male searched and dug for buried females, he would probably just serve as a female-finder for larger males. Instead, small males patrol potential pupation sites and wait for females that emerge without being discovered by large males. Intermediate-sized males sometimes dig and sometimes patrol.

Centris pallida

50.13 Size Can Affect Behavior The large male bee on the right is digging a female out of the soil, while the small male on the left waits for a female to emerge.

TABLE 50.1 Types of Social Ants

	Act benefits the recipient +	Act costs the recipient −
Act benefits the performer +	Cooperative	Selfish
Act costs the performer −	Altruistic	Spiteful

An almost universal cost associated with group living is higher exposure to diseases and parasites. Long before the causes of diseases were known, people knew that association with sick persons increased their chances of getting sick. Quarantine has been used to combat the spread of illness for as long as we have written records. The diseases of wild animals are not well known, but most of those that have been studied are spread by close contact.

Categories of Social Acts

Individuals living together perform many types of acts that are not performed by solitary animals. These acts can be grouped into four categories according to their effects on the interacting individuals (Table 50.1). An **altruistic act** benefits another individual at a cost to the performer. A **selfish act** benefits the performer but inflicts a cost on some other individual. A **cooperative act** benefits both the performer and the recipient. A **spiteful act** inflicts costs on both. These terms are purely descriptive; they do not imply conscious motivation or awareness on the part of the animal. If a genetic basis for a cooperative or selfish act exists, and if performing it increases the fitness of the performer, then the genes governing the act will increase in frequency in the lineage. In other words, cooperative or selfish behavior can evolve. We will now examine how altruistic behavior can evolve, both among close relatives and among unrelated individuals.

Altruism can evolve by means of natural selection

How could an act that *lowers* the performer's chances of survival evolve into a behavior pattern? One explanation for altruism lies in genetic relatedness: altruistic behaviors evolve most easily when performers and recipients are genetically related. Genetic relatedness, *r*, is the probability that an allele in one individual is a copy that is identical, by descent, to an allele in another individual. To calculate *r*, we construct a diagram showing the individuals and their common ancestors, linked across generations by arrows. Because meiosis takes place at each generational link, the probability that a copy of any allele will be passed on is 0.5. For *k* generational links, the probability is (0.5^k). To calculate *r*, we sum this value for all possible pathways between the two individuals. Some examples are diagrammed in Figure 50.14.

Genetic relatedness is important because, as we learned in Chapter 21, an individual may influence its fitness in two different ways. First, it may produce its own offspring, contributing to its individual fitness. Second, it may help relatives that bear the same alleles it does because they are descended from a common ancestor. This process is called kin selection. Together, individual fitness and kin selection determine the inclusive fitness of an individual. Occasional altruistic acts may eventually evolve into altruistic behavior patterns if the benefits of increasing the reproductive success of related individuals exceed the costs of decreasing the altruist's own reproductive success.

Many social groups consist of some individuals that are close relatives and others that are unrelated or distantly related. Individuals of some species recognize their relatives and adjust their behavior accordingly. White-fronted bee-eaters (Figure 50.15) are colonial African birds in which most breeding pairs are assisted by nonbreeding adults that help incubate eggs and feed nestlings. Both males and females help; individuals whose nests fail may help at other nests later in the same breeding season. Nearly all of these helpers assist close relatives. When helpers have a choice of two nests at which to help, about 95 percent of the time

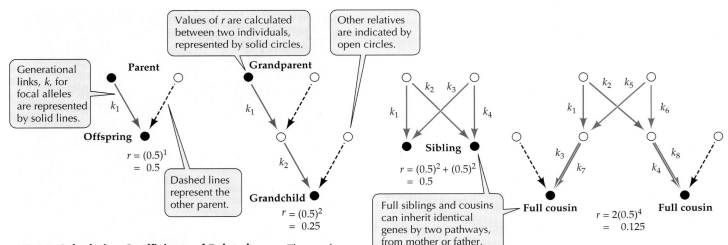

Generational links, k, for focal alleles are represented by solid lines.

Values of r are calculated between two individuals, represented by solid circles.

Other relatives are indicated by open circles.

Parent

Grandparent

k_1

Offspring
$r = (0.5)^1$
$= 0.5$

Dashed lines represent the other parent.

k_1

k_2

Grandchild
$r = (0.5)^2$
$= 0.25$

k_2 k_3

k_1 k_4

Sibling
$r = (0.5)^2 + (0.5)^2$
$= 0.5$

Full siblings and cousins can inherit identical genes by two pathways, from mother or father.

k_2 k_5

k_1 k_6

k_3 k_7

k_4 k_8

Full cousin

Full cousin
$r = 2(0.5)^4$
$= 0.125$

50.14 Calculating Coefficients of Relatedness The coefficient of relatedness, r, is the probability that an allele in one individual is identical, by descent, to an allele in another individual. To calculate r, we construct a diagram showing the individuals and their common ancestors, linked across generations by arrows.

they choose the nest with the young most closely related to them. Helping behavior among white-fronted bee-eaters is altruistic because individuals that help are not more successful when they become breeders than those that do not help, nor do they appear to gain any other advantage from helping. Extending help to nonrelatives does not improve fitness, which is why white-fronted bee-eaters rarely help nonrelatives.

Species whose social groups include sterile individuals are said to be **eusocial**. This extreme form of sociality has evolved in termites and some Hymenoptera (the ant, bee, and wasp order) in which worker females defend the group against predators or bring food to the colony, but do not reproduce. Some species have specialized soldiers with large defensive weapons (Figure 50.16). Workers are at risk of being killed while defending the colony.

How could eusociality evolve? The British evolutionist W. D. Hamilton first suggested that eusociality evolved because worker female hymenopterans benefit by helping to raise their sisters, to which they are more similar genetically ($r = 0.75$) than they would be to any offspring they produced on their own ($r = 0.50$). Sterile hymenopteran workers are more closely related to their sisters than they would be to their offspring

50.15 White-Fronted Bee-Eaters Are Altruists These African birds are perched near entrances to their nests. Tags (visible on one bird) enable investigators to identify individual birds; studies show bee-eaters often help raise young that are not their own, but that are closely related to them.

50.16 Sterile Individuals Are Extreme Altruists Eusocial insect species contain classes of sterile individuals. These individuals defend and provide food for the colony, but do not reproduce. These soldier army ants from Panama have evolved to contain powerful weaponry in the form of large jaws.

because members of the order Hymenoptera have an unusual sex determination system in which males are haploid, but females are diploid. A fertilized egg hatches into a female; an unfertilized egg hatches into a male. If a female mates with only one male, all the sperm she receives are identical because the haploid males have only one set of chromosomes, all of which are transmitted to each sperm cell. Therefore, a female's daughters share all of these genes. They also share, on average, half of the genes they receive from their mother. As a result, on average they share 75 percent of their alleles, rather than 50 percent as they would if both parents were diploid. Sisters are therefore genetically more similar to one another ($r = 0.75$) than a mother is to her daughters and sons ($r = 0.50$).

If Hamilton's hypothesis is correct, there should be a "genetic conflict" between workers and their queen mother. The queen, who is equally related ($r = 0.50$) to her sons and her daughters, would maximize her fitness by producing equal numbers of sons and daughters. Her daughters, however, which are more closely related to their sisters ($r = 0.75$) than they are to their brothers ($r = 0.25$), would maximize their inclusive fitness with an investment ratio of 75:25 (3:1) in favor of sisters.

The queen can control the sex of the eggs she lays, but the workers that care for and feed the larvae could skew the sex ratio of surviving offspring by giving more food to their sisters than to their brothers. In fact, among species in which the queen normally mates only once, workers invest much more in caring for females than in caring for males. If the founding queen is removed, one of her daughters typically becomes the new queen. She produces offspring that are nieces and nephews of the workers produced by the first queen ($r = 0.375$). As predicted by Hamilton's hypothesis, these manipulated colonies produce more males than colonies with their original queen.

However, Hamilton's hypothesis cannot account for many cases of eusociality. It cannot explain eusociality among the many hymenopteran species in which queens mate with many males. Nor can it explain eusociality in species, such as termites and naked mole-rats, in which both sexes are diploid. Naked mole-rats, the only eusocial mammals, live in underground colonies containing 70 to 80 individuals whose tunnel systems are maintained by sterile workers. Breeding is restricted to a single queen and several kings that live in a nest chamber in the center of the colony. Other females and males are sterile.

The inability of Hamilton's hypothesis to explain many aspects of eusocial behavior suggests that other environmental factors may also favor helping. One clue is provided by the fact that nearly all eusocial animals construct elaborate nests or burrow systems within which their offspring are reared (Figure 50.17).

50.17 Termite Mounds Are Large and Complex These immense Australian termite mounds are very costly to construct and maintain. Elaborate nests or burrows are a common characteristic of nearly all eusocial animals.

Forming a new colony and constructing these nests or tunnels is costly. Founding individuals are at high risk of being captured by predators, and most founding events fail. Thus, high predation rates, which favor cooperation among founding individuals, may facilitate the evolution of eusociality. Colonies founded by multiple queens grow more rapidly, and are more likely to be successful, than colonies founded by solitary queens.

Unrelated individuals may behave altruistically toward one another

It is easy to understand how cooperative behavior can evolve among related individuals. It is more difficult to understand the evolution of warnings of danger, sharing of food, and grooming among unrelated individuals of the same species or between members of different species. How can we explain the evolution of such cooperative behavior? The proposed answer is a model called **reciprocal altruism**. According to this model, reciprocal altruism evolves if a helper is in turn the recipient of beneficial acts by the individuals it has helped. If there is a genetic basis for the acts, natural selection may increase the frequency of alleles governing the cooperative behavior.

The Evolution of Animal Societies

The decisions animals make about where to settle, with whom to mate, what resources to invest in a reproductive effort, and when to terminate investment all help determine the type of social system that animals have. Today's social systems are the result of long periods of evolution, but there are few records of past social systems because behavior leaves few traces in the fossil record. Possible routes of the evolution of social systems must therefore be inferred primarily from current patterns of social organization. Fortunately, many stages of complexity exist among species, and the simpler systems provide clues about the stages through which the more complex ones may have passed. First we will examine the origins of all animal societies, which lie in the association of parents with their offspring. Then we will discuss how environmental factors affect social behavior. Finally, we will consider the influence of ancient evolutionary history on current social behaviors.

Parents of many species care for their offspring

Individuals of many species invest time and energy in caring for offspring. Parental care increases the chances of an offspring's survival, but it may reduce the ability of the parent to produce additional offspring. Parental care may also lower the chances of survival of the parent itself, because the parent could have used the time and energy to engage in other activities that would improve its own survival or reproductive success.

Males and females often differ strikingly in the kinds and amounts of parental investment they can and do make. Birds, mammals, and fishes illustrate these differences and why they exist. Only female mammals have functional mammary glands; males cannot produce milk. Males of most mammal species contribute nothing to offspring nutrition. Birds, on the other hand, do not produce milk. Among birds, all aspects of reproduction except production of eggs and sperm can be performed readily by both males and females. Not surprisingly, both males and females feed offspring in about 90 percent of bird species.

Sex roles among fishes differ from those of birds and mammals because most fish species do not feed their young. Parental care consists primarily of guarding eggs and young from predators (Figure 50.18). In many fish species, males are the primary guardians. A male can guard a clutch of eggs while attracting additional females to lay eggs in his nest. A female, on the other hand, can produce another clutch of eggs sooner if she resumes foraging immediately after mating than if she spends time guarding her eggs.

The most widespread form of social system is the family, an association of one or more adults and their

This dark area is a large brood of tiny young fish.

50.18 Cichlid Fish Guard Their Young Both parents vigorously defend their offspring for as long as 48 days.

dependent offspring. If parental care lasts a long time, or if the breeding season is longer than the time it takes for the young to mature, the adults may still be caring for younger offspring when older offspring reach the age at which they could help their parents. Many communal breeding systems probably evolved by this route. Florida scrub jays live all year on territories, each of which contains a breeding pair and up to six helpers (Figure 50.19). About three-fourths of the helpers are offspring from the previous breeding season that remain with their parents.

Most mammals evolved sociality via the extended family route. In simple mammalian social systems, solitary females or male–female pairs care for their young. As the period of parental care increases, older offspring are still present when the next generation is born, and they often help rear their younger siblings. In most social mammal species, female offspring remain in the group in which they were born, but males tend to leave, or are driven out, and must seek other social groups. Therefore, among mammals, most helpers are females.

The environment influences the evolution of animal societies

The type of social organization a species evolves is strongly related to the environment in which it lives. Among the African weaverbirds, species that live in forests eat insects, feed alone, and build well-hidden nests. Most of these species are monogamous, and males and females look alike. In marked contrast, weaverbirds that live in tree-studded grasslands called savannas eat primarily seeds, feed in large flocks, and nest in colonies, usually in isolated *Acacia* trees, where their nests are large and conspicuous (Figure 50.20). In most of these species, males have several mates—that

50.19 Cooperation among Florida Scrub Jays   Florida scrub jay helpers, most of which are offspring from the previous breeding season, are helping to feed nestlings and defend the nest against predators such as the approaching snake. By doing so, they are improving their inclusive fitness.

50.20 Savanna Weaverbirds Nest Colonially   Many African weaverbirds nest in colonies in isolated trees. Although these nests are highly conspicuous, it is difficult for most avian, mammalian, and reptilian predators to get to them at the tips of small branches.

is, they are polygynous—and are brightly colored, but females are not.

These striking differences probably evolved because nesting sites and food in forests are common and widely dispersed. Solitary pairs can use these resources more efficiently than animals in groups can. In savannas, good nesting trees are scarce and highly clumped, and nests are easy for predators to find. Males compete for these limited nest sites; the males that hold the best sites attract the most females. Males spend their time attempting to attract additional mates rather than helping to rear the offspring they already have, which explains the evolution of brighter plumage among males.

Among the herbivorous hoofed mammals of Africa, social organization and feeding ecology are correlated with the size of the animal (Table 50.2). Smaller animals have higher metabolic demands per unit of body weight than do larger ones (see Chapter 37). Therefore, smaller hoofed mammals are very selective in what they eat. They feed preferentially on high-protein foods such as buds, young leaves, and fruits. These foods are dispersed throughout forests, which also provide cover in which to hide from predators. Hiding is a tactic that is effective for solitary animals. The largest hoofed mammal species are able to eat lower-quality food, but they must process great quantities of it each day. They feed in grasslands with high standing crops of herbaceous vegetation, follow the rains to areas where grass growth is best, and live in large herds (Figure 50.21).

Living in herds makes it possible for males to compete among themselves for control of females, and for

TABLE 50.2 Social Organization and Ecology in African Hoofed Animals

SPECIES	BODY WEIGHT (KG)	FEEDING ECOLOGY	GROUP SIZE	SOCIAL ORGANIZATION
Dikdik, duiker	3–60	Selective browsing and grazing	1 or 2	Pair
Reedbuck, gerenuk	20–60	Selective browsing and grazing	2–12	Male with small harem
Impala, gazelles	20–250	Grazing, browsing	2–100	Territory-defending male with harem
Wildebeest, hartebeest	90–270	Grazing	Up to thousands	Herd in which males defend females
Eland, buffalo	300–600	Unselective grazing	Up to thousands	Herd with male dominance hierarchy

dominant males to defend medium to large harems. In these open environments, hiding is impossible, but it is also difficult for predators to approach the herds undetected. Species of intermediate sizes have feeding ecologies and social systems intermediate between those of smaller and larger species.

Among primates, nocturnal forest-dwelling insect eaters, such as lorises, some lemurs, and the owl monkey, live in pairs and are usually solitary foragers. Many diurnal species take insects and other animal food when they are available, but most of them eat fruits, seeds, and leaves. In Africa and Asia, group sizes are smallest among arboreal forest-dwelling species, whatever their diets, and largest among the ground-dwelling savanna species, such as baboons (Figure 50.22). In troops with more than one male,

Connochaetes sp.

50.21 Living in Large Herds East African wildebeest live in large herds that move from place to place to obtain the fresh green grass upon which they feed. Their major predators—lions—live on permanent territories and often have little to eat when the wildebeest herds are far away.

Papio sp.

50.22 Baboons in Groups Baboons forage in open savannas and travel in large groups. If a predator approaches, the formidable males cooperate in defending the group.

(a) Concentrated lek
(*Pipra serena*)

Several males dance, moving randomly in a small area.

(b) Dispersed lek
(*Masius chrysopterus*)

He lands…

…hops…

…and hops back.

(c) Cooperative lek
(*Chiroxiphia caudata*)

Two males cooperate in a "round dance."

50.23 Manakin Males Display in Leks   The three manakin species illustrated here display in different types of leks.

strong dominance hierarchies exist among the males, and one or two of the males do most of the copulating. Females may also have dominance relationships, and young females often assume the status of their mothers when they mature.

Social systems also reflect evolutionary history

Until now we have looked only at the current conditions, such as asymmetry in reproductive investment between the sexes and environmental conditions, that influence a species' social behavior. Until the development of new methods of determining the evolutionary relationships of species to one another, no other approach was possible. However, if a reliable phylogeny exists for an animal group, we can use it to tell whether differences in social behaviors among species in that lineage are determined solely by adaptation to current conditions or whether they also reflect longer-term history.

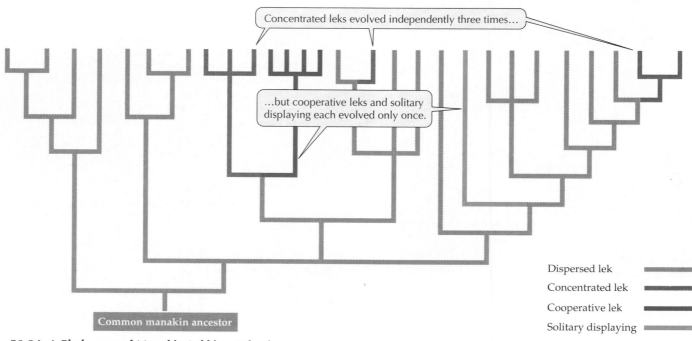

Concentrated leks evolved independently three times…

…but cooperative leks and solitary displaying each evolved only once.

Common manakin ancestor

Dispersed lek
Concentrated lek
Cooperative lek
Solitary displaying

50.24 A Phylogeny of Manakin Lekking Behaviors   This tree shows the evolutionary relationships of the manakins and the types of leks they use. Concentrated leks evolved several times.

This approach has been applied to the social systems of manakins, a group of about 40 species of small fruit-eating birds that live in tropical American forests. Male manakins form leks where females choose and copulate with males. Females then carry out all other reproductive activities unassisted by the males. Manakin leks differ in how close the displaying males are to one another and in whether the males coordinate and cooperate with one another as they display (Figure 50.23). The phylogeny of manakins indicates that lek-breeding behavior evolved in the common ancestor of all manakins, and that lekking has been lost only once during manakin social evolution. By contrast, leks in which the males display close to one another have evolved independently three times among manakins (Figure 50.24). Leks at which males coordinate their displays with one another have evolved independently five times. Cooperative leks and solitary displaying each evolved once.

Which manakin species uses which type of lek is not correlated with the type of environment in which it lives, but species that share recent common ancestors have similar lek types. This pattern indicates that once a particular type of lek evolved in a lineage, it was maintained among the descendants even if they came to live in different habitats. As more complete phylogenies become available, behavioral ecologists will be able to assess the roles of phylogenetic history and adaptation to current environments in the evolution of animal societies.

Summary of "Behavioral Ecology"

• Ecologists study the nature and consequences of interactions among organisms and their environments at all scales, from local to global.
• Behavioral ecology is the study of how animals decide where to carry out different activities, how to select the resources they need, how to respond to predators and competitors, and how to interact with other member of their own species.

Choosing Environments

• Selecting a habitat in which to live is one of the most important decisions an individual makes.
• The cues animals use to select habitats are good predictors of conditions suitable for future survival and reproduction. **Review Figure 50.1**
• Foraging theories were developed to understand how animals select prey items from the array present in the environment. **Review Figure 50.2**

Costs and Benefits of Behaviors

• Cost–benefit analyses of behavior are based on the principle that animals have only limited amounts of time and energy to devote to their activities.

Mating Tactics and Roles

• Individuals choose their associates, how to interact with them, and when to leave them. The most widespread choice of associates is the choice of a mate.
• Because males produce enough sperm to fertilize many more eggs than a single female can produce, males typically increase their reproductive success by mating with many females. The reproductive success of females, on the other hand, is typically limited by the cost of producing eggs. As a result, males usually initiate courtship and often fight for opportunities to mate with females. Females seldom fight over males and often reject courting males.
• Sexual selection often leads to exaggerated traits such as long tails. **Review Figure 50.4**
• During courtship, males signal their desirability as mating partners and may perform behaviors that increase the probability that their sperm will fertilize eggs. **Review Figure 50.5**
• By paying particular attention to those signals at which males cannot cheat, females have favored the evolution of "reliable" signals.
• Males of some species defend territories that contain food, nesting sites, or other resources. **Review Figures 50.6, 50.7, 50.8**

Social Behavior: Costs and Benefits

• Benefits of social living include better opportunities to capture prey and to avoid predators. **Review Figures 50.10, 50.11, 50.12**
• Costs of social living include competition for food, interference, and transmission of diseases.

Categories of Social Acts

• The acts performed by individuals living together can be grouped into four descriptive categories: altruistic, selfish, cooperative, and spiteful. **Review Table 50.1**
• Altruism among closely related individuals can evolve by means of natural selection because individuals that help close relatives can improve their inclusive fitness by kin selection. **Review Figure 50.14**
• Eusocial systems with sterile individuals have evolved among termites, hymenopterans (ants, bees, and wasps), and in one mammal, the naked mole-rat. **Review Figure 50.16**

The Evolution of Animal Societies

• The origin of most animal societies is the family, an association of one or more adults and their dependent offspring. **Review Figure 50.19**
• The type of social organization a species evolves is strongly related to the environment in which it lives. **Review Table 50.2 and Figures 50.20, 50.21, 50.22**
• Evolutionary history also influences a species' social organization. **Review Figures 50.23, 50.24**

Self-Quiz

1. Which of the following is *not* a component of the cost of performing a behavioral act?
   a. Its energetic cost
   b. The risk of being injured
   c. Its opportunity cost
   d. The risk of being attacked by a predator
   e. Its information cost

2. An almost universal cost associated with group living is
   a. increased risk of predation.
   b. interference with foraging.
   c. higher exposure to diseases and parasites.
   d. poorer access to mates.
   e. poorer access to sleeping sites.

3. An act is said to be altruistic if it
   a. confers a benefit on the performer by inflicting some cost on some other individual.
   b. confers a benefit both on the performer and on some other individual.
   c. inflicts a cost both on the performer and on some other individual.
   d. confers a benefit on another individual at some cost to the performer.
   e. imposes a cost on the performer without benefiting any other individual.

4. Which of the following statements about male and female roles in social systems is *not* correct?
   a. Females invest more in gamete production, but they may invest more or less than males in care of offspring.
   b. Biparental care is prevalent among birds.
   c. Males of most mammal species help feed offspring.
   d. Males with a high probability of parentage invest more in parental care than males that are less certainly related to the offspring of their mates.
   e. Among fishes, if there is unequal parental care by individuals of the two sexes, it is nearly always the male that does more.

5. Male and female mating tactics usually differ because
   a. males are typically larger than females.
   b. males do not contribute as many genes to their offspring as females do.
   c. males, but not females, usually can increase their fitness by mating with more than one partner.
   d. males can control copulations to their advantage.
   e. males and females occupy different positions when they copulate.

6. Choice of mating partner may be based on
   a. the inherent qualities of a potential mate.
   b. the resources held by a potential mate.
   c. both the inherent qualities of a potential mate and the resources it holds.
   d. the success of individuals of the opposite sex in courtship.
   e. all of the above

7. A lek is
   a. a territory held by a single male.
   b. a territory held by two or more males.
   c. a display ground at which a single male displays.
   d. a display ground at which two or more males display.
   e. a territory held by two or more females.

8. Among social birds, there are usually more male than female helpers in those species with helpers because
   a. males are better helpers than females.
   b. males typically receive greater benefits from helping.
   c. males survive better than females.
   d. mothers do not allow their daughters to help.
   e. males often need to wait for an unoccupied territory elsewhere.

9. Small African hoofed mammals are usually solitary because
   a. they feed on scattered high-quality foods in forested environments.
   b. the low quality of their food does not permit them to assemble in groups.
   c. they are too small to defend themselves against predators.
   d. they are too small to follow the rains to areas where grass growth is best.
   e. they are usually driven from their natal groups.

10. A phylogenetic analysis of lekking behavior in manakins shows that the type of lek a manakin species uses
    a. is correlated with the type of environment in which it lives.
    b. depends on how well males can coordinate their dancing.
    c. is similar among species that share a recent common ancestor.
    d. is unrelated to the phylogeny of manakins.
    e. evolves rapidly among closely related species.

Applying Concepts

1. Most hawks are solitary hunters. Swallows often hunt in groups. What are some plausible explanations for this difference? How could you test your ideas?

2. Because costs and benefits of behaviors can seldom be measured directly, behavioral ecologists often use indirect measures. What are the strengths and weaknesses of some of these indirect measures?

3. Among birds, males of polygamous species that display in leks are usually much larger and more brightly colored than females, whereas among species that form monogamous pairs, males are usually similar in size to females, whether or not they are more brightly colored. What hypotheses can be advanced to explain this difference?

4. Polyandry is a mating system in which one female has a "harem" of several males. Why is polyandry much rarer among both birds and mammals than polygyny, the situation in which one male mates with several females?

5. Many animals defend space, but the sizes of the territories they defend and the resources these areas provide vary enormously. Why don't all animals defend the same type and size of territory?

6. When frogs mate, a male clasps a gravid female behind her front legs and stays with her until she lays her eggs, at which time he fertilizes them. In most species of frogs, the male remains clasped to the female for a short time, usually no longer than a few hours. However, in some species, pairs may remain together for up to several weeks. In view of the fact that a male cannot court or mate with any other female while clasping one, and that

a female lays only a single clutch of eggs, why is it advantageous for males to behave this way? What can you guess about the breeding ecology of frogs that remain clasped for long periods? Why should females permit males to clasp them for so long? (Females do not struggle!)

7. Among vertebrates, helpers are individuals capable of reproducing, and most of them later breed on their own. Among eusocial insects, sterile castes have evolved repeatedly. What differences between vertebrates and insects might explain the failure of sterile castes to evolve in the former?

Readings

Alcock, J. 1998. *Animal Behavior*, 6th Edition. Sinauer Associates, Sunderland, MA. An excellent text of animal behavior from an evolutionary perspective. Contains additional material on all the topics discussed in this chapter.

Borgia, G. 1995. "Why Do Bowerbirds Build Bowers?" *American Scientist*, vol. 83, pages 542–547. Argues that courtship areas provide choosy females with easy avenues of escape.

Buss, D. M. 1994. "The Strategies of Human Mating." *American Scientist*, vol. 82, pages 238–248. Shows that people worldwide are attracted to the same qualities in the opposite sex.

De Waal, F. B. 1995. "Bonobo Sex and Society." *Scientific American*, March. Describes how bonobos, the most humanlike of the apes, live in peaceful societies in which females dominate the hierarchy and casual sex soothes all conflicts.

Emlen, S. T., P. H. Wrege and N. J. Demong. 1995. "Making Decisions in the Family: An Evolutionary Perspective." *American Scientist*, vol. 83, pages 148–157. Describes the complex social interactions in a family of white-fronted bee-eaters.

Gordon, D. M. 1995. "The Development of Organization in an Ant Colony." *American Scientist*, vol. 83, pages 50–57. Shows how simple, local decisions can generate a complex society.

Heinrich, B. and J. Marzluff. 1995. "Why Ravens Share." *American Scientist*, vol. 83, pages 428–437. Shows how by feeding in groups, ravens eat regularly even when food is scarce.

Krebs, J. R. and N. B. Davies. 1993. *An Introduction to Behavioral Ecology*, 3rd Edition. Blackwell Scientific Publications, Oxford. An excellent account of the methods and results of modern studies of behavioral ecology.

Narins, P. M. 1995. "Frog Communication." *Scientific American*, August. Discusses why loudly croaking frogs are not a chorus, but rather an assembly of males, each of which employs acoustic adaptations to make himself heard above the din.

Pfennig, D. W. and P. W. Sherman. 1995. "Kin Recognition." *Scientific American*, June. Describes how many animals can recognize close kin by assessing genetic similarities, by olfaction, or by their having grown up together.

Sherman, P. W. and J. Alcock. (Eds.). 1998. *Exploring Animal Behavior*, 2nd Edition. Sinauer Associatres, Sunderland, MA. An anthology of articles on behavior that have appeared in *American Scientist* over the last 20 years.

Thornhill, R. and J. Alcock. 1983. *The Evolution of Insect Mating Systems*. Harvard University Press, Cambridge, MA. A comprehensive account of the amazing variety of insect mating systems.

Wilson, E. O. 1975. *Sociobiology: The New Synthesis*. Harvard University Press, Cambridge, MA. A classic review of all aspects of social evolution and animal societies.

Chapter 51

Population Ecology

Loxodonta africana

Stages of Life
Female elephants groom young calves in Etosha National Park, Namibia. African elephants produce comparatively few offspring over their lifetime, and the females spend a great deal of energy caring for each one. Humans use a similar reproductive strategy.

*E*lephants reproduce at a slower rate than most animals. A female African elephant does not reach reproductive age until she is about 10 years old, after which she produces a single calf at a time. Because she nurses each calf for about 2 years and the gestation period of elephants is 22 months, a female produces only one calf every 4 years. Thus, although she may live for many decades, a female African elephant produces only 10 to 12 calves during her lifetime.

Female elephants live in small herds, each of which is led by an old, experienced female. Female calves remain in the herd with their mothers, but young males are expelled and live most of their lives in small bachelor herds. Males join family herds only when females are in breeding condition and they take no part in raising the offspring they sire.

The reproductive traits of all organisms have been molded by natural selection acting over many generations. In each lineage, those traits that maximized reproductive success were favored, but natural selection has not produced a single, dominant pattern of reproduction. Some organisms, like elephants and humans, usually give birth to a single offspring in each reproductive episode; others spawn thousands or millions of eggs in one bout. Some organisms begin to reproduce within days or weeks of being born; others live for many years before reproducing. The traits that an individual expresses during its life constitute its **life history**. Life history traits influence how many individuals of a species are found in an area and how dramatically their densities vary.

Why are life history traits so variable? What causes a species to be common or rare? Why do the sizes of populations fluctuate yearly, seasonally, or not at all? Why is a species common in some parts of its geographic range and rare in others? What determines the limits of the ranges of species? Answering such questions is the major task of population ecology.

A **population** consists of all the individuals of a species within a given area. The boundaries of the area are usually determined by the goals of a scientific study. To understand population dynamics, population ecologists count individuals in different places and try to determine the factors that influence birth, death, immigration, and emigration rates. In this chapter we will discuss how and why the sizes of populations of species vary over space and time, and show how this ecological knowledge is used to predict and manage the growth of populations.

Varied Life Histories

The complete life history of an organism consists of its birth, growth to maturity, reproduction, and death. During its life an individual organism ingests nutrients or food, grows, interacts with other individuals of the same and other species, reproduces, and usually moves or is moved so that it does not die exactly where it was born. Life histories describe how an organism divides its efforts among these activities.

All life histories are based on a certain set of traits: size at birth; how fast the individuals grow; how long they live; how many times they reproduce; the rate and pattern of growth and development; the ages at which they die; how much they move around; the number, size, and sex composition of their offspring; and the number and timing of reproductive events in an organism's life.

Many organisms, such as the red-spotted newt (*Notophthalmus viridescens*), a colorful salamander that breeds in small ponds throughout eastern North America, have complex life histories. The olive-colored, red-spotted adult newts feed and lay their eggs in ponds. The eggs hatch into larvae that grow and develop in the ponds for several months. Eventually the larvae lose their gills and change into *efts*, immature newts that leave the water to live on land for 4 to 9 years (Figure 51.1). The bright red efts are highly toxic to their natural predators.

Most efts return to the ponds in which they were born to reproduce, but some travel long distances and colonize new ponds. Newly created farm ponds in rural areas and wildlife management ponds in national forests are populated quickly by red-spotted newts, showing that they are good colonizers of new habitats. Thus, during its life, a red-spotted newt goes through three distinct stages—larva, eft, and adult—and lives in two different habitats. We will now describe in more general terms the life history stages illustrated by red-spotted newts.

Notophthalmus viridescens

51.1 A Red Eft The toxic red eft is the terrestrial dispersal stage in the life cycle of the red-spotted newt.

Life histories include stages for growth, change, and dispersal

For at least part of their lives, all organisms grow by gathering and assimilating energy and nutrients. Some organisms, such as red-spotted newts, gather energy and nutrients throughout their lives, even after they reach adult size and stop growing. Energy gathered after growth stops maintains organisms and supports reproduction. In many species, however, energy gathering is confined to a particular stage. Most moths, for example, feed only when they are larvae. The adults lack mouthparts and digestive tracts, live on energy gathered by the larvae, and survive only long enough to disperse, mate, and lay eggs. Having different stages specialized for different activities increases the efficiency of performing a particular activity.

Individuals of many species also change form during their lives. Human babies are unmistakably human, but newborns of many species differ dramatically from adults. Some of the most striking changes are found among insects such as beetles, flies, moths, butterflies, and bees, which undergo radical metamorphoses from their larval to their adult forms (see Chapter 29). These metamorphoses take place during the pupal stage, during which larval tissues and organs are broken down, and adult tissues and organs are constructed from the larval material. Many plants have resting stages, such as spores and seeds, that have low metabolic rates and are highly resistant to changes in the physical environment. Growth typically does not take place in these stages, but dispersal is common.

At some time in their lives, all organisms disperse. Some, such as plants and sessile animals, disperse as small eggs, larvae, spores, or seeds. Others, such as insects and birds, disperse primarily as adults. Still others may disperse during several different stages. Individuals of some species can change their location many times during their lives in response to environmental changes. Others must remain in the first place they settle.

Life histories embody trade-offs and constraints

Life history traits are molded by natural selection, but they also involve trade-offs. Changes that improve fitness by means of one life history trait often reduce fitness by means of another. For example, a higher number of births is often traded off for better survival of offspring already produced. What are the major trade-offs in life history traits?

A universal trade-off exists between number and size of offspring. Every newborn individual begins to grow with energy and nutrients from its maternal parent, but how much energy and nutrients individuals receive from their mothers varies greatly. Orchid seeds receive very little, grass seeds slightly more. Coconuts, birds, and placental mammals receive large amounts of maternal energy. The larger the amount of energy provided to each offspring, the larger it can grow before it must gather its own energy, but the fewer offspring a parent can produce for a given amount of energy—a major trade-off.

A trade-off also exists between number of offspring produced and the amount of care parents provide to their offspring after birth. Individuals of many species are completely independent of their parents during their growth periods, obtaining all their own energy and providing for their own protection. However, in some animal species, parents provide additional care and protection that may extend, as it does in many birds and mammals, until the offspring have reached adult size. The more parental care the parents provide, the fewer offspring they can produce for the same investment in reproduction.

Different species produce their offspring at different times, and produce different numbers of offspring in a given batch (known as a *clutch* or *litter* in animals or a *seed crop* in vascular plants). Some organisms reproduce only once and then die. A bacterium that forms two daughter cells may be considered to have died when it divided. Some plants (called *annuals*) invest so much of the energy they gain during their single growing season in seed production that they do not survive long after reproducing. Most insects and spiders also live for less than a year and die soon after reproducing.

Some longer-lived organisms also reproduce once in their lifetimes and die very soon afterward. Pacific salmon (genus *Oncorhynchus*) hatch in fresh water, spend a number of years at sea, return to fresh water, spawn, and die. Most agaves (century plants) of the American Southwest likewise store up energy for many years before producing a large flowering stalk, forming many seeds, and dying (Figure 51.2). Yucca plants, which grow in the same environments, appear

51.2 Big Bang Reproduction This century plant has mobilized the energy stored during its long life to produce a large flowering stalk with hundreds of flowers, literally reproducing itself to death.

similar, but they invest less in each reproduction and live to reproduce many times. If two individuals have the same amount of energy to invest in reproduction, and one reproduces only once while the other reproduces several times, the former can produce more offspring in a single episode than the latter because it reserves no energy for its own future survival.

A major trade-off exists between reproduction and survival. Studies of many species of plants, shrimp, snails, fishes, birds, and mammals show that engaging in reproduction reduces adult survival rates. Female red deer that produce fawns and nurse them die at higher rates than females that do not produce fawns. The more an adult invests in reproduction, the shorter its lifetime.

A conflict between reproduction and growth is a major life history trade-off. Members of many species do not begin to reproduce until they have reached full size, but others, such as most plants, mollusks, fishes, and reptiles, start to reproduce while they are still relatively small and continue to reproduce as they grow. Reproduction usually reduces growth because these two processes compete for the limited amount of energy an individual has at its disposal. Beech trees in

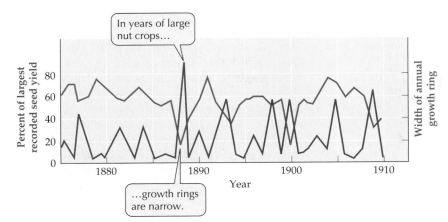

51.3 Reproduction Slows Growth Rates in Beech Trees The width of annual growth rings (plotted in red) reveals the growth rates of beech trees in different years. The trees grow slowly in years when they produce large crops of beechnuts.

Germany, for example, grew more slowly during years when they produced large seed crops than they did during years when their seed crops were small (Figure 51.3).

Offspring are like "money in the bank"

The potential contribution of an individual's offspring to future generations depends upon when they are produced. A useful analogy compares the production of offspring to earning interest on money deposited in a bank. It pays to deposit money in the bank as soon as possible so that it can begin earning interest. Offspring produced early in an adult's life likewise "yield interest" quickly—that is, they begin to reproduce sooner than offspring produced later. However, if juvenile survival is very poor and reproduction greatly reduces the life span of the parent, natural selection may favor delaying reproduction until the parent is older. For example, most birds begin to reproduce when they are 1 year old, but gulls, penguins, and albatrosses do not breed until they are 3 to 9 years old. Although individuals in these three groups reach full size within a year, they are not efficient foragers at this age and cannot gather enough energy to reproduce without jeopardizing their survival and future reproduction.

Reproductive value is the average number of offspring that remain to be born to individuals of a particular age. A newborn individual does not have the highest reproductive value, even though it has its full reproductive potential ahead of it, because many newborn individuals die before they have a chance to reproduce. Therefore we must discount the number of offspring an individual could produce if it survived by the chance that it will die before reaching reproductive age, or during reproduction. When we make the appropriate calculations, we find that the reproductive value of an individual steadily increases until it begins to reproduce, at which time it can no longer die before starting to produce offspring. After maturity, repro-

ductive value usually declines; in most species, it reaches zero when the individual has finished reproducing. However, individuals can still have positive reproductive value after they have stopped reproducing if they continue to assist the survival of their offspring and grandoffspring.

Because reproductive value declines with age in a mature individual, the power of natural selection acting on alleles that first produce their phenotypic effects at older ages grows increasingly weaker. Once reproductive value has dropped to zero, natural selection cannot influence phenotypic traits, even those that are highly detrimental to the individual's survival. As a result, increasing numbers of harmful alleles are expressed as individuals age, causing increased mortality rates, especially after reproduction has ceased. In this manner, **senescence**—an increased probability of dying per unit of time—has evolved.

Senescence poses serious social problems for people in modern industrial societies. As a result of improved hygiene and nutrition, most people in these societies are now spared the serious childhood infections that cause death rates to be high in nonindustrial societies. Most people live to the age when the so-called genetic diseases of old age begin to afflict them. Cancer and heart disease, the main killers in industrialized societies, are much more difficult to deal with than the contagious diseases that formerly caused most deaths. For this reason, despite the expenditure of enormous resources to extend life, the average age at death in the United States has changed very little during the past 30 years. As one source of mortality is eliminated, another takes its place (Figure 51.4). Life history theory suggests that this problem is likely to continue indefinitely.

Population Structure: Patterns in Space

At any given moment, an individual organism occupies only one spot and is of one particular age. The members of a population, however, are distributed over space and differ in age and size. These features determine **population structure**. Ecologists study population structure at different spatial scales ranging from local subpopulations to entire species. Spatial

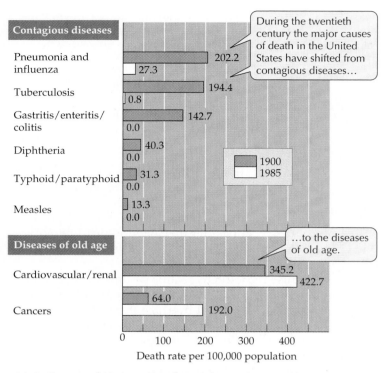

Contagious diseases

Pneumonia and influenza: 202.2 / 27.3

Tuberculosis: 194.4 / 0.8

Gastritis/enteritis/colitis: 142.7 / 0.0

Diphtheria: 40.3 / 0.0

Typhoid/paratyphoid: 31.3 / 0.0

Measles: 13.3 / 0.0

During the twentieth century the major causes of death in the United States have shifted from contagious diseases…

■ 1900
□ 1985

Diseases of old age

…to the diseases of old age.

Cardiovascular/renal: 345.2 / 422.7

Cancers: 64.0 / 192.0

Death rate per 100,000 population

51.4 Causes of Human Death   Today in the United States most people die of diseases of old age.

distributions of organisms influence the stability of populations and affect interactions among species. Geneticists and evolutionary biologists also study population structure, but they are interested primarily in distributions of genotypes and their degree of isolation from one another because that component of population structure influences how populations evolve (see Chapter 21).

Population density is an important feature

The number of individuals of a species per unit of area (or volume) is its **population density**. Ecologists are interested in population densities because dense populations often exert strong influences on their members and on populations of other species. Other scientists—those who work in agriculture, conservation, or medicine, for example—wish to manage species to raise their densities (crop plants, aesthetically attractive species, threatened or endangered species) or reduce their densities (agricultural pests, disease organisms). To manipulate population densities, we must know what factors make populations grow and shrink and how these factors work.

Because organisms and their environments differ, densities are measured in more than one way. Ecologists usually measure the densities of organisms in terrestrial environments as the number of individuals per unit of area, but number per unit of volume is

generally a more useful measure for organisms living in water. For species whose members differ markedly in size, as do most plants and some animals (such as mollusks, fishes, and reptiles), the total mass of individuals—their **biomass**—may be a more useful measure of density than the number of individuals.

Sometimes individuals can be counted directly without missing any of them or counting any of them twice, but this process is usually impossible or too laborious. Ecologists commonly estimate population densities by sampling a population in a representative area and extending their findings to a larger area. Estimates of population size can also be made by marking and recapturing individuals. For example, if we capture and mark 100 individuals in a population, we can take another sample later and determine what percentage of individuals captured in that sample were already marked. If 10 percent of the captured individuals were already marked, we would conclude that the population contained about 1,000 individuals.

Spacing patterns tell us much about populations

Ecologists studying population structure look at the way the individuals in a population are spaced. Spacing patterns often reveal why individuals settled and survived where they did. Individuals of a population may be tightly clumped together, evenly spaced,

51.5 Competitive Spacing   In the sand dunes of this Australian desert, plant spacing is regulated by the availability of water. Each plant removes so much water from the surrounding soil that young plants are unable to establish themselves in the bare areas.

or randomly scattered. Distributions can become clumped when young individuals settle close to their birthplaces, when suitable habitat patches are "islands" separated by unsuitable areas, or by chance. The spacing of many plants is a result of competition for light, water, and soil nutrients (Figure 51.5), or because they rub against one another when moved by wind or water currents. Among animals, defense of space is the most common cause of even distributions (see Figure 50.8). Random distributions may result when many factors interact to influence where individuals settle and survive.

Age distributions reflect past events

Populations are composed of individuals ranging from newborns to postreproductive adults. The proportions of individuals in each age group in a population make up its **age distribution**. The density and spacing of individuals are spatial attributes of a population; age distribution is a *temporal* (time-oriented) attribute. The timing and rates of births and deaths determine age distributions. If birth rates and death rates are both high, a population is dominated by young individuals. If birth rates and death rates are low, a relatively even distribution of individuals of different ages results. The age distribution of a population reveals much about its recent history of births and deaths.

The timing of births and deaths may influence age distributions for many years in populations of long-lived species. The human population of the United States is a good example. Between 1947 and 1964, the United States experienced what is called the post–World War II baby boom. During these years, average family size grew from 2.5 to 3.8 children; an unprecedented 4.3 million babies were born in 1957. Birth rates declined during the 1960s, but Americans born during the baby boom will constitute the dominant age class into the twenty-first century (Figure 51.6*a*). "Baby boomers" became parents in the 1980s, producing another bulge in the age distribution—a baby boom echo—but they had, on average, fewer children than their parents, so the bulge is not as large. Fish populations may also be dominated by individuals of one age group. In Lake Erie, 1944 was such an excellent year for reproduction and survival of whitefish that individuals of that age group dominated whitefish catches in the lake for several years (Figure 51.6*b*).

Population Dynamics: Changes over Time

At any moment in time, a population has a particular structure determined by the distribution of its mem-

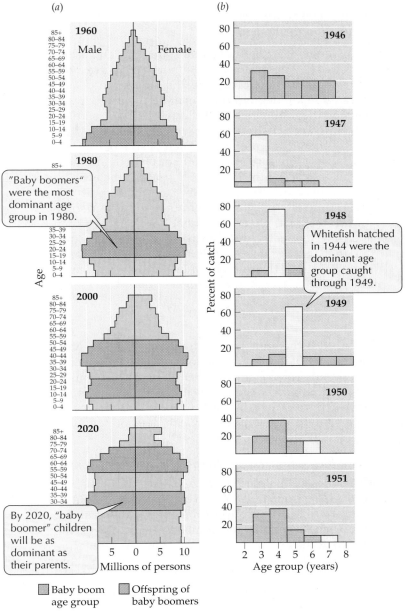

51.6 Age Distribution Can Be Influenced by the Timing of Births Individuals born during years of high birth rates may dominate age distributions for many years in populations of long-lived individuals; two examples are (*a*) humans in the United States and (*b*) whitefish in Lake Erie.

bers in space and their ages. However, as we have just seen, population structure is not static. Changes in population structure influence whether the population will increase or decrease; that is, they determine the *dynamics* of the population. We will now examine how ecologists measure changes in birth and death rates and use that information to understand changes in population densities.

Births, deaths, immigration, and emigration drive population dynamics

Knowledge of when individuals are born and when they die provides a surprising amount of information about a population. Births, deaths, immigration, and emigration are **demographic events**—events that determine the numbers of individuals in a population. Ecologists measure the *rates* at which these events take place—that is, the number of such events per unit of time. These rates are influenced both by environmental factors and by the life history traits of the species. The number of individuals in a population at any given time is equal to the number present at some time in the past, plus the number born between then and now, minus the number that died, plus the number that immigrated, minus the number that emigrated. That is, the number of individuals at a given time, N_1, is given by the equation

$$N_1 = N_0 + B - D + I - E$$

where N_1 is the number of individuals at time 1; N_0 is the number of individuals at time 0; B is the number of individuals born, D the number that died, I the number that immigrated, and E the number that emigrated between time 0 and time 1. If we measure these rates over many time intervals, we can determine why a population's density changes.

Life tables summarize patterns of births and deaths

Life tables can help us visualize patterns of births and deaths in a population. We can construct a life table by determining for a group of individuals born at the same time—a **cohort**—the number still alive at specific times and the number of offspring they produced during each time interval. An example is shown in Table 51.1.

As you can see from the table, members of a cohort of the grass *Poa annua* began producing seeds some time after they were 3 months old and continued to produce seeds for the rest of their lives. By the end of 2 years, all members of the cohort were dead. Note that the life table includes both numbers observed and rates calculated from those numbers. For example, we can calculate the probability of dying during the 6–9 month age interval by dividing the number of individuals that died by the number alive at the beginning of the interval: $211/527 = 0.4$. The data in Table 51.1 show that the number of seeds produced per individual peaks at 6 months of age, and that individuals produce very few seeds after they are a year old.

Ecologists often use graphs to highlight the most important changes in birth and death rates in populations. Graphs of survivorship—the mirror image of death rate—in relation to age show when individuals survive well and when they do not. To interpret survivorship data, ecologists have found it useful to compare real data with several hypothetical curves that illustrate a range of possible survivorship patterns. A useful type of graph plots the proportion of individuals of a cohort that are alive at different times during their total potential life span (Figure 51.7a). At one extreme, nearly all individuals survive for their entire potential life span and die almost simultaneously (hypothetical curve I). At the other extreme, the survivorship of young individuals is very low, but survivorship is high for the remainder of the life span (hypothetical curve III). An intermediate possibility is that survivorship is the same throughout the life span (hypothetical curve II).

Survivorship data from real populations often resemble one of these hypothetical curves. For example, survivorship of *Poa annua* seedlings is very high for the first 6 months, but then, as in hypothetical curve I, it declines significantly in older individuals (Figure

TABLE 51.1 Life Table for a Cohort of 843 Individuals of a Short-Lived Grass[a]

AGE INTERVAL (PERIOD BETWEEN TWO EXACT AGES, IN MONTHS)	NUMBER ALIVE AT BEGINNING OF AGE INTERVAL	PROPORTION ALIVE AT BEGINNING OF AGE INTERVAL	NUMBER DYING DURING AGE INTERVAL	DEATH RATE FOR AGE INTERVAL	NUMBER OF SEEDS PRODUCED DURING AGE INTERVAL
0–3	843	1.000	121	0.144	0
3–6	722	0.857	195	0.270	300
6–9	527	0.625	211	0.400	620
9–12	316	0.375	172	0.544	430
12–15	144	0.171	95	0.625	210
15–18	54	0.064	39	0.722	60
18–21	15	0.018	12	0.800	30
21–24	3	0.004	3	1.000	10
24	0	0.000	0	0.000	0

[a]*Poa annua*

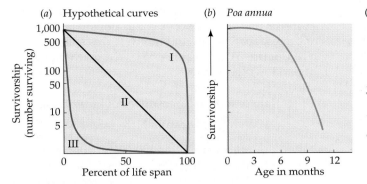

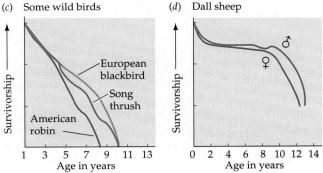

51.7 Survivorship Curves Survivorship curves show the proportion of individuals of a cohort still alive at different times over the life span. (*a*) The range of possible survivorship patterns. (*b*) *Poa annua*, a type of grass. (*c*) Three species of thrushes. (*d*) Male and female dall sheep.

51.7*b*). Many wild birds have survivorship curves similar to hypothetical curve II; the probability of their surviving is about the same over most of the life span once individuals are a few months old (Figure 51.7*c*).

A more common survivorship pattern, especially among organisms that produce large numbers of offspring, each of which receives little energy and no subsequent parental care, is one with low survivorship of young individuals, followed by high survivorship during the middle part of the life span, and then low survivorship toward the end of the life span. Dall sheep, although they do not have a high birth rate, have such a survivorship curve (Figure 51.7*d*), one that combines the first part of the type III curve with the middle and late parts of the type I curve. Survivorship curves help us understand how birth and death rates change over time, but the data ecologists use to estimate the curves are incomplete. None of them include deaths of zygotes prior to birth or seeds prior to germination.

Males and females of a single species may have different survivorship curves. For example, survivorship of adult female red deer is nearly constant once they reach reproductive age (2 to 3 years old), although females that reproduce do not survive as well as nonreproductive females. By contrast, the death rates of males rise sharply when they reach their breeding age of 7 to 8 years (Table 51.2) because males engage in heavy combat with each other during the breeding season. Many males are injured during these fights.

Individuals are not the most important units in modular organisms

For the kinds of organisms we have considered so far, called **unitary organisms**, individuals are easy to distinguish, and most adult members of a population are similar to one another in size and shape. For example, all populations of mollusks, echinoderms, insects, and vertebrates consist of unitary organisms. But not all organisms are unitary.

Modular organisms are organisms whose bodies consist of repeated units. The fertilized egg of a modular organism develops into a unit of construction—a module—which then produces additional modules much like itself. Many plants are modular, and there are many important groups of modular protists, fungi, and animals (sponges, corals, moss animals, colonial tunicates). Modular organisms may grow to large sizes, and it is often difficult to distinguish a modular organism from a cluster of genetically separate individuals (Figure 51.8).

The effect of modular organisms on their environment often depends primarily on the number and size of the modules. An aspen clone, which is a single genetic individual, affects a much larger area than a single tree does. The modules of a single organism may differ markedly in size and age. Therefore, students of modular organisms are often concerned primarily with the number, size, and shape of modules rather than with the number of genetically distinct individuals. But all population ecologists are interested in how

TABLE 51.2 Death Rates of Male and Female Red Deer on Rhum Island, Scotland

	DEATH RATE[a]	
AGE	FEMALES	MALES
0–6 months	12.4	12.4
6–12 months	15.0	20.8
Yearlings	7.4	13.0
2 years	1.1	1.8
3–4 years	3.6	1.7
5–6 years	3.8	2.2
7–8 years	2.3	6.1
9–10 years	2.8	16.3
11–12 years	8.7	37.0

[a] Percentage of those alive at beginning of year

51.8 A Single Modular Organism May Look Like a Population Each clump of these quaking aspens is a single genetic individual that has spread by underground roots and has sent up many stems, each of which appears to be a separate tree.

population densities change over time, the topic to which we now turn.

Exponential Population Growth

If a single bacterium selected at random from the surface of this book, and all its descendants, were able to grow and reproduce in an unlimited environment, an explosive population growth would result. In a month this bacterial colony would weigh more than the visible universe and would be expanding outward at the speed of light. Similarly, a single pair of Atlantic cod and their descendants, reproducing without hindrance, would fill the Atlantic Ocean in 6 years.

All populations have the potential for explosive growth because as the number of individuals in the population increases, the number of new individuals added per unit of time accelerates, even if the rate of growth per individual—called the per capita growth rate—remains constant. This form of explosive increase is called **exponential growth**. If for the moment

we ignore immigration and emigration and assume that births and deaths occur continuously and at constant rates, such a growth pattern forms a continuous curve (Figure 51.9a) that is expressed mathematically in the following way:

Rate of increase in number of individuals

$$= \left(\begin{array}{c} \text{Average per capita birth rate} \\ - \text{ Average per capita death rate} \end{array} \right)$$

$\times$ Number of individuals

or, more concisely,

$$\frac{\Delta N}{\Delta t} = (b - d)N$$

where $\Delta N / \Delta t$ is the rate of change in size of the population (ΔN = change in number of individuals; Δt = change in time). The difference between the average per capita birth rate (b) and the average per capita death rate (d) is called r. In these equations, b includes both births and immigration and d includes both deaths and emigration. When conditions are optimal for the population, r has its highest value, called r_{max}, the **intrinsic rate of increase**; r_{max} has a characteristic value for each species. Therefore, the rate of growth of a population under optimal conditions is $\Delta N / \Delta t = r_{max} N$.

However, as we have already seen, for many organisms, births and deaths are not continuous processes. Many insects and annual plants reproduce only once each year and then die. Many other species reproduce only during a short period of the year—the breeding season—but deaths occur continuously during the year. These populations have discrete generations, and they are modeled with a different equation:

$$N_{t+1} = \lambda N_1$$

51.9. Exponential Growth (a) A theoretical exponential growth curve. (b) The growth of a pheasant population introduced to Protection Island, Washington State.

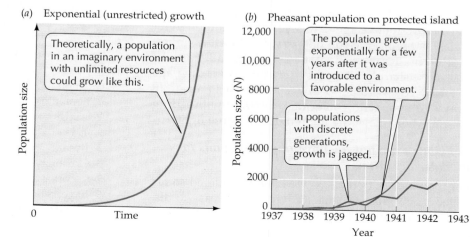

(a) Exponential (unrestricted) growth

Theoretically, a population in an imaginary environment with unlimited resources could grow like this.

Population size

0   Time

(b) Pheasant population on protected island

The population grew exponentially for a few years after it was introduced to a favorable environment.

In populations with discrete generations, growth is jagged.

Population size (N)

12,000
10,000
8000
6000
4000
2000
0
1937 1938 1939 1940 1941 1942 1943
Year

where λ, the **finite rate of increase**, is the ratio of the population size during the next time period to the population size during the current time period. The growth curve of a population with a single breeding season each year resembles a saw blade, with a sharp vertical increase during the breeding season followed by a gradual decline due to deaths during the rest of the year (see Figure 51.9*b*).

Populations may experience exponential growth when they are introduced to or colonize previously unoccupied but favorable environments. For example, in 1937, eight pheasants were introduced onto Protection Island off the coast of Washington State. The island, which had abundant food and no mammalian predators, was too far from the mainland for pheasants to fly to and from it. By 1942, the population had increased to nearly 2,000 birds. Because pheasants lay eggs only once each year, in spring, the growth curve is jagged. The rate of population growth initially was close to the theoretical maximum (Figure 51.9*b*). However, by 1942, the population was no longer growing exponentially, as it did during the first few years. In the next section we explore why growth rates in all populations eventually slow down, often very soon.

Population Growth in Limited Environments

No real population can maintain exponential growth for very long because environmental limitations cause birth rates to drop and death rates to rise. In fact, over long time periods, the size of most populations fluctuates around a relatively constant number. The simplest way to picture the limits imposed by the environment is to assume that an environment can support no more than a certain number of individuals of any particular species. This number, called the environmental **carrying capacity**, is determined by the availability of resources—food, nest sites, shelter—as well as by disease, predators, and perhaps social interactions.

Population growth is influenced by the carrying capacity

The limitations of the carrying capacity mean that, rather than being exponential, population growth follows a curve that flattens out as the population approaches the carrying capacity, so that the curve has an S-shape (Figure 51.10). This is what happened to the pheasant population on Protection Island.

The S-shaped growth pattern, which is characteristic of many populations growing in environments with limited resources, can be represented mathematically by adding to the equation for exponential growth a term, $(K - N)/K$, that slows the population's growth as it approaches the carrying capacity. The simplest such equation is that for **logistic growth**, in which

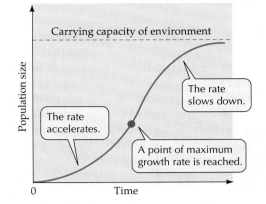

51.10 Logistic Growth Typically, a population in an environment with limited resources stops growing exponentially when it reaches the environmental carrying capacity.

each individual added to the population depresses population growth equally:

$$\frac{\Delta N}{\Delta t} = r \left(\frac{K - N}{K} \right) N$$

where K is the carrying capacity and the other symbols are the same as in the equation for exponential growth. The biological assumption in this equation is that each additional individual makes things slightly worse for the others because it competes with them for available resources, or for other reasons. Population growth stops when $N = K$ because then $(K - N) = 0$, so $(K - N)/K = 0$, and thus $\Delta N/\Delta t = 0$.

The logistic growth equation contains some important simplifications that are not true for most populations. Its most critical assumptions are that (1) each individual exerts its effects immediately at birth; (2) all individuals produce equal effects; and (3) births and deaths are continuous. However, in nature, organisms grow during their lives, and their effects on others normally increase with age, so there may be a delay between the birth of an individual and the time at which it begins to affect the other members of its population. A seedling tree exerts a much smaller effect on its neighbors than a large adult tree does, and it does not begin to reproduce until it reaches a relatively large size.

In addition, the logistic equation models a population in a single, uniform habitat patch. Individuals that live elsewhere may immigrate into the local population and some individuals may leave it, but we have modeled only the density of individuals in the local population. Next we consider the dynamics of an assembly of local populations, which is more complex than the growth of a single population.

Many species are divided into discrete subpopulations

Populations that are divided into discrete subpopulations, among which some exchange of individuals occurs, are called **metapopulations** (populations of populations). Each subpopulation of a metapopulation has a probability of birth (colonization) and death (extinction). Within each subpopulation, growth occurs in the ways we have just discussed, but because subpopulations are typically much smaller than the metapopulation as a whole, local disturbances and random fluctuations in numbers of individuals often cause the extinction of a subpopulation. However, if individuals move frequently between subpopulations, immigrants may prevent declining subpopulations from becoming extinct. This process is known as the **rescue effect**.

The bay checkerspot butterfly (*Euphydryas editha bayensis*) provides a good illustration of metapopulation dynamics. The larvae of this butterfly feed on only a few species of annual plants that are restricted to outcrops of a particular kind of rock on hills south of San Francisco (Figure 51.11). The bay checkerspot has been studied for many years by Stanford University biologists. During drought years, most host plant individuals die early in spring, before the butterfly larvae have completed their development. At least three butterfly subpopulations became extinct during a severe drought in 1975–1977. The largest patch of suitable habitat, Morgan Hill, typically supported thousands of butterflies. It probably served as a source

of individuals that dispersed to and colonized small patches where the butterflies had become extinct.

During the severe California drought of the 1990s, many subpopulations of the bay checkerspot became extinct. During the spring of 1996, no butterflies were found in any of the subpopulations. The metapopulation had become extinct. Drought was a major factor causing the extinction, but suburban expansion that destroyed or diminished the sizes of some of the patches probably also contributed to it.

The logistic growth equation is one way of expressing the fact that the births, deaths, and movements of organisms are influenced by the densities of individuals of the same and other species. The logistic growth equation incorporates only the influence of other individuals of the same species, but even this simple inclusion provides the foundation for considering the regulation of the density and distributions of individuals of a species, the topic to which we now turn.

Population Regulation

In a population that is growing logistically, growth slows down as density increases because the members increasingly affect one another adversely. As a result, a population above the environmental carrying capacity is more likely to decrease in density than one that is below the carrying capacity. In this section we discuss how populations may be regulated by interactions between their density and the carrying capacity of their environment, by disturbances, and by movements of individuals.

How does population density influence birth and death rates?

If the density of a population is determined primarily by changes in per capita birth or death rates in response to density, it is said to be regulated by **density-dependent** factors. Death or birth rates may be density-dependent for several reasons. First, as a species increases in abundance, it may deplete its food supply, reducing the amount of food that each individual gets. Poor nutrition may increase death rates and lower birth rates. Second, predators may be attracted to regions where densities of their prey have increased. If predators are able to capture a larger proportion of the prey than they did when the prey were scarce, the per capita death rate of the prey rises. Third, diseases, which may increase death rates, spread more easily in dense populations than in sparse populations.

If the per capita birth and death rates in a population are unrelated to its density, population regulation is said to be **density-independent**. A very cold spell in winter, for example, may kill a large proportion of the individuals in a population regardless of its density. However, even density-*independent* environmental fac-

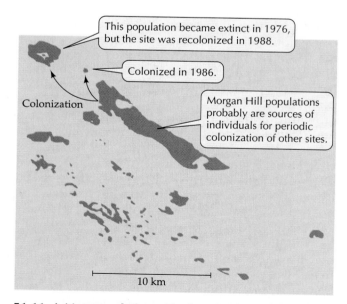

51.11 A Metapopulation The bay checkerspot butterfly metapopulation was divided into a number of subpopulations confined to habitats that contain the food plant of its larvae. (The entire metapopulation has since become extinct.)

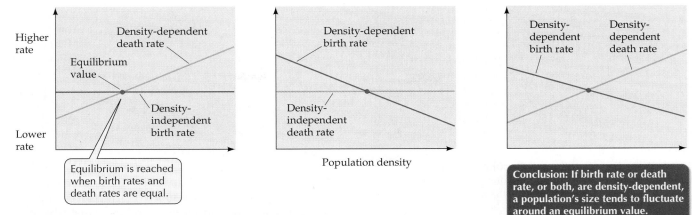

Equilibrium is reached when birth rates and death rates are equal.

Conclusion: If birth rate or death rate, or both, are density-dependent, a population's size tends to fluctuate around an equilibrium value.

51.12 Density-Dependent Factors Regulate Population Size Densities of all populations fluctuate, but fluctuations are diluted by density-dependent birth and death rates.

tors may result indirectly in density-*dependent* mortality. The cold weather, for example, may not kill organisms directly, but may increase the amount of food individuals need to eat each day. Individuals pushed by population density into poorer foraging areas may be more likely to die than those in better foraging areas. Or the death rate may be related to the quality of sleeping places. If population density is high, a larger proportion of individuals may be forced to sleep in places that expose them to the cold.

Various combinations of density-dependent and density-independent events can regulate the density of a population. The hypothetical graphs in Figure 51.12 show how birth and death rates can change in relation to population density. When birth and death rates are equal—the point at which the lines cross—the population neither grows nor shrinks. If birth or death rates or both are density-dependent, the population responds to increases or decreases in its density by returning toward an equilibrium density. If neither rate is density-dependent, there is no equilibrium. The abundance of a species is determined by the combined effects of all the factors and processes, density-dependent and density-independent, impinging upon its populations.

Ecologists often infer the operation of density-dependent processes by observing correlations between population density and birth and death rates. Stronger evidence is provided by experiments in which densities are manipulated. An experimental test of density-dependent population regulation was performed on a gall-forming fly (*Eurosta solidaginis*) that infests goldenrod (*Solidago altissima*), a common herbaceous plant found in fields in eastern North America. *Eurosta* is widespread, but it usually exists at moderately low and relatively constant densities. Adult flies emerge in early June from galls in which they have overwin-

tered. A female fly lays her eggs in a bud on a growing goldenrod stem. After it hatches, the larva bores into the stem and develops inside a gall, where it feeds on the plant's tissues. The gall forms rapidly, reaching full size in early July (Figure 51.13). In September, the fully developed larva pupates within the gall after excavating an exit tunnel through which it will emerge.

Eurosta larvae and pupae are attacked by a suite of predators that includes a parasitoid wasp, a beetle, and birds. To determine whether predation rates on *Eurosta* act in a density-dependent manner, an ecologist collected all the galls from a number of plots, kept them in an unheated room during the winter, and released the flies and their wasp predators into the fields

51.13 A Gall-Forming Fly This goldenrod was stimulated to form multiple galls when a female *Eurosta* laid her eggs on the stem. A larval fly will hatch from each gall.

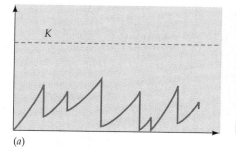
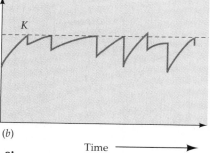
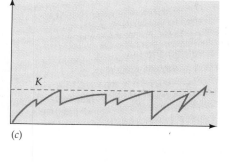

(a)

(b)

(c)

Time ⟶

51.14 Disturbances Influence Population Size
(a) Dynamics of a population dominated by phases of population growth after repeated disturbances. The population density is well below the environmental carrying capacity (K) most of the time. (b) Dynamics of a population dominated by limitations on environmental carrying capacity. The population density is close to K most of the time. (c) Same as in (b), but with a much lower carrying capacity.

at the appropriate time the following spring. On some plots, more flies were released than had been present the previous year; on others, fewer flies were released—in other words, population densities were augmented or reduced. At the end of the following winter all galls were collected, and survival rates of *Eurosta* were determined by dissecting the galls. Death rates during the year were strongly density-dependent; that is, they were highest in the experimental plots where the greatest density of flies had been released. Thus, populations of *Eurosta* in the study area appear to be maintained at relatively constant levels by density-dependent predation.

Disturbances affect population densities

Populations are repeatedly exposed to **disturbances**—short-term events that disrupt populations by changing their environment and, hence, its carrying capacity. Common physical disturbances are fires, hurricanes, ice storms, floods, landslides, and lava flows. Biological disturbances include tree falls, disease epidemics, and the burrowing and trampling activities of animals. Disturbances differ in their spatial distribution, frequency, predictability, and severity.

A disturbance may lower the environmental carrying capacity for only a short time. When it recovers, populations may grow rapidly. In general, smaller organisms are more strongly affected by environmental disturbances than larger ones are. Most insect populations in the temperate zone are constantly recovering from disturbances. In contrast, many bird species appear to be close to the environmental carrying capacity much of the time. Insect populations often behave according the pattern shown in Figure 51.14a, whereas birds generally follow a pattern that is similar to those in Figures 51.14b and c.

The responses of organisms to disturbances depend upon the frequency and severity of those disturbances. Organisms respond behaviorally and physiologically to disturbances that occur regularly and repeatedly during their lifetimes. Animals seek shelter in storms and go into hibernation in winter. Trees drop their leaves in winter and change physiologically so that they can tolerate frosts. However, if such a disturbance is unusually severe, the tolerances of individuals may be exceeded, and many may die. An organism is more likely to have adaptations for tolerating a particular kind of disturbance if the disturbance is frequent relative to the organism's life span.

Populations of organisms can also influence the frequency of some disturbances. Immediately after a fire, there is not enough combustible organic matter to carry another fire. However, as vegetation grows back, dead wood, branches, and leaves accumulate, gradually increasing the supply of fuel to support another fire. Thus the frequency of fires may be proportional to the rate at which fuel accumulates through the growth of plant populations, or the rate at which herbivores consume plant materials that would otherwise accumulate. Similarly, as many trees age, their roots become weakened by fungal infections. Old, large trees are thus susceptible to being toppled by high winds. Therefore, the likelihood of a major blowdown increases with forest age.

Organisms often cope with habitat changes by dispersing

A common response of animals to disturbances or other environmental changes is dispersal. If habitat quality declines greatly, individuals may be able to improve their survival and reproduction by going elsewhere. If regularly repeated seasons cause changes in a habitat, organisms may adjust their life cycles in equally regular ways, appearing to anticipate the changes.

One of the most spectacular responses to seasonal changes in habitat quality is **migration,** the regular seasonal movement of animals from one place to another. This behavior is most widespread among birds, but some insects, such as monarch butterflies, and

(a) *Danaus* sp.

(b) *Rangifer tarandus*

51.15 Animals Migrate to Remain in Suitable Environments (a) Most of North America's monarch butterflies migrate to central Mexico. They spend the winter in cool mountain valleys, where they can reduce their energy expenditure by keeping their metabolic rates low. (b) These caribou in the American Arctic are migrating from the open tundra to their winter feeding grounds at the edge of the boreal forest, where food, some of it in the form of lichens on the branches of trees, is more readily available when the ground is covered with snow.

some mammals also migrate (Figure 51.15). The primary function of migration is to keep the animals in good foraging areas at all times of the year. In arctic regions, caribou migrate each year between winter and summer ranges (see Figure 50.21). Most insectivorous birds leave high latitudes in autumn for more favorable wintering grounds at low latitudes.

The limits of species ranges are dynamic

Environmental conditions, births, deaths, and dispersal determine the density and distribution of a species. Within the geographic ranges of most species, population densities are higher toward the center of the range and decline toward the periphery, becoming zero at the limit of the range of the species. The scissor-tailed flycatcher (*Tyrannus forficatus*), a bird that winters in Central America and breeds in the southern Great Plains of the United States, illustrates this pattern of population density (Figure 51.16). The density pattern of scissor-tailed flycatchers is simple because the region where they breed is relatively flat. Species

that live in more mountainous regions typically have many centers of high population density.

Some species achieve their highest population densities at the margins of their ranges. This situation often arises when range boundaries are at the edges of land masses, where there is an abrupt transition from suitable terrestrial habitats to unsuitable marine habitats. Competition between ecologically similar species can also result in abrupt changes in population densities at the edges of ranges, as we will see in Chapter 52.

Humans Manage Populations

For many centuries, people have tried to decrease populations of species they consider undesirable and

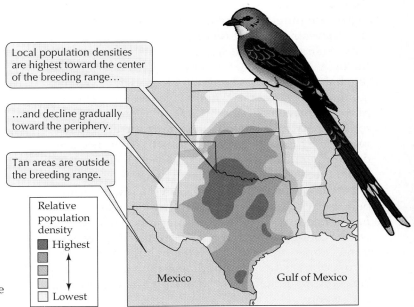

Local population densities are highest toward the center of the breeding range…

…and decline gradually toward the periphery.

Tan areas are outside the breeding range.

Relative population density

Highest
↕
Lowest

Mexico

Gulf of Mexico

51.16 Population Densities Are Greater toward the Center of a Range The population density pattern of the scissor-tailed flycatcher is simple because its breeding range is relatively flat.

increase populations of desirable species. Strategies for controlling and managing populations of organisms are based on our understanding of how populations grow and are regulated. A general principle of population dynamics is that the total number of births and the growth rates of individuals tend to be highest when a population is well below its carrying capacity. Therefore, if we wish to maximize the number of individuals that can be harvested, we should manage a population so that it is far enough below carrying capacity to have high birth and growth rates. Hunting seasons for birds and mammals are determined with this objective in mind.

Life history traits determine how heavily a population can be exploited

Populations of some organisms can sustain their growth despite a high rate of harvest. Such populations (many species of fish, for example) reproduce at high rates, with each female laying thousands or millions of eggs. Another characteristic of these high-yielding populations is that individual growth is density-dependent. If prereproductive individuals are harvested at a high rate, the remaining individuals may grow faster. Many fish populations can be harvested heavily for many years because only a modest number of females must survive to reproductive age to produce the eggs needed to maintain the population.

Fish can, of course, be overharvested. Many populations have been greatly reduced because so many individuals were harvested that too few reproductive adults survived to maintain the population. The Georges Bank off the coast of New England—a source of cod, halibut, and other prime food fishes—has been exploited so heavily that many fish stocks have been reduced to levels insufficient to support a commercial fishery.

The whaling industry has also engaged in excessive harvests. The blue whale, Earth's largest animal, was the first whale species to be hunted nearly to extinction. The industry then turned to smaller species of whales that were still numerous enough to support commercially viable whaling operations (Figure 51.17). Management of whale populations is difficult for two reasons. First, whales reproduce at very low rates—they have long prereproductive periods, produce only one offspring at a time, and have long intervals between births. Thus many whales are needed to produce even a small number of offspring. Second, because whales are distributed widely throughout Earth's oceans, they are an international resource whose conservation and wise management depends upon cooperative action by *all* whaling nations. The recovery of whale populations will require observance by all nations of a moratorium on commercial harvesting.

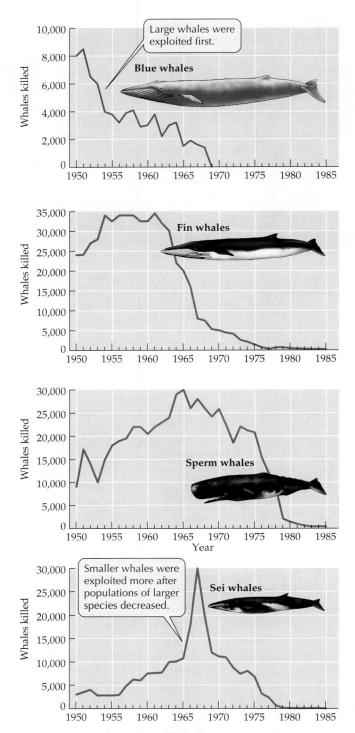

51.17 Overexploitation of Whales The graphs show the number of whales of four species killed each year from 1969 to 1985.

Life history information is used to control populations

The same principles apply if we wish to reduce the size of populations of undesirable species and keep

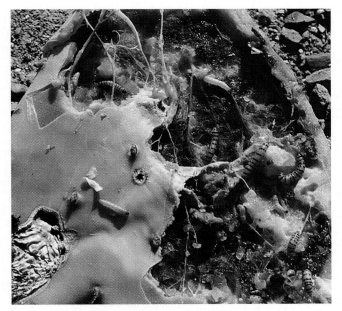

51.18 Biological Control of a Pest *Cactoblastis* caterpillars consume an *Opuntia* cactus, a pest in Australia.

ment. We can rid our dumps and cities of rats more easily by making garbage unavailable (reducing the carrying capacity of the rats' environment) than by poisoning the rats.

Similarly, if we wish to preserve a rare species, the most important step usually is to provide it with suitable habitat. If habitat is available, the species will usually reproduce at rates sufficient to maintain its population. If the habitat is insufficient, preserving the species usually requires expensive and continuing intervention, such as providing extra food.

Humans often introduce predators and parasites to regulate populations of undesirable species. For example, the cactus *Opuntia*, introduced into Australia from South America, spread rapidly and became a common pest species over vast expanses of valuable sheep-grazing land. It was controlled by introducing a moth species (*Cactoblastis cactorum*) whose larvae hatch in and completely destroy patches of *Opuntia* found by the egg-laying females (Figure 51.18). However, new patches of cactus arise in other places from seeds dispersed by birds. These new patches flourish until they are found and destroyed by *Cactoblastis*. Today, over a large region, the numbers of both *Opuntia* and *Cactoblastis* are fairly constant and low, but in the local areas that make up the whole, there are vigorous oscillations resulting from the extermination of first the plant and then the herbivore.

Can we manage our human population?

Managing our own population has become a matter of great concern because the size of the human population is responsible for most environmental problems,

them at low densities. At densities well below carrying capacity, populations have high birth rates, and can therefore withstand higher death rates than they could closer to carrying capacity. Killing part of a density-dependently regulated population only brings it back to the point at which it experiences the most rapid rate of growth. A far more effective approach to reducing the population of a species is to remove its resources, thereby lowering the carrying capacity of its environ-

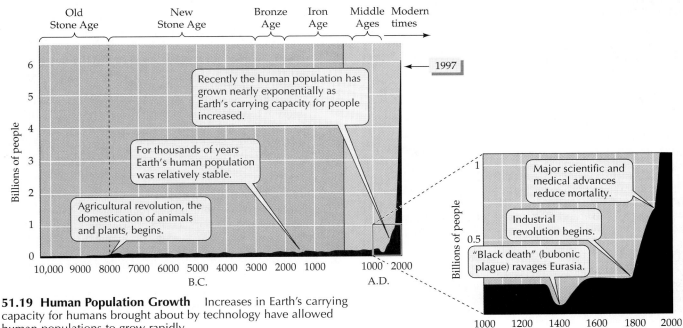

51.19 Human Population Growth Increases in Earth's carrying capacity for humans brought about by technology have allowed human populations to grow rapidly.

from pollution to extinctions of other species. For thousands of years, Earth's carrying capacity for human populations was set at a low level by food and water supplies and disease—that is, by the technology available to garner resources and combat diseases. Domestication of plants and animals and cultivation of the land enabled our ancestors to increase resources at their disposal dramatically. These developments stimulated rapid population growth up to the next carrying capacity limit, which was determined by the agricultural productivity possible with only human- and animal-powered tools. Agricultural machines and artificial fertilizers, made possible by the tapping of fossil fuels, greatly increased agricultural productivity, further raising Earth's carrying capacity for humans. The development of modern medicine reduced the effectiveness of disease as a limiting factor on human populations, raising the global carrying capacity still further (Figure 51.19). Medicine and better hygiene have allowed people to live in large numbers in areas where diseases formerly kept numbers very low.

What is Earth's present carrying capacity for people? Today's carrying capacity is set by Earth's ability to absorb the by-products of our enormous consumption of fossil fuel energy and by whether we are willing to cause the extinction of millions of other species to accommodate our increasing use of environmental resources. We will explore some of the consequences of high human population densities for the survival of other species in Chapter 55.

Summary of "Population Ecology"

Varied Life Histories

• The life history of a species consists of growth, dispersal, and reproductive stages. Many organisms have complex life histories.
• A population consists of all the individuals of a species within a given area.
• Trade-offs inevitably exist between number and size of offspring, between number of offspring and parental care, between survival and reproduction, and between growth and reproduction. **Review Figures 51.2, 51.3**
• Reproductive value is the average number of offspring that remain to be born to individuals of a particular age. Reproductive value rises to a peak when individuals first begin to reproduce and then declines to zero after reproduction ceases.

Population Structure: Patterns in Space

• The number of individuals of a species per unit of area (or volume) is its population density. Dense populations often exert strong influences on populations of other species.

• Its age distribution reveals much about the recent history of births and deaths in a population. The timing of births and deaths may influence age distributions for many years. **Review Figure 51.6**

Population Dynamics: Changes over Time

• Births, deaths, immigration, and emigration drive changes in population density and distribution.
• Life tables help us visualize patterns of births and deaths in a population. **Review Table 51.1**
• Graphs of survivorship in relation to age show when individuals survive well and when they do not. **Review Figure 51.7**

Exponential Population Growth

• All populations have the potential to grow exponentially when they colonize suitable environments. **Review Figure 51.9**

Population Growth in Limited Environments

• No population can maintain exponential growth for very long because environmental limitations cause birth rates to drop and death rates to rise.
• The number of individuals of a particular species that an environment can support—called the carrying capacity—is determined by the availability of resources and by disease and predators.
• A population in a constant but limited environment at first grows rapidly, but growth rates decrease as the carrying capacity is approached. **Review Figure 51.10**
• Metapopulation dynamics are determined by "births" (colonizations) and "deaths" (extinctions) of local subpopulations. Immigrants may prevent declining subpopulations from becoming extinct, a process known as the rescue effect. **Review Figure 51.11**

Population Regulation

• Regulation of a population by changes in per capita birth or death rates in response to density is said to be density-dependent.
• If per capita birth and death rates are unrelated to a population's density, population regulation is said to be density-independent.
• The abundance of a species is determined by the combined effects of all density-dependent and density-independent factors affecting it. **Review Figure 51.12**
• Populations of many organisms are below carrying capacity most of the time because they are recovering from disturbances. **Review Figure 51.14**
• Population densities of most species are high near the center of the species' range and decline toward the periphery. **Review Figure 51.16**

Humans Manage Populations

• Humans use the principles of population dynamics to control and manage populations of desirable and undesirable species. Nevertheless, many populations have been overexploited. **Review Figures 51.17, 51.18**
• Earth's carrying capacity for humans has been increased several times by technological developments. Whether the current human population exceeds Earth's carrying capacity is hotly debated. **Review Figure 51.19**

Self-Quiz

1. The number of individuals of a species per unit of area is known as its
   a. population size.
   b. population density.
   c. population structure.
   d. subpopulation.
   e. biomass.

2. The age distribution of a population is determined by
   a. the timing of births.
   b. the timing of deaths.
   c. the timing of both births and deaths.
   d. the rate at which the population is growing.
   e. all of the above

3. Which of the following is *not* a component of the life history of all organisms?
   a. Growth
   b. Dispersal
   c. Reproduction
   d. Reorganization
   e. Energy gathering

4. Which of the following is *not* a demographic event?
   a. Growth
   b. Birth
   c. Death
   d. Immigration
   e. Emigration

5. A group of individuals born at the same time is known as a
   a. deme.
   b. subpopulation.
   c. Mendelian population.
   d. cohort.
   e. taxon.

6. A population grows at a rate closest to its intrinsic rate of increase when
   a. its birth rates are the highest.
   b. its death rates are the lowest.
   c. environmental conditions are optimal.
   d. it is close to the environmental carrying capacity.
   e. it is well below the environmental carrying capacity.

7. Some organisms reproduce only once in their lifetimes because they
   a. invest so much in reproduction that they have insufficient reserves for survival.
   b. produce so many offspring at one time that they do not need to survive longer.
   c. don't have enough eggs to reproduce again.
   d. don't have enough sperm to reproduce again.
   e. have stopped growing.

8. Which of the following is *not* true of reproductive value?
   a. Reproductive value is the average number of offspring that remain to be born to an individual of a particular age.
   b. The reproductive value of an individual increases until it begins to reproduce.
   c. Reproductive value reaches its maximum when an individual completes reproduction.
   d. Reproductive value usually reaches its maximum when an individual begins to reproduce.
   e. Reproductive value usually declines during the reproductive life of an individual.

9. Density-dependent population regulation results when
   a. only birth rates change in response to density.
   b. only death rates change in response to density.
   c. diseases spread in populations at all densities.
   d. both birth and death rates change in response to density.
   e. population densities fluctuate very little.

10. The best way to reduce the population of an undesirable species in the long term is to
    a. reduce the carrying capacity of the environment for the species.
    b. selectively kill reproducing adults.
    c. selectively kill prereproductive individuals.
    d. attempt to kill individuals of all ages.
    e. sterilize individuals.

Applying Concepts

1. Huntington disease is a severe disorder of the human nervous system that generally results in death. It is caused by a dominant allele that does not usually express itself phenotypically until its bearer is 35 to 40 years old. How fast is the gene causing Huntington disease likely to be eliminated from the human population? How would your answer change if the gene expressed itself when its bearer was 20 years old? 10 years old?

2. Many people have improperly formed wisdom teeth and must spend considerable sums of money to have them removed. Assuming, as is probably the case, that the presence or absence of wisdom teeth and their mode of development are partly under genetic control, will we gradually lose our wisdom teeth by evolutionary processes?

3. Some organisms, such as oysters and elm trees, produce vast quantities of offspring, nearly all of which die before they reach adulthood. What fraction of such deaths are likely to be selective—that is, dependent on the genotypes of the individuals dying? If in fact most such deaths are nonselective, what does that imply for the rates of evolution of oysters and elms?

4. Most organisms whose populations we wish to manage for higher densities are long-lived and have low reproductive rates, whereas most organisms whose populations we attempt to reduce are short-lived but have high reproductive rates. What is the significance of this difference for management strategies and effectiveness of management practices?

5. In the mid-nineteenth century, the human population of Ireland was largely dependent upon a single food crop, the potato. When a disease caused the potato crop to fail, the Irish population declined drastically for three reasons: (1) a large percentage of the population emigrated to the United States and other countries; (2) the average age of a woman at marriage increased from about 20 to about 30 years; and (3) many families starved to death rather than accept food from Britain.

None of these social changes was planned at the national level, yet all contributed to adjusting population size to the new carrying capacity. Discuss the ecological strategies involved, using examples from other species. What would you have done had you been in charge of the national population policy for Ireland?

6. From a purely ecological standpoint, can the problem of world hunger ever be overcome by improved agriculture alone? What other components must a hunger-control policy include?

Readings

Begon, M., J. L. Harper and C. R. Townsend. 1996. *Ecology: Individuals, Populations and Communities*, 3rd Edition. Blackwell Scientific Publications, Oxford. A basic text for all aspects of contemporary ecology.

Begon, M. and M. Mortimer. 1996. *Population Ecology: A Unified Study of Animals and Plants*, 3rd Edition. Blackwell Scientific Publications, Oxford. This introduction to population dynamics stresses the differences between plants and animals.

Charnov, E. L. 1993. *Life History Invariants: Some Explorations of Symmetry in Evolutionary Ecology*. Oxford University Press, Oxford. An excellent but technical treatment of phylogenetic constraints and trade-offs in life histories.

Gotelli, N. J. 1995. *A Primer of Ecology*. Sinauer Associates, Sunderland, MA. A unique introduction to models of population growth, population structure, and metapopulation dynamics. Has problems to work at the end of each chapter.

Harper, J. L. 1977. *The Population Biology of Plants*. Academic Press, New York. The most complete review of the literature on plant populations; an excellent advanced reference for most topics in plant population biology.

Myers, J. H. 1993. "Population Outbreaks in Forest Lepidoptera." *American Scientist*, vol. 81, pages 240–251. Illustrates the role of viruses in causing the remarkable population cycles of tent caterpillars and other forest insects.

Ricklefs, R. E. 1997. *Ecology*, 4th Edition. W. H. Freeman and Company, New York. This text covers both the dynamic and the evolutionary aspects of ecology.

Stearns, S. C. 1992. *The Evolution of Life Histories*. Oxford University Press, Oxford. Thorough coverage of all aspects of life history evolution, including a discussion of the major analytic tools used in life history analyses.

Chapter 52

Community Ecology

Snow Triggers Community Interactions
Weather conditions in the mountains above the Four Corners desert affected the vegetation, which affected the insect and deer mouse populations, which resulted in the spread of hantaviruses carried by the mice. When predators controlled the deer mouse population, incidence of hantavirus-related respiratory illness among humans also declined.

During May and June, 1993, many people in the Four Corners region of Arizona, Colorado, New Mexico, and Utah contracted a rare respiratory disease. Scientists at the Center for Disease Control and Prevention (CDC) tested the blood of people stricken with the disease and found antibodies to hantaviruses in their tissues. Hantaviruses infest between 4,000 and 200,000 people each year in Europe and Asia; between 4,000 and 20,000 of them die. Because rodents are vectors of hantaviruses in the Old World, CDC scientists contacted ecologists in the Four Corners region to find out whether anything unusual had happened to rodent populations in the region. The ecologists, who had been studying rodents in the area for many years, knew that unusual events had indeed occurred.

Heavy snow and rain in the mountains during the previous winter and spring had resulted in an abundant crop of piñon pine cones and unusually large insect populations. Deer mice feasted on the abundant piñon nuts and grasshoppers; their population increased tenfold between May 1992 and May 1993. CDC scientists found that about 30 percent of the deer mice in the region were infected by hantaviruses, suggesting that they were the primary vectors of the disease. Fortunately, the population of deer mice declined rapidly during the summer of 1993 from 30 per hectare in June to 4 per hectare in August, probably because of heavy predation by foxes, owls, and snakes.

The rise and fall of populations of deer mice in the Four Corners region was the result of interactions between the mice, their food supply, and their predators. The populations of these organisms also interact with populations of other plants and animals, fungi, protists, bacteria, and archaebacteria. All of the organisms that live in a particular area constitute an **ecological community**. Each of the species interacts in unique ways with other species in its community. Some of these interactions are strong and important; others are weak and affect the functioning of the community very little. The study of such interactions, and how they determine which and how many species live in a place, is the focus of community ecology. In this chapter we consider the niches of organisms and the types of ecological interactions—predator–prey, competition, mutualism, and commensalism—and show how they determine the structure and functioning of ecological communities.

Ecological Interactions

Deer mice, piñon pines, hantaviruses, and the other organisms in the Four Corners region, and in all other ecological communities, interact with one another in four major ways (Table 52.1). One organism, by its activities, may benefit itself while harming the other, as when individuals of one species eat individuals of the other. The eater is called a predator or parasite and the eaten is its prey or host. These interactions are known as **predator–prey** or **host–parasite interactions** (a +/− interaction). Alternatively, two organisms may mutually harm one another. This type of interaction is common when two organisms use the same resources and those resources are insufficient to supply their combined needs. Such organisms are called competitors, and their interactions constitute **competition** (a −/− interaction). If both participants benefit from an interaction, we call them mutualists, and their +/+ interaction is a **mutualism**. If one participant benefits but the other is unaffected, the interaction is a **commensalism** (a +/0 interaction). If one participant is harmed but the other is unaffected, the interaction is an **amensalism** (a 0/− interaction).

Figs and fig wasps illustrate the complexity of ecological interactions

The categories of species interactions are not clear-cut, both because the strengths of interactions vary and because many cases do not fit the categories neatly. That

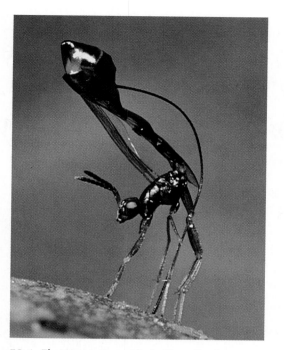

52.1 Fig Wasps Have Their Own Predators   A female parasitoid wasp bores through a fig with her long ovipositor to lay her eggs on the fig wasp larvae inside.

nature is more complex than this simple classification system is illustrated by interactions between figs and fig wasps. Most species of fig trees can reproduce only with the help of certain wasps. Fig wasps of most species visit only one fig species, and most fig species are visited by only one wasp species. A fig tree begins reproduction by producing a large number of closed flower clusters, within which many female flowers form. A female fig wasp bearing both her own fertilized eggs and fig pollen enters a cluster through a small hole, which soon seals. She pollinates receptive female flowers, lays her eggs in the ovaries of some of the flowers, and dies. Each wasp larva develops within—and eats—one seed. Not all of the larvae develop; some may be parasitized by another wasp species. A parasitic female wasp may puncture the fig with her long ovipositor and lay her eggs on the fig wasp larvae (Figure 52.1).

TABLE 52.1   Types of Ecological Interactions

		EFFECTS ON ORGANISM 2		
		BENEFIT	**HARM**	**NO EFFECT**
EFFECTS ON ORGANISM 1	**BENEFIT**	Mutualism	Predation or parasitism	Commensalism
	HARM	Predation or parasitism	Competition	Amensalism
	NO EFFECT	Commensalism	Amensalism	—

As the fig develops, pollen-bearing male flowers mature. At this point, the young male fig wasps chew their way out of the seeds in which they developed and crawl around inside the fig searching for seeds housing female wasps. The males chew open these seeds and mate with the females. The fertilized females emerge from their seeds and collect pollen from male flowers. The females leave the fig through a hole cut in its wall by the males, fly to another fig tree, and begin the reproductive cycle again. The fig finishes ripening, becoming a soft, sweet fig that may eventually be eaten by a bird or mammal.

Fig wasps depend completely on figs to complete their life cycles. To produce offspring, a female fig wasp must carry pollen to a receptive cluster and pollinate its flowers. Otherwise the ovaries of the flowers will not develop into seeds that can be eaten by her offspring. In some fig species, only one female wasp normally enters a cluster. If she is not carrying pollen, none of her offspring survive. A fig tree pays a price to get its flowers pollinated: Wasp larvae consume many potential fig seeds. From the perspective of figs, fig wasps are both mutualists and predators, and their interaction with figs does not fit cleanly into the scheme shown in Table 52.1.

Although there are many other exceptions to the simple scheme shown in Table 52.1, most interactions fit the categories well enough for us to use them as a guide for exploring interactions among species in this chapter.

Interactions between resources and consumers influence community dynamics

Many interactions between organisms within communities center on resources and their consumers. A **resource** is any substance directly used by an organism that can potentially lead to the growth of its population and whose availability is reduced when it is used. We usually think first of resources that can be consumed by being eaten, but space, including hiding places and nest sites, becomes unavailable if it is occupied. Factors such as temperature, humidity, salinity, and pH, even though they may strongly affect population growth, are not resources because they are not used up or monopolized. Some resources, such as nest sites, are not altered by being used and immediately become available for occupancy when the user leaves. Other resources must regenerate before they are again available to consumers.

Niches: The Conditions under which Species Persist

The set of environmental conditions under which a species can persist defines its **ecological niche**. If there were no competitors, predators, or disease organisms in its environment, a species would be able to persist under a broader array of conditions than it can in the presence of other species that negatively affect it. On the other hand, the presence of beneficial species may increase the range of environments in which a species can persist.

Both physical and biotic factors determine a species' niche

An experiment performed on two species of barnacles, *Balanus balanoides* and *Chthamalus stellatus*, demonstrated the importance of both physical and biotic factors in determining where the two species actually live. These barnacles live in the intertidal zone of rocky North Atlantic shores. Adult *Chthamalus* generally live higher in the intertidal zone than do adult *Balanus*, but young *Chthamalus* settle in large numbers in the *Balanus* zone. In the absence of *Balanus*, young *Chthamalus* survive and grow well in the *Balanus* zone, but if *Balanus* are present, the *Chthamalus* are eliminated by being smothered, crushed, or undercut by the larger, more rapidly growing *Balanus*. Young *Balanus* settle in the *Chthamalus* zone, but they grow poorly because they lose water rapidly when exposed to air; therefore *Chthamalus* can compete successfully with them there. However, *Balanus* would persist slightly higher in the intertidal zone in the absence of *Chthamalus*. The result is intertidal zonation, with *Chthamalus* growing above *Balanus* (Figure 52.2).

The actual distribution of *Chthamalus* is more restricted than its potential distribution because of competition with *Balanus* in the lower intertidal zone, and the distribution of *Balanus* is more restricted than its potential distribution because of the combined effects of desiccation and competition with *Chthamalus* in the higher intertidal zone. Experiments have shown that the vertical ranges of adults of both barnacles are greater if the other species is removed. Figure 52.2 shows the differences between the potential and actual distributions of the barnacles for only one dimension: height in the intertidal zone. Other important factors that influence barnacle distributions include the type of substrate, the amount of wave action, and water temperature.

Limiting resources determine the outcomes of interactions

Population ecologists typically direct their attention toward differences among species in their requirements for **limiting resources**—resources whose supply is less than the demand made upon them by organisms. Even though a species needs a resource, that resource may not be significant for understanding the species' population dynamics if it is not limiting. For example, most terrestrial animals have a strict but similar requirement for a certain minimum level of oxy-

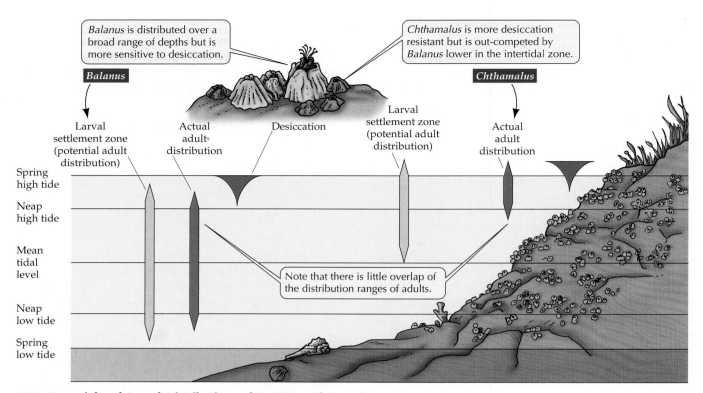

52.2 Potential and Actual Distributions of Two Barnacle Species
Because of interspecific competition, the zone each species occupies is different from the zone it could potentially occupy in the other's absence. (The width of the red and gold bars is proportional to the density of the populations.)

gen. However, studying the use of oxygen reveals very little about the structure of terrestrial communities because the supply of oxygen is nearly always above that minimum level. The primary resources that influence distributions and abundances of terrestrial species are those that are depletable and regenerate slowly, such as food. In some freshwater aquatic environments, however, organisms regularly deplete dissolved oxygen. Aquatic ecologists, unlike terrestrial ecologists, pay careful attention to oxygen levels.

Limiting resources differ among environments, but some resources, such as food supplies, are often limiting. Because of the importance of resources in the lives of all species, we first examine competition among organisms for scarce resources, and then consider predation.

Competition: Seeking Scarce Resources

If two or more organisms use the same resources, and those resources are insufficient to meet their demands, the individuals are competitors, whether they are members of the same or a different species. **Intraspecific competition,** competition among individuals of the same species, may result in reduced growth and reproductive rates for some individuals, may exclude some individuals from better habitats, and may cause the deaths of others. **Interspecific competition,** competition among individuals of different species, affects individuals in the same way, but in addition, an entire species may be excluded from habitats where it cannot compete successfully. In extreme cases, a competitor may cause the extinction of another species. In this section we will show how ecologists study competition in the laboratory and in the field. Then we will discuss how competition influences the composition of ecological communities.

Competition can be studied in the laboratory

We can model competition by adding to the equation for logistic growth (see Chapter 51) a term that incorporates the effects of members of a competing species on the growth of the focal species:

$$\Delta N_1 / \Delta t = r_1 N_1 (K_1 - N_1 - \alpha N_2 / K_1)$$

where N_1 is the number of individuals in species 1, N_2 is the number of individuals in species 2, K_1 is the environmental carrying capacity for species 1, and α is a competition coefficient that measures the *per individual* effect of species 2 on the growth rate of the population of species 1.

We can write a similar equation for the growth of species 2 in which β is the competition coefficient that measures the *per individual* effect of species 1 on the growth of the population of species 2. α and β are usually, but not always, less than 1; that is, the addition of an individual of another species has a smaller competitive effect than the addition of an individual of the same species.

To determine the effects of competition, ecologists can perform laboratory experiments. The first such experiments were performed by the Russian ecologist G. F. Gause and reported in his influential book, *The Struggle for Existence*, published in 1934. Gause began by conducting experiments with protozoans in containers within which the environment was homogeneous. In every experiment, the population of one species grew faster, monopolizing the food resource until the other died out, an outcome called **competitive exclusion**. The results of one of Gause's experiments with two species of *Paramecium*, in which *P. aurelia* was the winner, are shown in Figure 52.3. In simple laboratory environments such as these, competitive exclusion is common.

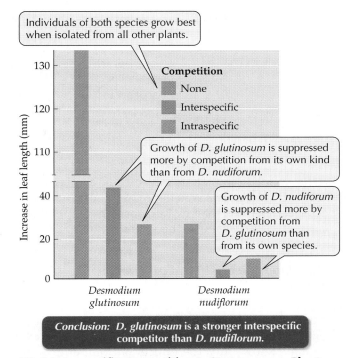

52.4 Interspecific Competition Is Strong among Plants
Plants compete with members of their own species and with members of other species for light, water, and nutrients, all of which can be manipulated in experiments.

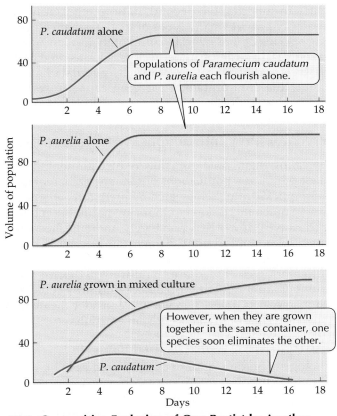

52.3 Competitive Exclusion of One Protist by Another
Competitive exclusion is common in simple laboratory environments.

In later experiments, Gause was able to prevent competitive exclusion by providing a heterogeneous environment containing some places where one species did better and other places where its competitor did better. For example, *Paramecium bursaria* and *P. aurelia* exclude one another in homogeneous environments, but they form stable mixtures if there is deoxygenated water at the bottom of their container. *P. bursaria*, which has mutualistic algae within its cell, can feed in deoxygenated water, whereas *P. aurelia* cannot.

Plants are good subjects for competition experiments because they compete for light, water, and nutrients, all of which can easily be manipulated. To measure the intensity of intraspecific and interspecific competition, investigators used two species of *Desmodium*, *D. glutinosum* and *D. nudiflorum*, both small herbaceous members of the pea family. To measure intraspecific competition, they planted small individuals of each species 10 cm from a large individual of the same species. To measure interspecific competition, they planted small individuals 10 cm from a large individual of the other species. Control individuals were planted at least 3 m from any other *Desmodium* plant.

Not surprisingly, individuals of both species grew best in the absence of competition (Figure 52.4). Growth of *D. nudiflorum* was depressed more by inter-

TABLE 52.2 Experiments Show How Ants and Rodents Interact with Their Food Supply

	RODENTS REMOVED	ANTS REMOVED	RODENTS AND ANTS REMOVED	CONTROL PLOTS
Number of ant colonies	543	0	0	318
Number of rodents	0	144	0	122
Density of seeds relative to control plots	1.0	1.0	5.5	1.0

specific competition than by intraspecific competition, whereas *D. glutinosum* was depressed more by interspecific competition than by intraspecific competition. Clearly *D. glutinosum* was the stronger competitor of the two.

Competition can be studied experimentally in nature

Although it is more difficult to do so, competition experiments can also be performed in nature. Experiments in nature are important because laboratory experiments show only the effects of competition in artificial environments. Field experiments are needed to reveal whether competition is influencing abundances and distributions of species in nature.

To determine whether the many coexisting species of seed-eating ants and rodents that live together in the Sonoran Desert of Arizona compete with one another, ecologists removed ants from some sites, rodents from other sites, and both ants and rodents from a third set of sites. If they removed either ants or rodents, population densities of the other group increased relative to densities in control plots (Table 52.2). These experimental results demonstrate that competition for food links the ants and the rodents. Further observations showed that ants and rodents both greatly reduce seed densities.

To determine whether different species of rodents also compete with one another, ecologists erected rodent-proof fences around 50 m by 50 m desert plots. The fences around the experimental plots had holes through which small rodents could pass, but too small to allow the passage of large kangaroo rats. The holes in the fences surrounding the control plots were large enough for all rodents to pass through. Within 2 years of the removal of kangaroo rats from the experimental plots, densities of small seed-eating rodents increased more than twofold, and the plots without kangaroo rats supported more rodent species than the control plots. Thus kangaroo rats reduce populations of some rodent species and eliminate others from places where they live. Kangaroo rats compete with other seed-eating rodents by reducing their food supply and by aggressively defending space.

Species with similar requirements often coexist in nature

Laboratory experiments show that competitors cannot coexist in simple, homogeneous environments. Yet many species with similar ecological requirements live together in most natural communities, as seed-eating ants and rodents do in the Sonoran Desert. How can so many similar species live together in natural environments? Part of the answer is that nature differs from simple laboratory environments in several important ways.

First, natural environments are variable in space and time. Therefore, competing species often eliminate one another from some parts of the environment, but not from others. This was the outcome of the competition among intertidal barnacles that we discussed early in this chapter. Second, natural environments typically provide many types of food, so that competing species do not overlap completely in their use of resources. In addition, other factors, such as predators, disease, and bad weather, may keep populations well below the environmental carrying capacity so that they rarely compete.

Predator–Prey Interactions

By eating them, predators may reduce populations of their prey. Local populations of some prey are replaced very quickly after being eaten by predators; others increase much more slowly. On rocky marine shorelines, each wave brings a new and barely undiminished supply of planktonic food to suspension-feeding animals such as barnacles. In this section we will describe the different types of predators, and we will discuss why numbers of interacting predators and prey typically fluctuate over time. Then we will consider the evolutionary results of predator–prey interactions.

Predators are classified by what they eat

Biologists find it useful to classify predators by what they eat. Because many predators eat organisms of many species, such dietary classifications are crude, but they are useful for describing interactions between predators and their prey.

Herbivores eat the tissues of plants. Plant tissues are abundant, but many of them offer poor nutrition for other organisms. Wood is primarily cellulose, itself very difficult to break down, impregnated with lignins, which are even more difficult to digest. Because wood is so hard to digest, few organisms attack branches and trunks unless the plant is already weak or dead. Plants also produce roots, leaves, flowers, nectar, pollen, fruits, and seeds, which differ in chemistry, size, structure, and pattern of production. Organisms specialized for eating different plant tissues are correspondingly diverse.

Carnivores are organisms that eat other animals. They are generally moderately larger than their prey, and they pursue, capture, and eat their prey one by one. Predators eat many individual prey items during their lives. For example, a small bird in a temperate forest during winter must eat several average-sized insects or seeds every minute during the day to maintain itself.

Suspension feeders eat prey much smaller than themselves that are suspended in the water or air. As we saw in Chapters 29 and 30, suspension feeders are found in many animal phyla (Brachiopoda, Mollusca, Annelida, the arthropod phyla, Phoronida, Ectoprocta, Echinodermata, Chordata). Every suspension feeder has a filtering apparatus whose structure determines the upper and lower size limits of prey that can be captured. Prey that are too small pass through the mesh of the structure; prey that are too large bounce off it. Most suspension feeders depend on the movement of the surrounding medium through their filtering apparatus, which can be accomplished by movement of either the medium or the animal.

The relative sizes of predators and prey influence their interactions

The relative sizes of predators and prey strongly influence their interactions because they determine how a predator captures and handles its prey. If the predator is much larger than its prey, prey are handled in bulk. The world's largest predators, baleen whales, feed on very small prey that they filter from the water. Predators that are only moderately larger than their prey usually pursue, capture, and eat their prey one at a time. Typical **parasites**, which are much smaller than their hosts, often maintain large populations on or in their hosts, only sometimes killing them (Figure 52.5). The parasitic larvae of many wasps and flies, called **parasitoids**, which are as small as or even smaller than their prey, can complete their development in a single host, eventually killing it.

A single prey individual may harbor hundreds or thousands of parasites without being killed by them. In addition, hosts have defenses against parasites, and

52.5 Most Parasites Are Smaller Than their Hosts This Caribbean soldierfish is host to the parasitic isopod attached to its head between its eyes. The fish has no way to remove the isopod, which feeds on its body tissues.

if a host is in good condition, it may be able to kill them. If a host is already weakened by stresses imposed by the physical environment or shortage of food, the parasites are more likely to succeed.

Pine trees, for example, defend themselves against parasites by exuding the sticky pitch for which pines are famous. Bark beetles attack pine trees by tunneling into the trunks of trees, but they find it difficult to chew through pitch. Weakened trees exude less pitch than healthy trees do, and are easier for the beetles to attack. The beetles lay eggs in their tunnels, and their larvae excavate more tunnels as they eat the nutritive layers just below the outer bark. Each colonizing beetle releases a powerful pheromone that attracts other individuals to the same tree. If enough beetles are attracted, their tunneling larvae may eventually kill the tree (Figure 52.6).

Numbers of predators and prey fluctuate in ecological time

When a typical predator captures and eats a prey individual, it reduces the size of the prey population by one, but the effects of predators on prey population dynamics cannot be determined simply by counting the number of prey eaten. We also need to know how prey densities influence the ease with which prey are captured and how rapidly they reproduce. To understand the direct and indirect interactions between predators and their prey, it is useful to consider the process of predation from the perspective of an individual predator.

Consider a predator that eats only one kind of prey. An individual predator can find enough to eat if the rate at which it encounters prey is above a certain threshold value. Below that threshold, it will lose weight and eventually starve. Nevertheless, the predator may continue to eat prey while slowly losing weight, driving the prey population down further. Eventually the number of predators may be reduced by starvation or emigration, which may allow the prey population to increase its numbers. This increase of prey may, in turn, permit the predator population to increase. Because of this pattern, the recovery of a predator population often lags behind the recovery of its prey population. Thus predator–prey interactions often change the population densities of both species, producing oscillations.

Population density changes among small mammals and their predators living at high latitudes are the best-known examples of predator–prey oscillations. Populations of Arctic lemmings and their chief predators—snowy owls, jaegers, and Arctic foxes—oscillate with a 3- to 4-year periodicity. Populations of Canadian lynx and their principal prey—snowshoe hares—oscillate on a 9- to 11-year cycle (Figure 52.7).

For many years, hare–lynx oscillations were believed to be driven only by interactions between hares and lynxes. Recently, ecologists performed experiments to find out whether any part of the lynx–hare oscillation could be explained by fluctuations in the hares' food supply in addition to predation by lynxes.

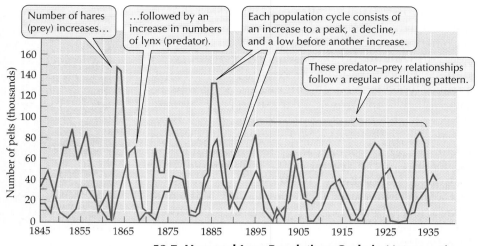

52.7 Hare and Lynx Populations Cycle in Nature The 9- to 11-year population cycles of the snowshoe hare and its major predator, the lynx, in Canada were revealed by the number of pelts sold by fur trappers to the Hudson Bay Company.

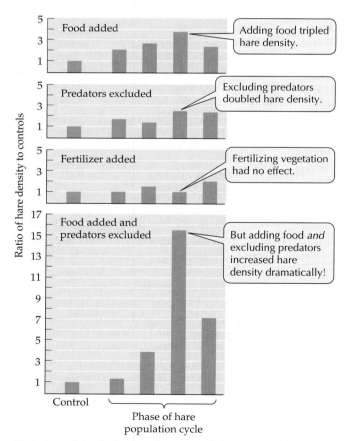

52.8 Prey Population Cycles May Have Multiple Causes Experiments showed that both food supply and predators (but not food quality) affect the population densities of snowshoe hares.

52.6 Attack by Bark Beetles Masses of egg-laying bark beetles have attacked this fallen Douglas fir tree. When the eggs hatched, the tunnels under the bark (which has been removed) were created by developing larvae burrowing through the tissues, eating as they went.

They selected nine 1 km² blocks of undisturbed coniferous forest in Yukon Territory, Canada. In two of the blocks, the hares were given supplemental food year-round. An electric fence with a grid large enough to allow hares, but not their mammalian predators, to pass through was erected around two other blocks. In one of these blocks, extra food was provided. In two other blocks, nitrogen-potassium-phosphorus fertilizer was added to increase plant growth. Three other blocks served as unmanipulated controls.

These experiments produced striking results. Excluding predators doubled, and adding food tripled, hare densities during the peak and decline phases of a cycle. Predator exclusion combined with food addition increased hare density 11-fold, but adding fertilizer had no effect on hare population density (Figure 52.8). Thus, the cycle is driven both by predation and by interactions between hares and their food supply.

The vegetation in the plots where these experiments on snowshoe hares were performed was relatively homogeneous. In nonuniform environments, predators may eliminate their prey in some places, but not in others. In ponds on islands in Lake Superior, chorus frogs (*Pseudacris triseriata*) are found in only some of the habitats that are suitable for them. Three major predators—larvae of a salamander, nymphs of a large dragonfly, and dytiscid beetles—eat chorus frog tadpoles. An ecologist noticed that the tadpoles were common in ponds with beetles, but were rare in ponds with salamander larvae and dragonfly nymphs. In laboratory experiments, he established that the salamander larvae could eat only small tadpoles, but that dragonfly nymphs could eat tadpoles of all sizes. Therefore, he hypothesized that dragonfly nymphs were responsible for eliminating chorus frogs from many ponds.

To test this hypothesis, he selected two large ponds that contained dragonfly nymphs but no tadpoles, and two ponds that contained tadpoles but no nymphs. So that all the tadpoles were handled equally, he removed the tadpoles from the two ponds that lacked dragonfly nymphs and then reintroduced them at the same density. He introduced dragonfly nymphs into one of the ponds at typical densities. He also removed nearly all dragonfly nymphs from one of the ponds that had them and then introduced tadpoles to both of those ponds. The dramatic results of the experiment supported his hypothesis (Figure 52.9). Tadpoles were eliminated from ponds with dragonfly nymphs, but survived well in ponds from which dragonfly nymphs were absent, or nearly so. This experiment shows that a particular predator may eliminate its prey in certain environments; however, it does not tell us why drag-

52.9 Nymphs Eliminate Tadpoles

The speed with which dragonfly nymphs can eliminate tadpoles of the chorus frog is illustrated by the results of two experiments, one in which dragonfly nymphs were added to pools with tadpoles, and the other in which tadpoles were added to pools with dragonfly nymphs.

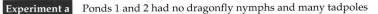

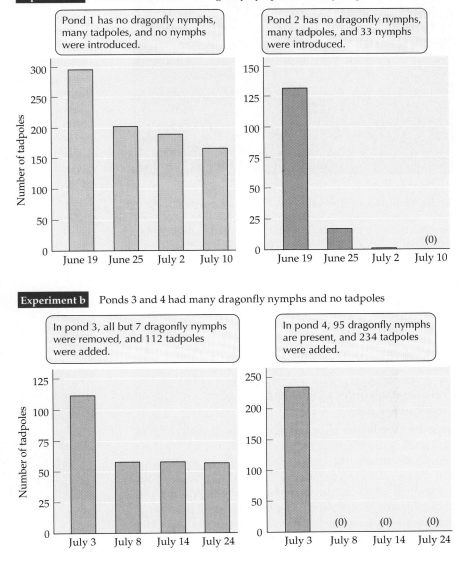

Experiment a Ponds 1 and 2 had no dragonfly nymphs and many tadpoles

Pond 1 has no dragonfly nymphs, many tadpoles, and no nymphs were introduced.

Pond 2 has no dragonfly nymphs, many tadpoles, and 33 nymphs were introduced.

Experiment b Ponds 3 and 4 had many dragonfly nymphs and no tadpoles

In pond 3, all but 7 dragonfly nymphs were removed, and 112 tadpoles were added.

In pond 4, 95 dragonfly nymphs are present, and 234 tadpoles were added.

(a) *Brachinus* sp.

(b) *Pterois volitans*

52.10 Defenses of Animal Prey (a) A bombardier beetle ejects a noxious spray at the temperature of boiling water in the direction of a predator. The spray is ejected in high-speed pulses more than 20 times in succession. (b) The Indo-Pacific lionfish is among the most toxic of all reef fishes. Glands at the base of its spines can inject poison into an attacker; its bright markings are thought to warn potential predators of this capability.

onfly nymphs were naturally absent from some of the ponds.

Predator–prey interactions change over evolutionary time

Because predators do not capture prey individuals randomly, they are agents of evolution as well as agents of mortality. As a consequence, prey have evolved a rich variety of adaptations that make them more difficult to capture, subdue, and eat. Among the evolutionary adaptations of prey are toxic hairs and bristles, tough spines, noxious chemicals and the means for ejecting them (Figure 52.10), camouflage,

and mimicry of inedible objects or of larger or more dangerous organisms. Predators, in turn, evolve to be more effective at overcoming these prey defenses.

MIMICRY. Among the best-studied adaptations to predation is **mimicry**: taking on the appearance of some inedible or unpalatable item. In **Batesian mimicry** a palatable species mimics a noxious or harmful species. Examples are the mimicry of ants by spiders and of bees and wasps by many different insects (Figure 52.11). Batesian mimicry works because a predator that captures an individual of an unpalatable species learns to avoid any prey of similar appearance. However, if a predator captures a palatable mimic, it is rewarded, and it learns to associate palatability with prey of that appearance. As a result, individuals of unpalatable species are attacked more often than they would be if there were no mimics. Because unpalatable individuals that differ from their mimics more than the average are less likely to be attacked by predators that have eaten a mimic, directional selection causes unpalatable species to evolve away from their mimics. Batesian mimicry systems are stable only if a mimic evolves toward a palatable species faster than the palatable species evolves away from it, which usually requires that the mimic be less common than the unpalatable species.

Another type of mimicry is **Müllerian mimicry**, the convergence over evolutionary time in the appearance of two or more unpalatable species. All species in a Müllerian mimicry system, including the predators, benefit when inexperienced predators eat individuals of any of the species because the predators learn rapidly that all similar species are unpalatable. Some of the most spectacular tropical butterflies are mem-

52.11 A Batesian Mimic Falsely Advertises Danger
By mimicking a wasp, this otenucid moth is protected from predators.

Highly unpalatable

Moderately unpalatable

Highly palatable (Batesian mimics)

Palatability not yet tested with birds

* Müllerian mimics of butterflies in the same group

52.12 Müllerian and Batesian Mimics By converging in appearance, the unpalatable Müllerian mimics among these different species of Costa Rican butterflies and moths reinforce each other in deterring predators; the palatable Batesian mimics benefit because predators learn to associate these color patterns with unpalatability.

diseases, rapidly growing parasite genotypes that may kill their hosts are often favored. This is why many waterborne human diseases, such as cholera, hepatitis, and dysentery, are so deadly.

PLANT DEFENSES. The leaves of many plants are defended physically against herbivores by being tough or having hairs or spines. Most leaves also contain chemicals called **secondary compounds** that have negative effects on herbivores. Defensive secondary compounds of one type—acute toxins—interfere with herbivore metabolism. Some of these toxins, such as nicotine, interfere with the transmission of nerve impulses to muscles. Other toxins are hallucinogens that cause individuals that ingest them to have a seriously distorted view of their environment. As a result, hallucinating herbivores are likely to ignore real environmental dangers. Some toxins imitate insect hormones and prevent insects from completing metamorphosis. Still other toxins are unusual amino acids that become incorporated into herbivore proteins and interfere with their functioning.

bers of Müllerian mimicry systems (Figure 52.12), as are many kinds of bees and wasps.

PARASITE VIRULENCE. Some parasites kill their hosts quickly; others live in or on their hosts without harming them very much. Why do parasites differ so much in virulence? To answer this question, we must consider the ability of a parasite to overcome the host's defenses, its rate of population growth within a host, and the length of time an infected host survives.

The faster parasites kill their hosts, the shorter the time during which they can be transferred to new hosts. Therefore, a parasite that does not kill its host, or kills it very slowly, is more likely to be transferred to another host than a parasite that kills its host quickly. On the other hand, a parasite that multiplies slowly within its host may be outcompeted by faster-multiplying individuals of its own species, with the result that more of the faster-multiplying individuals are transmitted to another host. However, if hosts are typically colonized by only one or a few genetically similar parasites, slow-growing genotypes may continue to dominate the parasite population. If parasites can continue to be transmitted to another host after the death of their host, as happens with many waterborne

Defensive chemicals of the second type make leaves difficult to digest, reducing their suitability as food for herbivores. The most common of these substances are tannins, which are present in the leaves of some herbaceous and most woody species. As most leaves age, their tannin concentrations increase, and the leaves also become tougher. Tannins may be present in such large quantities that waters draining from areas dominated by tanniferous plants are tea-colored. The most famous of such "blackwater rivers" is the Río Negro in Brazil (Figure 52.13).

Some plants respond to being eaten by increasing the concentrations of defensive chemicals in their leaves. Mountain birches in northern Finland are eaten by caterpillars of the moth *Oporinia autumnata*. When caterpillars attack a birch, the tree responds by increas-

52.13 The Black River The dark, tannin-laden waters of the Río Negro (top left) show up against the waters of the Amazon as the two rivers join.

ing the concentrations of defensive chemicals in its leaves, sometimes within a few days of the initial attack. The larvae of *Oporinia*, of another moth, and of two sawflies—all of which feed on birches—developed more slowly when they were fed on birch leaves that grew close to leaves heavily eaten by caterpillars during the *same* growing season. Similarly, caterpillars to which ecologists fed leaves from birch trees that had been severely damaged the previous year grew more slowly than caterpillars fed leaves from birches only lightly damaged the previous year. Slower growth lowers larval survival and reproduction, thereby reducing the damage to the trees from the next generation of caterpillars.

Beneficial Interspecific Interactions

During predator–prey and competitive interactions, one or both participants in the interaction are harmed. Amensalism causes harm to one of the partners without affecting the other (–/0). In the other two types of interspecific interactions—commensalism (+/0) and mutualism (+/+)—neither partner is harmed, and one or both may benefit. We will examine these interactions in the sections that follow.

In amensalism and commensalism one participant is unaffected

An individual may harm another organism without benefiting itself (a 0/– interaction). Mammals, for example, create bare spaces around waterholes. They benefit by drinking water, but not by trampling the plants they kill. Leaves and branches falling from trees damage smaller plants beneath them. The trees drop

52.14 Damage Comes from Above Shrubs and herbaceous plants are often destroyed by dead branches falling from tall trees.

old structures regardless of whether or not they damage other plants. Such interactions—amensalisms—are widespread and important. Herbs, shrubs, and small trees in tropical forests often are damaged more by falling objects than by herbivores (Figure 52.14).

Commensalism, a +/0 interaction, benefits one partner but has no effect on the other. An example is the relationship between cattle egrets and grazing mammals. Cattle egrets are found throughout the tropics and subtropics. They typically forage on the ground around cattle or other large mammals, concentrating their attention near the heads and feet of the mammals, where they catch insects flushed by their hooves and mouths (Figure 52.15). Cattle egrets foraging close to grazing mammals capture more food for less effort than egrets foraging away from grazing mammals. The benefit to the egrets is clear; the mammals neither gain nor lose.

52.15 Commensalism Is a +/0 Interaction Cattle egrets, such as these individuals foraging around elephants in East Africa, catch more insects with less work than do egrets foraging away from the larger beasts. The elephants are neither harmed nor helped by the egrets.

Mutualisms benefit both participants

Mutualisms—interactions that benefit both participants (+/+ interactions)—are important among virtually all groups of organisms. Mutualistic interactions exist between plants and microorganisms, protists and fungi, plants and insects, and among plants. Animals also have mutualistic interactions with protists and with one another. The evolution of eukaryotic organisms is believed to be the result of mutualistic interactions between previously free-living prokaryotes and the cells they originally infected (see Chapters 4 and 24).

Some of the most complex and ecologically important mutualisms are between members of different kingdoms. Nitrogen-fixing bacteria of the genus *Rhizobium* receive protection and nutrients from their host plant and provide their host with nitrogen (see Chapter 25). Lichens are compound organisms consisting of highly modified fungi that harbor either cyanobacteria or green algae among their hyphae (see Chapter 26). The fungi absorb water and nutrients from the environment and provide these as well as a supporting structure for the microorganisms, which conduct photosynthesis. This mutualistic combination is especially successful at occupying inhospitable habitats such as rock surfaces, tree bark, and bare, hard ground.

Animals have important mutualistic interactions with protists. Corals and some tunicates gain most of their energy from photosynthetic protists that live within their tissues. In exchange, they provide the pro-tists with nutrients from the small animals they capture. Termites have nitrogen-fixing protists in their guts that help them digest cellulose in the wood they eat. Young termites must acquire their protists by eating the feces of other termites; if prevented from doing so, they soon die. The protists are provided with a suitable environment in which to live and an abundant supply of cellulose.

ANIMAL–ANIMAL MUTUALISMS. Many species of ants have mutualistic relationships with aphids. Ants "milk" these small plant-sucking insects by stroking them with their forelegs and antennae. The aphids respond by secreting droplets of partly digested plant sap that has passed through their guts. In return, the ants protect the aphids from predatory wasps, beetles, and other natural enemies. The aphids lose nothing, because plant sap is high in sugar but low in amino acids, with the result that aphids ingest more sugar than they need.

Some coral reef fishes and shrimps obtain their energy by eating parasites from the scales and gills of larger fish (Figure 52.16). It presumably benefits the hosts to have their parasites removed, but this benefit has never been quantified. These mutualisms are particularly interesting because the cleaners are suitable prey that may actually enter the mouths of dangerous predators. The predators usually refrain from attacking the cleaners, but such restraint could not have been present when the interactions first began to evolve. Such cleaning probably began with the cleaners removing parasites from less dangerous locations.

PLANT–ANIMAL MUTUALISMS. Terrestrial plants have many mutualistic interactions with animals (Table

52.16 An Animal–Animal Mutualism This prawn (*Lysmata grabhami*) is removing parasites from a moray eel.

TABLE 52.3 Mutualistic Relationships of Plants with Animals

BENEFIT TO PLANTS	BENEFIT TO ANIMALS	SOME EXAMPLES
Animals disperse pollen	Animals feed on pollen or nectar	Most plants with brightly colored flowers
Animals disperse seeds	Animals feed on fleshy rewards surrounding or attached to seeds	Conifers such as junipers and yews
Animals disperse both pollen and seeds	Animals feed on both floral and fruit rewards	Most tropical trees, shrubs at all latitudes

52.3). A complex mutualism between trees and ants that live in Central America illustrates the benefits of such interactions. Trees of the species *Acacia cornigera* have large, hollow thorns in which ants of the genus *Pseudomyrmex* construct their nests and raise their young (Figure 52.17). These ants live only on acacias. The ants feed on nectar produced at the bases of the leaf petioles and on special nutritive bodies on the leaves. The ants attack and drive off leaf-eating insects, eat the eggs and larvae of herbivorous insects, and even bite and sting browsing mammals. They also cut back the tips of other plants, particularly vines that grow over their host tree. The ants get room and board; the plants get protection against both predators and competitors.

Many angiosperms depend on animals to move their pollen and seeds (Table 52.3). The plants benefit by hav-ing their pollen carried to other plants and by receiving pollen to fertilize their ovules. The animals benefit by obtaining food in the form of nectar and pollen. Plants provide animals with attractive rewards—nutrient-rich nectar. Movement to another flower of the same species is encouraged by the limited amount of nectar on any one plant, and by the existence of similar rewards on other individuals of the same species. As a result, the foraging animals transfer pollen to the stigmas of plants belonging to the same species. But there is a price: The energy and materials the plant spends to produce nectar and other rewards for animals cannot be used for growth or seed production.

Animals are induced to move seeds by the presence of nutritive rewards attached to or surrounding them. Many seeds are surrounded by fleshy fruits that are eaten by animals, which either regurgitate or defecate

52.17 A Plant–Animal Mutualism Some acacia trees have large, swollen, hollow thorns (*a*) that house ants. The ants patrol the trees, attacking herbivorous insects and cutting away vines and branches of neighboring plants that would otherwise smother the acacia. In an experiment, small acacia trees were cut down, and ants were allowed to recolonize some trees as they regrew, but not others. Those trees with ant colonies (*b*) grew back quickly, but those without ants (*c*) were heavily attacked by other insects and regained their leaves very slowly.

(a) *Bombycilla garrulus*

(b) *Formica* sp.

The ant-attracting elaiosome is the white tissue wrapped around the black seed.

52.18 Fruits Attract Different Frugivores (a) Bright red fruits are attractive to many birds, such as this waxwing. (b) A *Formica* ant removes a ripe seed from a pod of a golden snake plant in the Colorado Rockies.

the seeds some time later away from the plant (Figure 52.18a). A nutritive body called an elaiosome is attached to many seeds that are dispersed by bats or ants. Ants carry the elaiosome and the seed back to their nests, but do not eat the seed, and eventually discard it on a refuse pile (Figure 52.18b). Although ants carry seeds only short distances, seeds in their underground nests are protected from fires and predators.

Interactions between plants and their pollinators and seed dispersers are clearly mutualistic, but they are not purely mutualistic. As we saw earlier, fig wasps are both pollinators and seed predators. Many seed dispersers are also seed predators that destroy some of the seeds they remove from plants. Some organisms that collect rewards are not mutualists at all. Many animals visit flowers without transferring any pollen, sometimes cutting holes in them to get to the nectar-producing regions at the base of the flower. On the other hand, some plants exploit pollinators. The flowers of certain orchids, for example, mimic female insects, enticing male insects to copulate with them (Figure 52.19). The male insects neither sire any offspring nor obtain any reward, but they transfer pollen between flowers, benefiting the orchid.

Temporal Changes in Communities

The species that live together in ecological communities constantly change as environmental shifts alter interspecific interactions. The plants that first colonize a site after a disturbance, for example, differ from those that colonize the site later. The process by which the species composition of a community changes over time is called **ecological succession**. Patterns and causes of ecological succession are varied, but the early colonists always alter the conditions under which later-arriving species grow.

Succession may begin at sites that have never been modified by organisms. The retreat of a glacier in Glacier Bay, Alaska, over the last 200 years exposed unoccupied sites that were colonized by plants. The retreating glacier left a series of moraines—gravel deposits formed where the glacial front was stationary for a number of years. No scientist was present to measure changes over the 200-year period, but ecologists have inferred the temporal pattern of succession

52.19 Some Orchids Mimic Female Insects The flowers of some orchids so closely resemble female wasps that males are fooled into attempting to copulate with the flowers, as this male wasp is trying to do.

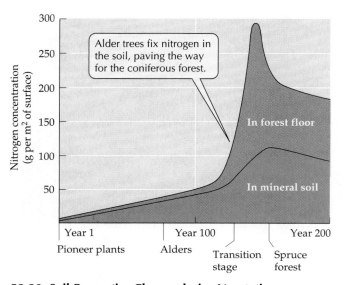

52.20 Soil Properties Change during Vegetation Succession As the plant community occupying an Alaskan glacial moraine changed from pioneering plants to a spruce forest, nitrogen accumulated in both the forest floor and the mineral soil.

by measuring plant communities on moraines of different ages. The youngest moraines, close to the current glacial front, are populated with small organisms such as bacteria, fungi, and algae. Slightly older moraines have lichens, mosses, and a few species of shallow-rooted herbs. Successively older moraines have shrubby willows, alders, and eventually conifers. By comparing moraines of different ages, the ecologists deduced the pattern of plant succession and of changes in soil nitrogen content at Glacier Bay, shown in Figure 52.20. Succession was caused in part by changes in the soil brought about by the plants themselves. Alder trees have nitrogen-fixing fungi in nodules on their roots. Because nitrogen is virtually absent from glacial moraines, nitrogen fixation by alders improved the soil for the growth of conifers. Conifers then outcompeted and displaced the alders.

Succession that takes place when all or part of the dead body of some plant or animal is decomposed is called **degradative succession**. The succession of fungal species that decompose pine needles in litter beneath Scots pines (*Pinus sylvestris*) is shown in Figure 52.21. New litter is continuously deposited under pines, so that the surface layer of litter is young and deeper layers are progressively older. Degradative succession begins when the first group of organisms starts consuming the needles as soon as they fall. Each group of organisms degrades certain compounds, converting them to other compounds that are attacked by the next successional group. This process continues over about 7 years, by which time the last group of organisms—basidiomycetes—has decomposed the remaining cellulose and lignin. By then, the remains are no longer recognizable as pine needles.

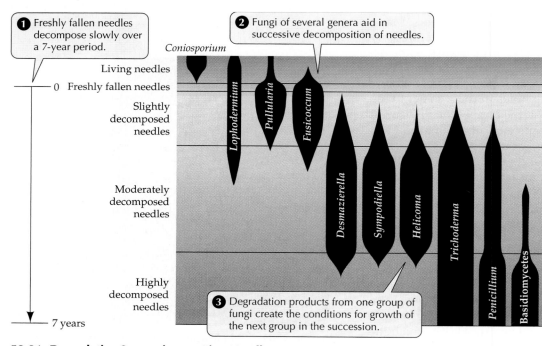

52.21 Degradative Succession on Pine Needles As indicated by the widths of the black bars, the abundances of ten types of fungi in pine litter change with time and with the depth of the layer.

(a) *Yucca brevifolia*

(b) *Tegeticula yuccasella*

52.22 Yucca–Yucca Moth Coevolution (a) The Joshua tree is pollinated only by (b) the yucca moth shown here on a flower.

Coevolution of Interacting Species

Over time, the evolution of many species traits has been influenced by interactions with other species. Species that have mutually influenced one another's evolution are said to have **coevolved**. Often coevolution is diffuse and general, but sometimes it is species-specific. The fig and fig wasp species discussed above have intimately coevolved—neither can reproduce without the other. So have yucca plants and the moths of the genus *Tegeticula* that pollinate them.

Female yucca moths lay their eggs only in the ovules of yucca flowers. A female *Tegeticula* lays no more than five eggs in one flower. After she has laid her eggs, she scrapes pollen from the flower's anthers, rolls it into a small ball, flies to another yucca plant, and places the pollen ball on the stigma of the flower before laying another batch of eggs. When the eggs hatch, the larvae burrow into the ovary and feed upon the developing seeds. *Yucca* has no other pollinators, *Tegeticula* larvae eat no other food, and each yucca species has a specific moth species associated with it (Figure 52.22).

One feature of the coevolved relationship between *Tegeticula* and *Yucca* is quite surprising: Why do female moths lay so few eggs per flower? Wouldn't a female moth that laid more than five eggs in a single flower produce more surviving offspring than moths that lay only the usual number? The evolutionary reason for their restraint is that *Yucca* plants selectively abort flowers in which more eggs are laid. As a result, fewer moth offspring are produced in flowers in which more than the normal number of eggs are laid. Thus, the mutualism is stabilized at a level that represents an "evolutionary compromise" between the fitness of both the moths and the yuccas.

Although species-specific coevolution is relatively rare, diffuse coevolution is widespread. In **diffuse coevolution**, species traits are influenced by interactions with a wide variety of predators, parasites, prey, and mutualists. Most flowers are pollinated by a number of animal species, and most pollinators visit many species of flowers. Most flowers adapted for bird pollination are red, a color that attracts most species of birds. Many flowers adapted for insect pollination have contrasting colors that form guidelines leading to the entrances to the flowers. These lines, which are conspicuous when viewed under ultraviolet light (Figure 52.23), are visible to bees and butterflies, but not to birds. Bat-pollinated flowers open at night and have wide openings into which a bat's head can enter, making the floral rewards accessible to many species of bats.

The traits of the fleshy fruits that surround many seeds are also the result of diffuse coevolution; very few fruits are adapted for dispersal by only a few species of animals. Most bird-dispersed fruits are red or some combination of red and another color. Fruits dispersed by nonflying mammals, many of which lack color vision, are typically purple and are not highly visible to birds. Bat-dispersed fruits are typically green and have a fruity odor when ripe. They are inconspicuous during the day but are easy for bats to detect at night. Many bird-dispersed fruits taste bitter to mammals, and vice versa.

(a) (b)

52.23 Bee Vision and Flower Colors Demonstrate Diffuse Coevolution
(a) Under normal sunlight, these black-eyed susans appear familiar to us.
(b) Bees, however, can see the patterns that appear in this ultraviolet photograph, in which part of the outer ray of petals blends with the central flowers in
the heads to make a larger visual target.

Most traits of flowers and fruits are the result of diffuse coevolution because most flower visitors and fruit
dispersers must use many different plant species to
survive throughout the year. Most plant species produce flowers and fruits for only a few weeks or
months each year. The animals must travel to where
flowers and fruits are available and must feed on
whichever plant species are flowering or bearing fruit.

Keystone Species: Major Influences on Community Composition

Organisms influence the communities in which they
live by altering microclimate, soil structure and chemistry, and water movement. These alterations change
the suitability of the physical environment for other
organisms. As we have just seen, organisms also
change the amount and distribution of resources, and
they consume one another. Species whose influences
on ecological communities are greater than would be
expected on the basis of their abundance are called
keystone species. Keystone species may influence the
species richness of communities, the flow of energy
and materials in ecosystems, or both. In this section
we will focus on how keystone species affect the numbers and kinds of species that live in a community. In
the next chapter, we will discuss the influence of keystone species on ecosystem processes.

Plants provide most of terrestrial ecosystem structure

In terrestrial communities, plants form most of the
structural environment, are the major modifiers of the
physical environment, and are the pathway through
which energy and nutrients enter communities. Any-

one who has walked into the shade of a tree on a hot,
sunny day knows that climate near the ground is
strongly influenced by plants. Temperatures fluctuate
less between day and night under trees than they do
in the open, and light levels are much lower there.
Leaves of trees intercept and evaporate much of the
rain that falls on them so that less reaches the ground
than in open areas. However, rain that does reach the
ground evaporates more slowly inside a forest than in
the open because temperatures are lower, humidities
are higher, and there is less wind.

Keystone animals may change vegetation structure

Animals that are able to change vegetation structure
can be powerful keystone species. Beavers, for example, cut trees and build dams; moose accelerate the
successional change from deciduous trees to conifers
by selectively browsing on deciduous trees. Ecologists
determined how moose alter plant succession by
building fences around four 100-m² plots of land on
Isle Royale in Lake Superior to keep moose out. They
also laid out, but did not fence, control plots outside
the exclosures.

Moose preferentially feed on deciduous trees, such
as mountain ash, mountain maple, aspen, and birch,
that colonize disturbed sites. They rarely eat white
spruce and balsam fir, species that replace deciduous
trees during succession, because the foliage of these
conifers has high concentrations of indigestible resins
and low concentrations of nitrogen. In the control plots,
deciduous species were so heavily eaten by moose that
spruce and fir were the only plants that grew above the
height at which moose feed. Inside the enclosures, on
the other hand, deciduous trees remained abundant,
and the succession to conifers was slower.

Predators may be keystone species

A predator, by consuming a prey that would dominate its environment if it were not eaten, may create openings for species that would otherwise be competitively excluded from the community. The sea star *Pisaster ochraceous*, an abundant predator in rocky intertidal communities on the Pacific coast of North America, functions as a keystone species in this way. In the absence of sea star predation, its preferred prey, the mussel *Mytilus californianus*, pushes out other competitors in a broad belt of the intertidal zone. By consuming mussels, *Pisaster* creates bare spaces that are taken over by a variety of other species (Figure 52.24).

The influence of *Pisaster* on community composition was demonstrated by experimentally removing them from selected parts of the intertidal zone repeatedly over a five-year period. The removals resulted in two major changes. First, the lower edge of the mussel bed extended farther down into the intertidal zone, showing that sea stars are able to eliminate the mussels completely where they are covered with water most of the time. Second, and more dramatically, 28 species of animals and algae disappeared from the removal zone, until only *Mytilus*, the competitive dominant, occupied the entire substratum. By altering competitive relationships, predation by *Pisaster*, in combination with physical factors such as desiccation and wave action, determines which species live in these rocky intertidal communities.

Some microorganisms are keystone species

Despite their small size, certain microorganisms strongly influence community structure. Wood is broken down primarily by microorganisms, especially

52.25 Nonregenerating Clear-Cuts in Oregon Because soil microorganisms were eliminated by burning and herbicides, no conifers are growing in these 15- to 20-year-old clear-cuts, even though each clearing has been planted with seedlings four times since the tree cover was cut.

fungi. Nitrogen fixation, the only source of biological nitrogen, is carried out only by prokaryotic microorganisms. Ignoring the importance of relationships between plants and microorganisms has sometimes led to the failure of reforestation attempts. For example, a 15-hectare plot in the Klamath Mountains of southern Oregon, clear-cut in 1968, has been replanted four times. All the plantings were failures, even though forests in this area regenerate readily after wildfires.

The reason for the failures is that the site was both burned and treated with herbicides to open it up for better growth of conifer seedlings. This treatment killed the early successional deciduous trees and shrubs that normally support soil organisms and ameliorate temperatures and moisture. When those plants were eliminated, most soil microorganisms, including those that form mutualistic associations with conifer roots, died, and conifers then were unable to grow. Many such nonregenerating clear-cuts dot high mountains in this region (Figure 52.25).

Indirect Effects of Interactions among Species

In the experiments we have described above, one member of a community was removed or excluded, and investigators measured the resulting changes. Such single-species removal experiments can demonstrate the direct effects of species on one another, but to detect their indirect effects, observations and manipulations of several species are needed.

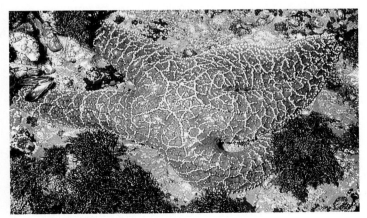

Pisaster ochraceous

52.24 A Sea Star Prevents Its Prey from Dominating Its Community This sea star is resting on rocks from which it has harvested all the mussels. Many other organisms, including the algae visible in the photograph, will now be able to colonize the site.

52.26 Direct and Indirect Interactions Control Populations

Several species, including mice, gypsy moths, and oak trees, interact to influence one another's population densities.

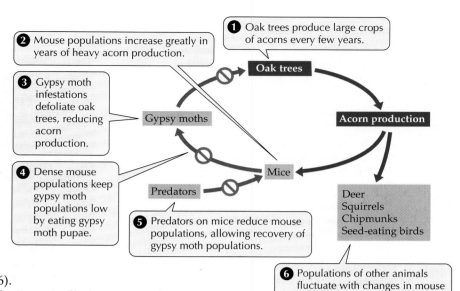

❶ Oak trees produce large crops of acorns every few years.

❷ Mouse populations increase greatly in years of heavy acorn production.

❸ Gypsy moth infestations defoliate oak trees, reducing acorn production.

❹ Dense mouse populations keep gypsy moth populations low by eating gypsy moth pupae.

❺ Predators on mice reduce mouse populations, allowing recovery of gypsy moth populations.

❻ Populations of other animals fluctuate with changes in mouse and gypsy moth densities.

Oak trees → Acorn production → Deer, Squirrels, Chipmunks, Seed-eating birds → Mice → Gypsy moths; Predators

Ecologists have assembled a variety of data to help them understand relationships between oak trees and the animals that eat their leaves and acorns. Most damage to leaves in the oak forests of eastern North America is caused by gypsy moths; most acorns are eaten by mice, chipmunks, deer, squirrels, and birds (Figure 52.26). The abundance of mice and chipmunks is controlled largely by acorn abundance, which varies greatly over the years. Deer consume large quantities of acorns during years of high acorn production, but during poor acorn years they shift to other foods.

Gypsy moths eat oak leaves, and during years when their populations are very large, they may defoliate large expanses of forest. Outbreaks of gypsy moths occur once every 6 to 10 years. Gypsy moth populations collapse after they defoliate a forest because most larvae die of starvation. In the year following defoliation, oak trees are full of leaves, but gypsy moth populations remain low for many years. Why is this?

To determine whether mice could prevent gypsy moth populations from recovering after a crash, ecologists measured predation rates by mice on gypsy moth pupae by attaching freeze-dried pupae to small squares of burlap. They placed the burlap panels on oak tree trunks at sites where gypsy moths typically pupate. During a year of moderately dense mouse populations, all of the introduced pupae were eaten within 8 days. During a year of low mouse density, half of the pupae survived more than 18 days, which is several days longer than it takes for gypsy moth caterpillars to complete metamorphosis and emerge from their pupae. Thus, in years when acorns are abundant, mice keep gypsy moth populations at low densities by eating most of their pupae. In so doing they allow the oak trees to recover from defoliation and accumulate enough energy reserves to produce another large crop of acorns.

Mouse populations typically drop precipitously about 1.5 years after a year of high acorn production, after which gypsy moth populations rebound and again defoliate the trees. If few mice were present, gypsy moths might rebound so quickly that oaks could never produce large crops of acorns. If the investigators had studied only the interactions between mice and acorns or between gypsy moths and oak trees, they would not have discovered the important influences of other species on the interactions.

Summary of "Community Ecology"

Ecological Interactions

• Species interact with one another in four major ways. **Review Table 52.1**

Niches: The Conditions under which Species Persist

• A species' niche is the range of environmental conditions under which it can persist.

• Interactions among species often restrict the range of a species to only part of its potential distribution. **Review Figure 52.2**

Competition: Seeking Scarce Resources

• If organisms use the same resources and those resources are in short supply, the individuals are competitors. Competition may be either intraspecific or interspecific.

• Competition is easily studied in the laboratory, where competitive exclusion is common in homogeneous environments, but not in heterogeneous ones. **Review Figure 52.3**

• Species that use similar resources commonly coexist in nature because nature is spatially and temporally complex, many resources are typically available, and other factors often keep populations below carrying capacity so that they do not compete strongly.

Predator–Prey Interactions

• Predators are classified as herbivores, carnivores, or suspension feeders depending upon what they eat. Herbivores consume large quantities of plant parts, which have low nutritional value. Carnivores eat other animals. Suspension feeders extract large numbers of small prey from water.

• Relative sizes of predators and prey influence their interactions. Parasites are typically much smaller than their prey and may live in or on their hosts without killing them.

• Because of time lags in the responses of both predators and prey, interactions dominated by one predator and one prey typically oscillate. **Review Figure 52.7**

• Experimental manipulation of predators in nature reveals that they are often important in determining both numbers and distributions of their prey. Predators may prevent prey from living in some environments that are otherwise suitable for them. **Review Figures 52.8, 52.9**

• Predators act as evolutionary agents against which prey evolve adaptations, such as toxic hairs and bristles, tough spines, noxious chemicals, and mimicry of inedible objects or dangerous organisms. **Review Figures 52.10, 52.11, 52.12**

Beneficial Interspecific Interactions

• Commensal interactions, in which one partner benefits while the other is unaffected, are common in nature. **Review Figure 52.15**

• Mutualistic interactions, in which both participants benefit, are also common in nature. Mutualistic interactions occur between members of different kingdoms (between plants and prokaryotes, between fungi and algae, and between animals and protists). Animals have mutualistic interactions with other animals and with plants (pollination, seed dispersal). **Review Figures 52.16, 52.17, 52.18**

Temporal Changes in Communities

• Ecological succession involves changes in the species composition of a community over time. Early colonists alter the conditions under which later-arriving species grow.

• Succession may begin at sites that have never been modified by organisms. **Review Figure 52.20**

• Succession may take place when all or part of the dead body of some organism is decomposed. **Review Figure 52.21**

Coevolution of Interacting Species

• Some mutualistic relationships, such as those between figs and fig wasps and yuccas and yucca moths, are tightly coevolved, but diffuse coevolution between many species is much more common. **Review Figures 52.22, 52.23**

Keystone Species: Major Influences on Community Composition

• Keystone species have influences on ecological communities that are greater than would be expected from their abundances.

• Plants, mammals that change vegetation structure, predators on dominant competitors, and microorganisms may function as keystone species. **Review Figures 52.24, 52.25**

Indirect Effects of Interactions among Species

• Indirect effects of species interactions affect many species populations. For example, mice prevent gypsy moth populations from recovering quickly after they defoliate oak trees, thereby allowing the trees to recover. **Review Figure 52.26**

Self-Quiz

1. Two organisms that use the same resources when those resources are in short supply are said to be
   a. predators.
   b. competitors.
   c. mutualists.
   d. commensalists.
   e. amensalists.

2. Which of the following is *not* a resource?
   a. Food
   b. Space
   c. Hiding places
   d. Nest sites
   e. Temperature

3. A species' potential distribution is the range of conditions under which it could survive if
   a. there were no predators or competitors.
   b. there were no predators or other negative influences.
   c. there were no competitors, predators, or disease-causing organisms.
   d. environmental conditions were ideal.
   e. the environment were fundamentally different.

4. An animal that is much smaller than its prey and which attacks it from the inside is called a
   a. predator.
   b. parasite.
   c. commensalist.
   d. competitor.
   e. parasitoid.

5. Which of the following factors tends to stabilize populations of predators and their prey?
   a. A high birth rate of the prey
   b. A high birth rate of the predator
   c. The ability of predators to further reduce prey when they are scarce
   d. The ability of predators to search widely for prey
   e. Environmental heterogeneity

6. The convergence over evolutionary time in the appearance of two or more unpalatable species is called
   a. cladism.
   b. mutual adaptation.
   c. Müllerian mimicry.
   d. Batesian mimicry.
   e. convergent mimicry.

7. Plants are good subjects for experiments to study competition because
   a. plants don't move around.
   b. the resources for which plants compete are easily measured.
   c. the resources for which plants compete are easily manipulated.
   d. plants often compete both in nature and in the laboratory.
   e. all of the above

8. Damage caused to shrubs by branches falling from overhead trees is an example of
   a. interference competition.
   b. partial predation.
   c. amensalism.
   d. commensalism.
   e. diffuse coevolution.

9. Ecological succession is
   a. the changes in species over time.
   b. the gradual process by which the species composition of a community changes.
   c. the changes in a forest as the trees grow larger.
   d. the process by which a species becomes abundant.
   e. the buildup of soil nutrients.

10. Keystone species
   a. influence the structure of the communities in which they live more than expected on the basis of their abundance.
   b. strongly influence the species composition of communities.
   c. may speed up the rate of vegetation succession.
   d. may be herbivores or carnivores.
   e. all of the above

Applying Concepts

1. A general rule commonly accepted by ecologists states that a "jack of all trades is master of none." Yet, most ecological communities are mixtures of jacks and masters—that is, generalists and specialists. Under what conditions would you expect jacks to be more successful? Masters? Why?

2. What features of predator–prey interactions tend to generate instabilities that lead to fluctuations in the densities of both species? Given that instabilities are expected, what keeps populations of either predator or prey from fluctuating to extinction?

3. Parasites usually have generation times much shorter than those of their hosts. Consequently, they should be able to evolve faster. What prevents them from evolving so fast that they completely overcome the resistances of their hosts and exterminate them?

4. Wind does not direct pollen toward conspecific stigmas. Given this inefficiency, why are there so many wind-pollinated plants? If seeds that land close to the parent plant survive less well than those that are carried farther away, why do so many plants produce seeds lacking dispersal devices?

5. On the eastern side of the Sierra Nevada in California, four species of chipmunks occupy adjacent habitats from which they exclude one another by direct aggressive interference. In the San Jacinto Mountains of southern California, three other chipmunk species similarly occupy adjacent habitats, but no interspecific aggression is observed. Each species simply remains in its own habitat. Which of these two assemblages do you think is the older one? Why?

6. Some direct interactions between two species benefit only one of those species. Give examples of such "one-way" benefits in each of the following cases:
   a. between two species of plants (give one example of energetic and another example of physical support)
   b. between a plant and an eater of its leaves
   c. between a predator and its prey

7. Wood is an abundant food source that has been available for millions of years. Why have so few animals evolved to be able to eat wood?

8. In the text, we showed how large mammals modify the communities in which they live. Give some examples of ways in which smaller animals modify their communities.

Readings

Barbour, M. G., H. J. Burk and W. D. Pitts. 1980. *Terrestrial Plant Ecology.* Benjamin Cummings, Menlo Park, CA. A general text on the interactions between plants and their surroundings, including several good chapters on communities and vegetation types.

Begon, M., J. L. Harper and C. R. Townsend. 1996. *Ecology: Individuals, Populations, and Communities,* 3rd Edition. Blackwell Scientific Publications, Oxford. A comprehensive treatment of all aspects of ecology.

Cox, P. A. and M. J. Balick. 1994. "The Ethnobotanical Approach to Drug Discovery." *Scientific American,* June. Discusses how biologists, by analyzing plants already used as drugs by indigenous cultures, can find new pharmaceutical compounds more rapidly than by randomly screening plants.

Fleming, T. H. 1993. "Plant-Visiting Bats." *American Scientist,* vol. 81, pages 460–467. Shows how the availability of fruits and flowers has resulted in evolutionary divergence in the behavior, ecology, and morphology of bats.

Fritz, R. S. and E. L. Simms (Eds.). 1992. *Plant Resistance to Herbivores and Pathogens.* University of Chicago Press, Chicago. Essays dealing with the ecology, evolution, and genetics of plant resistance mechanisms and how they work.

Futuyma, D. J. and M. Slatkin (Eds.). 1983. *Coevolution.* Sinauer Associates, Sunderland, MA. A collection of essays that summarize current knowledge of coevolutionary relationships among all types of living organisms.

Gotelli, N. J. 1995. *A Primer of Ecology.* Sinauer Associates, Sunderland, MA. An excellent introduction to population modeling.

Harborne, J. B. 1988. *Introduction to Ecological Biochemistry,* 3rd Edition. Academic Press, New York. An excellent presentation of the types of chemicals produced by living organisms and their roles in ecological interactions.

Ricklefs, R. E. 1997. *The Economy of Nature,* 4th Edition. W. H. Freeman, New York. A readable text covering all aspects of ecology.

Ricklefs, R. E. and D. Schluter (Eds). 1993. *Species Diversity in Ecological Communities: Historical and Geographical Perspectives.* University of Chicago Press, Chicago. A comprehensive multi-authored volume that explores species interactions at many different spatial and temporal scales.

Thompson, J. N. 1982. *Interaction and Coevolution.* Wiley, New York. A review of patterns in and conditions favoring the evolution of close interactions among species.

Thompson, J. N. 1994. *The Coevolutionary Process.* University of Chicago Press, Chicago. A thorough review of the processes by which coevolutionary relationships evolve.

Wickler, W. 1968. *Mimicry in Plants and Animals.* Wiedenfeld and Nicholson, London. A general and easily followed presentation of the evolution of close resemblances among organisms that are not closely related to one another.

Chapter 53

Ecosystems

*I*n 1976, the outlet of Southern Indian Lake in northern Manitoba, Canada, was dammed, raising the lake level 3 meters. Engineers then diverted the Churchill River so that rather than flowing into the lake, it flowed southward across a drainage divide and through a series of hydroelectric generating stations. Before the dam was built, ecologists studied the lake in detail to assess the likely consequences of raising its level and greatly reducing the flow of river water into it. They predicted that fewer nutrients would enter the lake, but that the reduction would be compensated by nutrients derived from increased soil erosion along the elevated shoreline. Based on their predictions, they saw no reason to believe that the Southern Indian Lake whitefish fishery, the most important commercial fishery in northern Manitoba, would be seriously adversely affected.

Dams: Some Surprising Aftereffects
Damming rivers and lakes in the expectation of creating benefits for the human population often results in unexpected and unpredictable detrimental effects to the ecosystem.

The ecologists' predictions of the future nutrient status of the lake and amounts of algal photosynthesis were correct. However, to everyone's surprise, the whitefish fishery collapsed. The greatly increased soil erosion on the new shoreline released large quantities of mercury into the lake. Mercury concentrations in fish in Southern Indian Lake now exceed Canadian safety standards and will probably remain above standard for many years. From 1977 to 1982, Manitoba Hydro, the builder of the dam, subsidized the commercial fishermen, and in 1982, it provided a one-time cash settlement of $2.5 million Canadian dollars for future losses to the fishermen.

Unexpected surprises commonly follow not only the damming of rivers and lakes, but any attempts to alter ecological systems. Surprises happen because the behavior of those systems is the result of interactions among many different processes, most of which are only incompletely understood. Ecologists now recognize that mercury pollution often results from the raising of lake levels, but they did not know that in the 1970s. As humans continue to alter Earth's ecological systems, new surprises confront us each year.

The organisms living in a particular area, such as Southern Indian Lake, together with the physical environment with which they interact, constitute an **ecosystem**. Ecosystems can be recognized and studied at many different spatial scales, ranging from local units, such as lakes, to the entire globe. At the global scale, Earth is a single ecosystem.

The dynamics of ecosystems are the result of the activities of myriad individual organisms, which are influenced by processes in the physical environment. Some of the processes are altered by organisms in turn, and some are not. Individuals of the many different species that interact in all ecosystems do so by capturing energy and materials, transforming and retaining them, and transferring them to other organisms that eat them.

The goal of ecosystem ecology, which we will discuss in this chapter, is to understand the factors that control the flow of energy and the cycling of materials through ecosystems. Ecologists use this knowledge to understand how and why ecosystems respond as they do to human-caused disturbances, and how society can best use the services that ecosystems provide for the benefit of humanity.

To set the stage for our study of energy flow and nutrient cycling in ecosystems, we will first describe climates on Earth and how they influence ecosystem functioning.

Climates on Earth

The sun drives the global circulation patterns of air and ocean waters and is the source of energy for photosynthesis and ecosystem energetics. The warming and cooling of moving masses of air and water explain much of Earth's climatic patterns. Climates, in turn, exert a powerful influence on the distributions, abundances, and evolution of species.

Climates vary greatly from place to place on Earth, primarily because different places receive different amounts of solar energy and because the monthly amount of incident solar energy is nearly constant at the equator, but varies dramatically at high latitudes. In this section we will examine how differences in solar energy input determine atmospheric and oceanic circulation.

Solar energy inputs drive global climates

Every place on Earth receives the same total number of hours of sunlight each year—an average of 12 hours per day—but not the same amount of *heat*. The rate at which heat arrives per unit of ground area depends primarily on the angle of sunlight. If the sun is low in the sky, a given amount of solar energy is spread over a larger area (and is thus less intense) than if the sun is directly overhead. In addition, when the sun is low in the sky, sunlight must pass through more atmosphere, with the result that more of its energy is absorbed and reflected before it reaches the ground. At higher latitudes (closer to the poles), there is more variation in both day length and the angle of arriving solar energy over the course of a year than at latitudes closer to the equator.

On average, the mean annual air temperature decreases about 0.4°C for every degree of latitude (about 110 kilometers) at sea level. Air temperature also decreases with elevation. The effect of elevation on temperature is due to the properties of gases. As a parcel of air rises, it expands, its pressure drops, and energy is expended in pushing molecules apart. With that loss of energy, the temperature of the air drops. When the parcel of air descends, it is compressed, its pressure rises, the same amount of energy is recovered, and its temperature increases.

When wind patterns bring air into contact with a mountain range, the air rises to pass over the mountains, cooling as it does so. Because cool air cannot hold as much moisture as warm air, clouds frequently form, and moisture is released as rain or snow. For this reason, the windward side of a mountain range generally receives more rainfall than the leeward side. On the leeward side, air descends and warms, and the dry air picks up moisture. Places on the leeward side of a mountain range that receive little rainfall for this reason are said to be in a **rain shadow** (Figure 53.1).

Global atmospheric circulation influences climates

Earth's climates are strongly influenced by global air circulation patterns. Air rises not only when it crosses mountains, but also when it is heated by the sun. Warm air rises in the tropics, which receive the greatest solar energy input. This air is replaced by air that flows toward the equator from the north and south. That air, in turn, is replaced by air from aloft that descends after having traveled away from the equator at great heights. At roughly 30° north and south latitudes, air that cooled and lost its moisture while rising at the equator descends and warms. Many of Earth's deserts, such as the Sahara and the Australian deserts, are located at these latitudes.

At about 60° north and south latitudes, air rises again, and cold, dense air descends at the poles, where there is little input of solar energy. The black arrows around the edge of Figure 53.2 show these vertical patterns, which are one component of Earth's winds.

The spinning of Earth on its axis influences surface winds because Earth's velocity is rapid at the equator,

53.1 A Rain Shadow Average annual rainfall tends to be lower on the leeward side of a mountain range than on the windward side.

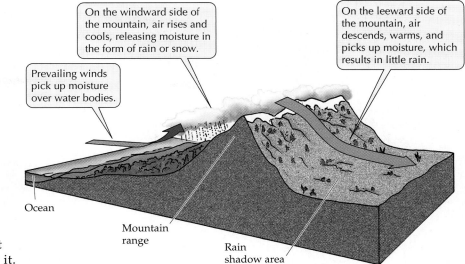

Prevailing winds pick up moisture over water bodies.

On the windward side of the mountain, air rises and cools, releasing moisture in the form of rain or snow.

On the leeward side of the mountain, air descends, warms, and picks up moisture, which results in little rain.

Ocean

Mountain range

Rain shadow area

but relatively slow close to the poles. An air mass at a particular latitude has the same velocity as Earth has at that latitude. As an air mass moves toward the equator, it confronts a faster and faster spin, and it slows down relative to Earth beneath it. As an air mass moves poleward, it confronts a slower and slower spin, and it speeds up relative to Earth beneath it. Therefore, air masses moving latitudinally are deflected to the right in the Northern Hemisphere and to the left in the Southern Hemisphere. Winds blowing toward the equator from the north and south veer to become the northeast and southeast trade winds, respectively. Winds blowing away from the equator also veer and become the westerlies that prevail at mid-latitudes. These surface winds are shown by the blue arrows in Figure 53.2.

Because Earth's axis is tilted, the amount of solar energy that reaches a given region varies seasonally as Earth orbits the sun. The amount of solar energy input is at its maximum at the time of year when the sun is closest to being overhead at noon. The **intertropical convergence zone**—the location of greatest solar energy input and the site where trade winds converge and air rises—thus shifts with the season. It shifts to

the north during the northern summer (southern winter) and to the south during the southern summer (northern winter).

However, the intertropical convergence zone lags behind the overhead passage of the sun by a bit more than a month because it takes that long to heat the surface mass of Earth. Seasonal changes in climate close to the equator are associated with the movement of the intertropical convergence zone because whenever an area is within the zone, air rises and heavy rains fall. When the zone is to the north or south of a tropical region, the prevailing winds are trade winds, which seldom yield rain unless forced to rise over mountains.

Global oceanic circulation is driven by the wind

The global pattern of wind circulation drives the circulation of ocean water. Ocean water generally moves in the direction of the prevailing winds (Figure 53.3). Winds blowing toward the equator from the northeast and southeast cause water to converge at the equator and move westward until it is blocked by a continental land mass. At that point

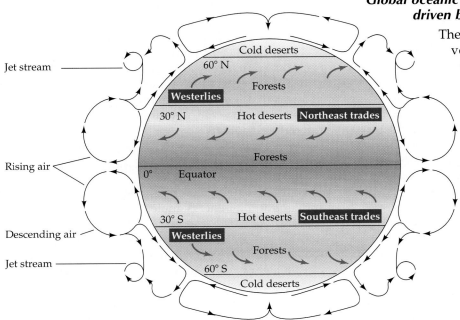

Jet stream

Cold deserts

60° N

Westerlies

Forests

30° N

Hot deserts

Northeast trades

Forests

Rising air

0° Equator

Forests

30° S

Hot deserts

Southeast trades

Westerlies

Descending air

Forests

Jet stream

60° S

Cold deserts

53.2 Circulation of Earth's Atmosphere If we could stand outside Earth and observe the movement of the air, we would see vertical movements like those indicated by the black arrows and surface winds like those shown by the blue arrows.

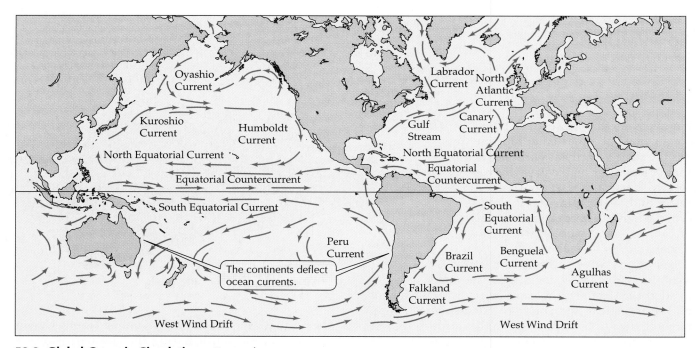

53.3 Global Oceanic Circulation To see that ocean currents are driven primarily by the wind, compare the surface currents shown here with the prevailing surface winds shown in Figure 53.2. Deep ocean currents differ strikingly from the surface ones shown here.

the water splits, some of it moving north and some of it moving south along continental shores. This poleward movement of ocean water is a major mechanism of heat transfer to high latitudes. As it moves toward the poles, the water veers right in the Northern Hemisphere and left in the Southern Hemisphere. Thus water turns eastward until it is blocked by another continent and is deflected laterally along its shores. In both hemispheres, water flows toward the equator along the west sides of continents, continuing to veer right or left until it meets at the equator and flows westward again.

The oceans play an important role in world climates, both because their waters move long distances and because water has a high specific heat. The **specific heat** of a substance is the amount of energy required to raise the temperature of 1 gram of the substance 1°C. For water, this value is 1 cal/g at 15°C. Similarly, 1 gram of water that cools 1°C gives off 1 cal/g. Air and land surfaces have a much lower specific heat. Consequently, in comparison with continents, oceans warm up more slowly in summer because it takes more heat to raise their temperature, and cool off more slowly in winter because more heat must be released to cool them.

At high latitudes, the temperatures of the interiors of large continents fluctuate greatly with the seasons, becoming very cold in winter and hot in summer, a pattern called a **continental climate**. The coasts of continents, particularly those on west sides at middle latitudes, where the prevailing winds blow from ocean to land, have **maritime climates**, with smaller differences between winter and summer temperatures. Seasonal temperatures change the most on the largest land mass, Asia, where strong winter high pressure (descending air) over Siberia causes winds to blow from the continent toward the coasts. In summer, however, strong low pressure (rising air) over Siberia draws great quantities of moist air over the land from the Indian Ocean, producing the great summer monsoons (rainy seasons) characteristic of southern Asia.

The amount and annual pattern of energy input into different ecosystems determines the rates at which those ecosystems function and the kinds of organisms that live there. Next we will discuss how climates influence the amount of energy that flows through ecosystems.

Energy Flow through Ecosystems

Organisms depend on inputs of energy (in the form of sunlight or high-energy molecules), water, and minerals for metabolism and growth. Except for a few limited ecosystems (caves, deep-sea thermal systems) in which solar energy is not the main energy source, almost all energy utilized by organisms comes (or once came) from the sun. Even the fossil fuels—coal, oil, and natural gas—upon which the economy of modern civilization is based are reserves of captured solar energy locked up in the remains of organisms that lived millions of years ago.

Only about 5 percent of the solar energy that arrives on Earth is captured by photosynthesis. The remaining energy is either radiated back into the atmosphere as heat (especially in places where Earth's surface is bare because there is too little water to support plant growth) or consumed by the evaporation of water from plants. The energy that *is* captured powers the "metabolism" of ecosystems. How that energy is captured by green plants and subsequently passes through a series of organisms is the topic of the next section.

Photosynthesis drives energy flow in ecosystems

Energy flow in most ecosystems originates with photosynthesis (see Chapter 8). The major factors influencing the rate of photosynthesis are the amount of solar radiation, the availability of water, the abundance of mineral nutrients and carbon dioxide, and temperature. The total amount of energy that plants assimilate by photosynthesis is called **gross primary production**; the production that remains after subtracting the energy that plants use for maintenance and for building tissues is called **net primary production.** The rate at which plants assimilate energy is called **primary pro-**ductivity. The distribution of primary production reflects the distribution of temperature and moisture on Earth (Figure 53.4).

Water availability and temperature are major determinants of gross primary productivity. To obtain minerals from the soil and to photosynthesize, plants must open their stomata, and when their stomata are open, they lose water. The rate of photosynthesis depends on temperature because most chemical reactions proceed faster as temperature increases, roughly doubling with every rise of 10°C. Temperature and moisture interact to determine primary productivity because the evaporative power of air, which affects transpiration and thus the flow of water within plants, is less at low temperatures than at high temperatures.

In many areas on Earth, the annual gross primary production is determined by available soil moisture and soil fertility. Shortage of water limits primary production during much of the year in most arid regions. Production in aquatic systems is limited by light, which decreases rapidly with depth; by nutrients, which sink and must be replaced by upwelling of water; and by temperature.

Close to the equator, temperatures are high throughout the year, and the water supply is adequate much of the time. In these climates, highly productive forests thrive. In lower- and mid-latitude deserts,

53.4 Primary Production in Different Ecosystems The (*a*) geographic extent, (*b*) annual production per unit of area, and (*c*) percentage of Earth's net primary production provided by different ecosystems.

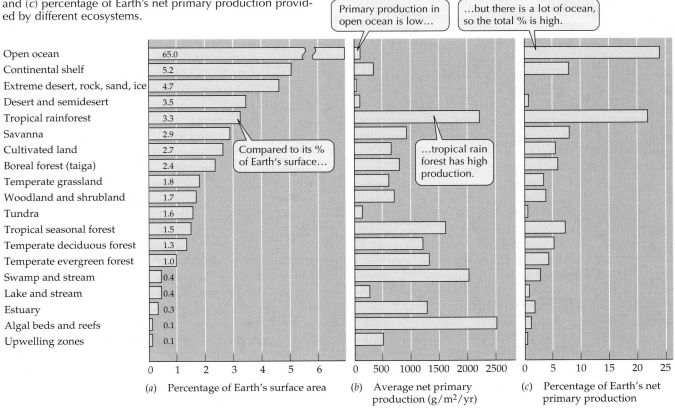

(*a*) Percentage of Earth's surface area

(*b*) Average net primary production (g/m²/yr)

(*c*) Percentage of Earth's net primary production

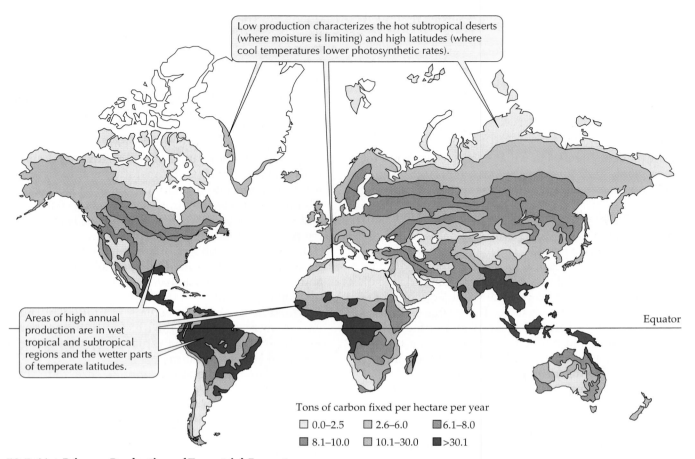

Low production characterizes the hot subtropical deserts (where moisture is limiting) and high latitudes (where cool temperatures lower photosynthetic rates).

Areas of high annual production are in wet tropical and subtropical regions and the wetter parts of temperate latitudes.

Equator

Tons of carbon fixed per hectare per year

☐ 0.0–2.5 ▨ 2.6–6.0 ▨ 6.1–8.0
▨ 8.1–10.0 ▨ 10.1–30.0 ■ >30.1

53.5 Net Primary Production of Terrestrial Ecosystems
Variations in temperature and water availability over Earth's land surface affect the productivity of its ecosystems.

where plant growth is limited by lack of moisture, primary production is low, and plants of low stature dominate the landscape. At still higher latitudes, where there is more moisture and trees grow well, primary production is limited by low temperatures during much of the year. The global distribution of net primary production is shown in Figure 53.5.

Plants use energy to maintain themselves

Plants use most of the energy they capture to maintain themselves and to grow and reproduce. Some of this energy produces new tissues that can be eaten by herbivores or used by organisms after the plants die. Because so much of the energy they capture goes to power their own metabolism, plants always contain much less energy than the total amount they have assimilated; only the energy plants do not use to maintain themselves is available to be harvested by animals.

Energy flows when organisms eat one another

Because energy flows through ecosystems when organisms eat one another, biologists find it useful to

group organisms according to their source of energy. The organisms that obtain their energy from a common source constitute a **trophic level** (Table 53.1). Organisms at a particular trophic level occupy a position in an ecosystem that is determined by the number of steps through which energy passes to reach them. Photosynthetic plants get their energy directly from sunlight. Collectively, they constitute the trophic level called photosynthesizers or **primary producers**. They produce the energy-rich organic molecules upon which nearly all other organisms feed.

All other organisms are called **consumers** because they consume, either directly or indirectly, the energy-rich organic molecules produced by photosynthetic organisms. Organisms that eat plants constitute the trophic level called herbivores. Organisms that eat herbivores are called primary carnivores. Those that eat primary carnivores are called secondary carnivores, and so on. Organisms that eat the dead bodies of organisms or their waste products are called detritivores or decomposers. The many organisms that obtain their food from more than one trophic level are called omnivores.

A sequence of linkages in which a plant is eaten by an herbivore, which is in turn eaten by a primary carni-

TABLE 53.1 The Major Trophic Levels

TROPHIC LEVEL	SOURCE OF ENERGY	EXAMPLES
Photosynthesizers (primary producers)	Solar energy	Green plants, photosynthetic bacteria, and protists
Herbivores	Tissues of primary producers	Termites, grasshoppers, water fleas, anchovies, deer, geese
Primary carnivores	Herbivores	Spiders, warblers, wolves, copepods
Secondary carnivores	Primary carnivores	Tuna, falcons, killer whales
Omnivores	Several trophic levels	Humans, opossums, crabs, robins
Detritivores	Dead bodies and waste products of other organisms	Fungi, many bacteria, vultures, earthworms

vore, and so on, is called a **food chain**. Food chains are usually interconnected to make a **food web**. Food webs result from the fact that most species in a community eat and are eaten by more than one other species.

The arrows in representations of food webs show who eats whom. A simplified food web, not including detritivores, for Gatun Lake, Panama, is shown in Figure 53.6. A food web is a useful summary of predator–prey interactions within a community. A complete food web, showing the position of every species in an ecosystem, would be confusingly complex because most biological communities contain so many species. Therefore, similar species, especially those at lower trophic levels, are usually lumped together, as they are in the diagram of the Gatun Lake food web.

Much energy is lost between trophic levels

Only a small portion of the energy captured at one trophic level is available to organisms at the next higher level because the energy that organisms use to

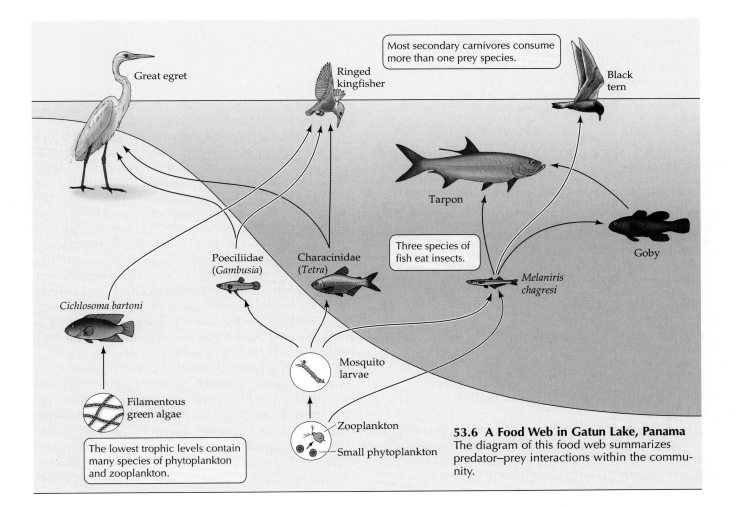

Great egret

Ringed kingfisher

Most secondary carnivores consume more than one prey species.

Black tern

Tarpon

Goby

Poeciliidae (*Gambusia*)

Characinidae (*Tetra*)

Three species of fish eat insects.

Melaniris chagresi

Cichlosoma bartoni

Mosquito larvae

Filamentous green algae

Zooplankton

Small phytoplankton

The lowest trophic levels contain many species of phytoplankton and zooplankton.

53.6 A Food Web in Gatun Lake, Panama
The diagram of this food web summarizes predator–prey interactions within the community.

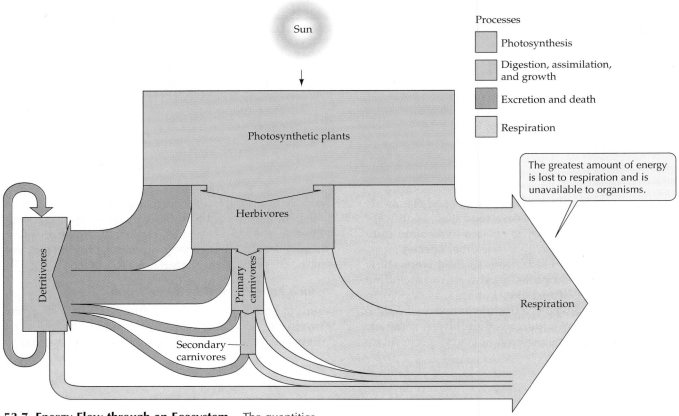

53.7 Energy Flow through an Ecosystem   The quantities of energy flowing through an ecosystem can be visualized using a diagram like this one, in which the width of the channels is roughly proportional to the amount of energy flowing through them. Trophic levels are indicated by blocks of green or blue; energy channels are in red or gold, and the direction of energy flow is shown by arrows.

maintain themselves is dissipated as heat, a form of energy that cannot be used by other organisms. The energy content of an organism's net production—its growth plus reproduction—is available to organisms at the next trophic level (Figure 53.7). The efficiency of energy transfer through food webs depends on the fraction of net production at one trophic level that is consumed by organisms at the next level, and how those organisms divide the ingested energy between production and maintenance (respiration). We can calculate the efficiency of energy transfer of a species or group of species as

$$E = P/(P + R)$$

where E is efficiency, P is net production, and R is respiration.

The E values for different animal taxa reveal two patterns (Table 53.2). First, birds and mammals have very low efficiencies because they expend so much energy maintaining constant high body temperatures.

Second, herbivores are less efficient than carnivores because plant tissues generally take more energy to digest than animal tissues do.

Even when efficiencies of energy transfer and consumption rates are both high, seldom is as much as 20 percent of the energy assimilated by a trophic level converted to production at the next trophic level. The amount of energy reaching a higher trophic level is determined by net primary production and by the efficiencies with which food energy is converted to biomass (the total weight of organisms) at the trophic levels below it. To show how energy decreases in moving from lower to higher trophic levels, ecologists construct diagrams called **pyramids of energy**. A **pyramid of biomass**, which shows the mass of organisms existing at different trophic levels, illustrates the amount of biomass that is available at a given moment in time for organisms at the next trophic level (Figure 53.8).

Pyramids of energy and biomass for the same ecosystem usually have similar shapes, but sometimes they do not. The shapes depend on the dominant organisms and how they allocate their energies. In most terrestrial ecosystems, the dominant photosynthetic plants are large and store energy for long periods, much of it in difficult-to-digest forms (cellulose, lignin). However, terrestrial ecosystems may differ

53.8 Pyramids of Biomass and Energy Ecosystems can be compared in terms of the amount of material present in organisms at different trophic levels (left), and in terms of energy flow (right).

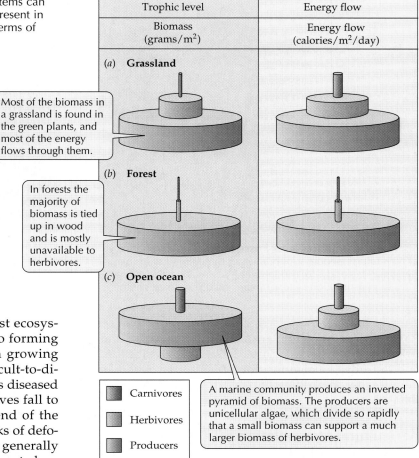

Most of the biomass in a grassland is found in the green plants, and most of the energy flows through them.

In forests the majority of biomass is tied up in wood and is mostly unavailable to herbivores.

A marine community produces an inverted pyramid of biomass. The producers are unicellular algae, which divide so rapidly that a small biomass can support a much larger biomass of herbivores.

strikingly in patterns of energy flow depending on the life forms of the dominant plants. In grassland ecosystems, because plants produce few hard-to-digest woody tissues, animals are able to consume most of the annual production of plant tissues each year. In grasslands, mammals—wild or domestic—may consume 30 to 40 percent of the annual aboveground net primary production. Insects may consume an additional 5 to 15 percent. Soil organisms, primarily nematodes, may consume 6 to 40 percent of the belowground biomass (Figure 53.8*a*).

By contrast, the dominant plants in forest ecosystems allocate a great deal of their energy to forming wood, which accumulates at high rates in growing forests. Wood, which is constructed of difficult-to-digest material, is rarely eaten unless a plant is diseased or otherwise weakened. In most forests leaves fall to the ground relatively undamaged at the end of the growing season. Although there are outbreaks of defoliating insects in forests, browsing rates are generally so low that forest ecologists often ignore losses to herbivores when calculating forest production (Figure 53.8*b*).

In most aquatic communities, on the other hand, the dominant photosynthesizers are bacteria and protists. They have such high rates of cell division that a small biomass of photosynthesizers can feed a much larger biomass of herbivores, which grow and reproduce much more slowly. This pattern can produce an inverted pyramid of biomass, even though the pyramid of energy for the same ecosystem has the typical shape (Figure 53.8*c*).

Much of the energy ingested by organisms is converted to biomass that is eventually consumed by detritivores (see Figure 53.7). Detritivores immobilize some nutrients, but they transform the remains and waste products of organisms (detritus) into carbon dioxide, water, and free mineral nutrients that can be taken up by plants again. If there were no detritivores, most nutrients would eventually be tied up in dead bodies, where they would be unavailable to plants. Therefore, continued ecosystem productivity depends on rapid decomposition of detritus.

Under the warm, wet conditions found in tropical forests, detritus is decomposed within a few weeks or months, and no litter accumulates on the soil surface.

TABLE 53.2 Average Production Efficiencies (*E*) for Various Groups of Animals

GROUP	PRODUCTION EFFICIENCY (%) *P*/(*P* + *R*)
Insectivores (mammals)	0.9
Birds	1.3
Small mammals	1.5
Large mammals	3.1
Fishes and social insects	10.0
Invertebrates other than insects	
Herbivores	21
Carnivores	28
Detritivores	36
Nonsocial insects	
Herbivores	39
Carnivores	56
Detritivores	47

(a)

(b)

53.9 Agriculture Requires Energy Inputs (a) In traditional agriculture, people supply most of the energy, as in this Bengali rice paddy. (b) Modern agriculture is based on high rates of consumption of fossil fuels and use of toxic chemicals.

Rates of decomposition are slower under colder and drier conditions (see Figure 52.22). At high altitudes and latitudes, decomposition of leaf litter may take decades; decomposition of tree trunks may take more than a century.

Agricultural Manipulation of Ecosystem Productivity

Humans exploit ecosystems by replacing species of low economic value with species of high value. By means of agriculture, we help some species compete with others and manipulate ecosystems so as to increase the yield of products useful to humans. Agriculture has several intricately intertwined components: We eliminate competition between crops and unwanted plants by cultivating and by applying herbicides; we reduce competing herbivores and disease-causing organisms, usually by applying toxic chemicals; we augment photosynthesis by fertilizing and irrigating; and we develop special high-yielding strains of plants that respond to additional fertilizer by increasing their growth rates. All these components must work together, because "miracle" strains of crops do not actually yield more than other strains unless they are provided with fertilizers and protected from competitors, herbivores, and pathogens. Agriculture also depends on energy from outside the system for cultivation and harvesting. In modern agriculture, this energy comes from fossil fuels (Figure 53.9).

Although human manipulations of agricultural systems have spectacularly increased food production per hectare, they have also created problems. Herbicides and insecticides have polluted lakes, rivers, and groundwater in most industrialized countries. Many agricultural pests have evolved resistances to pesticides. The usual response has been to increase the use of pesticides, creating even more severe pollution problems. Recently, however, agriculturists have developed new, less toxic methods of pest control in agricultural systems.

New methods of pest control, collectively known as **integrated pest management** (IPM), are growing in use. IPM combines chemical approaches to pest control with cultural practices—such as crop rotation, mixed plantings of crop plants, and mechanical tillage of the soil—and biological methods—such as development of pest-resistant strains of crops, use of natural predators and parasites, and use of chemical attractants. The reduced use of toxic chemicals avoids most pollution problems and reduces the chance that pests will evolve resistance to pesticides.

Cycles of Materials in Ecosystems

As we have just seen, energy is not fully recycled in ecosystems because at each transformation, much of it is dissipated as heat, a form that cannot be used by organisms to power their metabolism. Chemical elements, on the other hand, are not lost when they are transferred among organisms; they cycle through organisms and the physical environment. Carbon, nitrogen, phosphorus, calcium, sodium, sulfur, hydrogen, and oxygen, together with smaller amounts of other chemical elements, are the primary materials of which organisms are constructed. The quantities of these elements that are available to organisms are strongly influenced by how organisms get them, how long they hold onto them, and what they do with them while they have them.

To understand the cycling of elements, it is convenient to divide the global ecosystem into four com-

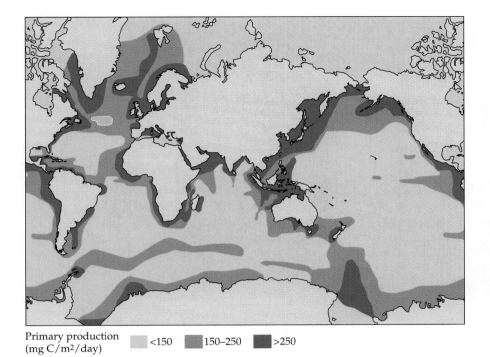

Primary production (mg C/m²/day) <150   150–250   >250

are very low over most of the oceans. Oxygen is usually present at all depths, because even slow mixing suffices to replenish the oxygen consumed by the respiration and decomposition of the few organisms that live in the nutrient-poor water. Most of the elements that enter the oceans settle to the bottom and remain there until bottom sediments are elevated above sea level by movements of Earth's crust, but this process may take many millions of years.

partments: oceans, fresh waters, atmosphere, and land. The physical environments in each compartment and the types of organisms living there are different, and therefore the amounts of elements found in the different compartments, what happens to those elements, and the rates at which they enter and leave the compartments differ strikingly. After we have described these compartments, we will consider them together to illustrate how elements cycle through the global ecosystem.

Oceans receive materials from the land and atmosphere

Oceans receive materials from land as runoff from rivers. On time scales of hundreds to thousands of years, oceans are the ultimate repository of most materials produced by human activity, even though the immediate receivers are often other compartments of the global ecosystem. Because of their huge size, and because they exchange materials with the atmosphere only at their surface, oceans respond very slowly to outside disturbances.

Except on continental shelves, ocean waters mix very slowly and are strongly stratified. Elements that enter the oceans from other compartments gradually sink to the seafloor, unless they are brought back to the surface by the cool bottom water that rises—**upwells**—near the coasts of continents (Figure 53.10). Waters in these zones of upwelling are rich in nutrients, and most of the world's great fisheries are concentrated there. Concentrations of mineral nutrients

Lakes and rivers contain only a small fraction of Earth's water

Lakes and rivers contain much less water than oceans do, and because these bodies of water are relatively small, most mineral nutrients entering them are not buried in bottom sediments for long periods of time. Some mineral nutrients enter fresh waters in rainfall, but most are released by the weathering of rocks and are carried to lakes and rivers via groundwater (the water that resides in the soil and in rocks) or by surface flow.

After entering rivers, mineral nutrients are usually carried rapidly to lakes or to the oceans. In lakes they are taken up by organisms and incorporated into their cells. These organisms eventually die and sink to the bottom, where decomposition of their tissues uses up the oxygen. Surface waters of lakes thus quickly become depleted of nutrients, while deeper waters become depleted of oxygen. However, this stratification process is countered by vertical movements of water—**turnover**—that bring nutrients to the surface and oxygen to deeper water. Wind is an important mixing agent in shallow lakes, but in deeper lakes it usually mixes only surface waters.

In temperate regions, lake waters turn over because water is most dense at 4°C; above and below that temperature it expands (Figure 53.11). In spring at mid-latitudes, the sun warms the surface layer of a lake. The depth of the warm layer gradually increases as spring and summer progress. However, there is still a well-defined zone—the **thermocline**—where the tempera-

53.11 Annual Temperature and Oxygen Cycles in a Temperate Lake These vertical temperature profiles are typical of temperate-zone lakes that freeze in winter.

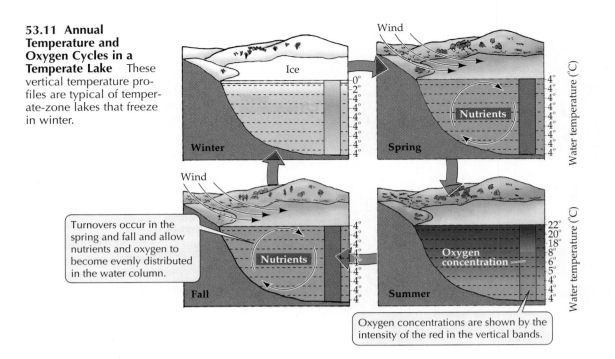

Turnovers occur in the spring and fall and allow nutrients and oxygen to become evenly distributed in the water column.

Oxygen concentrations are shown by the intensity of the red in the vertical bands.

ture drops abruptly to about 4°C. Only if the lake is shallow enough to warm to the bottom does the temperature of the deepest water rise above 4°C.

In autumn, as the surface of the lake cools, the cooler surface water, which is denser than the warmer water below it, sinks, and is replaced by warmer water from below. This process continues until the entire water column has reached 4°C. At this point, the density of the water is uniform throughout the lake, and even modest winds readily mix the entire water column. As colder weather then cools the surface water below 4°C, it becomes less dense than the 4°C water below it and floats at the top. Another turnover occurs in spring, when the surface layers above the thermocline warm to 4°C and the water column, again being of uniform density throughout, is easily mixed by wind.

Deep tropical and subtropical lakes may be permanently stratified because they never become cool enough to have uniformly dense water. Their bottom waters lack oxygen because decomposition quickly depletes any oxygen that reaches them. However, many tropical lakes are overturned at least periodically by strong winds so that their deeper waters are occasionally oxygenated. Arctic lakes turn over only once each year.

The atmosphere regulates temperatures close to Earth's surface

The atmosphere is a thin sphere of gases surrounding Earth. About 80 percent of the mass of the atmosphere lies in its lowest layer, the troposphere, which extends upward from Earth's surface about 17 km in the trop-

ics and subtropics, but only about 10 km at higher latitudes (Figure 53.12). Most global air circulation takes place within the troposphere, and virtually all atmospheric water vapor is located there.

Above the troposphere, the stratosphere, which extends upward to about 50 kilometers above Earth's surface, contains about 99 percent of the remaining atmospheric mass, but it is extremely dry. Materials enter the stratosphere from the troposphere near the equator, where air rises to high altitudes. These materials tend to remain in the stratosphere for a relatively long time because stratospheric air circulation is horizontal. Ozone (O^3) in the stratosphere absorbs most shorter wavelengths of biologically damaging ultraviolet radiation, which is why the development of the ozone hole, which we discussed in Chapter 19, is of great concern.

The atmosphere is 78.08 percent nitrogen as N_2, 20.95 percent oxygen, 0.93 percent argon, and 0.03 percent carbon dioxide. It also contains traces of hydrogen gas, neon, helium, krypton, xenon, ozone, and methane. The atmosphere contains Earth's biggest pool of nitrogen and large supplies of oxygen. Although carbon dioxide constitutes a very small fraction of the atmosphere, it is the source of the carbon used by terrestrial photosynthetic organisms. Concentrations of atmospheric water vapor are highly variable in space and time. The atmosphere is a transport medium for many gases, as well as for airborne particles containing carbon, nitrogen, sulfur, phosphorus, and other nutrient elements.

The atmosphere plays a decisive role in regulating temperatures at and close to Earth's surface. Without

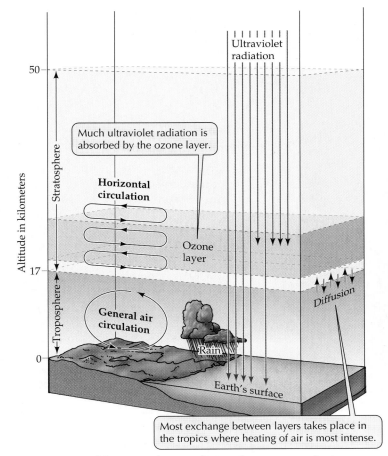

Much ultraviolet radiation is absorbed by the ozone layer.

Horizontal circulation

Ozone layer

General air circulation

Rain

Diffusion

Earth's surface

Ultraviolet radiation

Most exchange between layers takes place in the tropics where heating of air is most intense.

Altitude in kilometers

Stratosphere

Troposphere

53.12 Earth's Atmosphere The two lowest atmospheric layers, the troposphere and the stratosphere, differ in their circulation patterns, the amount of moisture they contain, and the amount of ultraviolet radiation they receive.

an atmosphere, the average surface temperature of Earth would be about –18°C rather than its actual +17°C. Earth remains at this warm temperature because the atmosphere is relatively transparent to visible light, but traps a large part of the outgoing infrared radiation (heat), the main radiation emitted by a cool body like Earth. Water vapor, carbon dioxide, and ozone are especially important trappers of infrared radiation. That is why, as we will see below, increased concentrations of atmospheric carbon dioxide may lead to important climatic changes.

Land covers about one-fourth of Earth's surface

About one-fourth of Earth's surface, most of it in the Northern Hemisphere, is currently above sea level. Most of this land is covered by a layer of soil of varying depths, weathered from parent rocks beneath it or carried to its present location by agents of erosion. Unlike their behavior in air and water, elements on land move slowly, and they usually move only short distances. For this reason, we will emphasize local

rather than global ecosystems in our discussion of the land compartment.

The land is connected to the atmospheric compartment by terrestrial organisms that take chemical elements from and release them to the air. Chemical elements in soils are carried in solution into the groundwater and eventually into rivers and oceans, where they are lost to organisms until an episode of uplifting raises marine sediments and a new cycle of erosion and weathering begins.

As you may recall from Chapter 34, the type of soil that forms in an area depends on the underlying rock, as well as on climate, topography, the organisms living there, and the length of time that soil-forming processes have been acting. As a soil weathers, its clay particles slowly decompose chemically. After hundreds of thousands of years, most nutrients needed by plants have weathered out and have been carried to the oceans. Therefore, very old soils are much less fertile than young soils are. Figure 34.6 shows a general picture of these changes over a period of 1 million years. Even though the global supply of nutrients is constant, regional and local deficiencies strongly affect ecosystem processes on land.

Biogeochemical Cycles

The chemical elements organisms need in large quantities—carbon, hydrogen, oxygen, nitrogen, phosphorus, and sulfur—cycle through organisms to the environment and back again. The pattern of movement of a chemical element through organisms and reservoirs in the physical environment is called its **biogeochemical cycle**. The carbon and nitrogen atoms of which life is composed today are the same atoms that made up dinosaurs, insects, and trees in the Mesozoic era. Some chemical elements circulate continually, but large quantities of other elements are temporarily lost from circulation through deposition in deep-sea sediments.

Each chemical element used by organisms has a distinctive biogeochemical cycle whose properties depend on the physical and chemical nature of the element and how organisms use it. All chemical elements cycle quickly through organisms because no individual, even of the longest-lived species, lives very long in geologic terms. Chemical elements, such as carbon and nitrogen, that exist in the atmosphere as a gas cycle faster than nongaseous elements. After discussing the movements of water, we'll discuss the cycling of the most abundant chemical elements in organisms.

Water cycles through the oceans, atmosphere, and land

The cycling of water through the oceans, atmosphere, and land is known as the **hydrological cycle**. Although water is a compound, not an element, we dis-

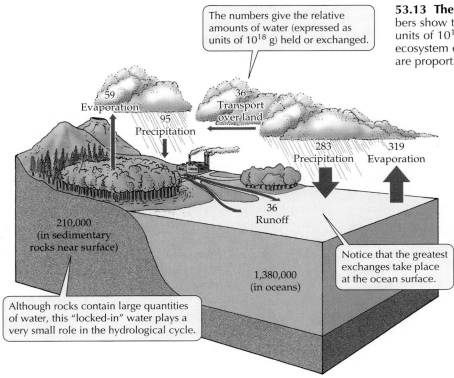

The numbers give the relative amounts of water (expressed as units of 10^{18} g) held or exchanged.

59
Evaporation

36
Transport over land

95
Precipitation

283
Precipitation

319
Evaporation

36
Runoff

210,000 (in sedimentary rocks near surface)

1,380,000 (in oceans)

Notice that the greatest exchanges take place at the ocean surface.

Although rocks contain large quantities of water, this "locked-in" water plays a very small role in the hydrological cycle.

53.13 The Global Hydrological Cycle The numbers show the relative amounts of water (expressed as units of 10^{18} g) held in or exchanged annually by ecosystem compartments. The widths of the arrows are proportional to the size of the fluxes.

Organisms profoundly influence the carbon cycle

Organisms are triumphs of carbon chemistry; to survive, they must have access to carbon atoms. Nearly all the carbon in organisms comes from carbon dioxide (CO_2) in the atmosphere or dissolved carbonate ions (HCO_3^-) in water. In the cells of some bacteria, algae, and leaves of plants, carbon is incorporated into organic molecules by photosynthesis. All organisms in other kingdoms get their carbon by consuming other organisms or their remains.

cuss its cycle here together with those of individual elements because of its importance to life. The hydrological cycle is driven by the evaporation of water, most of it from ocean surfaces. Some water returns to the oceans as precipitation, but much less falls back on the oceans than is evaporated from them. The remaining evaporated water is carried by winds over the land, where it falls as rain or snow.

Water also evaporates from soils, from freshwater lakes and rivers, and from the leaves of plants (transpiration), but the total amount evaporated is less than the amount that falls as precipitation. The excess water eventually returns to the oceans via rivers, coastal runoff, and subterranean flows (Figure 53.13).

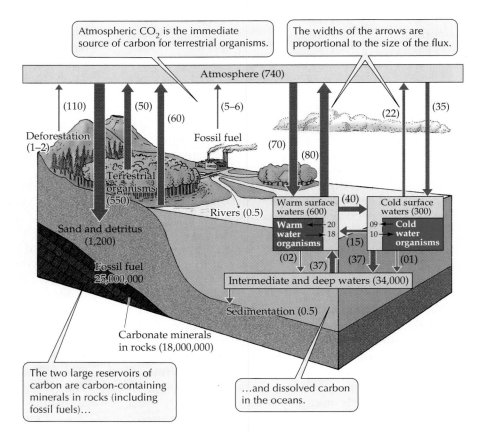

Atmospheric CO_2 is the immediate source of carbon for terrestrial organisms.

The widths of the arrows are proportional to the size of the flux.

Atmosphere (740)

(110) (50) (60) (5–6) (70) (80) (22) (35)

Deforestation (1–2)

Fossil fuel

Terrestrial organisms (550)

Rivers (0.5)

Sand and detritus (1,200)

Warm surface waters (600) (40) Cold surface waters (300)

Warm water organisms 20 18 09 10 Cold water organisms

(15)

(02) (37) (37) (01)

Fossil fuel 25,000,000

Intermediate and deep waters (34,000)

Sedimentation (0.5)

Carbonate minerals in rocks (18,000,000)

The two large reservoirs of carbon are carbon-containing minerals in rocks (including fossil fuels)...

...and dissolved carbon in the oceans.

53.14 The Global Carbon Cycle The numbers show the quantities of carbon (expressed as units of 10^{15} g) in organisms and in various carbon reservoirs and the amounts that move annually between the various ecosystem compartments. The widths of the arrows are proportional to the size of the fluxes.

Although atmospheric carbon dioxide is the immediate source of carbon for terrestrial organisms, only a small part of Earth's carbon is in the atmosphere. Most of it exists as nongaseous, dissolved carbon in oceans and carbon-containing minerals in rocks (Figure 53.14). Sedimentary rocks hold most of the carbon that is in rocks. Movement of carbon between rocks and other reservoirs of carbon is very slow.

Although marine organisms contain very little carbon, they have a profound influence on the distribution of carbon in the oceans. They convert soluble carbonate ions from seawater into insoluble ocean sediments by depositing carbon in their shells and skeletons, which eventually sink to the bottom.

Biological processes move carbon between the atmospheric and terrestrial compartments, removing it from the atmosphere during photosynthesis and returning it to the atmosphere during respiration. Growing plants at middle and high latitudes in the Northern Hemisphere incorporate so much carbon into their bodies during the summer that they reduce the concentration of atmospheric carbon dioxide from about 350 parts per million in winter to 335 parts per million in midsummer. This carbon is released back into the atmosphere by decomposition in autumn.

At times in the remote past, large quantities of carbon were removed from the global carbon cycle when organisms died in large numbers in environments without oxygen. In such environments, detritivores do not reduce organic carbon to carbon dioxide. Instead, organic molecules accumulate and eventually are transformed into oil, natural gas, coal, or peat. Humans have discovered and used these deposits, known as **fossil fuels**, at ever-increasing rates during the past 150 years. As a result, carbon dioxide, the final product of burning these fuels, is being released into the atmosphere faster than it is being transferred to the oceans and incorporated into terrestrial biomass (see Figure 53.18).

Elemental nitrogen can be used by few organisms

Nitrogen is an essential component of many organic molecules, such as nucleic acids and proteins. Although nitrogen (N_2) makes up 78 percent of the atmosphere, it cannot be used by most organisms in its gaseous form. It can be converted into biologically useful forms only by a few species of bacteria and cyanobacteria (see Chapter 23). Therefore, despite its abundance, usable nitrogen is often in short supply in ecosystems. This is why nearly all commercial fertilizers contain biologically useful compounds of nitrogen.

Just as organisms other than nitrogen fixers cannot take up nitrogen gas directly from the atmosphere, they do not respire nitrogen back to the atmosphere. Instead, organic molecules containing nitrogen are converted to inorganic molecules in several stages by different organisms. Most of the resulting nitrogen-containing compounds, such as nitrates or ammonia, are again taken up by plants. This movement of nitrogen among organisms accounts for about 95 percent of all nitrogen fluxes on Earth (Figure 53.15).

The phosphorus cycle has no gaseous phase

The phosphorus cycle differs from the other cycles discussed in this section in that it lacks a gaseous phase. Some phosphorus is transported on dust particles, but in general the atmosphere plays a very minor role in the phosphorus cycle. Phosphorus exists mostly as phosphate (PO_4^{3-}) or similar compounds.

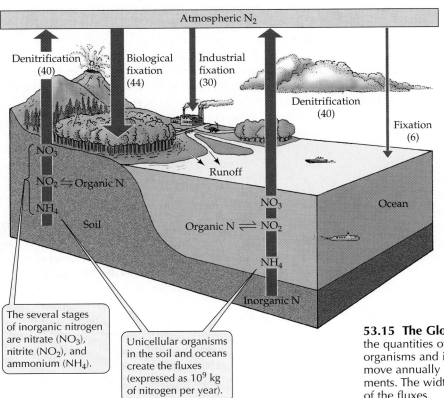

53.15 The Global Nitrogen Cycle The numbers show the quantities of nitrogen (expressed as units of 10^9 kg) in organisms and in various reservoirs and the amounts that move annually between the various ecosystem compartments. The widths of the arrows are proportional to the size of the fluxes.

TABLE 53.3 Reservoirs, Fluxes, and Residence Times of Phosphorus in the Global Ecosystem

RESERVOIR	AMOUNT IN RESERVOIR (10^6 METRIC TONS)	FLUX (10^6 METRIC TONS/YEAR)	RESIDENCE TIME (YEARS)
Atmosphere	0.0028	4.5	0.0006 (53 hours)
Land biota	3,000	63.5	47.2
Land	2,000,000	88–100	2,000
Shallow ocean	2,710	1,058	2.56
Ocean biota	138	1,040	0.1327 (48 days)
Deep ocean	87,100	60	1,45
Sediments	4×10^9	214	1.87×10^8
Total ocean ecosystem	89,810	1.9	47,270

Most phosphate deposits are of marine origin. In nature, phosphorus becomes available through the slow weathering and dissolution of rocks and minerals. Organisms need phosphorus as a component of the energy-rich molecules involved in cellular metabolism. Phosphorus is often a limiting nutrient in soils and lakes. That is why phosphate is a component of most fertilizers and why adding phosphate to lakes causes marked increases in their biological productivity.

Table 53.3 shows how much more rapidly phosphorus is cycled through marine organisms than through terrestrial organisms. Phosphorus also moves readily between the surface and the bottom of the oceans. On average, a phosphorus atom cycles about 50 times between deep waters and surface waters before it is deposited in ocean sediments. Each time a phosphorus atom reaches surface waters, it cycles between the marine biota and the dissolved phosphate in the water about 25 times before it returns to deep waters. As a result, the average phosphorus atom is incorporated into marine organisms about 1,250 times during its stay in the ocean!

Organisms drive the sulfur cycle

Sulfur is biologically important because it is a component of proteins. Emissions of sulfur dioxide (SO_2) and hydrogen sulfide (H_2S) from volcanoes and fumaroles (vents for hot gases) are the only significant natural nonbiological fluxes of sulfur. These emissions release, on average, between 10 and 20 percent of the total natural flux of gaseous sulfur to the atmosphere, but they vary greatly in time and space. Large eruptions spread great quantities of sulfur over broad areas, but they are rare events.

Volatile sulfur compounds are also emitted by both terrestrial and marine organisms. Certain marine algae produce large amounts of dimethyl sulfide (CH_3SCH_3), which accounts for about half of the biotic component of the sulfur cycle; the other half is produced by terrestrial organisms. On land, the break-down of organic sulfur compounds during fermentation is the most important mechanism of sulfur release (Figure 53.16).

Sulfur is apparently always abundant enough to meet the needs of living organisms. It also plays an important role in global climate. Even if air is moist, clouds do not form readily unless there are nuclei around which water can condense. Dimethyl sulfide is the major source of such nuclei. Therefore, increases or decreases in sulfur emissions can change cloud cover and hence climate.

Human Alterations of Biogeochemical Cycles

Human activity has greatly modified the quantities of elements being cycled as well as how and where they enter and exit ecosystems. These changes can increase metabolic rates if they increase the availability of nutrients, or decrease them if levels of elements become high enough to be toxic to organisms, or if high levels cause other detrimental environmental changes.

We will now consider several examples of consequences resulting from human modifications of biogeochemical cycles. These consequences range from local (lake eutrophication) to regional (acid precipitation) to global (climate change).

Lake eutrophication is a local effect

The most striking and best-studied example of a local effect of altered biogeochemical cycles is **eutrophication**—the addition of nutrients, especially phosphorus, to fresh water. Humans, who tend to live around water, dump large quantities of nutrients directly and indirectly into lakes and rivers. Most of these nutrients come from domestic and industrial sewage, but many come from leaching of fertilizers and pesticides from agricultural lands draining into rivers and lakes. Some nutrients also arrive naturally in precipitation, but human activities have greatly increased the quantities of nutrients that enter fresh water.

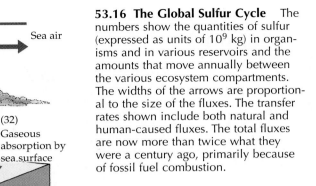

53.16 The Global Sulfur Cycle The numbers show the quantities of sulfur (expressed as units of 10^9 kg) in organisms and in various reservoirs and the amounts that move annually between the various ecosystem compartments. The widths of the arrows are proportional to the size of the fluxes. The transfer rates shown include both natural and human-caused fluxes. The total fluxes are now more than twice what they were a century ago, primarily because of fossil fuel combustion.

In fresh water, photosynthesis is most often limited by supplies of phosphorus. In eutrophic (enriched) lakes, the extra phosphorus provided by fertilizers and detergents allows algae and bacteria to multiply, forming blooms that turn the water green. The decomposition of dead cells produced by this increased biological activity uses up all the oxygen in the lake, and anaerobic organisms come to dominate the sediments. These anaerobic organisms do not break down organic compounds all the way to carbon dioxide, and many of the end products of their activities have unpleasant odors.

Lake Erie is a eutrophic lake, but two hundred years ago it had only moderate levels of photosynthesis and clear, oxygenated water. Today more than 15 million people live in the Lake Erie basin. Nearby cities pour more than 250 billion liters of domestic and industrial wastes into the lake annually. The entire basin is intensely farmed and heavily fertilized.

In the early part of the twentieth century, nutrients in the lake increased greatly, and algae proliferated. At the water filtration plant in Cleveland, algae increased from 81 per milliliter in 1929 to 2,423 per milliliter in 1962. Algal blooms and populations of bacteria also increased. The numbers of *Escherichia coli* increased enough to cause the closing of many of the lake's beaches as health hazards.

As oxygen levels dropped in deeper lake waters, many native species that thrive only in oxygenated water declined. Before 1900, the dominant fishes in Lake Erie were lake herring, blue pike, carp, yellow perch, sauger, whitefish, and walleye. Lake trout were common in deeper waters. By 1925, herring had become too scarce to support the herring fishery. After 1945, blue pike, sauger, and whitefish became very scarce and lake trout disappeared. Currently the lake's fishery depends upon yellow perch, smelt, sheepshead, white bass, carp, catfish, and walleye, most of which are less valuable commercially than the species that declined.

Since 1972, the United States and Canada have invested more than 8 billion dollars to improve municipal waste facilities and reduce discharges of phosphorus into Lake Erie. As a result, the amount of phosphate added to Lake Erie has decreased more than 80 percent from the maximum level, and phosphorus concentrations in the lake have declined substantially. Since 1985, estimated inputs of phosphorus to Lake Erie have been reduced to about the target goal of 11,000 metric tons per year. The deeper waters of Lake Erie still become poor in oxygen during the summer months, but the rate of oxygen depletion is declining. Algal blooms have decreased, as have populations of small fishes that feed on algae.

The rate at which a eutrophic lake recovers depends on the replacement rate of its waters. Because water flows slowly through Lake Erie, it will take many years for the lake to recover from the heavy pollutant loads it has received. By contrast, the molecules of water in Lake Washington, a smaller lake adjacent to the city of Seattle, are replaced every 3 years. When sewage was diverted away from Lake Washington, the lake returned to its former condition within a decade.

Acid precipitation is a regional effect

An important regional effect of human alteration of two major biogeochemical cycles is **acid precipitation**—rain or snow whose pH is lowered by the presence of sulfuric acid (H_2SO_4) and nitric acid (HNO_3), derived in large part from the burning of fossil fuels. These acids enter the atmosphere and may travel hundreds of kilometers before they settle to Earth in precipitation or as dry particles.

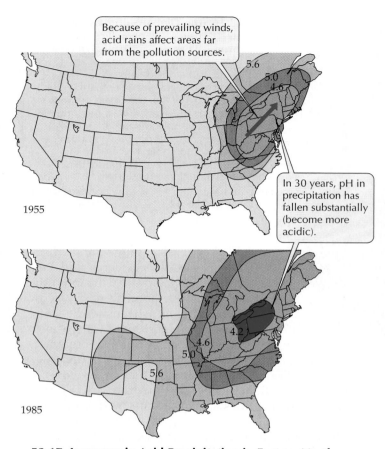

Because of prevailing winds, acid rains affect areas far from the pollution sources.

In 30 years, pH in precipitation has fallen substantially (become more acidic).

1955

1985

53.17 Increases in Acid Precipitation in Eastern North America The numbers represent the annual average pH of precipitation. Oxides of nitrogen and sulfur—the principal contributors to acid precipitation—travel far enough from their sources that the effects of many sources blend together to produce the pattern shown here.

Acid precipitation is now a phenomenon of all major industrial countries and is particularly widespread in eastern North America (Figure 53.17). The source of acid precipitation in New England is primarily the Ohio Valley; that of Scandinavia is primarily the industrial areas of England and Germany. The normal pH of precipitation in New England is about 5.6, but precipitation there now averages about pH 4.1, and there are occasional storms with a precipitation pH as low as 3.0. Precipitation with a pH of about 3.5 or lower causes direct damage to the leaves of plants and reduces photosynthetic rates. In central Europe, acid precipitation has contributed to moderate to severe damage to 15 to 20 percent of the growing stock of harvestable forest.

Ecologists in Canada studied the effects of acid precipitation on small lakes by adding enough sulfuric acid to two lakes to reduce their pH from about 6.6 to 5.2. In both lakes, nitrifying bacteria failed to adapt to these moderately acidic conditions, with the result that the nitrogen cycle was blocked and ammonium accu-

mulated in the water. When the ecologists stopped adding acid to one of the lakes, its pH increased to 5.4, and nitrification resumed after a lag of about 1 year. These experiments show that lakes are very sensitive to acidification, but can recover quickly when pH returns to normal values.

Acid precipitation also changes soil chemistry. Since 1963, scientists have monitored the chemistry of both precipitation and stream water at the Hubbard Brook Experimental Forest in New Hampshire. The pH of stream water leaving the forest has changed relatively little as precipitation has become more acid, but large amounts of calcium and magnesium have flowed out of the watershed. So much calcium has been lost from the soil that calcium supply has been limiting forest production at Hubbard Brook since 1987, and annual accumulation of biomass in the forest is now much lower than it was during the previous decades.

When pollution originates in one area but causes problems in another, as acid precipitation does, solving the problem is politically difficult. Acid precipitation is caused by the generation of the energy upon which modern societies depend. Oxides of nitrogen and sulfur can be removed from smokestack gases, but costs rise sharply as the percentage removed rises above 90 percent. The number of sources emitting these oxides is now so great that almost complete removal will be necessary to correct the problem, even if no new sources are added.

Alterations of the carbon cycle produce global effects

The biogeochemical cycle most seriously disturbed globally by human activity is the carbon cycle. Climatologists have measured atmospheric concentrations of carbon dioxide on top of Mauna Loa in Hawaii since 1958. Their measurements reveal a slow but steady increase in carbon dioxide concentrations (Figure 53.18). Based on a variety of calculations, atmospheric scientists believe that 150 years ago, before the Industrial Revolution, the concentration of atmospheric carbon dioxide was probably about 265 parts per million. Today it is 350 parts per million.

This increase has been caused primarily by combustion of fossil fuels and secondarily by the burning of forests. If current trends in both these activities continue, atmospheric carbon dioxide is expected to reach 580 parts per million by the middle of the twenty-first century. This carbon dioxide will eventually be transferred to the oceans and deposited in sediments as calcium carbonate ($CaCO_3$), but the rate of transfer is much slower than the rate at which humans are introducing carbon dioxide into the atmosphere.

Enough carbon is being released by the burning of fossil fuels to alter the heat balance of Earth, even though the absolute quantity is small relative to other

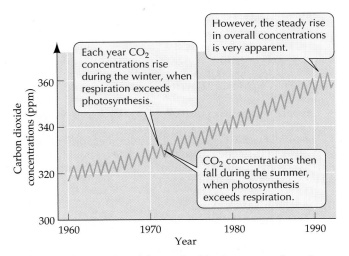

53.18 Atmospheric Carbon Dioxide Concentrations Are Increasing Carbon dioxide concentrations, expressed as parts per million by volume of dry air, are recorded on top of Mauna Loa, Hawaii.

components of the carbon cycle. The buildup of atmospheric carbon dioxide will warm Earth during the twenty-first century. Like the glass in a greenhouse, carbon dioxide is transparent to sunlight but opaque to radiated heat. Carbon dioxide thus permits sunlight to strike and warm Earth, but it traps some of the heat that Earth radiates back toward space.

The concentration of carbon dioxide in air trapped in the Antarctic and Greenland ice caps during the last Ice Age—between 15,000 and 30,000 years ago—was as low as 200 parts per million; during a warm interval 5,000 years ago it may have been slightly higher than it is today. This long-term record, which shows that Earth was warmer when CO_2 levels were higher than they are today, is a major reason why scientists expect Earth to warm as atmospheric CO_2 levels continue to increase.

In 1988, the United Nations Environment Programme and the World Meteorological Organization founded the Intergovernmental Panel on Climate Change (IPCC). The panel's mandate is to assess scientific and technical information about climate change. In its 1995 report, the work of 2,500 contributing scientists from more than 60 countries, the panel concluded that the average surface temperature over Earth has risen about 0.6°C since the late 1800s. More importantly, all of the 10 warmest years on record since 1860 have occurred during the last 15 years, despite the 3-year global cooling effect caused by the eruption of Mount Pinatubo in the Philippines in 1991, which ejected large amounts of dust into the atmosphere. The associated warming of ocean waters has resulted in a 3- to 6-centimeter rise in global sea level during the last 100 years.

Global climate models are used to predict the likely consequences of further increases in concentrations of atmospheric carbon dioxide. Predictions from these models are imprecise because nature is much more complicated than any model. Nonetheless, if the current concentration of atmospheric carbon dioxide doubles, the mean temperature of Earth is expected to increase 3 to 5°C, with greater increases at higher latitudes. A carbon dioxide doubling would also shift climatic patterns latitudinally, would probably cause droughts in the central regions of continents, and would increase precipitation in coastal areas. Global warming may result in the melting of the Greenland and Antarctic ice caps and will warm the oceans, which will thus expand, raising sea levels and flooding coastal cities and agricultural lands.

Because carbon dioxide is carried by air movements to places thousands of kilometers away from where it enters the atmosphere, the problem is global. Although it is very difficult for societies to decide how much they should invest today to avert potential future climatic problems, some nations have committed themselves to efforts at reducing their emissions of carbon dioxide. Such reductions will not be easy because carbon dioxide is the inevitable end product of fossil fuel combustion, and modern societies are powered by fossil fuels.

The amount of carbon dioxide emitted by a power plant burning fossil fuels cannot be reduced by cleansing of the gases leaving the smokestack. However, because so much energy is currently wasted, many steps can be taken to increase energy use efficiency so that we get more valuable services for the same amount of fuel burned. In addition, we can substitute energy sources that do not contain carbon (solar, wind, geothermal, nuclear) for fossil fuels.

Chemosynthesis-Powered Ecosystems

Most ecosystems are powered by sunlight falling directly on them, but some depend upon sunlight that falls elsewhere. For example, marine ecosystems below the level where enough light penetrates to permit photosynthesis depend on biomass produced in the well-lit zone above them. The productivity of most deep-sea ecosystems is low because only small amounts of detritus descend through the water column to reach them.

Some deep-sea ecosystems are totally independent of sunlight. The most striking are those around hot springs associated with seafloor spreading zones. The energy base of these ecosystems is chemoautotrophy by sulfur-oxidizing bacteria, which obtain energy by oxidizing hydrogen sulfide emitted from the vents. Most of the other organisms in these ecosystems, such as pogonophores, live directly or indirectly on these sulfur-oxidizing bacteria (see Figure 24.7).

Ecologists recently discovered a cave in southern Romania whose ecosystem is powered by bacteria that fix inorganic carbon by using hydrogen sulfide as an energy source. Chemoautotrophic production by these bacteria is the food base for 48 species of cave-adapted terrestrial and aquatic invertebrates. Air pockets in the submerged portions of the cave contain mats of bacteria and fungi that float on the water surface and grow on the limestone walls of the cave.

Summary of "Ecosystems"

• The organisms living in a particular area, together with the physical environment with which they interact, constitute an ecosystem. At a global scale, Earth is a single ecosystem.

Climates on Earth

• Biological processes on Earth are driven primarily by solar radiation.
• Climates determine the amount of heat, moisture, and sunlight available to living organisms in different places on Earth.
• Rising air expands and cools, releasing moisture. Descending air warms and dries and takes up moisture, creating rain shadows. **Review Figure 53.1**
• Global air circulation is driven by solar radiation and the spinning of Earth on its axis. **Review Figure 53.2**
• Oceanic currents are driven primarily by prevailing winds. **Review Figure 53.3**

Energy Flow through Ecosystems

• The capture of solar radiation by photosynthesis powers ecosystem productivity.
• The annual production of an area is determined primarily by temperature and moisture. **Review Figures 53.4, 53.5**
• Energy flows through ecosystems as organisms capture and store energy and transfer it to other organisms when they are eaten. Organisms are grouped into trophic levels according to the number of steps through which energy passes to get to them. **Review Table 53.1**
• Who eats whom in a ecosystem can be diagrammed as a food web. **Review Figure 53.6**
• The amount of energy flowing through an ecosystem depends on primary production and on the efficiency of transfer of energy from one trophic level to another. **Review Figures 53.7, 53.8**

Agricultural Manipulation of Ecosystem Productivity

• Humans manipulate ecosystem productivity by increasing rates of photosynthesis and reducing losses of crops to pests. In modern agriculture, the energy required to do this is provided by fossil fuels. **Review Figure 53.9**

Cycles of Materials in Ecosystems

• The main compartments of the global ecosystem are the oceans, fresh waters, land, and atmosphere, among which materials are constantly being exchanged.
• Primary production in oceans is highest adjacent to continents, where nutrient-rich waters rise to the surface. **Review Figure 53.10**
• Temperate-zone lakes turn over twice each year as water cools and warms. **Review Figure 53.11**
• The two lowest layers of Earth's atmosphere differ from each other in their circulation patterns, the amount of moisture they contain, and the amount of ultraviolet radiation they receive. **Review Figure 53.12**

Biogeochemical Cycles

• The elements organisms need in large quantities cycle though organisms to the environment and back again.
• The cycle of water—the hydrological cycle—is driven by evaporation of water, most of it from ocean surfaces. **Review Figure 53.13**
• Atmospheric carbon dioxide is the immediate source of carbon for terrestrial organisms, but only a small part of Earth's carbon is in the atmosphere. **Review Figure 53.14**
• Although nitrogen makes up 78 percent of Earth's atmosphere, nitrogen can be converted into biologically useful forms only by a few species of bacteria and cyanobacteria. **Review Figure 53.15**
• The phosphorus cycle differs from the cycles of carbon and nitrogen in that it lacks a gaseous phase. **Review Table 53.3**

Human Alterations of Biogeochemical Cycles

• The most striking example of a local effect of altered biogeochemical cycles is lake eutrophication.
• Acid precipitation is an important regional consequence of human modifications of the nitrogen and sulfur cycles. **Review Figure 53.17**
• Earth's climate is being changed as a result of increasing concentrations of carbon dioxide in the atmosphere. **Review Figure 53.18**

Chemosynthesis-Powered Ecosystems

• A few deep-sea and cave ecosystems are powered by chemosynthesis rather than photosynthesis.

Self-Quiz

1. Which of the following is true about the amount of sunlight and heat arriving on Earth?
   a. Every place on Earth receives the same annual number of hours of sunlight and the same amount of heat.
   b. Every place on Earth receives the same annual number of hours of sunlight, but not the same amount of heat.
   c. Every place on Earth receives the same annual amount of heat, but not the same number of hours of sunlight.
   d. Both the annual amount of sunlight and the amount of heat received vary over the surface of Earth.
   e. None of the above

2. When an area is within the intertropical convergence zone,
   a. the northeast trade winds blow steadily.
   b. the southeast trade winds blow steadily.
   c. air is descending and it seldom rains.
   d. air is rising and heavy rains fall frequently.
   e. westerly winds blow steadily.

3. Zones of marine upwelling are important because
   a. they help scientists measure the chemistry of deep ocean water.
   b. they bring to the surface organisms that are difficult to observe elsewhere.
   c. ships can sail faster in these zones.
   d. they increase marine productivity by bringing nutrients back to surface ocean waters.
   e. they bring oxygenated water to the surface.

4. Which of the following is *not* true of the troposphere?
   a. It contains nearly all atmospheric water vapor.
   b. Materials enter it primarily at the intertropical convergence zone.
   c. It is about 17 km deep in the tropics.
   d. Most global atmospheric circulation takes place there.
   e. It contains about 80 percent of the mass of the atmosphere.

5. Carbon dioxide is called a greenhouse gas because
   a. it is used in greenhouses to increase plant growth.
   b. it is transparent to heat radiation but opaque to sunlight.
   c. it is transparent to sunlight but opaque to heat radiation.
   d. it is transparent to both sunlight and heat radiation.
   e. it is opaque to both sunlight and heat radiation.

6. The phosphorus cycle differs from those of carbon and nitrogen in that
   a. it lacks a gaseous phase.
   b. it lacks a liquid phase.
   c. only phosphorus is cycled through marine organisms.
   d. living organisms do not need phosphorus.
   e. The phosphorus cycle does not differ importantly from the carbon and nitrogen cycles.

7. Acid precipitation results from human modifications of
   a. the carbon and nitrogen cycles.
   b. the carbon and sulfur cycles.
   c. the carbon and phosphorus cycles.
   d. the nitrogen and sulfur cycles.
   e. the nitrogen and phosphorus cycles.

8. The total amount of energy that plants assimilate by photosynthesis is called
   a. gross primary production.
   b. net primary production.
   c. biomass.
   d. a pyramid of energy.
   e. eutrophication.

9. The amount of energy reaching an upper trophic level is determined by
   a. net primary production.
   b. net primary production and the efficiencies with which food energy is converted to biomass.
   c. gross primary production.
   d. gross primary production and the efficiencies with which food energy is converted to biomass.
   e. gross primary production and net primary production.

10. Which of the following is *not* a component of integrated pest management?
    a. Use of cultural strategies such as crop rotation and mixed plantings
    b. Use of pest-resistant strains of crops
    c. Use of predators and parasites of crop pests
    d. Use of chemical attractants
    e. Use of chemical pesticides whenever pests are discovered

Applying Concepts

1. How would you expect temperature and oxygen profiles to appear in a broad, shallow tropical lake? In a very deep tropical lake? Why?

2. The waters of Lake Washington, adjacent to the city of Seattle, rapidly returned to their preindustrial condition when sewage was diverted from the lake to Puget Sound, an arm of the Pacific Ocean. Would all lakes being polluted with sewage clean themselves up as rapidly as Lake Washington if pollutant input were stopped? What characteristics of a lake are most important to its rate of recovery following removal of pollutant inputs? What is the diverted sewage likely to do to Puget Sound?

3. Tropical forests currently are being cut at a very rapid rate. Does this necessarily mean that deforestation is a major source of input of carbon dioxide to the atmosphere? If not, why not?

4. The two drawings below represent pyramids of biomass for (*a*) an old field in Georgia and (*b*) the English Channel. Explain the significance of the inversion of the second pyramid compared with the first.

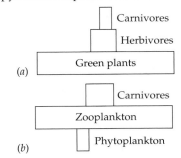

5. The amount of energy flowing through a food chain declines more or less rapidly depending upon the nature of the organisms in the chain. Which of the following simplified food chains is likely to be more efficient? Why? What criteria of efficiency are you using?
   a. phytoplankton → zooplankton → herring
   b. shrubs → deer → wolf

6. A government official authorizes the construction of a large power plant in a former wilderness area. Its smokestacks discharge great quantities of waste resulting from the combustion of coal. List and describe all *likely* ecological results at local, regional, and global levels. Now suppose the wastes were thoroughly scrubbed from the stack gases. Which of the ecological results you have just outlined would still happen?

Readings

Anderson, D. M. 1994. "Red Tides." *Scientific American,* August. Shows that the frequency of red tides, which can release potent toxins into the oceans, is increasing because pollution provides rich nutrients for the red tide organisms.

Bazzaz, F. A. and E. D. Fajer. 1992. "Plant Life in a CO_2-Rich World." *Scientific American*, January. Discusses the fact that even without considerations of global warming, increasing atmospheric levels of carbon dioxide may greatly alter the structure and functioning of ecosystems. These changes will not necessarily benefit plants.

Berner, R. A. and A. C. Lasaga. 1989. "Modeling the Geochemical Carbon Cycle." *Scientific American*, March. Discusses how natural geochemical processes that result in the slow buildup of atmospheric carbon dioxide may have caused past geologic intervals of global warming through the greenhouse effect.

Broeker, W. S. 1995. "Chaotic Climate." *Scientific American*, November. Describes the geologic record that shows that Earth's weather patterns have sometimes changed dramatically in a decade or less.

Butcher, S. S., R. J. Charlson, G. H. Orians and G. V. Wolfe (Eds). 1992. *Global Biochemical Cycles*. Academic Press, New York. A comprehensive review of all global biogeochemical cycles and the methods used to study them.

Gates, D. M. 1993. *Climate Change and Its Biological Consequences*. Sinauer Associates, Sunderland, MA. An introduction to Earth's climates, past and present, and the effect climate change has on organisms and ecosystems. An accessible explanation of the various computer models that predict massive climate warming in the twenty-first century.

Jordan, C. F. 1985. *Nutrient Cycling in Tropical Forest Ecosystems*. Wiley, New York. A useful summary of nutrient cycling patterns in tropical forests and how they differ from those of temperate forests.

Kusler, J. A., W. J. Mitsch and J. S. Larson. 1994. "Wetlands." *Scientific American*, January. Discusses how wetlands purify water, reduce floods, and serve as incubators for aquatic life. Assesses what happens when the protection of an essential natural resource must be weighed against society's demand for real estate, fossil fuels, and agricultural land.

Makarewicz, J. C. and P. Bertram. 1991. "Evidence for the Restoration of the Lake Erie Ecosystem." *BioScience,* vol. 41, pages 216–223. Describes changes in water quality, oxygen levels, and population dynamics of microorganisms, plants, and animals in Lake Erie.

Mohnen, V. A. 1988. "The Challenge of Acid Rain." *Scientific American*, August. Argues that acid rain's effects on soil and water leave no doubt that we need to control its causes, and shows how new advances in technology offer attractive solutions.

Nicol, S. and W. de la Mare. 1993. "Ecosystem Management and the Antarctic Krill." *American Scientist*, vol. 81, pages 36–47. Discusses how krill's unusual biology and ecological significance are driving an effort to apply systems theory to managing an ecosystem.

Rambler, M. B., L. Margulis and R. Fester (Eds.). 1989. *Global Ecology: Towards a Science of the Biosphere*. Academic Press, Boston. A collection of essays covering a wide variety of the interactions between organisms, global biogeochemical cycles, and global climate.

Rietzler, K. and I. C. Feller 1996. "Caribbean Mangrove Swamps." *Scientific American*, March. Describes the remarkable ecosystems that exist in shallow water on muddy tropical ocean shores.

Robison B. H. 1995. "Light in the Ocean's Midwater." *Scientific American*, July. Describes a rich and remarkable community of organisms below the photic zone that is illuminated only by the radiance of its residents.

Schlesinger, W. 1997. *Biogeochemistry*, 2nd Edition. Oxford University Press, Oxford. A thorough review of all aspects of biogeochemistry.

Schneider, S. H. 1989. "The Changing Climate." *Scientific American*, September. Argues that global warming should be unmistakable within a decade or two and shows how prompt emission cuts could slow the buildup of heat-trapping gases.

Whittaker, R. H. 1975. *Communities and Ecosystems*, 2nd Edition. Macmillan, New York. A good, short textbook with excellent coverage of theoretical topics; an excellent follow-up to the materials in this chapter.

Chapter 54

Biogeography

Phascolarctos cinerus

Strange Organisms in New Places
Animals like the koala were unfamiliar to the Europeans who colonized Australia.

When the first Europeans arrived in Australia, they saw plants and animals that differed in perplexing ways from the ones they had known at home, such as flowers pollinated by brush-tongued parrots and mammals that hopped around on their hind legs, carrying their young in pouches. On the other hand, the first Europeans to visit North America felt more at home because the plants and animals of North America were similar to those of Europe.

During their worldwide travels, European explorers found many vegetation types—tropical forests, mangrove forests, and deserts with tall cacti—that were unfamiliar to them, but they also found many areas where the vegetation was similar to what they knew back home, even though they seldom recognized any familiar species. The study of the diversity of organisms over space began when those eighteenth-century travelers who first noted intercontinental differences in biotas attempted to understand those differences.

Biogeography is the science that attempts to explain patterns in the variation of individuals, populations, species, and communities along geographic gradients. In this chapter, we will show how biogeographers determine how events in the remote past influenced distributional patterns today, and how biological differences among species in different geographic regions influence the species composition of ecological communities in different regions of Earth.

Why Are Species Found Where They Are?

Superficially, explaining species' distributions seems to be a simple matter because the question of why a species is or is not found in a certain location has only a few possible answers. If a species occupies a particular area, either it evolved there, or it evolved elsewhere and dispersed to the area. If a species is *not* found in a particular area, either it evolved elsewhere and never dispersed to the area, or it was once present in the area but no longer lives there. Unfortunately, determining which of these answers is correct turns out to be difficult.

To explain the distributions of organisms, biogeographers must draw upon and interpret a broad array of knowledge. Finding the answers to the questions listed above requires information about the evolutionary histories of species, which comes from fossils and from knowledge of phylogenetic relationships (see Chapter 22). In addition, it requires information about Earth's changes—continental drift, glacial advances and retreats, sea level changes, and mountain building—during the time when the organisms were evolving. Such geologic information can tell us whether organisms evolved where they are currently found or dispersed and colonized new areas from a distant area of origin. In this section we show how the acceptance of continental drift and the use of cladistic methods of analysis changed the science of biogeography.

Ancient events influence current distributions

Early biogeographers believed in a relatively constant Earth that was too young to account for the diversity and distribution of life by any means except divine creation. They assembled much valuable information, but their interpretations were constrained by their belief that organisms had been created in their current forms in their current ranges. Linnaeus, for example, believed that all organisms had been created in one place—which he called Paradise—from which they later dispersed. Indeed, because most people believed that the continents were fixed in their positions, the only way to account for current distributions was to invoke massive dispersal.

The notion that the continents might have moved was not seriously considered until 1912. Alfred Wegener, the German meteorologist who argued that the continents had drifted, based his ideas on the shapes of continents (the outlines of Africa and South America seem to fit together like pieces of a puzzle), the alignment of mountains chains and rock strata, coal beds, glacial deposits, and biogeography (the distributions of plants and animals in Africa and South America were hard to explain if one assumed that the continents had always been where they are now).

When Wegener proposed his ideas, few scientists took them seriously, primarily because there were no known mechanisms for continental movement and because no convincing geologic evidence of such movements existed. As we learned in Chapter 22, geologic evidence and plausible mechanisms were eventually discovered, and the broad pattern of continental movement is now clear.

About 280 million years ago, the continents were united to form a single land mass, Pangaea. By the early Mesozoic era (about 245 million years ago), when the continents were still very close to one another, many groups of nonmarine organisms, including insects, freshwater fishes, and frogs, had already evolved. Some organisms that live on widely separated continents today were probably present on those land masses when they were all part of Pangaea.

By 100 mya, Pangaea had separated into northern (Laurasia) and southern (Gondwanaland) land masses, and the southern continents were drifting away from each other (see Figure 19.16). Eventually, continental drift, which continues today, brought India into contact with southern Asia, Australia closer to Southeast Asia, and South America into contact with North America (see Figure 19.17). Continental drift has thus influenced biogeographic patterns throughout the history of life on Earth.

Modern biogeographic methods

As the great age of Earth and the fact of evolution began to be understood, two groups of investigators developed new methods for generating testable hypotheses about geographic distributions. One group consisted of ecologists who studied how current distributions were influenced by interactions among species and by interactions between species and their physical environments. Because local habitats always contain fewer species than the number living in the surrounding region, these ecologists believed that interactions of the types discussed in Chapter 52 could explain many aspects of the distributions of organisms.

The other group consisted of historical biogeographers who used new techniques to investigate ancient influences on distributions. An important technique they developed was the transformation of taxonomic cladograms into **area cladograms** by substituting the species' geographic distributions for their names. The distribution patterns identified by area cladograms may suggest routes of dispersal or point to the splitting of biota due to the appearance of major barriers (Figure 54.1). Comparisons of area cladograms of many evolutionary lineages can demonstrate similarities and differences in their distribution patterns. Similarities suggest common responses to physical events, such as continental drift, mountain building, and sea level changes. Differences suggest that organisms in different lineages responded in different ways,

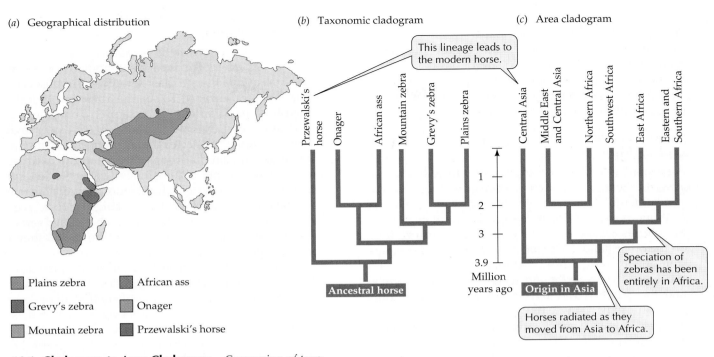

(a) Geographical distribution

Plains zebra
Grevy's zebra
Mountain zebra
African ass
Onager
Przewalski's horse

(b) Taxonomic cladogram

This lineage leads to the modern horse.

Przewalski's horse
Onager
African ass
Mountain zebra
Grevy's zebra
Plains zebra

Ancestral horse

(c) Area cladogram

Central Asia
Middle East and Central Asia
Northern Africa
Southwest Africa
East Africa
Eastern and Southern Africa

1
2
3
3.9
Million years ago

Origin in Asia

Speciation of zebras has been entirely in Africa.

Horses radiated as they moved from Asia to Africa.

54.1 Cladogram to Area Cladogram Conversion of taxonomic cladograms to area cladograms can aid in the understanding of distribution patterns.

or at different times, to past events, or that they have dispersed in unique ways.

These ecological and evolutionary approaches characterize the two main subdisciplines of biogeography: historical and ecological biogeography. *Historical biogeography* concerns itself primarily with the evolutionary histories of lineages of organisms: Where and when did they originate? How did they spread? What does their present-day distribution tell us about their past histories? *Ecological biogeography* concentrates on the current interactions of organisms with the physical environment and with one another. Ecological biogeographers seek to understand how ecological relationships influence where species and higher taxa are found today.

The names of these two subdisciplines are somewhat misleading. All biogeography is historical. Ecological biogeographers concentrate on recent history, current interactions, and changes within the past few thousand years. They also study patterns of distribution within local areas and regions. Historical biogeographers concentrate on longer time periods and larger spatial scales. Because time and space are continuous variables, these two subdisciplines blend together. The integration of knowledge from both of these subdisciplines is essential to a full understanding of the geographic distributions of organisms. We will first discuss the methods and findings of historical biogeographers.

Historical Biogeography

Historical biogeographers attempt to determine the influence of past events on today's patterns of distribution. We can never know past events with complete certainty, but by using a variety of types of evidence, historical biogeographers can develop, test, and adopt interpretations in which they have a high degree of confidence. As we have just seen, historical biogeographers often base their interpretations on phylogenies, which show the evolutionary relationships among organisms in a lineage. Phylogenies are most useful to biogeographers if the approximate times of evolutionary and geographic separations of lineages can be estimated.

Biogeographers use several approaches to infer the approximate times of separation of taxa within a lineage. First, if a "molecular clock" has been ticking, the degree of difference in the molecules of species is strongly correlated with the length of time their lineages have been separated (see Chapter 23). Second, fossils can show a biogeographer how long a taxon has been present in an area and whether its members formerly lived in areas where they are no longer found. The fossil record is helpful, but it is always incomplete. The first and last members of a taxon to live in an area are extremely unlikely to have become fossils that are discovered and described.

A third valuable source of information is the distributions of living species. Much more complete and extensive information can be gathered on such distributions than will ever be available from fossils. Much can be learned by examining the distribution patterns of *many different groups* of living organisms. Similarities in their distributions provide clues about past events that affected many of them.

Biogeographers wish to explain general patterns. To do so, they must compare the distributions of many different species, they must know the species' evolutionary relationships, and they must know the timing of geologic events that might have affected many species.

Vicariance and dispersal can both explain distributions

A species may be found in an area either because it evolved there or because it dispersed to the area. The appearance of a barrier that splits a species' previously continuous distribution is called a **vicariant event**, and the species is said to have a **vicariant distribution**. If, on the other hand, a barrier existed prior to the expansion of a species' range, and members of the species subsequently crossed it and established a new population, the species is said to have a **dispersal distribution**.

By studying a single lineage, a biogeographer may discover evidence suggesting that the distributions of ancestral species were influenced by some vicariant event, such as changes in sea level, mountain building, or continental movement. If that inference is correct, other lineages should also have been influenced by the same event—that is, they should have similar distribution patterns. Differences in distribution patterns among lineages indicate either that the lineages responded differently to the same vicariant events, or that the lineages separated at different times. By analyzing such similarities and differences, biogeographers can discover how vicariant events and dispersal may have influenced today's distribution patterns.

Species, genera, and families found in only one region are said to be **endemic** to that location. As far as we know, all species are endemic to Earth. Some species are endemic to one continent. Others are restricted to very small areas, such as tiny islands or single mountaintops. Because a species may disperse widely and then die out where it originated, biogeographers cannot assume that a species now endemic to a region actually evolved there. Endemic taxa can be very old ones that are in the process of becoming extinct, or very young taxa that have recently evolved in a restricted area.

The longer an area has been isolated from other areas by a vicariant event, such as continental drift, the more endemic taxa it is likely to have, because there has been more time for evolutionary divergence to take place. Australia, which has been separated the longest from the other continents (about 65 million years) has the most distinct biota. South America has the next most distinct biota, having been isolated from other continents for nearly 60 million years. North America and Eurasia, which were joined together for much of Earth's history, have very similar biotas. That is why the early European travelers felt more at home in North America than in Australia.

Biogeographers use parsimony to explain distributions

When several hypotheses can explain a pattern, scientists typically prefer the most parsimonious one—the one that requires the smallest number of unobserved events to account for it. To see the application of the **principle of parsimony** to biogeography, consider the distribution of the New Zealand flightless weevil *Lyperobius huttoni*, a species that is found in the mountains of South Island and on sea cliffs at the extreme southwestern corner of North Island (Figure 54.2). If you knew only its current distribution and the current positions of the two islands, you might guess, even though this weevil cannot fly, that *L. huttoni* had somehow managed to cross Cook Strait, the 25-kilometer body of water that separates the two islands.

However, more than 60 other animal and plant species, including other species of flightless insects, live on both sides of Cook Strait. It is unlikely that all of these species made the same ocean crossing. In fact, that assumption is unnecessary to explain the distribution patterns. Geologic evidence indicates that the present-day southwestern tip of North Island was formerly united with South Island. Therefore, none of the 60 species need have made a water crossing. A single vicariant event, the separation of the northern tip of South Island from the remainder of the island by the newly formed Cook Strait, could have split all of the distributions.

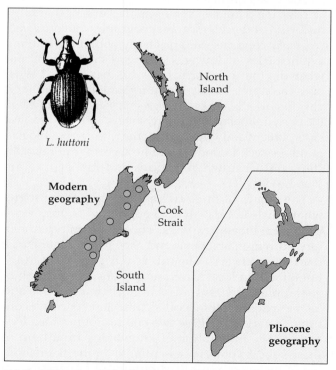

54.2 A Vicariant Distribution Explained The yellow circles indicate the current distribution of the weevil *Lyperobius huttoni*. Compare the present New Zealand geography with that in the Pliocene.

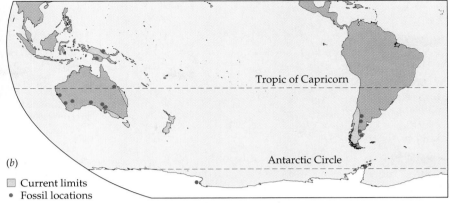

(b)

☐ Current limits
● Fossil locations

(a)

54.3 Southern Beeches Were Carried by Drifting Continents (a) *Nothofagus*, the southern beech tree, dominates the wet forests of southern Argentina. (b) Current and fossil distributions of *Nothofagus*.

Although organisms *do* cross major oceanic and terrestrial barriers, biogeographers often apply the rule of parsimony when interpreting distribution patterns, just as evolutionists do when reconstructing phylogenies (see Chapter 22). Although one vicariant event could have separated many formerly continuous distributions, some of the species found on both South and North Island, such as birds and flying insects, may indeed have dispersed across Cook Strait. However, in the absence of evidence indicating that they did, most biogeographers favor the more parsimonious interpretation. The principle of parsimony is a useful operating rule, but evolutionary history has not always been parsimonious. Other evidence may strongly support a less parsimonious explanation of current distributions of organisms.

Biogeographic histories are reconstructed from various kinds of data

Biogeographers use distribution maps, phylogenies, and knowledge of paleoclimates and paleogeography to reconstruct the biogeographic histories of taxa. This kind of information suggests, for example, that the current distributions of southern beeches (genus *Nothofagus*: family Nothofagaceae; Figure 54.3a) were influenced by continental drift many millions of years ago. Today southern beeches are found in South America, Australia, Tasmania, New Zealand, New Guinea, New Britain, and New Caledonia (Figure 54.3b).

An extensive fossil pollen record shows that southern beeches were distributed across Gondwanaland when South America, Antarctica, Australia, and the associated islands were still united. But they apparently arrived there after Africa and India had drifted away from the other southern continents, because no fossil pollen of *Nothofagus* has been found on those two land masses. *Nothofagus* subsequently became extinct in Antarctica when it became too cold to support trees.

Another taxon of woody plants whose current distribution was influenced by the breakup of Gondwanaland is the proteads (family Proteaceae; Figure 54.4). Unlike

(a)

Leucospermum conocarpodendron   *Banksia integrifolia*

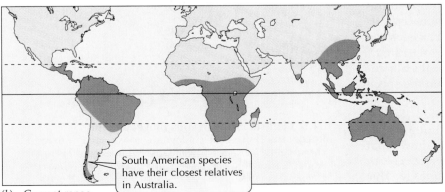

South American species have their closest relatives in Australia.

(b) Current range

54.4 Protead Distributions Reveal a Gondwanaland Ancestry (a) Proteads from South Africa (left) and Australia (right). (b) Current distribution of the family Proteaceae.

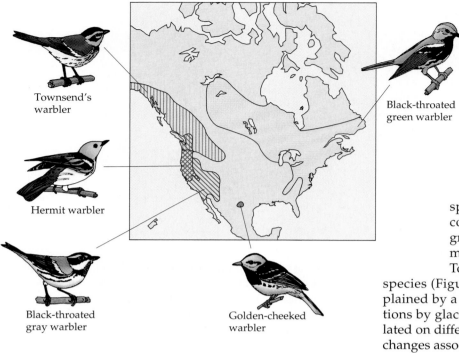

Townsend's warbler

Hermit warbler

Black-throated gray warbler

Golden-cheeked warbler

Black-throated green warbler

54.5 Why Are There Five Species? The present-day breeding ranges of five closely related species of warblers suggested that glaciers may have isolated the western and eastern populations of a widespread ancestral species.

Nothofagus, proteads are found in Africa, but the African species are highly specialized members of an endemic subfamily. The South American species are members of a different subfamily, whose closest relatives are found in the Australian region. Thus, the phylogeny and distribution of the proteads suggests that they had a broad distribution in Gondwanaland earlier than *Nothofagus* did, before that large land mass began to separate.

Similar combinations of geologic and phylogenetic information can be used to investigate distribution patterns that have been strongly influenced by more recent events. For example, biogeographers have investigated the possibility that recent advances and retreats of glaciers in North America initiated speciation events among songbirds. During times of glacial advance, populations of many forest-dwelling birds were divided into two segments, one in the western mountains and the other in the Appalachian mountains, both far south of the ice. One possibility is that populations isolated by glaciation evolved enough differences during their separation that they failed to interbreed when they again became sympatric as the glaciers retreated.

A biogeographer postulated that several such advances and retreats of glaciers could explain the speciation pattern and current distributions of species in the black-throated green warbler complex (Figure 54.5). No fossils of these warblers exist, but a cladogram, based on their mitochondrial DNA, shows that the black-throated gray warbler lineage was the earliest to separate from the remainder of the lineage. This

split preceded a glacial advance, which could have separated the black-throated green warbler from the ancestor of the hermit and Townsend's warblers. Hermit and Townsend's warblers, however, are sister species (Figure 54.6). Their separation cannot be explained by a splitting of eastern and western populations by glacial advances. They probably became isolated on different western mountain ranges by climate changes associated with glacial advances.

Earth can be divided into major terrestrial biogeographic regions

Although the drifting continents carried many kinds of organisms with them, the continents have been isolated from one another long enough to have evolved

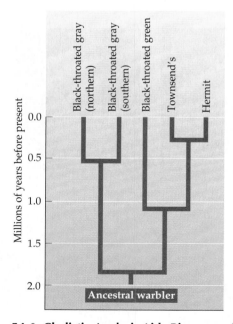

54.6 Cladistic Analysis Aids Biogeographic Reconstruction   This cladogram, constructed from mitochondrial DNA data, shows that some lineage splits in the warblers shown in Figure 54.5 preceded the Pleistocene glacial advances.

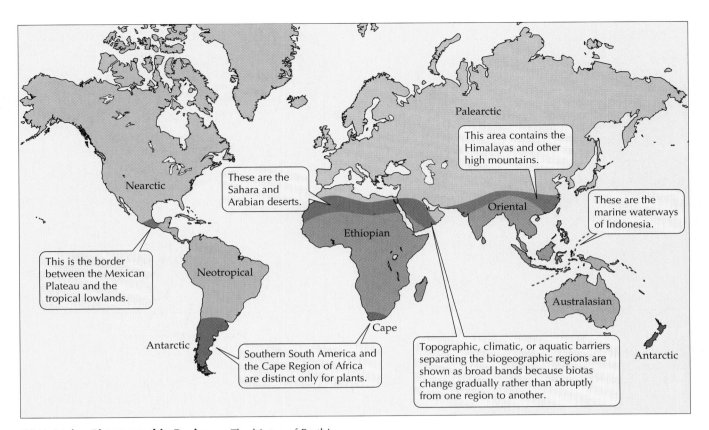

54.7 Major Biogeographic Regions The biotas of Earth's major biogeographic regions differ strikingly from one another, even though changes may be gradual.

distinct biotas. The differences among continental biotas were first recognized more than two centuries ago, and later formed the basis for dividing Earth into major biogeographic regions. Biogeo-graphers drew the boundaries of these regions where species compositions change dramatically over short distances (Figure 54.7).

All biogeographers agree on the boundaries of many of these regions, but plant biogeographers recognize two regions not used by zoogeographers: southern South America and the Cape Region of South Africa. The floras of these two regions are very distinct from those of adjacent areas on those continents, but the faunas of southern South America and the Cape Region are very similar to those of the remainder of those continents.

Except for the Australian region, the biogeographic regions are no longer separated from each other by water, as they were in the past (see Figure 19.16). The biological distinctness of these biogeographic regions is maintained today in part by mountain and desert barriers to dispersal and in part by major changes in climates over short distances.

Ecological Biogeography

Ecological biogeographers use the wealth of available information on current distributions of organisms to test theories that explain the numbers of species in different communities, the ways in which species disperse, and the effects of different types of barriers. They can also use experiments to test hypotheses, something that historical biogeographers cannot do.

First we will discuss a model that attempts to account for the number of species living in an area—its **species richness**—and then we will look at experiments conducted to test this model.

The species richness of an area is determined by rates of immigration and extinction

Over periods of a few thousand years (during which speciation is unlikely), the species richness of an area is influenced by the immigration of new species and the extinction of species already present. It is easiest to visualize the effects of these two processes if we consider, as did Robert MacArthur and Edward Wilson, an oceanic island that initially has no species.

Imagine a newly formed oceanic island that receives colonists from a mainland area. The list of species on the mainland that might possibly invade the island is called the *species pool*. The first colonists to arrive on the island are all "new" species because no

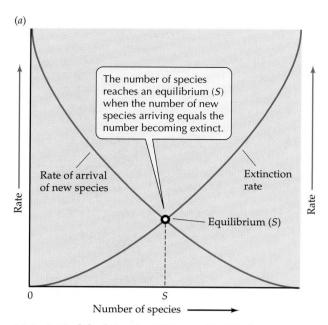

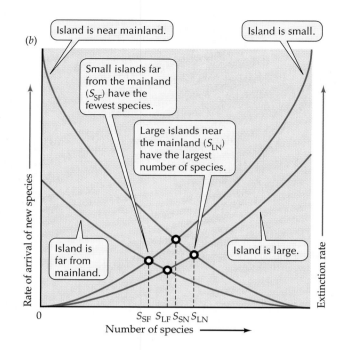

54.8 A Model of Species Richness Equilibrium Rates of arrival of new species and extinction of species already present determine the equilibrium number of species.

species live there initially. As the number of species on the island increases, a larger fraction of colonists will be members of species already present, so even if the same number of species arrive as before, the rate of arrival of *new* species decreases, until it reaches zero when the island has all the species in the species pool.

Now consider extinction rates. First there will be only a few species on the island, and their populations may grow large. As more species arrive and their numbers increase, the resources of the island will be divided among more species. We therefore expect the average population size of each species to become smaller as the number of species increases. The smaller a population, the more likely it is to become extinct. In addition, the number of species that can become extinct increases as species accumulate on the island. New arrivals to the island may include pathogens and predators that increase the probability of extinction of other species, further increasing the number of species becoming extinct per unit of time.

Because the rate of arrival of new species decreases and the extinction rate increases as the number of species increases, eventually the number of species should reach an equilibrium at which the rates of arrival and extinction are equal (Figure 54.8a). If there are more species than the equilibrium number, extinctions should exceed arrivals, and species richness should decline toward the equilibrium number. If there are fewer species than the equilibrium number, arrivals should exceed extinctions, and species rich-

ness should increase. Such an equilibrium is dynamic because even if species richness remains relatively constant, species composition may change as different species replace those that become extinct.

This equilibrium model does not predict which species will arrive and which will become extinct. It predicts only the equilibrium number of species if arrival and extinction rates are known and are constant. If either rate fluctuates, the equilibrium point shifts up and down.

If extinction and immigration rates are relatively constant, the equilibrium model can be used to predict how species richness should differ among islands of different sizes and different distances from the mainland. We expect extinction rates to be higher on small islands than on large islands because species' populations would, on average, be smaller there. Similarly, we expect fewer immigrants to reach islands more distant from the mainland. Figure 54.8b gives relative species richness equilibria for islands of different sizes and distances from the mainland. As you can see, the equilibrium number of species should be highest for islands that are relatively large and relatively close to the mainland.

The species richness equilibrium model has been tested

The species richness equilibrium model predicts that the equilibrium number of species should increase with island size and decrease with distance from the mainland. Bird species on islands in the Pacific Ocean exhibit these patterns (Figure 54.9). New Guinea supplies the mainland species pool for most of these small

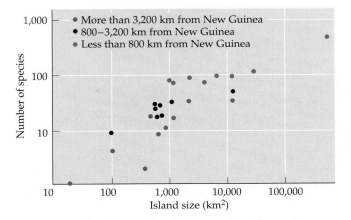

54.9 Small, Distant Islands Have Fewer Bird Species
The dots show the numbers of land and freshwater bird species on islands of different sizes in the Moluccas, Melanesia, Micronesia, and Polynesia. These islands have been divided into three groups according to their distance from the mainland, New Guinea.

islands. The species richness patterns of plants, insects, lizards, and mammals on Pacific islands are similar to those of birds.

The predictions of the equilibrium model are based on assumptions about rates of colonization and extinction. Major disturbances, which serve as "natural experiments," sometimes permit immigration and extinction rates to be estimated directly. The eruption of Krakatau in August 1883, described in Chapter 19, destroyed all life on the island's surface. After the lava cooled, Krakatau was colonized rapidly by plants and animals from Sumatra to the east and Java to the west. By 1933 Krakatau was again covered with a tropical evergreen forest, and 271 species of plants and 27 species of resident land birds were found there. During the 1920s, a forest canopy was developing, and there were high rates of immigration of both birds and plants to Krakatau (Table 54.1). Birds probably brought the seeds of many plants because, between 1908 and 1934, both the percentage (from 20 to 25) and the absolute number (from 21 to 54) of plant species with bird-dispersed seeds increased.

Today the numbers of species of plants and birds are not increasing as fast as they did during the 1920s, and they may be approaching equilibrium on Krakatau. Future biological censuses of Krakatau will continue to measure the arrival and extinction rates of species and will show if and when different taxa reach an equilibrium number of species.

A manipulative experiment testing the equilibrium model of species richness was carried out in the Florida Keys, a region dotted with thousands of small islands consisting entirely of red mangrove trees rooted in shallow water. Six tiny islands were fumigated with methyl bromide, which destroyed all arthropods living on them (Figure 54.10). Methyl bromide decomposes rapidly and does not inhibit recolonization. The design of this experiment permitted the investigators to measure arrival and extinction rates directly. They found that rates of recolonization of the islands by arthropods were very high. Within a year the fumigated islands had about their original number of species, a result supporting the equilibrium model.

Habitat islands influence species richness

Our discussion of the equilibrium species richness model has used as examples oceanic islands, where water provides barriers to dispersal. Mainland areas are full of **habitat islands**, which are patches of habitat separated from other similar patches by different types of habitat. For many species living in habitat islands, such as ponds or montane forests surrounded by deserts, the intervening areas may be as unsuitable to live in as if they were covered with water.

54.10 Testing the Equilibrium Species Richness Model Scaffolding is erected by scientists to enclose a small mangrove island in the Florida Keys. Methyl bromide introduced into the enclosure killed all arthropods inside it. When the enclosure was removed, arthropods quickly recolonized the island.

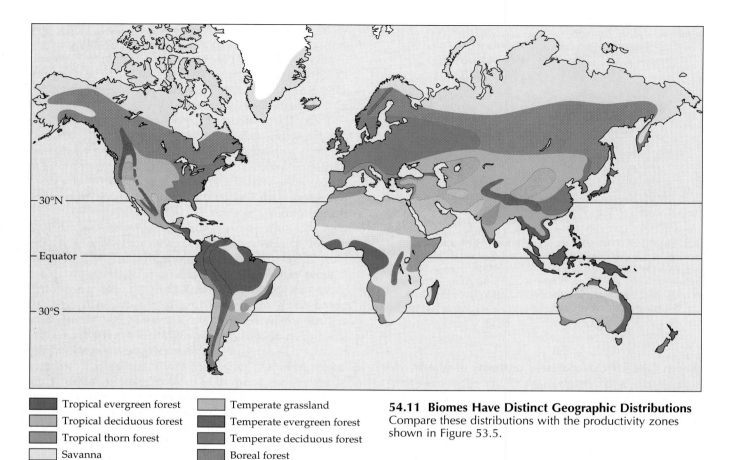

54.11 Biomes Have Distinct Geographic Distributions
Compare these distributions with the productivity zones shown in Figure 53.5.

Legend:
- Tropical evergreen forest
- Tropical deciduous forest
- Tropical thorn forest
- Savanna
- Hot desert
- Chaparral
- Cold desert
- Temperate grassland
- Temperate evergreen forest
- Temperate deciduous forest
- Boreal forest
- Tundra
- Alpine
- Polar ice cap

logenetically. Although biomes are named for and identified by their characteristic vegetation, sometimes supplemented by their location or climate, each biome contains many species of organisms in other kingdoms adapted to its physical environment and the physical structure provided by the plants.

Biomes are identified by their distinctive climates and dominant plants

Because climate plays a key role in determining which types of plants live in a given environment, the distri-

The equilibrium species richness model can be applied to habitat islands if we recognize that some intervening areas, though unsuitable for permanent occupancy, may nonetheless permit a brief stopover. Therefore, arrival rates are higher for most habitat islands than they are for similarly sized oceanic islands. We will discuss the importance of habitat islands to the conservation of species richness in the next chapter.

Terrestrial Biomes

Ecologists apply the name **biomes** to major ecosystem types that differ from one another in the structure of their dominant vegetation. The vegetation of a biome appears similar wherever that biome is found, but the plant species in these communities, despite their similarities, may not be closely related phy-

TABLE 54.1 Number of Species of Resident Land Birds on Krakatau

PERIOD	NUMBER OF SPECIES	EXTINCTIONS	COLONIZATIONS
1908	13		
1908–1919		2	17
1919–1921	28		
1921–1933		3	4
1933–1934	29		
1934–1951		3	7
1951	33		
1952–1984		4	7
1984–1996	36		

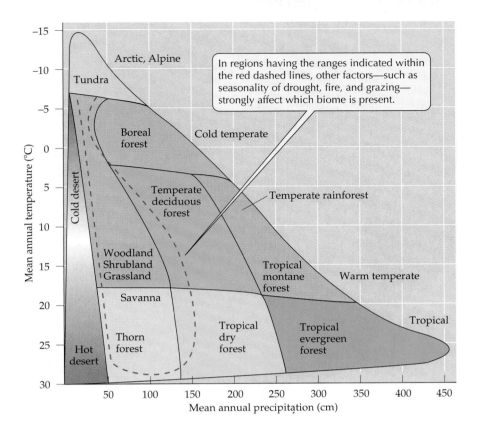

54.12 Most Biomes Have a Distinct Temperature and Precipitation Range Temperature and precipitation play key roles in determining which types of plants live in a given environment. The Mediterranean biome is not included because its distribution depends on the *seasonal* distribution of rainfall, a climate feature not shown here.

things, communities of streams, lakes, marshes, salt flats, dry slopes with shallow soils, moist valleys with deep soils, farmlands, pastures, and cities. Biomes usually blend into and intermingle with one another; sharp boundaries between them are rare. Nonetheless, by recognizing major biomes, we draw attention to the ecosystems that would predominate if natural processes had not been disturbed.

bution of biomes on Earth is strongly influenced by annual patterns of temperature and rainfall. The geographic distribution of biomes is shown in Figure 54.11; their distribution in relation to mean annual temperature and mean annual precipitation is shown in Figure 54.12. Taken together, these two figures show how each biome fits into global climate types.

In some biomes, such as temperate deciduous forest, precipitation is relatively constant throughout the year, but temperature varies strikingly between summer and winter. In other biomes, both temperature and precipitation change seasonally. In certain biomes, such as tropical rainforest, temperatures are nearly constant but rainfall varies seasonally. In the tropics, where seasonal temperature fluctuations are small, annual cycles are dominated by wet and dry seasons. In general, the length of time that a region is close to the intertropical convergence zone, and hence receives rainfall, increases toward the equator. The intertropical convergence zone shifts latitudinally in a seasonally predictable way (see Figure 53.2), resulting in a characteristic latitudinal pattern of distribution of rainy and dry seasons in tropical and subtropical regions (Figure 54.13).

Although terrestrial biomes are identified on the basis of their dominant plant communities, many other communities are found within each biome. For example, the deciduous forest biome contains, among other

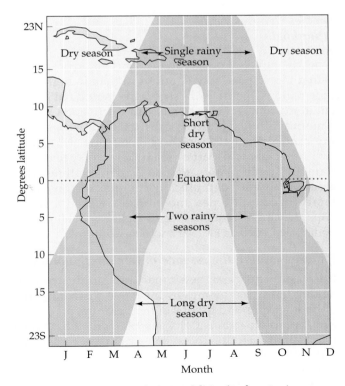

54.13 Rainy Seasons Change with Latitude In the tropics and subtropics, which months are rainy and which are dry is highly predictable based on the region's latitude.

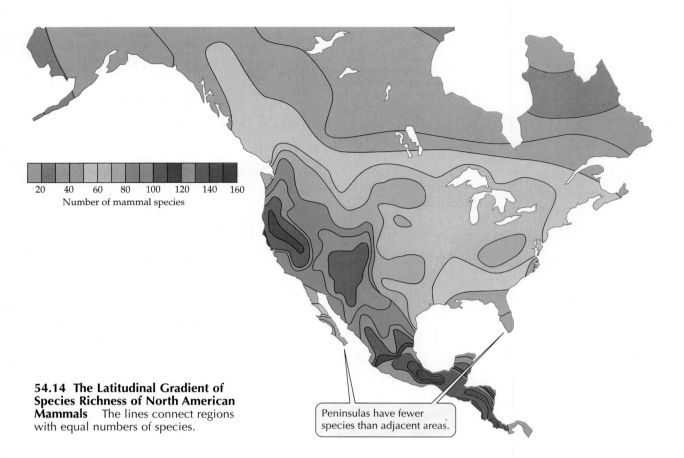

54.14 The Latitudinal Gradient of Species Richness of North American Mammals The lines connect regions with equal numbers of species.

Number of mammal species

Peninsulas have fewer species than adjacent areas.

Species richness varies latitudinally

A nearly universal pattern in the distribution of species is that more species live in tropical than in high-latitude regions. Figure 54.14 shows the latitudinal gradient in mammal species richness in North and Central America. Similar patterns exist for birds, frogs, and trees and for many marine taxa. The figure also shows two other general patterns. First, more species are found in mountainous regions than in relatively flat areas because topographically complex regions have more vegetation types and climates within a small area. Second, mammal species richness declines on peninsulas, such as Florida and Baja California.

Pictures capture the essence of terrestrial biomes

It is easiest to grasp the similarities and differences among terrestrial biomes by means of a combination of photographs and graphs of temperature, precipitation, and biological activity, supplemented by a few words that describe the species richness and other attributes of those biomes. We use this method in the following pages to describe the major terrestrial biomes of the world (Figures 54.15–54.24).

Each biome is illustrated by two photographs that show either the biome at different times of year or rep-resentatives of the biome in different parts of Earth. Below the photos are graphs that plot seasonal patterns of temperature and precipitation at a typical site in the biome, and graphs showing how active different kinds of organisms are during the year. Levels of biological activity, shown by the width of horizontal bars, change either because resident organisms become more active (produce leaves, come out of hibernation, hatch, or reproduce) or because organisms migrate into and out of the biome at different times of the year. A small box describes the kinds of plants that dominate vegetation in the biome and patterns of species richness there. These descriptions are very general and do not capture the variability that exists within each biome.

Tundra is found at high latitudes and in high mountains

The tundra biome is found in the Arctic and high in mountains at all latitudes. Because the climate is too cold for trees to grow, tundra vegetation is dominated by low-growing perennial plants (Figure 54.15).

In the Arctic, permanently frozen soil—**permafrost**—underlies the tundra's vegetation. The top few centimeters of soil thaw during the short (but often warm) summers, when the sun shines 24 hours a day. Even though there is little precipitation, lowland

54.15 Tundra   Arctic tundra: Denali National Park, Alaska (top). Tropical alpine tundra: Teleki Valley, Mt. Kenya (below).

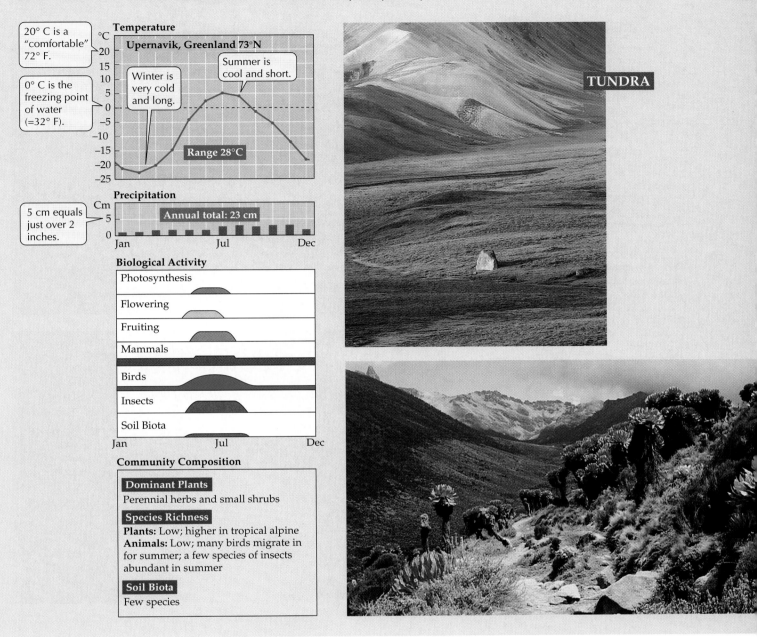

Temperature

20° C is a "comfortable" 72° F.

0° C is the freezing point of water (=32° F).

Upernavik, Greenland 73°N

Winter is very cold and long.

Summer is cool and short.

Range 28°C

Precipitation

5 cm equals just over 2 inches.

Annual total: 23 cm

Biological Activity

Photosynthesis

Flowering

Fruiting

Mammals

Birds

Insects

Soil Biota

Community Composition

Dominant Plants
Perennial herbs and small shrubs

Species Richness
Plants: Low; higher in tropical alpine
Animals: Low; many birds migrate in for summer; a few species of insects abundant in summer

Soil Biota
Few species

TUNDRA

Arctic tundra is very wet because water cannot drain down through the frozen soil. Upland, or montane, tundra in the Arctic is better drained than lowland tundra, and south-facing slopes may not be underlain by permafrost. Plants grow actively in the shallow, water-logged soils for a few months each year. Most Arctic tundra animals either migrate into the area for the summer and go elsewhere for the winter, or they are dormant for most of the year.

Tropical alpine tundra has days and nights of equal length (12 hours) all year. At these high altitudes the temperature is never very high, and it drops below freezing on most clear nights. Plants photosynthesize (although slowly) all year, and most animals are year-round residents.

Boreal forests are dominated by evergreen trees

Moving from the poles toward the equator, or to a lower elevation on temperate-zone mountains, we find the boreal forest (Figure 54.16). Over most of the boreal forest biome winters are long and very cold and summers are short (although often warm). The short-

54.16 Boreal Forest Northern spruce forest, Altay Mountains, Xinjiang, China (top). Bryophytes and lichens on southern evergreens, Tasmania, southern Australia (below).

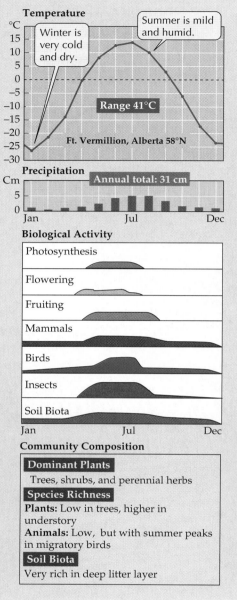

BOREAL FOREST

ness of the summers favors trees with evergreen leaves that live for several years. These trees are ready to photosynthesize as soon as temperatures become favorable in the spring.

The boreal forests of the Northern Hemisphere are dominated by coniferous evergreen gymnosperms, while in the Southern Hemisphere the dominant trees are southern beeches (*Nothofagus*), which are angiosperms with small leaves. Some species of *Nothofagus* are evergreen, others are deciduous. Coniferous evergreen forests also grow along the west coasts of continents at middle to high latitudes where winters are mild but very wet and summers are cool and dry;

these forests are home to Earth's tallest trees, and they support the highest standing biomasses of wood of all ecological communities.

Boreal forests have only a few tree species, nearly all of which are wind-pollinated and have wind-dispersed seeds. The dominant animals—such as insects, moose, and hares—eat leaves. The seeds in the cones of conifers also support a fauna of rodents and birds.

Temperate deciduous forests change with the seasons

The temperate deciduous forest biome (Figure 54.17) is found in eastern North America, eastern Asia, and in

54.17 Temperate Deciduous Forest   A Rhode Island forest in summer (left) and winter (right).

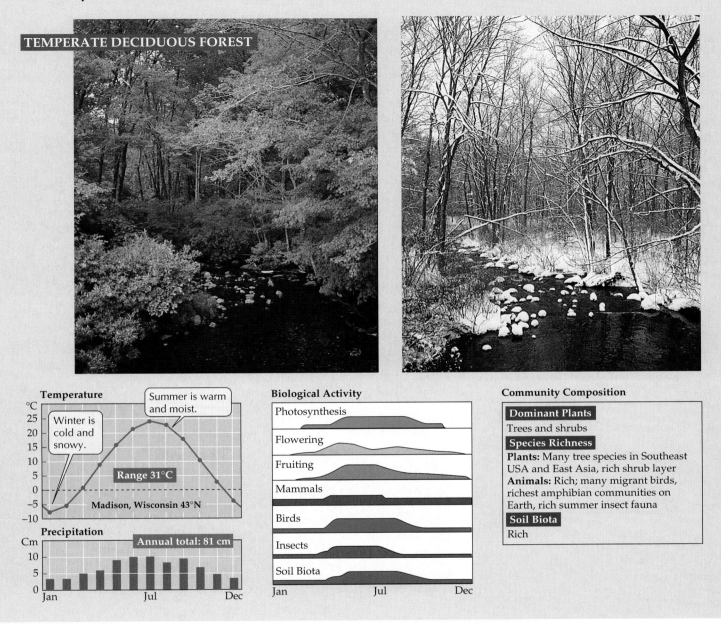

TEMPERATE DECIDUOUS FOREST

Temperature

°C

Winter is cold and snowy.

Summer is warm and moist.

Range 31°C

Madison, Wisconsin 43°N

Precipitation

Cm

Annual total: 81 cm

Biological Activity

Photosynthesis

Flowering

Fruiting

Mammals

Birds

Insects

Soil Biota

Community Composition

Dominant Plants
Trees and shrubs

Species Richness
Plants: Many tree species in Southeast USA and East Asia, rich shrub layer
Animals: Rich; many migrant birds, richest amphibian communities on Earth, rich summer insect fauna

Soil Biota
Rich

parts of western Europe. These regions have striking seasonal cycles of activity. Temperatures fluctuate dramatically between summer and winter and ample precipitation falls throughout the year.

Changes in the leaves are the most conspicuous sign of the seasons. Deciduous trees lose their leaves during the cold winters and produce leaves that photosynthesize rapidly during the warm, moist summers. There are many more tree species here than in boreal forests, and many deciduous trees have animal-dispersed pollen and fruits. Most deciduous trees and shrubs produce their fruits in autumn, at the end of the growing season.

Annual primary production in these forests may equal that of many tropical forests. Many birds migrate into this biome in summer, when insects are abundant. The temperate forests richest in species are in the southern Appalachian Mountains of the United States and in eastern China and Japan. Each of these is an area that was not disturbed by the glaciations of the Pleistocene (see Chapter 19).

Grasslands are ubiquitous

The temperate grassland biome is found in many parts of the world, all of which are relatively dry much of the year (Figure 54.18). Grasslands often experience

54.18 Temperate Grassland Nebraska prairie in spring (top). The Veldt, Natal, South Africa (below).

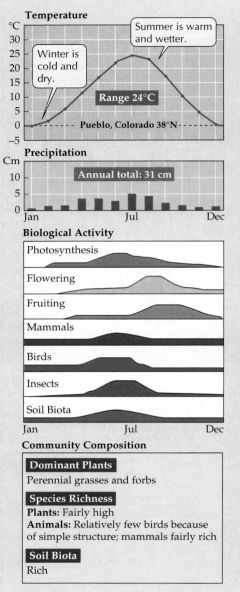

Temperature

°C

Winter is cold and dry.

Summer is warm and wetter.

Range 24°C

Pueblo, Colorado 38°N

Precipitation

Cm

Annual total: 31 cm

Jan   Jul   Dec

Biological Activity

Photosynthesis

Flowering

Fruiting

Mammals

Birds

Insects

Soil Biota

Jan   Jul   Dec

Community Composition

Dominant Plants
Perennial grasses and forbs

Species Richness
Plants: Fairly high
Animals: Relatively few birds because of simple structure; mammals fairly rich

Soil Biota
Rich

TEMPERATE GRASSLANDS

what is known as a continental climate, with hot summers and cold winters. Such regions as the pampas of Argentina, the veldt of South Africa, and the Great Plains of the United States are all part of this biome, much of which has been converted by humans for agricultural purposes.

Natural grasslands are structurally simple, but they are rich in species of perennial grasses, sedges, and forbs (broad-leaved herbaceous species). Many forbs have showy flowers, and grasslands are often riots of color when forbs are in bloom. Grasses are uniquely adapted to survive disturbances because they store much of their energy underground and quickly resprout after they are burned or heavily grazed. As we saw in Chapter 53, grasslands typically support large populations of grazing mammals, and fires are common. In some grasslands most of the precipitation falls in winter, while in others the majority of moisture falls in summer.

Cold deserts are high and dry

Deserts are characterized by low rainfall. The cold desert biome is found in dry regions of the middle to high latitudes, especially in the interiors of large conti-

54.19 Cold Desert Sagebrush steppe near Mono Lake, California (top). Los Glacieres National Park, Argentina (below).

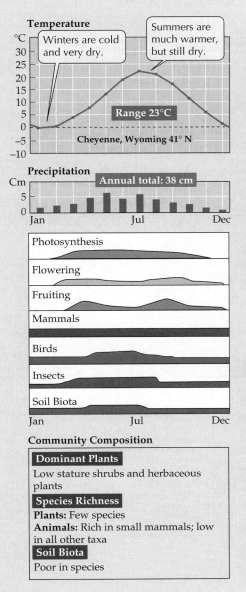

Temperature

°C

Winters are cold and very dry.

Summers are much warmer, but still dry.

Range 23°C

Cheyenne, Wyoming 41° N

Precipitation

Cm

Annual total: 38 cm

Jan   Jul   Dec

Photosynthesis

Flowering

Fruiting

Mammals

Birds

Insects

Soil Biota

Jan   Jul   Dec

Community Composition

Dominant Plants
Low stature shrubs and herbaceous plants

Species Richness
Plants: Few species
Animals: Rich in small mammals; low in all other taxa

Soil Biota
Poor in species

COLD DESERT

nents. Cold deserts also are found at fairly high latitudes in the rain shadows of mountain ranges (see Figure 53.1). Seasonal changes in temperature are great, and most of the small amount of rain that falls does so in the winter months.

Cold deserts are dominated by a few species of low-growing shrubs (Figure 54.19). The surface layers of the soil are recharged with moisture in winter, and plant growth is concentrated in spring. By early summer cold deserts are usually barren; so little rain falls that plants cannot conduct much photosynthesis.

Cold deserts are relatively poor in species in most

taxonomic groups, although the plants of this biome tend to produce a great many seeds, supporting a rich fauna of seed-eating birds, ants, and rodents.

Hot deserts form at 30° latitude

The hot desert biome is found in two belts, centered at 30° north and 30° south latitudes, respectively. These are the regions where air from aloft in the atmosphere descends, warms, and picks up moisture (see Figure 53.2). Hot deserts receive most of their rainfall in summer, when the intertropical convergence zone described in Chapter 53 moves toward the

54.20 Hot Desert Anza Borrego Desert, California (left). Rainbow Valley in the desert of central Australia (right).

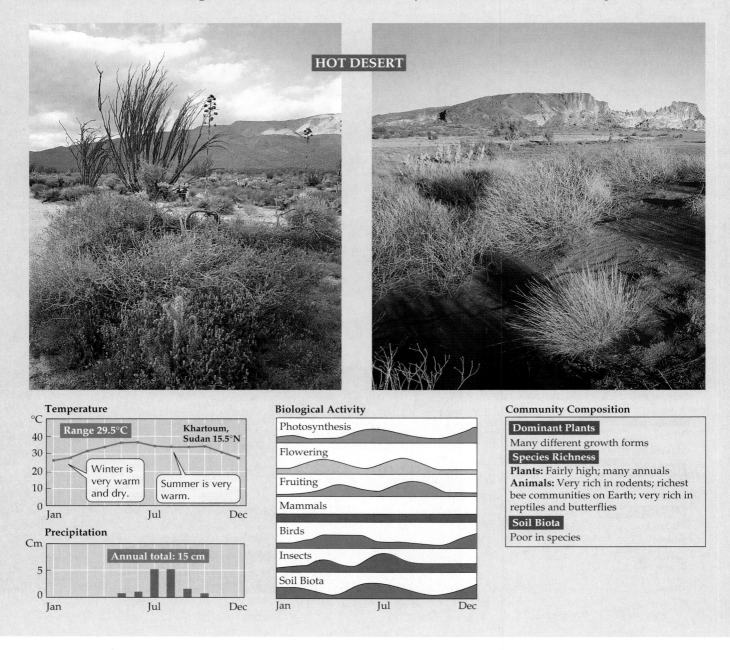

poles. However, they also receive winter rains from storms that form over the mid-latitude oceans. The driest regions, where summer and winter rains rarely penetrate, are in the center of Australia and the middle of the Sahara Desert of Africa.

Except in these driest regions, hot deserts have a richer and structurally more diverse vegetation than cold deserts do (Figure 54.20). Succulent plants such as cacti, which store large quantities of water in their expandable stems and that photosynthesize primarily with their stems rather than their leaves, are conspicu-

ous in many hot deserts. An abundance of annual plants springs suddenly into profusion when rain falls.

Desert plants produce great quantities of seeds. Pollination and dispersal of fruits by animals are typical. Population densities of rodents and ants are often remarkably high, and lizards and snakes typically are common.

The chaparral climate is dry and pleasant

The chaparral biome is found on the west sides of continents at moderate latitudes, where cool ocean

54.21 Chaparral Fynbos vegetation, Cape of Good Hope, South Africa (top). Mendocino, California (below).

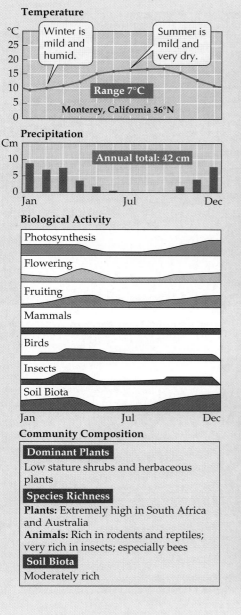

Temperature

°C

Winter is mild and humid.

Summer is mild and very dry.

Range 7°C

Monterey, California 36°N

Precipitation

Cm

Annual total: 42 cm

Jan   Jul   Dec

Biological Activity

Photosynthesis

Flowering

Fruiting

Mammals

Birds

Insects

Soil Biota

Jan   Jul   Dec

Community Composition

Dominant Plants
Low stature shrubs and herbaceous plants

Species Richness
Plants: Extremely high in South Africa and Australia
Animals: Rich in rodents and reptiles; very rich in insects; especially bees

Soil Biota
Moderately rich

CHAPARRAL

waters flow offshore. Winters in this biome are cool and wet and summers are hot and dry. Such climates are found in the Mediterranean region of Europe (hence the term "Mediterranean climate," which is sometimes applied to this biome), coastal California, central Chile, extreme southern Africa, and southwestern Australia.

Chaparral is dominated by low-growing shrubs that have tough, evergreen leaves (Figure 54.21). The shrubs carry out most of their growth and photosynthesis in early spring, which is when insects are active

and birds breed. Annual plants are abundant, providing seeds that store well during the hot, dry summers. This biome thus supports large populations of small rodents, most of which store seeds in underground burrows. The vegetation of the chaparral is naturally adapted to experience periodic fires.

Many shrubs of the Northern Hemisphere chaparral produce bird-dispersed fruits that ripen in the late fall or early spring, which is when large numbers of migrant birds arive from the north. One such fruit, the olive, has played a very important role in human his-

54.22 Thorn Forest and Savanna A thorn forest in Madagascar (top), and a grove of *Acacia* trees in Tanzania (below).

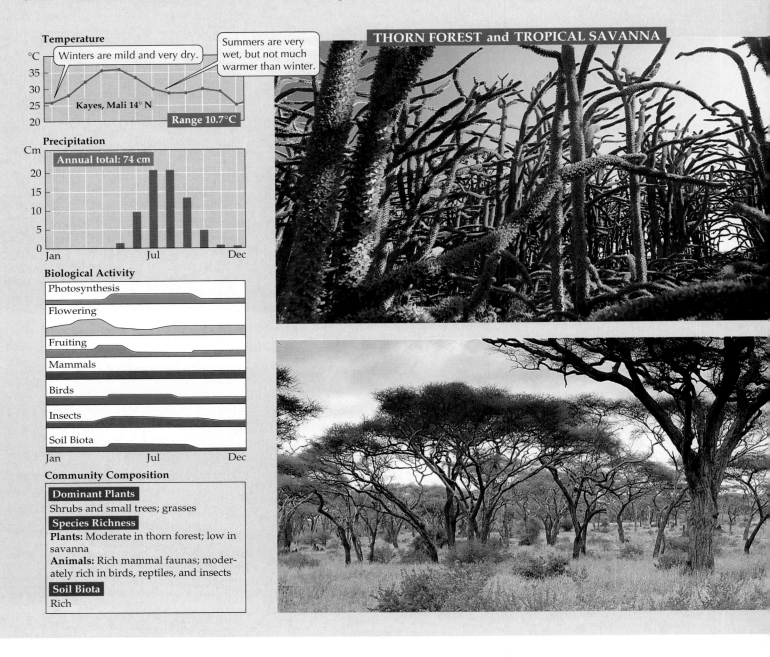

THORN FOREST and TROPICAL SAVANNA

Temperature

Winters are mild and very dry.

Summers are very wet, but not much warmer than winter.

Kayes, Mali 14° N

Range 10.7°C

Precipitation

Annual total: 74 cm

Biological Activity

Photosynthesis

Flowering

Fruiting

Mammals

Birds

Insects

Soil Biota

Community Composition

Dominant Plants
Shrubs and small trees; grasses

Species Richness
Plants: Moderate in thorn forest; low in savanna
Animals: Rich mammal faunas; moderately rich in birds, reptiles, and insects

Soil Biota
Rich

tory, providing a rich food source for people at a period of otherwise low food availability.

Thorn forests and savannas have similar climates

Thorn forests are found on the equatorial sides of hot deserts. The climate is semiarid; little or no rain falls during winter, but rainfall may be heavy during the summer wet season. Thorn forests contain many plants similar to those found in hot deserts. The dominant plants are small, spiny shrubs and trees (Figure 54.22). Members of the genus *Acacia* (see Figure 52.18) are common in thorn forests all over the world

The dry tropical and subtropical regions of Africa, South America, and Australia have extensive areas of savannas—expanses of perennial grasses and grasslike plants punctuated by scattered trees. The largest savannas are found in central and eastern Africa, where the biome supports huge numbers of grazing and browsing mammals, which in turn serve as prey for many large carnivores. Grazers and browsers maintain the savan-

54.23 Tropical Deciduous Forest Palo Verde National Park, Costa Rica, shown in the the rainy (top) and the dry season (below).

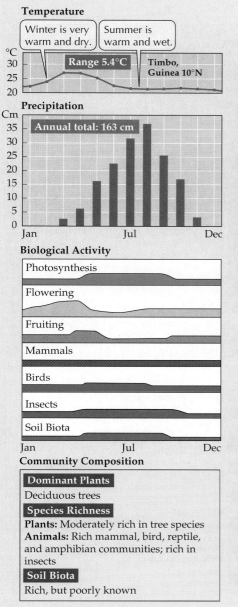

TROPICAL DECIDUOUS FOREST

nas; if savanna vegetation is not grazed, browsed, or burned, it reverts to dense thorn forest.

Tropical deciduous forests occur in hot lowlands

As the length of the rainy season increases toward the equator, thorn forests are replaced by tropical deciduous forests. These forests have taller trees and fewer succulent plants than the thorn forests, and they are much richer in species (Figure 54.23). The long dry season is very hot and often windy. Most of the trees, except for those growing along rivers, lose their leaves during the dry season; many of them flower while they are in this leafless state. The biome is very rich in species of both plants and animals.

Because the soils of the tropical deciduous forest biome are less leached of nutrients than are the soils of wetter areas, they are some of the best soils in the tropics for agriculture. As a result, most tropical deciduous forests have been cleared for grazing cattle and growing crops.

54.24 Tropical Evergreen Forest   Canopies of montane wet forest, Bwindi, Uganda (left) and lowland wet forest, Madagascar (right).

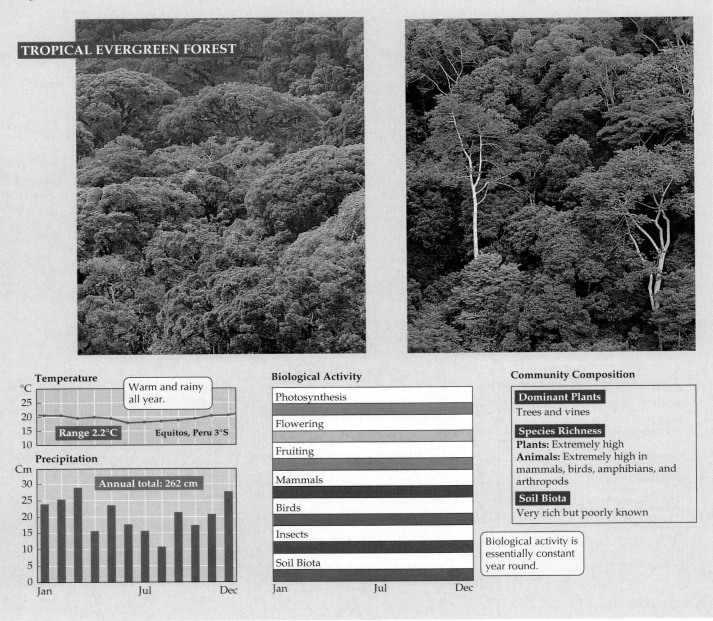

Tropical evergreen forests are species-rich

The tropical evergreen forest biome is found in equatorial regions where total rainfall exceeds 250 cm annually and the dry season lasts for no more than two or three months. It is the richest of all biomes in number of species of both plants and animals, with up to 500 species of trees per square kilometer. Many of these species are rare, and nearly all of them rely on animals to transport their pollen and disperse their fruits. Food webs in this community are extremely complex.

Along with their immense richness of species, tropical evergreen forests have the highest overall productivity of all ecological communities (Figure 54.24). However, most mineral nutrients are tied up in the vegetation; the soils are deeply weathered and usually cannot support agriculture without massive applications of fertilizers.

In the upland (montane) wet forests of the tropics, temperature decreases about 6° for each 1,000 m of elevation. Trees here are shorter than lowland tropical trees. Their leaves are smaller, and there are more epiphytes—plants that derive their nutrients and moisture from air and water rather than soil. Epiphytes, which grow on other plants, thrive in the high forests where clouds form regularly, bathing the forest in moisture. Photosynthetic rates in the mountain forests are slower than in the lowlands because the temperature is lower and because the leaves are wet most of the time.

Human activities are destroying tropical evergreen forests at a very high rate. Many of the species living here, especially the invertebrates, have not yet been described or named by scientists. Many will pass into extinction without our knowledge.

Aquatic Biomes

Three-fourths of Earth's surface is covered by water. For organisms that cannot survive out of water, terrestrial habitats are barriers to dispersal. However, some aquatic species have flying adults that can disperse widely among water bodies. Others have windborne, desiccation-resistant spores and seeds, and still others are small enough to be transported by means such as mud on the feet of birds. Many freshwater taxa that are capable of dispersing across terrestrial barriers are distributed widely over several continents.

Freshwater biomes have little water but many species

Although a minute fraction of Earth's water is found in ponds, lakes, and streams, about 10 percent of all aquatic species live in freshwater habitats. Prominent among these are the more than 25,000 species of insects that have at least one aquatic stage in their life cycle. Most commonly, eggs and larvae are aquatic and the winged adults are the primary dispersers. Adults of some of these insects, such as dragonflies, are powerful flyers, but adults of mayflies and some other species are weak flyers, rapidly desiccate in air, and live no longer than a few days. As you would expect, oceanic islands have no or very few species of mayflies because of their inability to survive long enough to disperse across wide expanses of salt water.

Similarly, fishes unable to tolerate salt water can disperse only within the connected rivers and lakes of a river basin. Dispersal between river basins can occur if their headwaters become joined when erosion removes the barrier between them. This happened, for example, when large amounts of water released by melting glaciers at the end of the Ice Age connected the headwaters of the Yukon River to the basin of the Mackenzie River (Figure 54.25). Today these rivers share a number of freshwater fish species, even though they are no longer connected.

Most families of freshwater fishes that cannot tolerate salt water are restricted to a single continent. Those families with species distributed on both sides of major saltwater barriers are believed to be ancient lineages whose ancestors were distributed widely in Pangaea or Gondwanaland.

Boundaries between marine biomes are determined primarily by changes in water temperature

All oceans are connected, and ocean water moves in great circular patterns—clockwise in the Northern Hemisphere and counterclockwise in the Southern Hemisphere (see Figure 53.3). These movements disperse organisms with limited swimming abilities. Nevertheless, most marine organisms have restricted ranges, indicating that important environmental limits to their distributions exist in the oceans.

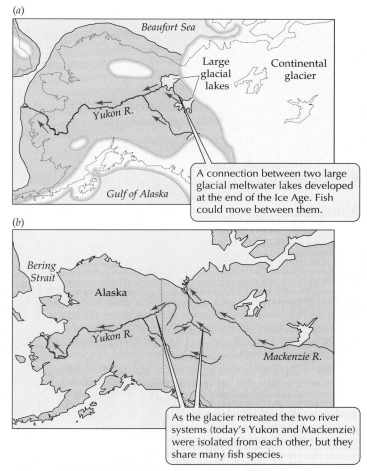

(a)

Beaufort Sea

Large glacial lakes

Continental glacier

Yukon R.

Gulf of Alaska

A connection between two large glacial meltwater lakes developed at the end of the Ice Age. Fish could move between them.

(b)

Bering Strait

Alaska

Yukon R.

Mackenzie R.

As the glacier retreated the two river systems (today's Yukon and Mackenzie) were isolated from each other, but they share many fish species.

54.25 Fish from One River System Enter Another The Yukon and Mackenzie Rivers of Canada share many freshwater fish species because they were connected by glacial melt at the end of the Ice Age. (a) The early stages of glacial retreat. Glaciers are shown in pale gray. (b) Today's drainage pattern.

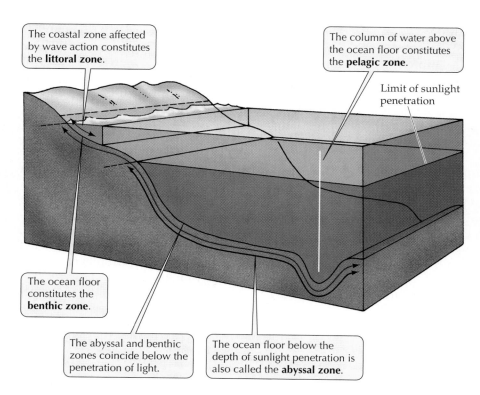

The coastal zone affected by wave action constitutes the **littoral zone**.

The column of water above the ocean floor constitutes the **pelagic zone**.

Limit of sunlight penetration

The ocean floor constitutes the **benthic zone**.

The abyssal and benthic zones coincide below the penetration of light.

The ocean floor below the depth of sunlight penetration is also called the **abyssal zone**.

54.26 Zones of the Ocean
Oceanic zones are shown schematically in relation to depth and sunlight penetration.

Horizontal and vertical gradients divide the oceans into zones with distinct physical conditions (Figure 54.26). Water temperatures, hydrostatic pressures, and food supplies all change with depth and influence biotic distributions. Food is scarce, for example, in the deep waters of the oceans, may arrive very infrequently, and cannot be hunted visually. Living successfully in different regions of the ocean requires different physiological tolerances and morphological attributes. Not surprisingly, even though organisms can disperse be-

tween these zones, organisms from one zone survive poorly if they attempt to live in another.

Ocean temperatures are barriers to colonization because many marine organisms are well adapted to only relatively narrow temperature ranges. The main biogeographic divisions of the pelagic zone coincide with regions where the temperature of surface waters changes relatively abruptly as a result of horizontal and vertical ocean currents (Figure 54.27). These temperature changes, in combination with seasonal changes in daylight, determine the seasons of maximum primary production. Marine algae tend to be adapted to photosynthesize either in summer or in winter, but not during both seasons.

Because nutrients gradually sink to the ocean bottom, high concentrations of nutrients in the pelagic zone are restricted to areas where upwelling currents bring nutrient-rich deep waters to the surface. Most marine organisms that grow and reproduce well in nutrient-rich waters perform relatively poorly in nutrient-poor waters, and vice versa. Therefore, nutrient-rich waters typically have biotas that differ considerably from those of nutrient-poor waters in the same general region.

Deep ocean waters are barriers to the dispersal of marine organisms that live only in shallow water. The distances that the eggs and larvae of marine organisms can be carried by ocean currents are determined in large part by the time it takes for the larvae to metamorphose into sedentary adults. Relatively few species have eggs and larvae that can dis-

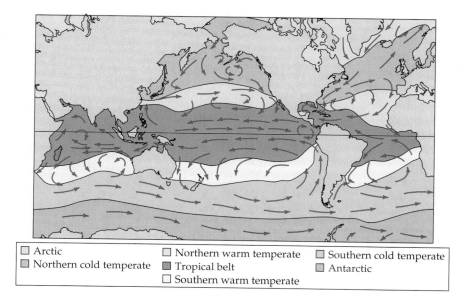

☐ Arctic
☐ Northern cold temperate
☐ Northern warm temperate
■ Tropical belt
☐ Southern warm temperate
☐ Southern cold temperate
☐ Antarctic

54.27 Pelagic Regions Are Determined by Ocean Currents The arrows represent ocean currents. Regions in which photosynthesis is maximized at different seasons are indicated by different colors.

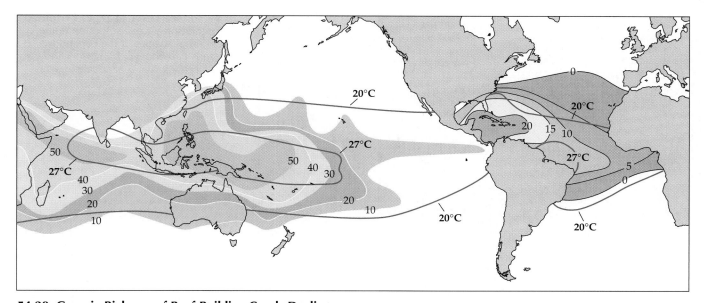

54.28 Generic Richness of Reef-Building Corals Declines with Distance from New Guinea The lines connect areas with equal numbers of genera. The 20° and 27° mean annual temperature isotherms are also shown.

perse across wide barriers of deep water. As a result, the richness of sedentary shallow-water species in the intertidal and subtidal zones of isolated islands in the Pacific Ocean decreases with distance from New Guinea (Figure 54.28).

Marine vicariant events influence species distributions

During the time of Pangaea, the seas were also united to form one world ocean, Panthalassa. Continental drift dramatically changed the sizes and shapes of the oceans, but ocean waters move so rapidly that ancient vicariant events may no longer influence distributions of marine organisms today. More recent marine vicariant events, however, have left their traces.

Relatively rapid changes in sea level are produced by major tectonic events and climate change. An important tectonic event was the elevation of the Panamanian Isthmus about 3 million years ago. This event separated the Pacific Ocean from the Caribbean Sea for the first time in more than 100 million years. Distinct marine biotas are now evolving on opposites sides of the isthmus, which also forms a barrier to the dispersal of Pacific species, such as sea snakes, that reached the west coast of the Americas after the barrier formed (Figure 54.29). If a sea level canal were constructed across the isthmus, poisonous sea snakes and other marine organisms would be able to disperse into the Caribbean. Currently the fresh waters of Gatun Lake form a barrier to the dispersal of marine organisms.

The total amount of water on Earth is roughly constant, so water that is tied up in continental glaciers is subtracted from the water that remains in the oceans. At numerous times during the Pleistocene, sea levels dropped more than 100 meters as continental glaciers formed and expanded, then rose when the glaciers re-

Pelamis platurus

54.29 A Block to Dispersal The presence of the Panamanian Isthmus means that poisonous sea snakes cannot enter the Caribbean Sea from the Pacific Ocean.

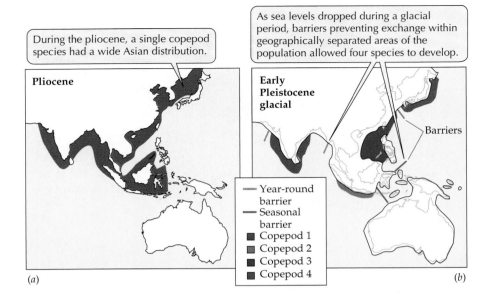

During the pliocene, a single copepod species had a wide Asian distribution.

As sea levels dropped during a glacial period, barriers preventing exchange within geographically separated areas of the population allowed four species to develop.

Pliocene

Early Pleistocene glacial

Barriers

— Year-round barrier
— Seasonal barrier
■ Copepod 1
■ Copepod 2
■ Copepod 3
■ Copepod 4

(a)   (b)

54.30 Sea Level Changes Permitted Speciation (a) Distribution of the ancestral species of the copepod genus *Labidocera*. (b) Barriers allowed the formation of four species.

ceded. The repeated formation and breakdown of seasonal and continental barriers as sea levels fluctuated in association with glacial cycles separated lineages of some marine organisms. These events resulted in repeated cycles of range expansion, contraction, and speciation in some taxa (Figure 54.30). As terrestrial animals we find it difficult to perceive marine barriers to dispersal, but they exist and strongly influence distributions of marine organisms.

Summary of "Biogeography"

Why Are Species Found Where They Are?

• If a species occupies an area, either it evolved there, or it evolved elsewhere and dispersed to the area. If a species is not found in a particular area, either it evolved elsewhere and never dispersed to the area, or it was once present in the area but no longer lives there.
• Continental drift has influenced the distributions of organisms throughout Earth's history.
• Biogeographers often analyze species distributions by converting taxonomic cladograms into area cladograms. **Review Figure 54.1**

Historical Biogeography

• Historical biogeographers attempt to determine the influence of past events on today's patterns of species distributions.
• Biogeographers use the principle of parsimony when they attempt to explain distribution patterns. **Review Figure 54.2**
• Vicariance and dispersal events have both influenced current distributions. **Review Figures 54.3, 54.4**
• Recent vicariant events, such as advances and retreats of glaciers, have affected current distributions of organisms.

Review Figures 54.5, 54.6
• Animal biogeographers divide Earth into six major biogeographic regions. Plant geographers recognize two additional regions. **Review Figure 54.7**

Ecological Biogeography

• Ecological biogeographers test theories that explain the numbers of species in different communities, how species disperse, and the effectiveness of barriers to movement.
• An equilibrium model of species richness, which predicts the number of species on islands, has been tested by examining patterns of distribution and by performing experiments. **Review Figures 54.8, 54.9, 54.10**

Terrestrial Biomes

• Terrestrial biomes are major ecosystem types that differ from one another in the structure of their dominant vegetation. These biomes are tundra, boreal forest, temperate deciduous forest, temperate grassland, cold desert, hot desert, chaparral, thorn forest and savanna, tropical deciduous forest, and tropical evergreen forest.
• The distribution of biomes on Earth is strongly influenced by annual patterns of temperature and rainfall. **Review Figures 54.11, 54.12, 54.13**
• The number of species in most lineages increases from polar to tropical regions. **Review Figure 54.14**
• Pictures and graphs capture the similarities and differences among Earth's terrestrial biomes. **Review Figures 54.15–54.24**

Aquatic Biomes

• No absolute barriers to the movement of marine organisms exist within the oceans, but most marine organisms have restricted ranges.
• Conditions in the oceans change dramatically with depth and sunlight penetration. **Review Figure 54.26**
• Boundaries between many pelagic regions are determined by ocean currents. **Review Figure 54.27**
• Species that live in shallow waters disperse with difficulty across wide deep-water barriers. **Review Figures 54.28**

- Changes in sea levels have resulted in speciation in some shallow-water lineages. **Review Figure 54.30**

Self-Quiz

1. Biogeography as a science began when
   a. eighteenth-century travelers first noted intercontinental differences in the distributions of organisms.
   b. Europeans went to the Middle East during the Crusades.
   c. cladistic methods were developed.
   d. the fact of continental drift was accepted.
   e. Charles Darwin proposed the theory of natural selection.

2. Historical and ecological biogeography differ in that
   a. only historical biogeography is concerned with history.
   b. historical biogeography is concerned with longer time periods and larger spatial scales.
   c. both are concerned with the same time scales, but historical biogeography deals with larger spatial scales.
   d. both are concerned with the same spatial scales, but historical biogeography deals with longer time scales.
   e. historical biogeography is not concerned with the current distributions of organisms.

3. Marine biogeographic regions are less distinct than terrestrial ones because
   a. the ocean biota is more poorly known than the terrestrial biota.
   b. there are currently fewer barriers to dispersal of marine organisms.
   c. most marine families and higher taxa evolved before the oceans were separated by continental drift.
   d. we know less about the distributions of marine organisms.
   e. oceanic circulation is faster than atmospheric circulation.

4. A parsimonious interpretation of a distribution pattern is one that
   a. requires the smallest number of undocumented vicariant events.
   b. requires the smallest number of undocumented dispersal events.
   c. requires the smallest total number of undocumented vicariant plus dispersal events.
   d. accords with the cladogram of a group.
   e. accounts for centers of endemism.

5. The only major biogeographic region that today is isolated by water from other regions is
   a. Greenland.
   b. Africa.
   c. South America.
   d. Australasia.
   e. North America.

6. Equilibrium species richness is reached in the MacArthur–Wilson model when
   a. immigration rates of new species and extinction rates of species are equal.
   b. immigration rates of all species and extinction rates of species are equal.
   c. the rate of vicariant events equals rates of dispersal.
   d. the rate of island formation equals the rate of island loss.
   e. No equilibrium number of species exists in that model.

7. Chaparral vegetation is dominated by
   a. deciduous trees.
   b. evergreen trees.
   c. deciduous shrubs.
   d. evergreen shrubs.
   e. grasses.

8. Which of the following is *not* true of tropical evergreen forests?
   a. They have large numbers of species of trees.
   b. Most plant species are animal-pollinated.
   c. Most plant species have animal-dispersed fruits.
   d. Biological energy flow is very high.
   e. Productivity depends on a rich supply of soil nutrients.

9. At all depths, the bottom of the ocean is known as the
   a. benthic zone.
   b. abyssal zone.
   c. pelagic zone.
   d. interoceanic convergence zone.
   e. subtidal zone.

10. Vicariant events that influence current distributions of marine organisms include
   a. the breakup of Gondwanaland.
   b. the breakup of Pangaea.
   c. sea level changes caused by the advance and retreat of Pleistocene glaciers.
   d. sea level changes caused by the advance and retreat of Permian glaciers.
   e. Ocean currents are too strong for past vicariant events to influence current distributions of marine organisms.

Applying Concepts

1. Horses evolved in North America but subsequently became extinct there. They survived to modern times only in Africa and Asia. In the absence of a fossil record we would probably infer that horses originated in the Old World. Today, the Hawaiian Islands have by far the greatest species richness of fruit flies (*Drosophila*). Would you conclude that the genus *Drosophila* originally evolved in Hawaii and spread to other regions? Under what circumstances do you think it is safe to conclude that a group of organisms evolved close to where the greatest number of species live today?

2. In nearly every ecological community, the number of species present is much smaller than the number potentially available to colonize it. Does this pattern constitute good evidence for species richness equilibrium? What do you consider the strongest evidence for species richness equilibrium?

3. A well-known legend states that Saint Patrick drove the snakes out of Ireland. Give some alternative explanations, based on sound biogeographic principles, for the absence of indigenous snakes in that country.

4. What are some significant present-day human problems whose solutions involve biogeographic considerations? What kinds of biogeographic knowledge are most important for each one?

5. Most of the world's flightless birds are either nocturnal and secretive (such as the kagu of New Caledonia) or large, swift, and well-armed (such as the ostrich of Africa). The exceptions are found primarily on islands,

and many of these island species have become extinct with the arrival of humans and their domestic animals. What special biogeographic conditions on islands might permit the survival of flightless birds? Why has human colonization so often resulted in the extinction of such birds? The power of flight has been lost secondarily in representatives of many groups of birds; what are some possible evolutionary advantages of flightlessness that might offset its obvious disadvantages?

Readings

Brown, J. H. and M. V. Lomolino. 1998. *Biogeography*, 2nd Edition. Sinauer Associates, Sunderland, MA. A comprehensive treatment of both ecological and historical biogeography.

Humphries, C. J. and L. R. Parenti. 1986. *Cladistic Biogeography*. Clarendon Press, Oxford. A concise treatment of the ways in which cladistic methods are used to determine the causes of current distributions of organisms.

MacArthur, R. H. and E. O. Wilson. 1967. *The Theory of Island Biogeography*. Princeton University Press, Princeton, NJ. The classic book that launched modern investigations of ecological biogeography.

Myers, A. A. and P. S. Giller. 1988. *Analytical Biogeography: An Integrated Approach to the Study of Animal and Plant Distributions.* Chapman & Hall, London. Contains chapters by different authors on many aspects of ecological and historical biogeography, including discussions of modern methods and their significance.

Terborgh, J. 1993. *Diversity and the Tropical Rain Forest*. W. H. Freeman, San Francisco. An exciting account of the richness of life in tropical evergreen forests.

Thornton, I. 1995. *Krakatau: The Destruction and Reassembly of an Island Ecosystem.* Harvard University Press, Cambridge, MA. The story of how scientists studied the recolonization of Krakatau.

Weiner, J. 1994. *The Beak of the Finch*. Alfred A. Knopf, New York. A stimulating account of field investigations of evolutionary changes in the beaks of Galapagos finches. The best book on modern evolutionary studies for the general reader.

Chapter 55

Conservation Biology

An Extinct Island Bird
This artist's reconstruction of a flightless Hawaiian goose shows one of the many bird species exterminated by the Polynesian settlers of the islands.

When Polynesian people settled in Hawaii about 2,000 years ago, they quickly exterminated, probably by overhunting, at least 39 species of endemic land birds, including 7 species of geese, 2 species of flightless ibises, a sea eagle, a small hawk, 7 flightless rails, 3 species of owls, 2 large crows, a honeyeater, and at least 15 finches.

No people lived in New Zealand until about 1,000 years ago, when the Polynesian ancestors of the Maori colonized the island. Hunting by the Maori caused the extinction of 13 species of flightless moas, some of which were larger than ostriches.

When humans arrived in North America over the Bering land bridge, about 20,000 years ago, they encountered a rich fauna of large mammals. Most of those species were exterminated within a few thousand years. A similar extermination of large animals followed the human colonization of Australia, about 40,000 years ago. At that time Australia had 13 genera of marsupials larger than 50 kg, a genus of gigantic lizards, and a genus of heavy flightless birds. All the species in 13 of those 15 genera had become extinct by 18,000 years ago.

The accelerating pace of human-caused extinctions of species, which raises serious concerns about the future of biological diversity on Earth, has led to the rapid development of **conservation biology**—the study of the diversity of life and how to preserve it. Conservation biology is an applied discipline that uses the best available science to preserve species and ecosystems. Conservation biologists study the causes of declines in species richness and develop methods by which genes, species, communities, and ecosystems can be preserved. The science of conservation biology draws heavily on concepts and knowledge from population genetics, evolution, ecology, biogeography, wildlife management, economics, and sociology. In turn, the needs of conservation are stimulating new research in those fields.

In this chapter we will see how biologists estimate rates of species extinctions. We will also explore why the accelerated extinction rates of other species are so important to humans, the causes of species endangerment, and how management plans that can reduce extinction rates and restore endangered species to a viable status are being developed.

Estimating Current Rates of Extinction

None of the human activities that are causing extinctions are new. It is just that many more of us are doing those things than ever before. We do not know how many species will become extinct during the next 100 years, in part because the number of extinctions will depend both on what we do and on unexpected events. Nevertheless, methods exist for estimating probable rates of extinction resulting from human actions. It is better to act using roughly correct estimates than to wait until precise estimates can be calculated. In this section we will discuss how conservation biologists estimate current rates of extinction and identify species at risk of extinction.

Species–area relationships are used to estimate extinction rates

Conservation biologists often use the well-established fact that the number of species present increases with the size of an area (see Figure 54.9) to estimate the number of species extinctions resulting from habitat destruction. The species–area relationship suggests that, on average, a 90 percent loss of habitat will result in the extirpation of half of the species living in that habitat. For example, if we assume that about half of existing terrestrial species live in tropical forests, and that about one-third of the remaining tropical forests will be logged during the next few decades, the species–area relationship suggests that about 1 million species will be extirpated during that period.

The rate at which tropical forests are being logged and converted to cropland and pastures is not precisely known, but it is currently very high (Figure 55.1). Moreover, even the lowest estimates of current extinction rates predict that at least 10 percent of Earth's species are likely to become extinct during the next two decades. Some estimates predict the extinction of 50 percent of Earth's species during the next 50 years.

Conservation biologists estimate risks of extinction

To estimate the risk of extinction of a particular species, conservation biologists develop models based on trends in its population size and changes in habitat availability to assess whether the species is at risk now or may become endangered in the future.

Although rarity itself is not always a cause for concern, species whose populations are shrinking rapidly usually are at risk. Species with only a few individuals confined to a small range are at risk of extinction because they can easily be eliminated by sudden local disturbances such as fires, unusual weather, disease, and predators.

Estimation of the risks faced by small populations is an important component of preservation analyses. The concept of a **minimum viable population** (MVP) is now widely used to estimate a population's risk of extinction. Its development was stimulated by the National Forest Management Act of 1976, which required the U.S. Forest Service to maintain viable popu-

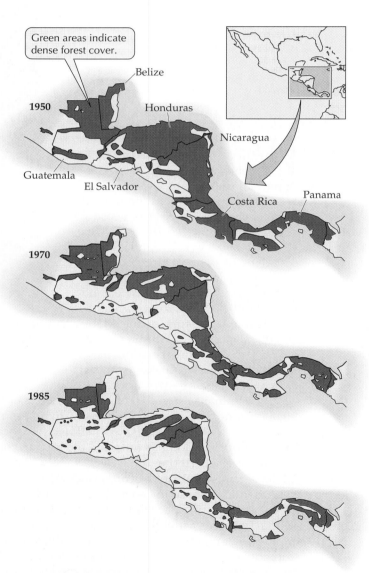

55.1 Deforestation Rates Are High in Tropical Forests
Central America provides an example of the high rate of destruction of tropical forests that has taken place in recent years. Less than half the forest that existed in 1950 remains, and much of that is in small patches.

lations of all native vertebrate species in each national forest. A minimum viable population is the estimated density or number of individuals necessary for the species to maintain or increase its numbers in a region. There is no sharp threshold above which a population is viable and below which it is not, but an MVP analysis can estimate a population's risk of extinction over decades and centuries, time frames that are appropriate for management plans.

A **population viability analysis** (PVA) is carried out to estimate how the size of a population influences its risk of becoming extinct within a specified time period—for example, 100 years. A PVA is based on knowledge of the interactions between the genetic variation, morphology, physiology, and behavior of a population and its environment, both physical and biological.

One component of a PVA is estimation of the extent and significance of *demographic stochasticity*—that is, the amount of random variation in birth and death rates. In a small population, extinction is likely when high death rates coincide with low birth rates. Estimates of the sizes of local populations at high risk of immediate extinction due to demographic stochasticity range from 10 individuals for microorganisms reproducing by fission to about 50 for sexually reproducing animals with lengthy prereproductive periods. Larger populations also may be at high risk because the same environmental conditions that cause low birth rates are likely to cause high death rates.

As we saw in Chapter 51, many species are distributed as metapopulations, collections of subpopulations each of which occupies a suitable patch in a landscape of otherwise unsuitable habitat (see Figure 51.11). The fraction of suitable habitat patches occupied by a metapopulation at any moment in time is determined by the rate at which local subpopulations become extinct and by the rate of colonization of empty patches by dispersing individuals. The prevention of extinction of a subpopulation by occasional immigrants from nearby patches is called the rescue effect.

The acorn woodpecker, a bird of the oak woodlands of western North America, exists as a metapopulation. The oak woodlands in New Mexico in which acorn woodpeckers live are found as small, isolated patches surrounded by open, more arid land. Conservation biologists performed a PVA for the acorn woodpecker by developing a computer simulation model to predict its survival and extinction. To run the model, they entered information on known birth and death rates of the woodpeckers.

The model predicted that, if birth and death rates remained constant at current levels, most of the acorn woodpecker subpopulations in isolated oak woodlands in New Mexico would become extinct within 20 years if no birds migrated between them. However, with only a small amount of migration, most simulated subpopulations survived more than 100 years (Figure 55.2). Many of these small populations of acorn woodpeckers are known to have survived for more than 70 years, suggesting that today birds occasionally fly between patches. However, if the patches of oak woodland were to become even more isolated, too few birds might disperse between patches to maintain the populations.

Another component of a PVA is analysis of **genetic stochasticity**, fluctuations in a population's level of ge-

55.2 A PVA Shows that Acorn Woodpeckers Occasionally Disperse between Woodland Patches
Acorn woodpeckers live in isolated oak woodlands in New Mexico. The graph shows population persistence as a function of the rate of dispersal between patches.

Acorn woodpecker

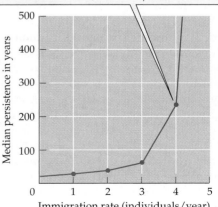

The immigration of as few as four individuals per year results in a dramatic increase in the persistence of the population—in this model, over 200 years.

netic variation. The level of inbreeding increases in small populations, leading to a lowering of fitness. Therefore, genetic information is important for planning the reintroduction of individuals from captivity or selecting individuals to reestablish extirpated populations.

Why Care about Species Extinctions?

The extinction of species is of great concern because, despite our increased ability to alter and restructure our surroundings, people depend on other species in many ways. We also derive enormous aesthetic pleasure from interacting with other organisms. Many people would consider a world with far fewer species a less desirable one in which to live.

More than half the medical prescriptions written in the United States contain a natural plant or animal product (Figure 55.3), yet the search for and exploitation of such products from the living world has barely begun. Many species may be eliminated by forest destruction before we find out if they might be sources of useful products.

Ecosystem processes produce many benefits to humanity, such as the generation and maintenance of fertile soils, prevention of soil erosion, detoxification and recycling of waste products, regulation of hydrological cycles and the composition of the atmosphere, control of agricultural pests, pollination, and the maintenance of the species richness upon which humanity depends for drugs, medicines, and aesthetic enjoyment. It is easy to list these benefits, but to justify the allocation of scarce public resources to maintain them, we need quantitative estimates of the value of ecosystem services.

A detailed study by economists, ecologists, and land managers in Western Cape Province, South Africa, has shown that an intensive program to eradicate invasive exotic plants in the mountains of the region is a highly cost effective way of maintaining a reliable regional supply of high-quality water. The native vegetation of the watersheds of the area is dominated by a species-rich community of shrubs known as fynbos (pronounced "fainbos") that can survive regular summer drought, nutrient-poor soils, and the fires that periodically sweep through the Cape mountains (Figure 55.4a). The fynbos-clad mountains provide about two-thirds of the Western Cape's water requirements. In addition, the species-rich endemic flora is widely harvested for cut and dried flowers and thatching grass. The combined value of these harvests in 1993 was about $19 million. Some of the income from tourism in the region comes from people who want to see the fynbos vegetation. About 400,000 people visit the Cape of Good Hope Nature Reserve each year, primarily to see plants.

Catharanthus roseus

55.3 Source of a Life-Saving Drug A drug for combating leukemia was derived from the Madagascar rosy periwinkle.

During recent decades a number of exotic plants, introduced into South Africa to provide a source of fast-growing timber and as hedge plants, have invaded the shrub-clad mountains. Because they are taller and faster-growing than the native plants, the exotics increase the intensity and severity of fires. By transpiring larger quantities of water, they decrease stream flows to less than half the amount flowing from mountains covered with native plants (Figure 55.4b). Removing the exotic plants by felling and digging out invasive trees and shrubs and managing fire is estimated to cost between $140 and $830 per hectare, depending on the densities of invasive plants. Annual follow-up operations cost about $8 per hectare.

The costs of alternative methods to replace the water lost from watersheds taken over by alien plants are much higher. A sewage purification plant that would deliver the same volume of water as a well-managed catchment of 10,000 hectares would cost $135 million to build and $2.6 million per year to operate. Desalination of seawater would cost four times as much.

Thus, the available alternatives would deliver water at a cost between 1.8 and 6.7 times more than the cost of maintaining natural vegetation in the watershed. Modern industrial societies often favor technologically sophisticated methods of substituting for lost ecosystem services. The study of water resources in the Western Cape Region shows that simple but labor-intensive methods—cutting and burning—can be cheaper and, in addition, preserve other valuable services, such as tourism and harvested plant products.

Some ecosystem services, such as aesthetic benefits, cannot be replaced with technological inventions.

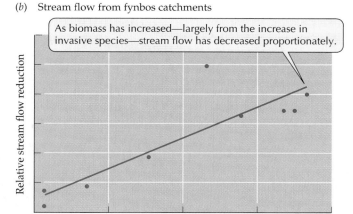

(b) Stream flow from fynbos catchments

As biomass has increased—largely from the increase in invasive species—stream flow has decreased proportionately.

Relative stream flow reduction

Relative biomass

(a) **55.4 Water Flows from Fynbos Vegetation** (a) The fynbos vegetation of South Africa. (b) Stream flow from fynbos watersheds in relation to plant biomass. (c) A computer simulations graphs water loss due to invasions of exotic trees.

Aesthetic benefits may contribute a great deal to a country's economy. One of the largest sources of foreign income in Kenya is nature tourism. The loss of a single species probably would not reduce the flow of tourists to Kenya, but if elephants, rhinoceroses, lions, leopards, and buffalo were all to disappear, few people would pay the high price of a Kenyan vacation (Figure 55.5). Populations of these species can be maintained only if large tracts of the ecosystems in which they live are preserved.

(c) Computer simulation

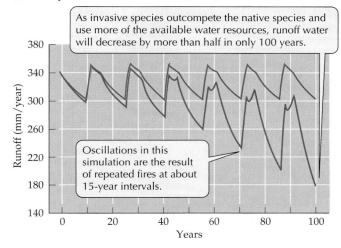

As invasive species outcompete the native species and use more of the available water resources, runoff water will decrease by more than half in only 100 years.

Runoff (mm/year)

Oscillations in this simulation are the result of repeated fires at about 15-year intervals.

Years

55.5 Large Mammals Support Ecotourism in Kenya
Without the wide array of large and impressive mammals that are present in Kenya's national parks, the country's flourishing ecotourism business would probably collapse.

Diceros bicornis

Panthera leo

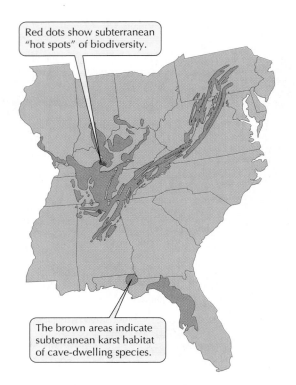

Red dots show subterranean "hot spots" of biodiversity.

The brown areas indicate subterranean karst habitat of cave-dwelling species.

55.6 Cave Animals Have Very Small Ranges   Cave animals tend to be rare because their habitat is rare. The map shows the distribution of cave habitat (subterranean karst) in the southeastern United States. This region has the highest biodiversity of cave animals in the Americas.

Determining Causes of Endangerment and Extinction

Species may be rare for any of several reasons. For example, they may live in a habitat that is rare. Cave-dwelling species typically have narrow ranges and small populations (Figure 55.6), as do many species that live only in desert lakes with high salt concentrations. Another reason is trophic—secondary carnivores are usually rare because so little energy is available to support their populations (see Chapter 53). Being rare may increase a species' chance of becoming extinct, but many rare species, especially those that have been rare for many years, are likely to persist for long time spans.

In this section we will examine the major causes of extinctions, which include overexploitation, the introduction of predators and diseases, the loss of mutualists, habitat destruction, and habitat fragmentation.

Overexploitation has driven many species to extinction

Until recently, humans caused extinctions primarily by overhunting. The passenger pigeon, the most abundant bird in North America in the early 1800s, became extinct by 1914, largely due to overhunting. Russian whalers exterminated the unusual Steller's sea cow of the North Pacific in the late 1800s, just 37 years after it was first described. The American bison was on the brink of extinction at the beginning of the twentieth century and might well be extinct today if its hunting had not been outlawed. Overexploitation continues today; for example, elephants and rhinoceroses are threatened in Africa because poachers kill them for their valuable tusks and horns.

Introduced pests, predators, and competitors have eliminated many species

Deliberately and accidentally, people move species from one continent to another. Pheasants and partridges were introduced into North America for hunting. European settlers took their crops and domesticated animals to Australia. They introduced other species, such as rabbits and foxes, for sport. Weed seeds were accidentally carried around the world in soil used as ballast in sailing ships and as contaminants in sacks of crop seeds. Despite quarantines, disease organisms spread rapidly, carried by infected plants, animals, and people.

A species that has evolved over time in a community with certain predators and competitors may be vulnerable to newly introduced predators and competitors. Introduced species have caused the extinctions of thousands of native species worldwide. Nearly half of the small to medium-sized marsupials and rodents of Australia have been exterminated during the last 100 years by a combination of competition with introduced rabbits and predation by introduced cats and foxes.

Black rats carried to remote oceanic islands on ships are especially destructive predators. On the Galapagos Islands, introduced pigs and rats exterminated several races of tortoises before Darwin explored the islands. Today on some islands they excavate all nests of the giant Galapagos tortoises and devour the eggs. Populations of tortoises on some islands are maintained today only because conservationists remove eggs and rear the young tortoises in captivity until they are large enough to defend themselves against pigs and rats (Figure 55.7).

Proliferation of pests with destructive consequences has quickly followed their introduction to new continents. Forest trees in eastern North America, for example, have been attacked by several European diseases. The chestnut blight, caused by a fungus originally from Europe, virtually eliminated the American chestnut, formerly a dominant tree in forests of the Appalachian Mountains. Some individuals still resprout, but the sprouts are soon infected by the fungus and killed. Nearly all American elms over large areas of the East and Midwest have been killed by Dutch elm disease, caused by the fungus *Ceratocystis ulmi*, which

55.7 Tortoises are Raised at Charles Darwin Station
Conservationists remove tortoise eggs from nests. When the eggs hatch, the young are reared in captivity until they are large enough to defend themselves.

55.8 Coevolved Mutualists   As the iiwi, a Hawaiian honeycreeper, inserts its bill into the corolla tube to extract nectar from a lobelia flower, it deposits pollen from flowers it visited previously. Declining populations of the honeycreeper also threaten the plant, which is left without a pollinator.

reached North America via Europe in 1930. Ecologists suspect that intercontinental movement of disease organisms caused extinctions in the past, but evidence of disease outbreaks is not usually preserved in the fossil record.

Loss of mutualists threatens some species

Many plants have mutualistic relationships with pollinators, but most of these mutualisms are not highly species-specific (see Chapter 52). On islands, however, where ecological communities contain relatively few species, plant–pollinator interactions often evolve to be highly specific. For example, a single species of *Lobelia* colonized the Hawaiian Islands, where it eventually gave rise to 110 of the 350 species in the genus. A single colonizing species of songbird gave rise to at least 47 species of honeycreepers, some of which have long, slender, curved bills. These nectar-feeding birds were the only pollinators of many species of Hawaiian lobelias, whose long, curved tubular flowers match the shapes of their bills (Figure 55.8).

Today, half of the nectar-feeding birds of Hawaii are extinct, leaving many lobelias without pollinators. Many of these lobelias still survive, but populations of some species have been reduced to only a few individuals. A few species reproduce now only because biologists artificially pollinate them.

Habitat destruction and fragmentation are important causes of extinctions today

The 6 billion people that exist on Earth today are fed, clothed, and housed by agricultural and forestry industries that convert natural ecological communities containing many species into highly modified communities dominated by one or a few species of plants. Within these communities, humans discourage the presence of other species by applying chemicals that kill competing plants, bacteria, fungi, nematodes, insects and other arthropods, and vertebrates.

Although agricultural ecosystems have long harbored fewer species than the complex ecosystems they replaced, only recently have farmers planted large tracts of land in single crops. Traditional farmers planted many different crops together, maintaining some of the diversity that is key to the success of natural communities (Figure 55.9). Many species that can-

55.9. Many Species of Plants Grow in Traditional Gardens   Traditional farmers plant many different crops together, as in this terrace garden in Honduras. The diversity maintained by these agricultural methods is vital to the continued well-being of the world's food crops.

TABLE 55.1 Co-option of Net Primary Production by Human Manipulations of Ecosystems

HABITAT CATEGORY	NET PRIMARY PRODUCTION CO-OPTED[a]
Cultivated land	15.0
Grazing land	11.6
Forest land	13.6
Human-occupied areas	0.4
Total	40.6
Total net primary production	132.1
Percent co-opted	30.7

[a] Values are in petagrams (one petagram = 10^{15} grams).

not survive in intensive modern agricultural systems thrive in traditional ones.

When ecosystems such as agricultural lands and plantation forests are managed so as to divert most of their primary production to certain species favored by people, we say that their production is **co-opted**. Agriculture and forestry today are so extensive that more than 30 percent of all terrestrial production is co-opted for human use (Table 55.1). All the other species on Earth have only two-thirds of the total global terrestrial production available for their use, and the fraction is steadily decreasing.

Because of increasing habitat modification and the co-option of primary production, habitat loss is certain to be the most important cause of species extinctions during the next century. The habitats required by some species are being completely destroyed. Other habitats, particularly old-growth forests, natural grasslands, and estuaries, are being reduced to widely separated patches that are too small to sustain populations of many of the species that live in them.

As habitats are progressively destroyed, the remaining patches become smaller and more isolated. Small habitat patches are qualitatively different from larger patches of the same habitat in ways that affect the survival of species. Small patches cannot maintain populations of species that require large areas, and they support only small populations of many of the species that can survive in them.

In addition, the fraction of a patch that is influenced by effects originating in adjacent habitats—**edge effects**—increases rapidly as patch size decreases (Figure 55.10). Close to the edges of forest patches, for example, winds are stronger, temperatures are higher, humidities are lower, and light levels are higher than they are farther inside the forest. Species from surrounding habitats often invade the edges of patches to compete with or prey upon the species living there.

More than 70 percent of the old-growth coniferous forests of the northwestern United States have been cut, mostly during the past 40 years. Second-growth forests in this area are being cut before they are 80 years old, so they do not acquire key characteristics of old-growth forests, such as trees of all ages (some of the conifers live more than 500 years), many standing dead trees, and large decomposing logs on the forest floor (Figure 55.11a). Among the species that require old growth to maintain viable populations are salamanders, which live inside rotting logs, the only microhabitat that retains moisture during the dry summers; the spotted owl, a species that hunts for rodents within mature forests; and the marbled murrelet, a seabird that nests on horizontal branches of large trees in coastal forests (Figure 55.11b).

As the fraction of the area remaining in old-growth forest is reduced, the home ranges of spotted owls must increase; hunting success, and hence reproductive success, thus decreases, and juvenile mortality during dispersal between patches increases. The spotted owl faces a high risk of being extirpated in most of Washington and Oregon within the next 50 years if clear-cutting of old-growth forests continues.

Usually we do not know which organisms lived in an area before its habitats became fragmented. To address this problem, a major research project near Manaus, Brazil, was launched before logging took place. The landowners agreed to preserve forest patches of certain sizes and locations, and censuses of those patches were conducted while the areas were still part of the continuous forest. Soon after the surrounding forest

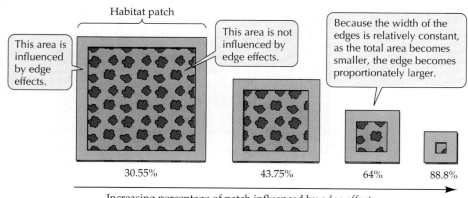

Habitat patch

This area is influenced by edge effects.

This area is not influenced by edge effects.

Because the width of the edges is relatively constant, as the total area becomes smaller, the edge becomes proportionately larger.

30.55%   43.75%   64%   88.8%

Increasing percentage of patch influenced by edge effects

55.10 Edge Effects The smaller the patch, the greater the proportion of the patch that is influenced by detrimental edge effects.

(a)

(b) *Strix occidentalis caurina*

55.11 Old-Growth Forests and Owls (a) Old-growth coniferous forests of the Pacific Northwest have trees of all ages, and there are many large logs on the forest floor. (b) The northern spotted owl reproduces at high rates only in old-growth forests.

was cut, species began to disappear from the isolated patches (Figure 55.12). The first species to be eliminated were monkeys with large home ranges, such as the black spider monkey, the tufted capuchin, and the bearded saki (Figure 55.13a).

Birds that follow swarms of army ants to capture insects flushed by the ants also disappeared quickly from small patches (Figure 55.13b). A particular colony of army ants is a useful resource for the birds only

(a) *Ateles* sp.

(b)

55.12 Brazilian Forest Fragments Studied for Species Loss Isolated patches (foreground) lose species much more quickly than patches connected to the main forest do, even if the isolated patches, such as the one in the foreground, are larger than the connected ones (in background).

55.13 Extinction in Patches (a) The spider monkey was rapidly displaced by the logging industry in South America. (b) The white-plumed antbird, a species common in Brazilian forests, has become extinct in isolated forest plots, but survives in connected patches.

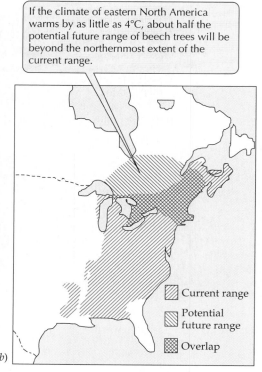

If the climate of eastern North America warms by as little as 4°C, about half the potential future range of beech trees will be beyond the northernmost extent of the current range.

(b)

Current range

Potential future range

Overlap

55.14 Beeches Are Threatened by Climate Warming
(a) Seedlings and saplings abound in this healthy beech forest. (b) If the climate of eastern North America warms by as little as 4°C, about half the potential future range of beech trees will be beyond the northernmost extent of their current range.

when it is raiding (about 27 days out of the 35-day period between colony moves). Therefore, the birds must have access to a number of colonies to be guaranteed that there is always at least one colony raiding. In small patches with few ant colonies, there are periods when none are raiding. During these days, ant-following birds have trouble finding enough to eat.

Climate change may cause species extinctions

Atmospheric scientists predict that, as a result of increasing concentrations of greenhouse gases (gases such as carbon dioxide and methane that are transparent to sunlight but opaque to heat radiated from Earth), average temperatures in North America will increase 2 to 5°C by the end of the twenty-first century. Conservation biologists are attempting to predict the effects of this warming trend on North American deciduous forest trees. If the climate warms by only 1°C, the average temperature formerly found at a particular location will shift 150 km north of that location. An organism that survives best at that average temperature will therefore need to shift its range 150 km north to remain in the same climate. Therefore, trees would need to shift their ranges as much as 500 to 800 km in a single century if the climate warmed 2 to 5°C.

Most deciduous forest trees grow for long periods before they begin to reproduce, and their seeds move only short distances. The American beech, like other large-seeded species, cannot shift its range rapidly. Beeches, whose seeds are dispersed primarily by jays,

advanced northward only about 20 km per century when glaciers retreated following the last glacial period. Beeches would thus have to migrate 40 times faster than they did in the past to keep up with the anticipated rate of climate change (Figure 55.14). Even though there might be areas of climate suitable for them, beeches probably could not reach them without human assistance. To maintain beech forests, we may need to intervene by moving seeds and by assisting seedling establishment.

In the past, global climate changed at a much slower rate than that predicted for the twenty-first century. Most organisms were able to shift their ranges rapidly enough to keep up with the changes. In addition, habitats were more continuous, and migration routes were not blocked by extensive areas of unsuitable human-modified habitats, as they are today. Organisms may therefore have much more difficulty dealing with climate changes during the twenty-first century than they did during glacial periods.

If Earth warms as predicted, climatic zones will not simply shift northward; new climates will develop and some existing climates will disappear. New climates are certain to develop at low elevations in the tropics. All models predict that the climate will warm less in tropical regions than at high latitudes, but a warming of even 2°C would result in climates near sea level that are hotter than those found anywhere in the humid tropics today. Adaptation to those climates may prove difficult for many tropical organisms.

Designing Recovery Plans

Once the causes of endangerment of species have been identified, appropriate remedies can be designed. If the cause of endangerment is overexploitation, harvesting can be controlled or banned. If habitat destruction is the cause, existing habitats can be protected and damaged habitats restored. If habitat isolation is the cause, individuals can be moved among subpopulations to increase dispersal rates. In this section we will describe how good analyses of population viability have been used to implement management actions designed to prevent species from becoming extinct by reducing mortality rates, reintroducing species to areas from which they have been extirpated, and raising endangered species in captivity.

Preserving habitats is always important

No population is viable without suitable habitat in which individuals can survive and reproduce. An excellent way to maintain viable populations is to set aside areas in which species and their habitats are protected. Protected areas may be established to preserve single species, but usually they are designed to main-

tain entire communities and ecosystems. Often single species of special economic, aesthetic, or ecological value serve as indicators of the viability of entire communities. The protection of single species that require large areas of suitable habitat—so-called "umbrella species" such as elephants, grizzly bears, and spotted owls—may promote the survival of most or all other species living in that ecosystem.

Demographic parameters may need to be manipulated

If the cause of a population decline is low reproductive success or unusually high mortality rates, managers may be able to intervene to change those rates. In the United States, losses of birds' nests to predators, such as crows, jays, native mammals, and domestic cats and dogs, are very high, especially in suburban areas. Many species of songbirds are declining in eastern North America because they cannot produce enough offspring to replace adult losses (Figure 55.15). Suburbanites need to recognize that their pets may be contributing to these declines.

Brown-headed cowbirds, which lay their eggs in other birds' nests, have increased dramatically in many parts of North America as a result of land clearing and associated high densities of domestic live-

55.15 Declining Songbird Populations in the Eastern United States   These graphs show the results of long-term counts in specific forest patches in the Middle Atlantic states. Brood parasitism and other predation on songbird nests is making it difficult for some populations to replace themselves.

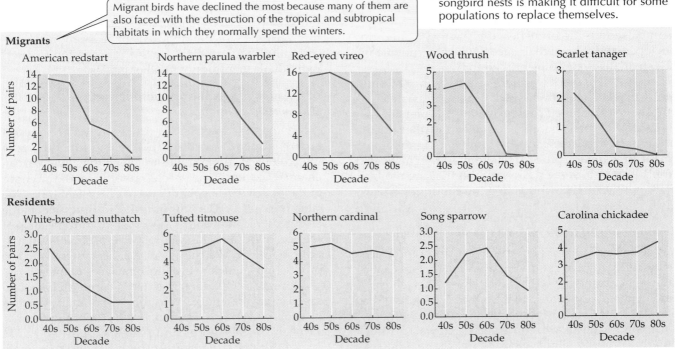

Migrant birds have declined the most because many of them are also faced with the destruction of the tropical and subtropical habitats in which they normally spend the winters.

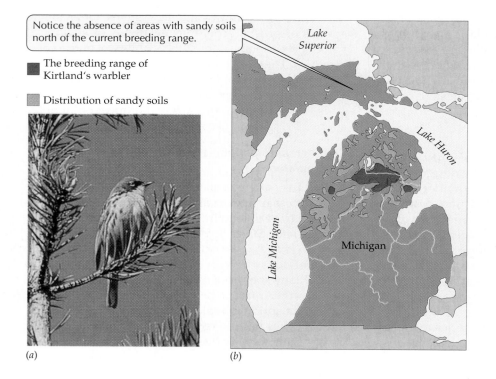

Notice the absence of areas with sandy soils north of the current breeding range.

■ The breeding range of Kirtland's warbler

■ Distribution of sandy soils

Lake Superior

Lake Huron

Lake Michigan

Michigan

(a)

(b)

55.16 The Kirtland's Warbler Is Threatened by Habitat Loss (*a*) A male Kirtland's warbler in a young jack pine. (*b*) The warbler's breeding range and the distribution of the sandy soils that support the stands of jack pines.

stock. Their avian hosts incubate the cowbird eggs and feed the cowbird nestlings in addition to, or instead of, their own offspring.

The Kirtland's warbler, an endangered species that nests only in young stands of jack pine on sandy soils in Michigan, is at risk from both heavy parasitism by cowbirds and loss of habitat (Figure 55.16). The current population of Kirtland's warbler is less than 1,000 individuals. Stands of jack pines depend on periodic fires for their persistence because jack pine cones remain closed on the branches. Only when heated do they open and release their seeds, which germinate in the ash on the floor of the burned forest.

Because Kirtland's warblers nest only in jack pine forests that are 8 to 18 years old, fire suppression measures have threatened to deprive the warblers of essential young forest habitat. To prevent further threat to the warblers, conservation biologists ignite controlled fires in jack pine forests to maintain a steady supply of forests of the right age. And they are removing brown-headed cowbirds to reduce nest parasitism rates.

Genotypes need to be matched to environments

Individuals are more likely to survive and reproduce successfully if their genotypes are appropriate to the environments in which they live. Thus, when conservation biologists prepared to introduce collared lizards to areas of the Ozark Mountains of southern Missouri from which the lizards had been extirpated, they thought carefully about where to get their colonists.

The open prairie vegetation required by collared lizards exists only in isolated, fire-prone glades on south-facing slopes with shallow soils, a habitat much reduced by agriculture and fire prevention (Figure 55.17). The Ozark populations of collared lizards, which have been isolated for about 2,000 lizard generations, are genetically distinct from those elsewhere in the range of the species. Therefore, the biologists decided to introduce lizards only from other glades in the Ozarks.

To obtain enough colonists, the founding lizards could not be taken from a single glade, because if fewer than ten mature lizards were introduced, the new population was unlikely to become established. And taking more than ten lizards from a single donor population would have threatened the future of the donor population. Therefore, lizards from at least five different glades were released together in a single new glade. Because individuals from different donor populations have distinct maternally inherited mitochondrial DNA markers, investigators will be able to determine which individuals have reproduced successfully. They can then use this information to select individuals to introduce to other glades.

Maintaining keystone species may prevent widespread extinctions

As we discussed in Chapter 52, keystone species exert strong influences on the structure and functioning of the ecological communities in which they live. We can characterize keystone species in terms of the strength of their effects on a community or ecosystem trait. One such measure, called **community importance**, is the change in a community or ecosystem trait per unit of change in the abundance of a species. A *trait* is a quantitative feature of a community or ecosystem, such as productivity, nutrient cycling, or species richness.

Mutualistic relationships that depend on keystone species may be vital for the survival of many species in a community. In Peruvian forests, for example, only a dozen species of figs and palms support an entire community of large fruit-eating birds and mammals

55.17 Both the Glades and the Lizards are Endangered (*a*) Open glades exist only as small patches on south-facing slopes in the Ozark Mountains. (*b*) The collared lizards of the Ozarks, which live only in these open glades, are genetically different from those elsewhere in the range of the species.

during the part of the year when fruits are least available. Loss of these few tree species would probably eliminate most of these birds and mammals, even if hundreds of other tree species remained. In turn, loss of the fruit-eaters might seriously impair the dispersal of the seeds of many other tree species. Thus, many mutualistic relationships are probably maintained by a few keystone tree species, which constitute only a small fraction of the 2,000 species of trees in the forest.

Because the extinction of keystone species could result in the extinction of many species in their communities, conservation biologists try to identify keystone species and take action to preserve them.

Captive propagation has a role in conservation

Species being threatened by overexploitation, loss of habitat, or environmental degradation through pollution can sometimes be maintained in captivity while the external threats to their existence are reduced or removed. Research on nutrition and the preparation of suitable diets, on the use of vaccinations and antibiotics, and on the control and enhancement of reproduction by both behavioral and technical means (such as artificial insemination and embryo transfers) supports captive propagation efforts.

Captive propagation is only a temporary measure that buys time. Existing zoos, aquariums, and botanical gardens do not have enough space to maintain adequate populations of more than a small fraction of Earth's rare and endangered species. Nonetheless, captive propagation can play an important role by maintaining species during critical periods and by providing a source of individuals for reintroduction into the wild. Captive propagation projects in zoos have also been very effective in raising public awareness of the rate at which species are threatened with extinction.

THE PEREGRINE FALCON. In 1942, about 350 pairs of peregrines bred in the United States east of the Mississippi River. This breeding population disappeared entirely by 1960, and no peregrines are known to have reproduced in the region during the next 20 years. The cause of their disappearance was the widespread use of organochlorine pesticides, such as DDT and dieldrin. These pesticides degrade very slowly in the environment and gradually accumulate in the prey of predators such as falcons. Their accumulation in the peregrines' bodies interfered with the deposition of calcium in eggshells. As a result, most of the falcons' eggs broke before they hatched. The successful reintroduction of the peregrine falcon to eastern North America depended on a strong captive propagation program.

Much of eastern North America became suitable habitat for peregrines again when the use of DDT was terminated by federal laws in the United States. Captive breeding of peregrines began at Cornell University in 1970, and by the end of 1986 more than 850 birds reared in captivity had been released in 13 eastern states, with spectacular success (Figure 55.18). In addition, some individuals adapted to live in urban environments, using ledges on buildings and other structures of cities as nesting sites (Figure 15.19). Peregrines probably would have recolonized the East by themselves, but they would have done so much more slowly without human assistance.

THE CALIFORNIA CONDOR. With its 9-foot wingspan, the California condor is North America's largest bird. Two hundred years ago, condors ranged from southern British Columbia to northern Mexico, but by the 1940s they were confined to a small region in the mountains and foothills north of Los Angeles. By 1978, the wild population was plunging toward extinction—only 25

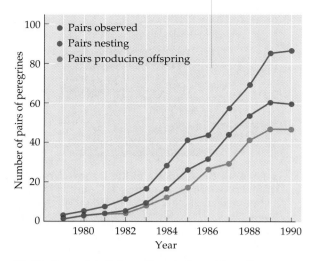

55.18 Peregrine Falcon Populations Have Been Reestablished Throughout the eastern United States, many pairs of peregrine falcons now attempt to reproduce, most of them successfully.

to 30 birds remained. In 1985 alone, 6 of the remaining 15 wild birds disappeared. To save the condor from extinction, biologists initiated a captive propagation program in 1983.

To maximize genetic variation in the captive population, all of the remaining wild birds were captured, the last one in April 1987. The first chick conceived in captivity hatched in 1988. By 1993, nine captive pairs were producing chicks, and the captive population had increased to more than 60 birds. The captive population was large enough to risk releasing 6 captive-bred birds in the mountains north of Los Angeles in 1992 (Figure 55.20).

55.19 A New Home in the City Peregrine falcons have responded well to captive propagation. Some individuals have adapted to urban life, nesting on the tall buildings of cities and feeding on pigeons.

Gymnogyps californianus

55.20 Back in the Wild Captive propagation of the California condor is providing the individuals that are being reintroduced into the species' former range.

The released condors are being provided with contaminant-free food in remote areas, and 3 of them were still alive in the spring of 1994. The released birds are using the same roosting sites, bathing pools, and mountain ridges that their predecessors did. More captive-reared birds also were released late in 1996 in northern Arizona. It is still too early to pronounce the program a success, but without captive propagation, the California condor would probably be extinct today.

CAPTIVE PROPAGATION HAS HIGH COSTS AND BENEFITS The California condor rehabilitation program costs about 1 million dollars a year. The Peregrine Fund at Cornell spent nearly 3 million dollars over the past 25 years; the expenses of other cooperating agencies add at least another half million to the total. These amounts may seem large, but they are small compared with the costs of other human activities, and compared with the cost of continued loss of species. The work needed to restore all of the world's threatened birds of prey probably could be accomplished with 5 million dollars per year, the approximate cost of one armored tank.

Establishing Priorities for Conservation Efforts

Many species and ecosystems are threatened, but the resources that can be allocated to preservation efforts are limited. How can those resources be allocated to achieve the most conservation benefits? Because many species can survive only in the ecological communities in which they evolved, preserving complete ecological communities and habitats is vital. Because only a small fraction of the landscape can be incorporated into parks and reserves, proper selection of those habitats is of great importance.

The primary function of parks, sanctuaries, and reserves is to maintain species and ecosystems relatively free of human disturbance. Parks are being established in many countries. Although their size and number will never be equal to the task of ecosystem and species preservation, more parks are urgently needed. Where should they be established?

Where should parks be established?

Two kinds of areas are high-priority sites for protection by parks and reserves—those that are home to unusually large numbers of species, and those that have many endemic species that live nowhere else. Because tropical ecosystems are generally richer in species than are ecosystems at higher latitudes (see Chapter 54), losses of tropical habitats threaten more species than do losses of comparable areas of temperate habitats.

The number of species that become extinct as a result of habitat destruction also depends on how many local species are found nowhere else. For example, nearly all the mammals and birds of Madagascar are found only on that island. Therefore, if the small fragments of tropical forests remaining on Madagascar are destroyed, the species dependent on them are certain to be exterminated in the wild (Figure 55.21).

Species with very small ranges are frequently found on islands, but some mainland regions also have many species found nowhere else. The Rift Valley lakes of Africa harbor more than 1,000 species of fish, most of which live in only one lake. The Atlantic coastal forests of southeastern Brazil are another center of endemism. Because only about 1 percent of the original extent of those forests remains, many species there have become extinct or are in danger of immediate extinction. Mountainous regions have many endemic species because temperature and rainfall change rapidly with elevation, creating many distinct habitats within a small area.

Centers of endemism are not the same for all groups of organisms. The Cape region at the southern tip of Africa, for example, has a flora of 8,500 species, 80 percent of which are endemic, but only 4 of the 187 species of birds found there are endemic.

Conservation biologists examine a region's conservation potential

To identify the places where conservation efforts are likely to provide the greatest benefits, conservation biologists compare countries with respect to estimates of the number of endemic species, the amount of land that is protected, and the amount of unprotected land that remains in a more or less natural state. Using these criteria, conservation biologists at the World Wildlife Fund divided the countries of the Indo-Pacific region into four categories (Figure 55.22). Category I countries have much land already protected and much land remaining in high-quality forests. Additional reserves can and should be established in these countries while the land is still available. Category III countries have little land already protected, but much land that could become reserves. They are high-priority countries for establishment of reserves that would transfer them to Category I. Countries in the other two categories have little land still available to be incorporated into protected reserves. Consequently, they are lower-priority targets for conservation efforts.

The Importance of Commercial Lands

In the majority of countries, most parks must be established in already settled areas because few pristine areas remain. The people living there cannot be evicted, nor is it appropriate, in most cases, to prevent hungry people from settling in or hunting in the parks. The high rates of human population growth in most tropical countries guarantee that pressures on parks from agricultural settlers will increase rather than decrease.

For these reasons, lands that are exploited for food, medicines, and fiber must play an important role in conservation. These lands are far more extensive than parks and reserves, and they include climates and ecosystems not represented in parks. Fortunately, many species can be preserved on lands that are being used in economically beneficial ways. Only a few species, such as predators on humans and domestic animals, or large, destructive herbivores, are incompatible with most human uses of land.

Some economic land uses are compatible with conservation

Forest reserves in which economically valuable products are harvested can support both species preservation and economic development. In Belize, about 75 percent of primary health care is provided by traditional healers using plant remedies. Persons known as *hierbateros* collect medicinal plants in the forests and sell them to the *curanderos* who actually provide the health care. A botanist and an economist determined that two 1-hectare plots in second-growth forest, one 30 years old, the other 50 years old, yielded 309 and 1,434 kg dry weight of medicines per year, which were sold at an average price of U.S.\$2.80 per kilogram. Thus, these 1-hectare plots yielded gross revenues of

Andansonia grandieri (giant baobob tree)

55.21 Madagascar Abounds with Endemic Species The majority of plant and animal species found on the island of Madagascar off Africa's east coast are found nowhere else in the world.

Southern spiny forest

Indri indri (indri)

Furcifer revocosus (warty chameleon)

Enlemur fulvus fulvus (brown lemur)

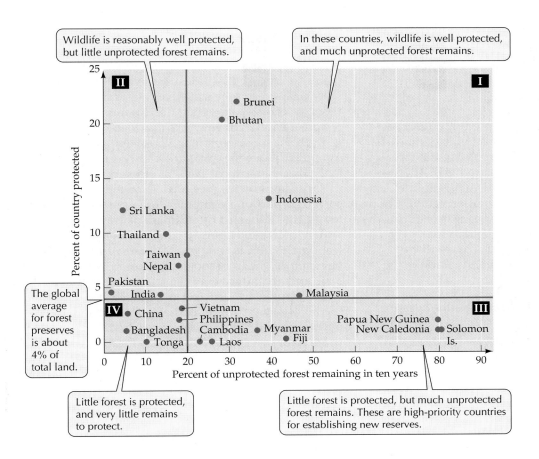

Wildlife is reasonably well protected, but little unprotected forest remains.

In these countries, wildlife is well protected, and much unprotected forest remains.

II

I

● Brunei
● Bhutan

● Sri Lanka
Thailand ●
Taiwan ●
Nepal ●
Pakistan ●
India ●

● Indonesia

● Malaysia

IV
● China
● Vietnam
● Philippines
● Cambodia
● Bangladesh
● Tonga
● Laos
Myanmar ●
● Fiji

Papua New Guinea ●
New Caledonia ●
● Solomon Is.

III

Percent of country protected

Percent of unprotected forest remaining in ten years

The global average for forest preserves is about 4% of total land.

Little forest is protected, and very little remains to protect.

Little forest is protected, but much unprotected forest remains. These are high-priority countries for establishing new reserves.

55.22 Identifying Conservation Priorities
In the World Wildlife Fund scheme, countries are grouped in four categories on the basis of the total area of their forest reserves and the area available to be incorporated into reserves.

areas managed for economically valuable products. Some of the reserves are the homes of indigenous people who continue to use the environment in their traditional ways.

The largest Costa Rican megareserve is La Amistad Biosphere Reserve, an area of more than 500,000 hectares that includes three national parks, a biological reserve, five Indian reservations, and two forest reserves. La Amistad contains the largest tract of highland vegetation in Central America, and it has considerable hydroelectric generating potential. All the native large predators still survive there. Managed properly, the reserve can provide drinking water, electricity, forest products, and nature tourism, protect indigenous cultures, and preserve

$865 and $4,017, which are greater than the incomes that would result from cultivating squash and corn on that land. Harvesting of medicinal plants can be compatible with agriculture if agricultural plots are allowed to regrow into forests for several decades before they are cleared again. These second-growth forests support many species that would otherwise not survive in the area.

Conservation requires large-scale planning

Knowing that large areas are required to preserve many species, conservation biologists promote the establishment of megareserves. A typical **megareserve** is a large area of land that has a central core of undisturbed habitat. Surrounding the core are buffer areas in which economic activity is permitted as long at it does not destroy the ecosystem. Appropriate activities may include sustainable harvesting of animal populations and plant products such as rubber, fruits, nuts, and wood. On the edges of a megareserve is a zone in which more intensive land uses, such as agriculture or plantation forestry, are permitted.

Costa Rica has pioneered the development of megareserves. It has consolidated its parks and reserves into eight megareserves that should maintain about 80 percent of the country's biodiversity (Figure 55.23). Each megareserve includes natural areas and

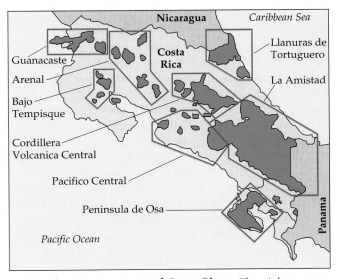

55.23 The Megareserves of Costa Rica The eight areas outlined in red are being managed for both biodiversity and economic activities. The green areas are the current reserves.

species. In combination, these benefits outweigh what could be gained by logging the steep slopes and converting them to low-productivity agricultural systems.

Restoring Degraded Ecosystems

Many areas that could be incorporated into reserves have been highly altered by human activities. Some of these areas can play their intended roles in biodiversity conservation only if they are restored to their original state. To accomplish this task, a subdiscipline of conservation biology, known as **restoration ecology**, is growing rapidly. Research on methods of restoring species and ecosystems is needed because many ecological communities will not recover, or will do so only very slowly, without creative intervention in the recovery process.

Tropical deciduous forests are being restored in Costa Rica

The world's largest restoration project is under way in Guanacaste National Park in northwestern Costa Rica. Its goal is to restore a large area of tropical deciduous forest, the most threatened ecosystem in Central America, from small fragments that remain in an area converted primarily to cattle pastures. One approach to restoration would be to exclude fire and domestic livestock from the park and let nature take its course. Grass patches of less than 120 hectares would be covered by woody vegetation within 20 years, but large expanses of pasture would require 50 to 200 years to regrow because tree seeds move into open pastures only very slowly.

Reforestation can be speeded up by manipulating the habitat, which is what Daniel Janzen, architect of the restoration project, is doing. To design his project, Janzen gathered and used basic ecological information about the abilities of different plant species to germinate and grow in open pastures. The single most important threat to Guanacaste National Park during the coming decades is fires, most of which are started by people. The introduced pasture grasses produce dense, highly flammable stands that spawn hot fires that penetrate far into surrounding forests. Domestic livestock keep these grasses under control by grazing, and they also disperse the seeds of native trees that are good at invading pastures. The restoration program is encouraging some grazing by domestic livestock in the park until plant succession has progressed to the point at which the grass no longer poses serious competition to the woody species and is no longer sufficiently dense to carry hot fires.

Restoring some habitats is difficult

Restoration of damaged and degraded habitats is an important activity, but ecologists still have limited ability to restore natural ecosystems. In the United States, the belief that existing ecosystems can be replaced has made it easy to get permits for development projects that destroy valuable habitats. Developers need only state that they will create substitutes for the areas they are destroying. Promising to restore a habitat, however, is much easier than doing it. Even the most experienced wetland ecologists, for example, are having great difficulty creating new wetlands that mimic those being destroyed.

An example is the "restored" wetland that was conceived as part of a compensation agreement that allowed the California Department of Transportation to widen Interstate Highway 5 near San Diego. The project damaged a marsh and jeopardized two endangered birds, the light-footed clapper rail and the least tern, and an endangered plant, the salt marsh bird's beak. Despite stringent, court-imposed standards and the involvement of wetland experts, the endangered birds were still not breeding in the "restored" marsh 9 years after it was created. The restoration ecologists have not given up, but the advice offered by a recent National Research Council committee on wetland restoration needs to be heeded: "Wetland restoration should not be used to mitigate avoidable destruction of other wetlands until it can be scientifically demonstrated that the replacement ecosystems are of equal or better functioning."

Markets and Conservation

Most species are common property resources—that is, they are "owned" by everybody. Because no individual or group of individuals typically has strong incentives to use common property resources sustainably, their preservation usually depends on central governments. Unfortunately, governments generally lack sufficient resources to do the job, and governmental actions sometimes are not well attuned to local situations. For these reasons, establishing property rights that allow owners to receive the economic benefits from managing biological resources can, under some conditions, assist conservation efforts. Such use of property rights and markets is most likely to be successful if the species being managed are relatively sedentary and if a reliable supply of the products derived from them is commercially important.

A good example of the use of property rights to preserve species is provided by butterfly farming in Papua New Guinea. Many species of butterflies in that country, and elsewhere in the tropics, are threatened by habitat destruction, environmental degradation, and overexploitation. By the mid-1960s, butterfly collecting and commercial harvesting, which provided butterflies for mounting in plastic and glass trays, tabletops, decorative screens, and even clear plastic

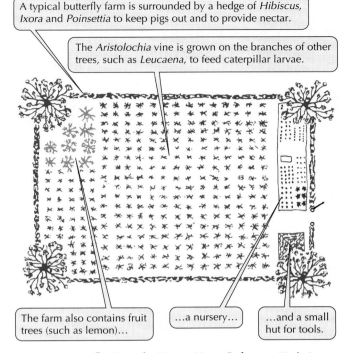

A typical butterfly farm is surrounded by a hedge of *Hibiscus*, *Ixora* and *Poinsettia* to keep pigs out and to provide nectar.

The *Aristolochia* vine is grown on the branches of other trees, such as *Leucaena*, to feed caterpillar larvae.

The farm also contains fruit trees (such as lemon)…

…a nursery…

…and a small hut for tools.

55.24 Butterfly Farm in Papua New Guinea Birdwing butterfly species are a valuable economic asset in Papua New Guinea.

toilet seats, had endangered some of the most striking and valuable birdwing butterflies.

In 1966, the government of Papua New Guinea prohibited collecting and trading in seven species of birdwing butterflies and established several butterfly reserves. In 1974, the government initiated a butterfly farming program designed to give it a monopoly on trade in its butterflies and to return all economic benefits to the country's people. By 1978, about 500 butterfly farms existed, operated by villagers who planted flowering hibiscus vines to attract adult butterflies and other plants on which the caterpillars feed (Figure 55.24). Local people tend the caterpillars and house the pupae so that the butterflies can be collected when they emerge. Many butterflies are still collected in the wild in Papua New Guinea, but the farms are a significant source of specimens, and they take pressure off the butterfly populations in the surrounding forest lands. Because farmed butterflies are generally in very good condition, they command high prices.

Investing to Preserve Biodiversity

Genes, species, and habitats are the sources of many economic benefits to people. However, most of these benefits will assist future generations, not the current one. We do not know how future generations will value these benefits or the biodiversity that generates them. We do not even know how the current generation values them. We also cannot predict which species will turn out to be sources of valuable foods, medicines, or drugs. Most species probably will not.

Between 1951 and 1981, the National Cancer Institute screened more than 100,000 extracts from 35,000 different species. To date, only one compound—taxol, derived from the Pacific yew tree—has received approval as a drug component. That drug is very valuable, but many species had to be searched to find it.

Although the chance that any given species will turn out to be the source of some valuable marketable product is small, biodiversity has another vital feature: extinction is forever. If we purposely or inadvertently exterminate a species, we have irreversibly destroyed a resource of unknown value. The irreversibility of ex-

tinction makes loss of biodiversity an urgent public issue.

How much should societies invest to preserve biodiversity? This question does not have an ecological answer. Economists and evolutionary biologists can contribute valuable information to the public debate, but the final decision is an ethical, spiritual, and political one that will depend on our beliefs about our responsibilities to the other organisms that share Earth with us.

The preservation of biological diversity and ecosystem services is one of the greatest challenges facing humankind. Many of the scientific tools needed for the task are already available, but appropriate use of these tools requires major changes in people's attitudes toward other species. If species are valued only because they are economically useful to us, increased losses of species are inevitable. Other uses of natural habitats are likely to be seen as more profitable, at least in the short run. Even though a wetland has great value, some people usually can make enough money in the short run by destroying it to be motivated to do so. Only when we value biological diversity and ecosystem functioning as the heritage of all humankind, a heritage to be passed on to our descendants as completely as possible, will we begin to reduce the current alarming rates of ecosystem destruction and species extinctions.

Summary of "Conservation Biology"

Estimating Current Rates of Extinction

• Estimates of current rates of extinction are based primarily on species–area relationships and rates of tropical deforestation. **Review Figure 55.1**

Why Care about Species Extinctions?

• Species provide the food, fiber, medicines, and aesthetic opportunities upon which human life depends.
• Ecosystems provide services that can be replaced only by expensive and continuing human effort. **Review Figure 55.4**

Determining Causes of Endangerment and Extinction

• Overexploitation, which historically resulted in most human-caused extinctions, is still an important cause of species extinctions today.
• Introduced species have caused many extinctions, particularly on islands. **Review Figure 55.7**
• Habitat destruction is the most important cause of species extinctions today. **Review Table 55.1**
• The fragmentation of habitat into patches that are too small to support populations is a major cause of extinctions.
• Edge effects that increase the detrimental effects of surrounding terrain on a habitat increase as habitat patch size decreases. **Review Figures 55.10, 55.12, 55.15**
• The predicted rapid warming of Earth's climate as a result of human activity could make survival difficult for many species that cannot shift their ranges quickly enough. **Review Figure 55.14**

Designing Recovery Plans

• The best way we know of to maintain populations is to set aside areas in which species and their habitats are protected. Human manipulation of these protected environments and the populations that inhabit them is sometimes needed.
• The identification and preservation of keystone species, whose loss could result in the extinction of many other species that depend on them, is a high priority of conservation biologists.
• Captive propagation can play a useful role in conservation. **Review Figures 55.18, 55.19**

Establishing Priorities for Conservation Efforts

• More parks and reserves are needed. High-priority areas for their establishment are regions of unusually high species richness and endemism and countries where substantial tracts of undisturbed habitats still remain. **Review Figure 55.22**

The Importance of Commercial Lands

• Megareserves combine areas under total protection with areas where commercial exploitation of the ecosystem is allowed as long as it does not damage ecosystem processes. **Review Figure 55.23**

Restoring Degraded Ecosystems

• Tropical deciduous forest, an endangered ecosystem in Central America, is being restored in Costa Rica. **Review Figure 55.24**

Markets and Conservation

• Properly employed, markets can help preserve biodiversity. **Review Figure 55.25**

Investing to Preserve Biodiversity

• Preserving biodiversity is not just a scientific issue. It raises serious moral and ethical concerns that define what it means to be a human being on Earth.

Self-Quiz

1. Which of the following is *not* currently a major cause of species extinctions?
   a. Habitat destruction
   b. Climate change
   c. Overexploitation
   d. Introduction of predators
   e. Introduction of diseases

2. When ecosystems are managed to favor strongly those species intended for human use, we say that their production is
   a. modified.
   b. diverted.
   c. co-opted.
   d. channeled.
   e. managed.

3. A minimum viable population is
   a. the estimated number of individuals necessary for the species to maintain genetic diversity.
   b. the estimated number of individuals necessary for the species to persist in all U.S. national forests.
   c. the estimated number of individuals necessary for the species to survive for several decades.
   d. the estimated number of individuals necessary for a species to maintain or increase its numbers in a region.
   e. the minimum density required for individuals to find mating partners.

4. Which of the following is *not* a component of a population vulnerability analysis?
   a. Spatial structure of a population
   b. Sex ratio within a population
   c. Amount of variation in birth and death rates
   d. Amount of heterozygosity and genetic variance
   e. Captive propagation of individuals

5. Which of the following is *not* an ecosystem service?
   a. Production of carbon dioxide
   b. Flood control
   c. Water purification
   d. Air purification
   e. Preservation of biological diversity

6. Conservation biologists are concerned about global warming because
   a. the rate of change in climate is projected to be faster than the rate at which many species can shift their ranges.

b. it is already too hot in the tropics.

c. climates have been so stable for thousands of years that many species lack the ability to tolerate variable temperatures.

d. climate change will be especially harmful to rare species.

e. none of the above

7. A species that is found only in a particular region is said to be

a. an indicator species for that region.

b. a restricted species.

c. a vulnerable species.

d. endemic to that region.

e. demographically constrained.

8. A keystone species is one that

a. preys heavily on a particular species.

b. is especially vulnerable to extinction.

c. is restricted to a small geographic area.

d. experiences considerable demographic stochasticity.

e. strongly influences the structure and functioning of its ecological community.

9. As a habitat patch gets smaller it

a. cannot support populations of species that require large areas.

b. supports only small populations of many species.

c. is influenced to an increasing degree by edge effects.

d. is invaded by species from surrounding habitats.

e. all of the above.

10. Restoration ecology is an important discipline because

a. many areas being incorporated into megareserves have been highly degraded.

b. many areas being incorporated into megareserves are vulnerable to global climate change.

c. many species suffer from demographic stochasticity.

d. many species are genetically impoverished.

e. fire is a threat to many reserves.

Applying Concepts

1. Most species driven to extinction by people in the past were large vertebrates. Do you expect this pattern to persist into the future? If not, why not?

2. Species endangered as a result of global climatic warming might be preserved if we move individuals from areas that are becoming unsuitable for them to areas likely to be better for them in the future. What are the major difficulties associated with such interventions? For what types of species would they work well? Poorly?

3. Conservation biologists have debated extensively which is better: many small reserves or a few large ones. What biological processes should be evaluated in making judgments about the size and location of reserves? To what extent should we be concerned with preserving the largest number of species rather than those species judged to be of unusual importance?

4. During World War I, French doctors adopted a "triage" system of dealing with wounded soldiers. The wounded were divided into three categories: those almost certain to die no matter what was done to help them, those likely to recover even if not assisted, and those whose prob-ability of survival would be greatly increased if they were given medical attention. The limited resources available to the doctors were directed primarily at the third category. What would be the implications of adopting a similar attitude toward species preservation?

5. Utilitarian arguments dominate discussions about the importance of preserving the biological richness of the planet. In your opinion, what role should moral arguments play?

Readings

Defenders of Wildlife. 1989. *Preserving Communities and Corridors.* Washington, D.C. A short collection of essays exploring the roles of corridors in preserving wildlife and how we can best implement important conservation legislation.

Flannery, T. 1994. *The Future Eaters*. Reed Books, Melbourne, Australia. A compelling examination of the consequences of unsustainable human activities. Focuses on the Australian region, but its message is applicable worldwide.

Gradwohl, J. and R. Greenberg. 1988. *Saving the Tropical Forests.* Island Press, Washington, D.C. A good account of the causes of tropical forest destruction and of successful projects throughout the world where local communities have averted forest destruction while reaping social and financial benefits.

Lawton, J. H. and R. M. May (Eds.). 1995. *Extinction Rates*. Oxford University Press, Oxford. Provides coverage of the quantitative and qualitative methods of estimating extinction rates and their ecological and evolutionary causes.

Meffe, G. K. and C. R. Carroll (Eds.). 1997. *Principles of Conservation Biology,* 2nd Edition. Sinauer Associates, Sunderland, MA. A multiauthored text that provides an extensive overview of conservation biology.

Perrings, C., K.-G. Mäler, C. Folke, C. S. Holling and B.-O. Jansson. 1995. *Biodiversity Loss: Economic and Ecological Issues*. Cambridge University Press, Cambridge. Reports findings of a research program that brought together economists and ecologists to consider the causes and consequences of biodiversity loss.

Primack, R. B. 1998. *Essentials of Conservation Biology*, 2nd Edition. Sinauer Associates, Sunderland, MA. An introductory text that combines theory with basic and applied research to explain the connections between conservation biology and other disciplines.

Repetto, R. 1990. "Deforestation in the Tropics." *Scientific American,* April. Argues that government policies that encourage exploitation (in particular, excessive logging and clearing for ranches and farms) are largely to blame for the accelerating destruction of tropical forests.

Tattersall, I. 1993. "Madagascar's Lemurs." *Scientific American,* January. Describes how diverse Madagascan habitats, and the endemic lemurs they house, are disappearing fast.

Terborgh, J. 1992. "Why American Songbirds Are Vanishing." *Scientific American,* May. Describes the sharp decline in the number of songbirds in eastern North America and tells why this trend will be difficult to reverse.

Western, D. and M. Pearl (Eds.). 1989. *Conservation for the Twenty-First Century*. Oxford University Press, New York. A set of essays by conservationists, governmental decision makers, and wildlife managers that identify gaps in knowledge and propose agendas for conservation action worldwide.

Wilson, E. O. 1992. *The Diversity of Life*. Belknap Press of Harvard University Press, Cambridge, MA. A readable book that outlines the processes that created the diversity of life, explains the threats to that diversity, and shows what we must do to preserve it.

Glossary

Abdomen (ab' duh mun) [L.: belly] In arthropods, the posterior portion of the body; in mammals, the part of the body containing the intestines and most other internal organs, posterior to the thorax.

Abomasum (ab' oh may' sum) The true stomach of ruminants (animals such as cattle, sheep, and goats).

Abscisic acid (ab sighs' ik) [L. *abscissio*: breaking off] A plant growth substance having growth-inhibiting action. Causes stomata to close.

Abscission (ab sizh' un) [L. *abscissio*: breaking off] The process by which leaves, petals, and fruits separate from a plant.

Absolute temperature scale A temperature scale in which the degree is the same size as in the Celsius (centigrade) scale, and zero is the state of no molecular motion. Absolute zero is −273° on the Celsius scale.

Absorption (1) Of light: complete retention, without reflection or transmission. (2) Of liquids: soaking up (taking in through pores or cracks).

Absorption spectrum A graph of light absorption versus wavelength of light; shows how much light is absorbed at each wavelength.

Abyssal zone (uh biss' ul) [Gr. *abyssos*: bottomless] That portion of the deep ocean where no light penetrates.

Accessory pigments Pigments that absorb light and transfer energy to chlorophylls for photosynthesis.

Acclimatization Changes in an organism that improve its ability to tolerate seasonal changes in its environment.

Acellular Not composed of cells.

Acetylcholine A neurotransmitter substance that carries information across vertebrate neuromuscular junctions and some other synapses. **Acetylcholinesterase** is an enzyme that breaks down acetylcholine.

Acetyl CoA (acetyl coenzyme A) Compound that reacts with oxaloacetate to produce citrate at the beginning of the citric acid cycle; a key metabolic intermediate in the formation of many compounds.

Acid [L. *acidus*: sharp, sour] A substance that can release a proton. (Contrast with base.)

Acid precipitation Precipitation that has a lower pH than normal as a result of acid-forming precursors introduced into the atmosphere by human activities.

Acidic Having a pH of less than 7.0 (a hydrogen ion concentration greater than 10^{-7} molar).

Acoelomate Lacking a coelom.

Acquired Immune Deficiency Syndrome See AIDS.

Acrosome (a' krow soam) [Gr. *akros*: highest or outermost + *soma*: body] The structure at the forward tip of an animal sperm which is the first to fuse with the egg membrane and enter the egg cell.

ACTH (adrenocorticotropin) A pituitary hormone that stimulates the adrenal cortex.

Actin [Gr. *aktis*: a ray] One of the two major proteins of muscle; it makes up the thin filaments. Forms the microfilaments found in most eukaryotic cells.

Action potential An impulse in a neuron taking the form of a wave of depolarization or hyperpolarization imposed on a polarized cell surface.

Action spectrum A graph of biological activity versus wavelength of light. It compares the effectiveness of light of different wavelengths.

Activating enzymes (also called aminoacyl-tRNA synthetases) These enzymes catalyze the addition of amino acids to their appropriate tRNAs.

Activation energy (E_a) The energy barrier that blocks the tendency for a set of chemical substances to react. A reaction is speeded up if this energy barrier is surmounted by adding heat, or if the barrier is lowered by providing a different reaction pathway with the aid of a catalyst.

Active site The region on the surface of an enzyme where the substrate binds, and where catalysis occurs.

Active transport The transport of a substance across a biological membrane against a concentration gradient—that is, from a region of low concentration (of that substance) to a region of high concentration. Active transport requires the expenditure of energy and is a saturable process. (Contrast with facilitated diffusion, free diffusion; see primary active transport, secondary active transport.)

Adaptation (a dap tay' shun) In evolutionary biology, a particular structure, physiological process, or behavior that makes an organism better able to survive and reproduce. Also, the evolutionary process that leads to the development or persistence of such a trait.

Adenine (a' den een) A nitrogen-containing base found in nucleic acids, ATP, NAD, etc.

Adenosine triphosphate See ATP.

Adenylate cyclase Enzyme catalyzing the formation of cyclic AMP from ATP.

Adhesion molecules See cell adhesion molecules.

Adrenal (a dree' nal) [L. *ad-*: toward + *renes*: kidneys] An endocrine gland located near the kidneys of vertebrates, consisting of two glandular parts, the cortex and medulla.

Adrenaline See epinephrine.

Adrenocorticotropin See ACTH.

Adsorption Binding of a gas or a solute to the surface of a solid.

Aerenchyma (air eng' kyma) [Gr. *aer*: air + *enchyma*: infusion] Modified parenchyma tissue, with many air spaces, found in shoots of some aquatic plants. (See parenchyma.)

Aerobic (air oh' bic) [Gr. *aer*: air + *bios*: life] In the presence of oxygen, or requiring oxygen.

Afferent (af' ur unt) [L. *ad*: to + *ferre*: to bear] To or toward, as in a neuron that carries impulses to the central nervous system, or a blood vessel that carries blood to a structure. (Contrast with efferents.)

Age distribution The proportion of individuals in a population belonging to each of the age categories into which the population has been divided. The number of divisions is arbitrary.

AIDS (Acquired immune deficiency syndrome) Condition in which the body's helper T lymphocytes are destroyed, leaving the victim subject to opportunistic diseases. Caused by the HIV-I virus.

Air sacs Structures in the avian respiratory system that facilitate unidirectional flow of air through the lungs.

Alcohol An organic compound with one or more hydroxyl (–OH) groups.

Aldehyde (al' duh hide)   A compound with a –CHO functional group. Many sugars are aldehydes. (Contrast with ketone.)

Aldosterone (al dahs' ter own)   A steroid hormone produced in the adrenal cortex of mammals. Promotes secretion of potassium and reabsorption of sodium in the kidney.

Aleurone layer (al' yur own) [Gr. *aleuron*: wheat flour]   In grass seeds, a specialized cell layer just between the seed coat and the endosperm, synthesizing hydrolytic enzymes under the influence of gibberellin, and thus helping mobilize reserves for the developing embryo.

Alga (al' gah) (plural: algae) [L.: seaweed]   Any one of a wide diversity of protists belonging to the phyla Pyrrophyta, Chrysophyta, Phaeophyta, Rhodophyta, and Chlorophyta (and, formerly, Cyanophyta—"blue-green algae"). Most live in the water, where they are the dominant autotrophs; most are unicellular, but a minority are multicellular ("seaweeds" and similar protists).

Allele (a leel') [Gr. *allos*: other]   The alternate forms of a genetic character found at a given locus on a chromosome.

Allele frequency   The relative proportion of a particular allele in a specific population.

Allergy [Ger. *allergie*: altered reaction]   An overreaction to an antigen in amounts that do not affect most people; often involves IgE antibodies.

Allometric growth   A pattern of growth in which some parts of the body of an organism grow faster than others, resulting in a change in body proportions as the organism grows.

Allopatric (al' lo pat' rick) [Gr. *allos*: other + *patria*: fatherland]   Pertaining to populations that occur in different places.

Allopatric speciation   See geographical speciation.

Allopolyploid   A polyploid in which the chromosome sets are derived from more than one species.

Allostery (al' lo steer' y) [Gr. *allos*: other + *stereos*: structure]   Regulation of the activity of an enzyme by binding, at a site other than the catalytic active site, of an effector molecule that does not have the same structure as any of the enzyme's substrates.

Alpha helix   Type of protein secondary structure; a right-handed spiral.

Alternation of generations   The succession of haploid and diploid phases in a sexually reproducing organism. In most animals (male wasps and honey bees are notable exceptions), the haploid phase consists only of the gametes. In fungi, algae, and plants, however, the haploid phase may be the more prominent phase (as in fungi and mosses) or may be as prominent as the diploid phase (see the life cycle of *Ulva*, for example). In vascular plants, the diploid phase is more prominent.

Altruistic act   A behavior whose performance harms the actor but benefits other individuals.

Alveolus (al ve' o lus) (plural: alveoli) [L. *alveus*: cavity]   A small, baglike cavity, especially the blind sacs of the lung.

Amensalism (a men' sul ism)   Interaction in which one animal is harmed and the other is unaffected. (Contrast with commensalism, mutualism.)

Ames test   A test for mutagens (and possible carcinogens) based on the ability of a test compound to cause mutations in the bacterium, *Salmonella*.

Amine   An organic compound with an amino group (see Amino acid).

Amino acid   An organic compound of the general formula $H_2N–CHR–COOH$, where R can be one of 20 or more different side groups. An amino acid is so named because it has both a basic amine group, $–NH_2$, and an acidic carboxyl group, $–COOH$. Proteins are polymers of amino acids.

Ammonotelic (am moan' o teel' ic) [Gr. *telos*: end]   Describes an organism in which the final product of breakdown of nitrogen-containing compounds (primarily proteins) is ammonia. (Contrast with ureotelic, uricotelic.)

Amniocentesis   A medical procedure in which cells from the fetus are obtained from the amniotic fluid. The genetic material of the cells is then examined. (Contrast with chorionic villus sampling.)

Amniotic egg   The eggs of birds and reptiles, which can be incubated in air because the embryo is enclosed by a fluid-filled sac.

Amoeba (a mee' bah) [Gr. *amoibe*: change]   Any one of a large number of different kinds of unicellular protists belonging to the phylum Rhizopoda, characterized among other features by its ability to change shape frequently through the protrusion and retraction of cytoplasmic extensions called pseudopods.

Amoeboid (a mee' boid)   Like an amoeba; constantly changing shape by the protrusion and retraction of pseudopodia.

Amphi- [Gr.: both]   Prefix used to denote a character or kind of organism that occupies two or more states. For example, amphibian (an animal that lives both on the land and in the water).

Amphipathic (am' fi path' ic) [Gr. *amphi*: both + *pathos*: emotion]   Of a molecule, having both hydrophilic and hydrophobic regions.

amu (atomic mass unit, or **dalton)**   The basic unit of mass on an atomic scale, defined as one-twelfth the mass of a carbon-12 atom. There are 6.023×10^{23} amu in one gram. This number is known as Avogadro's number.

Amylase (am' ill ase)   Any of a group of enzymes that digest starch.

Anabolism (an ab' uh liz' em) [Gr. *ana*: up, throughout + *ballein*: to throw]   Synthetic reactions of metabolism, in which complex molecules are formed from simpler ones. (Contrast with catabolism.)

Anaerobic (an ur row' bic) [Gr. *an*: not + *aer*: air + *bios*: life]   Occurring without the use of molecular oxygen, O_2.

Anagenesis   Evolutionary change in a single lineage over time.

Analogy (a nal' o jee) [Gr. *analogia*: resembling]   A resemblance in function, and often appearance as well, between two structures which is due to convergence in evolution rather than to common ancestry. (Contrast with homology.)

Anaphase (an' a phase) [Gr. *ana*: indicating upward progress]   The stage in nuclear division at which the first separation of sister chromatids (or, in the first meiotic division, of paired homologues) occurs. Anaphase lasts from the moment of first separation to the time at which the moving chromosomes converge at the poles of the spindle.

Anaphylactic shock   A precipitous drop in blood pressure caused by loss of fluid from capillaries because of an increase in their permeability stimulated by an allergic reaction.

Ancestral trait   Trait shared by a group of organisms as a result of descent from a common ancestor.

Androgens (an' dro jens)   The male sex steroids.

Aneuploidy (an' you ploy dee)   A condition in which one or more chromosomes or pieces of chromosomes are either lacking or present in excess.

Angiosperm (an' jee oh spurm) [Gr. *angion*: vessel + *sperma*: seed]   One of the flowering plants; literally, one whose seed is carried in a "vessel," which is the fruit. (See fruit.)

Angiotensin (an' jee oh ten' sin)   A peptide hormone that raises blood pressure by causing peripheral vessels to constrict; maintains glomerular filtration by constricting efferent glomerular vessels; stimulates thirst; and stimulates the release of aldosterone.

Animal [L. *animus*: breath, soul]   A member of the kingdom Animalia. In general, a multicellular eukaryote that obtains its food by ingestion.

Animal hemisphere   The metabolically active upper portion of some animal eggs, zygotes, and embryos, which does *not* contain the dense nutrient yolk. The **animal pole** refers to the very top of the egg or embryo. (Contrast with vegetal hemisphere.)

Anion (an' eye one)   An ion with one or more negative charges. (Contrast with cation.)

Anisogamy (an' eye sog' a mee) [Gr. *aniso*: unequal + *gamos*: marriage]   The existence of two dissimilar gametes (egg and sperm).

Annual   Referring to a plant whose life cycle is completed in one growing season. (Contrast with biennial, perennial.)

Anorexia nervosa (an or ex' ee ah) [Gr. *an*: not + *orexis*: appetite]   Severe malnutrition and body wasting brought on by a psychological aversion to food.

Antennapedia complex A group of homeotic genes that control the development of the head and anterior thorax of the fruit fly, *Drosophila melanogaster*.

Anterior Toward the front.

Anterior pituitary The portion of the vertebrate pituitary gland that derives from gut epithelium and produces tropic hormones.

Anther (an' thur) [Gr. *anthos*: flower] A pollen-bearing portion of the stamen of a flower.

Antheridium (an' thur id' ee um) (plural: antheridia) [Gr. *antheros*: blooming] The multicellular structure that produces the sperm in bryophytes and ferns.

Antibody One of millions of blood proteins, produced by the immune system, that specifically recognizes a foreign substance and initiates its removal from the body.

Anticodon A "triplet" of three nucleotides in transfer RNA that is able to pair with a complementary triplet (a codon) in messenger RNA, thus aligning the transfer RNA on the proper place on the messenger. The codon (and, reciprocally, the anticodon) codes for a specific amino acid.

Antidiuretic hormone A hormone that controls water reabsorption in the mammalian kidney. Also called vasopressin.

Antigen (an' ti jun) Any substance that stimulates the production of an antibody or antibodies upon introduction into the body of a vertebrate.

Antigen processing The breakdown of antigenic proteins into smaller fragments, which are then presented on the cell surface, along with MHC proteins, to T cells.

Antigenic determinant A specific region of an antigen, which is recognized by and binds to a specific antibody.

Antiparallel Parallel but running in opposite directions. The two strands of DNA are antiparallel.

Antipodals (an tip' o dulls) [Gr. *anti*: against + *podus*: foot] Cells (usually three) of the mature embryo sac of a flowering plant, located at the end opposite the egg (and micropyle).

Antiport A membrane transport process that carries one substance in one direction and another in the opposite direction. (Contrast with symport.)

Antisense nucleic acid A single-stranded RNA or DNA complementary to and thus targeted against the mRNA transcribed from a harmful gene such as an oncogene.

Anus (a' nus) Opening through which digestive wastes are expelled, located at the posterior end of the gut.

Aorta (a or' tuh) [Gr. *aorte*: aorta] The main trunk of the arteries leading to the systemic (as opposed to the pulmonary) circulation.

Apex (a' pecks) The tip or highest point of a structure, as the apex of a growing stem or root.

Apical (a' pi kul) Pertaining to the apex, as the apical meristem, which is the actively growing tissue at the tip of a stem or root.

Apomixis (ap oh mix' is) [Gr. *apo*: away from + *mixis*: sexual intercourse] The asexual production of seeds.

Apoplast (ap' oh plast) In plants, the continuous meshwork of cell walls and extracellular spaces through which material can pass without crossing a plasma membrane. (Contrast with symplast.)

Apoptosis (ay' pu toh sis) A series of genetically programmed events leading to cell death.

Appendix A vestigial portion of the human gut at the junction of the ileum with the colon.

Apterous Lacking wings.

Assisted reproductive technologies (ART) Any of a number of technological approaches to improve human fertility. They include in vitro fertilization of eggs, injection of sperm and eggs into the oviduct, and injection of individual sperm into individual eggs.

Archaebacteria (ark' ee bacteria) [Gr. *archaios*: ancient] One of the two kingdoms of prokaryotes; the archaebacteria possess distinctive lipids and lack peptidoglycan. Most live in extreme environments. (Contrast with eubacteria.)

Archegonium (ar' ke go' nee um) [Gr. *archegonos*: first of a kind] The multicellular structure that produces eggs in bryophytes, ferns, and gymnosperms.

Archenteron (ark en' ter on) [Gr. *archos*: beginning + *enteron*: bowel] The earliest primordial animal digestive tract.

Area cladogram A cladogram in which the geographic ranges of the taxa are substituted for the names of the taxa.

Arteriosclerosis See atherosclerosis.

Artery A muscular blood vessel carrying oxygenated blood away from the heart to other parts of the body. (Contrast with vein.)

Artifact [L. *ars, artis*: art + *facere*: to make] Something made by human effort or intervention. In biology, something that was not present in the living cell or organism, but was unintentionally produced by an experimental procedure.

Ascospore (ass' ko spor) A fungus spore produced within an ascus.

Ascus (ass' cuss) [Gr. *askos*: bladder] In fungi belonging to the class Ascomycetes (sac fungi), the club-shaped sporangium within which spores are produced by meiosis.

Asexual Without sex.

Associative learning "Pavlovian" learning, in which an animal comes to associate a previously neutral stimulus (such as the ringing of a bell) with a particular reward or punishment.

Assortative mating A breeding system under which mates are selected on the basis of a particular trait or group of traits.

Assortment (genetic) The random separation during meiosis of nonhomologous chromosomes and of genes carried on nonhomologous chromosomes. For example, if genes *A* and *B* are borne on nonhomologous chromosomes, meiosis of diploid cells of genotype *AaBb* will produce haploid cells of the following types in equal numbers: *AB, Ab, aB,* and *ab*.

Asymmetric The state of lacking any plane of symmetry.

Asymmetric carbon atom In a molecule, a carbon atom to which four different atoms or groups are bound.

Atherosclerosis (ath' er oh sklair oh' sis) A disease of the lining of the arteries characterized by fatty, cholesterol-rich deposits in the walls of the arteries. When fibroblasts infiltrate these deposits and calcium precipitates in them, the disease become arteriosclerosis, or "hardening of the arteries."

Atmosphere The gaseous mass surrounding our planet. Also: a unit of pressure, equal to the normal pressure of air at sea level.

Atom [Gr. *atomos*: indivisible] The smallest unit of a chemical element. Consists of a nucleus and one or more electrons.

Atomic mass (also called atomic weight) The average mass of an atom of an element on the amu scale. (The average depends upon the relative amounts of different isotopes of an element on Earth.)

Atomic mass unit See amu.

Atomic number The number of protons in the nucleus of an atom, also equal to the number of electrons around the neutral atom. Determines the chemical properties of the atom.

ATP (adenosine triphosphate) A compound containing adenine, ribose, and three phosphate groups. When it is formed, useful energy is stored; when it is broken down (to ADP or AMP), energy is released to drive endergonic reactions. ATP is a universal energy storage compound.

ATP synthase An integral membrane protein that couples the transport of proteins with the formation of ATP.

Atrium (a' tree um) A body cavity, as in the hearts of vertebrates. The thin-walled chamber(s) entered by blood on its way to the ventricle(s). Also, the outer ear.

Autocatalysis An enzymatic reaction in which the inactive form of an enzyme is converted into its active form by the enzyme itself.

Autoimmune disease A disorder in which the immune system attacks the animal's own body.

Autonomic nervous system The system (which in vertebrates comprises sympathetic and parasympathetic subsystems) that controls such involuntary functions as those of guts and glands.

Autopolyploid A polyploid in which the sets of chromosomes are derived from the same species.

Autoradiography The detection of a radioactive substance in a cell or organism by putting it in contact with a photographic emulsion and allowing the material to "take its own picture." The emulsion is developed, and the location of the radioactivity in the cell is seen by the presence of silver grains in the emulsion.

Autoregulatory mechanism A feedback mechanism that enables a structure to regulate its own function.

Autosome Any chromosome (in a eukaryote) other than a sex chromosome.

Autotroph (au′ tow trow′ fik) [Gr. *autos*: self + *trophe*: food] An organism that is capable of living exclusively on inorganic materials, water, and some energy source such as sunlight or chemically reduced matter. (Contrast with heterotroph.)

Auxin (awk′ sin) [Gr. *auxein*: increase] In plants, a substance (indoleacetic acid) that regulates growth and various aspects of development.

Auxotroph (awks′ o trofe) [Gr. *auxanein*: to grow + *trophe*: food] A mutant form of an organism that requires a nutrient or nutrients not required by the wild type, or reference, form of the organism. (Contrast with prototroph.)

Avogadro's number The conversion factor between atomic mass units and grams. More usefully, the number of atoms in that quantity of an element which, expressed in grams, is numerically equal to the atomic weight in amu; 6.023×10^{23} atoms. (See mole.)

Axon [Gr.: axle] Fiber of a neuron which can carry action potentials. Carries impulses away from the cell body of the neuron; releases a neurotransmitter substance.

Axon hillock The junction between an axon and its cell body; where action potentials are generated.

Axon terminals The endings of an axon; they form synapses and release neurotransmitter.

Axoneme (ax′ oh neem) The complex of microtubules and their crossbridges that forms the motile apparatus of a cilium.

Bacillus (buh sil′ us) [L.: little rod] Any of various rod-shaped bacteria.

Bacteriophage (bak teer′ ee o fayj) [Gr. *bakterion*: little rod + *phagein*: to eat] One of a group of viruses that infect bacteria and ultimately cause their disintegration.

Bacterium (bak teer′ ee um) (plural: bacteria) [Gr. *bakterion*: little rod] A prokaryote. An organism with chromosomes not contained in nuclear envelopes.

Balanced polymorphism [Gr. *polymorphos*: having many forms] The maintenance of more than one form, or the maintenance at a given locus of more than one allele, at frequencies of greater than one percent in a population. Often results when heterozygotes are superior to both homozygotes.

Baroreceptor [Gr. *baros*: weight] A pressure-sensing cell or organ.

Barr body In mammals, an inactivated X chromosome.

Basal body Centriole found at the base of a eukaryotic flagellum or cilium.

Basal metabolic rate The minimum rate of energy turnover in an awake (but resting) bird or mammal that is not expending energy for thermoregulation.

Base A substance which can accept a proton (H^+). (Contrast with acid.) In nucleic acids, a nitrogen-containing base (purine or pyrimidine) is attached to each sugar in the backbone.

Base pairing See complementary base pairing.

Basic Having a pH greater than 7.0 (having a hydrogen ion concentration lower than 10^{-7} molar).

Basidium (bass id′ ee yum) In fungi of the class Basidiomycetes, the characteristic sporangium in which four spores are formed by meiosis and then borne externally before being shed.

Batesian mimicry Mimicry by a relatively harmless kind of organism of a more dangerous one, by which the mimic enjoys protection from predators that mistake it for the dangerous model. (Contrast with Müllerian mimicry.)

B cell A type of lymphocyte involved in the humoral immune response of vertebrates. Upon recognizing an antigenic determinant, a B cell develops into a plasma cell, which secretes an antibody. (Contrast with a T cell.)

Benefit An improvement in survival and reproductive success resulting from a behavior. (Contrast with cost.)

Benign (be nine′) A tumor that grows to a certain size and then stops, uaually with a fibrous capsule surrounding the mass of cells. Benign tumors do not spread (metastasize) to other organs.

Benthic zone [Gr. *benthos*: bottom of the sea] The bottom of the ocean. (Contrast with pelagic zone.)

Beta-pleated sheet Type of protein secondary structure; results from hydrogen bonding between polypeptide regions running antiparallel to each other.

Biennial Referring to a plant whose life cycle includes vegetative growth in the first year and flowering and senescence in the second year. (Contrast with annual, perennial.)

Bilateral symmetry The condition in which only the right and left sides of an organism, divided exactly down the back, are mirror images of each other. (Contrast with radial symmetry.)

Bile A secretion of the liver delivered to the small intestine via the common bile duct. In the intestine, bile emulsifies fats.

Binocular cells Neurons in the visual cortex that respond to input from both retinas; involved in depth perception.

Binomial (bye nome′ ee al) Consisting of two names; for example, the binomial nomenclature of biology which gives the name of the genus followed by the name of the species.

Bioaccumulation The ever-increasing concentration of a chemical compound in the tissues of an organism as it is passed up the food chain.

Biodiversity crisis The current high rate of loss of species, caused primarily by human activities.

Biogenesis [Gr. *bios*: life + *genesis*: source] The origin of living things from other living things.

Biogeochemical cycles Movement of elements through living organisms and the physical environment.

Biogeography The scientific study of the geographic distribution of organisms. Ecological biogeography is concerned with the habitats in which organisms live, historical biogeography with the complete geographic ranges of organisms and the historical circumstances that determine the ranges.

Biological species concept The view that a species is most usefully defined as a population or series of populations within which there is a significant amount of gene flow under natural conditions, but which is genetically isolated from other populations.

Bioluminescence The production of light by biochemical processes in an organism.

Biomass The total weight of all the living organisms, or some designated group of living organisms, in a given area.

Biome (bye′ ome) A major division of the ecological communities of Earth; characterized by distinctive vegetation.

Biota (bye oh′ tah) All of the organisms, including animals, plants, fungi, and microorganisms, found in a given area.

Biotechnology The use of cells to make medicines, foods and other products useful to humans.

Biradial symmetry Radial symmetry modified so that only two planes can divide the animal into similar halves.

Bithorax complex A group of homeotic genes that control the development of the abdomen and posterior thorax of *Drosophila melanogaster*.

Blastocoel (blass′ toe seal) [Br. *blastos*: sprout + *koilos*: hollow] The central, hollow cavity of a blastula.

Blastodisc (blass′ toe disk) A disk of cells forming on the surface of a large yolk mass, comparable to a blastula, but occurring in animals such as birds and reptiles, in which the massive yolk restricts cleavage to one side of the egg only.

Blastomere A cell produced by the division of a fertilized egg.

Blastopore The opening from the archenteron to the exterior of a gastrula.

Blastula (blass′ chu luh) [Gr. *blastos*: sprout] An early stage in animal embryology; in many species, a hollow sphere of cells surrounding a central cavity, the blastocoel. (Contrast with blastodisc.)

Blood–brain barrier A property of the blood vessels of the brain that prevents most chemicals from diffusing from the blood into the brain.

Bohr effect (boar) The reduction in affinity of hemoglobin for oxygen caused by acidic conditions, usually as a result of increased CO_2.

Bottleneck A combination of environmental conditions that causes a serious reduction in the size of the population.

Bowman's capsule An elaboration of kidney tubule cells that surrounds a knot of capillaries (the glomerulus). Blood is filtered across the walls of these capillaries and the filtrate is collected into Bowman's capsule.

Brain stem The portion of the vertebrate brain between the spinal cord and the forebrain.

Bronchus (plural: bronchi) The major airway(s) branching off the trachea into the vertebrate lung.

Brown fat Fat tissue in mammals that is specialized to produce heat. It has many mitochondria and capillaries, and a protein that uncouples oxidative phosphorylation. It is frequently found in newborns, in mammals acclimated to cold, and in hibernators.

Browser An animal that feeds on the tissues of woody plants.

Bryophyte (bri' uh fite') [Gr. *bruon*: moss + *phyton*: plant] Any nonvascular plant, including mosses, liverworts, and hornworts.

Bud primordium [L. *primordium*: the beginning] In plants, a small mass of potentially meristematic tissue found in the angle between the leaf stalk and the shoot apex. Will give rise to a lateral branch under appropriate conditions.

Budding Asexual reproduction in which a more or less complete new organism simply grows from the body of the parent organism and eventually detaches itself.

Buffering A process by which a system resists change—particularly in pH, in which case added acid or base is partially converted to another form.

Bulb In plants, an underground storage organ composed principally of enlarged and fleshy leaf bases.

Bundle sheath In C$_4$ plants, a layer of photosynthetic cells between the mesophyll and a vascular bundle of a leaf.

C$_3$ photosynthesis The form of photosynthesis in which 3-phosphoglycerate is the first stable product, and ribulose bisphosphate is the CO$_2$ receptor.

C$_4$ photosynthesis The form of photosynthesis in which oxaloacetate is the first stable product, and phosphoenolpyruvate is the CO$_2$ acceptor. C$_4$ plants also perform the reactions of C$_3$ photosynthesis.

Calcitonin A hormone produced by the thyroid gland; it lowers blood calcium and promotes bone formation. (Contrast with parathormone.)

Calmodulin (cal mod' joo lin) A calcium-binding protein found in all animal and plant cells; mediates many calcium-regulated processes.

calorie [L. *calor*: heat] The amount of heat required to raise the temperature of one gram of water by one degree Celsius (1°C) from 14.5°C to 15.5°C. In nutrition studies, "Calorie" (spelled with a capital C) refers to the kilocalorie (1 kcal = 1,000 cal), the amount of heat required to raise the temperature of one kilogram of water by 1°C.

Calvin–Benson cycle The stage of photosynthesis in which CO$_2$ reacts with RuBP to form 3PG, 3PG is reduced to a sugar, and RuBP is regenerated, while other products are released to the rest of the plant.

Calyptra (kuh lip' tra) [Gr. *kalyptra*: covering for the head] A hood or cap found partially covering the apex of the sporophyte capsule in many moss species, formed from the expanded wall and neck of the archegonium.

Calyx (kay' licks) [Gr. *kalyx*: cup] All of the sepals of a flower, collectively.

CAM See crassulacean acid metabolism.

Cambium (kam' bee um) [L. *cambiare*: to exchange] A meristem that gives rise to radial rows of cells in stem and root, increasing them in girth; commonly applied to the vascular cambium which produces wood and phloem, and the cork cambium, which produces bark.

cAMP (cyclic AMP) A compound, formed from ATP, that mediates the effects of numerous animal hormones. Also needed for the transcription of catabolite-repressible operons in bacteria. Used for communication by cellular slime molds.

Canopy The leaf-bearing part of a tree. Collectively the aggregate of the leaves and branches of the larger woody plants of an ecological community.

Capacitance vessels Refers to veins because of their variable capacity to hold blood.

Capillaries [L. *capillaris*: hair] Very small tubes, especially the smallest blood-carrying vessels of animals between the termination of the arteries and the beginnings of the veins.

Capping In eukaryote RNA processing, the addition of a modified G at the 5' end of the molecule.

Capsid The protein coat of a virus.

Capsule In bryophytes, the spore case. In some bacteria, a gelatinous layer exterior to the cell wall.

Carbohydrates Organic compounds with the general formula $C_nH_{2m}O_m$. Common examples are sugars, starch, and cellulose.

Carboxylic acid (kar box sill' ik) An organic acid containing the carboxyl group, –COOH, which dissociates to the carboxylate ion, –COO$^-$.

Carcinogen (car sin' oh jen) A substance that causes cancer.

Cardiac (kar' dee ak) [Gr. *kardia*: heart] Pertaining to the heart and its functions.

Carnivore [L. *carn*: flesh + *vovare*: to devour] An organism that feeds on animal tissue. (Contrast with detritivore, herbivore, omnivore.)

Carotenoid (ka rah' tuh noid) [L. *carota*: carrot] A yellow, orange, or red lipid pigment commonly found as an accessory pigment in photosynthesis; also found in fungi.

Carpel (kar' pel) [Gr. *karpos*: fruit] The organ of the flower that contains one or more ovules.

Carrier In facilitated diffusion, a membrane protein that binds a specific molecule and transports it through the membrane. In genetics, a person heterozygous for a recessive trait. In respiratory and photosynthetic electron transport, a participating substance such as NAD that exists in both oxidized and reduced forms.

Carrying capacity In ecology, the largest number of organisms of a particular species that can be maintained indefinitely in a given part of the environment.

Cartilage In vertebrates, a tough connective tissue found in joints, the outer ear, and elsewhere. Forms the entire skeleton in some animal groups.

Casparian strip A band of cell wall containing suberin and lignin, found in the endodermis. Restricts the movement of water across the endodermis.

Catabolism [Ge. *kata*: down + *ballein*: to throw] Degradational reactions of metabolism, in which complex molecules are broken down. (Contrast with anabolism.)

Catabolite repression The decreased synthesis of many enzymes that tend to provide glucose for a cell; caused by the presence of excellent carbon sources, particularly glucose.

Catalyst (cat' a list) [Gr. *kata*-, implying the breaking down of a compound] A chemical substance that accelerates a reaction without itself being consumed in the overall course of the reaction. Catalysts lower the activation energy of a reaction. Enzymes are biological catalysts.

Cation (cat' eye on) An ion with one or more positive charges. (Contrast with anion.)

Caudal [L. *cauda*: tail] Pertaining to the tail, or to the posterior part of the body.

cDNA See complementary DNA.

Cecum (see' cum) [L. *caecus*: blind] A blind branch off the large intestine. In many nonruminant mammals, the cecum contains a colony of microorganisms that contribute to the digestion of food.

Cell adhesion molecules Molecules on animal cell surfaces that affect the selective association of cells during development of the embryo.

Cell cycle The stages through which a cell passes between one division and the next. Includes all stages of interphase and mitosis.

Cell division The reproduction of a cell to produce two new cells. In eukaryotes, this process involves nuclear division (mitosis) and cytoplasmic division (cytokinesis).

Cell theory The theory, well established, that organisms consist of cells, and that all cells come from preexisting cells.

Cell wall A relatively rigid structure that encloses cells of plants, fungi, many protists, and most bacteria. The cell wall gives these cells their shape and limits their expansion in hypotonic media.

Cellular immune system That part of the immune system that is based on the activities of T cells. Directed against parasites, fungi, intracellular viruses, and foreign tissues (grafts). (Contrast with humoral immune system.)

Cellular respiration   See respiration.

Cellulose (sell′ you lowss)   A straight-chain polymer of glucose molecules, used by plants as a structural supporting material. **Cellulase** is an enzyme that hydrolyzes cellulose.

Central dogma   The statement that information flows from DNA to RNA to polypeptide (in retroviruses, there is also information flow from RNA to cDNA).

Central nervous system   That part of the nervous system which is condensed and centrally located, e.g., the brain and spinal cord of vertebrates; the chain of cerebral, thoracic and abdominal ganglia of arthropods.

Centrifuge [L. *fugere*: to flee]   A device in which a sample can be spun around a central axis at high speed, creating a centrifugal force that mimics a very strong gravitational force. Used to separate mixtures of suspended materials.

Centriole (sen′ tree ole)   A paired organelle that helps organize the microtubules in animal and protist cells during nuclear division.

Centromere (sen′ tro meer) [Gr. *centron*: center + *meros*: part]   The region where sister chromatids join.

Centrosome (sen′ tro soam)   The major microtubule organizing center of an animal cell.

Cephalization (sef′ uh luh zay′ shun) [Gr. *kephale*: head]   The evolutionary trend toward increasing concentration of brain and sensory organs at the anterior end of the animal.

Cerebellum (sair′ uh bell′ um) [L.: diminutive of *cerebrum*: brain]   The brain region that controls muscular coordination; located at the anterior end of the hindbrain.

Cerebral cortex   The thin layer of gray matter (neuronal cell bodies) that overlays the cerebrum.

Cerebrum (su ree′ brum) [L.: brain]   The dorsal anterior portion of the forebrain, making up the largest part of the brain of mammals. In mammals, the chief coordination center of the nervous system; consists of two **cerebral hemispheres**.

Cervix (sir′ vix) [L.: neck]   The opening of the uterus into the vagina.

cGMP (cyclic guanosine monophosphate)   An intracellular messenger that is part of signal transmission pathways involving G proteins. (See G protein.)

Channel   A membrane protein that forms an aqueous passageway though which specific solutes may pass by simple diffusion; some channels are gated: they open and close in response to binding of specific molecules.

Chaperone protein   A protein that assists a newly forming protein in adopting its appropriate tertiary structure.

Chemical bond   An attractive force stably linking two atoms.

Chemiosmotic mechanism   According to this model, ATP formation in mitochondria and chloroplasts results from a pumping of protons across a membrane (against a gradient of electrical charge and of pH), followed by the return of the protons through a protein channel with ATPase activity.

Chemoautotroph   An organism that uses carbon dioxide as a carbon source and obtains energy by oxidizing inorganic substances from its environment. (Contrast with chemoheterotroph, photoautotroph, photoheterotroph.)

Chemoheterotroph   An organism that must obtain both carbon and energy from organic substances. (Contrast with chemoautotroph, photoautotroph, photoheterotroph.)

Chemosensor   A cell or tissue that senses specific substances in its environment.

Chemosynthesis   Synthesis of food substances, using the oxidation of reduced materials from the environment as a source of energy.

Chiasma (kie az′ muh) (plural: chiasmata) [Gr.: cross]   An X-shaped connection between paired homologous chromosomes in prophase I of meiosis. A chiasma is the visible manifestation of crossing over between homologous chromosomes.

Chitin (kye′ tin) [Gr. *chiton*: tunic]   The characteristic tough but flexible organic component of the exoskeleton of arthropods, consisting of a complex, nitrogen-containing polysaccharide. Also found in cell walls of fungi.

Chlorophyll (klor′ o fill) [Gr. *chloros*: green + *phyllon*: leaf]   Any of a few green pigments associated with chloroplasts or with certain bacterial membranes; responsible for trapping light energy for photosynthesis.

Chloroplast [Gr. *chloros*: green + *plast*: a particle]   An organelle bounded by a double membrane containing the enzymes and pigments that perform photosynthesis. Chloroplasts occur only in eukaryotes.

Choanocyte (cho′ an oh cite)   The collared, flagellated feeding cells of sponges.

Cholecystokinin (ko′ lee sis to kai nin)   A hormone produced and released by the lining of the duodenum when it is stimulated by undigested fats and proteins. It stimulates the gallbladder to release bile and slows stomach activity.

Chorion (kor′ ee on) [Gr. *khorion*: afterbirth]   The outermost of the membranes protecting mammal, bird, and reptile embryos; in mammals it forms part of the placenta.

Chorionic villus sampling   A medical procedure that extracts a portion of the chorion from a pregnant woman to enable genetic and biochemical analysis of the embryo. (Contrast with amniocentesis.)

Chromatid (kro′ ma tid)   Each of a pair of new sister chromosomes from the time at which the molecular duplication occurs until the time at which the centromeres separate at the anaphase of nuclear division.

Chromatin   The nucleic acid–protein complex found in eukaryotic chromosomes.

Chromatography   Any one of several techniques for the separation of chemical substances, based on differing relative tendencies of the substances to associate with a mobile phase or a stationary phase.

Chromatophore (krow mat′ o for) [Gr. *chroma*: color + *phoreus*: carrier]   A pigment-bearing cell that expands or contracts to change the color of the organism.

Chromosomal aberration   Any large change in the structure of a chromosome, including duplication or loss of chromosomes or parts thereof, usually gross enough to be detected with the light microscope.

Chromosome (krome′ o sowm) [Gr. *chroma*: color = *soma*: body]   In bacteria and viruses, the DNA molecule that contains most or all of the genetic information of the cell or virus. In eukaryotes, a structure composed of DNA and proteins that bears part of the genetic information of the cell.

Chromosome walking   A technique based on recognition of overlapping fragments; used as a step in DNA sequencing.

Chylomicron (ky low my′ cron)   Particles of lipid coated with protein, produced in the gut from dietary fats and secreted into the extracellular fluids.

Chyme (kime) [Gr. *chymus*, juice]   Created in the stomach; a mixture of ingested food with the digestive juices secreted by the salivary glands and the stomach lining.

Ciliate (sil′ ee ate)   A member of the protist phylum Ciliophora, unicellular organisms that propel themselves by means of cilia.

Cilium (sil′ ee um) (plural: cilia) [L. *cilium*: eyelash]   Hairlike organelle used for locomotion by many unicellular organisms and for moving water and mucus by many multicellular organisms. Generally shorter than a flagellum.

Circadian rhythm (sir kade′ ee an) [L. *circa*: approximately + *dies*: day]   A rhythm in behavior, growth, or some other activity that recurs about every 24 hours under constant conditions.

Circannual rhythm (sir can′ you al) [L. *circa*: approximately + *annus*: year]   A rhythm of behavior, growth, or some other activity that recurs on a yearly basis.

Citric acid cycle   A set of chemical reactions in cellular respiration, in which acetyl CoA reacts with oxaloacetate to form citric acid, and oxaloacetate is regenerated. Acetyl CoA is oxidized to carbon dioxide, and hydrogen atoms are stored as NADH and FADH$_2$.

Clade (clayd) [Gr. *klados*: branch]   All of the organisms, both living and fossil, descended from a particular common ancestor.

Cladistic classification   A classification based entirely on the phylogenetic relationships among organisms.

Cladogenesis (clay doh jen′ e sis) [Gr. *klados*: branch + *genesis*: source]   The formation of new species by the splitting of an evolutionary lineage.

Cladogram   The graphic representation of a clade.

Class   In taxonomy, the category below the phylum and above the order; a group of related, similar orders.

Class I MHC molecules These cell surface proteins participate in the cellular immune response directed against virus-infected cells.

Class II MHC molecules These cell surface proteins participate in the cell-cell interactions (of helper T cells, macrophages, and B cells) of the humoral immune response.

Class II MHC molecules These proteins do not present processed antigen, as do classes I and II, but instead include some proteins of the complement system and certain cytokines.

Class switching The process whereby a plasma cell changes the class of immunoglobulin that it synthesizes. This results from the deletion of part of the constant region of DNA, bringing in a new C segment. The variable region is the same as before, so that the new immunoglobulin has the same antigenic specificity.

Clathrin A fibrous protein on the inner surfaces of animal cell membranes that strengthens coated vesicles and thus participates in receptor-mediated endocytosis.

Clay A soil constituent comprising particles smaller than 2 micrometers in diameter.

Cleavages First divisions of the fertilized egg of an animal.

Climax In ecology, a community that terminates a succession and which tends to replace itself unless it is further disturbed or the physical environment changes.

Climograph (clime' o graf) Graph relating temperature and precipitation with time of year.

Cline A gradual change in the traits of a species over a geographical gradient.

Clitoris (klit' er us, klite' er us) A structure in the human female reproductive system that is homologous with the male penis and is involved in sexual stimulation.

Cloaca (klo ay' kuh) [L. *cloaca*: sewer] In some invertebrates, the posterior part of the gut; in many vertebrates, a cavity receiving material from the digestive, reproductive, and excretory systems.

Clonal anergy When a naive T cell encounters a self-antigen, the T cell may bind to the antigen but does not receive signals from an antigen-presenting cell. Instead of being activated, the T cell dies (becomes anergic). In this way, we avoid reacting to our own tissue-specific antigens.

Clonal deletion In immunology, the inactivation or destruction of lymphocyte clones that would produce immune reactions against the animal's own body.

Clonal selection The mechanism by which exposure to antigen results in the activation of selected T- or B-cell clones, resulting in an immune response.

Clone [Gr. *klon*: twig, shoot] Genetically identical cells or organisms produced from a common ancestor by asexual means.

Cnidocytes The feeding cells of cnidarians, within which nematocysts are housed.

Coacervate (ko as' er vate) [L. *coacervare*: to heap up] An aggregate of colloidal particles in suspension.

Coacervate drop Drops formed when a mixture of large proteins and polysaccharides is shaken in water. The interiors of these drops, which are often very stable, contain most of the proteins and polysaccharides.

Coated vesicle Vesicle, sometimes formed from a coated pit, with characteristic "bristly" surface; its membrane contains distinctive proteins, including clathrin.

Coccus (kock' us) [Gr. *kokkos*: berry, pit] Any of various spherical or spheroidal bacteria.

Cochlea (kock' lee uh) [Gr. *kokhlos*: a land snail] A spiral tube in the inner ear of vertebrates; it contains the sensory cells involved in hearing.

Codominance A condition in which two alleles at a locus produce different phenotypic effects and both effects appear in heterozygotes.

Codon A "triplet" of three nucleotides in messenger RNA that directs the placement of a particular amino acid into a polypeptide chain. (Contrast with anticodon.)

Coefficient of relatedness The probability that an allele in one individual is an identical copy, by descent, of an allele in another individual.

Coelom (see' lum) [Gr. *koiloma*: cavity] The body cavity of certain animals, which is lined with cells of mesodermal origin.

Coelomate Having a coelom.

Coenocyte (seen' a sight) [Gr.: common cell] A "cell" bounded by a single plasma membrane, but containing many nuclei.

Coenzyme A nonprotein molecule that plays a role in catalysis by an enzyme. The coenzyme may be part of the enzyme molecule or free in solution. Some coenzymes are oxidizing or reducing agents, others play different roles.

Coevolution Concurrent evolution of two or more species that are mutually affecting each other's evolution.

Cohort (co' hort) [L. *cohors*: company of soldiers] A group of similar-age organisms, considered as it passes through time.

Coleoptile (koe' lee op' til) [Gr. *koleos*: sheath + *ptilon*: feather] A pointed sheath covering the shoot of grass seedlings.

Collagen [Gr. *kolla*: glue] A fibrous protein found extensively in bone and connective tissue.

Collecting duct In vertebrates, a tubule that receives urine produced in the nephrons of the kidney and delivers that fluid to the ureter for excretion.

Collenchyma (cull eng' kyma) [Gr. *kolla*: glue + *enchyma*: infusion] A type of plant cell, living at functional maturity, which lends flexible support by virtue of primary cell walls thickened at the corners. (Contrast with parenchyma, sclerenchyma.)

Colon [Gr. *kolon*: large intestine] The large intestine.

Commensalism The form of symbiosis in which one species benefits from the association, while the other is neither harmed nor benefited.

Common bile duct A single duct that delivers bile from the gallbladder and secretions from the pancreas into the small intestine.

Communication A signal from one organism (or cell) that alters the pattern of behavior in another organism (or cell) in an adaptive fashion.

Community Any ecologically integrated group of species of microorganisms, plants, and animals inhabiting a given area.

Companion cell Specialized cell found adjacent to a sieve tube member in some flowering plants.

Comparative analysis An approach to studying evolution in which hypotheses are tested by measuring the distribution of states among a large number of species.

Compensation point The light intensity at which the rates of photosynthesis and of cellular respiration are equal.

Competitive inhibitor A substance, similar in structure to an enzyme's substrate, that binds the active site and thus inhibits a reaction.

Competition In ecology, use of the same resource by two or more species, when the resource is present in insufficient supply for the combined needs of the species.

Competitive exclusion A result of competition between species for a limiting resource in which one species completely eliminates the other.

Competitive inhibitor A substance, similar in structure to an enzyme's substrate, that binds the active site and inhibits a reaction.

Complement system A group of eleven proteins that play a role in some reactions of the immune system. The complement proteins are not immunoglobulins.

Complementary base pairing The A–T (or A–U), T–A (or U–A), C–G and G–C pairing of bases in double-stranded DNA, in transcription, and between tRNA and mRNA.

Complementary DNA (cDNA) DNA formed by reverse transcriptase acting on an RNA template; essential intermediate in the reproduction of retroviruses; used as a tool in recombinant DNA technology; lacks introns.

Complete metamorphosis A change of state during the life cycle of an organism in which the body is almost completely rebuilt to produce an individual with a very different body form. Characteristic of insects such as butterflies, moths, beetles, ants, wasps, and flies.

Compound (1) A substance made up of atoms of more than one element. (2) Made up of many units, as the compound eyes of arthropods (as opposed to the simple eyes of the same group of organisms).

Condensation reaction A reaction in which two molecules become connected by a covalent bond and a molecule of water is released. ($AH + BOH \rightarrow AB + H_2O$.)

Cones (1) In the vertebrate retina: photoreceptors responsible for color vision. (2) In gymnosperms: reproductive structures consisting of many sporophylls packed relatively tightly.

Conidium (ko nid' ee um) [Gr. *konis*: dust] An asexual fungus spore borne singly or in chains either apically or laterally on a hypha.

Conifer (kahn' e fer) [Gr. *konos*: cone + *phero*: carry] One of the cone-bearing gymnosperms, mostly trees, such as pines and firs.

Conjugation (kahn' jew gay' shun) [L. *conjugare*: yoke together] The close approximation of two cells during which they exchange genetic material, as in *Paramecium* and other ciliates, or during which DNA passes from one to the other through a tube, as in bacteria.

Connective tissue An animal tissue that connects or surrounds other tissues; its cells are embedded in a collagen-containing matrix.

Connexon In a gap junction, a protein channel linking adjacent animal cells.

Consensus sequences Short stretches of DNA that appear, with little variation, in many different genes.

Constant region The constant region in an immunoglobulin is encoded by a single exon and determines the function, but not the specificity, of the molecule. The constant region of the T cell receptor anchors the protein to the plasma membrane.

Constitutive enzyme An enzyme that is present in approximately constant amounts in a system, whether its substrates are present or absent. (Contrast with inducible enzyme.)

Consumer An organism that eats the tissues of some other organism.

Continental climate A pattern, typical of the interiors of large continents at high latitudes, in which bitterly cold winters alternate with hot summers. (Contrast with maritime climate.)

Continental drift The gradual drifting apart of the world's continents that has occurred over a period of billions of years.

Contractile vacuole An organelle, often found in protists, which pumps excess water out of the cell and keeps it from being "flooded" in hypotonic environments.

Convergent evolution The evolution of similar features independently in unrelated taxa from different ancestral structures.

Cooperative act Behavior in which two or more individuals interact to their mutual benefit. No conscious awareness by the actors of the effects of their behavior is implied.

Cooption The act of capturing something for a particular use. In ecology refers to the diversion of ecological production for human use. Such production is said to be coopted.

Copulation Reproductive behavior that results in a male depositing sperm in the reproductive tract of a female.

Corepressor A low molecular weight compound that unites with a protein (the repressor) to prevent transcription in a repressible operon.

Cork A waterproofing tissue in plants, with suberin-containing cell walls. Produced by a cork cambium.

Corm A conical, underground stem that gives rise to a new plant. (Contrast with bulb.)

Corolla (ko role' lah) [L.: diminutive of *corona*: wreath, crown] All of the petals of a flower, collectively.

Coronary (kor' oh nair ee) Referring to the blood vessels of the heart.

Corpus luteum (kor' pus loo' tee um) [L. *corpus*: body + *luteum*: yellow] A structure formed from a follicle after ovulation; it produces hormones important to the maintenance of pregnancy.

Cortex [L.: bark or rind] (1) In plants: the tissue between the epidermis and the vascular tissue of a stem or root. (2) In animals: the outer tissue of certain organs, such as the adrenal cortex and cerebral cortex.

Corticosteroids Steroid hormones produced and released by the cortex of the adrenal gland.

Cost See energetic cost, opportunity cost, risk cost.

Cotyledon (kot' ul lee' dun) [Gr. *kotyledon*: a hollow space] A "seed leaf." An embryonic organ which stores and digests reserve materials; may expand when seed germinates.

Countercurrent exchange An adaptation that promotes maximum exchange of heat or any diffusible substance between two fluids by the fluids flow in opposite directions through parallel tubes in close approximation to each other. An example is countercurrent heat exchange between arterioles and venules in the extremities of some animals.

Covalent bond A chemical bond that arises from the sharing of electrons between two atoms. Usually a strong bond.

Crassulacean acid metabolism (CAM) A metabolic pathway enabling the plants that possess it to store carbon dioxide at night and then perform photosynthesis during the day with stomata closed.

Crista (plural: cristae) A small, shelflike projection of the inner membrane of a mitochondrion; the site of oxidative phosphorylation.

Critical night length In the photoperiodic flowering response of short-day plants, the length of night above which flowering occurs and below which the plant remains vegetative. (The reverse applies in the case of long-day plants.)

Critical period The age during which some particular type of learning must take place or during which it occurs much more easily than at other times. Typical of song learning among birds.

Cross-pollination The pollination of one plant by pollen from another plant. (Contrast with self-pollination.)

Cross section (also called a transverse section) A section taken perpendicular to the longest axis of a structure.

Crossing over The mechanism by which linked markers undergo recombination. In general, the term refers to the reciprocal exchange of corresponding segments between two homologous chromatids. However, the reciprocity of crossing-over is problematical in prokaryotes and viruses; and even in eukaryotes, very closely linked markers often recombine by a nonreciprocal mechanism.

CRP The cAMP receptor protein that interacts with the promoter to enhance transcription; a lowered cAMP concentration results in catabolite repression.

Crustacean (crus tay' see an) A member of the phylum Crustacea, such as a crab, shrimp, or sowbug.

Cryptic appearance The resemblance of an animal to some part of its environment, which helps it to escape detection by predators.

Culture (1) A laboratory association of organisms under controlled conditions. (2) The collection of knowledge, tools, values, and rules that characterize a human society.

Cuticle A waxy layer on the outer surface of a plant or an insect, tending to retard water loss.

Cutin (cue' tin) [L. *cutis*: skin] A mixture of long, straight-chain hydrocarbons and waxes secreted by the plant epidermis, providing a water-impermeable coating on aerial plant parts.

Cyanobacteria (sigh an' o bacteria) [Gr. *kuanos*: the color blue] A division of photosynthetic bacteria, formerly referred to as blue-green algae; they lack sexual reproduction, and they use chlorophyll *a* in their photosynthesis.

Cyclic AMP See cAMP.

Cyclins Proteins that activate cyclin-dependent kinases, bringing about transitions in the cell cycle.

Cyclin-dependent kinase (cdk) A kinase is an enzyme that catalzyes the addition of phosphate groups from ATP to target molecules. Cdk's target proteins involved in transitions in the cell cycle and are active only when complexed to additional protein subunits, cyclins.

Cyst (sist) [Gr. *kystis*: pouch] (1) A resistant, thick-walled cell formed by some protists and other organisms. (2) An abnormal sac, containing a liquid or semisolid substance, produced in response to injury or illness.

Cytochromes (sy' toe chromes) [Gr. *kytos*: container + *chroma*: color] Iron-containing red proteins, components of the electron-transfer chains in photophosphorylation and respiration.

Cytokine A small protein that is made by one type of immune cell and stimulates a target cell which has a specific receptor for that cytokine.

Cytokinesis (sy' toe kine ee' sis) [Gr. *kytos*: container + *kinein*: to move] The division of the cytoplasm of a dividing cell. (Contrast with mitosis.)

Cytokinin (sy' toe kine' in) [Gr. *kytos*: container + *kinein*: to move]   A member of a class of plant growth substances playing roles in senescence, cell division, and other phenomena.

Cytoplasm   The contents of the cell, excluding the nucleus.

Cytoplasmic determinants   In animal development, gene products whose spatial distribution may determine such things as embryonic axes.

Cytosine (site' oh seen)   A nitrogen-containing base found in DNA and RNA.

Cytoskeleton   The network of microtubules and microfilaments that gives a eukaryotic cell its shape and its capacity to arrange its organelles and to move.

Cytosol   The fluid portion of the cytoplasm, excluding organelles and other solids.

Cytotoxic T cells   Cells of the cellular immune system that recognize and directly eliminate virus-infected cells. (Contrast with helper T cells, suppressor T cells.)

Dalton   See amu.

Deciduous (de sid' you us) [L. *decidere*: fall off]   Referring to a plant that sheds its leaves at certain seasons. (Contrast with evergreen.)

Degeneracy   The situation in which a single amino acid may be represented by any of two or more different codons in messenger RNA. Most of the amino acids can be represented by more than one codon.

Degradative succession   Ecological succession occuring on the dead remains of the bodies of plants and animals, as when leaves or animal bodies rot.

Dehydration   See condensation reaction.

Deletion (genetic)   A mutation resulting from the loss of a continuous segment of a gene or chromosome. Such mutations never revert to wild type. (Contrast with duplication, point mutation.)

Deme (deem) [Gr. *demos*: common people]   Any local population of individuals belonging to the same species that interbreed with one another.

Demographic processes   The events—such as births, deaths, immigration, and emigration—that determine the number of individuals in a population.

Demographic stochasticity   Random variations in the factors influencing the size, density, and distribution of a population.

Demography   The study of dynamical changes in the sizes, densities, and distributions of populations.

Denaturation   Loss of activity of an enzyme or nucleic acid molecule as a result of structural changes induced by heat or other means.

Dendrite [Gr. *dendron*: a tree]   A fiber of a neuron which often cannot carry action potentials. Usually much branched and relatively short compared with the axon, and commonly carries information to the cell body of the neuron.

Denitrification   Metabolic activity by which inorganic nitrogen-containing ions are reduced to form nitrogen gas and other products; carried on by certain soil bacteria.

Density dependence   Change in the severity of action of agents affecting birth and death rates within populations that are directly or inversely related to population density.

Density independence   The state where the severity of action of agents affecting birth and death rates within a population does not change with the density of the population.

Deoxyribonucleic acid   See DNA.

Depolarization   A change in the electric potential across a membrane from a condition in which the inside of the cell is more negative than the outside to a condition in which the inside is less negative, or even positive, with reference to the outside of the cell. (Contrast with hyperpolarization.)

Derived trait   A trait found among members of a lineage that was not present in the ancestors of that lineage.

Dermal tissue system   The outer covering of a plant, consisting of epidermis in the young plant and periderm in a plant with extensive secondary growth. (Contrast with ground tissue system and vascular tissue system.)

Desmosome (dez' mo sowm) [Gr. *desmos*: bond + *soma*: body]   An adhering junction between animal cells.

Determination   Process whereby an embryonic cell or group of cells becomes fixed into a predictable developmental pathway.

Detritivore (di try' ti vore) [L. *detritus*: worn away + *vorare*: to devour]   An organism that eats the dead remains of other organisms.

Deuterium   An isotope of hydrogen possessing one neutron in its nucleus. Deuterium oxide is called "heavy water."

Deuterostome   One of two major lines of evolution in animals, characterized by radial cleavage, enterocoelous development, and other traits.

Development   Progressive change, as in structure or metabolism; in most kinds of organisms, development continues throughout the life of the organism.

Dialysis (dye ahl' uh sis) [Gr. *dialyein*: separation]   The removal of ions or small molecules from a solution by their diffusion across a semipermeable membrane to a solvent where their concentration is lower.

Diaphragm (dye' uh fram) [Gr. *diaphrassein*, to barricade]   (1) A sheet of muscle that separates the thoracic and abdominal cavities in mammals; responsible for the action of breathing. (2) A method of birth control in which a sheet of rubber is fitted over the woman's cervix, blocking the entry of sperm.

Diastole (dye ahs' toll ee) [Gr.: dilation]   The portion of the cardiac cycle when the heart muscle relaxes. (Contrast with systole.)

Dicot (short for dicotyledon) [Gr. *dis*: two + *kotyledon*: a cup-shaped hollow]   Any member of the angiosperm class Dicotyledones, flowering plants in which the embryo produces two cotyledons prior to germination. Leaves of most dicots have major veins arranged in a branched or reticulate pattern.

Differentiation   Process whereby originally similar cells follow different developmental pathways. The actual expression of determination.

Diffuse coevolution   The situation in which the evolution of a lineage is influenced by its interactions with a number of species, most of which exert only a small influence on the evolution of the focal lineage.

Diffusion   Random movement of molecules or other particles, resulting in even distribution of the particles when no barriers are present.

Digestion   Enzyme-catalyzed process by which large, usually insoluble, molecules (foods) are hydrolyzed to form smaller molecules of soluble substances.

Dihybrid cross   A mating in which the parents differ with respect to the alleles of two loci of interest.

Dikaryon (di care' ee ahn) [Gr. *dis*: two + *karyon*: kernel]   A cell or organism carrying two genetically distinguishable nuclei. Common in fungi.

Dioecious (die eesh' us) [Gr.: two houses]   Organisms in which the two sexes are "housed" in two different individuals, so that eggs and sperm are not produced in the same individuals. Examples: humans, fruit flies, oak trees, date palms. (Contrast with monoecious.)

Diploblastic   Having two cell layers. (Contrast with triploblastic.)

Diploid (dip' loid) [Gr. *diploos*: double]   Having a chromosome complement consisting of two copies (homologues) of each chromosome. A diploid individual (or cell) usually arises as a result of the fusion of two gametes, each with just one copy of each chromosome. Thus, the two homologues in each chromosome pair in a diploid cell are of separate origin, one derived from the female parent and one from the male parent.

Diplontic life cycle   A life cycle in which every cell except the gametes is diploid.

Directional selection   Selection in which phenotypes at one extreme of the population distribution are favored. (Contrast with disruptive selection; stabilizing selection.)

Disaccharide   A carbohydrate made up of two monosaccharides (simple sugars).

Dispersal stage   Stage in its life history at which an organism moves from its birthplace to where it will live as an adult.

Displacement activity   Apparently irrelevant behavior performed by an animal under conflict situations, especially when tendencies to attack and escape are closely balanced.

Display A behavior that has evolved to influence the actions of other individuals.

Disruptive selection Selection in which phenotypes at both extremes of the population distribution are favored. (Contrast with directional selection; stabilizing selection.)

Distal Away from the point of attachment or other reference point. (Contrast with proximal.)

Disturbance A short-term event that disrupts populations, communities, or ecosystems by changing the environment.

Diverticulum (di ver tic' u lum) [L. *divertere*: turn away] A small cavity or tube that connects to a major cavity or tube.

Division A term used by some microbiologists and formerly by botanists, corresponding to the term phylum.

DNA (deoxyribonucleic acid) The fundamental hereditary material of all living organisms. In eukaryotes, stored primarily in the cell nucleus. A nucleic acid using deoxyribose rather than ribose.

DNA hybridization A process by which DNAs from two species are mixed and heated so that interspecific double helixes are formed.

DNA ligase Enzyme that unites Okazaki fragments of the lagging strand during DNA replication; also mends breaks in DNA strands. It connects pieces of a DNA strand and is used in recombinant DNA technology.

DNA methylation Addition of methyl groups to DNA; plays role in regulation of gene expression; protects a bacterium's DNA against its restriction endonucleases.

DNA polymerase Any of a group of enzymes that catalyze the formation of DNA strands from a DNA template.

Dominance In genetic terminology, the ability of one allelic form of a gene to determine the phenotype of a heterozygous individual, in which the homologous chromosome carries both it and a different allele. For example, if *A* and *a* are two allelic forms of a gene, *A* is said to be dominant to *a* if *AA* diploids and *Aa* diploids are phenotypically identical and are distinguishable from *aa* diploids. The *a* allele is said to be recessive.

Dominance hierarchy The set of relationships within a group of animals, usually established and maintained by aggression, in which one individual has precedence over all others in eating, mating, and other activities; a second individual has precedence over all but the highest-ranking individual, and so on down the line.

Dormancy A condition in which normal activity is suspended, as in some seeds and buds.

Dorsal [L. *dorsum*: back] Pertaining to the back or upper surface. (Contrast with ventral.)

Double fertilization Process virtually unique to angiosperms in which one sperm nucleus combines with the egg to produce a zygote, and the other sperm nucleus combines with the two polar nuclei to produce the first cell of the triploid endosperm.

Double helix Of DNA: molecular structure in which two complementary polynucleotide strands, antiparallel to each other, form a right-handed spiral.

Duodenum (doo' uh dee' num) The beginning portion of the vertebrate small intestine. (Contrast with ileum, jejunum.)

Duplication (genetic) A mutation resulting from the introduction into the genome of an extra copy of a segment of a gene or chromosome. (Contrast with deletion, point mutation.)

Dynein [Gr. *dunamis*: power] A protein that undergoes conformational changes and thus plays a part in the movement of eukaryotic flagella and cilia.

Ecdysone (eck die' sone) [Gr. *ek*: out of + *dyo*: to clothe] In insects, a hormone that induces molting.

Ecological biogeography The study of the distributions of organisms from an ecological perspective, usually concentrating on migration, dispersal, and species interactions.

Ecological community The species living together at a particular site.

Ecological niche (nitch) [L. *nidus*: nest] The functioning of a species in relation to other species and its physical environment.

Ecology [Gr. *oikos*: house + *logos*: discourse, study] The scientific study of the interaction of organisms with their environment, including both the physical environment and the other organisms that live in it.

Ecosystem (eek' oh sis tum) The organisms of a particular habitat, such as a pond or forest, together with the physical environment in which they live.

Ecto- (eck' toh) [Gr.: outer, outside] A prefix used to designate a structure on the outer surface of the body. For example, ectoderm. (Contrast with endo- and meso-.)

Ectoderm [Gr. *ektos*: outside + *derma*: skin] The outermost of the three embryonic tissue layers first delineated during gastrulation. Gives rise to the skin, sense organs, nervous system, etc.

Ectotherm [Gr. *ektos*: outside + *thermos*: heat] An animal unable to control its body temperature. (Contrast with endotherm.)

Edema (i dee' mah) [Gr. *oidema*: swelling] Tissue swelling caused by the accumulation of fluid.

Edge effect The changes in ecological processes in a community caused by physical and biological factors originating in an adjacent community.

Effector Any organ, cell, or organelle that moves the organism through the environment or else alters the environment to the organism's advantage. Examples include muscle, bone, and a wide variety of exocrine glands.

Effector cell A lymphocyte that performs a role in the immune system without further differentiation.

Effector phase In this phase of the immune response, effector T cells called cytotoxic T cells attack virus-infected cells, and effector helper T cells assist B cells to differentiate into plasma cells, which release antibodies.

Efferent [L. *ex*: out + *ferre*: to bear] Away from, as in neurons that conduct action potentials out from the central nervous system, or arterioles that conduct blood away from a structure. (Contrast with afferent.)

Egg In all sexually reproducing organisms, the female gamete; in birds, reptiles, and some other vertebrates, a structure witin which early embryonic development occurs.

Elasticity The property of returning quickly to a former state after a disturbance.

Electrocardiogram (EKG) A graphic recording of electrical potentials from the heart.

Electroencephalogram (EEG) A graphic recording of electrical potentials from the brain.

Electromyogram (EMG) A graphic recording of electrical potentials from muscle.

Electron (e lek' tron) [L. *electrum*: amber (associated with static electricity), from Gr. *slektor*: bright sun (color of amber)] One of the three most important fundamental particles of matter, with mass approximately 0.00055 amu and charge −1.

Electron microscope An instrument that uses an electron beam to form images of minute structures; the transmission electron microscope is useful for thinly-sliced material, and the scanning electron microscope gives surface views of cells and organisms.

Electrophoresis (e lek' tro fo ree' sis) [L. *electrum*: amber + Gr. *phorein*: to bear] A separation technique in which substances are separated from one another on the basis of their electric charges and molecular weights.

Electrotonic potential In neurons, a hyperpolarization or small depolarization of the membrane potential induced by the application of a small electric current. (Contrast with action potential, resting potential.)

Elemental substance A substance composed of only one type of atom.

Embolus (em' buh lus) [Gr. *embolos*: inserted object; stopper] A circulating blood clot. Blockage of a blood vessel by an embolus or by a bubble of gas is referred to as an **embolism**. (Contrast with thrombus.)

Embryo [Gr. *en-*: in + *bryein*: to grow] A young animal, or young plant sporophyte, while it is still contained within a protective structure such as a seed, egg, or uterus.

Embryo sac In angiosperms, the female gametophyte. Found within the ovule, it consists of eight or fewer cells, membrane bounded, but without cellulose walls between them.

Emergent property A property of a complex system that is not exhibited by its individual component parts.

Emigration The deliberate and usually oriented departure of an organism from the habitat in which it has been living.

3′ end (3-prime) The end of a DNA or RNA strand that has a free hydroxyl group at the 3′-carbon of the sugar (deoxyribose or ribose).

5′ end (5-prime) The end of a DNA or RNA strand that has a free phosphate group at the 5′-carbon of the sugar (deoxyribose or ribose).

Endemic (en dem′ ik) [Gr. *endemos*: dwelling in a place] Confined to a particular region, thus often having a comparatively restricted distribution.

Endergonic reaction One for which energy must be supplied. (Contrast with exergonic reaction.)

Endo- [Gr.: within, inside] A prefix used to designate an innermost structure. For example, endoderm, endocrine. (Contrast with ecto-, meso-.)

Endocrine gland (en′ doh krin) [Gr. *endon*: inside + *krinein*: to separate] Any gland, such as the adrenal or pituitary gland of vertebrates, that secretes certain substances, especially hormones, into the body through the blood.

Endocrinology The study of hormones and their actions.

Endocytosis A process by which liquids or solid particles are taken up by a cell through invagination of the plasma membrane. (Contrast with exocytosis.)

Endoderm [Gr. *endon*: within + *derma*: skin] The innermost of the three embryonic tissue layers first delineated during gastrulation. Gives rise to the digestive and respiratory tracts and structures associated with them.

Endodermis [Gr. *endon*: within + *derma*: skin] In plants, a specialized cell layer marking the inside of the cortex in roots and some stems. Frequently a barrier to free diffusion of solutes.

Endomembrane system Endoplasmic reticulum plus Golgi apparatus plus, when present, lysosomes; thus, a system of membranes that exchange material with one another.

Endometrium (en do mee′ tree um) [Gr. *endon*: within + *metrios*: womb] The epithelial cells lining the uterus of mammals.

Endoplasmic reticulum [Gr. *endon*: within + L. *plasma*: form; L. *reticulum*: little net] A system of membrane-bounded tubes and flattened sacs, often continuous with the nuclear envelope, found in the cytoplasm of eukaryotes. Exists as rough ER, studded with ribosomes, and smooth ER, lacking ribosomes.

Endorphins Naturally occurring, opiate-like substances in the mammalian brain.

Endoskeleton A skeleton covered by other, soft body tissues. (Contrast with exoskeleton.)

Endosperm [Gr. *endon*: within + *sperma*: seed] A specialized triploid seed tissue found only in angiosperms; contains stored food for the developing embryo.

Endosymbiosis [Gr. *endon*: within + *syn*: together + *bios*: life] The living together of two species, with one living inside the body (or even the cells) of the other.

Endosymbiotic theory Theory that the eukaryotic cell evolved from a prokaryote that contained other, endosymbiotic prokaryotes.

Endotherm [Gr. *endon*: within + *thermos*: hot] An animal that can control its body temperature by the expenditure of its own metabolic energy. (Contrast with ectotherm.)

Energetic cost The difference between the energy an animal would have expended had it rested, and that expended in performing a behavior.

Energy The capacity to do work.

Enhancer In eukaryotes, a DNA sequence, lying on either side of the gene it regulates, that stimulates a specific promoter.

Enterocoelous development A pattern of development in which the coelum is formed by an outpocketing of the embryonic gut (enteron).

Enterokinase (ent uh row kine′ ase) An enzyme secreted by the mucosa of the duodenum. It activates the zymogen trypsinogen to create the active digestive enzyme trypsin.

Entrainment With respect to circadian rhythms, the process whereby the period is adjusted to match the 24-hour environmental cycle.

Entropy (en′ tro pee) [Gr. *en*: in + *tropein*: to change] A measure of the degree of disorder in any system. A perfectly ordered system has zero entropy; increasing disorder is measured by positive entropy. Spontaneous reactions in a closed system are always accompanied by an increase in disorder and entropy.

Environment An organism's surroundings, both living and nonliving; includes temperature, light intensity, and all other species that influence the focal organism.

Environmental toxicology The study of the distribution and effects of toxic compounds in the environment.

Enzyme (en′ zime) [Gr. *en*: in + *zyme*: yeast] A protein, on the surface of which are chemical groups so arranged as to make the enzyme a catalyst for a chemical reaction.

Epi- [Gr.: upon, over] A prefix used to designate a structure located on top of another; for example: epidermis, epiphyte.

Epicotyl (epp′ i kot′ il) [Gr. *epi*: upon + *kotyle*: something hollow] That part of a plant embryo or seedling that is above the cotyledons.

Epidermis [Gr. *epi*: upon + *derma*: skin] In plants and animals, the outermost cell layers. (Only one cell layer thick in plants.)

Epididymis (epuh did′ uh mus) [Gr. *epi*: upon + *didymos*: testicle] Coiled tubules in the testes that store sperm and conduct sperm from the seiminiferous tubules to the vas deferens.

Epinephrine (ep i nef′ rin) [Gr. *epi*: upon + *nephros*: a kidney] The "fight or flight" hormone. Produced by the medulla of the adrenal gland, it also functions as a neurotransmitter. Also known as adrenaline.

Epiphyte (ep′ e fyte) [Gr. *epi*: upon + *phyton*: plant] A specialized plant that grows on the surface of other plants but does not parasitize them.

Episome A plasmid that may exist either free or integrated into a chromosome. (See plasmid.)

Epistasis An interaction between genes, in which the presence of a particular allele of one gene determines whether another gene will be expressed.

Epithelium In animals, a layer of cells covering or lining an external surface or a cavity.

Equilibrium (1) In biochemistry, a state in which forward and reverse reactions are proceeding at counterbalancing rates, so there is no observable change in the concentrations of reactants and products. (2) In evolutionary genetics, a condition in which allele and genotype frequencies in a population are constant from generation to generation.

Error signal In physiology, the difference between a set point and a feedback signal that results in a corrective response.

Erythrocyte (ur rith′ row sight) [Gr. *erythros*: red + *kytos*: hollow vessel] A red blood cell.

Esophagus (i soff′ i gus) [Gr. *oisophagos*: gullet] That part of the gut between the pharynx and the stomach.

Essential amino acid An amino acid an animal cannot synthesize for itself and must obtain from its diet.

Essential element An irreplaceable mineral element without which normal growth and reproduction cannot proceed.

Estivation (ess tuh vay′ shun) [L. *aestivalis*: summer] A state of dormancy and hypometabolism that occurs during the summer; usually a means of surviving drought and/or intense heat. Contrast with hibernation.

Estrogen Any of several steroid sex hormones, produced chiefly by the ovaries in mammals.

Estrous cycle The cyclical changes in reproductive physiology and behavior in female mammals (other than some primates), culminating in estrus.

Estrus (es′ truss) [L. *oestrus*: frenzy] The period of heat, or maximum sexual receptivity, in some female mammals. Ordinarily, the estrus is also the time of release of eggs in the female.

Ethology (ee thol′ o jee) [Gr. *ethos*: habit, custom + *logos*: discourse] The study of whole patterns of animal behavior in natural environments, stressing the analysis of adaptation and evolution of the patterns.

Ethylene One of the plant hormones, the gas $H_2C=2CH_2$.

Etiolation Plant growth in the absence of light.

Eubacteria (yew bacteria) Kingdom including the great majority of bacteria, such as the gram negative bacteria, gram positive bacteria, mycoplasmas, etc. (Contrast with Archaebacteria.)

Euchromatin Chromatin that is diffuse and non-staining during interphase; may be transcribed. (Contrast with heterochromatin.)

Eukaryotes (yew car' ry otes) [Gr. *eu*: true + *karyon*: kernel or nucleus] Organisms whose cells contain their genetic material inside a nucleus. Includes all life other than the viruses, Archaebacteria, and Eubacteria.

Eusocial Term applied to insects, such as termites, ants, and many bees and wasps, in which individuals cooperate in the care of offspring, there are sterile castes, and generations overlap.

Eutrophication (yoo trofe' ik ay' shun) [Gr. *eu*-: well + *trephein*: to flourish] The addition of nutrient materials to a body of water, resulting in changes to species composition therein.

Evergreen A plant that retains its leaves through all seasons. (Contrast with deciduous.)

Evolution Any gradual change. Organic evolution, often referred to as evolution, is any genetic and resulting phenotypic change in organisms from generation to generation.

Evolutionary agent Any factor that influences the direction and rate of evolutionary changes.

Evolutionary biology The collective branches of biology that study evolutionary process and their products—the diversity and history of living things.

Evolutionarily conservative Traits of organisms that evolve very slowly.

Evolutionary innovations Major changes in body plans of organisms; these have been very rare during evolutionary history.

Evolutionary radiation The proliferation of species within a single evolutionary lineage.

Excision repair The removal and damaged DNA and its replacement by the appropriate nucleotides. Often, several bases on either side of the damaged base are removed by the action of an endonuclease. Then a DNA polymerase adds the correct bases according to the template still present on the other strand of DNA. DNA ligase catalyzes the sealing up of the repaired strand.

Excitatory postsynaptic potential (EPSP) A change in the resting potential of a postsynaptic membrane in a positive (depolarizing) direction. (Contrast with inhibitory postsynaptic potential.)

Excretion Release of metabolic wastes by an organism.

Exergonic reaction A reaction in which free energy is released. (Contrast with endergonic reaction.)

Exo- (eks' oh) Same as ecto-.

Exocrine gland (eks' oh krin) [Gr. *exo*: outside + *krinein*: to separate] Any gland, such as a salivary gland, that secretes to the outside of the body or into the gut.

Exocytosis A process by which a vesicle within a cell fuses with the plasma membrane and releases its contents to the outside. (Contrast with endocytosis.)

Exon A portion of a DNA molecule, in eukaryotes, that codes for part of a polypeptide. (Contrast with intron.)

Exoskeleton (eks' oh skel' e ton) A hard covering on the outside of the body to which muscles are attached. (Contrast with endoskeleton.)

Experiment A scientific method in which particular factors are manipulated while other factors are held constant so that the potential influences of the manipulated factors can be determined.

Exploitation competition Competition that occurs because resources are depleted. (Contrast with interference competition.)

Exponential growth Growth, especially in the number of organisms in a population, which is a simple function of the size of the growing entity: the larger the entity, the faster it grows. (Contrast with logistic growth.)

Expression vector A DNA vector, such as a plasmid, that carries a DNA sequence that includes the adjacent sequences for its expression into mRNA and protein in a host cell.

Expressivity The degree to which a genotype is expressed in the phenotype— may be affected by the environment.

Extensor A muscle that extends an appendage.

Extinction The termination of a lineage of organisms.

Extrinsic protein A membrane protein found only on the surface of the membrane. (Contrast with intrinsic protein.)

F-duction Transfer of genes from one bacterium to another, using the F-factor as a vehicle.

F-factor In some bacteria, the fertility factor; a plasmid conferring "maleness" on the cell that contains it.

F_1 generation The immediate progeny of a parental (P) mating; the first filial generation.

F_2 generation The immediate progeny of a mating between members of the F_1 generation.

Facilitated diffusion Passive movement through a membrane involving a specific carrier protein; does not proceed against a concentration gradient. (Contrast with active transport, free diffusion.)

Facultative Capable of occurring or not occurring, as in facultative aerobes. (Contrast with obligate.)

Family In taxonomy, the category below the order and above the genus; a group of related, similar genera.

Fat A triglyceride that is solid at room temperature. (Contrast with oil.)

Fatty acid A molecule with a long hydrocarbon tail and a carboxyl group at the other end. Found in many lipids.

Fauna (faw' nah) All of the animals found in a given area. (Contrast with flora.)

Feces [L. *faeces*: dregs] Waste excreted from the digestive system.

Feedback control Control of a particular step of a multistep process, induced by the presence or absence of a product of one of the later steps. A thermostat regulating the flow of heating oil to a furnace in a home is a negative feedback control device.

Fermentation (fur men tay' shun) [L. *fermentum*: yeast] The degradation of a substance such as glucose to smaller molecules with the extraction of energy, without the use of oxygen (i.e., anaerobically). Involves the glycolytic pathway.

Fertilization Union of gametes. Also known as syngamy.

Fertilization membrane A membrane surrounding an animal egg which becomes rapidly raised above the egg surface within seconds after fertilization, serving to prevent entry of a second sperm.

Fetus The latter stages of an embryo that is still contained in an egg or uterus; in humans, the unborn young from the eighth week of pregnancy to the moment of birth.

Fiber An elongated and tapering cell of vascular plants, usually with a thick cell wall. Serves a support function.

Fibrin A protein that polymerizes to form long threads that provide structure to a blood clot.

Filter feeder An organism that feeds upon much smaller organisms, that are suspended in water or air, by means of a straining device.

Filtration In the excretory physiology of some animals, the process by which the initial urine is formed; water and most solutes are transferred into the excretory tract, while proteins are retained in the blood or hemolymph.

First law of thermodynamics Energy can be neither created nor destroyed.

Fission Reproduction of a prokaryote by division of a cell into two comparable progeny cells.

Fitness The contribution of a genotype or phenotype to the composition of subsequent generations, relative to the contribution of other genotypes or phenotypes. (See inclusive fitness.)

Fixed action pattern A behavior that is genetically programmed.

Flagellin (fla jell' in) The protein from which prokaryotic (but not eukaryotic) flagella are constructed.

Flagellum (fla jell' um) (plural: flagella) [L. *flagellum*: whip] Long, whiplike appendage that propels cells. Prokaryotic flagella differ sharply from those found in eukaryotes.

Flexor A muscle that flexes an appendage.

Flora (flore' ah) All of the plants found in a given area. (Contrast with fauna.)

Florigen A plant hormone (not yet isolated) involved in the conversion of a vegetative shoot apex to a flower.

Flower The total reproductive structure of an angiosperm; its basic parts include the calyx, corolla, stamens, and carpels.

Fluorescence The emission of a photon of visible light by an excited atom or molecule.

Follicle [L. *folliculus*: little bag] In female mammals, an immature egg surrounded by nutritive cells.

Follicle-stimulating hormone A gonadotropic hormone produced by the anterior pituitary.

Food chain A portion of a food web, most commonly a simple sequence of prey species and the predators that consume them.

Food web The complete set of food links between species in a community; a diagram indicating which ones are the eaters and which are consumed.

Forb Any broad-leaved (dicotyledonous), herbaceous plant. Especially applied to such plants growing in grasslands.

Fossil Any recognizable structure originating from an organism, or any impression from such a structure, that has been preserved over geological time.

Founder effect Random changes in allele frequencies resulting from establishment of a population by a very small number of individuals.

Fovea [L. *fovea*; a small pit] The area, in the vertebrate retina, of most distinct vision.

Frame-shift mutation A mutation resulting from the addition or deletion of a single base pair in the DNA sequence of a gene. As a result of this, mRNA transcribed from such a gene is translated normally until the ribosome reaches the point at which the mutation has occurred. From that point on, codons are read out of proper register and the amino acid sequence bears no resemblance to the normal sequence. (Contrast with missense mutation, nonsense mutation.)

Free diffusion Diffusion directly across a membrane without the involvement of carrier molecules. Free diffusion is not saturable and cannot cause the net transport from a region of low concentration to a region of higher concentration. (Contrast with facilitated diffusion and active transport.)

Free energy That energy which is available for doing useful work, after allowance has been made for the increase or decrease of disorder. Designated by the symbol G (for Gibbs free energy), and defined by: $G = H - TS$, where H = heat, S = entropy, and T = absolute (Kelvin) temperature.

Frequency-dependent selection Selection that changes in intensity with the proportion of individuals having the trait.

Fruit In angiosperms, a ripened and mature ovary (or group of ovaries) containing the seeds. Sometimes applied to reproductive structures of other groups of plants, and includes any adjacent parts which may be fused with the reproductive structures.

Fruiting body A structure that bears spores.

Fundamental niche The range of condition under which an organism could survive if it were the only one in the environment. (Contrast with realized niche.)

Fungus (fung' gus) A member of the kingdom Fungi, a (usually) multicellular eukaryote with absorptive nutrition.

G_1 phase In the cell cycle, the gap between the end of mitosis and the onset of the S phase.

G_2 phase In the cell cycle, the gap between the S (synthesis) phase and the onset of mitosis.

G protein A membrane protein involved in signal transduction; characterized by binding guanyl nucleotides. The activation of certain receptors activates the G protein, which in turn activates adenylate cyclase. G protein activation involves binding a GTP molecule in place of a GDP molecule.

Gametangium (gam i tan' gee um) [Gr. *gamos*: marriage + *angeion*: vessel or reservoir] Any plant or fungal structure within which a gamete is formed.

Gamete (gam' eet) [Gr. *gamete*: wife, *gametes*: husband] The mature sexual reproductive cell: the egg or the sperm.

Gametocyte (ga meet' oh site) [Gr. *gamete*: wife, *gametes*: husband + *kytos*: cell] The cell that gives rise to sex cells, either the eggs or the sperm. (See oocyte and spermatocyte.)

Gametogenesis (ga meet' oh jen' e sis) [Gr. *gamete*: wife, *gametes*: husband + *genesis*: source] The specialized series of cellular divisions that leads to the production of sex cells (gametes). (Contrast with oogenesis and spermatogenesis.)

Gametophyte (ga meet' oh fyte) In plants with alternation of generations, the haploid phase that produces the gametes. (Contrast with sporophyte.)

Ganglion (gang' glee un) [Gr.: tumor] A group or concentration of neuron cell bodies.

Gap junction A 2.7-nanometer gap between plasma membranes of two animal cells, spanned by protein channels. Gap junctions allow chemical substances or electrical signals to pass from cell to cell.

Gas exchange In animals, the process of taking up oxygen from the environment and releasing carbon dioxide to the environment.

Gastrovascular cavity Serving for both digestion (gastro) and circulation (vascular); in particular, the central cavity of the body of jellyfish and other cnidarians.

Gastrula (gas' true luh) [Gr. *gaster*: stomach] An embryo forming the characteristic three cell layers (ectoderm, endoderm, and mesoderm) which will give rise to all of the major tissue systems of the adult animal.

Gastrulation Development of a blastula into a gastrula.

Gated channel A channel (membrane protein) that opens and closes in response to binding of specific molecules or to changes in membrane potential.

Gel electrophoresis (jel ul lec tro for' eesis) A semisolid matrix suspended in a salty buffer in which molecules can be separated on the basis of their size and change when current is passed through the gel.

Gene [Gr. *gen*: to produce] A unit of heredity. Used here as the unit of genetic function which carries the information for a single polypeptide.

Gene amplification Creation of multiple copies of a particular gene, allowing the production of large amounts of the RNA transcript (as in rRNA synthesis in oocytes).

Gene cloning Formation of a clone of bacteria or yeast cells containing a particular foreign gene.

Gene family A set of identical, or once-identical, genes, derived from a single parent gene; need not be on the same chromosomes; classic example is the globin family in vertebrates.

Gene flow The exchange of genes between different species (an extreme case referred to as hybridization) or between different populations of the same species caused by migration following breeding.

Gene pool All of the genes in a population.

Gene therapy Treatment of a genetic disease by providing patients with cells containing wild type alleles for the genes that are nonfunctional in their bodies.

Generalized transduction The transfer of any bacterial host gene in a virus particle to another bacterium.

Generative nucleus In a pollen tube, a haploid nucleus that undergoes mitosis to produce the two sperm nuclei that participate in double fertilization. (Contrast with tube nucleus.)

Generator potential A stimulus-induced change in membrane resting potential in the direction of threshold for generating action potentials.

Genet The genetic individual of a plant that is composed of a number of nearly identical but repeated units.

Genetic drift Changes in gene frequencies from generation to generation in a small population as a result of random processes.

Genetic stochasticity Variation in the frequencies of alleles and genotypes in a population over time.

Genetics The study of heredity.

Genetic structure The frequencies of alleles and genotypes in a population.

Genome (jee' nome) The genes in a complete haploid set of chromosomes.

Genotype (jean' oh type) [Gr. *gen*: to produce + *typos*: impression] An exact description of the genetic constitution of an individual, either with respect to a single trait or with respect to a larger set of traits. (Contrast with phenotype.)

Genus (jean' us) (plural: genera) [Gr. *genos*: stock, kind] A group of related, similar species.

Geographical (allopatric) speciation Formation of two species from one by the interposition of (or crossing of) a physical barrier. (Contrast with parapatric, sympatric speciation.)

Geotropism See gravitropism.

Germ cell A reproductive cell or gamete of a multicellular organism.

Germination The sprouting of a seed or spore.

Gestation (jes tay' shun) [L. *gestare*: to bear] The period during which the embryo of a mammal develops within the uterus. Also known as **pregnancy**.

Gibberellin (jib er el' lin) [L. *gibberella*: hunchback (refers to shape of a reproductive structure of a fungus that produces gibberellins)] One of a class of plant growth substances playing roles in stem elongation, seed germination, flowering of certain plants, etc. Named for the fungus *Gibberella*.

Gill An organ for gas exchange in aquatic organisms.

Gill arch A skeletal structure that supports gill filaments and the blood vessels that supply them.

Gizzard (giz' erd) [L. *gigeria*: cooked chicken parts] A very muscular port of the stomach of birds that grinds up food, sometimes with the aid of fragments of stone.

Gland An organ or group of cells that produces and secretes one or more substances.

Glans penis Sexually sensitive tissue at the tip of the penis.

Glia (glee' uh) [Gr.: glue] Cells, found only in the nervous system, which do not conduct action potentials.

Glomerulus (glo mare' yew lus) [L. *glomus*: ball] Sites in the kidney where blood filtration takes place. Each glomerulus consists of a knot of capillaries served by afferent and efferent arterioles.

Glucocorticoids Steroid hormones produced by the adrenal cortex. Secreted in response to ACTH, they inhibit glucose uptake by many tissues in addition to mediating other stress responses.

Glucagon A hormone produced and released by cells in the islets of Langerhans of the pancreas. It stimulates the breakdown of glycogen in liver cells.

Gluconeogenesis The biochemical synthesis of glucose from other substances, such as amino acids, lactate, and glycerol.

Glucose (glue' kose) [Gr. *gleukos*: sweet wine mash for fermentation] The most common sugar, one of several monosaccharides with the formula $C_6H_{12}O_6$.

Glycerol (gliss' er ole) A three-carbon alcohol with three hydroxyl groups, the linking component of phospholipids and triglycerides.

Glycogen (gly' ko jen) A branched-chain polymer of glucose, similar to starch (which is less branched and may be of lower molecular weight). Exists mostly in liver and muscle; the principal storage carbohydrate of most animals and fungi.

Glycolysis (gly kol' li sis) [from glucose + Gr. *lysis*: loosening] The enzymatic breakdown of glucose to pyruvic acid. One of the oldest energy-yielding mechanisms in living organisms.

Glycosidic linkage The connection in an oligosaccharide or polysaccharide chain, formed by removal of water during the linking of monosaccharides.by root pressure.

Glyoxysome (gly ox' ee soam) An organelle found in plants, in which stored lipids are converted to carbohydrates.

Golgi apparatus (goal' jee) A system of concentrically folded membranes found in the cytoplasm of eukaryotic cells. Plays a role in the production and release of secretory materials such as the digestive enzymes manufactured in the pancreas. First described by Camillo Golgi (1844–1926).

Gonad (go' nad) [Gr. *gone*: seed, that which produces seed] An organ that produces sex cells in animals: either an ovary (female gonad) or testis (male gonad).

Gonadotropin A hormone that stimulates the gonads.

Grade The level of complexity found in an animal's body plan.

Gram stain A differential stain useful in characterizing bacteria.

Granum Within a chloroplast, a stack of thylakoids.

Gravitropism A directed plant growth response to gravity.

Grazer An animal that eats the vegetative tissues of herbaceous plants.

Green gland An excretory organ of crustaceans.

Gross morphology The sizes and shapes of the major body parts of a plant or animal.

Gross primary production The total energy captured by plants growing in a particular area.

Ground meristem That part of an apical meristem that gives rise to the ground tissue system of the primary plant body.

Ground tissue system Those parts of the plant body not included in the dermal or vascular tissue systems. Ground tissues function in storage, photosynthesis, and support.

Groundwater Water present deep in soils and rocks; may be stationary or flow slowly eventually to discharge into lakes, rivers, or oceans.

Group transfer The exchange of atoms between molecules.

Growth Irreversible increase in volume (probably the most accurate definition, but at best a dangerous oversimplification).

Growth factors A group of proteins that circulate in the blood and trigger the normal growth of cells. Each growth factor acts only on certain target cells.

Growth stage That stage in the life history of an organism in which it grows to its adult size.

Guanine (gwan'een) A nitrogen-containing base found in DNA, RNA, and GTP.

Guard cells In plants, paired epidermal cells which surround and control the opening of a stoma (pore).

Gut An animal's digestive tract.

Guttation The extrusion of liquid water through openings in leaves, caused by root pressure.

Gymnosperm (jim' no sperm) [Gr. *gymnos*: naked + *sperma*: seed] A plant, such as a pine or other conifer, whose seeds do not develop within an ovary (hence, the seeds are "naked").

Gyrus (plural: gyri) The raised or ridged portion of the convoluted surface of the brain. (Contrast to sulcus.)

Habit The form or pattern of growth characteristic of an organism.

Habitat The environment in which an organism lives.

Habituation (ha bich' oo ay shun) The simplest form of learning, in which an animal presented with a stimulus without reward or punishment eventually ceases to respond.

Hair cell A type of mechanosensor in animals.

Half-life The time required for half of a sample of a radioactive isotope to decay to its stable, nonradioactive form.

Halophyte (hal' oh fyte) [Gr. *halos*: salt + *phyton*: plant] A plant that grows in a saline (salty) environment.

Haploid (hap' loid) [Gr. *haploeides*: single] Having a chromosome complement consisting of just one copy of each chromosome. This is the normal "ploidy" of gametes or of asexual spores produced by meiosis or of organisms (such as the gametophyte generation of plants) that grow from such spores without fertilization.

Haplontic life cycle A life cycle in which the zygote is the only diploid cell.

Hardy–Weinberg rule The rule that the basic processes of Mendelian heredity (meiosis and recombination) do not alter either the frequencies of genes or their diploid combinations. The Law also states how the percentages of diploid combinations can be predicted from a knowledge of the proportions of alleles in the population.

Haustorium (haw stor' ee um) [L. *haustus*: draw up] A specialized hypha or other structure by which fungi and some parasitic plants draw food from a host plant.

Haversian systems Units of organization in compact bone that reflect the action of intercommunicating osteoblasts.

Helicase (heel' uh case) An enzyme that unwinds the DNA double helix for DNA relplication.

Helper T cells T cells that participate in the activation of B cells and of other T cells; targets of the HIV-I virus, the agent of AIDS. (Contrast with cytotoxic T cells, suppressor T cells.)

Hematocrit (heme at o krit) [Gr. *haima*: blood + *krites*: judge]   The proportion of 100 cc of blood that consists of red blood cells.

Hemizygous (hem' ee zie' gus) [Gr. *hemi*: half + *zygotos*: joined]   In a diploid organism, having only one allele for a given trait, typically the case for X-linked genes in male mammals and Z-linked genes in female birds. (Contrast with homozygous, heterozygous.)

Hemoglobin (hee' mo glow' bin) [Gr. *haima*: blood + L. *globus*: globe]   The colored protein of vertebrate blood (and blood of some invertebrates) which transports oxygen.

Hepatic (heh pat' ik) [Gr. *hepar*: liver]   Pertaining to the liver.

Hepatic duct   The duct that conveys bile from the liver to the gallbladder.

Herbicide (ur' bis ide)   A chemical substance that kills plants.

Herbivore [L. *herba*: plant + *vorare*: to devour]   An animal which eats the tissues of plants. (Contrast with carnivore, detritivore, omnivore.)

Heritable   Able to be inherited; in biology usually refers to genetically determined traits.

Hermaphroditism (her maf' row dite' ism) [Gr. *hermaphroditos*: a person with both male and female traits]   The coexistence of both female and male sex organs in the same organism.

Hertz (abbreviated as Hz)   Cycles per second.

Hetero- [Gr.: other, different]   A prefix used in biology to mean that two or more different conditions are involved; for example, heterotroph, heterozygous.

Heterochromatin   Chromatin that retains its coiling during interphase; generally not transcribed. (Contrast with euchromatin.)

Heterocyst   A large, thick-walled cell in the filaments of certain cyanobacteria; performs nitrogen fixation.

Heterogeneous nuclear RNA (hnRNA)   The product of transcription of a eukaryotic gene, including transcripts of introns.

Heterokaryon (het' er oh care' ee ahn) [Gr. *heteros*: different + *karyon*: kernel]   A cell or organism carrying a mixture of genetically distinguishable nuclei. A heterokaryon is usually the result of the fusion of two cells without fusion of their nuclei.

Heteromorphic (het' er oh more' fik) [Gr. *heteros*: different + *morphe*: form]   having a different form or appearance, as two heteromorphic life stages of a plant. (Contrast with isomorphic.)

Heterosporous (het' er os' por us)   Producing two types of spores, one of which gives rise to a female megaspore and the other to a male microspore. Heterosporous plants produce distinct female and male gametophytes. (Contrast with homosporous.)

Heterotherm   An animal that regulates its body temperature at a constant level at some times but not others, such as a hibernator.

Heterotroph (het' er oh trof) [Gr. *heteros*: different + *trophe*: food]   An organism that requires preformed organic molecules as food. (Contrast with autotroph.)

Heterozygous (het' er oh zie' gus) [Gr. *heteros*: different + *zygotos*: joined]   Of a diploid organism having different alleles of a given gene on the pair of homologues carrying that gene. (Contrast with homozygous.)

Hfr (for "high frequency of recombination")   Donor bacterium in which the F-factor has been integrated into the chromosome. This produces a bacterium that transfers its chromosomal markers at a very high frequency to recipient (F⁻) cells.

Hibernation [L. *hibernus*: winter]   The state of inactivity of some animals during winter; marked by a drop in body temperature and metabolic rate.

Highly repetitive DNA   Short DNA sequences present in millions of copies in the genome, next to each other (in tandem). In a In a reassociation experiment, denatured highly repetitive DNA reanneals very quickly.

Hippocampus   A part of the forebrain that takes part in long-term memory formation.

Histamine (hiss' tah meen)   A substance released within a damaged tissue by a type of white blood cell. Histamines are responsible for aspects of allergice reactions, including the increased vascular permeability that leads to edema (swelling).

Histology   The study of tissues.

Histone   Any one of a group of basic proteins forming the core of a nucleosome, the structural unit of a eukaryotic chromosome. (See nucleosome.)

Historical biogeography   The study of the distributions of organisms from a long-term, historical perspective.

hnRNA   See heterogeneous nuclear RNA.

Holdfast   In many large attached algae, specialized tissue attaching the plant to its substratum.

Homeobox   A segment of DNA, found in a few genes, perhaps regulating the expression of other genes and thus controlling large-scale developmental processes.

Homeostasis (home' ee o sta' sis) [Gr. *homos*: same + *stasis*: position]   The maintenance of a steady state, such as a constant temperature or a stable social structure, by means of physiological or behavioral feedback responses.

Homeotherm (home' ee o therm) [Gr. *homos*: same + *therme*: heat]   An animal which maintains a constant body temperature by virtue of its own heating and cooling mechanisms. (Contrast with heterotherm, poikilotherm.)

Homeotic genes (home' ee ott' ic)   Genes that determine what entire segments of an animal become.

Homeotic mutation   A drastic mutation causing the transformation of body parts in *Drosophila* metamorphosis. Examples include the *Antennapedia* and *ophthalmoptera* mutants.

Homolog (home' o log') [Gr. *homos*: same + *logos*: word]   One of a pair, or larger set, of chromosomes having the same overall genetic composition and sequence. In diploid organisms, each chromosome inherited from one parent is matched by an identical (except for mutational changes) chromosome—its homolog—from the other parent.

Homology (ho mol' o jee) [Gr. *homologi(a)*: agreement]   A similarity between two structures that is due to inheritance from a common ancestor. The structures are said to be homologous. (Contrast with analogy.)

Homoplasy (home' uh play zee) [Gr. *homos*: same + *plastikos*: to mold]   The presence in several species of a trait not present in their most common ancestor. Can result from convergent evolution, reverse evolution, or parallel evolution.

Homosporous   Producing a single type of spore that gives rise to a single type of gametophyte, bearing both female and male reproductive organs. (Contrast with heterosporous.)

Homozygous (home' o zie' gus) [Gr. *homos*: same + *zygotos*: joined]   Of a diploid organism having identical alleles of a given gene on both homologous chromosomes. An organism may be a "homozygote" with respect to one gene and, at the same time, a "heterozygote" with respect to another. (Contrast with heterozygous.)

Hormone (hore' mone) [Gr. *hormon*: excite, stimulate]   A substance produced in one part of a multicellular organism and transported to another part where it exerts its specific effect on the physiology or biochemistry of the target cells.

Host   An organism that harbors a parasite and provides it with nourishment.

Host–parasite interaction   The dynamic interaction between populations of a host and the parasites that attack it.

Humoral immune system   The part of the immune system mediated by B cells; it is mediated by circulating antibodies and is active against extracellular bacterial and viral infections.

Humus (hew' muss)   The partly decomposed remains of plants and animals on the surface of a soil. Its characteristics depend primarily upon climate and the species of plants growing on the site.

Hyaluronidase (hill yew ron' uh dase)   An enzyme that digests proteoglycans. Found in sperm cells, it helps digest the coatings surrounding an egg so the sperm can penetrate the egg cell membrane.

Hybrid (high' brid) [L. *hybrida*: mongrel]   The offspring of genetically dissimilar parents. In molecular biology, a double helix formed of nucleic acids from different sources.

Hybridoma   A cell produced by the fusion of an antibody-producing cell with a myeloma cell; it produces monoclonal antibodies.

Hydrocarbon   A compound containing only carbon and hydrogen atoms.

Hydrogen bond A chemical bond which arises from the attraction between the slight positive charge on a hydrogen atom and a slight negative charge on a nearby fluorine, oxygen, or nitrogen atom. Weak bonds, but found in great quantities in proteins, nucleic acids, and other biological macromolecules.

Hydrological cycle The sum total of movement of water from the oceans to the atmosphere, to the soil, and back to the oceans. Some water is cycled many times within compartments of the system before completing one full circuit.

Hydrolyze (hi' dro lize) [Gr. *hydro*: water + *lysis*: cleavage] To break a chemical bond, as in a peptide linkage, with the insertion of the components of water, –H and –OH, at the cleaved ends of a chain. The digestion of proteins is a hydrolysis.

Hydrophilic [Gr. *hydro*: water + *philia*: love] Having an affinity for water. (Contrast with hydrophobic.)

Hydrophobic [Gr. *hydro*: water + *phobia*: fear] Molecules and amino acid side chains, which are mainly hydrocarbons (compounds of C and H with no charged groups or polar groups), have a lower energy when they are clustered together than when they are distributed through an aqueous solution. Because of their attraction for one another and their reluctance to mix with water they are called "hydrophobic." Oil is a hydrophobic substance; phenylalanine is a hydrophobic amino acid. (Contrast with hydrophilic.)

Hydrophobic interaction A weak attraction between highly nonpolar molecules or parts of molecules suspended in water.

Hydrostatic skeleton The incompressible internal liquids of some animals that transfer forces from one part of the body to another when acted upon by the surrounding muscles.

Hydroxyl group The —OH group, characteristic of alcohols.

Hyperosmotic Having a more negative osmotic potential, as a result of having a higher concentration of osmotically active particles. Said of one solution as compared with another. (Contrast with hypoosmotic, isosmotic.)

Hyperpolarization A change in the resting potential of a membrane so the inside of a cell becomes more electronegative. (Contrast with depolarization.)

Hypersensitive response A defensive response of plants to microbial infection; it results in a "dead spot."

Hypertension High blood pressure.

Hypha (high' fuh) (plural: hyphae) [Gr. *hyphe*: web] In the fungi, any single filament. May be multinucleate (zygomycetes, ascomycetes) or multicellular (basidiomycetes).

Hypocotyl That part of the embryonic or seedling plant shoot that is below the cotyledons.

Hypoosmotic Having a less negative osmotic potential, as a result of having a lower concentration of osmotically active particles. Said of one solution as compared

with another. (Contrast with hyperosmotic, isosmotic.)

Hypothalamus The part of the brain lying below the thalamus; it coordinates water balance, reproduction, temperature regulation, and metabolism.

Hypothetico-deductive method A method of science in which hypotheses are erected, predictions are made from them, and experiments and observations are performed to test the predictions. The process may be repeated many times in the course of answering a question.

Icosahedron (eye kos a heed' ron) A 20-sided crystal. Some viruses have coat proteins which form a icosahedron.

Imaginal disc In insect larvae, groups of cells that develop into specific adult organs.

Imbibition [L. *imbibo*: to drink] The binding of a solvent to another molecule. Dry starch and protein will imbibe water.

Immune system A system in mammals that recognizes and eliminates or neutralizes either foreign substances or self substances that have been altered to appear foreign.

Immunization The deliberate introduction of antigen to bring about an immune response.

Immunoglobulins A class of proteins, with a characteristic structure, active as receptors and effectors in the immune system.

Immunological memory Certain clones of immune system cells made to respond to an antigen persist. This leads to a more rapid and massive response of the immune system to any subsequent exposure to that antigen.

Immunological tolerance A mechanism by which an animal does not mount an immune response to the antigenic determinants of its own macromolecules.

Imprinting A rapid form of learning, in which an animal comes to make a particular response, which is maintained for life, to some object or other organism.

Inclusive fitness The sum of an individual's own fitness (the effect of producing its own offspring: the individual selection component) plus its influence on fitness in relatives other than direct descendants (the kin selection component).

Incomplete dominance Condition in which the heterozygous phenotype is intermediate between the two homozygous phenotypes.

Incomplete metamorphosis Insect development in which changes between instars are gradual.

Incus (in' kus) [L. *incus*: anvil] The middle of the three bones that conduct movements of the eardrum to the oval window of the inner ear. (See malleus, stapes.)

Individual fitness That component of inclusive fitness that results from an organism producing its own offspring. (Contrast with kin selection component.)

Indoleacetic acid See auxin.

Induced fit A change in the tertiary structures of some enzymes, caused by binding of substrate to the active site.

Inducer (1) In enzyme systems, a small molecule which, when added to a growth medium, causes a large increase in the level of some enzyme. (2) In embryology, a substance that causes a group of target cells to differentiate in a particular way.

Inducible enzyme An enzyme that is present in much larger amounts when a particular compound (the inducer) has been added to the system. (Contrast with constitutive enzyme.)

Inflammation A nonspecific defense against pathogens; characterized by redness, swelling, pain, and increased temperature.

Inflorescence A structure composed of several flowers.

Inhibitor A substance which binds to the surface of an enzyme and interferes with its action on its substrates.

Inhibitory postsynaptic potential A change in the resting potential of a postsynaptic membrane in the hyperpolarizing (negative) direction.

Initiation complex Combination of a ribosomal light subunit, an mRNA molecule, and the tRNA charged with the first amino acid coded for by the mRNA; formed at the onset of translation.

Initiation factors Proteins that assist in forming the translation initiation complex at the ribosome.

Inositol triphosphate (IP3) An intracellular second messenger derived from membrane phospholipids.

Insertion sequence A large piece of DNA that can give rise to copies at other loci; a type of transposable genetic element.

Instar (in' star) [L.: image, form] An immature stage of an insect between molts.

Instinct Behavior that is relatively highly steretyped and self-differentiating, that develops in individuals unable to observe other individuals performing the behavior or to practice the behavior in the presence of the objects toward which it is usually directed.

Insulin (in' su lin) [L. *insula*: island] A hormone, synthesized in islet cells of the pancreas, that promotes the conversion of glucose to the storage material, glycogen.

Integrase An enzyme that integrates retroviral cDNA into the genome of the host cell.

Integrated pest management A method of control of pests in which natural predators and parasites are used in conjunction with sparing use of chemical methods to achieve control of a pest without causing serious adverse environmental side effects.

Integument [L. *integumentum*: covering] A protective surface structure. In gymnosperms and angiosperms, a layer of tissue around the ovule which will become the seed coat. Gymnosperm ovules have one integument, angiosperm ovules two.

Intention movement   The preparatory motions that animals go through prior to a complete behavior response; for example, the crouch before flying, the snarl before biting, etc.

Intercalary meristem   A meristematic region in plants which occurs not apically, but between two regions of mature tissue. Intercalary meristems occur in the nodes of grass stems, for example.

Intercostal muscles   Muscles between the ribs that can augment breathing movements by elevating and suppressing the rib cage.

Interference competition   Competition resulting from direct behavioral interactions between organisms. (Contrast with exploitation competition.)

Interferon   A glycoprotein produced by virus-infected animal cells; increases the resistance of neighboring cells to the virus.

Interkinesis   The phase between the first and second meiotic divisions.

Interleukins   Regulatory proteins, produced by macrophages and lymphocytes, that act upon other lymphocytes and direct their development.

Intermediate filaments   Fibrous proteins that stabilize cell structure and resist tension.

Internode   Section between two nodes of a plant stem.

Interphase   The period between successive nuclear divisions during which the chromosomes are diffuse and the nuclear envelope is intact. It is during this period that the cell is most active in transcribing and translating genetic information.

Interspecific competition   Competition between members of two or more species.

Intertropical convergence zone   The tropical region where the air rises most strongly; moves north and south with the passage of the sun overhead.

Intraspecific competition   Competition among members of a single species.

Intrinsic protein   A membrane protein that is embedded in the phospholipid bilayer of the membrane. (Contrast with extrinsic protein.)

Intrinsic rate of increase   The rate at which a population can grow when its density is low and environmental conditions are highly favorable.

Intron   A portion of a DNA molecule that, because of RNA splicing, is not involved in coding for part of a polypeptide molecule. (Contrast with exon.)

Invagination   An infolding.

Inversion (genetic)   A rare mutational event that leads to the reversal of the order of genes within a segment of a chromosome, as if that segment had been removed from the chromosome, turned 180°, and then reattached.

Invertebrate   Any animal that is not a vertebrate, that is, whose nerve cord is not enclosed in a backbone of bony segments.

In vitro [L.: in glass]   In a test tube, rather than in a living organism. (Contrast with in vivo.)

In vivo [L.: in the living state]   In a living organism. Many processes that occur in vivo can be reproduced in vitro with the right selection of cellular components. (Contrast with in vitro.)

Ion (eye' on) [Gr.: wanderer]   An atom or group of atoms with electrons added or removed, giving it a negative or positive electrical charge.

Ionic channel   A membrane protein that can let ions pass across the membrane. The channel can be ion-selective, and it can be voltage-gated or ligand-gated.

Ionic bond   A chemical bond which arises from the electrostatic attraction between positively and negatively charged ions. Usually a strong bond.

Iris (eye' ris) [Gr. *iris*: rainbow]   The round, pigmented membrane that surrounds the pupil of the eye and adjusts its aperture to regulate the amount of light entering the eye.

Irruption   A rapid increase in the density of a population. Often followed by massive emigration.

Islets of Langerhans   Clusters of hormone-producing cells in the pancreas.

Isogamy (eye sog' ah mee) [Gr. *isos*: equal + *gamos*: marriage]   A kind of sexual reproduction in which the gametes (or gametangia) are not distinguishable on the basis of size or morphology.

Isolating mechanism   Geographical, physiological, ecological, or behavioral mechanisms that lead to a reduction in the frequency of hybrid matings.

Isomers   Molecules consisting of the same numbers and kinds of atoms, but differing in the way in which the atoms are combined.

Isomorphic (eye' so more' fik) [Gr. *isos*: equal + *morphe*: form]   having the same form or appearance, as two isomorphic life stages. (Contrast with heteromorphic.)

Isosmotic   Having the same osmotic potential. Said of two solutions. (Contrast with hyperosmotic, hypoosmotic.)

Isotope (eye' so tope) [Gr. *isos*: equal + *topos*: place]   Two isotopes of the same chemical element have the same number of protons in their nuclei, but differ in the number of neutrons.

Isozymes   Chemically different enzymes that catalyze the same reaction.

Jejunum (jih jew' num)   The middle division of the small intestine, where most absorption of nutrients occurs. (See duodenum, ileum.)

Joule (jool, or jowl)   A unit of energy, equal to 0.24 calories.

Juvenile hormone   In insects, a hormone maintaining larval growth and preventing maturation or pupation.

Karyotype   The number, forms, and types of chromosomes in a cell.

Kelvin temperature scale   See absolute temperature scale.

Keratin (ker' a tin) [Gr. *keras*: horn]   A protein which contains sulfur and is part of such hard tissues as horn, nail, and the outermost cells of the skin.

Ketone (key' tone)   A compound with a C=O group attached to two other groups, neither of which is an H atom. Many sugars are ketones. (Contrast with aldehyde.)

Keystone species   A species that exerts a major influence on the composition and dynamics of the community in which it lives.

Kidneys   A pair of excretory organs in vertebrates.

Kin selection component   The component of inclusive fitness resulting from helping the survival of relatives containing the same alleles by descent from a common ancestor.

Kinase (kye' nase)   An enzyme that transfers a phosphate group from ATP to another molecule. Protein kinases transfer phosphate from ATP to specific proteins, playing important roles in cell regulation.

Kinesis (ki nee' sis) [Gr.: movement]   Orientation behavior in which the organism does not move in a particular direction with reference to a stimulus but instead simply moves at an increasing or decreasing rate until it ends up farther from the object or closer to it. (Contrast with taxis.)

Kinetochore (kin net' oh core) [Gr. *kinetos*: moving + *khorein*: to move]   Specialized structure on a centromere to which microtubules attach.

Kingdom   The highest taxonomic category in the Linnaean system.

Knockout mouse   A genetically engineered mouse in which one or more functioning alleles have been replaced by defective alleles.

***Lac* operon**   A region of DNA in *E. coli* that contains a single promoter and operator controlling the expression of three adjacent genes involved in the utilization of the sugar, lactose.

Lactic acid   The end product of fermentation in vertebrate muscle and some microorganisms.

Lagging strand   In DNA replication, the daughter strand that is synthesized discontinuously.

Lamella   Layer.

Larynx (lar' inks)   A structure between the pharynx and the trachea that includes the vocal cords.

Larva (plural: larvae) [L.: ghost, early stage]   An immature stage of any invertebrate animal that differs dramatically in appearance from the adult.

Lateral   Pertaining to the side.

Lateral inhibition   In visual information processing in the arthropod eye, the mutual inhibition of optic nerve cells; results in enhanced detection of edges.

Laterization (lat' ur iz ay shun)   The formation of a nutrient-poor soil that is rich in nsoluble iron and aluminum compounds.

Law of independent assortment Alleles of different, unlinked genes assort independently of one another during gamete formation, Mendel's second law.

Law of segregation Alleles segregate from one another during gamete formation, Mendel's first law.

Leader sequence A sequence of amino acids at the N-terminal end of a newly synthesized protein, determining where the protein will be placed in the cell.

Leading strand In DNA replication, the daughter strand that is synthesized continuously.

Leaf axil The upper angle between a leaf and the stem, site of lateral buds which under appropriate circumstances become activated to form lateral branches.

Leaf primordium [L.: the beginning] A small mound of cells on the flank of a shoot apical meristem that will give rise to a leaf.

Lek A traditional courtship display ground, where males display to females.

Lenticel Spongy region in a plant's periderm, allowing gas exchange.

Leukocyte (loo' ko sight) [Gr. *leukos*: clear + *kutos*: hollow vessel] A white blood cell.

Leuteinizing hormone A peptide hormone produced by pituitary cells that stimulates follicle maturation in females.

Lichen (lie' kun) [Gr. *leikhen*: licker] An organism resulting from the symbiotic association of a true fungus and either a cyanobacterium or a unicellular alga.

Life cycle The entire span of the life of an organism from the moment of fertilization (or asexual generation) to the time it reproduces in turn.

Life history The stages an individual goes through during its life.

Life table A table showing, for a group of equal-aged individuals, the proportion still alive at different times in the future and the number of offspring they produce during each time interval.

Ligament A band of connective tissue linking two bones in a joint.

Ligand (lig' and) A molecule that binds to a receptor site of another molecule.

Lignin The principal noncarbohydrate component of wood, a polymer that binds together cellulose fibrils in some plant cell walls.

Limbic system A group of primitive vertebrate forebrain nuclei that form a network and are involved in emotions, drives, instinctive behaviors, learning, and memory.

Limiting resource The required resource whose supply most strongly influences the size of a population.

Linkage Association between genetic markers on the same chromosome such that they do not show random assortment and seldom recombine; the closer the markers, the lower the frequency of recombination.

Lipase (lip' ase; lye' pase) An enzyme that digests fats.

Lipids (lip' ids) [Gr. *lipos*: fat] Substances in a cell which are easily extracted by organic solvents; fats, oils, waxes, steroids, and other large organic molecules, including those which, with proteins, make up the cell membranes. (See phospholipids.)

Litter The partly decomposed remains of plants on the surface and in the upper layers of the soil.

Littoral zone The coastal zone from the upper limits of tidal action down to the depths where the water is thoroughly stirred by wave action.

Liver A large digestive gland. In vertebrates, it secretes bile and is involved in the formation of blood.

Lobes Regions of the human cerebral hemispheres; includes the temporal, frontal, parietal, and occipital lobes.

Locus In genetics, a specific location on a chromosome. May be considered to be synonymous with "gene."

Logistic growth Growth, especially in the size of an organism or in the number of organisms that constitute a population, which slows steadily as the entity approaches its maximum size. (Contrast with exponential growth.)

Loop of Henle (hen' lee) Long, hairpin loop of the mammalian renal tubule that runs from the cortex down into the medulla, and back to the cortex. Creates a concentration gradient in the interstitial fluids in the medulla.

Lophophore A U-shaped fold of the body wall with hollow, ciliated tentacles that encircles the mouth of animals in several different phyla. Used for filtering prey from the surrounding water.

Lordosis (lor doe' sis) [Gk. *lordosis*: curving forward] A posture assumed by females of some mammalian species (especially rodents) to signal sexual receptivity.

Lumen (loo' men) [L.: light] The cavity inside any tubular part of an organ, such as a piece of gut or a kidney tubule.

Lungs A pair of saclike chambers within the bodies of some animals, functioning in gas exchange.

Luteinizing hormone A gonadotropin produced by the anterior pituitary. It stimulates the gonads to produce sex hormones.

Lymph [L. *lympha*: water] A clear, watery fluid that is formed as a filtrate of blood; it contains white blood cells; it collects in a series of special vessels and is returned to the bloodstream.

Lymph nodes Specialized tissue regions that act as filters for cells, bacteria and foreign matter.

Lymphocyte A major class of white blood cells. Includes T cells, B cells, and other cell types important in the immune response.

Lysis (lie' sis) [Gr.: a loosening] Bursting of a cell.

Lysogenic The condition of a bacterium that carries the genome of a virus in a relatively stable form. (Contrast with lytic.)

Lysosome (lie' so soam) [Gr. *lysis*: a loosening + *soma*: body] A membrane-bounded inclusion found in eukaryotic cells (other than plants). Lysosomes contain a mixture of enzymes that can digest most of the macromolecules found in the rest of the cell.

Lysozyme (lie' so zyme) An enzyme in saliva, tears, and nasal secretions that attacks bacterial cell walls, as one of the body's nonspecific defense mechanisms.

Lytic Condition in which a bacterium lyses shortly after infection by a virus; the viral genome does not become stabilized within the bacterial cell. (Contrast with lysogenic.)

Macro- (mack' roh) [Gr. *makros*: large, long] A prefix commonly used to denote something large. (Contrast with micro-.)

Macroevolution Evolutionary changes occurring over long time spans and usually involving changes in many traits. (Contrast with microevolution.)

Macroevolutionary time The time required for macroveolutionary changes in a lineage.

Macromolecule A giant polymeric molecule. The macromolecules are proteins, polysaccharides, and nucleic acids.

Macronutrient A mineral element required by plant tissues in concentrations of at least 1 milligram per gram of their dry matter.

Macrophage (mac' roh faj) A type of white blood cell that endocytoses bacteria and other cells.

Major histocompatibility complex (MHC) A complex of linked genes, with multiple alleles, that control a number of immunological phenomena; it is important in graft rejection.

Malignant tumor A tumor whose cells can invade surrounding tissues and spread to other organs.

Malleus (mal' ee us) [L. *malleus*: hammer] The first of the three bones that conduct movements of the eardrum to the oval window of the inner ear. (See incus, stapes.)

Malpighian tubule (mal pee' gy un) A type of protonephridium found in insects.

Mammal [L. *mamma*: breast, teat] Any animal of the class Mammalia, characterized by the production of milk by the female mammary glands and the possession of hair for body covering.

Mantle A sheet of specialized tissues that covers most of the viscera of mollusks; provides protection to internal organs and secretes the shell.

Map unit In eukaryotic genetics, one map unit corresponds to a recombinant frequency of 0.01.

Mapping In genetics, determining the order of genes on a chromosome and the distances between them.

Marine [L. *mare*: sea, ocean] Pertaining to or living in the ocean. (Contrast with aquatic, terrestrial.)

Maritime climate   Weather pattern typical of coasts of continents, particularly those on the western sides at mid latitudes, in which the difference between summer and winter is relatively small. (Contrast with continental climate.)

Marsupial (mar soo' pee al)   A mammal belonging to the subclass Metatheria, such as opossums and kangaroos. Most have a pouch (marsupium) that contains the milk glands and serves as a receptacle for the young.

Mass extinctions   Geological periods during which rates of extinction were much higher than during intervening times.

Mass number   The sum of the number of protons and neutrons in an atom's nucleus.

Mast cells   Typically found in connective tissue, mast cells can be provoked by antigens or inflammation to release histamine.

Maternal effect genes   These genes code for morphogens that determine the polarity of the egg and larva in the fruit fly, *Drosophila melanogaster*.

Maternal inheritance (cytoplasmic inheritance)   Inheritance in which the phenotype of the offspring depends on factors, such as mitochondria or chloroplasts, that are inherited from the female parent through the cytoplasm of the female gamete.

Mating type   In some bacteria, fungi, and protists, sexual reproduction can occur only between partners of a different mating type. "Mating type" is not the same as "sex," since some species have as many as 8 mating types; mating may also be between hermaphroditic partners of opposite mating type, with both partners acting as both "male" and "female" in terms of donating and receiving genetic information.

Maturation   The automatic development of a pattern of behavior, which becomes increasingly complex or precise as the animal matures. Unlike learning, the development does not require experience to occur.

Mechanosensor   A cell that is sensitive to physical movement and generates action potentials in response.

Medulla (meh dull' luh) [L.: narrow]   (1) The inner, core region of an organ, as in the adrenal medulla (adrenal gland) or the renal medulla (kidneys). (2) The portion of the brain stem that connects to the spinal cord.

Medusa (meh doo' suh)   The tentacle-bearing, jellyfish-like, free-swimming sexual stage in the life cycle of a cnidarian.

Mega- [Gr. *megas*: large, great]   A prefix often used to denote something large. (Contrast with micro-.)

Megareserve   A large park or reserve; usually has associated buffer areas in which human use of the environment is restricted to activities that do not destroy the functioning of the ecosystem.

Megasporangium   The special structure (sporangium) that produces the megaspores.

Megaspore [Gr. *megas*: large + *spora*:seed]   In plants, a haploid spore that produces a female gametophyte. In many cases the megaspore is larger than the male-producing microspore.

Meiosis (my oh' sis) [Gr.: diminution]   Division of a diploid nucleus to produce four haploid daughter cells. The process consists of two successive nuclear divisions with only one cycle of chromosome replication.

Membrane potential   The difference in electrical charge between the inside and the outside of a cell, caused by a difference in the distribution of ions.

Memory cells   Long-lived lymphocytes produced by exposure to antigen. They persist in the body and are able to mount a rapid response to subsequent exposures to the antigen.

Mendelian population   A local population of individuals belonging to the same species and exchanging genes with one another.

Menopause   The time in a human female's life when the ovarian and menstrual cycles cease.

Menstrual cycle   The monthly sloughing off of the uterine lining if fertilization does not occur in the female. Occurs between puberty and menopause.

Meristem [Gr. *meristos*: divided]   Plant tissue made up of actively dividing cells.

Mesenchyme (mez' en kyme) [Gr. *mesos*: middle + *enchyma*: infusion]   Embryonic or unspecialized cells derived from the mesoderm.

Meso- (mez' oh) [Gr.: middle]   A prefix often used to designate a structure located in the middle, or a stage that appears at some intermediate time. For example, mesoderm, Mesozoic.

Mesoderm [Gr. *mesos*: middle + *derma*: skin]   The middle of the three embryonic tissue layers first delineated during gastrulation. Gives rise to skeleton, circulatory system, muscles, excretory system, and most of the reproductive system.

Mesoglea   The jelly-like middle layer that constitutes the bulk of the bodies of the medusae of many cnidarians; not a true cell layer.

Mesophyll (mez' a fill) [Gr. *mesos*: middle + *phyllon*: leaf]   Chloroplast-containing, photosynthetic cells in the interior of leaves.

Mesosome (mez' o soam') [Gr. *mesos*: middle + *soma*: body]   A localized infolding of the plasma membrane of a bacterium.

Messenger RNA (mRNA)   A transcript of one of the strands of DNA, it carries information (as a sequence of codons) for the synthesis of one or more proteins.

Meta- [Gr.: between, along with, beyond]   A prefix used in biology to denote a change or a shift to a new form or level; for example, as used in metamorphosis.

Metabolic compensation   Changes in biochemical properties of an organism that render it less sensitive to temperature changes.

Metabolic pathway   A series of enzyme-catalyzed reactions so arranged that the product of one reaction is the substrate of the next.

Metabolism (meh tab' a lizm) [Gr. *metabole*: to change]   The sum total of the chemical reactions that occur in an organism, or some subset of that total (as in "respiratory metabolism").

Metamorphosis (met' a mor' fo sis) [Gr. *meta*: between + *morphe*: form, shape]   A radical change occurring between one developmental stage and another, as for example from a tadpole to a frog or an insect larva to the adult.

Metaphase (met' a phase) [Gr. *meta*: between]   The stage in nuclear division at which the centromeres of the highly supercoiled chromosomes are all lying on a plane (the metaphase plane or plate) perpendicular to a line connecting the division poles.

Metapopulation   A population divided into subpopulations, among which there are occasional exchanges of individuals.

Metastasis (meh tass' tuh sis)   The spread of cancer cells from their original site to other parts of the body.

Methanogen   Any member of a group of Archaebacteria that release methane as a metabolic product. This group is considered to be an extremely ancient one.

MHC   See major histocompatibility complex.

Micelles (my sells') [L. *mica*: grain, crumb]   The small particles of fat in the small intestine, resulting from the emulsification of dietary fat by bile.

Micro- (mike' roh) [Gr. *mikros*: small]   A prefix often used to denote something small. (Contrast with macro-, mega-.)

Microbiology [Gr. *mikros*: small + *bios*: life + *logos*: discourse]   The scientific study of microscopic organisms, particularly bacteria, unicellular algae, protists, and viruses.

Microevolution   The small evolutionary changes typically occurring over short time spans; generally involving a small number of traits and minor genetic changes. (Contrast with macroevolution.)

Microevolutionary time   The time required for microevolutionary changes within a lineage of organisms.

Microfilament   Minute fibrous structure generally composed of actin found in the cytoplasm of eukaryotic cells. They play a role in the motion of cells.

Micromorphology   The structure of the macromolecules of an organism.

Micronutrient   A mineral element required by plant tissues in concentrations of less than 100 micrograms per gram of their dry matter.

Microorganism   Any microscopic organism, such as a bacterium or one-celled alga.

Micropyle (mike' roh pile) [Gr. *mikros*: small + *pyle*: gate]   Opening in the integument(s) of a seed plant ovule through which pollen grows to reach the female gametophyte within.

Microsporangium   The special structure (sporangium) that produces the microspores.

Microspores [Gr. *mikros*: small + *spora*: seed] In plants, a haploid spore that produces a male gametophyte. In many cases the microspore is smaller than the female-producing megaspore.

Microtubules Minute tubular structures found in centrioles, spindle apparatus, cilia, flagella, and other places in the cytoplasm of eukaryotic cells. These tubules play roles in the motion and maintenance of shape of eukaryotic cells.

Microvilli (singular: microvillus) The projections of epithelial cells, such as the cells lining the small intestine, that increase their surface area.

Middle lamella A layer of derivative polysaccharides that separates plant cells; a common middle lamella lies outside the primary walls of the two cells.

Migration The regular, seasonal movements of animals between breeding and nonbreeding ranges.

Mimicry (mim' ik ree) The resemblance of one kind of organism to another, or to some inanimate object; serves the function of making the organism difficult to find, of discouraging potential enemies or of attracting potential prey. (See Batesian mimicry and Müllerian mimicry.)

Mineral An inorganic substance other than water.

Mineralocorticoid A hormone produced by the adrenal cortex that influences mineral ion balance; aldosterone.

Minimal medium A medium for the growth of bacteria, fungi, or tissue cultures, containing only those nutrients absolutely required for the growth of wild type cells.

Minimum viable population. The smallest number of individuals required for a population to persist in a region.

Mismatch repair When a single base in DNA is changed into a different base, or the wrong base inserted during DNA replication, there is a mismatch in base pairing with the base on the opposite strand. A repair system removes the incorrect base and inserts the proper one for pairing with the opposite strand.

Missense mutation A mutation that changes a codon for one amino acid to a codon for a different amino acid. (Contrast with frame-shift mutation, nonsense mutation.)

Mitochondrial matrix The fluid interior of the mitochondrion, enclosed by the inner mitochondrial membrane.

Mitochondrion (my' toe kon' dree un) (plural: mitochondria) [Gr. *mitos*: thread + *chondros*: cartilage, or grain] An organelle that occurs in eukaryotic cells and contains the enzymes of the ctric acid cycle, the respiratory chain, and oxidative phosphorylation. A mitochondrion is bounded by a double membrane.

Mitosis (my toe' sis) [Gr. *mitos*: thread] Nuclear division in eukaryotes leading to the formation of two daughter nuclei each with a chromosome complement identical to that of the original nucleus.

Mitotic center Cellular region that organizes the microtubules for mitosis. In animals a centrosome serves as the mitotic center.

Mobbing Gathering of calling animals around a predator; their calls and the confusion they create reduce the probability that the predator can hunt successfully in the area.

Moderately repetitive DNA DNA sequences that appear hundreds to thousands of times in the genome. They include the DNA sequences coding for rRNAs and tRNAs, as well as the DNA at telomeres.

Modular organism An organism which grows by producing additional units of body construction that are very similar to the units of which it is already composed.

Mole A quantity of a compound whose weight in grams is numerically equal to its molecular weight expressed in atomic mass units. Avogadro's number of molecules: 6.023×10^{23} molecules.

Molecular formula A representation that shows how many atoms of each element are present in a molecule.

Molecular weight The sum of the atomic weights of the atoms in a molecule.

Molecule A particle made up of two or more atoms joined by covalent bonds or ionic attractions.

Molting The process of shedding part or all of an outer covering, as the shedding of feathers by birds or of the entire exoskeleton by arthropods.

Monoecious (mo nee' shus) [Gr.: one house] Organisms in which both sexes are "housed" in a single individual, which produces both eggs and sperm. (In some plants, these are found in different flowers within the same plant.) Examples: corn, peas, earthworms, hydras. (Contrast with dioecious, perfect flower.)

Moneran (moh neer' un) A bacterium. This term was coined when both archaebacteria and eubacteria were considered to be members of a single kingdom, Monera.

Mono- [Gr. *monos*: one] Prefix denoting a single entity. (Contrast with poly.)

Monoclonal antibody Antibody produced in the laboratory from a clone of hybridoma cells, each of which produces the same specific antibody.

Monocot (short for monocotyledon) [Gr. *monos*: one + *kotyledon*: a cup-shaped hollow] Any member of the angiosperm class Monocotyledones, plants in which the embryo produces but a single cotyledon (seed leaf). Leaves of most monocots have their major veins arranged parallel to each other.

Monocytes White blood cells that produce macrophages.

Monohybrid cross A mating in which the parents differ with respect to the alleles of only one locus of interest.

Monomer A small molecule, two or more of which can be combined to form oligomers (consisting of a few monomers) or polymers (consisting of many monomers).

Monophyletic (mon' oh fih leht' ik) [Gk. *monos*: single + *phylon*: tribe] Being descended from a single ancestral stock.

Monosaccharide A simple sugar. Oligosaccharides and polysaccharides are made up of monosaccharides.

Monosynaptic reflex A neural reflex that begins in a sensory neuron and makes a single synapse before activating a motor neuron.

Morphogens Diffusible substances whose concentration gradients determine patterns of development in animals and plants.

Morphogenesis (more' fo jen' e sis) [Gr. *morphe*: form + *genesis*: origin] The development of form. Morphogenesis is the overall consequence of determination, differentiation, and growth.

Morphology (more fol' o jee) [Gr. *morphe*: form + *logos*: discourse] The scientific study of organic form, including both its development and function.

Mosaic development Pattern of animal embryonic development in which each blastomere contributes a specific part of the adult body. (Contrast with regulative development.)

Motor end plate The modified area on a muscle cell membrane where a synapse is formed with a motor neuron.

Motor neuron A neuron carrying information from the central nervous system to an effector such as a muscle fiber.

Motor unit A motor neuron and the set of muscle fibers it controls.

mRNA (See messenger RNA.)

Mucosa (mew koh' sah) An epithelial membrane containing cells that secrete mucus. The inner cell layers of the digestive and respiratory tracts.

Müllerian mimicry The resemblance of two or more unpleasant or dangerous kinds of organisms to each other.

Multicellular [L. *multus*: much + *cella*: chamber] Consisting of more than one cell, as for example a multicellular organism. (Contrast with unicellular.)

Muscle Contractile tissue containing actin and myosin organized into polymeric chains called microfilaments. In vertebrates, the tissues are either cardiac muscle, smooth muscle, or striated (skeletal) muscle.

Muscle fiber A single muscle cell. In the case of striated muscle, a syncitial, multinucleate cell.

Muscle spindle Modified muscle fibers encased in a connective sheat and functioning as stretch sensors.

Mutagen (mute' ah jen) [L. *mutare*: change + Gr. *genesis*: source] An agent, especially a chemical, that increases the mutation rate.

Mutation In the broad sense, any discontinuous change in the genetic constitution of an organism. In the narrow sense, the word usually refers to a "point mutation," a change along a very narrow portion of the nucleic acid sequence.

Mutation pressure   Evolution (change in gene proportions) by different mutation rates alone.

Mutualism   The type of symbiosis, such as that exhibited by fungi and algae or cyanobacteria in forming lichens, in which both species profit from the association.

Mycelium (my seel' ee yum) [Gr. *mykes*: fungus]   In the fungi, a mass of hyphae.

Mycorrhiza (my' ka rye' za) [Gr. *mykes*: fungus + *rhiza*: root]   An association of the root of a plant with the mycelium of a fungus.

Myelin (my' a lin)   A material forming a sheath around some axons. It is formed by Schwann cells that wrap themselves about the axon. It serves to insulate the axon electrically and to increase the rate of transmission of a nervous impulse.

Myofibril (my' oh fy' bril) [Gr. *mys*: muscle + L. *fibrilla*: small fiber]   A polymeric unit of actin or myosin in a muscle.

Myogenic (my oh jen' ik) [Gr. *mys*: muscle + *genesis*: source]   Originating in muscle.

Myoglobin (my' oh globe' in) [Gr. *mys*: muscle + L. *globus*: sphere]   An oxygen-binding molecule found in muscle. Consists of a heme unit and a single globiin chain, and carrys less oxygen than hemoglobin.

Myosin [Gr. *mys*: muscle]   One of the two major proteins of muscle, it makes up the thick filaments. (See actin.)

NAD (nicotinamide adenine dinucleotide)   A compound found in all living cells, existing in two interconvertible forms: the oxidizing agent NAD^+ and the reducing agent NADH.

NADP (nicotinamide adenine dinucleotide phosphate)   Like NAD, but possessing another phosphate group; plays similar roles but is used by different enzymes.

Natal group   The group into which an individual was born.

Natural selection   The differential contribution of offspring to the next generation by various genetic types belonging to the same population. The mechanism of evolution proposed by Charles Darwin.

Nauplius (no' plee us) [Gk. *nauplios*: shellfish]   The typical larva of crustaceans. Has three pairs of appendages and a median compound eye.

Necrosis (nec roh' sis)   Tissue damage resulting from cell death.

Negative control   The situation in which a regulatory macromolecule (generally a repressor) functions to turn off transcription. In the absence of a regulatory macromolecule, the structural genes are turned on.

Negative feedback   A pattern of regulation in which a change in a sensed variable results in a correction that opposes the change.

Nekton [Gr. *nekhein*: to swim]   Animals, such as fish, that can swim against currents of water. (Contrast with plankton.)

Nematocyst (ne mat' o sist) [Gr. *nema*: thread + *kystis*: cell]   An elaborate, thread-like structure produced by cells of jellyfish and other cnidarians, used chiefly to paralyze and capture prey.

Nephridium (nef rid' ee um) [Gr. *nephros*: kidney]   An organ which is involved in excretion, and often in water balance, involving a tube that opens to the exterior at one end.

Nephron (nef' ron) [Gr. *nephros*: kidney]   The basic component of the kidney, which is made up of numerous nephrons. Its form varies in detail, but it always has at one end a device for receiving a filtrate of blood, and then a tubule that absorbs selected parts of the filtrate back into the bloodstream.

Nephrostome (nef' ro stome) [Gr. *nephros*: kidney + *stoma*: opening]   An opening in a nephridium through which body fluids can enter.

Nerve   A structure consisting of many neuronal axons and connective tissue.

Net primary production   Total photosynthesis minus respiration by plants.

Neural plate   A thickened strip of ectoderm along the dorsal side of the early vertebrate embryo; gives rise to the central nervous system.

Neural tube   An early stage in the development of the vertebrate nervous system consisting of a hollow tube created by two opposing folds of the dorsal ectoderm along the anterior–posterior body axis.

Neuromuscular junction   The region where a motor neuron contacts a muscle fiber, creating a synapse.

Neuron (noor' on) [Gr. *neuron*: nerve, sinew]   A cell derived from embryonic ectoderm and characterized by a membrane potential that can change in response to stimuli, generating action potentials. Action potentials are generated along an extension of the cell (the axon), which makes junctions (synapses) with other neurons, muscle cells, or gland cells.

Neurotransmitter   A substance, produced in and released by one neuron, that diffuses across a synapse and excites or inhibits the postsynaptic neuron.

Neurula (nure' you la) [Gr. *neuron*: nerve]   Embryonic stage during formation of the dorsal nerve cord by two ectodermal ridges.

Neutral alleles   Alleles that differ so slightly that the proteins for which they code function identically.

Neutron (new' tron) [E.: neutral]   One of the three most fundamental particles of matter, with mass approximately 1 amu and no electrical charge.

Nicotinamide adenine dinucleotide   (See NAD.)

Nicotinamide adenine dinucleotide phosphate   (See NADP.)

Nitrification   The oxidation of ammonia to nitrite and nitrate ions, performed by certain soil bacteria.

Nitrogenase   In nitrogen-fixing organisms, an enzyme complex that mediates the stepwise reduction of atmospheric N_2 to ammonia.

Nitrogen fixation   Conversion of nitrogen gas to ammonia, which makes nitrogen available to living things. Carried out by certain prokaryotes, some of them free-living and others living within plant roots.

Node [L. *nodus*: knob, knot]   In plants, a (sometimes enlarged) point on a stem where a leaf or bud is or was attached.

Node of Ranvier   A gap in the myelin sheath covering an axons, where the axonal membrane can fire action potentials.

Noncompetitive inhibitor   An inhibitor that binds the enzyme at a site other than the active site. (Contrast with competitive inhibitor.)

Nondisjunction   Failure of sister chromatids to separate in meiosis II or mitosis, or failure of homologous chromosomes to separate in meiosis I. Results in aneuploidy.

Nonpolar molecule   A molecule whose electric charge is evenly balanced from one end of the molecule to the other.

Nonsense (chain-terminating) mutation   Mutations that change a codon for an amino acid to one of the codons (UAG, UAA, or UGA) that signal termination of translation. The resulting gene product is a shortened polypeptide that begins normally at the amino-terminal end and ends at the position of the altered codon. (Contrast with frame-shift mutation, missense mutation.)

Nonspecific defenses   Immunologic responses directed against most or all pathogens, generally without reference to the pathogens' antigens. These defenses include the skin, normal flora, lysozyme, the acidic stomach, interferon, and the inflammatory response.

Nontracheophytes   Those plants lacking well-developed vascular tissue; the liverworts, hornworts, and mosses. (Contrast with tracheophytes.)

Normal flora   The bacteria and fungi that live on animal body surfaces without causing disease.

Norepinephrine   A neurotransmitter found in the central nervous system and also at the postganglionic nerve endings of the sympathetic nervous system. Also called noradrenaline.

Notochord (no' tow kord) [Gr. *notos*: back + *chorde*: string]   A flexible rod of gelatinous material serving as a support in the embryos of all chordates and in the adults of tunicates and lancelets.

Nuclear envelope   The surface, consisting of two layers of membrane, that encloses the nucleus of eukaryotic cells.

Nucleic acid (new klay' ik) [E.: nucleus of a cell]   A long-chain alternating polymer of deoxyribose or ribose and phosphate groups, with nitrogenous bases—adenine, thymine, uracil, guanine, or cytosine (A, T, U, G, or C)—as side chains. DNA and RNA are nucleic acids.

Nucleoid (new' klee oid)   The region that harbors the chromosomes of a prokaryotic cell. Unlike the eukaryotic nucleus, it is not bounded by a membrane.

Nucleolar organizer (new klee' o lar)   A region on a chromosome that is associated with the formation of a new nucleolus following nuclear division. The site of the genes that code for ribosomal RNA.

Nucleolus (new klee' oh lus) [from L. diminutive of *nux*: little kernel or little nut] A small, generally spherical body found within the nucleus of eukaryotic cells. The site of synthesis of ribosomal RNA.

Nucleoplasm (new' klee o plazm)   The fluid material within the nuclear envelope of a cell, as opposed to the chromosomes, nucleoli, and other particulate constituents.

Nucleosome   A portion of a eukaryotic chromosome, consisting of part of the DNA molecule wrapped around a group of histone molecules, and held together by another type of histone molecule. The chromosome is made up of many nucleosomes.

Nucleotide   The basic chemical unit (monomer) in a nucleic acid. A nucleotide in RNA consists of one of four nitrogenous bases linked to ribose, which in turn is linked to phosphate. In DNA, deoxyribose is present instead of ribose.

Nucleus (new' klee us) [from L. diminutive of *nux*: kernel or nut]   (1) In chemistry, the dense central portion of an atom, made up of protons and neutrons, with a positive charge. Surrounded by a cloud of negatively charged electrons. (2) In cells, the centrally located chamber of eukaryotic cells that is bounded by a double membrane and contains the chromosomes. The information center of the cell.

Nutrient   A food substance; or, in the case of mineral nutrients, an inorganic element required for completion of the life cycle of an organism.

Obligate (ob' li gut)   Necessary, as in obligate anaerobe. (Contrast with facultative.)

Obligate anaerobe   An animal that can live only in oxygenated environments.

Oil   A triglyceride that is liquid at room temperature. (Contrast with fat.)

Okazaki fragments   Newly formed DNA strands making up the lagging strand in DNA replication. DNA ligase links the Okazaki fragments to give a continuous strand.

Olfactory   Having to do with the sense of smell.

Oligomer   A compound molecule of intermediate size, made up of two to a few monomers. (Contrast with monomer, polymer.)

Ommatidium [Gr. *omma*: an eye]   One of the units which, collected into groups of up to 20,000, make up the compound eye of arthropods.

Omnivore [L. *omnis*: all, everything + *vorare*: to devour]   An organism that eats both animal and plant material. (Contrast with carnivore, detritivore, herbivore.)

Oncogenic (ong' co jen' ik) [Gr. *onkos*: mass, tumor + *genes*: born]   Causing cancer.

Oocyte (oh' eh site) [Gr. *oon*: egg + *kytos*: cell]   The cell that gives rise to eggs in animals.

Oogenesis (oh' eh jen e sis) [Gr. *oon*: egg + *genesis*: source]   Female gametogenesis, leading to production of the egg.

Oogonium (oh' eh go' nee um)   In some algae and fungi, a cell in which an egg is produced.

Operator   The region of an operon that acts as the binding site for the repressor.

Operon   A genetic unit of transcription, typically consisting of several structural genes that are transcribed together; the operon contains at least two control regions: the promoter and the operator.

Opportunity cost   The sum of the benefits an animal forfeits by not being able to perform some other behavior during the time when it is performing a given behavior.

Opsin (op' sin) [Gr. *opsis*: sight]   The protein protion of the visual pigment rhodopsin. (See rhodopsin.)

Optic chiasm   Stucture on the lower surface of the vertebrate brain where the two optic nerves come together.

Optical isomers   Isomers that differ in the configuration of the four different groups attached to a single carbon atom; so named because solutions of the two isomers rotate the plane of polarized light in opposite directions. The two isomers are mirror images of one another.

Optimality models   Models developed to determine the structures or behaviors that best solve particular problems faced by organisms.

Order   In taxonomy, the category below the class and above the family; a group of related, similar families.

Organ   A body part, such as the heart, liver, brain, root, or leaf, composed of different tissues integrated to perform a distinct function for the body as a whole.

Organ identity genes   These plant genes specify the various parts of the flower.

Organ of Corti   Structure in the inner ear that transforms mechanical forces produced from pressure waves ("sound waves") into action potentials that are sensed as sound.

Organelles (or' gan els') [L.: little organ] Organized structures that are found in or on cells. Examples: ribosomes, nuclei, mitochrondria, chloroplasts, cilia, and contractile vacuoles.

Organic   Pertaining to any aspect of living matter, e.g., to its evolution, structure, or chemistry. The term is also applied to any chemical compound that contains carbon.

Organism   Any living creature.

Organizer, embryonic   A region of an embryo which directs the development of nearby regions. In amphibian early gastrulas, the dorsal lip of the blastopore.

Origin of replication   A DNA sequence at which helicase unwinds the DNA double helix and DNA polymerase binds to initiate DNA replication.

Osmoregulation   Regulation of the chemical composition of the body fluids of an organism.

Osmosensor   A neuron that converts changes in the osmotic potential of interstial fluids into action potentials.

Osmosis (oz mo' sis) [Gr. *osmos*: to push] The movement of water through a differentially permeable membrane from one region to another where the water potential is more negative. This is often a region in which the concentration of dissolved molecules or ions is higher, although the effect of dissolved substances may be offset by hydrostatic pressure in cells with semi-rigid walls.

Osmotic potential   A property of any solution, resulting from its solute content; it may be zero or have a negative value. A negative osmotic potential tends to cause water to move into the solution; it may be offset by a positive pressure potential in the solution or by a more negative water potential in a neighboring solution. (Contrast with turgor pressure.)

Ossicle (ah' sick ul) [L. *os*: bone]   The calcified construction unit of echinoderm skeletons.

Osteoblasts   Cells that lay down the protein matrix of bone.

Osteoclasts   Cells that dissolve bone.

Otolith (oh' tuh lith) [Gk.*otikos*: ear + *lithos*: stone]   Structures in the vertebrate vestibular apparatus that mechanically stimulate hair cells when the head moves or changes position.

Outgroup   A taxon that separated from another taxon, whose lineage is to be inferred, before the latter underwent evolutionary radiation.

Oval window   The flexible membrane which, when moved by the bones of the middle ear, produces pressure waves in the inner ear

Ovary (oh' var ee)   Any female organ, in plants or animals, that produces an egg.

Oviduct [L. *ovum*: egg + *ducere*: to lead]   In mammals, the tube serving to transport eggs to the uterus or to outside of the body.

Oviparous (oh vip' uh rus)   Reproduction in which eggs are released by the female and development is external to the mother's body. (Contrast with viviparous.)

Ovulation   The release of an egg from an ovary.

Ovule (oh' vule) [L. *ovulum*: little egg]   In plants, an organ that contains a gametophyte and, within the gametophyte, an egg; when it matures, an ovule becomes a seed.

Ovum (oh' vum) [L.: egg]   The egg, the female sex cell.

Oxidation (ox i day' shun)   Relative loss of electrons in a chemical reaction; either outright removal to form an ion, or the sharing of electrons with substances having a greater affinity for them, such as oxygen. Most oxidation, including biological ones, are associated with the liberation of energy. (Contrast with reduction.)

Oxidative phosphorylation ATP formation in the mitochondrion, associated with flow of electrons through the respiratory chain.

Oxidizing agent A substance that can accept electrons from another. The oxidizing agent becomes reduced; its partner becomes oxidized.

P generation Also called the parental generation. The individuals that mate in a genetic cross. Their immediate offspring are the F_1 generation.

Pacemaker That part of the heart which undergoes most rapid spontaneous contraction, thus setting the pace for the beat of the entire heart. In mammals, the sinoatrial (SA) node. Also, an artificial device, implanted in the heart, that initiates rhythmic contraction of the organ.

Pacinian corpuscle A sensory neuron surrounded by sheaths of connective tissue. Found in the deep layers of the skin, where it senses touch and vibration.

Pair rule genes Segmentation genes that divide the *Drosophila* larva into two segments each.

Paleontology (pale′ ee on tol′ oh jee) [Gr. *palaios*: ancient, old + *logos*: discourse] The scientific study of fossils and all aspects of extinct life.

Palisade parenchyma In leaves, one or several layers of tightly packed, columnar photosynthetic cells, frequently found just below the upper epidermis.

Pancreas (pan′ cree us) A gland, located near the stomach of vertebrates, that secretes digestive enzymes into the small intestine and releases insulin into the bloodstream.

Pangaea (pan jee′ uh) [Gk. *pan*: all, every] The single land mass formed when all the continents came together in the Permian period.

Parabronchi Passages in the lungs of birds through which air flows.

Paradigm A general framework within which some scientific discipline (or even the whole Earth) is viewed and within which questions are asked and hypotheses are developed. Scientific revolutions usually involve major paradigm changes.

Parapatric speciation Development of reproductive isolation among members of a continuous population in the absence of a geographical barrier. (Contrast with geographic, sympatric speciation.)

Paraphyletic taxon A taxon that includes some, but not all, of the descendants of a single ancestor.

Parasite An organism that attacks and consumes parts of an organism much larger than itself. Parasites sometimes, but not always, kill the host.

Parasitoid A parasite that is so large relative to its host that only one individual or at

most a few individuals can live within a single host.

Parasympathetic nervous system A portion of the autonomic (involuntary) nervous system. Activity in the parasympathetic nervous system produces effects such as decreased blood pressure and decelerated heart beat. (Contrast with sympathetic nervous system.)

Parathormone Hormone secreted by the parathyroid glands. Stimulates osteoclast activity and raises blood calcium levels.

Parathyroids Four glands on the posterior surface of the thyroid that produce and release parathormone.

Parenchyma (pair eng′ kyma) [Gr. *para*: beside + *enchyma*: infusion] A plant tissue composed of relatively unspecialized cells without secondary walls.

Parental investment Investment in one offspring or group of offspring that reduces the ability of the parent to assist other offspring.

Parsimony The principle of preferring the simplest among a set of plausible explanations of a phenomenon. Commonly employed in evolutionary and biogeographic studies.

Parthenocarpy Formation of fruit from a flower without fertilization.

Parthenogenesis (par′ then oh jen′ e sis) [Gr. *parthenos*: virgin + *genesis*: source] The production of an organism from an unfertilized egg.

Partial pressure The portion of the barometric pressure of a mixture of gases that is due to one component of that mixture. For example, the partial pressure of oxygen at sea level is 20.9% of barometric pressure.

Pasteur effect The sharp decrease in rate of glucose utilization when conditions become aerobic.

Pastoralism A nomadic form of human culture based on the tending of herds of domestic animals.

Patch clamping A technique for isolating a tiny patch of membrane to allow the study of ion movement through a particular channel.

Pathogen (path′ o jen) [Gr. *pathos*: suffering + *gignomai*: causing] An organism that causes disease.

Pattern formation In animal embryonic development, the organization of differentiated tissues into specific structures such as wings.

Pedigree The pattern of transmission of a genetic trait in a family.

Pelagic zone (puh ladj′ ik) [Gr. *pelagos*: the sea] The open waters of the ocean.

Penetrance Of a genotype, the proportion of individuals with that genotype who show the expected phenotype.

Penis (pee′ nis) [L.: tail] The male organ inserted into the female during sexual intercourse.

PEP carboxylase The enzyme that combines carbon dioxide with PEP to form a 4-carbon dicarboxylic acid at the start of C_4

photosynthesis or of Crassulacean acid metabolism (CAM).

Pepsin [Gr. *pepsis*: digestion] An enzyme, in gastric juice, that digests protein.

Peptide linkage The connecting group in a protein chain, –CO–NH–, formed by removal of water during the linking of amino acids, –COOH to –NH$_2$. Also called an amide linkage.

Peptidoglycan The cell wall material of many prokaryotes, consisting of a single enormous molecule that surrounds the entire cell.

Perennial (per ren′ ee al) [L. *per*: through + *annus*: a year] Referring to a plant that lives from year to year. (Contrast with annual, biennial.)

Perfect flower A flower with both stamens and carpels, therefore hermaphroditic.

Pericycle [Gr. *peri*: around + *kyklos*: ring or circle] In plant roots, tissue just within the endodermis, but outside of the root vascular tissue. Meristematic activity of pericycle cells produces lateral root primordia.

Periderm The outer tissue of the secondary plant body, consisting primarily of cork.

Period (1) A minor category in the geological time scale. (2) The duration of a cyclical event, such as a circadian rhythm.

Peripheral nervous system Neurons that transmit information to and from the central nervous system and whose cell bodies reside outside the brain or spinal cord.

Peristalsis (pair′ i stall′ sis) [Gr. *peri*: around + *stellein*: place] Wavelike muscular contractions proceeding along a tubular organ, propelling the contents along the tube.

Peritoneum The mesodermal lining of the coelom among coelomate animals.

Permease A protein in membranes that specifically transports a compound or family of compounds across the membrane.

Peroxisome An organelle that houses reactions in which toxic peroxides are formed. The peroxisome isolates these peroxides from the rest of the cell.

Petal In an angiosperm flower, a sterile modified leaf, nonphotosynthetic, frequently brightly colored, and often serving to attract pollinating insects.

Petiole (pet′ ee ole) [L. *petiolus*: small foot] The stalk of a leaf.

pH The negative logarithm of the hydrogen ion concentration; a measure of the acidity of a solution. A solution with pH = 7 is said to be neutral; pH values higher than 7 characterize basic solutions, while acidic solutions have pH values less than 7.

Phage (fayj) Short for bacteriophage.

Phagocyte A white blood cell that ingests microorganisms by endocytosis.

Phagocytosis [Gr.: *phagein* to eat; cell-eating] A form of endocytosis, the uptake of a solid particle by forming a pocket of plasma membrane around the particle and pinching off the pocket to form an intracellular particle bounded by membrane. (Contrast with pinocytosis.)

Pharynx [Gr.: throat] The part of the gut between the mouth and the esophagus.

Phenotype (fee' no type) [Gr. *phanein*: to show] The observable properties of an individual as they have developed under the combined influences of the genetic constitution of the individual and the effects of environmental factors. (Contrast with genotype.)

Pheromone (feer' o mone) [Gr. *phero*: carry + *hormon*: excite, arouse] A chemical substance used in communication between organisms of the same species.

Phloem (flo' um) [Gr. *phloos*: bark] In vascular plants, the food-conducting tissue. It consists of sieve cells or sieve tubes, fibers, and other specialized cells.

Phosphate group The functional group $-OPO_3H_2$; the transfer of energy from one compound to another is often accomplished by the transfer of a phosphate group.

Phosphodiester linkage The connection in a nucleic acid strand, formed by linking two nucleotides.

3-Phosphoglycerate The first product of photosynthesis, produced by the reaction of ribulose bisphosphate with carbon dioxide.

Phospholipids Cellular materials that contain phosphorus and are soluble in organic solvents. An example is lecithin (phosphatidyl choline). Phospholipids are important constituents of cellular membranes. (See lipids.)

Phosphorylation The addition of a phosphate group.

Photoautotroph An organism that obtains energy from light and carbon from carbon dioxide. (Contrast with chemoautotroph, chemoheterotroph, photoheterotroph.)

Photoheterotroph An organism that obtains energy from light but must obtain its carbon from organic compounds. (Contrast with chemoautotroph, chemoheterotroph, photoautotroph.)

Photon (foe' tohn) [Gr. *photos*: light] A quantum of visible radiation; a "packet" of light energy.

Photoperiod (foe' tow peer' ee ud) The duration of a period of light, such as the length of time in a 24-hour cycle in which daylight is present. The regulation of processes such as flowering by the changing length of day (or of night) is known as photoperiodism.

Photophosphorylation Photosynthetic reactions in which light energy trapped by chlorophyll is used to produce ATP and, in noncyclic photophosphorylation, is used to reduce $NADP^+$ to NADPH.

Photorespiration Light-driven uptake of oxygen and release of carbon dioxide, the carbon being derived from the early reactions of photosynthesis.

Photosensor A cell that senses and responds to light energy. Also called a **photoreceptor**.

Photosynthesis (foe tow sin' the sis) [literally, "synthesis out of light"] Metabolic processes, carried out by green plants, by which visible light is trapped and the energy used to synthesize compounds such as ATP and glucose.

Phototropism [Gr. *photos*: light + *trope*: a turning] A directed plant growth response to light.

Phylogenetic tree Graphic representation of lines of descent among organisms.

Phylogeny (fy loj' e nee) [Gr. *phylon*: tribe, race + *genesis*: source] The evolutionary history of a particular group of organisms; also, the diagram of the "family tree" that shows genetic linkages between ancestors and descendants.

Phylum (plural: phyla) [Gr. *phylon*: tribe, stock] In taxonomy, a high-level category just beneath kingdom and above the class; a group of related, similar classes.

Physiology (fiz' ee ol' o jee) [Gr. *physis*: natural form + *logos*: discourse, study] The scientific study of the functions of living organisms and the individual organs, tissues, and cells of which they are composed.

Phytoalexins Substances toxic to fungi, produced by plants in response to fungal infection.

Phytochrome (fy' tow krome) [Gr. *phyton*: plant + *chroma*: color] A plant pigment regulating a large number of developmental and other phenomena in plants; can exist in two different forms, one of which is active and the other is not. Different wavelengths of light can drive it from one form to the other.

Phytoplankton (fy' tow plangk' ton) [Gr. *phyton*: plant + *planktos*: wandering] The autotrophic portion of the plankton, consisting mostly of algae.

Pigment A substance that absorbs visible light.

Pilus (pill' us) [Lat. *pilus*: hair] A surface appendage by which some bacteria adhere to one another during conjugation.

Pinocytosis [Gr.: drinking cell] A form of endocytosis; the uptake of liquids by engulfing a sample of the external medium into a pocket of the plasma membrane followed by pinching off the pocket to form an intracellular vesicle. (Contrast with phagocytosis and endocytosis.)

Pistil [L. *pistillum*: pestle] The female structure of an angiosperm flower, within which the ovules are borne. May consist of a single carpel, or of several carpels fused into a single structure. Usually differentiated into ovary, style, and stigma.

Pith In plants, relatively unspecialized tissue found within a cylinder of vascular tissue.

Pituitary A small gland attached to the base of the brain in vertebrates. Its hormones control the activities of other glands. Also known as the hypophysis.

Placenta (pla sen' ta) [Gr. *plax*: flat surface] The organ found in most mammals that provides for the nourishment of the fetus and elimination of the fetal waste products.

Placental (pla sen' tal) Pertaining to mammals of the subclass Eutheria, a group characterized by the presence of a placenta; contains the majority of living species of mammals.

Plankton [Gr. *planktos*: wandering] The free-floating organisms of the sea and fresh water that for the most part move passively with the water currents. Consisting mostly of microorganisms and small plants and animals. (Contrast with nekton.)

Plant A member of the kingdom Plantae. Multicellular, gaining its nutrition by photosynthesis.

Planula (plan' yew la) [L. *planum*: something flat] The free-swimming, ciliated larva of the cnidarians.

Plaque (plack) [Fr.: a metal plate or coin] (1) A circular clearing in a turbid layer (lawn) of bacteria growing on the surface of a nutrient agar gel. Produced by successive rounds of infection initiated by a single bacteriophage. (2) An accumulation of prokaryotic organisms on tooth enamel. Acids produced by the metabolism of these microorganisms can cause tooth decay.

Plasma (plaz' muh) [Gr. *plassein*: to mold] The liquid portion of blood, in which blood cells and other particulates are suspended.

Plasma cell An antibody-secreting cell that developed from a B cell. The effector cell of the humoral immune system.

Plasma membrane The membrane that surrounds the cell, regulating the entry and exit of molecules and ions. Every cell has a plasma membrane.

Plasmid A DNA molecule distinct from the chromosome(s); that is, an extrachromosomal element. May replicate independently of the chromosome.

Plasmodesma (plural: plasmodesmata) [Gr. *plasma*: formed or molded + *desmos*: band] A cytoplasmic strand connecting two adjacent plant cells.

Plasmodium In the noncellular slime molds, a multinucleate mass of protoplasm surrounded by a membrane; characteristic of the vegetative feeding stage.

Plasmolysis (plaz mol' i sis) Shrinking of the cytoplasm and plasma membrane away from the cell wall, resulting from the osmotic outflow of water. Occurs only in cells with rigid cell walls.

Plastid Organelle in plants that serves for food manufacture (by photosynthesis) or food storage; bounded by a double membrane.

Platelet A membrane-bounded body without a nucleus, arising as a fragment of a cell in the bone marrow of mammals. Important to blood-clotting action.

Pleiotropy (plee' a tro pee) [Gr. *pleion*: more] The determination of more than one character by a single gene.

Pleural membrane [Gk. *pleuras*: rib, side] The membrane lining the outside of the lungs and the walls of the thoracic cavity. Inflammation of these membranes is a condition known as *pleurisy*.

Podocytes Cells of Bowman's capsule of the nephron that cover the capillaries of the glomerulus, forming filtration slits.

Poikilotherm (poy' kill o therm) [Gr. *poikilos*: varied + *therme*: heat] An animal whose body temperature tends to vary with the surrounding environment. (Contrast with homeotherm, heterotherm.)

Point mutation A mutation that results from a small, localized alteration in the chemical structure of a gene. Such mutations can give rise to wild-type revertants as a result of reverse mutation. In genetic crosses, a point mutation behaves as if it resided at a single point on the genetic map. (Contrast with deletion.)

Polar body A nonfunctional nucleus produced by meiosis, accompanied by very little cytoplasm. The meiosis which produces the mammalian egg produces in addition three polar bodies.

Polar molecule A molecule in which the electric charge is not distributed evenly in the covalent bonds.

Polar nucleus One of two nuclei derived from each end of the angiosperm embryo sac, both of which become centrally located. They fuse with a male nucleus to form the primary triploid nucleus that will prduce the endosperm tissue of the angiosperm seed.

Polarity In development, the difference between one end and the other. In chemistry, the property that makes a polar molecule.

Pollen [L.: fine powder, dust] The fertilizing element of seed plants, containing the male gametophyte and the gamete, at the stage in which it is shed.

Pollination Process of transferring pollen from the anther to the receptive surface (stigma) of the ovary in plants.

Poly- [Gr. *poly*: many] A prefix denoting multiple entities.

Polygamy [Gr. *poly*: many + *gamos*: marriage] A breeding system in which an individual acquires more than one mate. In polyandry, a female mates with more than one male, in polygyny, a male mates with more than one female.

Polygenes Multiple loci whose alleles increase or decrease a continuously variable phenotypic trait.

Polymer A large molecule made up of similar or identical subunits called monomers. (Contrast with monomer, oligomer.)

Polymerase chain reaction (PCR) A technique for the rapid production of millions of copies of a particular stretch of DNA.

Polymerization reactions Chemical reactions that generate polymers by means of condensation reactions.

Polymorphism (pol' lee mor' fiz um) [Gr. *poly*: many + *morphe*: form, shape] (1) In genetics, the coexistence in the same population of two distinct hereditary types based on different alleles. (2) In social organisms such as colonial cnidarians and social insects, the coexistence of two or more functionally different castes within the same colony.

Polyp The sessile, asexual stage in the life cycle of most cnidarians.

Polypeptide A large molecule made up of many amino acids joined by peptide linkages. Large polypeptides are called proteins.

Polyphyletic group A group containing taxa, not all of which share the most recent common ancestor.

Polyploid (pol' lee ploid) A cell or an organism in which the number of complete sets of chromosomes is greater than two.

Polysaccharide A macromolecule composed of many monosaccharides (simple sugars). Common examples are cellulose and starch.

Polysome A complex consisting of a threadlike molecule of messenger RNA and several (or many) ribosomes. The ribosomes move along the mRNA, synthesizing polypeptide chains as they proceed.

Polytene (pol' lee teen) [Gr. *poly*: many + *taenia*: ribbon] An adjective describing giant interphase chromosomes, such as those found in the salivary glands of fly larvae. The characteristic, reproducible pattern of bands and bulges seen on these chromosomes has provided a method for preparing detailed chromosome maps of several organisms.

Pons [L. *pons*: bridge] Region of the brain stem anterior to the medulla.

Population Any group of organisms coexisting at the same time and in the same place and capable of interbreeding with one another.

Population density The number of individuals (or modules) of a population in a unit of area or volume.

Population dynamics Changes in the distribution and abundance of individuals in a population.

Population structure The proportions of individuals in a population belonging to different age classes (age structure). Also, the distribution of the population in space.

Population vulnerability analysis (PVA) A determination of the risk of extinction of a population given its current size and distribution.

Portal vein A vein connecting two capillary beds, as in the hepatic portal system.

Positive control The situation in which a regulatory macromolecule is needed to turn transcription of structural genes on. In its absence, transcription will not occur.

Positive cooperativity Occurs when a molecule can bind several ligands and each one that binds alters the conformation of the molecule so that it can bind the next ligand more easily. The binding of four molecules of O_2 by hemoglobin is an example of positive cooperativity.

Positive feedback A regulatory system in which an error signal stimulates responses that increase the error.

Postabsorptive period When there is no food in the gut and no nutrients are being absorbed.

Posterior Toward or pertaining to the rear.

Postsynaptic cell The cell whose membranes receive the neurotransmitter released at a synapse.

Postzygotic isolating mechanism Any factor that reduces the viability of zygotes resulting from matings between individuals of different species.

Predator An organism that kills and eats other organisms. Predation is usually thought of as involving the consumption of animals by animals, but it can also mean the eating of plants.

Presynaptic excitation/inhibition Occurs when a neuron modifies activity at a synapse by releasing a neurotransmitter onto the presynaptic nerve terminal.

Prey [L. *praeda*: booty] An organism consumed as an energy source.

Prezygotic isolating mechanism A mechanism that reduces the probability that individuals of different species will mate.

Primary active transport Form of active transport in which ATP is hydrolyzed, yielding the energy required to transport ions against their concentration gradients. (Contrast with secondary active transport.)

Primary growth In plants, growth produced by the apical meristems. (Contrast with secondary growth.)

Primary producer A photosynthetic or chemosynthetic organism that synthesizes complex organic molecules from simple inorganic ones.

Primary succession Succession that begins in an areas initially devoid of life, such as on recently exposed glacial till or lava flows.

Primary structure The specific sequence of amino acids in a protein.

Primary wall Cellulose-rich cell wall layers laid down by a growing plant cell.

Primate (pry' mate) A member of the order Primates, such as a lemur, monkey, ape, or human.

Primer A short, single-stranded segment of DNA serving as the necessary starting material for the synthesis of a new DNA strand, which is synthesized from the 3' end of the primer.

Primitive streak A line running axially along the blastodisc, the site of inward cell migration during formation of the three-layered embryo. Formed in the embryos of birds and fish.

Primordium [L. *primordium*: origin] The most rudimentary stage of an organ or other part.

Pro- [L.: first, before, favoring] A prefix often used in biology to denote a developmental stage that comes first or an evolutionary form that appeared earlier than another. For example, prokaryote, prophase.

Probe A segment of single stranded nucleic acid used to identify DNA molecules containing the complementary sequence.

Procambium Primary meristem that produces the vascular tissue.

Progesterone [L. *pro*: favoring + *gestare*: to bear] A vertebrate female sex hormone that maintains pregnancy.

Prokaryotes (pro kar' ry otes) [L. *pro*: before + Gk. *karyon*: kernel, nucleus] Organisms whose genetic material is not contained within a nucleus. The bacteria. Considered an earlier stage in the evolution of life than the eukaryotes.

Prometaphase The phase of nuclear division that begins with the disintegration of the nuclear envelope.

Promoter The region of an operon that acts as the initial binding site for RNA polymerase.

Proofreading The correction of an error in DNA replication just after an incorrectly paired base is added to the growing polynucleotide chain.

Prophage (pro' fayj) The noninfectious units that are linked with the chromosomes of the host bacteria and multiply with them but do not cause dissolution of the cell. Prophage can later enter into the lytic phase to complete the virus life cycle.

Prophase (pro' phase) The first stage of nuclear division, during which chromosomes condense from diffuse, threadlike material to discrete, compact bodies.

Proplastid [Gr. *pro*: before + *plastos*: molded] A plant cell organelle which under appropriate conditions will develop into a plastid, usually the photosynthetic chloroplast. If plants are kept in the dark, proplastids may become quite large and complex.

Prostaglandin Any one of a group of specialized lipids with hormone-like functions. It is not clear that they act at any considerable distance from the site of their production.

Prosthetic group Any nonprotein portion of an enzyme.

Protease (pro' tee ase) See proteolytic enzyme.

Protein (pro' teen) [Gr. *protos*: first] One of the most fundamental building substances of living organisms. A long-chain polymer of amino acids with twenty different common side chains. Occurs with its polymer chain extended in fibrous proteins, or coiled into a compact macromolecule in enzymes and other globular proteins.

Proteolytic enzyme An enzyme whose main catalytic function is the digestion of a protein or polypeptide chain. The digestive enzymes trypsin, pepsin, and carboxypeptidase are all proteolytic enzymes (proteases).

Protist A member of the kingdom Protista, which consists of those eukaryotes not included in the kingdoms Animalia, Fungi, or Plantae. Many protists are unicellular. The kingdom Protista includes protozoa, algae, and fungus-like protists.

Protoderm Primary meristem that gives rise to epidermis.

Proton (pro' ton) [Gr. *protos*: first] One of the three most fundamental particles of matter, with mass approximately 1 amu and an electrical charge of +1.

Proton motive force The proton gradient and electric charge difference produced by chemiosmotic proton pumping. It drives protons back across the membrane, with the concomitant formation of ATP.

Protonema (pro' tow nee' mah) [Gr. *protos*: first + *nema*: thread] The hairlike growth form that constitutes an early stage in the development of a moss gametophyte.

Proto-oncogenes The normal alleles of genes possessing oncogenes (cancer-causing genes) as mutant alleles. Proto-oncogenes encode growth factors and receptor proteins.

Protoplast A cell that would normally have a cell wall, from which the wall has been removed by enzymatic digestion or by special growth conditions.

Protostome One of two major lines of animal evolution, characterized by spiral, determinate cleavage of the egg, and by schizocoelous development. (Contrast with deuterostome.)

Prototroph (pro' tow trofe') [Gr. *protos*: first + *trophein*: to nourish] The nutritional wild type, or reference form, of an organism. Any deviant form that requires growth nutrients not required by the prototrophic form is said to be a nutritional mutant, or auxotroph.

Protozoa A group of single-celled organisms classified by some biologists as a single phylum; includes the flagellates, amoebas, and ciliates. This textbook follows most modern classifications in elevating the protozoans to a distinct kingdom (Protista) and each of their major subgroups to the rank of phylum.

Provincialized A biogeographic term referring to the separation, by environmental barriers, of the biota into units with distinct species compositions.

Provirus See prophage.

Proximal Near the point of attachment or other reference point. (Contrast with distal.)

Pseudocoelom A body cavity not surrounded by a peritoneum. Characteristic of nematodes and rotifers.

Pseudogene A DNA segment that is homologous to a functional gene but contains a nucleotide change that prevents its expression.

Pseudoplasmodium [Gr. *pseudes*: false + *plasma*: mold or form] In the cellular slime molds such as *Dictyostelium*, an aggregation of single amoeboid cells. Occurs prior to formation of a fruiting structure.

Pseudopod (soo' do pod) [Gr. *pseudes*: false + *podos*: foot] A temporary, soft extension of the cell body that is used in location, attachment to surfaces, or engulfing particles.

Pulmonary Pertaining to the lungs.

Punctuated equilibrium An evolutionary pattern in which periods of rapid change are separated by longer periods of little or no change.

Pupa (pew' pa) [L.: doll, puppet] In certain insects (the Holometabola), the encased developmental stage that intervenes between the larva and the adult.

Pupil The opening in the vertebrate eye through which light passes.

Purine (pure' een) A type of nitrogenous base. The purines adenine and guanine are found in nucleic acids.

Purkinje fibers Specialized heart muscle cells that conduct excitation throughout the ventricular muscle.

Pyramid of biomass Graphical representation of the total masses at different trophic levels in an ecosystem.

Pyramid of energy Graphical representation of the total energy contents at different trophic levels in an ecosystem.

Pyrimidine (peer im' a deen) A type of nitrogenous base. The pyrimidines cytosine, thymine, and uracil are found in nucleic acids.

Pyrogen A substance that causes fever.

Pyruvate A three-carbon acid; the end product of glycolysis and the raw material for the citric acid cycle.

Q_{10} A value that compares the rate of a biochemical process or reaction over a 10°C range of temperature. A process that is not temperature-sensitive has a Q_{10} of 1. Values of 2 or 3 mean the reaction speeds up as temperature increases.

Quantum (kwon' tum) [L. *quantus*: how great] An indivisible unit of energy.

Quaternary structure Of aggregating proteins, the arrangement of polypeptide subunits.

R factor (resistance factor) A plasmid that contains one or more genes that encode resistance to antibiotics.

Radial symmetry The condition in which two halves of a body are mirror images of each other regardless of the angle of the cut, providing the cut is made along the center line. Thus, a cylinder cut lengthwise down its center displays this form of symmetry. (Contrast with bilateral symmetry.)

Radioisotope A radioactive isotope of an element. Examples are carbon-14 (^{14}C) and hydrogen-3, or tritium (^{3}H).

Radiotherapy Treatment, as of cancer, with X- or gamma rays.

Rain shadow A region of low precipitation on the leeward side of a mountain range.

Ramet The repeated morphological units of sessile, modular organisms. (Contrast with genet.)

Random drift Evolution (change in gene proportions) by chance processes alone.

Rate constant Of a particular chemical reaction, a constant which, when multiplied by the concentration(s) of reactant(s), gives the rate of the reaction.

Reactant A chemical substance that enters into a chemical reaction with another substance.

Reaction, chemical A process in which atoms combine or change bonding partners.

Reaction wood Modified wood produced in branches in response to gravitational stimulation. Gymnosperms produce compression wood that tends to push the branch up; angiosperms produce tension wood that tends to pull the branch up.

Realized niche The actual niche occupied by an organism; it differs from the fundamental niche because of the presence of other species.

Receptacle [L. *receptaculum*: reservoir] In an angiosperm flower, the end of the stem to which all of the various flower parts are attached.

Receptive field Of a neuron, the area on the retina from which the activity of that neuron can be influenced.

Receptor-mediated endocytosis A form of endocytosis in which macromolecules in the environment bind specific receptor proteins in the plasma membrane and are brought into the cell interior in coated vesicles.

Receptor potential The change in the resting potential of a sensory cell when it is stimulated.

Recessive See dominance.

Reciprocal altruism The exchange of altruistic acts between two or more individuals. The acts may be separated considerably in time.

Reciprocal crosses A pair of crosses, in one of which a female of genotype A mates with a male of genotype B and in the other of which a female of genotype B mates with a male of genotype A.

Recognition site (also called a restriction site) A sequence of nucleotides in DNA to which a restriction enzyme binds and then cuts the DNA.

Recombinant An individual, meiotic product, or single chromosome in which genetic materials originally present in two individuals end up in the same haploid complement of genes. The reshuffling of genes can be either by independent segragation, or by crossing over between homologous chromosomes. For example, a human may pass on genes from both parents in a single haploid gamete.

Recombinant DNA technology The application of genetic tools (restriction endonucleases, plasmids, and transformation) to the production of specific proteins by biological "factories" such as bacteria.

Rectum The terminal portion of the gut, ending at the anus.

Redox reaction A chemical reaction in which one reactant becomes oxidized and the other becomes reduced.

Reducing agent A substance that can donate electrons to another substance. The reducing agent becomes oxidized, and its partner becomes reduced.

Reduction (re duk' shun) Gain of electrons; the reverse of oxidation. Most reductions lead to the storage of chemical energy, which can be released later by an oxidation reaction. Energy storage compounds such as sugars and fats are highly reduced compounds. (Contrast with oxidation.)

Reflex An automatic action, involving only a few neurons (in vertebrates, often in the spinal cord), in which a motor response swiftly follows a sensory stimulus.

Refractory period Of a neuron, the time interval after an action potential, during which another action potential cannot be elicited.

Regulative development A pattern of animal embryonic development in which the fates of the first blastomeres are not absolutely fixed. (Contrast with mosaic development.)

Regulatory gene A gene that contains the information for making a regulatory macromolecule, often a repressor protein.

Releaser A sensory stimulus that triggers a fixed action pattern.

Releasing hormone One of several hypothalamic hormones that stimulates the secretion of anterior pituitary hormone.

REM sleep A sleep state characterized by dreaming, skeletal muscle relaxation, and rapid eye movements.

Renal [L. *renes*: kidneys] Relating to the kidneys.

Replica plating A technique used in the selection of colonies of cells with a desired genotype.

Replication fork A point at which a DNA molecule is replicating. The fork forms by the unwinding of the parent molecule.

Repressible enzyme An enzyme whose synthesis can be decreased or prevented by the presence of a particular compound. A repressible opren often controls the sythesis of such an enzyme.

Repressor A protein coded by the regulatory gene. The repressor can bind to a specific operator and prevent transcription of the operon.

Reproductive isolating mechanism Any trait that prevents individuals from two different populations from producing fertile hybrids.

Reproductive isolation The condition in which a population is not exchanging genes with other populations of the same species.

Reproductive value The expected contribution of an individual of a particular age to the future growth of the population to which it belongs.

Rescue effect The avoidance of extinction by immigration of individuals from other populations.

Resolving power Of an optical device such as a microscope, the smallest distance between two lines that allows the lines to be seen as separate from one another.

Resource Something in the environment required by an organism for its maintenance and growth that is consumed in the process of being used.

Resource defense polygamy A breeding system in which individuals of one sex (usually males) defend resources that are attractive to individuals of the other sex (usually females); individuals holding better resources attract more mates.

Respiration (res pi ra' shun) [L. *spirare*: to breathe] (1) Cellular respiration; the oxidation of the end products of glycolysis with the storage of much energy in ATP. The oxidant in the respiration of eukaryotes is oxygen gas. Some bacteria can use nitrate or sulfate instead of O_2. (2) Breathing.

Respiratory chain The terminal reactions of cellular respiration, in which electrons are passed from NAD or FAD, through a series of intermediate carriers, to molecular oxygen, with the concomitant production of ATP.

Respiratory uncoupler A substance that allows protons to cross the inner mitochondrial membrane without the concomitant formation of ATP, thus uncoupling respiration from phosphorylation.

Resting potential The membrane potential of a living cell at rest. In cells at rest, the interior is negative to the exterior. (Contrast with action potential, electrotonic potential.)

Restoration ecology The science and practice of restoring damaged or degraded ecosystems.

Restriction endonuclease Any one of several enzymes, produced by bacteria, that break foreign DNA molecules at very specific sites. Some produce "sticky ends." Extensively used in recombinant DNA technology.

Restriction map A partial genetic map of a DNA molecule, showing the points at which particular restriction endonuclease recognition sites reside.

Reticular system A central region of the vertebrate brain stem that includes complex fiber tracts conveying neural signals between the forebrain and the spinal cord, with collateral fibers to a variety of nuclei that are involved in autonomic functions, including arousal from sleep.

Retina (rett' in uh) [L. *rete*: net] The light-sensitive layer of cells in the vertebrate or cephalopod eye.

Retinal The light-absorbing portion of visual pigment molecules. Derived from β-carotene.

Retrovirus An RNA virus that contains reverse transcriptase. Its RNA serves as a template for cDNA production, and the cDNA is integrated into a chromosome of the mammalian host cell.

Reverse transcriptase An enzyme that catalyzes the production of DNA (cDNA), using RNA as a template; essential to the reproduction of retroviruses.

Reversion (genetic) A mutational event that restores wild type phenotype to a mutant.

RFLP (Restriction fragment length polymorphism) Coexistence of two or more patterns of restriction fragments (patterns produced by restriction enzymes), as revealed by a probe. The polymorphism reflects a difference in DNA sequence on homologous chromosomes.

Rhizoids (rye' zoids) [Gr. *rhiza*: root] Hairlike extensions of cells in mosses, liverworts, and a few vascular plants that serve the same function as roots and root hairs in vascular plants. The term is also applied to branched, rootlike extensions of some fungi and algae.

Rhizome (rye' zome) [Gr. *rhizoma*: mass of roots]   A special underground stem (as opposed to root) that runs horizontally beneath the ground.

Rhodopsin   A photopigment used in the visual process of transducing photons of light into changes in the membrane potential of photosensory cells.

Ribonucleic acid   See RNA.

Ribose (rye' bose)   A sugar of chemical formula $C_5H_{10}O_5$, one of the building blocks of ribonucleic acids.

Ribosomal RNA (rRNA)   Several species of RNA that are incorporated into the ribosome.

Ribosome   A small organelle that is the site of protein synthesis.

Ribozyme   An RNA molecule with catalytic activity.

Ribulose 1,5-bisphosphate (RuBP)   The compound in chloroplasts which reacts with carbon dioxide in the first reaction of the Calvin-Benson cycle.

Risk cost   The increased chance of being injured or killed as a result of performing a behavior, compared to resting.

RNA (ribonucleic acid)   A nucleic acid using ribose. Various classes of RNA are involved in the transcription and translation of genetic information. RNA serves as the genetic storage material in some viruses.

RNA polymerase   An enzyme that catalyzes the formation of RNA from a DNA template.

RNA splicing   The last stage of RNA processing in eukaryotes, in which the transcripts of introns are excised through the action of small nuclear ribonucleoprotein particles (snRNP).

Rods   Light-sensitive cells (photosensors) in the retina. (Contrast with cones.)

Root cap   A thimble-shaped mass of cells, produced by the root apical meristem, that protects the meristem and that is the organ that perceives the gravitational stimulus in root gravitropism.

Root hair   A specialized epidermal cell with a long, thin process that absorbs water and minerals from the soil solution.

Round dance   The dance performed on the vertical surface of a honeycomb by a returning honeybee forager when she has discovered a food source less than 100 meters from the hive.

Round window   A flexible membrane between the middle and inner ear that distributes pressure waves in the fluid of the inner ear.

rRNA   See ribosomal RNA.

Rubisco (RuBP carboxylase)   Enzyme that combines carbon dioxide with ribulose bisphosphate to produce 3-phosphoglycerate, the first product of C_3 photosynthesis. The most abundant protein on Earth.

Rumen (rew' mun)   The first division of the ruminant stomach. It stores and initiates bacterial fermentation of food. Food is regurgitated from the rumen for further chewing.

Ruminant   An herbivorous, cud-chewing mammal such as a cow, sheep, or deer, having a stomach consisting of four compartments.

S phase   In the cell cycle, the stage of interphase during which DNA is replicated. (Contrast with G_1 phase, G_2 phase.)

Sap   An aqueous solution of nutrients, minerals, and other substances that passes through the xylem of plants.

Saprobe [Gr. *sapros*:rotten + *bios*: life]   An organism (usually a bacterium or fungus) that obtains its carbon and energy directly from dead organic matter.

Sarcomere (sark' o meer) [Gr. *sark*: flesh + *meros*: a part]   The contractile unit of a skeletal muscle.

Saturated hydrocarbon   A compound consisting only of carbon and hydrogen, with the hydrogen atoms connected by single bonds.

Schizocoelous development   Formation of a coelom during embryological development by a splitting of mesodermal masses.

Schwann cell   A glial cell that wraps around part of the axon of a peripheral neuron, creating a myelin sheath.

Sclereid   A type of sclerenchyma cell, commonly found in nutshells, that is not elongated.

Sclerenchyma (skler eng' kyma)   A plant tissue composed of cells with heavily thickened cell walls, dead at functional maturity. The principal types of sclerenchyma cells are fibers and sclereids.

Second messenger   A signaling molecule that is created or actived inside the cell in response to activation of a receptor on the cell surface. The second messenger molecule then triggers the cell's response. An example is cyclic AMP.

Secondary active transport   Form of active transport in which ions or molecules are transported against their concentration gradient using energy obtained by relaxation of a gradient of sodium ion concentration rather than directly from ATP. (Contrast with primary active transport.)

Secondary compound   A compound synthesized by a plant that is not needed for basic cellular metabolism. Typically has an antiherbivore or antiparasite function.

Secondary growth   In plants, growth produced by vascular and cork cambia, contributing to an increase in girth. (Contrast with primary growth.)

Secondary structure   Of a protein, localized regularities of structure, such as the α helix and the β pleated sheet.

Secondary wall   Wall layers laid down by a plant cell that has ceased growing; often impregnated with lignin or suberin.

Second law of thermodynamics   States that in any real (irreversible) process, there is a decrease in free energy and an increase in entropy.

Second messenger   A compound, such as cyclic AMP, that is released within a target cell after a hormone or other "first messenger" has bound to a surface receptor on a cell; the second messenger triggers further reactions within the cell.

Secretin (si kreet' in)   A peptide hormone secreted by the upper region of the small intestine when acidic chyme is present. Stimulates the pancreatic duct to secrete bicarbonate ions.

Section   A thin slice, usually for microscopy, as a tangential section or a transverse section.

Seed   A fertilized, ripened ovule of a gymnosperm or angiosperm. Consists of the embryo, nutritive tissue, and a seed coat.

Seed crop   The number of seeds produced by a plant during a particular bout of reproduction.

Seedling   A young plant that has grown from a seed (rather than by grafting or by other means).

Segmentation genes   In insect larvae, genes that determine the number and polarity of larval segments.

Segment polarity genes   Genes that determine the boundaries and front-to-back organization of the segments in the *Drosophila* larva.

Segregation (genetic)   The separation of alleles, or of homologous chromosomes, from one another during meiosis so that each of the haploid daughter nuclei produced by meiosis contains one or the other member of the pair found in the diploid mother cell, but never both.

Selective permeability   A characteristic of a membrane, allowing certain substances to pass through while other substances are excluded.

Self-differentiating   Behavior that develops without experience with the normal objects toward which it is usually directed and without any practice. (See also instinct.)

Selfish act   A behavioral act that benefits its performer but harms the recipients.

Self-pollination   The fertilization of a plant by its own pollen. (Contrast with cross-pollination.)

Semelparous organism   An organism that reproduces only once in its lifetime. (Contrast with iteroparous.)

Semen (see' men) [L.: seed]   The thick, whitish liquid produced by the male reproductive organ in mammals, containing the sperm.

Semicircular canals   Part of the vestibular system of mammals.

Semiconservative replication   The common way in which DNA is synthesized. Each of the two partner strands in a double helix acts as a template for a new partner strand. Hence, after replication, each double helix consists of one old and one new strand.

Seminiferous tubules The tubules within the testes within which sperm production occurs.

Senescence [L. *senescere*: to grow old] Aging; deteriorative changes with aging.

Sensor A sensory cell; a cell transduces a physical or chemical stimulus into a membrane potential change.

Sensory neuron A neuron leading from a sensory cell to the central nervous system. (Contrast with motor neuron.)

Sepal (see' pul) One of the outermost structures of the flower, usually protective in function and enclosing the rest of the flower in the bud stage.

Septum [L.: partition] A membrane or wall between two cavities.

Sertoli cells Cells in the seminiferous tubules that nuture the developing sperm.

Serum That part of the blood plasma that remains after clots have formed and been removed.

Sessile (sess' ul) [L. *sedere*: to sit] Permanently attached; not moving.

Sertoli cells Cells in the seminiferous tubules that nuture the developing sperm.

Set point In a regulatory system, the threshold sensitivity to the feedback stimulus.

Sex chromosome In organisms with a chromosomal mechanism of sex determination, one of the chromosomes involved in sex determination. One sex chromosome, the X chromosome, is present in two copies in one sex and only one copy in the other sex. The autosomes, as opposed to the sex chromosomes, are present in two copies in both sexes. In many organisms, there is a second sex chromosome, the Y chromosome, that is found in only one sex—the sex having only one copy of the X.

Sexduction See F-duction.

Sex linkage The pattern of inheritance characteristic of genes located on the sex chromosomes of organisms having a chromosomal mechanism for sex determination. The sex that is diploid with respect to sex chromosomes can assume three genotypes: homozygous wild type, homozygous mutant, or heterozygous carrier. The other sex, haploid for sex chromosomes, is either hemizygous wild type or hemizygous mutant.

Sexuality The ability, by any of a multitude of mechanisms, to bring together in one individual genes that were originally carried by two different individuals. The capacity for genetic recombination.

Sexual selection Selection by one sex of characteristics in individuals of the opposite sex. Also, the favoring of characteristics in one sex as a result of competition among individuals of that sex for mates.

Shoot The aerial part of a vascular plant, consisting of the leaves, stem(s), and flowers.

Sieve plate In sieve tubes, the highly specialized end walls in which are concentrated the clusters of pores through which the protoplasts of adjacent sieve tube members are interconnected.

Sieve tube A column of specialized cells found in the phloem, specialized to conduct organic matter from sources (such as photosynthesizing leaves) to sinks (such as roots). Found principally in flowering plants.

Sieve tube member A single cell of a sieve tube, containing cytoplasm but relatively few organelles, with highly specialized perforated end walls leading to elements above and below.

Sign stimulus The single stimulus, or one out of a very few stimuli, by which an animal distinguishes key objects, such as an enemy, or a mate, or a place to nest, etc.

Signal sequence The sequence of a protein that directs the protein through a particular cellular membrane.

Signal transduction pathway The series of biochemical steps whereby a stimulus to a cell (such as a hormone or neurotransmitter binding to a receptor) is translated into a response of the cell.

Silencer A sequence of eukaryotic DNA that binds proteins that inhibit the transcription of an associated gene.

Silent mutations Genetic changes that do not lead to a phenotypic change. At the molecular level, these are DNA sequence changes that, because of the redundancy of the genetic code, result in the same amino acids in the resulting protein.

Similarity matrix A matrix to compare the structures of two molecules constructed by adding the number of their amino acids that are identical or different

Sinoatrial node (sigh' no ay' tree al) The pacemaker of the mammalian heart.

Sinus (sigh' nus) [L. *sinus*: a bend, hollow] A cavity in a bone, a tissue space, or an enlargement in a blood vessel.

Skeletal muscle See striated muscle.

Sliding filament theory A proposed mechanism of muscle contraction based on formation and breaking of crossbridges between actin and myosin filaments, causing them to slide together.

Small intestine The portion of the gut between the stomach and the colon, consisting of the duodenum, the jejunum, and the ileum.

Small nuclear ribonucleoprotein particle (snRNP) A complex of an enzyme and a small nuclear RNA molecule, functioning in RNA splicing.

Smooth muscle One of three types of muscle tissue. Usually consists of sheets of mononucleated cells innervated by the autonomic nervous system.

Society A group of individuals belonging to the same species and organized in a cooperative manner; in the broadest sense, includes parents and their offspring.

Sodium cotransport Carrier-mediated transport of molecules across membranes driven by sodium ions binding to the same carrier and moving down their concentration gradient.

Sodium–potassium pump The complex protein in plasma membranes that is responsible for primary active transport; it pumps sodium ions out of the cell and potassium ions into the cell, both against their concentration gradients.

Solute A substance that is dissolved in a liquid (solvent).

Solution A liquid (solvent) and its dissolved solutes.

Solvent A liquid that has dissolved or can dissolve one or more solutes.

Somatic [Gr. *soma*: body] Pertaining to the body, or body cells (rather than to germ cells).

Somite (so' might) One of the segments into which an embryo becomes divided longitudinally, leading to the eventual segmentation of the animal as illustrated by the spinal column, ribs, and associated muscles.

Southern blotting Transfer of DNA fragments from an electrophoretic gel to a sheet of paper or other absorbent material for analysis with a probe.

Spatial summation In the production or inhibition of action potentials in a postsynaptic neuron, the interaction of depolarizations and hyperpolarizations produced by several terminal boutons.

Spawning The direct release of sex cells into the water.

Specialized transduction In some types of bacteriophage (e.g., lambda), a prophage inserts at a specific location in the genome. When the prophage is induced to become lytic, it leaves the host chromosome and may take only the adjacent bacterial genes along with its phage DNA.

Speciation (spee' shee ay' shun) The process of splitting one population into two populations that are reproductively isolated from one another.

Species (spee' shees) [L.: kind] The basic lower unit of classification, consisting of a population or series of populations of closely related and similar organisms. The more narrowly defined "biological species" consists of individuals capable of interbreeding freely with each other but not with members of other species.

Species diversity A weighted representation of the species of organisms living in a region; large and common species are given greater weight than are small and rare ones. (Contrast with species richness.)

Species pool All the species potentially available to colonize a particular habitat.

Species richness The number of species of organisms living in a region. (Contrast with species diversity.)

Specific heat The amount of energy that must be absorbed by a gram of a substance to raise its temperature by one degree centigrade. By convention, water is assigned a specific heat of one.

Sperm [Gr. *sperma*: seed] A male reproductive cell.

Spermatocyte (spur mat' oh site) [Gr. *sperma*: seed + *kytos*: cell] The cell that gives rise to the sperm in animals.

Spermatogenesis (spur mat' oh jen' e sis) [Gr. *sperma*: seed + *genesis*: source] Male gametogenesis, leading to the production of sperm.

Spermatogonia Undifferentiated germ cells that give rise to primary spermatocytes and hence to sperm.

Sphincter (sfingk' ter) [Gr. *sphinkter*: that which binds tight] A ring of muscle that can close an orifice, for example at the anus.

Spindle apparatus An array of microtubules stretching from pole to pole of a dividing nucleus and playing a role in the movement of chromosomes at nuclear division. Named for its shape.

Spiracle (spy' rih kel) [L. *spirare*: to breathe] An opening of the treacheal respiratory system of terrestrial arthorpods.

Spiteful act A behavioral act that harms both the actor and the recipient of the act.

Spliceosome An RNA–protein complex that splices out introns from eukaryotic pre-mRNAs.

Splicing The removal of introns and connecting of exons in eukaryotic pre-mRNAs.

Spongy parenchyma In leaves, a layer of loosely packed photosynthetic cells with extensive intercellular spaces for gas diffusion. Frequently found between the palisade parenchyma and the lower epidermis.

Spontaneous generation The idea that life is generated continually from nonliving matter. Usually distinguished from the current idea that life evolved from nonliving matter under primordial conditions at an early stage in the history of earth.

Spontaneous reaction A chemical reaction which will proceed on its own, without any outside influence. A spontaneous reaction need not be rapid.

Sporangiophore [Gr. *phore*: to bear] Any branch bearing one or more sporangia.

Sporangium (spor an' gee um) [Gr. *spora*: seed + *angeion*: vessel or reservoir] In plants and fungi, any specialized stucture within which one or more spores are formed.

Spore [Gr. *spora*: seed] Any asexual reproductive cell capable of developing into an adult plant without gametic fusion. Haploid spores develop into gametophytes, diploid spores into sporophytes. In prokaryotes, a resistant cell capable of surviving unfavorable periods.

Sporophyll (spor' o fill) [Gr. *spora*: seed + *phyllon*: leaf] Any leaf or leaflike structure that bears sporangia; refers to carpels and stamens of angiosperms and to sporangium-bearing leaves on ferns, for example.

Sporophyte (spor' o fyte) [Gr. *spora*: seed + *phyton*: plant] In plants with alternation of generations, the diploid phase that produces the spores. (Contrast with gametophyte.)

Stabilizing selection Selection against the extreme phenotypes in a population, so that the intermediate types are favored. (Contrast with disruptive selection.)

Stamen (stay' men) [L.: thread] A male (pollen-producing) unit of a flower, usually composed of an anther, which bears the pollen, and a filament, which is a stalk supporting the anther.

Starch [O.E. *stearc*: stiff] An α-linked polymer of glucose; used by plants as a means of storing energy and carbon atoms.

Start codon The mRNA triplet (AUG) that acts as signals for the beginning of translation at the ribosome. (Compare with stop codons. There are a few mnior exceptions to these codons.)

Stasis Period during which little or no evolutionary change takes place within a lineage or groups of lineages.

Statocyst (stat' oh sist) [Gk. *statos*: stationary + *kystos*: pouch] An organ of equilibrium in some invertebrates.

Statolith (stat' oh lith) [Gk. *statos*: stationary + *lithos*: stone] A solid object that responds to gravity or movement and stimulates the mechanosensors of a statocyst.

Stele (steel) [Gr. *stele*: pillar] The central cylinder of vascular tissue in a plant stem.

Stem cell A cell capable of extensive proliferation, generating more stem cells and a large clone of differentiated progeny cells, as in the formation of red blood cells.

Step cline A sudden change in one or more traits of a species along a geographical gradient.

Steroid Any of numerous lipids based on a 17-carbon atom ring system.

Sticky ends On a piece of two-stranded DNA, short, complementary, one-stranded regions produced by the action of a restriction endonuclease. Sticky ends allow the joining of segments of DNA from different sources.

Stigma [L.: mark, brand] The part of the pistil at the apex of the style, which is receptive to pollen, and on which pollen germinates.

Stimulus Something causing a response; something in the environment detected by a receptor.

Stolon A horizontal stem that forms roots at intervals.

Stoma (plural: stomata) [Gr. *stoma*: mouth, opening] Small opening in the plant epidermis that permits gas exchange; bounded by a pair of guard cells whose osmotic status regulates the size of the opening.

Stop codons Triplets (UAG, UGA, UAA) in mRNA that act as signals for the end of translation at the ribosome. (See also start codon. There are a few mnior exceptions to these codons.)

Stratosphere The part of the atmosphere above the troposphere; extends upward to approximately 50 kilometers above the surface of the earth; contains very little water.

Stratum (plural strata) A layer or sedimentary rock laid down at a particular time in a past.

Striated muscle Contractile tissue characterized by multinucleated cells containing highly ordered arrangements of actin and myosin microfilaments. Also known as **skeletal muscle**.

Strobilus (strobe' a lus) [Gr. *strobilos*: a cone] The cone, or characteristic fruit, of the pine and other gymnosperms. Also, a cone-shaped mass of sprophylls found in club mosses.

Stroma The fluid contents of an organelle, such as a chloroplast.

Stromatolite A composite, flat-to-domed structure composed of successive mineral layers. Some are known to be produced by the action of bacteria in salt or fresh water, and some ancient ones are considered to be evidence for early life on the earth.

Structural formula A representation of the positions of atoms and bonds in a molecule.

Structural gene A gene that encodes the primary structure of a protein.

Style [Gr. *stylos*: pillar or column] In flowering plants, a column of tissue extending from the tip of the ovary, and bearing the stigma or receptive surface for pollen at its apex.

Sub- [L.: under] A prefix often used to designate a structure that lies beneath another or is less than another. For example, subcutaneous, subspecies.

Suberin A waxy material serving as a waterproofing agent in cork and in the Casparian strips of the endodermis in plants.

Submucosa (sub mew koe' sah) The tissue layer just under the epithelial lining of the lumen of the digestive tract. (Contrast with mucosa.)

Substrate (sub' strayte) The molecule or molecules on which an enzyme exerts catalytic action.

Substrate level phosphorylation ATP formation resulting from direct transfer of a phosphate group to ADP from an intermediate in glycolysis. (Contrast with oxidative phosphorylation.)

Succession In ecology, the gradual, sequential series of changes in species composition of a community following a disturbance.

Sulcus (plural: sulci) The valleys or creases between the raised portions of the convoluted surface of the brain. (Contrast to gyrus.)

Sulfhydryl group The —SH group.

Summation The ability of a neuron to fire action potentials in response to numerous subthreshold postsynaptic potentials arriving simultaneously at differentiated places on the cell, or arriving at the same site in rapid succession.

Suppressor T cells T cells that inhibit the responses of B cells and other T cells to antigens. (Contrast with cytotoxic T cells, helper T cells.)

Surface-to-volume ratio For any cell, organism, or geometrical solid, the ratio of surface area to volume; this is an important factor in setting an upper limit on the size a cell or organism can attain.

Surfactant A substance that decreases the surface tension of a liquid. Lung surfactant, secreted by cells of the alveoli, is mostly phospholipid and decreases the amount of work necessary to inflate the lungs.

Survivorship curve A plot of the logarithm of the fraction of individuals still alive, as a function of time.

Suspensor In plants, a cell or group of cells derived from the zygote, but not actually part of the embryo proper, which in some seed plants pushes the young embryo deeper into nutritive gametophyte tissue or endosperm by its growth.

Swim bladder An internal gas-filled organ that helps fishes maintain their position in the water column; later evolved into an organ for gas exchange in some lineages.

Symbiosis (sim′ bee oh′ sis) [Gr.: to live together] The living together of two or more species in a prolonged and intimate ecological relationship. (See parasitism, commensalism, mutualism.)

Symmetry In biology, the property that two halves of an object are mirror images of each other. (See bilateral symmetry and radial symmetry.)

Sympathetic nervous system A division of the autonomic (involuntary) nervous system. Its activities include increasing blood pressure and acceleration of the heartbeat. The neurotransmitter at the sympathetic terminals is epinephrine or norepinephrine. (Contrast with parasympathetic nervous system.)

Sympatric (sim pat′ rik) [Gr. *syn*: together + *patria*: homeland] Referring to populations whose geographic regions overlap at least in part.

Sympatric speciation Formation of new species even though members of the daughter species overlap in their distribution during the speciation process. (Contrast with geographic, parapatric speciation.)

Symplast The continuous meshwork of the interiors of living cells in the plant body, resulting from the presence of plasmodesmata. (Contrast with apoplast.)

Symport A membrane transport process that carries two substances in the same direction across the membrane. (Contrast with antiport.)

Synapse (sin′ aps) [Gr. *syn*: together + *haptein*: to fasten] The narrow gap between the terminal bouton of one neutron and the dendrite or cell body of another.

Synapsis (sin ap′ sis) The highly specific parallel alignment (pairing) of homologous chromosomes during the first division of meiosis.

Synaptic vesicle A membrane-bounded vesicle, containing neurotransmitter, which is produced in and discharged by the presynaptic neuron.

Synergids (sin nur′ jids) Two cells found close to the egg cell in the angiosperm embryo sac; they disappear shortly after fertilization.

Syngamy (sing′ guh mee) [Gr. *sun-*: together + *gamos*: marriage] Union of gametes. Also known as fertilization.

Syrinx (sear′ inks) [Gr.: pipe, cavity] A specialized structure at the junction of the trachea and the primary bronchi leading to the lungs. The vocal organ of birds.

Systematics The scientific study of the diversity of organisms.

Systemic circulation The part of the circulatory system serving those parts of the body other than the lungs or gills.

Systole (sis′ tuh lee) [Gr.: contraction] Contraction of a chamber of the heart, driving blood forward in the circulatory system.

T cell A type of lymphocyte, involved in the cellular immune response. The final stages of its development occur in the thymus gland. (Contrast with B cell; see also cytotoxic T cell, helper T cell, suppressor T cell.)

T cell receptor A protein on the surface of a T cell that recognizes the antigenic determinant for which the cell is specific.

T tubules A system of tubules that runs throughout the cytoplasm of muscle fibers, through which action potentials spread.

Target cell A cell with the appropriate receptors to bind and respond to a particular hormone or other chemical mediator.

Taste bud A structure in the epithelium of the tongue that includes a cluster of chemosensors innervated by sensory neurons.

TATA box An eight-base-pair sequence, found about 25 base pairs before the starting point for transcription in many eukaryotic promoters, that binds a transcription factor and thus helps initiate transcription.

Taxis (tak′ sis) [Gr. *taxis*: arrange, put in order] The movement of an organism in a particular direction with reference to a stimulus. A taxis usually involves the employment of one sense and a movement directly toward or away from the stimulus, or else the maintenance of a constant angle to it. Thus a positive phototaxis is movement toward a light source, negative geotaxis is movement upward (away from gravity), and so on.

Taxon A unit in a taxonomic system.

Taxonomy (taks on′ oh me) [Gr. *taxis*: arrange, classify] The science of classification of organisms.

Telomeres (tee′ lo merz) [Gr. *telos*: end] Repeated DNA sequences at the ends of eukaryotic chromosomes.

Telophase (tee′ lo phase) [Gr. *telos*: end] The final phase of mitosis or meiosis during which chromosomes became diffuse, nuclear envelopes reform, and nucleoli begin to reappear in the daughter nuclei.

Template In biochemistry, a molecule or surface upon which another molecule is synthesized in complementary fashion, as in the replication of DNA. In the brain, a pattern that responds to a normal input but not to incorrect inputs.

Template strand In a stretch of double-stranded DNA, the strand that is transcribed.

Temporal summation In the production or inhibition of action potentials in a postsynaptic neuron, the interaction of depolarizations or hyperpolarizations produced by rapidly repeated stimulation of a single point.

Tendon A collagen-containing band of tissue that connects a muscle with a bone.

Tepal In an angiosperm flower, a sterile modified leaf. This term is used to refer to such flower parts when one is unable to distinguish between petals and sepals.

Terminal transferase An enzyme that adds nucleotides to free ends of DNA, without reference to a template strand.

Terrestrial (ter res′ tree al) [L. *terra*: earth] Pertaining to the land.

Territory A fixed area from which an animal or group of animals excludes other members of the same species by aggressive behavior or display.

Tertiary structure In reference to a protein, the relative locations in three-dimensional space of all the atoms in the molecule. The overall shape of a protein. (Contrast with primary, secondary, and quaternary structures.)

Test cross A cross of a dominant-phenotype individual (which may be either heterozygous or homozygous) with a homozygous-recessive individual.

Testis (tes′ tis) (plural: testes) [L.: witness] The male gonad; that is, the organ that produces the male sex cells.

Testosterone (tes toss′ tuhr own) A male sex steroid hormone.

Tetanus [Gr. *tetanos*: stretched] (1) In physiology, a state of sustained, maximal muscular contraction caused by rapidly repeated stimulation. (2) In medicine, an often-fatal disease ("lockjaw") caused by the bacterium *Clostridium tetani*.

Thalamus A region of the vertebrate forebrain; involved in integration of sensory input.

Thallus (thal′ us) [Gr.: sprout] Any algal body which is not differentiated into root, stem, and leaf.

Thermocline In a body of water, the zone where the temperatures change abruptly to about 4°C.

Thermoneutral zone The range of temperatures over which an endotherm does not have to expend extra energy to thermoregulate.

Thermosensor A cell or structure that responds to changes in temperature.

Thoracic cavity The portion of the mammalian body cavity bounded by the ribs, shoulders, and diaphragm. Contains the heart and the lungs.

Thorax In an insect, the middle region of the body, between the head and abdomen. In mammals, the part of the body between the neck and the diaphragm.

Thrombin An enzyme that converts fibrinogen to fibrin, thus triggering the formation of blood clots.

Thrombus (throm′ bus) [Gk. *thrombos*: clot] A blood clot that forms within a blood vessel and remains attached to the wall of the vessel. (Contrast with embolus.)

Thylakoid A flattened sac within a chloroplast. The membranes of the numerous thylakoids contain all of the chlorophyll in a plant, in addition to the electron carriers of photophosphorylation. Thylakoids stack to form grana.

Thymine A nitrogen-containing base found in DNA.

Thymus A ductless, glandular portion of the lymphoid system, involved in development of the immune system of vertebrates.

Thyroid [Gr. *thyreos*: door-shaped] A two-lobed gland in vertebrates. Produces the hormone thyroxin.

Thyrotropic hormone A hormone that is produced in the pituitary gland of amphibia such as frogs and transported in the bloodstream to the thyroid gland, inducing the thyroid gland to produce the thyroid hormone that regulates metamorphosis from tadpole to adult frog.

Tight junction A junction between epithelial cells, in which there is no gap whatever between the adjacent cells. Materials may get through a tight junction only by entering the epithelial cells themselves.

Tissue A group of similar cells organized into a functional unit and usually integrated with other tissues to form part of an organ such as a heart or leaf.

Tonus A low level of muscular tension that is maintained even when the body is at rest.

Tornaria (tor nare' e ah) [L. *tornus*: lathe] The free-swimming ciliated larva of certain echinoderms and hemichordates; its existence indicates the evolutionary relationship of these two groups.

Totipotency In a cell, the condition of possessing all the genetic information and other capacities necessary to form an entire individual.

Toxigenicity The ability of a bacterium to produce chemical substances injurious to the tissues of the host organism.

Trachea (tray' kee ah) [Gr. *trakhoia*: a small tube] A tube that carries air to the bronchi of the lungs of vertebrates, or to the cells of arthropods.

Tracheid (tray' kee id) A distinctive conducting and supporting cell found in the xylem of nearly all vascular plants, characterized by tapering ends and walls that are pitted but not perforated.

Tracheophytes [Gr. *trakhoia*: a small tube + *phyton*: plant] Those plants with xylem and phloem, including psilophytes, club mosses, horsetails, ferns, gymnosperms, and angiosperms. (Contrast with nontrachoephytes.)

Trade winds The winds that blow toward the intertropical convergence zone from the northeast and southeast.

Trait One form of a character: Eye color is a character; brown eyes and blue eyes are traits.

Transcription The synthesis of RNA, using one strand of DNA as the template.

Transcription factors Proteins that assemble on a eukaryotic chromosome, allowing RNA polymerase II to perform transcription.

Transdetermination Alteration of the developmental fate of an imaginal disc in *Drosophila*.

Transduction (1) Transfer of genes from one bacterium to another, with a bacterial virus acting as the carrier of the genes. (2) In sensory cells, the transformation of a stimulus (e.g., light energy, sound pressure waves, chemical or electrical stimulants) into action potentials.

Transfection Uptake, incorporation, and expression of recombinant DNA.

Transfer cells A modified parenchyma cell that transports mineral ions from its cytoplasm into its cell wall, thus moving the ions from the symplast into the apoplast.

Transfer RNA (tRNA) A category of relatively small RNA molecules (about 75 nucleotides). Each kind of transfer RNA is able to accept a particular activated amino acid from its specific activating enzyme, after which the amino acid is added to a growing polypeptide chain.

Transformation Mechanism for transfer of genetic information in bacteria in which pure DNA extracted from bacteria of one genotype is taken in through the cell surface of bacteria of a different genotype and incorporated into the chromosome of the recipient cell. By extension, the term has come to be applied to phenomena in other organisms in which specific genetic alterations have been produced by treatment with purified DNA from genetically marked donors.

Transgenic Containing recombinant DNA incorporated into its genetic material.

Translation The synthesis of a protein (polypeptide). This occurs on ribosomes, using the information encoded in messenger RNA.

Translocation (1) In genetics, a rare mutational event that moves a portion of a chromosome to a new location, generally on a nonhomologous chromosome. (2) In vascular plants, movement of solutes in the phloem.

Transpiration [L. *spirare*: to breathe] The evaporation of water from plant leaves and stem, driven by heat from the sun, and providing the motive force to raise water (plus ions) from the roots.

Transposable element A segment of DNA that can move to, or give rise to copies at, another locus on the same or a different chromosome. May be a single insertion sequence or a more complex structure (transposon) consisting of two insertion sequences and one or more intervening genes.

Trichocyst (trick' o sist) [Gr. *trichos*: hair + *kystis*: cell] A threadlike organelle ejected from the surface of ciliates, used both as a weapon and as an anchoring device.

Triglyceride A simple lipid in which three fatty acids are combined with one molecule of glycerol.

Triplet See codon.

Triplet repeat Occurrence of repeated triplet of bases in a gene, often leading to genetic disease, as does excessive repetition of CGG in the gene responsible for fragile-X syndrome.

Triploblastic Having three cell layers. (Contrast with diploblastic.)

Trisomic Containing three, rather than two members of a chromosome pair.

tRNA See transfer RNA.

Trochophore (troke' o fore) [Gr. *trochos*: wheel + *phoreus*: bearer] The free-swimming larva of some annelids and mollusks, distinguished by a wheel-like band of cilia around the middle, and indicating an evolutionary relationship between these two groups.

Trophic level A group of organisms united by obtaining their energy from the same part of the food web of a biological community.

Tropic hormones Hormones of the anterior pituitary that control the secretion of hormones by other endocrine glands.

Tropism [Gr. *tropos*: to turn] In plants, growth toward or away from a stimulus such as light (phototropism) or gravity (gravitropism).

Tropomyosin (troe poe my' oh sin) A protein that, along with actin, constitutes the thin filaments of myofibrils. It controls the interactions of actin and myosin necessary for muscle contraction.

Troposphere The atmospheric zone reaching upward approximately 17 km in the tropics and subtropics but only to about 10 km at higher latitudes. The zone in which virtually all the water vapor in the atmosphere is located.

Trypsin A protein-digesting enzyme. Secreted by the pancreas in its inactive form (trypsinogen), it becomes active in the duodenum of the small intestine.

T-tubules A set of transverse tubes that penetrates skeletal muscle fibers and terminates in the sarcoplasmic reticulum. The T-system transmits impulses to the sacs, which then release Ca^{2+} to initiate muscle contraction.

Tube foot In echinoderms, a part of the water vascular system. It grasps the substratum, prey, or other solid objects.

Tube nucleus In a pollen tube, the haploid nucleus that does not participate in double fertilization. (Contrast with generative nucleus.)

Tuber [L.: swelling] A short, fleshy underground stem, usually much enlarged, and serving a storage function, as in the case of the potato.

Tubulin A protein that polymerizes to form microtubules.

Tumor A disorganized mass of cells, often growing out of control. Malignant tumors spread to other parts of the body.

Tumor suppressor genes Genes which, when homozygous mutant, result in cancer. Such genes code for protein products that inhibit cell proliferation.

Turgor pressure   The actual physical (hydrostatic) pressure within a cell. (Contrast with osmotic potential, water potential.)

Twitch   A single unit of muscle contraction.

Tympanic membrane [Gr. *tympanum*: drum] The eardrum.

Umbilical cord   Tissue made up of embryonic membranes and blood vessels that connects the embryo to the placenta in eutherian mammals.

Uncoupler   See respiratory uncoupler.

Understory   The aggregate of smaller plants growing beneath the canopy of dominant plants in a forest.

Unicellular (yoon' e sell' yer ler) [L. *unus*: one + *cella*: chamber]   Consisting of a single cell; as for example a unicellular organism. (Contrast with multicellular.)

Uniport   A membrane transport process that carries a single substance. (Contrast with antiport, symport.)

Unitary organism   An organism that consists of only one module.

Unsaturated hydrocarbon   A compound containing only carbon and hydrogen atoms. One or more pairs of carbon atoms are connected by double bonds.

Upwelling   The upward movement of nutrient-rich, cooler water from deeper layers of the ocean.

Urea   A compound serving as the main excreted form of nitrogen by many animals, including mammals.

Ureotelic   Describes an organism in which the final product of the breakdown of nitrogen-containing compounds (primarily proteins) is urea. (Contrast with ammonotelic, uricotelic.)

Ureter (your' uh tur) [Gr. *ouron*: urine]   A long duct leading from the vertebrate kidney to the urinary bladder or the cloaca.

Urethra (you ree' thra) [Gr. *ouron*: urine] In most mammals, the canal through which urine is discharged from the bladder and which serves as the genital duct in males.

Uric acid   A compound that serves as the main excreted form of nitrogen in some animals, particularly those which must conserve water, such as birds, insects, and reptiles.

Uricotelic   Describes an organism in which the final product of the breakdown of nitrogen-containing compounds (primarily proteins) is uric acid. (Contrast with ammonotelic, ureotelic.)

Urinary bladder   A structure structure that receives urine from the kidneys via the ureter, stores it, and expels it periodically through the urethra.

Urine (you' rin) [Gk. *ouron*: urine]   In vertebrates, the fluid waste product containing the toxic nitrogenous by-products of protein and amino acid metabolism.

Uterus (yoo' ter us) [L.: womb]   The uterus or womb is a specialized portion of the female reproductive tract in certain mammals. It receives the fertilized egg and nurtures the embryo in its early development.

Vaccination   Injection of virus or bacteria or their proteins into the body, to induce immunization. The injected material is usually attenuated (weakened) before injection.

Vacuole (vac' yew ole) [Fr.: small vacuum]   A liquid-filled cavity in a cell, enclosed within a single membrane. Vacuoles play a wide variety of roles in cellular metabolism, some being digestive chambers, some storage chambers, some waste bins, and so forth.

Vagina (vuh jine' uh) [L.: sheath]   In female mammals, the passage leading from the external genital orifice to the uterus; receives the copulatory organ of the male in mating.

Van der Waals interaction   A weak attraction between atoms resulting from the interaction of the electrons of one atom with the nucleus of the other atom. This attraction is about one-fourth as strong as a hydrogen bond.

Variable regions   The part of an immunoglobulin molecule or T-cell receptor that includes the antigen-binding site.

Vascular (vas' kew lar)   Pertaining to organs and tissues that conduct fluid, such as blood vessels in animals and phloem and xylem in plants.

Vascular bundle   In vascular plants, a strand of vascular tissue, including conducting cells of xylem and phloem as well as thick-walled fibers.

Vascular ray   In vascular plants, radially oriented sheets of cells produced by the vascular cambium, carrying materials laterally between the wood and the phloem.

Vascular tissue system   The conductive system of the plant, consisting primarily of xylem and phloem. (Contrast with dermal tissue system, ground tissue system.)

Vasopressin   See antidiuretic hormone.

Vector   (1) An agent, such as an insect, that carries a pathogen affecting another species. (2) A plasmid or virus that carries an inserted piece of DNA into a bacterium for cloning purposes in recombinant DNA technology.

Vegetal hemisphere   The lower portion of some animal eggs, zygotes, and embryos, in which the dense nutrient yolk settles. The **vegetal pole** refers to the very bottom of the egg or embryo. (Contrast with animal hemisphere.)

Vegetative   Nonreproductive, or nonflowering, or asexual.

Vein [L. *vena*: channel]   A blood vessel that returns blood to the heart. (Contrast with artery.)

Vena cava [L.: hollow vein]   One of a pair of large veins that carry blood from the systemic circulatory system into the heart.

Ventral [L. *venter*: belly, womb]   Toward or pertaining to the belly or lower side. (Contrast with dorsal.)

Ventricle   A muscular heart chamber that pumps blood through the body.

Vernalization [L. *vernalis*: belonging to spring]   Events occurring during a required chilling period, leading eventually to flowering. Vernalization may require many weeks of below-freezing temperatures.

Vertebral column   The jointed, dorsal column that is the primary support structure of vertebrates.

Vertebrate   An animal whose nerve cord is enclosed in a backbone of bony segments, called vertebrae. The principal groups of vertebrate animals are the fishes, amphibians, reptiles, birds, and mammals.

Vessel [L. *vasculum*: a small vessel]   In botany, a tube-shaped portion of the xylem consisting of hollow cells (vessel elements) placed end to end and connected by perforations. Together with tracheids, vessel elements conduct water and minerals in the plant.

Vestibular apparatus (ves tib' yew lar) [L. *vestibulum*: an enclosed passage]   Structures associated with the vertebrate ear; these structures sense changes in position or momentum of the head, affecing balance and motor skills.

Vestigial (ves tij' ee al) [L. *vestigium*: footprint, track]   The remains of body structures that are no longer of adaptive value to the organism and therefore are not maintained by selection.

Vicariance (vye care' ee unce) [L. *vicus*: change]   The splitting of the range of a taxon by the imposition of some barrier to dispersal of its members. May lead to cladogenesis.

Vicariant distribution   A distribution resulting from the disruption of a formerly continuous range by a vicariant event.

Villus (vil' lus) (plural: villi) [L.: shaggy hair]   A hairlike projection from a membrane; for example, from many gut walls.

Virion (veer' e on)   The virus particle, the minimum unit capable of infecting a cell.

Viroid (vye' roid)   An infectious agent consisting of a single-stranded RNA molecule with no protein coat; produces diseases in plants.

Virus [L.: poison, slimy liquid]   Any of a group of ultramicroscopic infectious particles constructed of nucleic acid and protein (and, sometimes, lipid) that can reproduce only in living cells.

Visceral mass   The major internal organs of a mollusk.

Vitamin [L. *vita*: life]   Any one of several structurally unrelated organic compounds that an organism cannot synthesize itself, but nevertheless requires in small quantity for normal growth and metabolism.

Viviparous (vye vip' uh rus) [L. *vivus*: alive]   Reproduction in which fertilization of the egg and development of the embryo occur inside the mother's body. (Contrast with oviparous.)

Waggle dance The running movement of a working honey bee on the hive, during which the worker traces out a repeated figure eight. The dance contains elements that transmit to other bees the location of the food.

Water potential In osmosis, the tendency for a system (a cell or solution) to take up water from pure water, through a differentially permeable membrane. Water flows toward the system with a more negative water potential. (Contrast with osmotic potential, turgor pressure.)

Water vascular system The array of canals and tubelike appendages that serves as the circulatory system, locomotory system, and food-capturing system of many echinoderms; is in direct connection with the surrounding sea water.

Wavelength The distance between successive peaks of a wave train, such as electromagnetic radiation.

Wild type Geneticists' term for standard or reference type. Deviants from this standard, even if the deviants are found in the wild, are said to be mutant.

Xanthophyll (zan' tho fill) [Gr. *xanthos*: yellowish-brown + *phyllon*: leaf] A yellow or orange pigment commonly found as an accessory pigment in photosynthesis, but found elsewhere as well. An oxygen-containing carotenoid.

X chromosome See sex chromosome.

X-linked (also called sex-linked) A character that is coded for by a gene on the X chromosome.

Xerophyte (zee' row fyte) [Gr. *xerox*: dry + *phyton*: plant] A plant adapted to an environment with a limited water supply.

Xylem (zy' lum) [Gr. *xylon*: wood] In vascular plants, the woody tissue that conducts water and minerals; xylem consists, in various plants, of tracheids, vessel elements, fibers, and other highly specialized cells.

Y chromosome See sex chromosome.

Yeast artificial chromosome A laboratory-made DNA molecule containing sequences of yeast chromosomes (origin or replication, telomeres, centromere, and selectable markers) so that it can be used as a vector in yeast.

Yolk The stored food material in animal eggs, usually rich in protein and lipid.

Z-DNA A form of DNA in which the molecule spirals to the left rather than to the right.

Zooplankton (zoe' o plang ton) [Gr. *zoon*: animal + *planktos*: wandering] The animal portion of the plankton.

Zoospore (zoe' o spore) [Gr. *zoon*: animal + *spora*: seed] In algae and fungi, any swimming spore. May be diploid or haploid.

Zygospore A highly resistant type of fungal spore produced by the zygomycetes (conjugating fungi).

Zygote (zye' gote) [Gr. *zygotos*: yoked] The cell created by the union of two gametes, in which the gamete nuclei are also fused. The earliest stage of the diploid generation.

Zymogen An inactive precursor of a digestive enzyme secreted into the lumen of the gut, where a protease cleaves it to form the active enzyme.

Answers to Self-Quizzes

Chapter 2
1. b   6. a
2. e   7. c
3. c   8. b
4. c   9. e
5. d   10. d

Chapter 3
1. e   6. a
2. d   7. c
3. c   8. e
4. d   9. a
5. b   10. d

Chapter 4
1. a   6. e
2. e   7. a
3. c   8. d
4. e   9. b
5. c   10. d

Chapter 5
1. e   6. b
2. d   7. c
3. a   8. b
4. d   9. e
5. c   10. c

Chapter 6
1. c   6. a
2. e   7. e
3. b   8. b
4. c   9. d
5. c   10. e

Chapter 7
1. a   6. d
2. d   7. a
3. c   8. e
4. e   9. c
5. c   10. e

Chapter 8
1. c   6. d
2. b   7. c
3. d   8. d
4. b   9. b
5. e   10. b

Chapter 9
1. e   6. a
2. c   7. e
3. b   8. d
4. d   9. b
5. c   10. a

Chapter 10*
1. d   6. d
2. a   7. b
3. e   8. a
4. d   9. b
5. d   10. c

Chapter 11
1. c   6. b
2. a   7. d
3. c   8. d
4. b   9. a
5. e   10. c

Chapter 12
1. c   6. d
2. d   7. b
3. c   8. d
4. d   9. d
5. b   10. a

Chapter 13
1. c   6. d
2. d   7. c
3. b   8. a
4. b   9. b
5. e   10. d

Chapter 14
1. c   6. c
2. d   7. c
3. c   8. b
4. a   9. e
5. c   10. d

Chapter 15
1. c   6. c
2. a   7. d
3. d   8. b
4. a   9. a
5. b   10. b

Chapter 16
1. b   6. b
2. d   7. c
3. a   8. a
4. c   9. c
5. b   10. e

Chapter 17
1. a   6. b
2. c   7. e
3. b   8. d
4. b   9. c
5. e   10. b

Chapter 18
1. d   6. a
2. b   7. d
3. e   8. d
4. e   9. a
5. c   10. d

Chapter 19
1. d   6. b
2. e   7. c
3. a   8. e
4. a   9. e
5. c   10. b

Chapter 20
1. d   6. e
2. c   7. b
3. d   8. e
4. b   9. d
5. d   10. e

Chapter 21
1. c   6. a
2. a   7. b
3. e   8. a
4. d   9. c
5. c   10. a

Chapter 22
1. e   6. d
2. c   7. a
3. a   8. d
4. c   9. b
5. a   10. d

Chapter 23
1. b   6. a
2. e   7. a
3. c   8. c
4. a   9. a
5. a   10. c

Chapter 24
1. e   6. c
2. d   7. a
3. e   8. a
4. c   9. d
5. e   10. b

Chapter 25
1. e   6. b
2. e   7. d
3. b   8. a
4. c   9. c
5. e   10. b

Chapter 26
1. a   6. d
2. e   7. c
3. c   8. b
4. d   9. b
5. a   10. d

Chapter 27
1. d   6. b
2. c   7. b
3. e   8. c
4. b   9. d
5. c   10. c

Chapter 28
1. b   6. a
2. d   7. e
3. e   8. a
4. c   9. c
5. d   10. c

Chapter 29
1. c   6. d
2. d   7. d
3. b   8. e
4. e   9. d
5. b   10. a

Chapter 30
1. b   7. d
2. c   8. e
3. a   9. a
4. d   10. e
5. c   11. c
6. b   12. c

Chapter 31
1. d   6. b
2. b   7. b
3. e   8. c
4. e   9. a
5. a   10. d

*Answers to the "Applying Concepts" questions in Chapter 10 appear at the end of this section.

Chapter 32

1. c	6. d
2. d	7. e
3. b	8. a
4. e	9. d
5. b	10. d

Chapter 33

1. e	6. a
2. b	7. b
3. c	8. c
4. c	9. d
5. d	10. a

Chapter 34

1. d	6. e
2. c	7. a
3. a	8. b
4. e	9. d
5. c	10. e

Chapter 35

1. a	6. c
2. e	7. e
3. c	8. c
4. d	9. a
5. b	10. b

Chapter 36

1. d	6. e
2. b	7. a
3. e	8. b
4. b	9. c
5. d	10. d

Chapter 37

1. c	6. e
2. a	7. a
3. d	8. e
4. b	9. d
5. b	10. c

Chapter 38

1. c	6. b
2. b	7. a
3. d	8. e
4. b	9. c
5. a	10. e

Chapter 39

1. (i) b	4. a
(ii) a	5. d
(iii) c	6. d
(iv) a,b,c	7. d
(v) a,b,c	8. d
2. c, e	9. c
3. e	10. d

Chapter 40

1. a	6. c
2. c	7. b
3. e	8. d
4. c	9. b
5. d	10. c

Chapter 41

1. d	6. e
2. a	7. e
3. d	8. c
4. c	9. d
5. c	10. a

Chapter 42

1. d	6. e
2. d	7. e
3. a	8. c
4. b	9. c
5. e	10. c

Chapter 43

1. c	6. c
2. a	7. a
3. e	8. b
4. d	9. a
5. d	10. b

Chapter 44

1. d	6. d
2. e	7. b
3. a	8. a
4. b	9. a
5. b	10. e

Chapter 45

1. e	6. b
2. d	7. c
3. a	8. c
4. b	9. a
5. c	10. d

Chapter 46

1. d	6. d
2. a	7. b
3. c	8. d
4. d	9. c
5. c	10. e

Chapter 47

1. b	6. d
2. e	7. a
3. c	8. b
4. a	9. d
5. b	10. d

Chapter 48

1. b	6. b
2. a	7. e
3. d	8. a
4. c	9. c
5. d	10. e

Chapter 49

1. a	6. c
2. e	7. d
3. b	8. c
4. b	9. d
5. a	10. a

Chapter 50

1. e	6. e
2. c	7. d
3. d	8. e
4. c	9. a
5. c	10. c

Chapter 51

1. b	6. e
2. e	7. a
3. d	8. c
4. a	9. d
5. d	10. a

Chapter 52

1. b	6. c
2. e	7. e
3. c	8. c
4. b	9. b
5. e	10. e

Chapter 53

1. b	6. a
2. d	7. e
3. d	8. a
4. b	9. d
5. c	10. e

Chapter 54

1. a	6. a
2. d	7. d
3. b	8. e
4. c	9. a
5. d	10. c

Chapter 55

1. b	6. a
2. c	7. d
3. c	8. e
4. e	9. e
5. a	10. a

Answers to "Applying Concepts" for Chapter 10, "Transmission Genetics: Mendel and Beyond"

1. Each of the eight boxes in the Punnet squares should contain the genotype Tt, regardless of which parent was tall and which dwarf.

2. Yellow parent = $s^Y s^b$; offspring 3 yellow (s^Y–): 1 black ($s^b s^b$). Black parent = $s^b s^b$; offspring all black ($s^b s^b$). Orange parent = $s^O s^b$; offspring 3 orange (s^O–): 1 black ($s^b s^b$). Both s^O and s^Y are dominant to s^b.

3. See Figure 10.5.

4. The trait is autosomal. Mother $dp\ dp$, father $Dp\ dp$. If the trait were sex-linked, all daughters would be wild-type and sons would be $dumpy$.

5. All females wild-type; all males spotted.

6. F_1 all wild-type, $PpSwsw$; F_2 9:3:3:1 in phenotypes. See Figure 10.8 for analogous genotypes.

7a. Ratio of phenotypes in F_2 is 3:1 (double dominant to double recessive).

7b. The F_1 are $Pby\ pB^Y$; they produce just two kinds of gametes (Pby and pBy). Combine them carefully and see the 1:2:1 phenotypic ratio fall out in the F_2.

7c. Pink-blistery.

7d. See Figures 9.15 and 9.17. Crossing over took place in the F_1 generation.

8. The genotypes are $PpSwsw$, $Ppswsw$, and $ppswsw$ in a ratio of 1:1:1:1.

9a. 1 black:2 blue:1 splashed white

9b. Always cross black with splashed white.

10a. $w^+ > w^e > w$

10b. Parents $w^e w$ and $w^+ Y$. Progeny $w+w^e$, $w+w$, $w^e Y$, and wY.

11. All will have normal vision because they inherit Dad's wild-type X chromosome, but half of them will be carriers.

12. Agouti parent $AaBb$. Albino offspring $aaBb$ and $aabb$; black offspring $Aabb$; agouti offspring $AaBb$.

13. Because the gene is carried on mitochondrial DNA, it is passed through the mother only. Thus if the woman does not have the disease but her husband does, their child will not be affected. On the other hand, if the woman has the disease but her husband does not, their child $will$ have the disease.

Illustration Credits

Authors' Photographs
William K. Purves by Mark Cameron
Gordon H. Orians by Elizabeth N. Orians
H. Craig Heller by Meera Heller
David Sadava by Mark Cameron

Table of Contents
Cell division (mitosis) in a bean: Spike Walker/Tony Stone Images
Galapagos iguana: Frans Lanting/Minden Pictures
Star coral: Linda Pitkin
Poppies: Jan Tove Hohansson
Frost-covered plants: Darrell Gulin/Tony Stone Images
Rainbow lorikeet: Hans Christian Heap/TCL
Iceberg: B. and C. Alexander/Photo Researchers, Inc.

Part-Opener Photographs
Part One: Dr. Kari Lounatmaa/Photo Researchers, Inc.
Part Two: C. L. Rieder
Part Three: Staffan Widstrand
Part Four: Ted Mead/Woodfall Wild Images
Part Five: Adam Jones
Part Six: James Beveridge/Visuals Unlimited
Part Seven: Carr Clifton/Minden Pictures

Chapter 1 *Opener*: William M. Smithey Jr./Planet Earth Pictures. 1.1: T. Stevens and P. McKinley/Photo Researchers, Inc. 1.2: B. Dowsett/Photo Researchers, Inc. 1.3: Fred Marsik/Visuals Unlimited. 1.4: Michael Abbey/Visuals Unlimited. 1.5 *larva*: Valorie Hodgson/Visuals Unlimited. 1.5 *pupa*: Dick Poe/Visuals Unlimited. 1.5 *butterfly*: Bill Beatty/Visuals Unlimited. 1.6: K. and K. Amman/Planet Earth Pictures. 1.7a: Staffan Widstrand. 1.7b: David Kjaer/Masterfile. 1.7c: Art Wolfe. 1.7d: Jack Fields/Photo Researchers, Inc. *C. Darwin*: American Philosophical Society. 1.10: Marty Snyderman/Planet Earth Pictures. 1.11: Bruce S. Cushing/Visuals Unlimited. 1.12: Levi, W. 1965. *Encyclopedia of Pigeon Breeds*. T. F. H. Publications, Jersey City, NJ. (*a,b*: photos by R. L. Kienlen, courtesy of Ralston Purina Company; *c,d*: photos by Stauber.). 1.13: Doug Perrine/Planet Earth Pictures. 1.14a: D. Cavanaugh/Visuals Unlimited. 1.14b: J. H. Robinson/Photo Researchers, Inc. 1.15a: Gregory G. Dimijian/Photo Researchers, Inc. 1.15b: Sinclair Stammers/Photo Researchers, Inc.

Chapter 2 *Opener*: Alex Williams/Masterfile. 2.3: Courtesy of Walter Gehring. 2.17: Art Wolfe. 2.20: P. Armstrong/Visuals Unlimited.

Chapter 3 *Opener*: Dan Richardson. 3.1: Christopher Small. 3.5: Gavriel Jecan/Art Wolfe Inc. 3.13a: Biophoto Associates/Photo Researchers, Inc. 3.13b: W. F. Schadel, BPS*. 3.13c: CNRI, Science World Enterprises/BPS. 3.18a,b,c; 3.19; 3.23: Dan Richardson.

Chapter 4 *Opener*: Jeremy Burgess/Photo Researchers, Inc. 4.3: J. J. Cardamone Jr. & B. K. Pugashetti/BPS. 4.4a,b: Stanley C. Holt/BPS. 4.5a: J. J. Cardamone Jr./BPS. 4.5c: S. Abraham & E. H. Beachey, VA Medical Center, Memphis, TN. Table 4.1 *upper row*: David M. Phillips/Visuals Unlimited. Table 4.1 *lower left*: Conly L. Rieder/BPS. Table 4.1 *lower center*: David Albertini, Tufts Univ. School of Medicine. Table 4.1 *lower right*: M. Abbey/Photo Researchers, Inc. 4.7 *centrioles*: Barry F. King/BPS. 4.7 *mitochondrion*: K. Porter, D. Fawcett/Visuals Unlimited. 4.7 *rough ER*: Fred E. Hossler/Visuals Unlimited. 4.7 *plasma membrane*: J. David Robertson, Duke Univ. Medical Center. 4.7 *nucleolus*: Richard Rodewald/BPS. 4.7 *golgi apparatus*: Gary T. Cole/BPS. 4.7 *smooth ER*: David M. Phillips/Visuals Unlimited. 4.7 *cell wall*: David M. Phillips/Visuals Unlimited. 4.7 *chloroplast*: W. P. Wergin, courtesy of E. H. Newcomb/BPS. 4.8 *upper*: Richard Rodewald/BPS. 4.8 *lower*, 4.9: Larry Gerace, Scripps Research Institute. 4.10: Jim Solliday/BPS. 4.11: K. Porter, D. Fawcett/Visuals Unlimited. 4.12: Alfred Owezarzak/BPS. 4.13: W. P. Wergin, courtesy of E. H. Newcomb/BPS. 4.14: Chuck Davis/Tony Stone Images. 4.16: Don Fawcett/Visuals Unlimited. 4.17a: Gary T. Cole/BPS. 4.18a: K. G. Murti/Visuals Unlimited. 4.19: E. H. Newcomb & S. E. Frederick/BPS. 4.20: M. C. Ledbetter, Brookhaven National Laboratory. 4.22a: H. W. Beams and R. G. Kessel. 1976. *Am. Sci.* 64: 279–290. 4.22b: Stanley Flegler/Visuals Unlimited. 4.23: Fred E. Hossler/Visuals Unlimited. 4.25a,b: W. L. Dentler/BPS. 4.27c: Barry F. King/BPS. 4.28: David M. Phillips/Visuals Unlimited.

*BPS = Biological Photo Service

Chapter 5 *Opener* and 5.3: L. Andrew Staehelin, Univ. of Colorado. 5.6 *top*: D. S. Friend, Univ. of California, SF. 5.6 *center*: Darcy E. Kelly, Univ. of Washington. 5.6 *bottom*: Courtesy of C. Peracchia. 5.15: M. M. Perry. 1979. *J. Cell Sci.* 39, p. 26.

Chapter 6 *Opener*: Greg Epperson/Adventure Photo & Film. 6.2: Jonathan Scott/Masterfile. 6.8: E. R. Degginger/Photo Researchers, Inc. 6.17: Clive Freeman, The Royal Institution/Photo Researchers, Inc. 6.19: Dan Richardson.

Chapter 7 *Opener*: Antonia Deutsch/Tony Stone Images. 7.9b: Francis Leroy/Photo Researchers, Inc. 7.14: Ephraim Racker/BPS.

Chapter 8 *Opener*: Susan McCartney/Photo Researchers, Inc. 8.1: C. G. Van Dyke/Visuals Unlimited. 8.16a: Lawrence Radiation Lab., Univ. of California. 8.16b: J. A. Bassham, Lawrence Berkeley Lab., Univ. of California. 8.22; 8.23b: E. H. Newcomb & S. E. Frederick/BPS. 8.25 *left*: Arthur R. Hill/Visuals Unlimited. 8.26 *right*: David Matherly/Visuals Unlimited.

Chapter 9 *Opener*: Steve Rogers & Vladimir Gelfand. 9.1a: John D. Cunningham/Visuals Unlimited. 9.1b: David M. Phillips/Visuals Unlimited. 9.1c: John D. Cunningham/Visuals Unlimited. 9.2: Ruth Kavenoff, Designergenes Ltd., P.O. Box 100, Del Mar, CA 90214. 9.3: John J. Cardamone Jr./BPS. 9.6: G. F. Bahr/BPS. 9.8 *upper inset*: A. L. Olins/BPS. 9.8 *lower inset*: David Ward, Yale Univ. School of Medicine. 9.9: Andrew S. Bajer, Univ. of Oregon. 9.10b,c: C. L. Rieder/BPS. 9.11a: T. E. Schroeder/BPS. 9.11b: B. A. Palevitz & E. H. Newcomb/BPS. 9.12: Gary T. Cole/BPS. 9.14: Courtesy of Applied Spectral Imaging. 9.15: C. A. Hasenkampf/BPS. 9.16: B. John Cabisco. 9.20b: Gopal Murti/Photo Researchers, Inc.

Chapter 10 *Opener*: Russell & Sons, ca. 1894. Gernsheim Collection, courtesy of the Harry Ransom Humanities Research Center, Univ. of Texas, Austin. *G. Mendel*: Leslie Holzer/Photo Researchers, Inc. 10.3: R. W. Van Norman/Visuals Unlimited. 10.14: NCI/Photo Researchers, Inc. 10.16: MERO/JACANA/Photo Researchers, Inc. *T. H. Morgan*: Corbis Bettmann. 10.26: Science

VU/Visuals Unlimited. *Bay scallops*: Barbara J. Miller/BPS.

Chapter 11 *Opener*: Dan Richardson. 11.2: Lee D. Simon/Photo Researchers, Inc. 11.4: Courtesy of Prof. M. H. F. Wilkins, Dept. of Biophysics, King's College, Univ. of London. 11.6b *left*: Dan Richardson. 11.6b *right*: A. Barrington Brown/Photo Researchers, Inc.

Chapter 12 *Opener* and 12.6: Dan Richardson. 12.12b: Courtesy of J. E. Edstrom and *EMBO J.* 12.14: Michael Abbey/Photo Researchers, Inc.

Chapter 13 *Opener*: Rosenfeld Images LTD/Photo Researchers, Inc. 13.1a: D. L. D. Caspar, Brandeis Univ. 13.1b: Omikron/Photo Researchers, Inc. 13.1c: A. B Dowsett/Photo Researchers, Inc. 13.7 *upper*: Hans Gelderblom/Visuals Unlimited. 13.7 *lower*: Science VU/Visuals Unlimited. 13.9: Rich Hambert/BPS. 13.11: Courtesy of L. Caro and R. Curtiss.

Chapter 14 *Opener*: Andrew Syred/Tony Stone Images. 14.9: Tiemeier et al., *Cell* 14:237–246, 1978. 14.19: Karen Dyer, Vivigen. 14.20: O. L. Miller, Jr.

Chapter 15 *Opener*: Courtesy of E. B. Lewis. 15.1: AP Photo/PA Files/Wide World Photo. 15.7: J. E. Sulston and H. R. Horvitz. 1977. *Dev. Bio.* 56, p. 100. 15.8: Courtesy of S. Strome. 15.17: Courtesy of W. Driever and C. Nüsslein-Volhard. 15.18: Courtesy of C. Rushlow and M. Levine.

Chapter 16 *Opener*: James Holmes/Fulmer Research/Photo Researchers, Inc. 16.3: Philippe Plailly/Photo Researchers, Inc. 16.6b *upper*: N. Y. State Agricultural Experiment Station, Cornell Univ. 16.6b *lower*: J. S. Yun & T. E. Wagner, Ohio Univ. 16.8: Courtesy of In Vitrogen Corporation. 16.16: Courtesy of Novartis Seeds.

Chapter 17 *Opener*: UPI/Corbis-Bettmann. 17.4: C. Harrison et al. 1983. *J. Med. Genet.* 20, p. 280. 17.9: Courtesy of Harvey Levy and Cecelia Walraven, New England Newborn Screening Program. 17.12: P. P. H. DeBruyn, Univ. of Chicago. 17.19: James King-Holmes/Photo Researchers, Inc.

Chapter 18 *Opener*: Andrew Syred/Photo Researchers, Inc. 18.2: G. W. Willis/Tony Stone Images. 18.3: Ziedonis Skobe/BPS. 18.5: Courtesy of Lennart Nilsson/Boehringer Ingelheim GmbH. 18.11: R. Rodewald/BPS. 18.17: A. Liepins, Sloan-Kettering Research Inst.

Chapter 19 *Opener*: National Institutes of Health/Photo Researchers, Inc. 19.5: Peter Ward, Univ. of Washington. 19.6: W. B. Saunders/BPS. 19.9 *left*: Ken Lucas/BPS. 19.9 *right*: Stanley M. Awramik/BPS. 19.10b: S. Conway Morris. 19.12b: Tom McHugh/Field Museum, Chicago/Photo Researchers, Inc. 19.13: B. Miller/BPS. 19.15: Ludek Pesek/Photo Researchers, Inc.

Chapter 20 *Opener*: Jon Riley/Tony Stone Images. 20.7: Frank S. Balthis. 20.12a,b,c: Tom Vezo. 20.13: From R. Dawkins. 1969. *The Blind Watchmaker*. W. W. Norton, New York.

Chapter 21 *Opener*: Ray Coleman/Photo Researchers, Inc. 21.1a,b: Tom Vezo. 21.7: Anthony D. Bradshaw, Univ. of Liverpool. 21.8a: Virginia P. Weinland/Photo Researchers, Inc. 21.8b: Heather Angel/BIOFOTOS. 21.10: M. Patterson/Photo Researchers, Inc. 21.11: Tom Vezo. 21.12 *left*, *right*: Peter J. Bryant/BPS. 21.12 *center*: Kenneth Y. Kaneshiro, Univ. of Hawaii. 21.15 *left*: Elizabeth Orians. 21.15 *center*, *right*: Jim Denny.

Chapter 22 *Opener*: Gary Brettnacher/Adventure Photo & Film. 22.1a: Michael Giannechini/Photo Researchers, Inc. 22.1b: Helen Carr/BPS. 22.1c: Nigel Downer/Planet Earth Pictures. 22.3 *left*: Adam Jones. 22.3 *right*: Staffan Widstand. 22.7 *left*, *right*: Peter J. Bryant/BPS. 22.9 *upper*, *lower*: Art Wolfe.

Chapter 23 *Opener*; 23.1: Richard Alexander, Univ. of Pennsylvania. 23.5: Courtesy of E. B. Lewis.

Chapter 24 *Opener*: NASA/Photo Researchers, Inc. 24.8a: Sinclair Stammers/Photo Researchers, Inc. 24.8b: Stanley M. Awramik/BPS.

Chapter 25 *Opener*: Kari Lounatmaa/Photo Researchers, Inc. 25.2: ASM/Visuals Unlimited. 25.3a: David Phillips/Photo Researchers, Inc. 25.3b: R. Kessel-G. Shih/Visuals Unlimited. 25.3c: Stanley Flegler/Visuals Unlimited. 25.4: T. J. Beveridge/BPS. 25.5a: J. A. Breznak and H. S. Pankratz/BPS. 25.5b: J. Robert Waaland/BPS. 25.6: George Musil/Visuals Unlimited. 25.7a *left*: S. C. Holt/BPS. 25.7a *center*: David M. Phillips/Visuals Unlimited. 25.7b *left*: Leon J. LeBeau/BPS. 25.7b *center*: A. J. J. Cardamone, Jr./BPS. 25.8: Alfred Pasieka/Photo Researchers, Inc. 25.9: H. W. Jannasch/BPS. 25.12: S. C. Holt/BPS. 25.13a: Paul W. Johnson/BPS. 25.13b: H. S. Pankratz/BPS. 25.13c: Bill Kamin/Visuals Unlimited. 25.14a: Paul W. Johnson/BPS. 25.14b: K. Stephens/BPS. 25.15: Science VU/Visuals Unlimited. 25.16: G. W. Willis/BPS. 25.17: D. A. Glawe/BPS. 25.18: Randall C. Cutlip/BPS. 25.19: T. J. Beveridge/BPS. 25.20: G. W. Willis/BPS. 25.21: Science VU/Visuals Unlimited. 25.22: Michael Gabridge/Visuals Unlimited. 25.24: Krafft-Explorer. 25.25: Martin G. Miller/Visuals Unlimited.

Chapter 26 *Opener*: M. Abbey/Visuals Unlimited. 26.1a: David Phillips/Visuals Unlimited. 26.1b: J. Paulin/Visuals Unlimited. 26.1c: Hal Beral/Visuals Unlimited. 26.7a: E. R. Degginger/Photo Researchers, Inc. 26.7b: Cabisco/Visuals Unlimited. 26.7c: M. Abbey/Photo Researchers, Inc.

26.8: David M. Phillips/Visuals Unlimited. 26.9: G. W. Willis/BPS. 26.11a: Robert Brons/BPS. 26.11b: A. M. Siegelman/Visuals Unlimited. 26.13a: Mike Abbey/Visuals Unlimited. 26.13b: M. Abbey/Photo Researchers, Inc. 26.13c,d: Paul W. Johnson/BPS. 26.14b: M. A. Jakus, NIH. 26.16: John D. Cunningham/Visuals Unlimited. 26.18a: Barbara J. Miller/BPS. 26.18b: Cabisco/Visuals Unlimited. 26.19a: D. W. Francis, Univ. of Delaware. 26.19b: David Scharf. 26.20: James W. Richardson/Visuals Unlimited. 26.21: Sanford Berry/Visuals Unlimited. 26.22a: Jan Hinsch/Photo Researchers, Inc. 26.22b: Biophoto Assoc./Photo Researchers, Inc. 26.24a: J. N. A. Lott/BPS. 26.24b: J. Robert Waaland/BPS. 26.25a: Paul Gier/Visuals Unlimited. 26.25b: J. N. A. Lott/BPS. 26.27a: Maria Schefter/BPS. 26.27b: J. N. A. Lott/BPS. 26.28a: Cabisco/Visuals Unlimited. 26.28b: Andrew J. Martinez/Photo Researchers, Inc. 26.28c: M. Abbey/Visuals Unlimited.

Chapter 27 *Opener upper*: Staffan Widstrand. *Opener lower*: Michael Busselle/Masterfile. 27.1: Peter F. Zika/Visuals Unlimited. 27.2: Ron Dengler/Visuals Unlimited. 27.4: Brian Enting/Photo Researchers, Inc. 27.5a,b: J. Robert Waaland/BPS. 27.6a: John D. Cunningham/Visuals Unlimited. 27.6b: William Harlow/Photo Researchers, Inc. 27.6c: Science VU/Visuals Unlimited. 27.8b: J. H. Troughton. 27.9: Farrell Grehan/Photo Researchers, Inc. 27.11: Figure information provided by Hermann Pfefferkorn, Dept. of Geology, Univ. of Pennsylvania. Original oil painting by John Woolsey. 27.16a: Bill Beatty/Visuals Unlimited. 27.16b: Cabisco/Visuals Unlimited. 27.17a: J. N. A. Lott/BPS. 27.17b: David Sieren/Visuals Unlimited. 27.18: W. Ormerod/Visuals Unlimited. 27.19a: Adam Jones. 27.19b: Nuridsany et Perennou/Photo Researchers, Inc. 27.19c: Dick Keen/Visuals Unlimited. 27.20: L. West/Photo Researchers, Inc. 27.23: Phil Gates/BPS. 27.24a: Richard Shiell. 27.24b: Bernd Wittich/Visuals Unlimited. 27.24c: M. Graybill/J. Hodder/BPS. 27.24d: Louisa Preston/Photo Researchers, Inc. 27.27a: Dick Poe/Visuals Unlimited. 27.27b,c; 27.28a: Richard Shiell. 27.28b: Noboru Komine/Photo Researchers, Inc. 27.30a: Ross Frid/Visuals Unlimited. 27.30b: Holt Studios/Photo Researchers, Inc. 27.30c: Catherine M. Pringle/BPS. 27.30d: Richard Shiell. 27.32a: Dora Lambrecht/Visuals Unlimited. 27.32b: Helen Carr/BPS. 27.32c; 27.33a: Richard Shiell. 27.33b: Adam Jones. 27.33c: Richard Shiell. 27.34: Rod Planck/Photo Researchers, Inc.

Chapter 28 *Opener*: Adam Jones. 28.1a: Jim W. Grace/Photo Researchers, Inc. 28.1b: L. E. Gilbert/BPS. 28.1c: G. L. Barron/BPS. 28.2: G. T. Cole/BPS. 28.3: N. Allin and G. L. Barron/BPS. 28.4: Milton H. Tierney, Jr./Visuals Unlimited. 28.5 *upper*: R. Calentine/Visuals Unlimited. 28.5 *lower*: John D. Cunningham/Visuals Unlimited. 28.7: J. Robert Waaland/BPS. 28.8: Gary R. Robinson/Visuals Unlimited. 28.9: James W. Richardson/Visuals Unlimited. 28.10: John D. Cunningham/Visuals Unlimited. 28.11a: Jim Solliday/BPS. 28.11b: L. West, National Audubon Society/Photo Researchers, Inc. 28.11c: Ray Coleman/Photo Researchers, Inc. 28.12: Centers for Disease Control, Atlanta. 28.14a: Angelina Lax/Photo Researchers, Inc. 28.14b: Manfred Danegger/Photo Researchers, Inc. 28.14c: Stan Flegler/Visuals Unlimited. 28.15: Biophoto Associates. 28.16: R. L. Peterson/BPS. 28.17: Paul A. Zahl/Photo Researchers, Inc. 28.18a: Gregory G. Dimijian/Photo Researchers, Inc. 28.18b: George Herben Photo/Visuals Unlimited. 28.18c: David Sieren/Visuals Unlimited. 28.19a: J. N. A. Lott/BPS.

Chapter 29 *Opener:* Rod Salm/Planet Earth Pictures. 29.1: Courtesy of R. L. Trelsted. 29.6 *left:* Gillian Lythgoe/Planet Earth Pictures. 29.6 *right:* Christian Petron/Planet Earth Pictures. 29.13a: Georgette Douwma/Planet Earth Pictures. 29.13b: Rod Salm/Planet Earth Pictures. 29.14a: Robert Brons/BPS. 29.14b,c: Chuck Davis Photo. 29.15: David J. Wrobel/BPS. 29.17b, 29.19b: James Solliday/BPS. 29.19c: Jim Solliday/BPS. 29.21a: Fred McConnaughey/Photo Researchers, Inc. 29.21c: Robert Brons/BPS. 29.25c: R. R. Hessler, Scripps Institute of Oceanography. 29.27a: Georgette Douwma/Planet Earth Pictures. 29.27c: Roger K. Burnard/BPS. 29.27d: Roger & Linda Mitchell. 29.29a: Ken Lucas/Planet Earth Pictures. 29.29b: Pete Atkinson/Masterfile. 29.29c: Chuck Davis Photo. 29.29d: Richard Humbert/BPS. 29.29e: Alex Kerstitch/Planet Earth Pictures. 29.29f: David J. Wrobel/BPS. 29.32: J. N. A. Lott/BPS. 29.33a: Joel Simon. 29.33b: Barbara Miller/BPS. 29.34a: Peter J. Bryant/BPS. 29.34b: David Maitland/Masterfile. 29.34c: L. E. Gilbert/BPS. 29.34d: Robert Brons/BPS. 29.36a: Gary Bell/Masterfile. 29.36b: Peter J. Byrant/BPS. 29.36c: David Wrobel/BPS. 29.36d: Geoff du Feu/Planet Earth Pictures. 29.37a: Charles R. Wyttenbach/BPS. 29.37b: Roger K. Burnard/BPS. 29.38a: R. F. Ashley/Visuals Unlimited. 29.38b: From Kristensen and Hallas, 1980. 29.39a: Richard Humbert/BPS. 29.39b,c: Peter J. Bryant/BPS. 29.39d,g: David Maitland/Masterfile. 29.39e: Steve Nicholls/Planet Earth Pictures. 29.39f: Brian Kenney/Planet Earth Pictures. 29.39h: Steve Hopkin/Planet Earth Pictures.

Chapter 30 *Opener:* Tim Davis/Tony Stone Images. 30.3; 30.4: David Wrobel/BPS. 30.5a: Ken Lucas/Planet Earth Pictures. 30.9a: Doug Perrine/DRK PHOTO. 30.9b–e: David Wrobel/BPS. 30.10: C. R. Wyttenbach/BPS. 30.11: Gary Bell/Masterfile. 30.12: M. Laverack/Planet Earth Pictures. 30.15: H. W. Pratt/BPS. 30.17a: David Wrobel/BPS. 30.17b: Marty Snyderman/Masterfile. 30.18a,d: David Wrobel/BPS. 30.18b: Doug Perrine/Masterfile. 30.18c: Ken Lucas/Planet Earth Pictures. 30.19: Peter Scoones/Planet Earth Pictures. 30.21a: Ken Lucas/BPS. 30.21b: Nick Garbutt/Indri Images. 30.21c: Art Wolfe. 30.24a: David J.

Wrobel/BPS. 30.24b: Carl Gans/BPS. 30.24c,d: Brian Kenney/Masterfile. 30.25a: Gavriel Jecan/Art Wolfe Inc. 30.25b: Art Wolfe. 30.26: Courtesy of Carnegie Museum of Natural History, Pittsburgh. 30.28a: Peter Scoones/Masterfile. 30.28b: Larry Tackett/Masterfile. 30.28c,d: Adam Jones. 30.30: Fritz Prenzel/Animals Animals. 30.31a: Art Wolfe. 30.31b: Jany Sauvanet/Photo Researchers, Inc. 30.32a: Staffan Widstrand. 30.32b: Merlin D. Tuttle, Bat Conservation International. 30.32c: Carol Farneti/Masterfile. 30.32d: Theo Allofs. 30.34a,b,c: Art Wolfe. 30.35a: Steve Kaufman/DRK PHOTO. 30.35b: John Bracegirdle/Masterfile. 30.36a: Art Wolfe. 30.36b: Kennan Ward/DRK Photo. 30.36c: K. & K. Ammann/Masterfile. 30.36d: Brian Kenney/Planet Earth Pictures. 30.38a: The American Museum of Natural History. 30.38b,c: Terraphotographics/BPS.

Chapter 31 *Opener:* Rod Planck/Photo Researchers, Inc. 31.1a: Hubertus Kanus/Photo Researchers, Inc. 31.2a: Michael P. Gadomski/Bruce Coleman Inc. 31.3a: D. Waugh/Tony Stone Images. 31.4: Nigel Cattlin/Holt Studios International/Photo Researchers, Inc. 31.7a: John Kaprielian/Photo Researchers, Inc. 31.7b, 31.8: John D. Cunningham/Visuals Unlimited. 31.9a: J. Robert Waaland/BPS. 31.9b: Joyce Photographics/Photo Researchers, Inc. 31.9c: Renee Lynn/Photo Researchers, Inc. 31.13a,d: Phil Gates/BPS. 31.13b: Biophoto Associates/Photo Researchers, Inc. 31.13c: Jack M. Bostrack/Visuals Unlimited. 31.13e: John D. Cunningham/Visuals Unlimited. 31.13f; 31.15b; 31.16 *upper, lower:* J. Robert Waaland/BPS. 31.18a: Jim Solliday/BPS. 31.18b: Microfield Scientific LTD/Photo Researchers, Inc. 31.18c: Alfred Owczarzak/BPS. 31.18d: John D. Cunningham/Visuals Unlimited. 31.20a *upper:* J. Robert Waaland/BPS. 31.20a *lower:* Cabisco/Visuals Unlimited. 31.20b *upper:* J. Robert Waaland/BPS. 31.20b *lower:* Cabisco/Visuals Unlimited. 31.22: J. N. A. Lott/BPS. 31.23: Jim Solliday/BPS. 31.24a,b: Phil Gates/BPS. 31.25b: Jeff Lepore/Photo Researchers, Inc. 31.25c: C. G. Van Dyke/Visuals Unlimited.

Chapter 32 *Opener:* Peter K. Ziminski/Visuals Unlimited. 32.2: Biophoto Associates/Photo Researchers, Inc. 32.7: John D. Cunningham/Visuals Unlimited. 32.10a: David M. Phillips/Visuals Unlimited. 32.12 *right:* Derrick Ditchburn/Visuals Unlimited. 32.13: Thomas Eisner, Cornell Univ.

Chapter 33 *Opener* and 33.2: Nigel Cattlin/Holt Studios International/Photo Researchers, Inc. 33.4: Jess R. Lee/Photo Researchers, Inc. 33.6: Stephen Parker/Photo Researchers, Inc. 33.8: Thomas Eisner, Cornell Univ. 33.9: Jon Mark Stewart/BPS. 33.10: J. N. A. Lott/BPS. 33.11, 33.12: Richard Shiell. 33.13: Janine Pestel/Visuals Unlimited. 33.14: Barbara J. Miller/BPS. 33.15: J. N. A. Lott/BPS. 33.16: Robert & Linda Mitchell. 33.17: Budd Titlow/Visuals Unlimited.

Chapter 34 *Opener:* Cabisco/Visuals Unlimited. 34.1: The Photo Works/Photo Researchers, Inc. 34.4: Kathleen Blanchard/Visuals Unlimited. 34.7: Jeremy Burgess/Photo Researchers, Inc. 34.8: Barbara J. O' Donnell/BPS. 34.10: E. H. Newcomb and S. R. Tandon/BPS. 34.12: Richard C. Johnson/Visuals Unlimited. 34.13: Nuridsany et Pérennou/Photo Researchers, Inc.

Chapter 35 *Opener:* Mary Clay/Masterfile. 35.5: Tom J. Ulrich/Visuals Unlimited. 35.6: Barbara J. Miller/BPS. 35.7: J. N. A. Lott/BPS. 35.11: J. A. D. Zeevaart, Michigan State Univ. 35.19a: Biophoto Associates/Photo Researchers, Inc. 35.21: Cabisco/Visuals Unlimited. 35.22: Adam Jones. 35.23; 35.25: J. N. A. Lott/BPS. 35.29: R. Last, Cornell Univ. Courtesy of the Society for Plant Physiology.

Chapter 36 *Opener:* David Woodfall/Woodfall Wild Images. 36:2: Stan Flegler/Visuals Unlimited. 36.3: Nigel Cattlin, Holt Studios International/Photo Researchers, Inc. 36.4: Bowman, J. (ed.). 1994. Arabidopsis: *An Atlas of Morphology and Development.* Springer-Verlag, New York. Photos by S. Craig & A. Chaudhury, Plate 6.2. 36.8a: John Kaprielian/Photo Researchers, Inc. 36.8b: Renee Lynn/Photo Researchers, Inc. 36.17a: Nigel Cattlin, Holt Studios International/Photo Researchers, Inc. 36.17b: James W. Richardson/Visuals Unlimited.

Chapter 37 *Opener:* Daniel J. Cox/Natural Exposures. 37.15a: B. & C. Alexander/Photo Researchers, Inc. 37.15b: Timothy Ransom/BPS. 37.20 *left, center:* G. W. Willis/BPS. 37.20 *right:* Fran Thomas, Stanford Univ. 37.21a: Barbara Gerlach/Visuals Unlimited. 37.21b: Art Wolfe.

Chapter 38 *Opener:* R. D. Fernald, Stanford Univ. 38.6a: Associated Press Photo. 38.6b: The Bettmann Archive, Inc.

Chapter 39 *Opener:* Ron Austing/Photo Researchers, Inc. 39.1a: Biophoto Associates/Photo Researchers, Inc. 39.1b: Andrew J. Martinez/Photo Researchers, Inc. 39.1c: Peter J. Bryant/BPS. 39.2: David M. Phillips/Photo Researchers, Inc. 39.4: Paul W. Johnson/BPS. 39.5: Fletcher & Baylis/Photo Researchers, Inc. 39.6: James Beveridge/Visuals Unlimited. 39.7a: Jim Merli/Visuals Unlimited. 39.7b: Robert W. Hernandez. 39.8: Renee Lynn/Photo Researchers, Inc. 39.13 *inset:* P. Bagavandoss/Photo Researchers, Inc. 39.17: Lara Hartley/BPS. 39.19: CC Studio/Photo Researchers, Inc. 39.21: C. Eldeman/Photo Researchers, Inc. 39.22: Nestle/Photo Researchers, Inc. 39.24: S. I. U. School of Med./Photo Researchers, Inc.

Chapter 40 *Opener:* Courtesy of John Morrill, New College. 40.5 *inset:* Courtesy of Richard Elinson, Univ. of Toronto.

Chapter 41 *Opener:* Stephen Krasemann, Nature Conservancy/Photo Researchers, Inc. 41.4: C. Raines/Visuals Unlimited.

Index

Numbers in **boldface italic** refer to information in an illustration, caption, or table.

About the Book

Editor: Andrew D. Sinauer
Project Editor: Carol J. Wigg
Developmental Editor: James Funston
Copy Editor: Stephanie Hiebert
Production Manager: Christopher Small
Book Layout and Production: Janice Holabird and Jefferson Johnson
Art Editing and Illustration Program: J/B Woolsey Associates
Design: Jefferson Johnson and Christopher Small
Book Cover Design: MBDesign
Photo Research: Jane Potter
Color Separations: Vision Graphics, Inc.
Cover Manufacture: Henry N. Sawyer Company, Inc.
Book Manufacture: R. R. Donnelley & Sons Company